Grzimek's
Animal Life Encyclopedia

Second Edition

●●●●

Grzimek's
Animal Life Encyclopedia

Second Edition

●●●●

Volume 10
Birds III

Jerome A. Jackson, Advisory Editor
Walter J. Bock, Taxonomic Editor
Donna Olendorf, Project Editor

Joseph E. Trumpey, Chief Scientific Illustrator

Michael Hutchins, Series Editor
In association with the American Zoo and Aquarium Association

GALE®

THOMSON
★
™
GALE

Detroit • New York • San Diego • San Francisco • Cleveland • New Haven, Conn. • Waterville, Maine • London • Munich

Grzimek's Animal Life Encyclopedia, Second Edition
Volume 10: Birds III

Project Editor
Donna Olendorf

Editorial
Deirdre Blanchfield, Madeline Harris, Christine Jeryan, Kristine M. Krapp, Kate Kretschmann, Melissa C. McDade, Mark Springer

Permissions
Kim Davis

Imaging and Multimedia
Mary K. Grimes, Lezlie Light, Christine O'Bryan, Barbara Yarrow, Robyn V. Young

Product Design
Tracey Rowens, Jennifer Wahi

Manufacturing
Dorothy Maki, Evi Seoud, Mary Beth Trimper

LIBRARY OF CONGRESS CATALOGING-IN-PUBLICATION DATA

Grzimek, Bernhard.
  [Tierleben. English]
  Grzimek's animal life encyclopedia.— 2nd ed.
    v. cm.
  Includes bibliographical references.
  Contents: v. 1. Lower metazoans and lesser deuterosomes / Neil Schlager, editor — v. 2. Protostomes / Neil Schlager, editor — v. 3. Insects / Neil Schlager, editor — v. 4-5. Fishes I-II / Neil Schlager, editor — v. 6. Amphibians / Neil Schlager, editor — v. 7. Reptiles / Neil Schlager, editor — v. 8-11. Birds I-IV / Donna Olendorf, editor — v. 12-16. Mammals I-V / Melissa C. McDade, editor — v. 17. Cumulative index / Melissa C. McDade, editor.
  ISBN 0-7876-5362-4 (set hardcover : alk. paper)
   1. Zoology—Encyclopedias. I. Title: Animal life encyclopedia. II. Schlager, Neil, 1966- III. Olendorf, Donna IV. McDade, Melissa C. V. American Zoo and Aquarium Association. VI. Title.
  QL7 .G7813 2004

                                                  590′.3—dc21
                                                  2002003351

Printed in Canada
10 9 8 7 6 5 4 3 2 1

Recommended citation: *Grzimek's Animal Life Encyclopedia*, 2nd edition. Volumes 8–11, *Birds I–IV*, edited by Michael Hutchins, Jerome A. Jackson, Walter J. Bock, and Donna Olendorf. Farmington Hills, MI: Gale Group, 2002.

Contents

Contents

Foreword

Earth is teeming with life. No one knows exactly how many distinct organisms inhabit our planet, but more than 5 million different species of animals and plants could exist, ranging from microscopic algae and bacteria to gigantic elephants, redwood trees and blue whales. Yet, throughout this wonderful tapestry of living creatures, there runs a single thread: Deoxyribonucleic acid or DNA. The existence of DNA, an elegant, twisted organic molecule that is the building block of all life, is perhaps the best evidence that all living organisms on this planet share a common ancestry. Our ancient connection to the living world may drive our curiosity, and perhaps also explain our seemingly insatiable desire for information about animals and nature. Noted zoologist, E.O. Wilson, recently coined the term "biophilia" to describe this phenomenon. The term is derived from the Greek *bios* meaning "life" and *philos* meaning "love." Wilson argues that we are human because of our innate affinity to and interest in the other organisms with which we share our planet. They are, as he says, "the matrix in which the human mind originated and is permanently rooted." To put it simply and metaphorically, our love for nature flows in our blood and is deeply engrained in both our psyche and cultural traditions.

Our own personal awakenings to the natural world are as diverse as humanity itself. I spent my early childhood in rural Iowa where nature was an integral part of my life. My father and I spent many hours collecting, identifying and studying local insects, amphibians and reptiles. These experiences had a significant impact on my early intellectual and even spiritual development. One event I can recall most vividly. I had collected a cocoon in a field near my home in early spring. The large, silky capsule was attached to a stick. I brought the cocoon back to my room and placed it in a jar on top of my dresser. I remember waking one morning and, there, perched on the tip of the stick was a large moth, slowly moving its delicate, light green wings in the early morning sunlight. It took my breath away. To my inexperienced eyes, it was one of the most beautiful things I had ever seen. I knew it was a moth, but did not know which species. Upon closer examination, I noticed two moon-like markings on the wings and also noted that the wings had long "tails", much like the ubiquitous tiger swallow-tail butterflies that visited the lilac bush in our backyard. Not wanting to suffer my ignorance any longer, I reached immediately for my *Golden Guide to North American Insects* and searched through the section on moths and butterflies. It was a luna moth! My heart was pounding with the excitement of new knowledge as I ran to share the discovery with my parents.

I consider myself very fortunate to have made a living as a professional biologist and conservationist for the past 20 years. I've traveled to over 30 countries and six continents to study and photograph wildlife or to attend related conferences and meetings. Yet, each time I encounter a new and unusual animal or habitat my heart still races with the same excitement of my youth. If this is biophilia, then I certainly possess it, and it is my hope that others will experience it too. I am therefore extremely proud to have served as the series editor for the Gale Group's rewrite of *Grzimek's Animal Life Encyclopedia*, one of the best known and widely used reference works on the animal world. *Grzimek's* is a celebration of animals, a snapshot of our current knowledge of the Earth's incredible range of biological diversity. Although many other animal encyclopedias exist, *Grzimek's Animal Life Encyclopedia* remains unparalleled in its size and in the breadth of topics and organisms it covers.

The revision of these volumes could not come at a more opportune time. In fact, there is a desperate need for a deeper understanding and appreciation of our natural world. Many species are classified as threatened or endangered, and the situation is expected to get much worse before it gets better. Species extinction has always been part of the evolutionary history of life; some organisms adapt to changing circumstances and some do not. However, the current rate of species loss is now estimated to be 1,000–10,000 times the normal "background" rate of extinction since life began on Earth some 4 billion years ago. The primary factor responsible for this decline in biological diversity is the exponential growth of human populations, combined with peoples' unsustainable appetite for natural resources, such as land, water, minerals, oil, and timber. The world's human population now exceeds 6 billion, and even though the average birth rate has begun to decline, most demographers believe that the global human population will reach 8–10 billion in the next 50 years. Much of this projected growth will occur in developing countries in Central and South America, Asia and Africa—regions that are rich in unique biological diversity.

Finding solutions to conservation challenges will not be easy in today's human-dominated world. A growing number of people live in urban settings and are becoming increasingly isolated from nature. They "hunt" in super markets and malls, live in apartments and houses, spend their time watching television and searching the World Wide Web. Children and adults must be taught to value biological diversity and the habitats that support it. Education is of prime importance now while we still have time to respond to the impending crisis. There still exist in many parts of the world large numbers of biological "hotspots"—places that are relatively unaffected by humans and which still contain a rich store of their original animal and plant life. These living repositories, along with selected populations of animals and plants held in professionally managed zoos, aquariums and botanical gardens, could provide the basis for restoring the planet's biological wealth and ecological health. This encyclopedia and the collective knowledge it represents can assist in educating people about animals and their ecological and cultural significance. Perhaps it will also assist others in making deeper connections to nature and spreading biophilia. Information on the conservation status, threats and efforts to preserve various species have been integrated into this revision. We have also included information on the cultural significance of animals, including their roles in art and religion.

It was over 30 years ago that Dr. Bernhard Grzimek, then director of the Frankfurt Zoo in Frankfurt, Germany, edited the first edition of *Grzimek's Animal Life Encyclopedia*. Dr. Grzimek was among the world's best known zoo directors and conservationists. He was a prolific author, publishing nine books. Among his contributions were: *Serengeti Shall Not Die*, *Rhinos Belong to Everybody* and *He and I and the Elephants*. Dr. Grzimek's career was remarkable. He was one of the first modern zoo or aquarium directors to understand the importance of zoo involvement in *in situ* conservation, that is, of their role in preserving wildlife in nature. During his tenure, Frankfurt Zoo became one of the leading western advocates and supporters of wildlife conservation in East Africa. Dr. Grzimek served as a Trustee of the National Parks Board of Uganda and Tanzania and assisted in the development of several protected areas. The film he made with his son Michael, *Serengeti Shall Not Die*, won the 1959 Oscar for best documentary.

Professor Grzimek has recently been criticized by some for his failure to consider the human element in wildlife conservation. He once wrote: "A national park must remain a primordial wilderness to be effective. No men, not even native ones, should live inside its borders." Such ideas, although considered politically incorrect by many, may in retrospect actually prove to be true. Human populations throughout Africa continue to grow exponentially, forcing wildlife into small islands of natural habitat surrounded by a sea of humanity. The illegal commercial bushmeat trade—the hunting of endangered wild animals for large scale human consumption—is pushing many species, including our closest relatives, the gorillas, bonobos, and chimpanzees, to the brink of extinction. The trade is driven by widespread poverty and lack of economic alternatives. In order for some species to survive it will be necessary, as Grzimek suggested, to establish and enforce a system of protected areas where wildlife can roam free from exploitation of any kind.

While it is clear that modern conservation must take the needs of both wildlife and people into consideration, what will the quality of human life be if the collective impact of short-term economic decisions is allowed to drive wildlife populations into irreversible extinction? Many rural populations living in areas of high biodiversity are dependent on wild animals as their major source of protein. In addition, wildlife tourism is the primary source of foreign currency in many developing countries and is critical to their financial and social stability. When this source of protein and income is gone, what will become of the local people? The loss of species is not only a conservation disaster; it also has the potential to be a human tragedy of immense proportions. Protected areas, such as national parks, and regulated hunting in areas outside of parks are the only solutions. What critics do not realize is that the fate of wildlife and people in developing countries is closely intertwined. Forests and savannas emptied of wildlife will result in hungry, desperate people, and will, in the long-term lead to extreme poverty and social instability. Dr. Grzimek's early contributions to conservation should be recognized, not only as benefiting wildlife, but as benefiting local people as well.

Dr. Grzimek's hope in publishing his *Animal Life Encyclopedia* was that it would "...disseminate knowledge of the animals and love for them", so that future generations would "...have an opportunity to live together with the great diversity of these magnificent creatures." As stated above, our goals in producing this updated and revised edition are similar. However, our challenges in producing this encyclopedia were more formidable. The volume of knowledge to be summarized is certainly much greater in the twenty-first century than it was in the 1970's and 80's. Scientists, both professional and amateur, have learned and published a great deal about the animal kingdom in the past three decades, and our understanding of biological and ecological theory has also progressed. Perhaps our greatest hurdle in producing this revision was to include the new information, while at the same time retaining some of the characteristics that have made *Grzimek's Animal Life Encyclopedia* so popular. We have therefore strived to retain the series' narrative style, while giving the information more organizational structure. Unlike the original *Grzimek's*, this updated version organizes information under specific topic areas, such as reproduction, behavior, ecology and so forth. In addition, the basic organizational structure is generally consistent from one volume to the next, regardless of the animal groups covered. This should make it easier for users to locate information more quickly and efficiently. Like the original Grzimek's, we have done our best to avoid any overly technical language that would make the work difficult to understand by non-biologists. When certain technical expressions were necessary, we have included explanations or clarifications.

Considering the vast array of knowledge that such a work represents, it would be impossible for any one zoologist to have completed these volumes. We have therefore sought specialists from various disciplines to write the sections with which they are most familiar. As with the original *Grzimek's*, we have engaged the best scholars available to serve as topic editors, writers, and consultants. There were some complaints about inaccuracies in the original English version that may have been due to mistakes or misinterpretation during the complicated translation process. However, unlike the original *Grzimek's*, which was translated from German, this revision has been completely re-written by English-speaking scientists. This work was truly a cooperative endeavor, and I thank all of those dedicated individuals who have written, edited, consulted, drawn, photographed, or contributed to its production in any way. The names of the topic editors, authors, and illustrators are presented in the list of contributors in each individual volume.

The overall structure of this reference work is based on the classification of animals into naturally related groups, a discipline known as taxonomy or biosystematics. Taxonomy is the science through which various organisms are discovered, identified, described, named, classified and catalogued. It should be noted that in preparing this volume we adopted what might be termed a conservative approach, relying primarily on traditional animal classification schemes. Taxonomy has always been a volatile field, with frequent arguments over the naming of or evolutionary relationships between various organisms. The advent of DNA fingerprinting and other advanced biochemical techniques has revolutionized the field and, not unexpectedly, has produced both advances and confusion. In producing these volumes, we have consulted with specialists to obtain the most up-to-date information possible, but knowing that new findings may result in changes at any time. When scientific controversy over the classification of a particular animal or group of animals existed, we did our best to point this out in the text.

Readers should note that it was impossible to include as much detail on some animal groups as was provided on others. For example, the marine and freshwater fish, with vast numbers of orders, families, and species, did not receive as

detailed a treatment as did the birds and mammals. Due to practical and financial considerations, the publishers could provide only so much space for each animal group. In such cases, it was impossible to provide more than a broad overview and to feature a few selected examples for the purposes of illustration. To help compensate, we have provided a few key bibliographic references in each section to aid those interested in learning more. This is a common limitation in all reference works, but *Grzimek's Encyclopedia of Animal Life* is still the most comprehensive work of its kind.

I am indebted to the Gale Group, Inc. and Senior Editor Donna Olendorf for selecting me as Series Editor for this project. It was an honor to follow in the footsteps of Dr. Grzimek and to play a key role in the revision that still bears his name. *Grzimek's Animal Life Encyclopedia* is being published by the Gale Group, Inc. in affiliation with my employer, the American Zoo and Aquarium Association (AZA), and I would like to thank AZA Executive Director, Sydney J. Butler; AZA Past-President Ted Beattie (John G. Shedd Aquarium, Chicago, IL); and current AZA President, John Lewis (John Ball Zoological Garden, Grand Rapids, MI), for approving my participation. I would also like to thank AZA Conservation and Science Department Program Assistant, Michael Souza, for his assistance during the project. The AZA is a professional membership association, representing 205 accredited zoological parks and aquariums in North America. As Director/William Conway Chair, AZA Department of Conservation and Science, I feel that I am a philosophical descendant of Dr. Grzimek, whose many works I have collected and read. The zoo and aquarium profession has come a long way since the 1970s, due, in part, to innovative thinkers such as Dr. Grzimek. I hope this latest revision of his work will continue his extraordinary legacy.

Silver Spring, Maryland, 2001
Michael Hutchins
Series Editor

• • • • •

How to use this book

Gzimek's Animal Life Encyclopedia is an internationally prominent scientific reference compilation, first published in German in the late 1960s, under the editorship of zoologist Bernhard Grzimek (1909–1987). In a cooperative effort between Gale and the American Zoo and Aquarium Association, the series is being completely revised and updated for the first time in over 30 years. Gale is expanding the series from 13 to 17 volumes, commissioning new color images, and updating the information while also making the set easier to use. The order of revisions is:

Vol 8–11: Birds I–IV
Vol 6: Amphibians
Vol 7: Reptiles
Vol 4–5: Fishes I–II
Vol 12–16: Mammals I–V
Vol 1: Lower Metazoans and Lesser Deuterostomes
Vol 2: Protostomes
Vol 3: Insects
Vol 17: Cumulative Index

Organized by order and family

The overall structure of this reference work is based on the classification of animals into naturally related groups, a discipline known as taxonomy—the science through which various organisms are discovered, identified, described, named, classified, and catalogued. Starting with the simplest life forms, the protostomes, in Vol. 1, the series progresses through the more complex animal classes, culminating with the mammals in Vols. 12–16. Volume 17 is a stand-alone cumulative index.

Organization of chapters within each volume reinforces the taxonomic hierarchy. Opening chapters introduce the class of animal, followed by chapters dedicated to order and family. Species accounts appear at the end of family chapters. To help the reader grasp the scientific arrangement, each type of chapter has a distinctive color and symbol:

▲ = Family Chapter (yellow background)

● = Order Chapter (blue background)

◐ = Monotypic Order Chapter (green background)

As chapters narrow in focus, they become more tightly formatted. General chapters have a loose structure, reminiscent of the first edition. While not strictly formatted, order chapters are carefully structured to cover basic information about member families. Monotypic orders, comprised of a single family, utilize family chapter organization. Family chapters are most tightly structured, following a prescribed format of standard rubrics that make information easy to find and understand. Family chapters typically include:

Thumbnail introduction
    Common name
    Scientific name
    Class
    Order
    Suborder
    Family
    Thumbnail description
    Size
    Number of genera, species
    Habitat
    Conservation status
Main essay
    Evolution and systematics
    Physical characteristics
    Distribution
    Habitat
    Behavior
    Feeding ecology and diet
    Reproductive biology
    Conservation status
    Significance to humans
Species accounts
    Common name
    Scientific name
    Subfamily
    Taxonomy
    Other common names
    Physical characteristics
    Distribution
    Habitat
    Behavior
    Feeding ecology and diet
    Reproductive biology

Conservation status
Significance to humans
Resources
Books
Periodicals
Organizations
Other

Color graphics enhance understanding

Grzimek's features approximately 3,500 color photos, including approximately 480 in four Birds volumes; 3,500 total color maps, including almost 1,500 in the four Birds volumes; and approximately 5,500 total color illustrations, including 1,385 in four Birds volumes. Each featured species of animal is accompanied by both a distribution map and an illustration.

All maps in *Grzimek's* were created specifically for the project by XNR Productions. Distribution information was provided by expert contributors and, if necessary, further researched at the University of Michigan Zoological Museum library. Maps are intended to show broad distribution, not definitive ranges, and are color coded to show resident, breeding, and nonbreeding locations (where appropriate).

All the color illustrations in *Grzimek's* were created specifically for the project by Michigan Science Art. Expert contributors recommended the species to be illustrated and provided feedback to the artists, who supplemented this information with authoritative references and animal skins from University of Michgan Zoological Museum library. In addition to species illustrations, *Grzimek's* features conceptual drawings that illustrate characteristic traits and behaviors.

About the contributors

The essays were written by expert contributors, including ornithologists, curators, professors, zookeepers, and other reputable professionals. *Grzimek's* subject advisors reviewed the completed essays to insure that they are appropriate, accurate, and up-to-date.

Standards employed

In preparing these volumes, the editors adopted a conservative approach to taxonomy, relying primarily on Peters Checklist (1934–1986)—a traditional classification scheme. Taxonomy has always been a volatile field, with frequent arguments over the naming of or evolutionary relationships between various organisms. The advent of DNA fingerprinting and other advanced biochemical techniques has revolutionized the field and, not unexpectedly, has produced both advances and confusion. In producing these volumes, Gale consulted with noted taxonomist Professor Walter J. Bock as well as other specialists to obtain the most up-to-date information possible. When scientific controversy over the classification of a particular animal or group of animals existed, the text makes this clear.

Grzimek's has been designed with ready reference in mind and the editors have standardized information wherever fea-

sible. For **Conservation status,** *Grzimek's* follows the IUCN Red List system, developed by its Species Survival Commission. The Red List provides the world's most comprehensive inventory of the global conservation status of plants and animals. Using a set of criteria to evaluate extinction risk, the IUCN recognizes the following categories: Extinct, Extinct in the Wild, Critically Endangered, Endangered, Vulnerable, Conservation Dependent, Near Threatened, Least Concern, and Data Deficient. For a complete explanation of each category, visit the IUCN web page at http://www.iucn.org/themes/ssc/redlists/categor.htm

In addition to IUCN ratings, essays may contain other conservation information, such as a species' inclusion on one of three Convention on International Trade in Endangered Species (CITES) appendices. Adopted in 1975, CITES is a global treaty whose focus is the protection of plant and animal species from unregulated international trade.

Grzimek's provides the following standard information on avian lineage in **Taxonomy** rubric of each Species account: [First described as] *Muscicapa rufifrons* [by] Latham, [in] 1801, [based on a specimen from] Sydney, New South Wales, Australia. The person's name and date refer to earliest identification of a species, although the species name may have changed since first identification. However, the organism described is the same.

Other common names in English, French, German, and Spanish are given when an accepted common name is available.

Appendices and index

For further reading directs readers to additional sources of information about birds. Valuable contact information for **Organizations** is also included in an appendix. While the encyclopedia minimizes scientific jargon, it also provides a **Glossary** at the back of the book to define unfamiliar terms. An exhaustive **Aves species list** records all known species of birds, categorized according to Peters Checklist (1934–1986). And a full-color **Geologic time scale** helps readers understand prehistoric time periods. Additionally, each of the four volumes contains a full **Subject index** for the Birds subset.

Acknowledgements

Gale would like to thank several individuals for their important contributions to the series. Michael Souza, Program Assistant, Department of Conservation and Science, American Zoo and Aquarium Association, provided valuable behind-the-scenes research and reliable support at every juncture of the project. Also deserving of recognition are Christine Sheppard, Curator of Ornithology at Bronx Zoo, and Barry Taylor, professor at the University of Natal, in Pietermaritzburg, South Africa, who assisted subject advisors in reviewing manuscripts for accuracy and currency. And, last but not least, Janet Hinshaw, Bird Division Collection Manager at the University of Michigan Museum of Zoology, who opened her collections to *Grzimek's* artists and staff and also compiled the "For Further Reading" bibliography at the back of the book.

• • • • •

Advisory boards

Series advisor

Michael Hutchins, PhD
Director of Conservation and William Conway Chair
American Zoo and Aquarium Association
Silver Spring, Maryland

Subject advisors

Volume 1: Lower Metazoans and Lesser Deuterostomes
Dennis Thoney, PhD
Director, Marine Laboratory & Facilities
Humboldt State University
Arcata, California

Volume 2: Protostomes
Dennis Thoney, PhD
Director, Marine Laboratory & Facilities
Humboldt State University
Arcata, California

Sean F. Craig, PhD
Assistant Professor, Department of Biological Sciences
Humboldt State University
Arcata, California

Volume 3: Insects
Art Evans, PhD
Entomologist
Richmond, Virginia

Rosser W. Garrison, PhD
Systematic Entomologist, Los Angeles County
Los Angeles, California

Volumes 4–5: Fishes I–II
Paul Loiselle, PhD
Curator, Freshwater Fishes
New York Aquarium
Brooklyn, New York

Dennis Thoney, PhD

Director, Marine Laboratory & Facilities
Humboldt State University
Arcata, California

Volume 6: Amphibians
William E. Duellman, PhD
Curator of Herpetology Emeritus
Natural History Museum and Biodiversity Research Center
University of Kansas
Lawrence, Kansas

Volume 7: Reptiles
James B. Murphy, PhD
Smithsonian Research Associate
Department of Herpetology
National Zoological Park
Washington, DC

Volumes 8–11: Birds I–IV
Walter J. Bock, PhD
Permanent secretary, International Ornithological Congress
Professor of Evolutionary Biology
Department of Biological Sciences,
Columbia University
New York, New York

Jerome A. Jackson, PhD
Program Director, Whitaker Center for Science,
Mathematics, and Technology Education
Florida Gulf Coast University
Ft. Myers, Florida

Volumes 12–16: Mammals I–V
Valerius Geist, PhD
Professor Emeritus of Environmental Science
University of Calgary
Calgary, Alberta
Canada

Devra Gail Kleiman, PhD
Smithsonian Research Associate
National Zoological Park
Washington, DC

Library advisors

• • • • •

Contributing writers

Birds I–IV

Michael Abs, Dr. rer. nat.
Berlin, Germany

George William Archibald, PhD
International Crane Foundation
Baraboo, Wisconsin

Helen Baker, PhD
Joint Nature Conservation Committee
Peterborough, Cambridgeshire
United Kingdom

Cynthia Ann Berger, MS
Pennsylvania State University
State College, Pennsylvania

Matthew A. Bille, MSc
Colorado Springs, Colorado

Walter E. Boles, PhD
Australian Museum
Sydney, New South Wales
Australia

Carlos Bosque, PhD
Universidad Simón Bolívar
Caracas, Venezuela

David Brewer, PhD
Research Associate
Royal Ontario Museum
Toronto, Ontario
Canada

Daniel M. Brooks, PhD
Houston Museum of Natural Science
Houston, Texas

Donald F. Bruning, PhD
Wildlife Conservation Society
Bronx, New York

Joanna Burger, PhD
Rutgers University
Piscataway, New Jersey

Carles Carboneras
SEO/BirdLife
Barcelona, Spain

John Patrick Carroll, PhD
University of Georgia
Athens, Georgia

Robert Alexander Cheke, PhD
Natural Resources Institute
University of Greenwich
Chatham, Kent
United Kingdom

Jay Robert Christie, MBA
Racine Zoological Gardens
Racine, Wisconsin

Charles T. Collins, PhD
California State University
Long Beach, California

Malcolm C. Coulter, PhD
IUCN Specialist Group on Storks,
Ibises and Spoonbills
Chocorua, New Hampshire

Adrian Craig, PhD
Rhodes University
Grahamstown, South Africa

Francis Hugh John Crome, BSc
Consultant
Atheron, Queensland
Australia

Timothy Michael Crowe, PhD
University of Cape Town
Rondebosch, South Africa

H. Sydney Curtis, BSc
Queensland National Parks &
Wildlife Service (Retired)
Brisbane, Queensland
Australia

S. J. J. F. Davies, ScD
Curtin University of Technology
Department of Environmental Biology
Perth, Western Australia
Australia

Gregory J. Davis, PhD
University of Wisconsin-Green Bay
Green Bay, Wisconsin

William E. Davis, Jr., PhD
Boston University
Boston, Massachusetts

Stephen Debus, MSc
University of New England
Armidale, New South Wales
Australia

Michael Colin Double, PhD
Australian National University
Canberra, A.C.T.
Australia

Rachel Ehrenberg, MS
University of Michigan
Ann Arbor, Michigan

Contributing writers

Eladio M. Fernandez
Santo Domingo
Dominican Republic

Simon Ferrier, PhD
New South Wales National Parks and
Wildlife Service
Armidale, New South Wales
Australia

Kevin F. Fitzgerald, BS
South Windsor, Connecticut

Hugh Alastair Ford, PhD
University of New England
Armidale, New South Wales
Australia

Joseph M. Forshaw
Australian Museum
Sydney, New South Wales
Australia

Bill Freedman, PhD
Department of Biology
Dalhousie University
Halifax, Nova Scotia
Canada

Clifford B. Frith, PhD
Honorary research fellow
Queensland Museum
Brisbane, Australia

Dawn W. Frith, PhD
Honorary research fellow
Queensland Museum
Brisbane, Australia

Peter Jeffery Garson, DPhil
University of Newcastle
Newcastle upon Tyne
United Kingdom

Michael Gochfeld, PhD, MD
UMDNJ-Robert Wood Johnson
Medical School
Piscataway, New Jersey

Michelle L. Hall, PhD
Australian National University
School of Botany and Zoology
Canberra, A.C.T.
Australia

Frank Hawkins, PhD
Conservation International
Antananarivo, Madagascar

David G. Hoccom, BSc
Royal Society for the Protection of
Birds
Sandy, Bedfordshire
United Kingdom

Peter Andrew Hosner
Cornell University
Ithaca, New York

Brian Douglas Hoyle PhD
Bedford, Nova Scotia
Canada

Julian Hughes
Royal Society for the Protection of
Birds
Sandy, Bedfordshire
United Kingdom

Robert Arthur Hume, BA
Royal Society for the Protection of
Birds
Sandy, Bedfordshire
United Kingdom

Gavin Raymond Hunt, PhD
University of Auckland
Auckland, New Zealand

Jerome A. Jackson, PhD
Florida Gulf Coast University
Ft. Myers, Florida

Bette J. S. Jackson, PhD
Florida Gulf Coast University
Ft. Myers, Florida

Darryl N. Jones, PhD
Griffith University
Queensland, Australia

Alan C. Kemp, PhD
Naturalists & Nomads
Pretoria, South Africa

Angela Kay Kepler, PhD
Pan-Pacific Ecological Consulting
Maui, Hawaii

Jiro Kikkawa, DSc
Professor Emeritus
University of Queensland,
Brisbane, Queensland
Australia

Margaret Field Kinnaird, PhD
Wildlife Conservation Society
Bronx, New York

Guy M. Kirwan, BA
Ornithological Society of the Middle
East
Sandy, Bedfordshire
United Kingdom

Melissa Knopper, MS
Denver Colorado

Niels K. Krabbe, PhD
University of Copenhagen
Copenhagen, Denmark

James A. Kushlan, PhD
U.S. Geological Survey
Smithsonian Environmental Research
Center
Edgewater, Maryland

Norbert Lefranc, PhD
Ministère de l'Environnement,
Direction Régionale
Metz, France

P. D. Lewis, BS
Jacksonville Zoological Gardens
Jacksonville, Florida

Josef H. Lindholm III, BA
Cameron Park Zoo
Waco, Texas

Peter E. Lowther, PhD
Field Museum
Chicago, Illinois

Gordon Lindsay Maclean, PhD, DSc
Rosetta, South Africa

Steve Madge
Downderry, Torpoint
Cornwall
United Kingdom

Albrecht Manegold
Institut für Biologie/Zoologie
Berlin, Germany

Jeffrey S. Marks, PhD
University of Montana
Missoula, Montana

Juan Gabriel Martínez, PhD
Universidad de Granada
Departamento de Biologia
Animal y Ecologia
Granada, Spain

Barbara Jean Maynard, PhD
Laporte, Colorado

Cherie A. McCollough, MS
PhD candidate, University of Texas
Austin, Texas

Leslie Ann Mertz, PhD
Fish Lake Biological Program
Wayne State University
Biological Station
Lapeer, Michigan

Derek William Niemann, BA
Royal Society for the Protection of
Birds
Sandy, Bedfordshire
United Kingdom

Malcolm Ogilvie, PhD
Glencairn, Bruichladdich
Isle of Islay
United Kingdom

Penny Olsen, PhD
Australian National University
Canberra, A.C.T.
Australia

Jemima Parry-Jones, MBE
National Birds of Prey Centre
Newent, Gloucestershire
United Kingdom

Colin Pennycuick, PhD, FRS
University of Bristol
Bristol, United Kingdom

James David Rising, PhD
University of Toronto
Department of Zoology
Toronto, Ontario
Canada

Christopher John Rutherford Robertson
Wellington, New Zealand

Peter Martin Sanzenbacher, MS
USGS Forest & Rangeland Ecosystem
Science Center
Corvallis, Oregon

Matthew J. Sarver, BS
Ithaca, New York

Herbert K. Schifter, PhD
Naturhistorisches Museum
Vienna, Austria

Richard Schodde PhD, CFAOU
Australian National Wildlife
Collection, CSIRO
Canberra, A.C.T.
Australia

Karl-L. Schuchmann, PhD
Alexander Koenig Zoological Research
Institute and Zoological Museum
Bonn, Germany

Tamara Schuyler, MA
Santa Cruz, California

Nathaniel E. Seavy, MS
Department of Zoology
University of Florida
Gainesville, Florida

Charles E. Siegel, MS
Dallas Zoo
Dallas, Texas

Julian Smith, MS
Katonah, New York

Joseph Allen Smith
Baton Rouge, Louisiana

Walter Sudhaus, PhD
Institut für Zoologie
Berlin, Germany

J. Denis Summers-Smith, PhD
Cleveland, North England
United Kingdom

Barry Taylor, PhD
University of Natal
Pietermaritzburg, South Africa

Markus Patricio Tellkamp, MS
University of Florida
Gainesville, Florida

Joseph Andrew Tobias, PhD
BirdLife International
Cambridge
United Kingdom

Susan L. Tomlinson, PhD
Texas Tech University
Lubbock, Texas

Donald Arthur Turner, PhD
East African Natural History Society
Nairobi, Kenya

Michael Phillip Wallace, PhD
Zoological Society of San Diego
San Diego, California

John Warham, PhD, DSc
University of Canterbury
Christchurch, New Zealand

Tony Whitehead, BSc
Ipplepen, Devon
United Kingdom

Peter H. Wrege, PhD
Cornell University
Ithaca, New York

• • • • •
Contributing illustrators

Drawings by Michigan Science Art

Joseph E. Trumpey, Director, AB, MFA
Science Illustration, School of Art and Design, University of Michigan

Wendy Baker, ADN, BFA

Brian Cressman, BFA, MFA

Emily S. Damstra, BFA, MFA

Maggie Dongvillo, BFA

Barbara Duperron, BFA, MFA

Dan Erickson, BA, MS

Patricia Ferrer, AB, BFA, MFA

Maps by XNR Productions

Paul Exner, Chief cartographer
XNR Productions, Madison, WI

Tanya Buckingham

Jon Daugherity

Gillian Harris, BA

Jonathan Higgins, BFA, MFA

Amanda Humphrey, BFA

Jacqueline Mahannah, BFA, MFA

John Megahan, BA, BS, MS

Michelle L. Meneghini, BFA, MFA

Bruce D. Worden, BFA

Thanks are due to the University of Michigan, Museum of Zoology, which provided specimens that served as models for the images.

Laura Exner

Andy Grosvold

Cory Johnson

Paula Robbins

Coraciiformes

(Kingfishers, todies, hoopoes, and relatives)

Class
Aves

Order
Coraciiformes

Number of families 8

Number of genera, species 49 genera; 211 species

Photo: Blue-bellied rollers (*Coracias cyanogaster*) share a branch. (Photo by Daniel Zupanc. Bruce Coleman Inc. Reproduced by permission.)

Systematics

The order Coraciiformes includes many species that represent some of the most colorful and unusual bird families in the world. The order is named after the rollers of the family Coraciidae, whose members have the least specialized or most basic design within the order. While each family within the Coraciiformes can be defined rather clearly, relationships between the families (and which families to include in the order) are more difficult to ascertain. This also means that the exact criteria for membership in the order are difficult to define, due to the diversity of form and behavior spread across such variable families. Ten families are usually incorporated into the Coraciiformes and these can be divided into four main groups. This approach will be adopted here; each group is sometimes designated as a sub-order:

1. Kingfishers (family Alcedinidae) and the allied families of todies (family Todidae) and motmots (family Momotidae) in the suborder Alcedines.

2. Bee-eaters (family Meropidae) alone in the suborder Meropes.

3. Rollers (family Coraciidae) alone in the suborder Coracii. The allied families of ground-rollers (family Brachypteraciidae) and the anomalous cuckoo-roller or courol (family Leptosomidae) are commonly included as Coracii in other treatments.

4. Hornbills (family Bucerotidae) and the closely related common hoopoe (family Upupidae) and woodhoopoes (family Phoeniculidae) in the suborder Bucerotes.

Given the diversity of the order, other arrangements of the Coraciiformes have been proposed. On the one hand, some

other families have been proposed for membership to a wider and more inclusive order Coraciiformes, such as the trogons (family Trogonidae), jacamars (family Galbulidae), and puffbirds (family Bucconidae). Of these, the trogons, with their worldwide and tropical distribution, are prime contenders for inclusion; although they differ in the unique heterodactyl arrangement of their toes, they are similar to other Coraciiformes in many other features. On the other hand, several coraciiform families have been elevated to the level of an order, such as the todies, hornbills, and hoopoes, while several subfamilies have been elevated to the level of families, such as the kingfishers, hoopoes, and hornbills. More distant relatives of Coraciiformes have been proposed to occur mainly among the woodpeckers and barbets, and especially the jacamars and puffbirds, of the Piciformes, but also among the trogons of the Trogoniformes, the cuckoos of the Cuculiformes, and the mousebirds of the Coliiformes. The rollers *sensu lato* have also been proposed as the primitive evolutionary template that gave rise to the sub-oscines, such as broadbills and pittas. All of these orders have been proposed to be precursors to the great avian order of the oscine Passeriformes, a suggestion based in part on the different and possibly relict forms that are now isolated on Madagascar.

Inter-family relationships

Many studies have attempted to unravel the relationships among families that might be included in the Coraciiformes. These range from anatomical studies, popular in the eighteenth century, to modern molecular studies of nuclear and mitochondrial DNA, such as the pioneering work of Charles Sibley and Jon Ahlquist (1990) on proteins from the whites of the eggs published in 1972 and on the hybridization of nuclear DNA in 1990. There is now some consensus on how the natural family groups cluster together, based wherever possible on unique shared features, reflected partly in the sub-

A belted kingfisher (*Megaceryle alcyon*) feeding. (Photo by Scott Nielsen. Bruce Coleman Inc. Reproduced by permission.)

ordinal divisions above and supported by some of the characteristics listed above and below. The common hoopoe and woodhoopoes are related by the unique anvil-shaped bone of the inner ear bone, while their most similar relatives, also with oval eggs and pitted shells, appear to be hornbills, which are defined by their uniquely fused neck vertebrae, the atlas and axis. The two special New World families of todies and motmots, as might be expected from their distribution, appear to be each other's closest relatives. These families are linked by species of intermediate characteristics, such as the tody motmot (*Hylomanes momotula*), and the connection is supported by a useful fossil record that shows an earlier and wider diversity in the Northern Hemisphere. Their biology also supports a more distant link with either the kingfishers, reflected in the diminutive kingfisher-like form of some todies and inclusion in the suborder Alcidines, or the bee-eaters, alone in their suborder Meropes but linked by the heavy bee-eater-like bill and the behavior of the larger motmots. It is notable that most members of these last four families excavate their own nests as tunnels into the ground, termite nests, or epiphyte roots, while only a minority of members use natural tree or ground holes. They all share the stirrup-shaped inner ear bone, together with trogons, and they all have chicks with well-developed pads on the "heels."

Physical characteristics

Coraciiformes are generally recognized by the large and long bill, large head, short neck, short legs, and weak feet with short toes. The single feature shared by almost all families in the Coraciiformes is that the front three toes are fused partially at the base. The middle toe is fused to the inner toe at its base and to the outer toe for most of its length. This toe-fusion is a condition termed syndactyly and is the main criterion by which the order is defined. Coraciiformes also share, but are not unique, in the design of the palate bones (classi-

fied as desmognathous), in the lack of an ambiens muscle in the leg, and in the rather small feet.

Other features are widespread in the order but not present in or unique to each family. The wing has 10 primary feathers, often with a vestigial eleventh feather; the tail has 12 feathers in all families other than the motmots, hoopoes, and hornbills, for whom the tail has 10 feathers. The breast bone has two notches on the sternum in most families, but only one notch in common hoopoes (*Upupa epops*) and hornbills. The inner ear bone, or columella (stapes), is of a simple reptilian design in rollers and hornbills, but of a unique anvil-like shape in hoopoes and woodhoopoes, and of a stirrup-like shape in kingfishers, bee-eaters, todies, motmots, and trogons. The eggs are generally white, rounded, and shiny, except for being oval in hoopoes and hornbills, and tinted light blue-green in hoopoes. The chicks hatch blind and with the upper mandible noticeably shorter than the lower. The chicks are naked in all families, except for hoopoes, which have patches of fine down. However, the chicks always develop later through a spiky "pin-cushion" stage, when their emerging feathers are retained in their quills for several days. Kingfishers and bee-eaters excavate and nest in earthen burrows, and their chicks have well-developed papillae on their "heel" joint, similar to those of honeyguides and woodpeckers (Piciformes). Sexes are similar in most species, except for most hornbills and some kingfishers. Many species have a brilliant plumage, often with a large colorful bill, long tail, or tall crest. While there is no single character that is unique to and defines any combination of families that might comprise the Coraciiformes, there is an overlap in shared characters that links at least the ten main families as presented here.

Behavior

Most coraciiform species are arboreal in their feeding, breeding, and roosting habits, though a minority of species spend much time on the ground. Most species feed on small animals, especially small vertebrates and large arthropods, and they catch their prey mainly by dropping down from a perch to the ground (e.g., true rollers) or into water (e.g., kingfishers). More aerial species may hover in search of prey (e.g., kingfishers), or they may take most food by hawking it on the wing (e.g., bee-eaters and broad-billed rollers). Many species, such the todies and motmots, combine terrestrial and aerial capture of prey into their foraging repertoire, often in quite different proportions. A few species are specialized in their foraging habits or diet; for example, bee-eaters de-venom their prey, cuckoo-rollers concentrate on chameleons, and shovel-billed kingfishers (*Clytoceyx rex*) specialize on earthworms. A few species collect most of their food while they walk or run about on the ground, such as the common hoopoes and some African hornbills. Other species consume fruit as their main diet and only add animal food secondarily when rearing chicks, as seen for many forest hornbills.

Most species are territorial when breeding, usually as single pairs, but in several families there are species that live and breed as cooperative groups or even some species that nest in large colonies. Most species lay their eggs inside a cavity (often sparsely lined with plant materials), which can

Carmine bee-eaters (*Merops nubicoides*) and white-fronted bee-eaters (*Melittophagus bullockoides*) in Okavango, Botswana. (Photo by Nigel Dennis. Photo Researchers, Inc. Reproduced by permission.)

be either a natural hole in a tree, a rock face, a building, or the ground, or an excavated tunnel in the ground with a nest chamber at the end. Interestingly, some kingfisher species excavate nest cavities in arboreal termite nests, rotten wood, or even sawdust piles. Most hornbills exhibit the unique habit of sealing the entrance of the nest to form a narrow slit. In all species, both members of a pair generally take part in nesting activities, including defense, construction, and delivery of food. In most species, the female does most or all of the incubation of eggs and the brooding of young chicks, while the male delivers food to the female and the chicks. Only later, when the demands of the growing chicks rise, are both sexes involved in provisioning at the nest. The nest-tunnels of bee-eaters, motmots, todies, rollers, and especially kingfishers become quite smelly as nesting progresses due to the accumulation of feces and the remains of food in the chamber. The chicks of hoopoes also produce a noxious odor in the nest, apparently a mixture of preen oil, copious feces, and possibly with the aid of special bacterial products. Only the hornbills practice good nest sanitation, which is accomplished by squirting out feces and tossing out food remains through the partly sealed nest entrance.

Distribution and evolution

Even though the order Coraciiformes is cosmopolitan, it is only the kingfishers that occur widely across every ice-free continent. Many other families have very restricted ranges.

The only other families that occur in the New World are the few species of motmots, confined to the tropics of Central and South America, and of todies, confined to the few islands of the Greater Antilles in the West Indies. Most families and species occur in the Old World. Even though bee-eaters, rollers, hoopoes, and hornbills range widely across Africa, Eurasia, and even into Australasia, all of the woodhoopoes and the majority of bee-eaters and rollers occur in Africa. Hornbills are divided mainly between Africa and Asia with only a single species extending to Australasia. The kingfishers, despite their comopolitan distribution, have only six of their 91 species in the Americas. To the north of this mainly tropical distribution, only one species of kingfisher, bee-eater, and roller and the hoopoe breed in Europe. To the south, only one species of bee-eater and roller breeds in Australia; however, these species are joined there by at least ten species of kingfishers in what is the major and Australasian center for the diversity of that family.

The distribution of the fossilized remains of coraciiform birds suggests an ancient and somewhat different geographic and evolutionary history than what might be inferred from the current range of each family. Fossil evidence suggests that coraciiform birds were the dominant radiation of arboreal perching birds in North America and Europe by the early Tertiary, about 60 million years before present. This evidence includes the remains of a roller-like bird, *Halcyornis*, from England and a motmot-like bird, *Protornis*, from Switzerland. Despite a currently wide distrib-

Hoopoe (*Upupa epops*) with an insect in its beak in South Africa. (Photo by John A. Snyder. Bruce Coleman Inc. Reproduced by permission.)

ution, no examples of kingfishers are yet recorded from this original radiation. More recently, from 25 to 40 million years ago, the remains of kingfisher-like birds are known from Wyoming, Germany, and France; of roller-like birds from France, Germany, and Wyoming; of hoopoe-like birds from France; and of tody-like birds from Wyoming and France. Even more recently, fossil kingfishers are also known from Australia. The Northern Hemisphere origin of motmots, and probably of todies, is supported even further by a 20-million-year-old motmot-like fossil from Florida. This species must have existed well before North and South America were finally connected (as recently as 2.5 million years ago), and is consistent with motmots and todies being the only avian families with their main diversity in Central America.

The Coraciiformes provide an ideal example to caution us against making assumptions about the history of the range and diversity of any group of birds based only on the present distribution and design of its members.

Resources

Books

Campbell, B., and E. Lack. *A Dictionary of Birds.* Calton, United Kingdom: T & AD Poyser, 1985.

del Hoyo, Josep, Andrew Elliott, and Jordi Sargatal, eds. "Order Coraciiformes." In *Handbook of Birds of the World.* Vol. 6, *Mousebirds to Hornbills.* Barcelona: Lynx Edicions, 2001.

Fry, C.H., and K. Fry. *Kingfishers, Bee-eaters & Rollers.* Princeton, New Jersey: Princeton University Press, 1992.

Kemp, A.C. *Birds Families of the World. Hornbills: Bucerotiformes.* Oxford: Oxford University Press, 1995.

Sibley, C.G., and J.E. Ahlquist. *Phylogeny and Classification of Birds: A Study in Molecular Evolution.* New Haven and London: Yale University Press, 1990.

Periodicals

Burton, P.K.J. "Anatomy and Evolution of the Feeding Apparatus in the Avian Order Coraciiformes and Piciformes." *Bulletin of the British Museum (Natural History)* 47, no. 6 (1984): 1–113.

Johansson, U.S., T.J. Parsons, M. Irestedt, and P.G.P. Ericson. "Clades within the 'Higher Land Birds,' Evaluated by Nuclear DNA Sequences." *Journal of Zoological Systematics and Evolutionary Research* 39 (2001): 37–51.

Organizations

Coraciiformes Taxon Advisory Group. Web site: <http://www.coraciiformestag.com>

Alan C. Kemp, PhD

Kingfishers

(Alcedinidae)

Class Aves
Order Coraciiformes
Suborder Alcidines
Family Alcedinidae

Thumbnail description
Small to medium-sized birds with a large head, long pointed bill, compact body, short neck, and small weak feet; plumage often black, white, or reddish, with areas of iridescent blue or green; bill and feet often black or bright red, orange, or yellow when adult; iris usually dark brown; flight fast and direct on rounded wings with short tail, but central tail feathers elongated in some species

Size
4–18 in (10–46 cm); 0.3–16.4 oz (9–465 g)

Number of genera, species
14 genera; 91 species

Habitat
Wide range of wooded or aquatic habitats, from arid savanna to dense rainforest or from sea coast to high mountain streams

Conservation status
Endangered: 1 species; Vulnerable: 11 species; Near Threatened: 12 species. Most populations face local threats to their habitat from logging of tropical forests, pollution of waterways, and development of oceanic islands

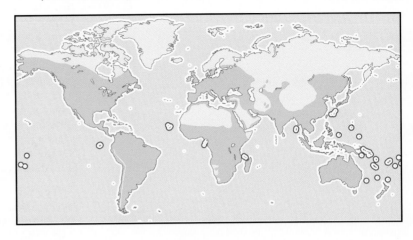

Distribution
Cosmopolitan; on all continents except Antarctica

Evolution and systematics

Kingfishers are a clearly defined group of birds, usually classified as the family Alcedinidae within the avian order Coraciiformes. They are most often placed in the suborder Alcidines, along with two small groups of birds from Central America, todies (family Todidae) and motmots (family Momotidae). Their other near relatives are bee-eaters (family Meropidae), while they appear to be more distantly related to rollers (family Coraciidae), hoopoes (families Upupidae and Phoeniculidae), hornbills (family Bucerotidae), and possibly even trogons (family Trogonidae).

The earliest fossil kingfishers are known from deposits that date to the relatively recent Lower Eocene, about 40 million years ago. These deposits in Wyoming are complemented by even more recent deposits from Germany and France, and also by material less than 25 million years old from Australia. This suggests that kingfishers have always been widely distributed across the world, including during the last two million years, when fossils similar to or identical with modern species have also been recorded from Australia, New Caledonia, Israel, Europe, North America, and Brazil. However, kingfishers form part of the radiation of coraciiform perching birds that was already well-established by about 60 mil-lion years ago, soon after the end of the Cretaceous era, so even more ancient fossils can be expected.

The 91 species of modern kingfisher can be divided easily into three groups. These are usually recognized as the subfamilies Halcyoninae, Alcedininae, and Cerylinae. However, each group is so distinct that each is sometimes elevated to the level of a family. Despite the clarity of these divisions, the relationships among the subfamilies and of the species within them remain unresolved. Traditional evidence, from morphology and behavior, suggests that the halcyonines are the least advanced kingfishers and that the cerylines and alcedinines are more advanced and more closely related to one another. More recent molecular studies, based on the technique of DNA-DNA hybridization, suggest that the alcedinines are least advanced and that the halcyonines and cerylines are most advanced and closely related. The main analysis by Hilary Fry in 1980 is based on certain assumptions. First, the inhabitants of stable habitats, such as rainforest, are more likely to be ancestral than species of recently habitable land, such as post-glacial Europe and North America or newly emerged islands. Second, species with unspecialized hunting techniques, such as simple sit-and-wait hunting from a perch to the ground, will precede

A ringed kingfisher (*Ceryle torquata*) in flight. (Photo by A. Papadatos/ VIREO. Reproduced by permission.)

more specialized modes of foraging, such as hawking insects or hovering over fish. Third, that singular species, such as the shovel-billed kookaburra (*Clytoceyx rex*), are more likely to be ancestral than larger groups of similar species, such as the collared (*Halcyon chloris*), sacred (*Halcyon sancta*), and chattering (*Halcyon tuta*) kingfishers.

Halcyoninae is the best match to these assumptions as the ancestral group. It has a number of species in primitive rainforest habitats, especially in Indonesia and New Guinea. It has many species that hunt using generalized techniques, yet it has a number of specialized and distinctive species. It includes the largest of all kingfishers, the kookaburras (genus *Dacelo*), of which the well-known laughing kookaburra (*Dacelo novaeguineae*) of Australia is the biggest. A number of other large species, all in Australasia or nearby Indonesia, show such affinities to kookaburras as similar call structure, raising the tail when calling, or blue color on the rump. These include the shovel-billed kookaburra, the striking white-rumped kingfisher (*Caridonax fulgidus*), and maybe even the smaller banded kingfisher (*Lacedo pulchella*). The unusual hook-billed kingfisher (*Melidora macrorrhina*), and the spiky-eared lilac-cheeked kingfisher (*Cittura cyanotis*) also show some affinities with kookaburras. However, they show other similarities to paradise kingfishers (genus *Tanysiptera*), named for their handsome blue-and-white plumage, red bill, and long racquet-tipped central tail feathers. The kookaburras are similarly linked, via the white-rumped *Halcyon fulgida* and moustached

Halcyon bougainvillei kingfishers, to the genus *Actenoides* (considered part of *Halycon* in Peters). The remaining species of this subfamily are not obviously specialized, other than having juveniles with faint barring on the breast. However, because some smaller groupings are evident among them and because there are so many species, they are usually separated into four genera for convenience. *Pelargopsis* (three species) have large stork-like bills, while *Syma* (two species, considered part of *Halycon* in Peters) have serrated edges to their yellow bills. The remainder are all extremely similar but can be divided into *Halcyon* species of Asia and Africa, and the various forms of blue-green and white *Todiramphus* (20 species, considered part of *Halycon* in Peters) of Asia and Australasia with their white collar, dark eye patch and, in some species, blue or reddish breast.

The subfamily Alcedininae offers a much simpler arrangement, with only two genera to contain these small, mainly piscivorous kingfishers of Africa and Asia, several of which have the second toe reduced or absent. They include the smallest of the dwarf or pygmy kingfishers of the genus *Ceyx* (11 species), several with reddish upperparts, all inhabitants of forest and woodland, feeding mainly on small insects, and most with a red, dorsally flattened bill when adult. The remainder are combined in the genus *Alcedo* (9 species), most of which feed predominately on fish, have blue upperparts and a blue breast band, and a long, black, laterally flattened bill. The common kingfisher (*Alcedo atthis*) of Eurasia is probably the best-known member of this genus. Some African species form a link between these two genera, with blue backs but red bills, which sometimes leads to their separation as a third genus *Corythornis* in other treatments.

The third subfamily, Cerylinae, is the only one with members in the Americas, including all members of the green-backed genus *Chloroceryle* (four species). All species feed mainly on fish and appear closely related to the smaller alcedinines. The cerylines include the largest piscivorous species in the genus *Megaceryle* (four species, considered part of *Ceryle* in Peters), each with pied-and-reddish plumage and inhabiting one of the continents of the Americas, Africa, or Eurasia. Finally, the pied kingfisher (*Ceryle rudis*) of Africa and mainland Asia, also with sexually dimorphic bands across the breast, is sometimes placed in its own genus *Ceryle* but only because of its smaller size and specialized hover-hunting behavior.

Physical characteristics

Kingfishers are a uniform and distinctive group of birds, all immediately recognizable as members of the Alcedinidae. They are small to medium-sized birds with a large head, long pointed bill, compact body, short neck, small legs, and weak feet. Kingfishers' feet have three front toes that are fused at their bases. Most have a fast direct flight on rounded wings and a short tail. The greatest differences are in overall form; the shape of the bill, from narrow and dagger-like to broad and shovel-like; or the development of long central tail feathers. Species range in mass from the 0.3–0.4 oz (9–11 g) African dwarf kingfisher (*Myioceyx lecontei*) to the 6.7–16 oz (190–465 g) laughing kookaburra (*Dacelo novaeguineae*). The sexes of

A pink-breasted paradise-kingfisher (*Tanysiptera nympha*) in flight. (Photo by C.H. Greenewalt/VIREO. Reproduced by permission.)

most species are similar in size. In a few species, one or the other sex is slightly larger, but only in the two largest species of kookaburra are the females markedly larger, while in some paradise kingfishers the males have longer streamers to the central tail feathers.

The sexes of most species are also similar in plumage, bill, and foot color, while juveniles are generally similar to adults except that the bill is often a dull black. Most species have at least some iridescent blue or green in the plumage, offset by large areas of black, white, or brown. The bill and feet are black or brown in many species, but in others one or other appendage may be bright yellow, orange, or red. The iris is dark in most species, with only three exceptions. In many of the cerylines and halcyonines, the sexes are distinguishable through differences in the color of the breast bands or back, but only in two species of the alcedinines is there obvious sexual dichromatism.

The bill shape is generally suggestive of feeding habits, being laterally flattened and dagger-like in species that regularly dive into water after slippery aquatic prey, but dorsoventrally flattened and more scoop-like in species that catch small animals on the ground, and especially wide in those forest species that dig in soil or leaf litter for their prey. One species has a hook and another has serrations at the tip of the bill, but both are of unknown function.

The eyes of kingfishers are also specialized for sighting prey. Ganglion cells that connect the light-sensitive cone cells on the retina are especially dense across a horizontal streak, at each end of which is a depression or fovea packed with cone cells and, by its shape, especially sensitive to movement across its surface. The outer or temporal fovea includes the area of binocular vision, while the inner or nasal fovea covers monocular vision and is also especially densely packed with ganglion cells. The angles of the streak and the well-connected nasal fovea coincide with what would be predicted for birds that search below them for prey and are especially sensitive to movement in their peripheral monocular vision. The birds' ability to turn the head through a wide angle allows fixation of the object with the binocular vision of the temporal fovea. The cone cells of kingfishers are also especially rich in the

droplets of red oil that signal excellent color vision. One species has already tested positive for vision near the ultraviolet range.

Species that dive into water in pursuit of prey also have to cope with the problem of the different refractive indices of air and water, and the effect that this has on the apparent location of an object due to bending of light rays at the surface. Tests have shown that pied kingfishers are able to compensate for this, mainly by increasing the dive angle and speed as the depth of prey below the water increases. This species, the most specialized piscivore of all kingfishers, also has a bony plate on the prefrontal area that slides across and screens the eyes as the head strikes the water.

Distribution

Kingfishers are cosmopolitan as a family, occurring on all ice-free continents, but with an uneven distribution of species. Only one subfamily, Cerylinae, occurs in the New World, with a few species in continental North America and a few more tropical species in Central and South America. The remaining species of the subfamily are virtually restricted to sub-Saharan Africa and the Asian mainland. The other two subfamilies, Halcyoninae and Alcedininae, occur across Africa, Asia, and Australasia, with a few species that extend north into the Paleoarctic regions of Europe, the Middle East, and continental Asia. Only members of the halcyonine genus *Todiramphus* extend east of Australia into the oceanic islands of the Pacific.

A rufous-backed kingfisher (*Ceyx rufidorsum*) with its insect prey. (Photo by Doug Wechsler/VIREO. Reproduced by permission.)

A European kingfisher (*Alcedo atthis*) inside its nest with newly hatched chicks in northern France. (Photo by J.C. Carton. Bruce Coleman Inc. Reproduced by permission.)

Most species of kingfisher are found in the Australasian region of Australia, New Guinea, and Indonesia east of Bali and Sulawesi, some of these on the oceanic islands of the Pacific. Many species are found in the adjacent areas of western Indonesia and the Sunda or Malesian region of the Malay peninsula, the islands of Borneo, Java, and Sumatra, and also the Philippines. Fewer species are found on the Asian mainland and in India and the Middle East, with only a few more in the Afrotropical region of sub-Saharan Africa and Madagascar. Only one, the common kingfisher, extends north into Europe. Species have been recorded for more than one region where appropriate, depending on the extent of their known breeding and non-breeding distributions.

Habitat

Habitats that provide both food and nest sites are essential to all kingfishers. Most kingfishers have the ability to excavate their own nests in soft earth, wood, or termite nests, besides the use of natural cavities, yet nest sites often remain the most limiting resource. Species that feed mainly on aquatic animals extend from arid seashores to small mountain streams, provided that there are earth banks or termite nests into which most species will excavate their nest tunnels. Species that feed on terrestrial prey occur from arid savanna, provided that there are banks or natural tree holes in which to nest, to dense rainforest, with its greater abundance of nest sites. A subjective analysis of the main habitat requirements suggests that 31 species are primarily aquatic, whether they occupy forest or not; 44 species feed mainly in closed-canopy forests; and 17 species are most abundant in wooded savanna. Only aquatic species occur in the New World, while forest-dependant species predominate in Asia and Australasia and savanna species in Africa.

Behavior

Most kingfishers are territorial as breeding pairs, but a few species, such as the laughing kookaburra and pied kingfisher, live as cooperative groups that consist of a pair with several non-breeding helpers. Most species are sedentary, but a few, such as the belted kingfisher (*Megaceryle alcyon*) of North America and the gray-headed kingfisher (*Halcyon leucocephala*) of Africa, perform regular nocturnal migrations between breeding and non-breeding areas. All but one species are diurnal, although the hook-billed kingfisher is probably joined by the shovel-billed kookaburra in the nights of the New Guinea forests. Many species bathe by diving repeatedly into water, especially after becoming soiled in the smelly nest cavity. Most species roost alone on a perch within vegetation, rarely in an old nest cavity. Sometimes juveniles roost with adults, and a few species, especially the pied kingfisher, gather at communal roosts for part of the year. All species are highly vocal, with a variety of distinct calls that assist in their location and identification. Loud calls are used to advertise territories, while communication between mates or with offspring is often quieter.

Feeding ecology and diet

Kingfishers eat a wide range of small animals and are capable of taking prey from the ground, water, air, or foliage. Most species spend much of their time perched on the lookout for prey, and only a few expend energy to hover or hawk after prey. Despite their name, none of the kingfishers feed exclusively on fish, and ignore aquatic animals for their diet. Most are adaptable and consume a range of relatively large invertebrates, especially grasshoppers in savanna, earthworms in forest, and crustacea in water; as well as small vertebrates, especially reptiles, fish, and amphibia. Only three species have been reported eating fruit: two eating fruit during winter at

A kookaburra beats its prey against a tree. (Illustration by Brian Cressman)

Kingfisher courting display. (Illustration by Brian Cressman)

high northern latitudes, and the other eating the nutritious fruits of oil palms in the African tropics. Where several species occur together, each has a preferred habitat, such as open or closed forest; each prefers a particular size of prey; and each employs a predominant foraging technique, such as hovering, digging, or exploiting the forest canopy. A species may also alter its feeding patterns in different areas of its range, depending on the other species with which it overlaps. After capture, prey is usually carried back to and beaten against a perch with the bill until it is soft enough to swallow. A few species follow otters, platypus, cormorants, egrets, cattle, or army ants for any prey they might disturb. Others attend grass fires for the insects they disturb.

Reproductive biology

Males of most species call frequently to advertise and defend their territory. Aerial pursuit, exposure of plumage patterns in special joint displays, and courtship feeding of the female by the male are all reported prior to copulation and nesting. Food is always held head-out in the bill during breeding to allow its passage to the female or chicks. Both sexes take some part in nest excavation and cavity choice, usually in an earth bank, less often in rotten wood or in a terrestrial or arboreal termite's nest, and occasionally in a natural tree

hole. Excavation is started by flying bill-first into the surface, continued later by pecking and scraping out debris with the bill or feet. The entrance tunnel, 3–26 ft (1–8 m) long depending on site and species, usually leads into a larger nest cavity, but no special lining is added. In most species, each pair nests alone, but a few species breed cooperatively, whether they are attended by helpers or nest together in a colony. Kingfisher eggs are white, round, and shiny. An egg is normally laid daily, but the size of an average clutch, ranging from two to seven, depends on the species. Both sexes usually take part in incubation and care of the young, although the female usually remains at the nest overnight. Incubation takes two to four weeks, and the nestling period three to eight weeks, related to the size of the species. Chicks hatch naked and blind, with the upper mandible of the bill notably shorter than the lower. Later, when the feathers emerge, they are retained in their quills initially, giving the chicks a prickly porcupine-like appearance. There is no nest sanitation, other than that chicks may loosen soil from the chamber walls to partly cover their droppings. Nests, especially those in earthen tunnels, often become smelly and full of maggots as feces and food remains accumulate. The chicks continue to be fed by the parents after fledging. They become independent within a few days or weeks and are sexually mature within a year.

Belted kingfisher (*Megaceryle alcyon*) chick swallowing fish in nest. (Photo by Anthony Mercieca. Photo Researchers, Inc. Reproduced by permission.)

Conservation status

The main threats to kingfishers are the clearing of their rainforest habitats and the draining or pollution of their aquatic habitats. These problems are exacerbated for species with a small total range or population, such as the Endangered Marquesas kingfisher (*Halcyon godeffroyi*), which lives on an island. A different threat comes from the lack of biological information about many species, so that it is difficult to plan for their conservation. Twelve species are considered threatened in some way and at least two subspecies have become extinct within historic times. All occur in Southeast Asia and Oceania; all but one inhabit forest; all but one is restricted to islands; and most occur in the Philippine center of endemic species.

Significance to humans

Kingfishers were featured in Greek mythology and on Egyptian friezes. Skulls of yellow-billed kingfishers (*Syma torotoro*) were worn as hair decorations in New Guinea, while the calls or sightings of some species were observed as omens, good or bad, by people of New Guinea and Borneo. Victorians added kingfishers to their collections of stuffed birds, drawn by the royal blue of the common kingfisher that gives the group its name. Kingfishers form part of legends among Arawak and Arikana tribes of Guyana and the Missouri River, respectively. Early in the twentieth century, the laughing kookaburra became an important symbol of Australia. Many other examples of human-kingfisher interaction probably exist. Currently, several species are persecuted for eating fish stocks bred for angling or farming.

1. Striped kingfisher (*Halcyon chelicuti*); 2. Common kingfisher (*Alcedo atthis*); 3. Pied kingfisher (*Ceryle rudis*); 4. Collared kingfisher (*Todiram-phus chloris*); 5. African pygmy-kingfisher (*Ceyx pictus*); 6. Belted kingfisher (*Megaceryle alcyon*); 7. Yellow-billed kingfisher (*Syma torotoro*); 8. Amazon kingfisher (*Chloroceryle amazona*). (Illustration by Brian Cressman)

1. Laughing kookaburra (*Dacelo novaeguineae*); 2. Common paradise kingfisher (*Tanysiptera galatea*); 3. Lilac-cheeked kingfisher (*Cittura cyanotis*); 4. Rufous-collared kingfisher (*Actenoides concretus*); 5. Hook-billed kingfisher (*Melidora macrorrhina*); 6. White-rumped kingfisher (*Caridonax fulgidus*); 7. Banded kingfisher (*Lacedo pulchella*); 8. Stork-billed kingfisher (*Pelargopsis capensis*); 9. Shovel-billed kookaburra (*Clytoceyx rex*). (Illustration by Brian Cressman)

Species accounts

Laughing kookaburra
Dacelo novaeguineae

SUBFAMILY
Halcyoninae

TAXONOMY
Alcedo novaeguineae Hermann, 1783, New South Wales. Two subspecies.

OTHER COMMON NAMES
English: Kookaburra, jackass, brown, giant, or laughing kingfisher; French: Martin chasseur géant; German: Jägerliest; Spanish: Cucaburra Comin.

PHYSICAL CHARACTERISTICS
15–17 in (39–42 cm), 7–16 oz (190–465 g). Largest of the kingfishers, dark brown and white with blue rump and barred reddish tail. Has a dark mask through the eye. The bill is black above and horn (or horn-colored) below.

DISTRIBUTION
Eastern Australia, introduced to southwest since 1897, also to Tasmania in 1905 and New Zealand since 1866.

HABITAT
Eucalyptus forest and woodlands, extending into parks and gardens.

BEHAVIOR
Group starts the day with a loud cackling laughing chorus, led by the pair and accompanied by their mature offspring. They spend long periods perched motionless and on the lookout for prey. Are generally sedentary and inactive. Group members roost together in dense foliage.

FEEDING ECOLOGY AND DIET
Swoops down from low perch to pick up small animals as food, mainly arthropods, such as grasshoppers, beetles, and spiders. Also small vertebrates, such as snakes, lizards, mice, and small birds. Members of group usually feed separately.

REPRODUCTIVE BIOLOGY
Monogamous pair breeds cooperatively with help of previous offspring. Nests are usually in natural cavity, less often excavated in termite nests or soft dead wood. Lays one to five eggs, usually two or three. Incubation period is 24–29 days, mainly by female but assisted by the group. Nestling period is 32–40 days; chicks are fed by whole group.

CONSERVATION STATUS
Not threatened. Widespread and common, the species even benefits from most human developments of bush clearance and gardens. The only kingfisher to have its range extended by human introductions.

SIGNIFICANCE TO HUMANS
Well-known emblem of Australia and its bird-life. ◆

Shovel-billed kookaburra
Clytoceyx rex

SUBFAMILY
Halcyoninae

TAXONOMY
Clytoceyx rex Sharpe, 1880, East Cape, New Guinea. Two subspecies.

OTHER COMMON NAMES
English: Shovel-billed kingfisher, emperor or crab-eating kingfisher; French: Martin-chasseur bec-en-cuillièr; German: Froschschnabel; Spanish: Martin Cazador Picopala.

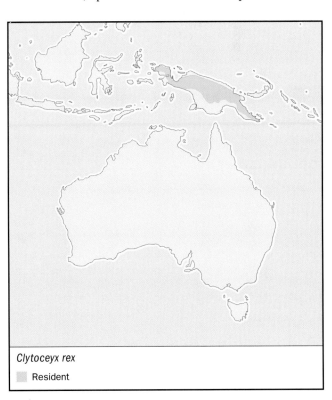

Clytoceyx rex

■ Resident

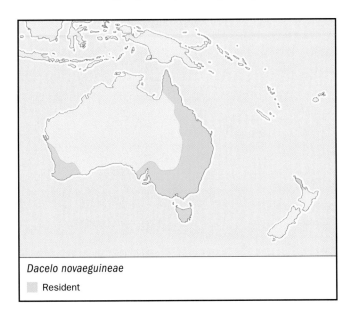

Dacelo novaeguineae

■ Resident

PHYSICAL CHARACTERISTICS
12–13 in (30–34 cm), 8.6–11.5 oz (245–325 g). Large king-fisher with dark brown above with blue rump, reddish below; tail is blue (male) or reddish (female). Unique broad, deep stubby bill, with dark brown above, pale horn below.

DISTRIBUTION
New Guinea.

HABITAT
Rainforest in lowlands, but especially on foothills up to 8,000 ft (2,400 m) above sea level.

BEHAVIOR
Calls at dawn from tree top, three to four long liquid notes each accompanied by tail flicking. Often perches low above forest floor, on the lookout for prey. Bill and breast often are caked with mud.

FEEDING ECOLOGY AND DIET
Feeds on forest floor, picking up prey or ploughing through soil to a depth of 3 in (8 cm), often at the base of tree or bush. Pulls out earthworms, insects, and small reptiles. Crab eating is unconfirmed. Forages at night, maybe predominately.

REPRODUCTIVE BIOLOGY
Almost unknown. Adults are in breeding condition in January. A chick on sale at a market was said to be one of two taken from a tree hole.

CONSERVATION STATUS
Not threatened, but poorly known. Does use forest edge and large gardens.

SIGNIFICANCE TO HUMANS
None known. Some nest-robbing for markets; are attractive to bird-watching tourists. ◆

Melidora macrorrhina
▨ Resident

Hook-billed kingfisher
Melidora macrorrhina

SUBFAMILY
Halcyoninae

TAXONOMY
Dacelo macrorrhinus Lesson, 1827, Manokwari, New Guinea. Three subspecies.

OTHER COMMON NAMES
French: Martin-chasseur d'Euphrosine; German: Hakenliest; Spanish: Martin Cazador Ganchudo.

PHYSICAL CHARACTERISTICS
11 in (27 cm), 3.1–3.9 oz (90–110 g). Large kingfisher, brown above and white (male) or buff (female) below. Feathers of crown are black with blue (male) or green (female) edges. Long bill has hooked tip with dark brown above and pale below.

DISTRIBUTION
New Guinea and some adjacent small islands.

HABITAT
Lowland rainforest, both primary and secondary, and also agri-cultural plantations.

BEHAVIOR
Calls at dusk, dawn, and throughout moonlit nights; one to three whistles followed by one to four short, high-pitched notes. Bill often is caked with mud.

FEEDING ECOLOGY AND DIET
Feeds on large insects and frogs, probably by digging in soil. Main activity is at twilight and during the night.

REPRODUCTIVE BIOLOGY
5 in (12 cm) wide nest chamber dug into active nests of arboreal termites. Lays two to three eggs; male incubates eggs and broods chicks by day. Collect food for chicks by day and night.

CONSERVATION STATUS
Not threatened, but little known due to its nocturnal habits.

SIGNIFICANCE TO HUMANS
None known. Attractive to bird-watching tourists. ◆

Common paradise kingfisher
Tanysiptera galatea

SUBFAMILY
Halcyoninae

TAXONOMY
Tanysiptera galatea G. R. Gray, 1859, Manokwari, New Guinea. At least 15 subspecies, some elevated to species.

OTHER COMMON NAMES
English: Galatea racquet-tail; French: Martin-chasseur à longs brins; German: Spatelliest; Spanish: Alción Colilargo Común.

Tanysiptera galatea
□ Resident

PHYSICAL CHARACTERISTICS
13–17 in (33–43 cm), 1.9–2.4 oz (55–69 g). Spectacular medium-sized kingfisher, dark blue above, white below with shining blue crown, red bill and long blue central tail feathers with white racquets at the tip.

DISTRIBUTION
New Guinea, west to main islands of Halmahera and Buru in Indonesia and several smaller islands in between.

HABITAT
Lowland rainforest, even small patches, up to 980 ft (300 m) above sea level, but also more open areas of monsoon and riparian forest, even extending into secondary forest.

BEHAVIOR
Calls with one to four long whistles, ending with a loud trill. Very sedentary and spends much time perched low down, deep within favorite small area of forest.

FEEDING ECOLOGY AND DIET
Flies down to forest floor to catch prey, less often to snatch insects off foliage. Eats wide range of small animals, mainly earthworms, but also snails, centipedes, beetles, grasshoppers, caterpillars, and lizards.

REPRODUCTIVE BIOLOGY
Both sexes of a monogamous pair excavate nest chamber in active nest of arboreal termite *Microtermes biroi*, an essential component of their habitat. Lays up to five eggs and both sexes care for eggs and chicks. Good breeding success.

CONSERVATION STATUS
Common and not threatened. Some threat from forest clearance, especially to small populations of subspecies on isolated islands.

SIGNIFICANCE TO HUMANS
None known. ◆

Rufous-collared kingfisher
Actenoides concretus

SUBFAMILY
Halcyoninae

TAXONOMY
Dacelo concreta Temminck, 1825, Sumatra. Three subspecies.

OTHER COMMON NAMES
English: Chestnut-collared kingfisher; French: Martin-chasseur trapu; German: Malaienliest; Spanish: Alción Malayo.

PHYSICAL CHARACTERISTICS
9–9.5 in (23–24 cm), 2.1–3.2 oz (60–90 g). Dumpy, medium-sized kingfisher, with green crown, blue (male), or buff-spotted green (female) back, rufous below and on collar. Bill black above and yellow below.

DISTRIBUTION
Sunda region of Malay Peninsula, Borneo, and Sumatra.

HABITAT
Dense lowland rainforest, even secondary forest in which canopy regenerated, up to 5,600 ft (1,700 m) above sea level.

BEHAVIOR
Calls with loud, long whistle that rises in tone. Perches mainly in middle and lower levels of dense forest, often with slow head bobbing and tail pumping.

FEEDING ECOLOGY AND DIET
Drops from low perch to snatch prey at water surface or pick up from the ground, sometimes turning over leaves in its search. Feeds on various arthropods, including insects and large scorpions, also snails, fish, small snakes and lizards.

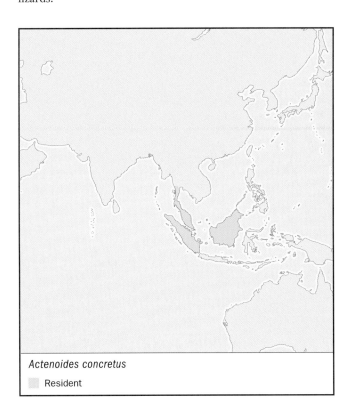

Actenoides concretus
□ Resident

REPRODUCTIVE BIOLOGY
Monogamous pair excavates nest burrow in earth bank, rarely in rotten tree trunk, ending in 8 in (20 cm) diameter nest chamber. Lay two eggs that are incubated for about 22 days.

CONSERVATION STATUS
Considered Near Threatened due to extensive removal of lowland forest, but survives in hill forest and in conserved tracts.

SIGNIFICANCE TO HUMANS
None known, though most widespread species in genus of six species spread across Southeast Asia. ◆

Lilac-cheeked kingfisher
Cittura cyanotis

SUBFAMILY
Halcyoninae

TAXONOMY
Dacelo cyanotis Temminck, 1824, Sulawesi (Sumatra in error). Three subspecies.

OTHER COMMON NAMES
English: Blue-eared, lilac kingfisher; French: Martin-chasseur oreillard; German: Blauohrliest; Spanish: Martin Cazador de Célebes.

PHYSICAL CHARACTERISTICS
11 in (28 cm), no mass data. Large kingfisher with unique, spiky, lilac ear coverts, brown above, lilac and white below with black (male) or blue (female) shoulder to dark wings. Black mask, bright red bill and feet. Red iris distinctive.

Cittura cyanotis

▨ Resident

DISTRIBUTION
Sulawesi in Indonesia, and adjacent larger islands.

HABITAT
Tall primary or secondary rainforest, up to 3,300 ft (1,000 m) above sea level.

BEHAVIOR
Known to roost on bare branch of low tree. Calls with rapid descending series of four notes, repeated every few minutes. Perches for long period in dark forest at middle to lower levels, watching for prey or for a mate to which it may display with raised bill and fanned tail.

FEEDING ECOLOGY AND DIET
Swoops down to capture prey on ground, mainly large insects, such as mantids, cicadas, grasshoppers and beetles, together with millipedes and small reptiles. Sometimes also hunts along the edge of clearings.

REPRODUCTIVE BIOLOGY
Nest reported from burrow in sloping ground, no further data.

CONSERVATION STATUS
Considered Near Threatened due to restricted range, patchy distribution and steady removal of its forest habitat. Nowhere common and biology not well understood.

SIGNIFICANCE TO HUMANS
None known. ◆

Banded kingfisher
Lacedo pulchella

SUBFAMILY
Halcyoninae

TAXONOMY
Dacelo pulchella Horsfield, 1821, Java. Three subspecies.

OTHER COMMON NAMES
French: Martin-chasseur mignon; German: Wellenliest; Spanish: Martin Cazador Chico.

PHYSICAL CHARACTERISTICS
8 in (20 cm), no mass data. Medium-sized kingfisher with back and tail banded in black. Sexes differently colored, male with crown and back metallic blue, forehead and mask rufous or black and pale buff below, female reddish brown above and white below with fine black banding extending over head and across breast. Both sexes with red bill, one of few kingfishers with pale, yellow-brown iris.

DISTRIBUTION
Myanmar, Thailand, and Sunda region of Borneo, Sumatra, and Java.

HABITAT
Lowland evergreen and bamboo forest, extending to 5,600 ft (1,700 m) above sea level on hill forests of Borneo.

BEHAVIOR
Perches motionless for long periods, except to slowly raise and lower long crown feathers, at various heights in the forest.

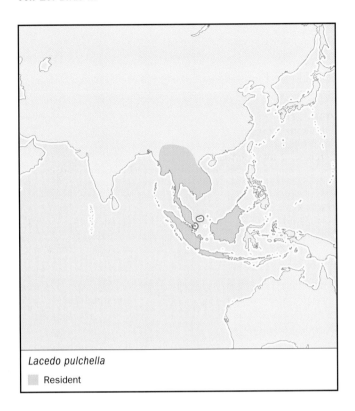

Lacedo pulchella
Resident

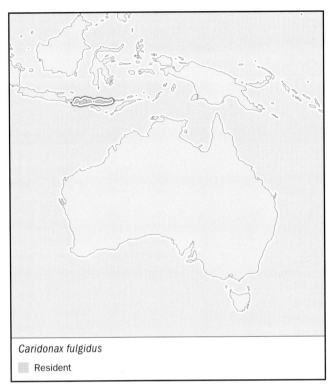

Caridonax fulgidus
Resident

Members of pair territorial, often perched in close proximity to one another. Call with single long whistle, followed by long series of short tri- and di-syllabic whistles.

FEEDING ECOLOGY AND DIET
Forages in diverse ways, swooping to the ground, onto fallen trees or into water, probing in loose soil, or snatching off foliage. Diet also diverse, a wide variety of arthropods and small vertebrates.

REPRODUCTIVE BIOLOGY
Excavate nest cavity in rotten wood, earth bank or arboreal termite nest. Lays two to five eggs. No further details recorded.

CONSERVATION STATUS
Not threatened. Widespread and common in many areas of unlogged forest, including large conservation reserves. Easily overlooked unless calling.

SIGNIFICANCE TO HUMANS
Regarded as an omen bird by Iban people of Borneo. ◆

White-rumped kingfisher
Caridonax fulgidus

SUBFAMILY
Halcyoninae

TAXONOMY
Halcyon fuligidus Gould, 1857, Lombok. Two subspecies. Looks superficially like a *Halcyon* species, but shows more similarity to paradise kingfishers, kookaburras, and *Actenoides* species.

OTHER COMMON NAMES
English: Blue-and-white kingfisher, glittering kingfisher; French: Martin-chasseur étincelant; German: Glitzerliest; Spanish: Alción Culiblanco.

PHYSICAL CHARACTERISTICS
12 in (30 cm), no mass data. Large kingfisher, blue-black above, white below and on rump, with bright red bill, feet and eye ring. Deep orange iris distinctive.

DISTRIBUTION
Main islands of Lesser Sundas archipelago in Indonesia, Lombok, Sumbawa, Flores, and Besar.

HABITAT
Primary and secondary forests, including wooded areas with cultivation.

BEHAVIOR
Calls with long rapid series of yapping notes, one per second, sometimes for over half a minute. Cocks the tail when calling, like a kookaburra.

FEEDING ECOLOGY AND DIET
Known to eat insects and their larvae but no description of how they are captured.

REPRODUCTIVE BIOLOGY
Each pair is territorial and excavates a nest tunnel in an earth bank. One nest had two eggs, another a single chick.

CONSERVATION STATUS
Not threatened. Still considered widespread and fairly common, despite restricted range and alteration of favored habitats in primary evergreen and deciduous forests. Biology poorly known.

SIGNIFICANCE TO HUMANS
None known, though attractive to bird-watching tourists with handsome colors and unique taxonomic status. ◆

Stork-billed kingfisher
Pelargopsis capensis

SUBFAMILY
Halcyoninae

TAXONOMY
Alcedo capensis Linnaeus, 1766, Chandernagor, West Bengal. At least 16 subspecies.

OTHER COMMON NAMES
English: Brown-headed stork-billed kingfisher; French: Marin-chasseur gurial; German: Storchsnabelliest; Spanish: Alción Picocigüeña.

PHYSICAL CHARACTERISTICS
14 in (35 cm), 5.0–7.9 oz (143–225 g). Large kingfisher with red bill and feet, varies from dark blue to pale turquoise above, rufous to white below, on head and on neck.

DISTRIBUTION
India east to China, Vietnam, Philippines, Borneo, and Lesser Sundas archipelago, including many intervening islands such as Java, Sumatra, Andamans, and Sri Lanka.

HABITAT
Stream and river banks up to 3,900 ft (1,200 m) above sea level, in forest, open woodland, and mangroves on the seashore, even among plantations and paddy fields.

BEHAVIOR
Perches on snags close to or over water. Main call a series of harsh cackling notes, uttered at perch or in flight. Spends long periods watching for prey, sometimes slowly bobbing head or wagging tail.

FEEDING ECOLOGY AND DIET
Dives into water or to the ground to capture food, then returns to perch to soften prey. Diet mainly aquatic, such as fish, crabs, and prawns, but also frogs, lizards, mice, and insects. Known to catch prey flushed by otters.

REPRODUCTIVE BIOLOGY
Monogamous breeding pairs are aggressively territorial. Excavate nest cavity in river bank, flat ground, rotten wood or arboreal termitaria, rarely use natural tree hole. Nest tunnel about 4 in (10 cm) wide, 3 ft (1 m) long, ending in 9–12 in (23–30 cm) diameter chamber in which two to five eggs are laid. Details of nesting cycle unrecorded.

CONSERVATION STATUS
Not threatened. Widespread, locally common, and occupies wide range of habitats. Human disturbance may cause local disappearance.

SIGNIFICANCE TO HUMANS
None known. ◆

Striped kingfisher
Halcyon chelicuti

SUBFAMILY
Halcyoninae

TAXONOMY
Alaudo chelicuti Stanley, 1814, Chelicut, Ethiopia. Two subspecies.

OTHER COMMON NAMES
French: Martin-chasseur strié; German: Streifenliest; Spanish: Alción Estriado.

PHYSICAL CHARACTERISTICS
6 in (17 cm), 1.1–1.8 oz (30–50 g). Small dull kingfisher, brown on crown and back, white on face, collar, and breast,

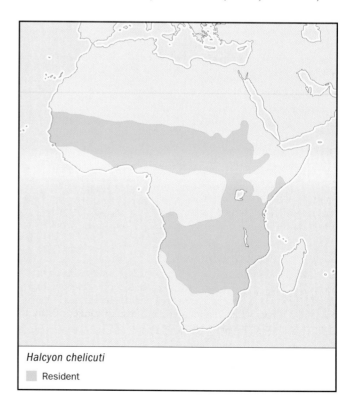

Pelargopsis capensis
 Resident

Halcyon chelicuti
 Resident

dark brown streaks on crown and flanks, blue-green back, rump, and flight feathers. Black mask, bill black above and red below, feet red.

DISTRIBUTION
Wooded savannas of sub-Saharan Africa.

HABITAT
Wide tolerance, from clearings in forest to riverine trees within scrublands. Extends into most arid habitats, along with red-backed kingfisher (*Todiramphus pyrrhopygius*) of Australia, which also has a striped crown.

BEHAVIOR
Perches 6.5–13 ft (2–4 m) up, usually on dry twigs on lookout for food. Often perches higher when territorial calling, a high disyllabic trill repeated up to 10 times, often a pair in duet. Calls accompanied by alternate flashing of blue and white patterns on upper and underside of wings, while swiveling back and forth on perch with tail cocked.

FEEDING ECOLOGY AND DIET
Feeds mainly on insects, especially grasshoppers, beetles and larvae, but will eat wide range of small invertebrates and a few vertebrates. Takes most prey on the ground, sometimes in the air, with 80% capture success.

REPRODUCTIVE BIOLOGY
Monogamous pair nests in old hole excavated by barbet or woodpecker, less often in natural cavity or old swallow nest. Often assisted by a second male. Lays two to six eggs, duration of nesting cycle unrecorded, sometimes double-brooded.

CONSERVATION STATUS
Not threatened. Common across wide range of extensive habitats, including in many large reserves and in areas of shifting cultivation.

SIGNIFICANCE TO HUMANS
None known. ◆

Collared kingfisher
Todiramphus chloris

SUBFAMILY
Halcyoninae

TAXONOMY
Alcedo chloris Boddaert, 1783, Buru. Exact taxonomy incomplete, at least 49 subspecies described.

OTHER COMMON NAMES
English: Black-masked/white-collared kingfisher; French: Martin-chasseur à collier blanc; German: Halsbandliest; Spanish: Alción Acollarado.

PHYSICAL CHARACTERISTICS
9–10 in (23–25 cm), 1.8–3.5 oz (51–100 g). Small kingfisher, generally blue-green above, white or buff below and on collar. Mask, feet and bill black, latter with yellow base. Much local variation across wide, fragmented, and insular range, mainly in intensity of colors and of white or rufous on head.

DISTRIBUTION
Red Sea east to China, Philippines, Indonesia, New Guinea, Australia, and many adjacent oceanic islands, reaching Andamans, Marianas, Fiji, and Tonga, among others.

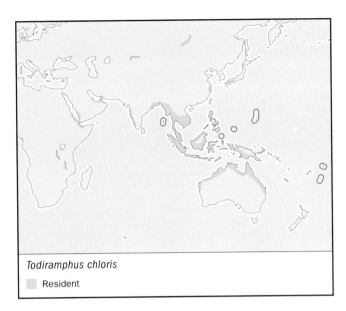

Todiramphus chloris
☐ Resident

HABITAT
Mainly coastal in mangroves and estuaries, but extends inland along major rivers and into adjacent forests and croplands. Widest habitat tolerance on some islands, up to 4,900 ft (1,500 m) above sea level on Java and Sumatra.

BEHAVIOR
Calls with two to five shrill notes from perch or in flight. Perches for long periods in the open, usually below 9 ft (3 m). Nocturnal migrant in some areas.

FEEDING ECOLOGY AND DIET
Dives from perch to ground, mud, or into water after prey. Eats wide range of small animals, along coasts mainly fish and crustacea, inland includes more insects in diet. Known to follow otters for any prey they might disturb.

REPRODUCTIVE BIOLOGY
Lays two to five eggs in cavity excavated by both members of monogamous pair in dead wood, earth bank, arboreal termitarium or even fern roots, less often natural tree hole or old woodpecker nest. Incubation about 14 days, nestling period 29–30 days.

CONSERVATION STATUS
Not threatened. Wide distribution, common or even abundant in many areas, range extends into areas of cultivation. Suffers from local habitat destruction in some areas, especially on small islands with distinctive subspecies.

SIGNIFICANCE TO HUMANS
None known. ◆

Yellow-billed kingfisher
Syma torotoro

SUBFAMILY
Halcyonine

TAXONOMY
Syma torotoro Lesson, 1827, Manokwari, New Guinea. Three subspecies.

Syma torotoro
■ Resident

OTHER COMMON NAMES
English: Lesser/lowland yellow-billed kingfisher, saw-billed kingfisher; French: Martin-chasseur torotoro; German: Gelb-schnabelliest; Spanish: Alción Torotoro.

PHYSICAL CHARACTERISTICS
8 in (20 cm), 1.1–1.8 oz (30–52 g). Small rufous kingfisher, with green back and tail and blue rump. Black patch on nape, sometimes on crown. Only kingfishers with yellow bill and feet, and with serrated tip to upper mandible of bill.

DISTRIBUTION
New Guinea, northern Australia, and adjacent islands.

HABITAT
Primary and secondary forest, and wooded areas of cultivation.

BEHAVIOR
Usually perches below 26 ft (8 m), but at any height including forest canopy. Calls with either short abrupt or longer fading loud trill. Sways from side to side while perched. May raise crown in threat, to display black eye-like spots on nape.

FEEDING ECOLOGY AND DIET
Captures most prey from ground, some off foliage or in air, rarely from water's edge or under leaf. Diet mainly insects, also earthworms and few small lizards, geckos, and snakes. May follow columns of ants for any insects they disturb.

REPRODUCTIVE BIOLOGY
Both members of monogamous, territorial pair excavate nest chamber in arboreal termite nest or soft dead wood. Lay one to four eggs, incubated and later brooding by both sexes.

CONSERVATION STATUS
Not threatened. Widespread, common and at densities of pair per 2.5–5 acres (1–2 ha) in good forest habitat.

SIGNIFICANCE TO HUMANS
Skulls valued as ornaments for hair by people of Middle Sepik River in New Guinea. ◆

African pygmy-kingfisher
Ceyx pictus

SUBFAMILY
Alcedininae

TAXONOMY
Todus pictus Boddaert, 1783, St. Louis, Senegal. Three subspecies.

OTHER COMMON NAMES
English: Pygmy/miniature kingfisher; French: Martin-pêcheur pygmée; German: Natalzwergfischer; Spanish: Martin Pigmeo Africano.

PHYSICAL CHARACTERISTICS
5 in (12 cm), 0.3–0.6 oz (9–16 g). Very small rufous kingfisher, with blue on crown, wings and rump, white on throat and ear spot, and lilac sides to face. Bill red, black in juvenile.

DISTRIBUTION
Resident in tropical Africa, expanding to breed in subtropics on either side during respective summers. Most widespread small African kingfisher, genus of six other species also widespread across southeastern Asia and New Guinea.

HABITAT
Resident in dense forest and woodland, migrating to more open woodland and borders of cultivated areas.

BEHAVIOR
Usually perches within 3 ft (1 m) of ground, searching for prey, often bobbing head or pumping tail. Utters soft high-

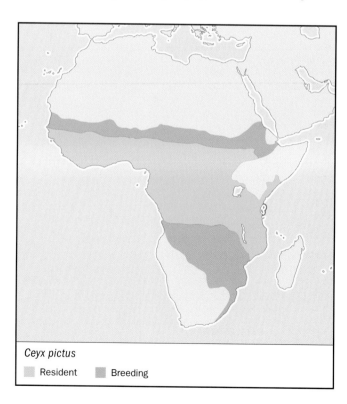

Ceyx pictus
■ Resident   ■ Breeding

pitched song, more often calls single, high, insect-like squeak during rapid flight. Migrates at night.

FEEDING ECOLOGY AND DIET
Pounces on prey mainly on ground, also in air or from water. Eats mainly insects, also spiders, myriapods, and even small frogs and lizards.

REPRODUCTIVE BIOLOGY
Monogamous pair excavates nest tunnel about 16 in (40 cm) long in earth bank, termitarium or aardvark burrow, sometimes in small colonies but usually spaced within territories. Lay three to six eggs, incubation and nestling periods each 18 days, attended by both parents. Often raise second or third broods during summer breeding season.

CONSERVATION STATUS
Not threatened. Widespread and common in most areas of range.

SIGNIFICANCE TO HUMANS
None known. Numbers killed during night migration by collision with windows. ◆

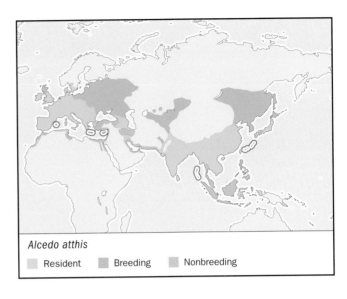

Alcedo atthis

■ Resident ■ Breeding ■ Nonbreeding

Common kingfisher
Alcedo atthis

SUBFAMILY
Alcedininae

TAXONOMY
Gracula atthis Linnaeus, 1758, Egypt. Seven subspecies.

OTHER COMMON NAMES
English: European/Eurasian kingfisher; French: Martin-pêcheur d'Europe; German: Eisvogel; Spanish: Martin Pescador Común.

PHYSICAL CHARACTERISTICS
6 in (16 cm), 0.7–1.6 oz (20–46 g). Small kingfisher, typical of genus, blue-green above, rufous below and on cheeks and forehead, white throat and spot on neck. Bill black (male) or red below (female), feet red.

DISTRIBUTION
Resident across mainland Europe, northern Africa, Asia, New Guinea and adjacent islands, breeding summer migrant to about 55° N across Palearctic, non-breeding winter migrant mainly to Middle East and islands of tropical southeastern Asia.

HABITAT
Rivers with vegetation along banks, less often lakes and dams, more coastal during non-breeding season.

BEHAVIOR
Largely aquatic. Perches inconspicuously, usually low above the water. Often turns around on perch to extend search area, bobs head when sights prey. Sings with whistles and warbles, shrill two-note screech in flight.

FEEDING ECOLOGY AND DIET
Diet mainly small fish, augmented with some aquatic crustacea, insects and frogs. Dives underwater from perch, to depths up to 3 ft (1 m). Rarely hovers, hawks insects, or follows otters to obtain food.

REPRODUCTIVE BIOLOGY
Both sexes territorial, and monogamous pair excavates nest tunnel into earth bank, 22–54 in (15–137 cm) long depending on soil conditions. Rarely use rotten wood or disused burrow of another animal. Breed during austral summer. Lay 3–10 eggs, incubation 19–21 days, shared by day female by night. Nestling period 23–27 days, chicks fed by both parents at maximum average rate of three to four items per hour.

CONSERVATION STATUS
Not threatened. Widespread and common, at densities up to four pairs/0.6 mi (four pair/km) on river, but some local problems with polluted or altered river courses.

SIGNIFICANCE TO HUMANS
Depletes stock on angling rivers or in fish ponds, sometimes persecuted. ◆

Amazon kingfisher
Chloroceyle amazona

SUBFAMILY
Cerylinae

TAXONOMY
Alcedo amazona Latham, 1790, Cayenne. Monotypic.

OTHER COMMON NAMES
French: Martin-pêcheur d'Amazonie; German: Amazonasfischer; Spanish: Martin Pescador Amazónico.

PHYSICAL CHARACTERISTICS
12 in (30 cm), 3.5–4.9 oz (98–140 g). Very large, metallic-green kingfisher, with white underparts and collar. Flanks streaked with green, breast band rufous (male) or green (female). Long heavy black bill.

DISTRIBUTION
Mexico, central and South America, east of Andes and south to northern Argentina.

HABITAT
Large rivers, lakes and estuaries, especially along more open shores, up to 8,202 ft (2,500 m) above sea level.

Chloroceyle amazona
Resident

BEHAVIOR
Most often perched about 16 ft (5 m) up in large tree over-looking water. Sometimes bobs head or pumps tail. Utters loud harsh barks, singly or in rapid series.

FEEDING ECOLOGY AND DIET
Aquatic diet, mainly of 0.4–6.7 in (10–170 mm) long fish, especially characid species, and some crustacea. Dives into water after prey from perch, rarely after hovering briefly.

REPRODUCTIVE BIOLOGY
Both members of monogamous pair excavate nest tunnel in earth back near water. Lays two to four eggs, incubated for 22 days by female at night and mainly by male by day. Nestling period 29–30 days.

CONSERVATION STATUS
Not threatened. Widespread and common, at densities of up to 1/km (large rivers), 4/km (streams) and even 5–6/km (lakes) at center of range. Occurs alongside four other species, two of similar size, two smaller, that may affect abundance, ecology, and success.

SIGNIFICANCE TO HUMANS
None known. ◆

Belted kingfisher
Megaceryle alcyon

SUBFAMILY
Cerylinae

TAXONOMY
Alcedo alcyon Linnaeus, 1758, South Carolina. Monotypic.

OTHER COMMON NAMES
French: Martin-pêcheur d'Amérique; German: Gütelfischer; Spanish: Martin Gigante Norteamericano.

PHYSICAL CHARACTERISTICS
11–13 in (28–33 cm), 4.0–6.3 oz (113–178 g). Very large, blue-gray kingfisher, with white breast and collar. Breast band plain blue-gray (male) or with rufous below (female). Juveniles of both sexes resemble adult female.

DISTRIBUTION
Resident across central United States of America and southern Canada, breeding summer migrant almost to Arctic Circle (to about 65°N), non-breeding winter migrant to southern USA and central America, south to Galapagos Islands and Guyana.

HABITAT
Lakes, rivers, streams, ponds and estuaries, from seashore to 8,200 ft (2,500 m) above sea level. Uses mainly mangroves, coasts, and offshore islands during non-breeding season.

BEHAVIOR
Usually perched in large tree overlooking water. Main call a harsh series of rattling notes. Visually sensitive to near-ultraviolet wavelengths, but behavioral significance unknown.

FEEDING ECOLOGY AND DIET
Aquatic diet, mainly fish, but also crustacea, amphibians, mollusks, and insects. Some fruit taken in winter. Hunts from perch or by hovering about 49 ft (15 m) above water, sometimes 0.6 mi (1 km) out from shore. Rarely submerges, catches most prey within 24 in (60 cm) of surface. Hunts mainly in late morning and afternoon, sometimes following egrets for any prey they disturb.

REPRODUCTIVE BIOLOGY
Monogamous, breeding during northern summer. Both parents excavate nest tunnel in earth bank, within range of but not always close to water. Burrow usually 3.3–6.6 ft (1–2 m) deep

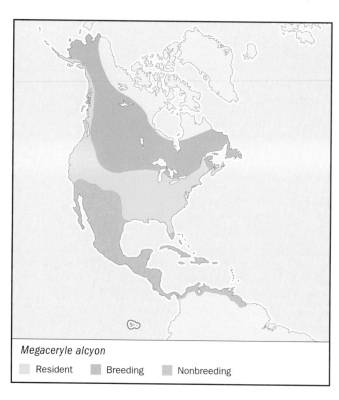

Megaceryle alcyon
Resident   Breeding   Nonbreeding

with 8–12 in (20–30 cm) diameter cavity at end. Lays five to eight eggs, incubation 22–24 days, nestling period 27–35 days. Sexes share duties of incubation, brooding, and provisioning.

CONSERVATION STATUS
Not threatened. Widespread and common in many areas, more resistant to pollution than most kingfishers. Some local disturbance at nest sites.

SIGNIFICANCE TO HUMANS
Feeds on some fish stocks and so persecuted locally. ◆

Pied kingfisher
Ceryle rudis

SUBFAMILY
Cerylinae

TAXONOMY
Alcedo rudis Linnaeus, 1758, Egypt. Four subspecies.

OTHER COMMON NAMES
English: Lesser/small/Indian pied kingfisher; French: Martin-pêcheur pie; German: Graufischer; Spanish: Martin Pescador Pie.

PHYSICAL CHARACTERISTICS
10 in (25 cm), 2.4–3.9 oz (68–110 g). Medium-sized kingfisher patterned in black and white. Black crown and broad mask distinctive, with double (male) or single (female) black band across white underparts. Very long black bill.

DISTRIBUTION
Sub-Saharan Africa, through Middle East, India, and Asian mainland to southern China.

HABITAT
Mainly large rivers, estuaries, and lakes, but from seashores to 8,200 ft (2,500 m) above sea level, also streams, ponds, and irrigation ditches. Absent from center of large swamps.

BEHAVIOR
Often perched on waterside vegetation or lookouts, rarely on the backs of hippos. Regularly bobs head or pumps tail. Noisy, with variety of shrill trills and chirps, uttered at perch or in flight.

FEEDING ECOLOGY AND DIET
Perches wherever possible, to save energy, but hovers in search of prey more than any other kingfisher, especially under windy

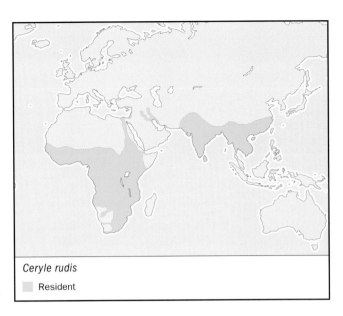

Ceryle rudis

▮ Resident

conditions. May then dive to 18 in (45 cm) below the surface and forage up to 2 mi (3 km) from shore, where it swallows prey in flight rather than return to a perch. Diet mainly small 1–2.4 in (25–60 mm) fish, supplemented by aquatic insects and crustacea. Eats few amphibians and mollusks, even insects taken ashore or in the air.

REPRODUCTIVE BIOLOGY
Monogamous pair excavates a nest tunnel in an earth bank, alone, or in colony of up to 100 pairs where nest sites limited. Normal clutch four to five eggs, incubation 18 days, nestling period 23–26 days. Sexes share nest duties, often assisted by a son from a previous brood and, especially in feeding chicks, by unrelated males.

CONSERVATION STATUS
Not threatened. Widespread and common, locally even abundant. The most numerous kingfisher in the world. Benefited in many areas from artificial dams and fish farming or stocking activities. Suffers locally from water pollution and use of pesticides.

SIGNIFICANCE TO HUMANS
None known. ◆

Resources

Books
Fry, C.H., K. Fry, and A. Harris. *Kingfishers, Bee-eaters and Rollers.* Halfway House, South Africa: Russel Friedman Books, 1992.

Sibley, C.G., and J.E. Ahlquist. *Phylogeny and Classification of Birds: A Study in Molecular Evolution.* New Haven and London: Yale University Press, 1990.

Woodall, P.F. "Family Alcedinidae (Kingfishers)." In *Handbook of Birds of the World.* Vol. 6, *Mousebirds to Hornbills,* edited by J. del Hoyo, A. Elliott, and J. Sargatal. Barcelona: Lynx Edicions, 2001.

Periodicals
Fry, C.H. "The Evolutionary Biology of Kingfishers (Alcedinidae)." *Living Bird* 18 (1980): 113–60.

Fry, C.H. "The Origin of Afrotropical Kingfishers." *Ibis* 122 (1980): 57–64.

Woodall, P.F. "Morphometry, Diet and Habitat in the Kingfishers (Aves: Alcedinidae)." *Journal of Zoology, London* 233, no. 1 (1991): 79–90.

Other
Coraciiformes Taxon Advisory Group. <http://www.coraciiformestag.com>.

Alan C. Kemp, PhD

Todies
(Todidae)

Class Aves
Order Coraciiformes
Suborder Alcedines
Family Todidae

Thumbnail description
Very small, rather plump kingfisher-like birds with long, somewhat flattened red bills, crimson gorget, stubby tail, shining green wings, and dorsal plumage

Size
4.0–4.6 in (10–11.5 cm); 0.19–0.27 oz (5.2–7.5 g)

Number of genera, species
1 genus; 5 species

Habitat
Tropical forests and woodlands, primary and secondary, ranging from arid lowlands to lushly forested highlands

Conservation status
Not threatened; locally common, although in 2001 population densities decreasing due to habitat destruction.

Distribution
Greater Antilles, West Indies

Evolution and systematics

Five species comprise one of the most uniform families in the ornithological world. Early taxonomists, analyzing museum skins tagged with exceedingly sparse field notes and puzzling over their relationships, allied them in a mixed bag: nightjars, trogons, jacamars, puffbirds, barbets, motmots, kingfishers, broadbills, cotingas, manakins, flowerpeckers, and tyrant-flycatchers. Until about 1900, all were regarded as variations of the Jamaican tody (*Todus todus*), then *Todus viridis*. Ultimately, these endearing Caribbean birds entered the order Coraciiformes, reflecting close kinships with motmots and kingfishers.

Fossils are unfortunately extremely scant: ancestral fossils (*Palaeotodus emryi*), are known from the Oligocene of Wyoming (37–24 million years ago), France, and Switzerland. *Palaeotodus*, although fragmentary, suggests close affinities with today's tody motmot (*Hylomanes momotula*) and the Swiss fossil *Protornis glarniensis*, a missing link between contemporary todies and motmots.

Geographical, paleontological, behavioral, morphological, and genetic data aid in the construction of a hypothetical evolutionary scenario for tody evolution. Approximately 30 million years ago, a primitive motmot/tody-like ancestor inhabited Northern Hemisphere forests. Twenty million years ago, climates began cooling, and by seven million years ago, only relict (from an earlier geological period) tody-motmot populations survived. Eventually, Central American birds similar to today's todies flew eastward, colonizing large Caribbean islands and

evolving into five species. This was possible because of Ice Age glaciations (one million years ago), when polar icecaps froze gigantic volumes of sea water, lowering sea levels worldwide by about 300 ft (90 m), thus reducing distances between continents and islands.

Physical characteristics

Todies are tiny, plump-bodied, large-headed, stub-tailed birds that characteristically perch with bill uptilted. Shining emerald-green above and creamy below, with a prominent crimson bib and a long, flat, bicolored red-and-black bill, each bird bears some resemblance to a miniature kingfisher or hummingbird. Visible similarities are accented by behaviors such as hovering and zooming courtship displays. Todies resemble kingfishers so much that taxonomist Linnaeus mistakenly applied a kingfisher genus (*Alcedo*) to the first tody described—the Jamaican. Conversely, locals in Hispaniola call todies *colibri*, which means hummingbird in Spanish.

Individual species are distinguished by flank and belly color (pink and yellow, yellow, greenish yellow), by cheeks (sky blue, gray), and by unique vocalizations. A family trademark, the scarlet bib, elicited curiosity from earliest times. For example: "As it sits on a twig in the verdure of spring, its grass-green coat is sometimes indistinguishable from the leaves in which it is embowered, itself looking like a leaf; but a little change in position, bringing its throat into the sun's rays, the light suddenly gleams as from a glowing coal" (Gosse, 1847).

Like most tropical forest birds, adult todies show no obvious seasonal differences in plumage. However, the dull attire of a summer adult, frazzled from raising three or four ravenous nestlings, can hardly compare with its shiny, semi-iridescent, feathered attire of early spring.

Flat, narrow, and shallow, the bill is ideally shaped for feeding—snapping up insects from the undersides of leaves in short, sweeping forays. Todies' short, rounded wings and loosely fluffed plumage are perfectly designed for their short flight paths and non-migratory lifestyles. Species flying the greatest distances on feeding forays have the longest wings.

Distribution

The family's geographical distribution is restricted to the Greater Antilles, West Indies. Cuba, Jamaica, and Puerto Rico each have one species, while Hispaniola supports two (narrow-billed tody [*Todus angustirostris*] and broad-billed tody [*Todus subulatus*]). The latter species are usually separated by habitat and elevation; where overlap occurs, foraging behavior provides classic insights into ecological isolating mechanisms.

Habitat

Three major factors limit tody distribution: vegetation, insect abundance, and territory requirements—especially the availability of suitable nesting banks. They occupy diverse habitats, ranging from 160 ft (50 m) below sea level to eleva-

tions above 9,800 ft (3,000 m). They favor brush and forest with interlacing foliage, epiphytes, and vines. Specific habitats include lowland or mountain rainforests (very dry to extremely lush, including elfin woodlands), pine groves, second growth, streamside vegetation, pasture borders, limestone karst, cactus deserts, and shaded coffee plantations. These little avian jewels are never garden birds; favoring natural forests, they adapt poorly to human incursion into their domains, except for roadcuts, which are welcomed for nesting.

Behavior

The first glimpses of todies are invariably of diminutive, vivid green, rapidly bobbing birds uttering loud nasal *beeps* that are quite disproportionate to their size. Adults and children consider them cute, joining the company of old-time naturalists. For example, here is a quote from esteemed ornithologist Dr. Alexander Wetmore: "If there be gnomes and elves in our world of birds, among them are the tiny todies, whose long, spade-like bills, light eyes, brilliant plumage and peculiar mannerisms make them the dwarfs and hobgoblins of the West Indian forests...their acquaintance is one of the greatest pleasures that comes to a foreign ornithologist travelling in their haunts" (1927).

Although strictly territorial, todies temporarily join mixed species feeding flocks passing through their territories. This behavior is pronounced during autumn and spring, when migrating warblers (Parulidae) visit West Indian forests. Here, the avifauna is impoverished compared to that of the continental tropics, so flocks are small, often averaging only six species.

Because todies are among the smallest and most active feeders of all birds, it is only natural that they have evolved effective modes of conserving energy. They do not employ typical methods of keeping warm, such as group roosting and huddling. Instead they rely on internal physiological mechanisms. The Puerto Rican tody, for instance, exhibits a very low normal body temperature of 98.1° F (36.7° C), rather than the 104° F (40° C) typical of its relatives. This enables it to decrease its expected energy expenditure by 33%, reducing the body's demand for additional heat production.

Most birds are homeotherms, just like people. This means they maintain a constant body temperature with little fluctuation. The Puerto Rican tody exhibits a rare thermoregulatory pattern in which its basic temperature varies widely. It can consciously control its normal body temperature by 27° F (15° C) from 82 to 109° F (28–43° C).

True torpor is relatively rare among birds. Todies not only exhibit torpor, but a controlled, sex-dependent, low metabolic rate that saves females approximately 70% of their daily energy expenditure.

Feeding ecology and diet

Voracious tody appetites require super-abundant annual food supplies. For example, for three years, feeding rates of Puerto Rican todies averaged 1.9 insects per minute in arid

This Cuban tody (*Todus multicolor*) shows off the colorful feathers under its bill. (Photo by Doug Wechsler/VIREO. Reproduced by permission.)

scrub and 1.1 per minute in rainforest. One of the highest chick feeding rates ever recorded for birds was when 420 feeds per day were delivered to a brood of three tody nestlings.

Fifty insect families have been identified as prey, chiefly grasshoppers, bugs, moths, butterflies, flies, cockroaches, damselflies, mantises, mayflies, and ants. Todies also consume seeds, lizards, and spiders. Fearsome creatures (scorpions, whipscorpions, snakes) are wisely avoided.

Tody feeding behavior resembles that of tody-flycatchers (tropical America). All capture insects using a distinctive aerial feeding technique, the underleaf-sally. With head directed upward, the bird scans undersides of leaves. Alert, jerking head and eyes rapidly, it darts upward at a shallow angle and, winging a short parabolic flight path, snatches an insect and continues in an unbroken arc to another perch. Occasionally, todies also swoop downward to snap insects from leaf surfaces, or hover in mid-air.

Todies prefer low-to-middle forest strata, often at eye level. Broad-billed todies feed higher than narrow-billed todies, a survival tactic that has evolved in the only two tody species whose geographical ranges overlap.

Reproductive biology

Loud courtship displays involve male-female chases and vigorous wing-rattling and wing-cracking. They pursue each other, often at lightning speed, weaving around foliage in parabolic arcs and circles. Another stunning component of courtship is the flank display, best developed in the pink-thighed species. For example, when the Cuban tody's display is at its height, its tiny body inflates into a green, neckless fuzzball with bright rosy flank tufts touching mid-dorsally. Simultaneously, this fluffy avian ball hops and bobs rapidly, uttering loud vocalizations to attract the perfect mate. Once paired, a mutual gift exchange of fresh insects occurs, like a bridal couple feeding each other slices of wedding cake.

The first (and rather quaint) published description of tody nesting habits comes from Moritz of Puerto Rico (1836): "In shady trees is seen once in a while the lovely green San Pedrito, rattling hoarsely...The locals believe that it nests in holes in the earth." It is true; todies are burrow-nesters, like their kingfisher cousins. They excavate cylindrical, angled tunnels in vertical soil embankments that are typically low, amphitheater-shaped slippages, roadside cuts, or natural inclines. The most successful burrows are those hewn from moderately overgrown embankments providing soil stabilization and partial concealment from predators.

Fresh tunnels are dug annually, primarily February to May. Each requires about eight weeks. Long, strong tody bills act as chisels to gouge out soil, while their tiny feet scrape away underneath. They tunnel energetically, initially visiting up to 60 times per hour.

Tody eggs are exceptional in the avian world in that they are huge compared to the bird's size. Average egg weights are 26% of the adult's body weight, comparable to the well-known

case of New Zealand kiwis. For comparison, egg-to-body-weight ratios of most birds range from 1.8–11%. Normally todies lay only one clutch (average 2.4 eggs), but will re-lay if it is destroyed. Eggs are tiny, white, glossy, and ovate.

Known incubation periods are 21–22 days, while nestling periods are 19–20 days. Each parent spends only two to three daylight hours incubating, a stark contrast to the assiduous kingfishers, which incubate up to seven continuous hours daily. Hatching occurs principally in the late afternoon, with one attendant adult in the nest chamber. Nestlings are naked, bearing conspicuous cushioned heels that cover the feet and legs with thick pads of swollen skin and leathery tubercles, like a baby born with leather boots.

Tody parents may not be over-attentive incubators, but once the chicks hatch they become highly diligent, supplying enormous quantities of insects to the offspring.

Notable also is nest-helping. As of 2001, nest assistance by one or two other adults during incubation and nestling periods is only known in the Puerto Rican tody. Two independent studies concluded that at least 50% of breeding pairs were given assistance. Nests with helpers contained significantly larger clutches (2.9 eggs) than did those without (2.3); chicks also grew quicker and fledged earlier than normal. Nest-helping is especially favorable in rainforests, where torrential rain often limits tody foraging rates. Banding studies suggest that helpers are not young from the previous year, and are likely adults from nearby territories whose breeding activities were curtailed or that did not breed at all. Nest cooperation in todies is unique because there is no apparent genetic relationship between helpers and recipients.

Adult todies use many innovations in teaching independence to newly fledged chicks. Weaning is not easy. Parents sometimes force them to fly by pushing them off perches or hovering in front of them with food in the bill, then pulling away at the last moment. Females are more likely to give in to a hungry chick than males. At times males physically prevent females from feeding fledglings. For the first six weeks, young todies have short black bills and gray bibs whose feathers gradually turn crimson. The entire repertoire of adult behavior is not achieved for several months.

Conservation status

As a family, todies seem reasonably secure. Overall, they remain common in natural habitats with high insect abundance. Only the narrow-billed tody is considered Near Threatened. Todies have partly benefited from human activities, excavating burrows in road cuttings and trailside banks. Clear-cutting and urbanization spell doom. They also cannot adapt to gardens, orchards, or pastures.

Little information is available concerning increasing pesticide use. On Cuba's Cayo Coco, currently undergoing rapid tourism expansion, malathion is sprayed aerially in wetlands and forests; hand-held fogging sprayers are also commonly used around hotels to combat mosquitoes. Pesticides decimate tody populations.

Coffee plantations formerly thrived under shady, indigenous trees. Here, todies enjoyed healthy populations and controlled insect numbers. From the 1980s on, international corporations discovered that growing coffee in direct sunlight was more profitable, although the beans were more bitter. Todies cannot adapt to modern coffee plantations, where pesticides are used liberally.

Although natural predators on wildlife are inevitable, introduced predators present greater threats. Todies are no exception; their numbers are everywhere seriously reduced by Indian mongooses (*Herpestes auropunctatus*), voracious mammals that destroy approximately 80% of tody burrows in Puerto Rico's rainforests. As the world's human population increases exponentially, human predation also increases. Many West Indians are sufficiently poor that they seek supplemental protein, however scant. Because tody nest holes are common along roads and trails, they succumb to such predation.

Significance to humans

Despite their tameness and abundance in varied habitats, todies were never culturally significant beyond their role as an ecotourism draw. Locals bestowed many endearing names on them: their bank-nesting habits prompted the French *perroquet de terre* (parakeet of the earth), and the Spanish *barrancoli*, *barranco-rio*, and *barranquero* (from *la barranca*, the Spanish word for bank). Confusion with hummingbirds inspired the Spanish *colibri* (hummingbird) in Hispaniola. A common vernacular in Puerto Rico is *papagayo* (loud, noisy), and bird enthusiasts in Jamaica often use robin red-breast.

Species accounts

Cuban tody
Todus multicolor

TAXONOMY
Todus multicolor John Gould, 1837, Cuba.

OTHER COMMON NAMES
French: Todier de Cuba; German: Vielfarbentodi; Spanish: Barrancolí Cubano.

PHYSICAL CHARACTERISTICS
4.3 in (10.8 cm), wing chord 1.7 in (4.4 cm), estimated weight 0.21–0.23 oz (6–6.5 g). Most brilliantly colored tody, with smallest bill. Rosy flanks, yellow undertail coverts. Sky-blue cheek patch and wrists; yellow base of bill, whitish belly. Yellow-green, almost iridescent eyebrow.

Todus multicolor

DISTRIBUTION
Cuba, including Isle of Pines (Isle of Youth) and larger cays off Cuba's north coast.

HABITAT
Ecologically adaptable. Locally common in xeric (extremely dry), moist, and wet forests; mountains; and lowlands, especially in gullies. Only tody inhabiting shoreline vegetation. Highest elevation recorded 8,184 ft (2,494 m) (Sierra Maestra).

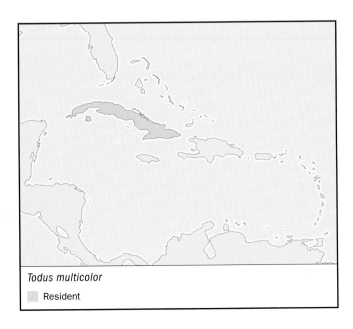

Todus multicolor

▨ Resident

BEHAVIOR
No recorded movements.

FEEDING ECOLOGY AND DIET
Primarily insectivorous, plus spiders and lizards. Mean foraging height 9 ft (2.6 m) in arid scrub.

REPRODUCTIVE BIOLOGY
Monogamous with striking courtship, exhibiting bright pink flanks. Smallest eggs in family. Breeds April to June. Excavates burrows in earth banks, rotten logs, natural limestone cavities, and (rarely) cave entrances. On Cayo Coco, uses sand at entrances of crab burrows.

CONSERVATION STATUS
Not threatened. In 1970, common in protected Guantanamo Naval Base. Cuba's poverty and unstable economy may affect tody populations. Recent pesticide use has reduced tody populations.

SIGNIFICANCE TO HUMANS
May be eaten in economically depressed areas and, like all todies, a delight to young and old. ◆

Puerto Rican tody
Todus mexicanus

TAXONOMY
Todus mexicanus Lesson, 1838, Tampico and Vera Cruz, Mexico. Clearly an error: todies do not occur in Mexico.

OTHER COMMON NAMES
French: Todier de Puerto Rico; German: Gelbflankentodi; Spanish: Barrancolí Puertorriqueño.

PHYSICAL CHARACTERISTICS
4.4 in (11 cm), wing chord 1.7 in (4.3 cm), weight male 0.22 oz (6.3 g), female 0.21 oz (5.9 g). The least colorful, smallest tody. Flanks yellow, no pink, no blue cheek patch, belly whitish. Sexual dimorphism in eye color: slate (males), white (females). Juveniles have four matu-

Todus mexicanus

ration stages (three weeks) when bill lengthens to adult size, grayish bib brightens to crimson, and yellow flanks develop.

DISTRIBUTION
Puerto Rico.

HABITAT
Rainforests, arid scrub, coffee plantations, moderately wet forests, karst (limestone) topography, often near streams.

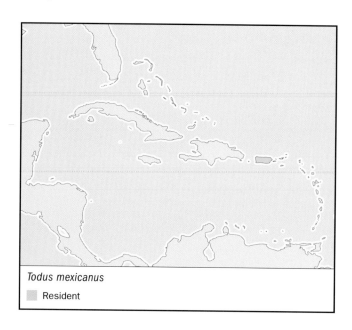

Todus mexicanus

■ Resident

BEHAVIOR

Maintain year-round home ranges and breeding territories. Usually single or paired, never in pure flocks although may temporarily join mixed feeding flocks.

FEEDING ECOLOGY AND DIET

Typical tody. Insectivorous, 50% of diet is flies and beetles. Also spiders, nematodes, millipedes, lizards, and fruits from six families (principally Ficus, Chenopodium, Rubus, Xanthoxylum, Psychotria, Clusia). Adults select sizes and variety of nestling foods.

REPRODUCTIVE BIOLOGY

Burrow excavation primarily February to May. Tunnel horizontal, almost always with right-angled curves and enlarged, depressed, unlined terminal chambers. Average burrow length in rainforest 13.9 in (30.5 cm). Nest helping common. Clutch size averages 2-4 eggs.

CONSERVATION STATUS

Not threatened. No recent information on status but populations diminishing due to habitat destruction and non-shade coffee.

SIGNIFICANCE TO HUMANS

None known. ◆

Resources

Books

Gosse, P. H. *The Birds of Jamaica.* London: John Van Voorst, 1847.

Kepler, A. K. *Comparative Study of Todies (Todidae), with Emphasis on the Puerto Rican Tody, Todus mexicanus.* No. 16. Cambridge, MA: Nuttall Ornithology Club, 1977.

Kepler, A. K. "Family Todidae." In *Handbook of the Birds of the World*, edited by J. del Hoyo, A. Elliott, and J. Sargatal. Vol. 6, *Mousebirds to Hornbills.* Barcelona: Lynx Edicions, 2001.

Sibley, C. G., and B. L. Monroe. *Distribution and Taxonomy of Birds of the World.* New Haven and London: Yale University Press, 1990.

Periodicals

Bond, J. "Origin of the Bird Fauna of the West Indies." *Wilson Bulletin* 60 (1948): 207–229.

Latta, S. C., and J. M. Wunderle, Jr. "Ecological Relationships of Two Todies in Hispaniola: Effects of Habitat and Flocking." *Condor* 98 (1996a): 769–779.

Latta, S. C., and J. M. Wunderle. "The Composition and Foraging Ecology of Mixed-species Flocks in Pine Forests of Hispaniola." *Condor* 98 (1996): 595–607.

Merola-Zwartjes, M., and J.D. Ligon. "The Ecological Energetics of the Puerto Rican Tody: Heterothermy, Torpor, and Intra-island Variation." *Ecology* (in press)

Olson, S. L. "Oligocene Fossils Bearing on the Origins of the Todidae and Momotidae (Aves: Coraciiformes)." *Smithsonian Contributions to Palaeobiology* No. 27 (1976): 111–119.

Oniki, Y. "Temperatures of Some Puerto Rican Birds, with Note of Low Temperatures in Todies." *Condor* 77 (1975): 344.

Ricklefs, R. E., and E. Bermingham. "Molecular Phylogenetics and Conservation of Caribbean Birds." *El Pitirre* 10, no. 3 (1997): 85–92.

Wetmore, A. "Birds of Porto Rico." *Bulletin of the U.S. Department of Agriculture.* 326 (1916): 1–140.

Wetmore, A. "Birds of Porto Rico and the Virgin Islands. Scientific Survey of Porto Rico and the Virgin Islands." *New York Academy of Sciences* 9, no. 4 (1927): 245–598.

Wetmore, A., and B. H. Swales. "The Birds of Haiti and the Dominican Republic." *Bulletin of the United States National Museum* 155 (1931): 1–483.

Wunderle, J. M., and S. C. Latta. "Avian Abundance in Sun and Shade Coffee Plantations and Remnant Pine Forest in the Cordillera Central, Dominican Republic." *Ornitologica Neotropical* 7 (1996): 19–24.

Angela Kay Kepler, PhD

Motmots

(Momotidae)

Class Aves
Order Coraciiformes
Suborder Alcedines
Family Momotidae

Thumbnail description
Spectacular birds that range in size from starling to pigeon. Characterized by green and blue hues, a black mask, and a long racquet-tipped tail in most species

Size
6–19 in (16–47 cm); 0.9–6.2 oz (25–175 g)

Number of genera, species
6 genera; 9 species

Habitat
Predominantly tropical forest and woodland

Conservation status
Vulnerable: 1 species

Distribution
Northeastern Mexico to northern Argentina

Evolution and systematics

Motmots (Momotidae) are a striking group. An Oligocene (30–40 million years ago) fossil (*Protornis*) from Switzerland suggests that ancestors of contemporary motmots were from the Old World; climatic events and competition probably led to their absence. A fossil was found in Florida from the late Miocene (25–30 million years ago), before North and South America were joined by the Panamanian land bridge. More recent Pleistocene (20,000 years ago) fossils were found in South America, in several fossil quarries in Minas Gerais, Brazil. Climatic changes probably restricted motmots to Central America, and they later dispersed into South America (where little radiation occurred) about 2.5 million years ago during the late Pliocene. These events suggest that motmots are the only avian family with a center of origin and diversity in middle America.

Motmots are closely related to bee eaters (Meropidae), Kingfishers (Alcedinidae), and todies (Todidae), with closest kinship to todies. In fact, the tody motmot (*Hylomanes momotula*) is considered the most primitive family member, and perhaps links the motmot and tody families. Shared characteristics of motmots, bee eaters, kingfishers, and todies include a unique

middle-ear ossicle, aspects of limb musculature, DNA, bright plumage, little or no dimorphism between sexes, predation by sallying (short flights from a perch to seize prey and return to the perch), and earth-excavated nest burrows. All four families are Coraciiformes, which likely radiated in the Eocene, and include rollers, hoopoes, and hornbills.

There are six genera with nine species. The turquoise-browed motmot (*Eumomota superciliosa*) represented by a single species, is more closely related to members of the genera *Electron* and *Baryphthengus*, each represented by two species, and the genus *Momotus*, represented by three species. These include the broad-billed motmot (*Electron platyrhynchum*), keel-billed motmot (*E. carinatum*), rufous motmot (*Baryphthengus martii*), rufous-capped motmot (*B. ruficapillus*), russet-crowned motmot (*Momotus mexicanus*), blue-crowned motmot (*M. motmota*), and highland motmot (*M. aequatorialis*). The rufous-capped motmot, though superficially similar to other species, lacks a racquet-tipped tail, as do some populations of broad-billed and rufous motmots. This suggests relatively recent divergence in these closely related genera. Supporting this hypothesis, while conducting playback experiments between motmots occurring together in Amazonian Peru, field ornithologist D. F. Lane discovered that rufous, blue-crowned, and broad-billed motmots seemed to respond territorially to each other's voices.

Physical characteristics

Motmots resemble kingfishers and have similar habits, although they are not found near water. They range from 6 to 20 in (16–50 cm), and 0.9–7.4 oz (25–210 g). The bill curves downward at the tip and in most species has serrated edges along the tomia (cutting edges of the bill). The tongue is relatively long. The tarsus is short and the middle toe is almost completely joined to the inner toe; there is only one rear toe. The wings are short and rounded. Plumage is soft blue or reddish brown; some species have blue or emerald stripes at the side of the head. A group of black feathers at chin and throat is characteristic of all motmots.

The tail is spatulate. The central pair of feathers is elongated, and barbs near the tail fall off readily, leaving part of the shaft of these feathers bare and resembling a thin wire. This barren area gives way to an oval disk at the feather tip where the barbs are retained, forming the spatulate-shaped racquet tip. The tody motmot is the smallest family member, characterized by more drab coloration and a shorter tail with no racquet tip. The blue-throated motmot (*Aspatha gularis*) also lacks a racquet-tipped tail.

These birds exhibit jerky tail twitching when disturbed. Male and female are similar in all species.

Distribution

Motmots are restricted to the neotropics, distributed from northeastern Mexico through most of tropical South America as far as northern Argentina. Although most species are lowland forms, there are two exceptions: the blue-throated motmot ranges 4,900–10,000 ft (1,500–3,100 m) in middle America, and the highland motmot ranges 4,100–7,200 ft (1,250–2,200 m) in the South American Andes.

The country harboring the highest diversity of motmots is Honduras, with seven species; Mexico, Guatemala, and Nicaragua each contain six species. In contrast, countries forming the Guiana shield (Venezuela, the Guianas, and Suriname) harbor a single species, the blue-crowned motmot.

About half the motmots are regionally restricted (four to middle America, one to the Andes). The species with widest distribution is perhaps the blue-crowned motmot, which ranges from Mexico through Argentina. The most range-restricted species is perhaps the blue-throated motmot, restricted to the central highlands of middle America.

Habitat

Most species are found in tropical or montane rainforest. Riverine gallery forest may be inhabited by blue-crowned, russet-crowned, and turquoise-browed motmots. Blue-crowned motmots will inhabit flooded forest, and blue-throated motmots will live in highland pine-oak forest.

Several motmot species are found in secondary forests, often visually inconspicuous and widely distributed. Most motmots inhabit the midstory or understory of forest or woodland.

Behavior

Motmots appear solitary, but seem to maintain pair bonds during and between years.

Many motmots have a subtle or soft "hooting" call, but there are exceptions. The call of the tody motmot has a ring to it, and the call of the broad-billed motmot is a louder, resonant "honk".

Motmots are not very active and often go undetected. The tail often pendulates, sometimes jerkily. They are inactive at night, active during twilight at dawn and dusk. Calling is most active during early morning light.

Some short-distance migration patterns are probable. For example, turquoise-browed motmots are often absent from breeding grounds for most of February, but return in March. Such seasonal movements are likely associated with changes in habitat association, but the family is mostly non-migratory.

Feeding ecology and diet

Motmots are omnivorous, taking invertebrates, small animals, and fruits. Invertebrates include beetles, butterflies (*Morphos*) and caterpillars, dragonflies, mantises, cicadas, spiders, centipedes, millipedes, scorpions, snails, earthworms, and crabs. Small animals include anole and gecko lizards, small snakes, frogs, small fish, an occasional nestling bird, and, in one recorded case, a blue-crowned motmot took a mouse. Fruits include those of palms, heliconia (*Heliconia*), nutmegs (*Compsoneura*, *Virola*), incense (*Bursera*), figs (*Ficus*), and other fruits. Frugivory (fruit consumption) seems to increase with size. For example, Remsen and colleagues found broad-billed motmots to be largely insectivorous, but rufous and blue-crowned motmots were more frugivorous. There are no records of the smallest species, the tody motmot, taking fruit.

Different motmot species obtain their prey in different ways, although patterns overlap. Smaller species appear to sally more as sit-and-wait strategists; larger species often perform long, broadcasting flights while continuously searching for prey. Smaller species also seem to catch more prey on the wing than larger species. Prey too large to swallow whole are often seized with the bill and clubbed against a perch. Pellets may be regurgitated. Several motmot species are "ant-following" birds that take insects turned up by long trains of army ants.

Smaller seeds of consumed fruits are passed through and dispersed; larger seeds are regurgitated on the spot. Species such as rufous motmots are important dispersers of the nutmeg *Virola surinamensis*, accounting for approximately 17% of dispersed fruits. Seed dispersal helps regenerate tropical forests.

A blue-crowned motmot (*Momotus momota*) carrying a rain frog to its nestlings. (Photo by M.P.L. Fogden. Bruce Coleman Inc. Reproduced by permission.)

Reproductive biology

Vocal duetting between male and female occurs to some degree and may serve to strengthen the monogamous pair bond during non-breeding season and maintain the territory. Nests are typically built solitarily, but there are exceptions involving colonial nesting that ranges from a few nests to more than 100 in species such as the turquoise-browed motmot.

Motmots nest in an underground chamber dug by both sexes. They take turns loosening the soil and kicking dirt out toward the opening. The chamber may be up to 16 ft (5 m) long in larger species. Eggs remain on bare soil, but hard insect parts regurgitated by incubating parents may be added underneath.

Eggs are rounded, shiny, and white. The clutch typically ranges from three to five eggs. In middle America, eggs are typically laid every other day between March and June; April and May are peak laying months. Typically a single clutch is laid each season, but if a clutch is lost replacement clutches are laid after 10–21 days. Both sexes incubate eggs during long shifts, perhaps changing duties once in 24 hours. Incubation is 17–22 days, depending on the species.

Chicks hatch blind, featherless, and dependent on parents. Skutch provides information on development from studying broad-billed motmots: partly feathered at 11 days, eyes begin opening at 12 days, soft calls at 13 days, taking food at burrow entrance at 15 days, and leave nest at 25 days. Young blue-throated motmots have soft down that appears soon after hatching. Both sexes care for the brood and feed the chicks lepidopterans and other insects, vertebrate innards, and protein-rich fruits. Young generally leave the nest at 24–32 days, though one record has a blue-crowned motmot leaving the nest at 38 days.

Conservation status

Of the nine species, only the keel-billed motmot is considered Vulnerable. The main threat is a rapid rate of habitat destruction.

Significance to humans

Indigenous tribes may use motmot tail feathers or wings in ornamentation. Folklore of the Brazilian Pareci Tribe of Mato Grosso attributes gaps without barbs along the tail of the rufous motmot to carrying fire embers. Motmots were well known in the Mayan culture.

1. Blue-crowned motmot (*Momotus momota*); 2. Blue-throated motmot (*Aspatha gularis*); 3. Broad-billed motmot (*Electron platyrhynchum*); 4. Keel-billed motmot (*Electron carinatum*). (Illustration by Jacqueline Mahannah)

Species accounts

Blue-throated motmot
Aspatha gularis

TAXONOMY
Prionites gularis Lafresnaye, 1840, Guatemala. Monotypic.

OTHER COMMON NAMES
French: Motmot à gorge bleue; German: Blaukehlmotmot; Spanish: Momoto Gorgiazul.

PHYSICAL CHARACTERISTICS
2.0–2.3 oz (56–65 g), 11 in (28 cm). Side of head is ochre-colored, with a black spot behind eye. Green above and paler green below, with dark green tail. Blue throat with black spot on chest and blackish bill.

DISTRIBUTION
Mountains of southern Mexico to El Salvador. Overlaps the same geographic range as several other motmots but lives at higher altitudes.

HABITAT
Montane evergreen forest; 4,900–10,000 ft (1,500–3,100 m).

BEHAVIOR
Appears solitary, but seems to maintain pair bonds during and between years. They are not very active, often go undetected. Tail often pendulates, sometimes jerkily. They are inactive at night, active at twilight at dawn and dusk. The blue-throated motmot sings at daybreak after leaving its earth hole; its song consists of pure full tones that rise and fall.

FEEDING ECOLOGY AND DIET
Insects are seized in flight by sallying; beetles make up a high proportion of the diet. Fruits are also consumed.

REPRODUCTIVE BIOLOGY
In Guatemalan highlands, motmots dig holes soon after young are fledged in late June or July. Pair spends nights in the hole during rainy season and dry winter months. In April the female lays 3–4 white eggs. After an incubation of 21–22 days, parents keep young warm for a considerable period. Young do not return to the nest hole at night after fledging.

CONSERVATION STATUS
Not threatened.

SIGNIFICANCE TO HUMANS
None known. ◆

Broad-billed motmot
Electron platyrhynchum

TAXONOMY
Momotus platyrhynchus Leadbeater, 1829, Brazil. Six subspecies.

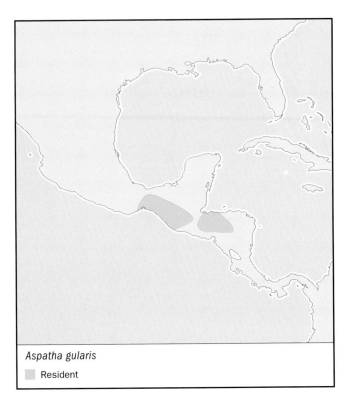

Aspatha gularis
  Resident

Electron platyrhynchum
  Resident

OTHER COMMON NAMES
English: Plain-tailed motmot; French: Motmot à bec large;
German: Plattschnabelmotmot; Spanish: Momoto Picoancho.

PHYSICAL CHARACTERISTICS
2–2.3 oz (56–66 g); 12–15 in (31–39 cm), with an unusual
widened bill form. Head, neck, and chest are rufous with a
black mask on face. Blue-green underbill and black spot on
chest. Green upperparts, blue-green belly, and green bluish
flight feathers.

DISTRIBUTION
Honduras to northern Bolivia and Mato Grosso, and eastward
to Paraguay; overlaps geographic distribution of several other
motmots.

HABITAT
Tropical evergreen rainforest and secondary vegetation; may
range to 3,600 ft (1,100 m).

BEHAVIOR
Appear solitary but seem to maintain pair bonds during and
between years. They are not very active and often go undetected. Tail often pendulates, sometimes jerkily. Inactive at
night, but active during twilight at dawn and dusk. Calling
most active at early morning light.

FEEDING ECOLOGY AND DIET
Consumes mostly insect adults and larvae, including butterflies,
dragonflies, and *Panaponera* ants as well as spiders, centipedes,
scorpions, small lizards, and frogs. Takes fruit minimally. Takes
prey on the wing during sallying, or gleans off the ground. Follows trains of army ants to consume displaced insects.

REPRODUCTIVE BIOLOGY
Nest holes are 3.2–6.6 ft (1–2 m) long and may change direction suddenly. Conspicuous opening is in vertical earth banks,
on river banks, on steep slopes beside roads or railways, or in
cave or well fissures. Partners relieve each other during incubation twice in each 24-hour period. Young broad-billed motmots hatch naked and blind; they rest on rumps and "ankles,"
which are protected against friction by calluses. Both parents
feed young, initially with squashed arthropods and soon with
adult-sized pieces of food.

CONSERVATION STATUS
Not threatened.

SIGNIFICANCE TO HUMANS
None known. ◆

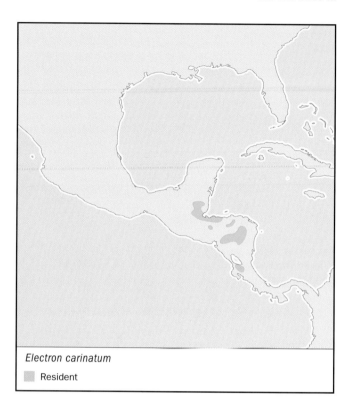

Electron carinatum
█ Resident

Keel-billed motmot

Electron carinatum

TAXONOMY
Prionites carinatus Bernard Du Bus de Gisignies, 1847,
Guatemala. Monotypic.

OTHER COMMON NAMES
French: Motmot à bec caréné; German: Kielschnabelmotmot;
Spanish: Momoto Carenado.

PHYSICAL CHARACTERISTICS
2.3 oz (65 g); 12–15 in (30–38 cm). Broad and flattened bill,
with a pronounced ridge. Rufous forehead and black mask,

with light turquoise streak above mask. Upperparts green, underparts greenish with light turquoise chin and black spot on
chest.

DISTRIBUTION
Patchily distributed from western Belize to northern Costa
Rica.

HABITAT
Tropical lowland and some montane rainforest; may range up
to 5,100 ft (1,550 m).

BEHAVIOR
Appear solitary but seem to maintain pair bonds during and
between years. They are not very active and often go undetected. The tail often pendulates, sometimes jerkily. Inactive at
night, active during twilight at dawn and dusk. Calling most
active at early morning light.

FEEDING ECOLOGY AND DIET
Prey is apparently taken on the wing during sallying. Little information on diet is available.

REPRODUCTIVE BIOLOGY
Males maintain territories through much vocal activity between
January and March. Excavated nests are in steep banks along
seasonal streams.

CONSERVATION STATUS
The keel-billed motmot is considered Vulnerable, with habitat
fragmentation due to agrarian conversion and banana plantations as principal threats. Geographic range is estimated at
14,000 mi² (36,000 km²). Its numbers are estimated at less than
10,000, with populations declining.

SIGNIFICANCE TO HUMANS
None known. ◆

Blue-crowned motmot
Momotus momota

TAXONOMY
Ramphastos momota Linnaeus, 1766, Cayenne. Twenty subspecies.

OTHER COMMON NAMES
English: Lesson's motmot, Caribbean motmot, tawny-bellied motmot, blue-diademed motmot; French: Motmot houtouc; German: Blauscheitelmotmot; Spanish: Momoto Común.

PHYSICAL CHARACTERISTICS
2.7–5.2 oz (77–148 g); 15–17 in (38–44 cm). Black crown bordered with violet and turquoise. Black mask with turquoise above and below. Back of neck is rufous, back is green. Greenish underparts and black spot(s) on chest.

DISTRIBUTION
Widest distribution of any motmot, northeastern Mexico to northern Argentina.

HABITAT
Occupies a variety of habitats, including tropical evergreen and deciduous forest, flooded and riverine gallery forest, montane and elfin forest, deciduous forest, and secondary vegetation. May range to 4,300 ft (1,300 m).

BEHAVIOR
Appear solitary but seem to maintain pair bonds during and between years. They are not very active and often go undetected. Tail often pendulates, sometimes jerkily. Inactive at night, active during twilight at dawn and dusk. Calling most active at early morning light.

FEEDING ECOLOGY AND DIET
Consumes insects and other invertebrates, including snails, earthworms, and centipedes. Also small reptiles, mice, and some fruits. Food is obtained by sallying and taken on the wing or off the ground. Fruits are often plucked while hovering on the wing. Blue-crowned motmots consume insects disturbed by trains of army ants.

REPRODUCTIVE BIOLOGY
In southern Costa Rica, the blue-crowned motmot digs its hole during rainy months (Aug.–Oct.) when soil is soft. Birds do not reappear until breeding season (March or April). One adult incubates from early afternoon to dawn, then the partner takes its

Momotus momota

☐ Resident

place. Incubation lasts 21 days. Lowland motmots stop covering their young at night when they are a week old. The nest is not clean, yet at fledging juvenal plumage looks fresh and clean. Young resemble adults in coloration but lack long racket-like tail feathers.

CONSERVATION STATUS
Not threatened.

SIGNIFICANCE TO HUMANS
Only motmot to have successfully bred in captivity. ◆

Resources

Books
BirdLife International. *Threatened Birds of the World.* Barcelona: Lynx Edicions, 2000.

del Hoyo, J., A. Elliot, and J. Sargatal, eds. *Handbook of the Birds of the World.* Vol. 6, *Mousebirds to Hornbills.* Barcelona: Lynx Edicions, 2001.

Feduccia, A. R. *The Origin and Evolution of Birds.* New Haven: Yale University Press, 1999.

Forshaw, Joseph M., and W. T. Cooper. *Kingfishers and Related Birds.* Vol. 3, *Todidae, Momotidae, Meropidae.* Sydney: Lansdowne Editions, 1987.

Sick, Helmut. *Birds in Brazil: A Natural History.* New Jersey: Princeton University Press, 1993.

Stiles, F. G., and A. F. Skutch. *A Guide to the Birds of Costa Rica.* London: Christopher Helm, 1989.

Stotz, D. F., et al. *Neotropical Birds: Ecology and Conservation.* Chicago: University of Chicago Press, 1996.

Periodicals
Becker, J. J. "A Fossil Motmot (Momotidae) from Late Miocene of Florida." *Condor* 88 (1986): 478–482.

Delgado-V., Carlos A., and D. M. Brooks. "Aberrant Vertebrate Prey taken by Neotropical Birds." *Cotinga* (2002) (in press).

Olson, S. L. "Oligocene Fossils Bearing on the Relationships of the Todidae and Momotidae (Aves: Coraciiformes)." *Smithsonian Contributions to Paleobiology.* 27 (1976): 111–119.

Resources

Orejuela, J. E. "Comparative Biology of Turquoise Browed and Blue Crowned Motmots in the Yucatan Peninsula, Mexico." *Living Bird* 16 (1977): 193–208.

Remsen, J. V. et al. "The Diets of Neotropical Trogons, Motmots, Barbets, and Toucans." *Condor* 95 (1993): 178–192.

Organizations

Neotropical Bird Club. c/o The Lodge, Sandy, Bedfordshire SG19 2DL United Kingdom. E-mail: secretary@ neotropicalbirdclub.org

Daniel M. Brooks, PhD

Bee-eaters
(Meropidae)

Class Aves

Order Coraciiformes

Suborder Alcedines

Family Meropidae

Thumbnail description
Small to medium-sized, active, colorful, sociable birds with long, slender, decurved bills and a characteristic upright, alert posture when perched

Size
6–10.5 in (15–27 cm)

Number of genera, species
3 genera; 24 species (22 species recognized by some taxonomists)

Habitat
Savanna, open woodlands, and desert-scrub, rarely forests

Conservation status
Not threatened

Distribution
Paleotropics

Evolution and systematics

Taxa in the bee-eater family are morphologically rather homogeneous, and their affinities with the other five families in the Alcedines suborder are more distant than the affinities those families share with each other. C. Hilary Fry, a recognized authority on bee-eater biology and evolution, recognizes three genera and 24 species. Analyses by Charles Sibley and Burt Monroe Jr. in the 1990s, using the technique of DNA-DNA hybridization, suggested only minor adjustments; they demote two species recognized by Fry to subspecies.

The genus *Nyctyornis* seems the most primitive and may point to the evolutionary roots of bee-eaters. Although the family is most diverse in Africa, the restricted distribution of *Nyctyornis*, in forested southeast Asia, suggests that the family originated in this region.

Physical characteristics

Bee-eaters are a family of small to medium-sized active birds, primarily of open habitats. They are easy to see in many parts of the paleotropics because of their striking color patterns and acrobatic flights to capture prey. Nearly all bee-eaters have beautiful plumages, typically with bright gourgets of red or yellow that contrast with blues and ochres of the breast and belly. Eye stripes bordered with contrasting colors are nearly ubiquitous. Two of the largest species, the carmine (*Merops nubicoides*) and rosy bee-eaters (*Merops malimbicus*), are a brilliant magenta or pink over most of the body.

All bee-eaters share the characteristics of a strong, slightly decurved bill, and a foot structure that includes the partial fusion of the three forward-facing toes. In most species the sexes are very similar or identical in appearance. Juveniles may have distinguishing plumage for a short time.

Distribution

Southern and southeastern Europe, sub-Saharan Africa, Arabian Peninsula, south and Southeast Asia, Indonesia, Philippines, Papua New Guinea, and Australia.

Habitat

Most bee-eaters favor open grasslands, dry scrub habitats, and large openings in woodlands, but there are a handful of little-known forest dwellers scattered from western Africa to southeast Asia and the island of Sulawesi, Indonesia. With few exceptions, even the forest species are usually found near openings, along river courses, and at road cuts.

Behavior

With few exceptions, bee-eaters are gregarious and social. Only about five species are likely completely solitary. Fifteen species are colonial or loosely colonial. Physical body contact is not common among birds, but with bee-eaters it is characteristic. A branch catching the early morning sun will attract first one, then another, and finally a whole family of bee-eaters

Bee-eaters exhibiting huddling behavior on a branch. (Illustration by Barbara Duperron)

that will huddle closely together, all facing into the sun. Colony sites are centers of high activity, and there is a cacophony of sound as family members greet each other, fuss over the proximity of intruders to their nesting hole, or fight to remove an unwanted visitor.

Complex social relationships might be quite common among bee-eaters, particularly those with sedentary populations, but few species have been studied in enough detail to know for certain. Detailed studies of the white-fronted bee-eater (*Merops bullockoides*) in Kenya by Stephen Emlen, Robert Hegner, and Peter Wrege revealed one of the most complex societies known for any species of bird.

Populations of white-fronted bee-eaters spend the entire year roosting together at a colony that may contain from 30 to 450 birds, and the population is structured into extended family groups or "clans" that include several generations. During the non-breeding season, all members of a clan, as many as 15 individuals, may roost together in one burrow at the colony. For breeding, clans often split up into nuclear family sub-groups. Members of a clan interact frequently, joining and greeting one another on perches near the colony, roosting together, or visiting each other's roosting burrows.

Feeding ecology and diet

All bee-eaters appear to be specialists in eating bees and other venomous hymenoptera. Studies of the diet of more than 15 species show that 60–80% of the diet is honeybees, wasps, and ants. But they will also pursue nearly any insect of suitable size, provided it is flying. A few species forage occasionally for large insects and small lizards on the ground, and there are even observations of bee-eaters catching small fish. A few larger species forage mainly on the wing, but most bee-

Little green bee-eater (*Merops orientalis*) with insect. (Illustration by Barbara Duperron)

eaters are "sit-and-wait" hunters, scanning the habitat from a perch with their keen eyes, chasing down likely prey with a flight that may take them even 150–180 ft (50–60 m) from their perch, then returning in a flash of color to subdue the prey for consumption.

Bee-eaters have stereotyped behaviors for dealing with prey—and they clearly recognize those with potentially dangerous venom and sting. After repeatedly smacking the head of the prey item against a perch, the bee-eater juggles the insect's body in its bill, biting along the abdomen from near the middle (thorax) toward the tip. This behavior often expresses a droplet of venom from the bee or wasp. Then the bee-eater wipes the tip of the insect's abdomen back and forth across the perch in a behavior known as "bee-rubbing." Many times this behavior pulls the stinger and poison gland out of the prey item.

Reproductive biology

Reproduction in bee-eaters depends on the species. Some have solitary nesting by unaided pairs, while others have extremely dense colonies with complex social structures and cooperative breeding. Cooperative breeding, which involves assistance by non-breeding adults, may be the rule rather than the exception. Seventeen species are known or suspected to be cooperative breeders.

All bee-eaters dig their own nesting burrows into soil or sand, and burrows may be 1.5–9 ft (0.5–3 m) in length, depending on the species and soil type. Thirteen species dig nests into nearly flat ground or into banks along rivers, irrigation ditches, erosion gullies, and even into roofs of aardvark or warthog dens.

The family structure and helping behaviors of the white-fronted bee-eater were studied in great detail and will serve as a complex example of cooperative breeding. The basic elements likely apply to all bee-eater species with cooperative breeding, although in some species family groups may be less complex. Helpers are often associated with all aspects of the reproductive attempt, from helping to dig the nesting burrow to incubating eggs and feeding nestlings. The most important function of helpers is in bringing food to the nestlings and recently fledged offspring. In Kenya, one helper could nearly double the number of offspring produced by a pair. Thus the ability to attract a helper could make a significant contribution to the reproductive output of a breeder.

Many helpers are young unpaired sons (and sometimes daughters) of the breeding pair, but if an older pair fails in

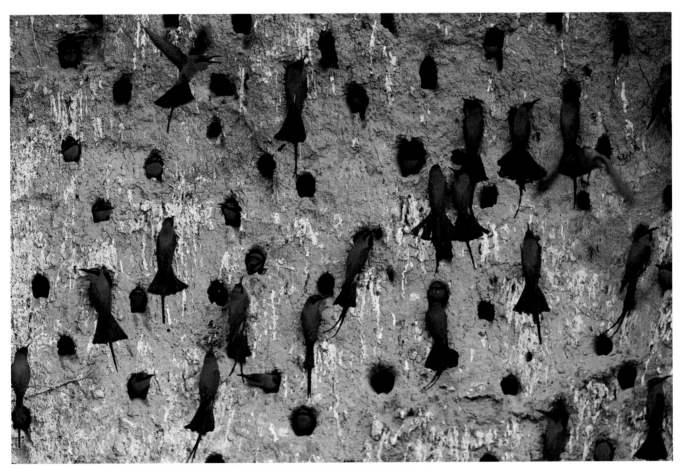

Carmine bee-eater (*Merops nubicoides*) nests on the Okavango River in Botswana. (Photo by Art Wolfe. Photo Researchers, Inc. Reproduced by permission.)

A European bee-eater (*Merops apiaster*) holds a bee in its mouth. Bee-eaters have particular behaviors for dealing with prey that have a potentially dangerous venom and sting. (Photo by R. Tipper/VIREO. Reproduced by permission.)

their nesting attempt, they may become helpers at the nest of a son or grandson.

In addition to benefits that a helper gains by increasing the production of close relatives, sometimes a bit of selfish reproduction is possible as well. It is common for a young female helper, nearly always the daughter of the breeding pair, to slip one of her own eggs into the nest of her parents.

Conservation status
Not threatened.

Significance to humans
In Australia, the arrival of the rainbow bee-eater (*Merops ornatus*) heralds the arrival of the rainy season, and in Africa huge colonies of rosy bee-eaters are a source of great pride to people in nearby villages. However, bee-eaters hold little significance to humans apart from the enjoyment that comes from observing their beauty and aerial antics. On all shores of the Mediterranean, those who make a living from the honey of the bee *Apis mellifera* have considered the European bee-eater (*Merops apiaster*) a pest, despite little scientific evidence that honey production is ever seriously compromised by bee-eater predation.

1. Black bee-eater (*Merops gularis*); 2. White-fronted bee-eater (*Merops bullockoides*); 3. Blue-bearded bee-eater (*Nyctyornis athertoni*); 4. European bee-eater (*Merops apiaster*); 5. Purple-bearded bee-eater (*Meropogon forsteni*); 6. Carmine bee-eater (*Merops nubicoides*); 7. White-throated bee-eater (*Merops albicollis*); 8. Female rainbow bee-eater (*Merops ornatus*). (Illustration by Barbara Duperron)

Species accounts

Blue-bearded bee-eater

Nyctyornis athertoni

TAXONOMY

Merops athertoni Jardine and Selby, 1830, India. Two subspecies.

OTHER COMMON NAMES

French: Guêpier à barbe bleue; German: Blaubartspint; Spanish: Abejaruco Barbiazul.

PHYSICAL CHARACTERISTICS

12.5–13.5 in (31–34 cm); 2.5–3.3 oz (70–93 g). The largest bee-eater and, along with its congener, clearly differentiated from the other genera by a stouter bill and a lax beard of feathers running from the base of the bill to the lower breast. Upperparts mainly green; belly is buff with broad green streaks. Forehead is pale azure-blue; long, broad throat feathers are dark blue grading to azure on the tips.

DISTRIBUTION

India to Indochina, but absent from the Malay Peninsula.

HABITAT

Clearings and more open areas of deep forest, occurring at elevations up to 7,150 ft (2,200 m), where it inhabits moss and deciduous forests.

BEHAVIOR

Birds spend most of their time in pairs, perched at the top of or at the outer side of trees, in search of food.

FEEDING ECOLOGY AND DIET

Relatively little is known, but honeybees, wasps, large beetles, and dragonflies are commonly taken. The blue beard may act as a flower mimic, attracting honeybees close enough to be snapped up by birds without moving from the perch.

REPRODUCTIVE BIOLOGY

Solitary nester, digging nesting burrows up to 9.75 ft (3 m) into stream or road banks. In India and Nepal, there are two breeding periods, February to May and August to October. In Indochina, egg laying is most common in April to May, but breeding may occur anytime until October. Clutch size is six eggs.

CONSERVATION STATUS

Not threatened.

SIGNIFICANCE TO HUMANS

None known. ◆

Purple-bearded bee-eater

Meropogon forsteni

TAXONOMY

Meropogon forsteni Bonaparte, 1850, Celebes. Monotypic.

OTHER COMMON NAMES

English: Celebes bee-eater; French: Guêpier des Célèbes; German: Celebesspint; Spanish: Abejaruco del Célebes.

PHYSICAL CHARACTERISTICS

10 in (25–26 cm), excluding tail streamers up to 2.5 in (6 cm). No weights available. Upperparts and wings are green; lower belly is dark brown; tail feathers are green and russet. Forehead and crown are blackish; ear-coverts, nape, and sides of

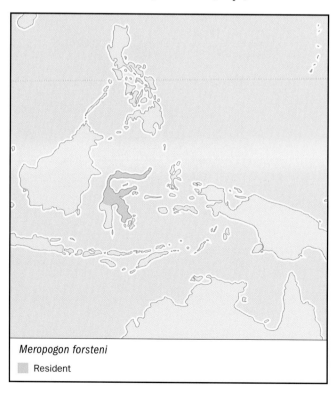

Nyctyornis athertoni
Resident

Meropogon forsteni
Resident

neck are chocolate to dark vinous-brown; long, broad throat feathers are purple and overlap the breast.

DISTRIBUTION
The most restricted species in the family, occurring only on the island of Sulawesi, Indonesia.

HABITAT
Open areas of rainforest.

BEHAVIOR
Sedentary or near-migrant. May travel to coasts for rainy seasons, and return inland for dry seasons.

FEEDING ECOLOGY AND DIET
Forages from mid- and upper-canopy perches for bees, wasps, beetles, and dragonflies.

REPRODUCTIVE BIOLOGY
Few records available, but active nests have been found in April, July, September, and December, so breeding may occur in nearly any month.

CONSERVATION STATUS
Not threatened.

SIGNIFICANCE TO HUMANS
None known. ◆

Black bee-eater
Merops gularis

TAXONOMY
Merops gularis Shaw, 1798, Sierra Leone. Two subspecies.

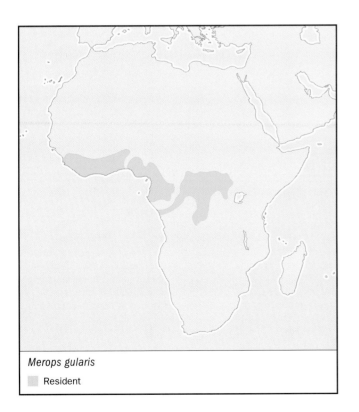

Merops gularis
▨ Resident

OTHER COMMON NAMES
French: Guêpier noir; German: Purpurspint; Spanish: Abejaruco Negro.

PHYSICAL CHARACTERISTICS
8 in (20 cm); 0.85–1.0 oz (24–30 g). Particularly striking and distinctive plumage, with nearly entirely black head and black back with contrasting scarlet throat and azure-blue rump.

DISTRIBUTION
West Africa from Sierra Leone west through Central Africa to eastern Zaire.

HABITAT
Clearings and stream edges in rainforest, secondary forest, wooded farmland, and gallery forest, usually high above the ground.

BEHAVIOR
Sedentary or partial migrant. Appears to be some movement corresponding to wet and dry seasons.

FEEDING ECOLOGY AND DIET
Usually forages high in the tree canopy.

REPRODUCTIVE BIOLOGY
Poorly known. Usually a solitary breeder; small colonies have been observed in Sierra Leone and Liberia. Breeding occurs from March to May.

CONSERVATION STATUS
Not threatened.

SIGNIFICANCE TO HUMANS
None known. ◆

White-fronted bee-eater
Merops bullockoides

TAXONOMY
Merops bullockoides A. Smith, 1834, Marico River, South Africa. Sometimes considered a subspecies of *M. bullocki*. Monotypic.

OTHER COMMON NAMES
French: Guêpier à front blanc; German: Weissstirnspint; Spanish: Abejaruco Frentiblanco.

PHYSICAL CHARACTERISTICS
8.5–9.5 in (22–24 cm); 1–1.4 oz (28–38 g). Upperparts and wings are blue-tinged green; underparts are buff; thighs and undertail coverts are blue; spread tail is green above and blackish below. White forehead, cheeks, and chin are sharply defined from the black mask and red throat.

DISTRIBUTION
Occurs south of the forested Congo basin across the breadth of Africa. Locally common north along the Rift Valley in Kenya to Lake Turkana, and on the west side of Lake Tanganyika north to Rwanda.

HABITAT
Occupies wooded savannas.

BEHAVIOR
Among the most social of all bee-eaters, roosting and breeding in large colonies (up to 400 nests) and interacting in extended family groups throughout its life. Sedentary in Kenya, but may

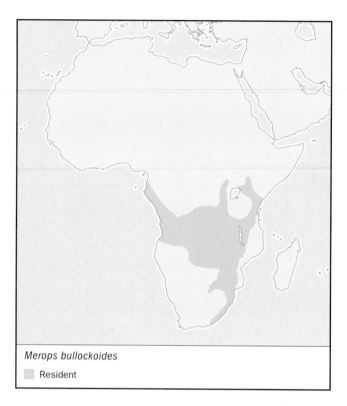

Merops bullockoides
  Resident

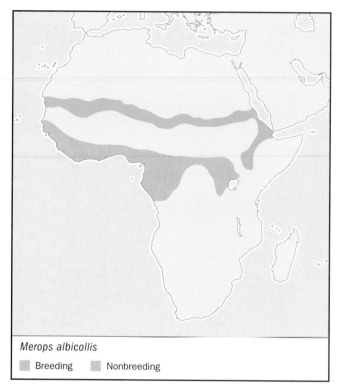

Merops albicollis
  Breeding   Nonbreeding

move widely during the non-breeding season in the southern range.

FEEDING ECOLOGY AND DIET
Family groups or clans defend foraging territories up to 4.5 mi (7 km) from the roosting/breeding colony. Within territories, members of the clan spend most of the time spaced apart on favorite perches, from which they make sallies for insect prey.

REPRODUCTIVE BIOLOGY
Throughout most of the range, breeding begins at the end of the dry season, August to October. In Kenya, where there are two somewhat unpredictable rainy periods, egg-laying may begin in October to November, or April to May, but any given population breeds during only one season. Clutch size is two to five eggs.

Cooperative breeding is common. Sixty percent of nests have one or more helpers (up to five), usually males.

CONSERVATION STATUS
Not threatened.

SIGNIFICANCE TO HUMANS
None known. ◆

White-throated bee-eater
Merops albicollis

TAXONOMY
Merops albicollis Vieillot, 1817, Senegal. Monotypic.

OTHER COMMON NAMES
French: Guêpier à gorge blanche; German: Weisskehlspint: Spanish: Abejaruco Gorgiblanco.

PHYSICAL CHARACTERISTICS
8 in (19–21 cm), excluding tail-streamers which can exceed 4.75 in (12 cm); 0.7–1 oz (20–28 g). The black crown and mask, separated by white supercilliary, cheeks, and throat make this species unmistakable. Hindneck is ochre; back is green; rump and tail are bluish; breast is pale green; belly is white. Longest tail streamers in the family.

DISTRIBUTION
Northern tropics, breeding across sub-Saharan Africa in very dry habitats, wintering in forested areas to the south, across the continent.

HABITAT
Occupies thorn scrub, open sandy dunes, and river washes during breeding, but rainforest canopy, woodlands, and orchards during the winter.

BEHAVIOR
Gregarious and vocal, this species is a conspicuous daytime migrant between desert breeding grounds and wet forests of tropical Africa, where it spends the non-breeding season flycatching from the canopy. Individuals wintering near the Zaire River must migrate nearly 1,400 mi (2,200 km) to the nearest breeding locations.

FEEDING ECOLOGY AND DIET
The diet is unusual, with a high proportion of flying ants, especially in forested habitats. White-throated bee-eaters will take ground prey such as lizards, tenebrionid beetles, and grasshoppers, and also forage in continuous flight like many of the larger bee-eaters. Most peculiar is an association with squirrels feeding on the oil palm *Elaeis guineensis*. Squirrels strip and discard the oily pericarp from the fruits, and bee-eaters snatch these nutritious pieces as they fall from high in the palm crown.

REPRODUCTIVE BIOLOGY
Breeds in loose colonies on flat or tiered ground surfaces from February to October (the later months in Chad and Nigeria).

Merops ornatus
■ Breeding   ■ Nonbreeding

Clutch size averages six eggs. Helping behavior is well developed. In one study, 90% of nests were attended by one or more non-breeding adults—the highest frequency of helping known for any bee-eater species.

CONSERVATION STATUS
Not threatened.

SIGNIFICANCE TO HUMANS
None known. ◆

Rainbow bee-eater
Merops ornatus

TAXONOMY
Merops ornatus Latham, 1801, New South Wales, Australia. Monotypic.

OTHER COMMON NAMES
English: Rainbow bird; French: Guêpier arc-en-ciel; German: Regenbogenspint; Spanish: Abejaruco Australiano.

PHYSICAL CHARACTERISTICS
7.5–8 in (19–21 cm), excluding tail streamers, which are 0.8 in (2 cm) in females and up to 2.8 in (7 cm) in males; 0.7–1.2 oz (20–33 g). Males are mainly glossy green with azure rump and uniquely black tail. Crown and nape are bronze; broad, black eyestripe is bordered below by pale blue band; chin and cheeks are yellow; throat is rufous; triangular gorget is black. Females are similar, but hind crown is less bronzy and tail streamers are shorter and wider.

DISTRIBUTION
Occurs throughout Australia, except most of the central arid region, and locally in Papua New Guinea. Winters in northern

Australia, New Guinea, as well as the Indonesian islands west to Lombok and north to Sulawesi.

HABITAT
Open habitats of almost all descriptions, perhaps determined mostly by availability of nesting sites.

BEHAVIOR
Gregarious outside of the breeding season, roosting and feeding in flocks, but aggregated only in loose colonies during breeding.

FEEDING ECOLOGY AND DIET
Primarily bees, wasps, and other hymenopterans. Also takes beetles, flies, moths, dragonflies, damselflies, and an occasional spider. Prey is captured on wing after being spotted from a perch.

REPRODUCTIVE BIOLOGY
Breeding in northern areas is underway by late August or September, but in southern Australia not until November and December. Nest burrows are usually dug into flat or gently sloping ground, with occasional use of low banks. Clutch size averages five eggs.

Roughly 15% of pairs are helped by one or more male helpers.

CONSERVATION STATUS
Not threatened. Populations may be increasing as a consequence of open habitats being created by human activity.

SIGNIFICANCE TO HUMANS
None known. ◆

European bee-eater
Merops apiaster

TAXONOMY
Merops apiaster Linnaeus, 1758, Southern Europe. Monotypic.

OTHER COMMON NAMES
French: Guêpier d'Europe; German: Bienenfresser; Spanish: Abejaruco Europeo.

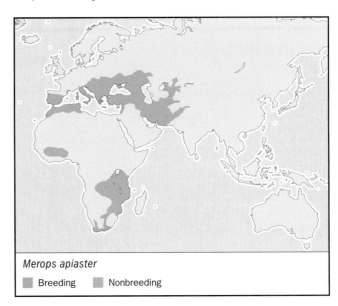

Merops apiaster
■ Breeding   ■ Nonbreeding

PHYSICAL CHARACTERISTICS
9–10 in (23–25 cm), excluding tail streamers of 0.8 in (2 cm); 1.6–2.8 oz (44–78 g). Very distinctive multi-colored plumage. Sexes distinguishable in breeding plumage, with females paler in coloration overall.

DISTRIBUTION
Northwest coast of Africa from Morocco to Libya, Mediterranean islands, countries of the northern Mediterranean east through the Middle East to Pakistan, northern India and Afghanistan. A small breeding population in South Africa and Namibia is largely disjunct from the wintering distribution, which extends from Lake Victoria in Kenya, south to the Transvaal, and west to Angola.

HABITAT
Grasslands, open woodlands, pasturelands with scattered trees, and gallery forests in drier habitats.

BEHAVIOR
Gregarious at all times of year, breeding in colonies and remaining in flocks on wintering grounds.

FEEDING ECOLOGY AND DIET
Forages primarily from a perch, as is typical of most bee-eaters, but also feeds for considerable periods in continuous flight.

REPRODUCTIVE BIOLOGY
Sometimes a solitary nester, it is more commonly found breeding in colonies, some with as many as 400 nests. Egg-laying occurs during May in the southern part of the range, and June and early July in Russia. South African populations begin breeding in October. Clutch sizes are the largest of any bee-eater, with up to 10 eggs, generally five or six.

Cooperative breeding is common, with about 20% of nests having a helper.

CONSERVATION STATUS
Not threatened.

SIGNIFICANCE TO HUMANS
Probably the most persecuted bee-eater, particularly wherever apiculture (bee-keeping) is an important industry. It is considered a pest in much of its range. ◆

Carmine bee-eater
Merops nubicoides

TAXONOMY
Merops nubicus Gmelin, 1788, Nubia. Two subspecies.

OTHER COMMON NAMES
English: Southern carmine bee-eater: French: Guêpier carmin; German: Karminspint; Spanish: Abejaruco Carmesí.

PHYSICAL CHARACTERISTICS
9.5–10.5 in (24–27 cm), excluding tail streamers of up to 4.75 in (12 cm); 1.2–2.1 oz (34–59 g). Upperparts are predominantly carmine with contrasting beryl-green crown, chin, cheeks, and rump. Black mask. Throat is olive-green but appears azure to greenish blue against the light. Belly is carmine shading to azure on undertail-coverts.

DISTRIBUTION
Disjunct distribution in Africa. Occurs from Senegal to Somalia, north of the Congo Basin. In winter, these northern birds move south into Sierra Leone, Nigeria, Cameroon, or across

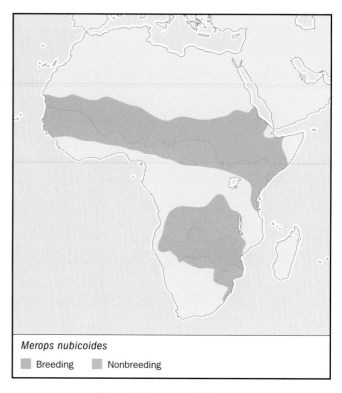

Merops nubicoides
▨ Breeding ▨ Nonbreeding

the continent to Uganda, Kenya, and Tanzania. In the south, they occur from the Okavango headwaters in Angola east to Mozambique, south into northern Botswana and southern Zimbabwe. Non-breeding birds spread into northeast South Africa and the Transvaal and north into Angola, southern Zaire, Burundi, and west Tanzania.

HABITAT
Savannas, open woodlands, lakeshores, and mangroves in coastal regions.

BEHAVIOR
Has three-stage migration pattern in northern and southern tropics. In the southern tropics, from breeding grounds, travels south from December onward, and back north in March to August, until breeding season. In the northern tropics, the birds move north following breeding.

FEEDING ECOLOGY AND DIET
Nearly all food is taken in continuous flight. The bird is common at bush fires, where it swoops above the leading edge to capture fleeing insects. In northeastern savannas, it sometimes associates with large ungulates and large cursorial birds (e.g., ostrich, bustards), using their backs as a perch from which to sally after insects disturbed by grazing animals.

REPRODUCTIVE BIOLOGY
Large breeding colonies of 100–1,000 nests are common. Nests occur in cliffs or on flat ground, where colonies are often larger—2,000–3,000 nests. North of the equator, egg-laying occurs from February to May (later farther north), and in the southern tropics from September to November.

Cooperative breeding has not been confirmed, although casual observations suggest it is likely.

CONSERVATION STATUS
Not threatened.

SIGNIFICANCE TO HUMANS
None known. ◆

Resources

Books

Fry, C. Hilary. *The Bee-Eaters*. Vermillion: Buteo Books, 1984.

Fry, C. Hilary, et al. *Kingfishers, Bee-eaters and Rollers: A Handbook*. Princeton: Princeton University Press, 1992.

Sibley, Charles G., and Burt L. Monroe Jr. *Distribution and Taxonomy of Birds of the World*. New Haven, CT: Yale University Press, 1990.

Periodicals

Bell, H. L. "Abundance and Seasonality of the Savanna Avifauna at Port Moresby Papua New-Guinea." *Ibis* 124 (1982): 252–274.

Crick, H. Q. P., and C. H. Fry. "Effects of Helpers on Parental Condition in Red-throated Bee-eaters *Merops bullocki*." *Journal of Animal Ecology* 55 (1986): 893–906.

Dyer, M., and H. G. P. Crick. "Observations on White-throated Bee-eaters *Merops albicollis* Breeding in Nigeria." *Ostrich* 54 (1983): 52–55.

Emlen, Stephen T., and Peter H. Wrege. "Breeding Biology of White-fronted Bee-eaters at Nakuru Kenya: The Influence of Helpers on Breeder Fitness." *Journal of Animal Ecology* 60 (1991): 309–326.

Emlen, Stephen T., and Peter H. Wrege. "Gender, Status and Family Fortunes in the White-fronted Bee-eater." *Nature* 367, no. 6459 (1994): 129–132.

Garnett, S. "Mortality and Group Cohesion in Migrating Rainbow Bee-eaters (*Merops ornatus*)." *Emu* 85 (1985): 267–268.

Hegner, R. E. et al. "Spacial Organization of the White-fronted Bee-eater (*Merops bullockoides*)." *Nature* 298, no. 5871 (1982): 264–266.

Karthikeyan, S., and J. N. Prasad. "Recent Sighting of Bluebearded Bee-eater (*Nyctyornis athertoni*) (Jardine and Selby)." *Journal of the Bombay Natural History Society* 90, no. 2 (1993): 290–291.

Lessells, C. M. et al. "Why Do European Bee-eater (*Merops apiaster*) Brothers Nest Close Together?" *Behavioral Ecology* 5 (1994): 105–113.

Lessells, C. M. et al. "Individual and Sex Differences in the Provisioning Calls of European Bee-eaters." *Animal Behaviour* 49 (1995): 244–247.

Lill, Alan. "Breeding of Rainbow Bee-eaters in Southern Victoria." *Corella* 17 (1993): 100–106.

Saffer, V. M., and M. C. Calver. "The Size and Type of Prey Taken by Adult Rainbow Bee-eaters in the South-west of Australia." *Emu* 97 (1997): 329–332.

Salewski, Volker, and Mark O. Roedel. "Fish Eating by Red-throated Bee-eaters, *Merops bullocki*." *Ostrich* 71 (2000): 425.

Sridhar, S., and Karanth K. Praveen. "Helpers in Cooperatively Breeding Small Green Bee-eater (*Merops orientalis*)." *Current Science Bangalore* 65 (1993): 489–490.

Wrege, Peter H., and Stephen T. Emlen. "Breeding Seasonality and Reproductive Success of White-fronted Bee-eaters in Kenya." *Auk* 108 (1991): 673–687.

Wrege, Peter H., and Stephen T. Emlen. "Family Structure Influences Mate Choice in White-fronted Bee-eaters" *Behavioral Ecology and Sociobiology* 35 (1994): 185–191.

Peter H. Wrege, PhD

Rollers
(*Coraciidae*)

Class Aves

Order Coraciiformes

Suborder Coracii

Family Coraciidae

Thumbnail description
Medium-sized, brightly colored birds with stout, hook-tipped bills and small, strong syndactylous (toes merged with no intermediate web) feet

Size
9–20 in (22–50 cm); 0.16–0.5 lb (80–250 g)

Number of genera, species
5 genera; 17 species

Habitat
Forests, woodlands, savanna, urban

Conservation status
Vulnerable: 4 species

Distribution
Africa, southern Europe, and southern Asia to northeastern and southeastern Asia, and Australasia, east to the Solomon Islands

Evolution and systematics

Although common throughout much of their range, rollers are poorly known. This scarcity of information is reflected in a sketchy understanding of their evolutionary history. The present predominantly Afrotropical distribution suggests that they originated in ancient Africa, but in *The Origin and Evolution of Birds* (1999), Alan Feduccia points out that the few fossils identified confidently as belonging to the family come from Europe. Dating from the late Eocene and Oligocene (35 million years ago), the fossil *Geranopterus alatus* was unearthed from deposits in France, and is similar to modern *Coracias* species. Three unnamed species date from the middle Eocene (45 million years ago), and were taken from deposits in Germany. Older roller-like birds, which may be nearer to ground rollers, are known from deposits in Britain and the United States (Wyoming).

Three groups of birds known collectively as rollers traditionally are placed in the suborder Coracii, and share a series of skull characteristics that tend to separate them from other groups in the diverse Coraciiformes. Although some anatomical features denote affinities with kingfishers (Alcedinidae), motmots (Momotidae), and bee-eaters (Meropidae), a lack of many derived muscle characteristics found in other Coraciiformes and a primitive condition of the middle-ear bone suggest that rollers form a relatively primitive group. It is speculated, though apparently with little supporting evidence, that they represent generalized, primitive perching birds, and from ancient roller-like birds the now-prevalent passerines or perching birds may have evolved.

Ornithological debate has focused on relationships among "true" rollers, ground-rollers, and the strikingly aberrant courol or cuckoo rollers (*Leptosomus discolor*), with three separate families—Coraciidae, Brachyteraciidae, and Leptosomatidae—as the commonly adopted arrangement. Morphological evidence argues that true rollers and ground rollers had a more recent common ancestor than either had with the courol, so a more relevant arrangement could retain Leptosomatidae, but include within Coraciidae two separate subfamilies—Brachypteraciinae for the five species of ground-rollers and Coraciinae for the 12 species of true rollers.

Physical characteristics

True rollers constitute a fairly homogeneous assemblage of medium-sized, heavy-bodied birds characterized by a proportionately large head, short neck, and stout robust bill. Syndactylous feet are proportionately small but strong. Broad, long wings reflect strong powers of flight. The rather broad tail may be squarish, slightly rounded, or somewhat forked, sometimes with markedly elongated outermost feathers. Variation in size is not great, and ranges from 10 in (26 cm) and 3 oz (90 g) for the blue-throated roller (*Eurystomus gularis*), to 14.5 in (37 cm) and 5.3 oz (150 g) for the blue-bellied roller (*Coracias cyanogaster*). All species are handsomely plumaged in tones of blue or lilac, with olive, chestnut, or pink markings; young birds resemble adults. Modifications in bill structure that evolved in response to different foraging techniques are the key distinguishing feature separating the two genera.

In *Coracias*, the shrike-like bill is strong, arched, and hook-tipped, and is suited to grasping prey captured on the ground. *Eurystomus* species catch flying insects, and the short, wide bill is well adapted to aerial feeding.

Ground-rollers are medium-sized terrestrial birds resembling *Coracias* rollers, but readily distinguished by long legs, distinctive plumage patterns, and short, rounded wings. All are stout-bodied, with a proportionately large head and a strong, robust bill. The large eyes possibly enable these birds to forage effectively in heavily shaded undergrowth, at dusk, or even during the night. Total length for four species with broad, rounded tails is 8.7–13 in (22–33 cm); a very long, graduated tail gives the long-tailed ground roller (*Uratelornis chimaera*) a disproportionate total length of 18 in (46 cm). Bold markings are a feature of the rich plumage coloration, with rufous or dark-green upperparts contrasting with finely patterned underparts and striking facial patterns. Sexes are alike, and young birds resemble adults.

Distribution

In 1971, when commenting on restriction of the courol to Madagascar and the nearby Comoro Islands, Joel Cracraft postulated that the first invasion of Madagascar was by primitive rollers that probably were terrestrial and arboreal. These gave rise to a lineage that became predominantly arboreal, eventually resulting in the evolution of *Leptosomus*. A second invasion by a more coraciid-like stock gave rise to ground-rollers, which now survive only on Madagascar. The only true roller to reach Madagascar is a recent colonizer from Africa.

Africa is the home of true rollers; eight of the 12 species occur on the continent as breeding residents. Outside Africa, they are dispersed widely through much of the Old World, from Ireland and Britain in the west to the Solomon Islands and eastern Australia in the east, and from about 55° N in the western Palaearctic to approximately 38° S in southeastern Australia. Nowhere do more than two species occur together. Western and eastern sectors of the world range each are occupied by a widespread species from different genera: the European roller (*Cocacias garrulus*) in the west, and the dollarbird (*Eurystomus orientalis*) in the east. Restricted insular ranges in the Indonesian archipelago are occupied by Temminck's roller (*Coracias temminckii*) and the azure roller (*Eurystomus azureus*).

Habitat

A strong preference for evergreen forest is shown by four species of ground-rollers, which keep to the heavily shaded undergrowth or forest floor. Occurring only in a narrow coastal strip of arid southwestern Madagascar, the long-tailed ground-roller prefers low, fairly dense deciduous woodland with a sparse groundcover on sandy soil.

Although occurring at high altitudes when crossing major mountain ranges during migration, true rollers are lowland birds, frequenting habitats that vary from tropical or subtropical woodland to most types of open country, including grassy hillsides with scattered trees, plains, scrublands, cultivated fields, and urban parks or gardens. In West Africa, the blue-bellied roller is very much an inhabitant of *Isoberlinia* woodlands, and the racket-tailed roller (*Coracias spatulata*) from southern Africa shows a preference for woodlands in which *Brachystegia*, *Colophospermum*, or *Baikaea* trees dominate. Other species appear to exhibit no marked habitat preferences.

Behavior

Little is known of the life and habits of shy, secretive ground-rollers, but available information indicates they are essentially terrestrial, especially when feeding, and usually encountered singly or in pairs. Instead of flying, they prefer to escape danger by running or standing motionless in a well-concealed position. The brief, guttural call notes are heard mainly during breeding season.

While on migration, some true rollers gather into loose flocks that may come together in large aggregations. However, for most of the year they are strongly territorial, normally encountered in pairs or post-breeding family parties. In the breeding territories, they are noisy and conspicuous, with loud cackling call notes that constantly accompany spectacular acrobatics. This contrasts with quiet, often lethargic behavior in wintering areas. *Eurystomus* species advertise territorial occupation by flying high above treetops and calling loudly, but the spectacular rolling flights undertaken by *Coracias* species give the birds their collective name. The flights feature prominently in territorial and courtship displays. They are made with powerful wingbeats as the bird flies

A dollarbird (*Eurystomus orientalis*) emerges from its nest in a tree hollow in Australia. (Photo by J. Warham/VIREO. Reproduced by permission.)

up at a steep angle, then suddenly tips forward and plunges down with rapid wingbeats while rolling the body from side to side before leveling out and moving away to repeat the sequence. In fine warm weather, true rollers are active throughout the day, spending much time sitting on vantage perches awaiting the appearance of prey. Activity levels decline, often significantly, during rain. Movements undertaken by true rollers vary from regular, seasonal migration, sometimes over vast distances, to local, irregular wanderings. Toward the tropics, seasonal migration tends to be replaced by irregular local movements or post-breeding dispersals.

Feeding ecology and diet

Courols are arboreal foragers, searching amid tree or shrub branches for large insects and small reptiles, especially chameleons. Captured prey is struck repeatedly against a stout limb before being swallowed. Ground-rollers are almost exclusively ground-foraging insectivores, capturing prey by searching amid leaf litter or probing with the bill into soft soil. They take small vertebrates such as frogs or lizards. The more arboreal short-legged ground-roller (*Brachypteracias leptosomus*) prefers to forage from perches in low-to-mid levels of the forest.

Vantage perches are of the utmost importance to true rollers, for it is from these that birds pounce to capture prey on the ground or sally forth to catch flying insects. *Coracias* species take mainly large arthropods, almost always from the ground; *Eurystomus* rollers are almost exclusively aerial feeders, taking flying insects after sallying forth from a high, exposed perch. The latter method is considered to be the more advanced technique. Small, soft-bodied prey are swallowed whole; larger prey are brought back and struck repeatedly against the perch before being swallowed.

Reproductive biology

A summer breeding season has been recorded for ground-rollers, with egg-laying occurring mostly in December. Pairs defend nesting territories, and courtship feeding of females by males has been observed. Little is known of nesting behavior, and the nest of the short-legged ground-roller—the only species known to nest in tree hollows—was not discovered until 1996. Other species nest in a chamber at the end of a burrow excavated by the birds in the ground, and usually two, but up to four, white eggs are laid. Incubation seems to be undertaken only by the female, and nestlings are fed by both parents.

Loud vocalization during flight features prominently in courtship displays of true rollers, but spectacular aerobatics are performed only by *Coracias* species. Bowing displays are performed by paired birds while perched facing each other. Copulation may occur after display flights, or more frequently after the bowing display. For long-distance migrants, courtship begins during spring migration, and nesting gets underway soon after pairs arrive in breeding territories. Nests mostly are in holes in trees, but sometimes in crevices in cliff faces or building walls. Clutches are 3–5 white eggs, and for 18–20 days incubation is undertaken by both sexes, though mostly by the

Newly hatched European rollers (*Coracias garrulus*). (Photo by IFA. Bruce Coleman Inc. Reproduced by permission.)

female. Newly hatched chicks are naked, with pin feathers first appearing at about seven days and full feathering being acquired between 17 and 22 days. Both parents feed nestlings for some 30 days and for up to 20 days after fledging.

Conservation status

Although generally common and apparently able to survive in relatively small, fragmented forest patches, the courol is adversely affected by widespread land clearing. The subspecies *gracilis* is restricted to mountain slopes on Grand Comoro Island, and is estimated to number about 100 pairs. Deforestation poses a more serious threat to ground-rollers, with four species listed as Vulnerable. The long-tailed ground-roller has one of the most restricted ranges of all Madagascar birds. As a group, true rollers have fared well, and most remain common throughout all or part of their ranges. Only the azure roller (*Eurystomus azureus*) is considered Vulnerable because of loss of forest habitat within its restricted insular range. Since the 1960s, significant local declines of European rollers have occurred in central and eastern Europe, but the species remains numerous elsewhere, with counts made in the 1970s of an estimated two to three million overwintering birds in eastern Kenya.

Significance to humans

There is little evidence of the courol being subjected to human interference, though in some districts body parts are used in love potions. Traditionally, it is seen as a bird of good omen or as representing the bond between loving persons. Only hunters show an interest in ground-rollers, which are considered easy targets. All species are killed for food. Hunting for sport, food, or taxidermy affects migrating European rollers passing through the Mediterranean region, especially on Cyprus and Malta where significant numbers are taken each year. The species is widely admired because of its colorful plumage; in agricultural districts it is considered beneficial as a destroyer of insect pests. Elsewhere, there appears to be little interest in rollers among local people.

1. Blue-bellied roller (*Coracias cyanogaster*); 2. European roller (*Coracias garrulus*); 3. Dollarbird (*Eurystomus orientalis*); 4. Long-tailed ground-roller (*Uratelornis chimaera*); 5. Scaly ground-roller (*Brachypteracias squamigera*); 6. Courol (*Leptosomus discolor*). (Illustration by Brian Cressman)

Species accounts

European roller
Coracias garrulus

TAXONOMY
Coracias garrulus Linnaeus, 1758, Sweden. Two subspecies.

OTHER COMMON NAMES
French: Rollier d'Europe; German: Blauracke; Spanish: Carraca Europea.

PHYSICAL CHARACTERISTICS
12.5 in (32 cm); 0.24–0.4 lb (110–190 g). Large roller with no tail-streamers. Head, neck, and underparts light blue; upperparts rufous-brown.

DISTRIBUTION
C. g. garrulus breeds northwest Africa, central and southern Europe to northwest Iran and southwest Siberia; *C. g. semenowi* breeds Middle East to northwest India and southwest China; all winter in sub-Saharan Africa.

HABITAT
Lowlands in open woodlands, wooded grasslands, cultivated fields, urban parks, or gardens.

BEHAVIOR
Seasonal migrant. Noisy, conspicuous in breeding territory; quiet, lethargic at wintering sites. Found in pairs when breeding, but groups or loose flocks on long-distance migration. Active on fine warm days, less active in rain.

FEEDING ECOLOGY AND DIET
Hunts from vantage perch, taking mainly large insects on ground. Prey struck repeatedly against ground or perch before swallowing; regurgitates undigested remains in pellets.

REPRODUCTIVE BIOLOGY
Monogamous; pairs vigorously defend nest. Spectacular rolling flight features in courtship display. Breeds May–July, laying 4–5 white eggs in a tree hollow, crevice in a rock face, or a hole in the wall of building. Incubation, mainly by female, 17–19 days; chicks fed by both parents, fledging at 25–30 days.

CONSERVATION STATUS
Not threatened. Generally common; in 1990s estimated 20,000 pairs in Europe; in 1970s estimated 2–3 million birds wintering in eastern Africa.

SIGNIFICANCE TO HUMANS
Admired because of beauty; beneficial as destroyer of insect pests; hunted for food, sport, and taxidermy. ◆

Blue-bellied roller
Coracias cyanogaster

TAXONOMY
Coracias cyanogaster Cuvier, 1817, Senegal. Monotypic.

OTHER COMMON NAMES
French: Rollier à ventre bleu; German: Opalracke; Spanish: Carraca Blanquiazul.

PHYSICAL CHARACTERISTICS
15 in (37 cm); 0.24–0.38 lb (110–190 g). Dark blue with black back and creamy head and breast.

DISTRIBUTION
Western Africa, from Senegal to southernmost Sudan.

HABITAT
Favors *Isoberlinia* woodland, also forest clearings and plantations.

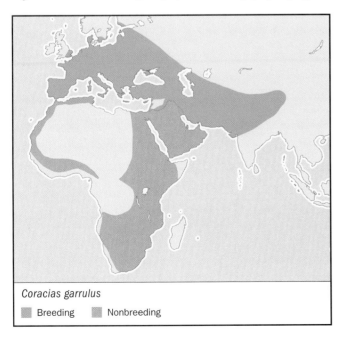

Coracias garrulus

■ Breeding   ■ Nonbreeding

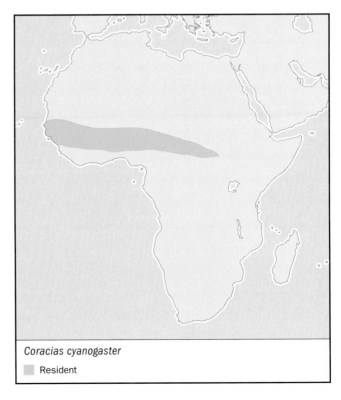

Coracias cyanogaster

■ Resident

BEHAVIOR
Partial migrant. In small groups, roosting together and occupying favored haunts. Conspicuous when sitting on exposed hunting perch or in overhead flight, but noisy only in breeding season.

FEEDING ECOLOGY AND DIET
From high-vantage perch flies down to take large invertebrates and small reptiles on ground below; some flying insects captured in flight. Prey repeatedly struck against ground or perch before swallowing; regurgitates undigested remains in pellets.

REPRODUCTIVE BIOLOGY
Poorly known. Monogamous or polygamous, males being more numerous. Defends nesting territory. Aerial chases and rolling flights in courtship. Breeding during wet season, mainly April–July; nest in tree hollow.

CONSERVATION STATUS
Not threatened. Common in west, less numerous in east.

SIGNIFICANCE TO HUMANS
Admired as most colorful roller. ◆

Dollarbird
Eurystomus orientalis

TAXONOMY
Coracias orientalis by Linnaeus, 1766, Java. Ten subspecies.

OTHER COMMON NAMES
English: Red-billed roller, eastern broad-billed roller; French: Rolle oriental; German: Dollarvogel; Spanish: Carraca Oriental.

PHYSICAL CHARACTERISTICS
10.2 in (26 cm); 0.3–0.35 lb (115–160 g). The white dollar-like circles on outspread wings are diagnostic.

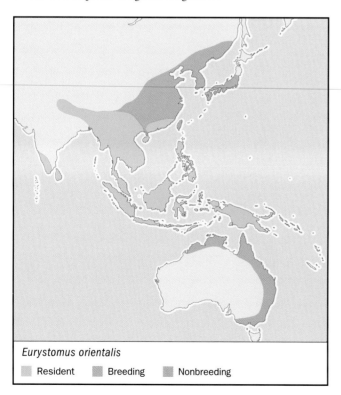

Eurystomus orientalis
■ Resident   ■ Breeding   ■ Nonbreeding

DISTRIBUTION
E. o. orientalis: breeds northwestern India to Indochina, Philippines, and Greater Sunda Islands; winters south and east to Sulawesi and Moluccas. *E. o. calonyx*: breeds northern India to eastern China, southeastern Russia, and Japan; winters south to southern Asia and Greater Sunda Islands. *E. o. laetior*: southwestern India. *E. o. irisi*: Sri Lanka. *E. o. gigas*: Southern Andaman Islands. *E. o. oberholseri*: Simeulue Island, off Sumatra. *E. o. waigiouensis*: New Guinea and adjacent islands. *E. o. crassirostris*: Bismarck archipelago. *E. o. solomonensis*: Solomon Islands. *E. o. pacificus*: Breeds northern and eastern Australia and Lesser Sunda Islands; winters New Guinea and Indonesian archipelago.

HABITAT
Woodlands, forest margins, savanna, farmland, urban parks, or gardens up to 4,900 ft (1,500 m); favors denser woodland where coexisting with *Coracias*.

BEHAVIOR
Migrant at higher latitudes, resident in tropics. Noisy, conspicuous in high, wheeling flight or perched atop a high tree. Occurs as pairs or family parties when breeding, and as flocks when migrating or at swarms of flying insects.

FEEDING ECOLOGY AND DIET
Large insects are captured in flight and crushed in the bill before swallowed.

REPRODUCTIVE BIOLOGY
Monogamous; breeds during summer. Pair vigorously defends nesting territory. Loud calling and aerobatics in courtship. Three or four eggs laid in high tree hollow; incubation 22–23 days; both parents feed chicks. Departs to wintering areas soon after chicks fledge.

CONSERVATION STATUS
Not threatened.

SIGNIFICANCE TO HUMANS
None known. ◆

Scaly ground-roller
Brachypteracias squamigera

SUBFAMILY
Brachypteraciinae

TAXONOMY
Brachypteracias squamiger Lafresnaye, 1838, Madagascar. Monotypic.

OTHER COMMON NAMES
English: Scaled ground-roller; French: Brachyptérolle écaillé; German: Schuppenedracke; Spanish: Carraca-Terrestre Escamosa.

PHYSICAL CHARACTERISTICS
12 in (30 cm). Cryptic plumage blends with foliage. White head and underparts with black and brown scaly pattern; upperparts bronze and green.

DISTRIBUTION
Eastern Madagascar.

HABITAT
Humid forest below 4,900 ft (1,800 m); prefers shaded areas with sparse groundcover.

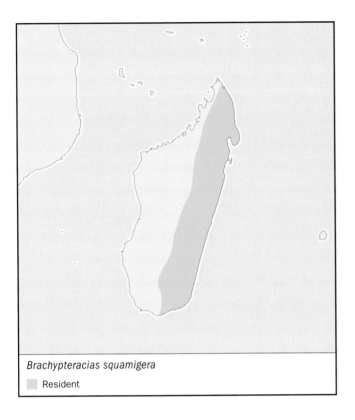

Brachypteracias squamigera

▢ Resident

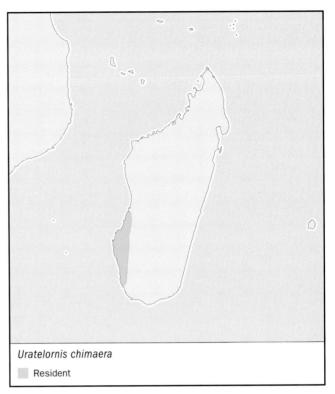

Uratelornis chimaera

▢ Resident

BEHAVIOR
Terrestrial, sedentary; shy, secretive, quiet. Seldom flies, escaping danger by running or standing motionless in concealed position.

FEEDING ECOLOGY AND DIET
Forages mainly in leaf litter on ground; takes insects, snails, small frogs.

REPRODUCTIVE BIOLOGY
Little known; breeds September–January; 1–2 white eggs laid at end of burrow excavated in earth bank; incubation by female at least 18 days; chicks fed by both parents, fledging at 18 days.

CONSERVATION STATUS
Vulnerable; threatened by deforestation and hunting.

SIGNIFICANCE TO HUMANS
Hunted for food. ◆

Long-tailed ground-roller
Uratelornis chimaera

SUBFAMILY
Brachypteraciinae

TAXONOMY
Uratelornis chimaera Rothschild, 1895, Madagascar. Monotypic.

OTHER COMMON NAMES
French: Brachyptérolle à longue queue; German: Langschwanzerdracke; Spanish: Carraca-Terrestre Colilarga.

PHYSICAL CHARACTERISTICS
18 in (46 cm) including tail. Sandy-brown upperparts with dark brown streaks; white underparts; and light blue outer wing feathers.

DISTRIBUTION
Southwestern Madagascar.

HABITAT
Dry *Didierea* woodland on sandy soil up to 330 ft (100 m).

BEHAVIOR
Terrestrial; inconspicuous, mottled plumage blending with leaf litter in shade. Pairs or family parties when breeding, singly at other times. Stands motionless awaiting prey, long tail held horizontal or cocked up. Calls from low branch, often at night.

FEEDING ECOLOGY AND DIET
Forages in leaf litter, taking terrestrial invertebrates.

REPRODUCTIVE BIOLOGY
Monogamous; breeds during summer monsoons; 3–4 white eggs laid in burrow excavated in sandy soil. Courtship features neck stretching and vertical spring up to perch; feeding of female by male. Pair always near nest, but nesting behavior not known.

CONSERVATION STATUS
Vulnerable in very restricted range.

SIGNIFICANCE TO HUMANS
Birds and eggs taken for food. ◆

Courol
Leptosomus discolor

TAXONOMY
Cuculus discolor Hermann, 1783, Madagascar. Three subspecies.

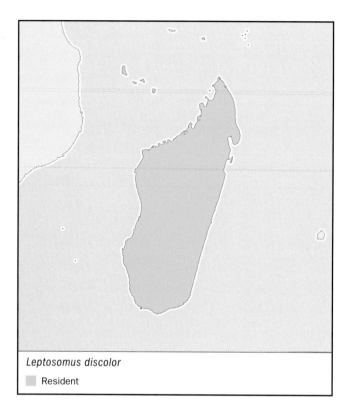

Leptosomus discolor

■ Resident

OTHER COMMON NAMES
English: Cuckoo-roller; French: Courol; German: Kurol; Spanish: Carraca Curol.

PHYSICAL CHARACTERISTICS
20 in (50 cm); 0.48–0.53 lb (219–240 g). The courol or cuckoo roller is a large bird characterized by a massive head with a stout bill; long, slightly rounded wings; a moderately long, broad tail; and very short legs. Back and wings grayish black, strongly glossed with metallic green and mauve-red, underparts ash gray, and crown to nape black (male). In the adult female and young, upperparts are dark brown, barred black and rufous

on the crown; underparts are pale rufous, boldly spotted with black.

DISTRIBUTION
Madagascar and Comoro Islands.

HABITAT
Forest and woodlands, especially at margins. Up to about 6,400 ft (2,000 m), the courol is widely dispersed in forests and wooded areas of Madagascar, being encountered more frequently at or near forest margins, where a preference is shown for large evergreen trees with leafy crowns. On the Comoro Islands, there seems to be a wider habitat tolerance, with the birds less dependent on tall trees.

BEHAVIOR
Arboreal; singly or pairs when breeding, groups at other times. Noisy, conspicuous in flight, inconspicuous at rest in leafy treetops. During the breeding season, courols seem to be highly territorial and usually are encountered singly or in pairs, but at other times are seen in groups of a dozen or more, often with males outnumbering females. When silently feeding or resting amid inner branches of leafy trees, they can escape detection, but in the air they are noisy and conspicuous, the peculiarly undulating flight accompanied by loud, whistling call notes.

FEEDING ECOLOGY AND DIET
Large insects and small reptiles, mainly chameleons, captured amid foliage and struck repeatedly against branch before swallowing.

REPRODUCTIVE BIOLOGY
Monogamous; breeds October–January; 2–4 white eggs laid in tree hollow; incubation by female probably 20 days; chicks fed by both parents, fledging at 30 days. Breeding by courols also takes place during summer monsoon season, when a clutch of up to four white eggs is laid in a hollow limb or tree hole, and incubation by the female lasts at least 20 days. Newly hatched chicks are down covered.

CONSERVATION STATUS
Not threatened, widespread and common.

SIGNIFICANCE TO HUMANS
Largely untouched, but sometimes killed to make love potions. ◆

Resources

Books
del Hoyo, J., A. Elliott, and J. Sargatal, eds. *Handbook of the Birds of the World.* Vol. 6, *Mousebirds to Hornbills.* Barcelona: Lynx Edicions, 2001.

Feduccia, A. *The Origin and Evolution of Birds.* 2nd ed. New Haven: Yale University Press. 1999.

Forshaw J. M., and W. T. Cooper. *Kingfishers and Related Birds.* Vol. 3, *Leptosomatidae, Coraciidae, Upupidae, Phoeniculidae.* Sydney: Lansdowne Editions, 1993.

Fry, C.H., K. Fry, and A. Harris. *Kingfishers, Bee-eaters and Rollers.* London: Christopher-Helm, 1992.

Periodicals
Balanca, G., and M. N. de Visscher. "Observations sur la Reproduction et les Déplacements du Rollier d'Abyssinie *Coracias abyssinica* du Rollier Varié *C. naevia* et du Rollier Africaine *Eurystomus glaucurus* au Nord du Burkina Faso." *Malimbus* 18 (1996): 44–57.

Cassola, F., and S. Lovari. "Food Habits of Rollers During the Nesting Season." *Bollettino di Zoologia* 46 (1979): 87–90.

Forbes-Watson, A. D. "Observations at a Nest of the Cuckoo-roller *Leptosomus discolor.*" *Ibis* 109 (1967): 425–430.

Garbutt, N. "Madagascar's Ground Rollers. Jewels among the Shadows." *Africa Birds and Birding* 5 (2000): 52–57.

Robel, D., and S. Robel. "Zum Verhalten der Blauracke (*Coracias garrulus*) Gegenüber Anderen Vogelarten im Brutgebiet." *Beitr. Vogelkd.* 30 (1984): 361–382.

Sosnowski, J., and S. Chmielewski. "Breeding Biology of the Roller *Coracias garrulus* in Puszeza Pilicka Forest

Resources

(Central Poland)." *Acta Ornithologica* 31 (1996): 119–131.

Thorstrom, R., and J. Lind. "First Nest Description, Breeding, Ranging and Foraging Behavior of the Short-legged Ground-Roller *Brachypteracias leptosomus* in Madagascar." *Ibis* 141 (1999): 569–576.

Organizations
Coraciiformes Taxon Advisory Group. Web site: <http://www .coraciiformestag.com>

Joseph M. Forshaw

Hoopoes
(Upupidae)

Class Aves

Order Coraciiformes

Suborder Bucerotes

Family Upupidae

Thumbnail description
Medium-sized bird with long, thin, decurved bill, prominent erectile crest, broad, rounded wings, and short legs

Size
10.2–12.6 in (26–32 cm); 0.08–0.2 lb (38–89 g)

Number of genera, species
1 genus; 1 species

Habitat
Open country with bare earth or short grass, usually with some trees; needs cavities for nesting

Conservation status
Not threatened; one species probably extinct since 1600

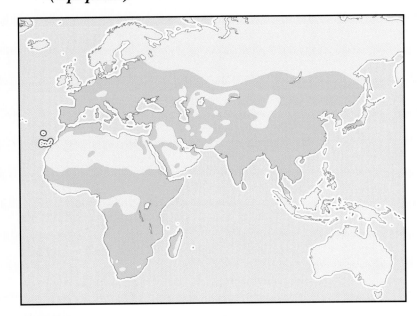

Distribution
Africa, Europe, and Asia

Evolution and systematics

The hoopoe (*Upupa epops*) has long been considered to be related to the hornbills (Bucerotidae), and Tertiary fossil evidence supports this view. Its closest relatives are the wood-hoopoes and scimitarbills (Phoeniculidae), with which the hoopoe shares characters such as feather structure, pterylosis, tongue structure, bill morphology, skeletal features, and egg-white proteins. DNA-DNA hybridization studies suggest that the hoopoe diverged from the hornbills, and the wood-hoopoes and scimitarbills from the hoopoe.

The hoopoe differs from all other Coraciiform birds in having no expansor secondarium muscle and in the newly hatched chick being downy. The bird also has a tufted uropygial gland that produces a foul smell to repel nest intruders. Such characteristics have led some authors to place it in a separate order, the Upupiformes.

Although it is usually regarded as a polytypic single species, some authors regard the Madagascar hoopoe (subspecies *marginata*) as a full species on the basis of its distinctive advertising call and large size. The nine subspecies are separated mainly on size, color, and wing pattern. Subfossils of a large, probably flightless hoopoe (*Upupa antaois*) are known from St. Helena island in the South Atlantic.

Physical characteristics

Several striking plumage and structural features make this medium-sized bird unmistakable. The plumage on the chest varies from pinkish to chestnut, while the broad, rounded wings, the back, and the tail are boldly barred black and white. The spectacular erectile crest is the same color as the head and tipped with black. The bill is long, slender, and decurved; modified musculature allows it to be opened when the bird probes for food. The tongue is reduced. Hoopoes have shorter legs, better suited to their ground-foraging habits.

The sexes are extremely similar. Juveniles are duller than adults, the white in the wings is tinged with cream and the crest and bill are relatively short. Nestlings have long, fluffy white down covering their entire bodies.

Distribution

The hoopoe ranges throughout Africa except in deserts and forested regions, in the drier west of Madagascar, and through Asia and Europe from the Iberian Peninsula north to the Gulf of Finland, the Sakmara River, the southern Lake Baikal region, and the middle Amur and Khungari Rivers. In Asia the species occurs throughout Sri Lanka, Indochina, and Taiwan, east to Japan, and south to the Malay Peninsula. It is a rare straggler south to northern Sumatra, Borneo, and the Philippines. The distribution of the nine subspecies is as follows: *U. e. epops*: northwestern Africa, Canary Islands, and Europe east to south central Russia, northwestern China, and northwestern India; *U. e. major*: Egypt, northern Sudan, and eastern Chad; *U. e. senegalensis*: southern Algeria and Senegal east to Ethiopia and Somalia; *U. e. waibeli*: Cameroon and northern DRC east to Uganda and northern Kenya; *U. e. africana*: central DRC east to central Kenya and south to Cape;

U. e. marginata: Madagascar; *U. e. saturata*: south central Russia east to Japan and south to central China and Tibet; *U. e. ceylonensis*: Pakistan and northern India south to Sri Lanka; *U. e. longirostris*: Assam and Bangladesh east to southern China, south to northern Malay Peninsula and Indochina.

Habitat

Hoopoes occur from boreal to temperate and tropical zones, preferring warm, dry regions with level or undulating terrain. They require bare or short-grassed open land for foraging and cavities in trees, walls, rocks, banks, or termite mounds for nesting. Habitats include pastures, parkland, orchards, vineyards, woodland edges and clearings, steppes, plains, dry and wooded savannas, river valleys, foothills, scrub, semi-desert, and (in Southeast Asia) coastal dune scrub. Hoopoes avoid areas with damp substrates. Only in Madagascar are they associated with the forest, being most common at the margins of heavily wooded forest and even penetrating primary forest. Hoopoes range up to 10,170 ft (3,100 m) above sea level, normally occurring below 6,560 ft (2,000 m).

Hoopoe (*Upupa epops*). (Illustration by Joseph E. Trumpey)

Behavior

The hoopoe is a monogamous species and strongly territorial when breeding. Males begin calling several weeks

Two hoopoes (*Upupa epops*) share an elder (*Sambucus nigra*) branch. (Photo by H. Reinhard/OKAPIA. Photo Researchers, Inc. Reproduced by permission.)

before breeding commences. The song is series of two to five far-carrying "hoop" notes, except in the Madagascar race, which has a soft, rolling purr. Territorial birds use song posts.

The hoopoe is a relatively confiding bird and in some areas has become a commensal of humans. The crest is usually held flat but raised when the bird alights or is excited. The flight is distinctive, with erratic, butterfly-like flapping. Hoopoes perch readily and can climb rough surfaces. They are diurnal, roosting in cavities at night.

Hoopoes are migratory over much of their range. Most Palearctic birds migrate to Africa and southern Asia after breeding. Races breeding in Asia make shorter-range movements to southern Asia, where the Siberian race (*saturatus*) also winters. Local populations in Africa and southern Asia are migratory, resident, or nomadic.

Feeding ecology and diet

The food is mainly insects, particularly larvae and pupae. Prey includes beetles, bugs, grasshoppers, butterflies, dragonflies, ants, termites, and flies. Spiders, earthworms, woodlice, and centipedes are also taken, while lizards, frogs, and small snakes have been recorded.

Hoopoes forage mainly in short grass and on bare soil. They walk about, constantly making short probes into the ground and sometimes pausing to insert the bill fully, opening and closing it to test or seize objects encountered. Hoopoes sometimes dig small holes with the bill to extract prey. They make short runs to catch prey, hawk flying insects, and search below refuse and dry dung, which are turned over with the bill.

Reproductive biology

Hoopoes nest in holes in trees, walls, cliffs, banks, termite mounds, flat ground, and crevices between rocks. Little or no nest material is used, and the nest cavity is often fetid. A nest site may be used for several years.

The male selects the nest site and establishes the territory. Intruders are chased in the air and on the ground. The male courtship-feeds the female, after which activity copulation usually takes place. The monogamous pair often fly slowly round the territory, one behind the other, raising and lowering their crests.

Eggs are produced at a rate of one per day. The clutch size is four to seven in the tropics and five to nine (maximum 12) in temperate regions. The incubation period is 15–18 days, only the female incubates, and hatching is asynchronous. The nestling period is 25–32 days. Fledged young start self-feeding after six days, thereafter remaining with the parents for some weeks. Hoopoes are normally single-brooded, although two or three broods are recorded. Of 172 eggs laid in 24 nests (Europe), 74% hatched and 58% fledged.

Nestlings defend themselves by hissing, jabbing upwards with the bill, producing an evil-smelling secretion from the uropygial gland, and spraying feces. Adults have a striking defense posture, with wings and tail spread on the ground, head thrown back, and bill raised.

Conservation status

This widespread species is locally common in many areas but has suffered obvious losses, especially at the edges of its range. In Europe, the hoopoe's range has been contracting since the late nineteenth century, with an especially strong decline since 1955–60. This decline is often attributed to climatic change, but may be due more to changes in farming practices and land use. In 2001, its European population was estimated at 700,000–900,000, with five to 10 million birds worldwide.

In Africa, Madagascar, and Asia desertification and habitat loss through high-intensity farming practices have adversely affected the hoopoe's numbers. Migrants are still persecuted by hunters in southern Europe and parts of Asia.

Significance to humans

This striking and unmistakable bird has a long pedigree in human culture. It was used as a hieroglyphic and revered in ancient Egypt. The hoopoe figures prominently in Aristophanes' play *The Birds*, features widely in folklore, and has long been celebrated in literature. Its scientific name and vernacular names in several languages, are onomatopoeic.

The hoopoe's diet includes insect pests of agriculture and forestry, and its usefulness in controlling such pests has been recognized in many areas. The hoopoe is consequently very widely protected by national laws.

Resources

Books

Cramp, S., ed. *The Birds of the Western Palearctic.* Vol. 4, *Terns to Woodpeckers.* Oxford: Oxford University Press, 1985.

Fry, C. H., S. Keith, and E. K. Urban, eds. *The Birds of Africa.* Vol. 3. London: Academic Press, 1988.

Glutz von Blotzheim, U. N., and K. M. Bauer. *Handbuch der Vögel Mitteleuropas.* Vol. 9. Frankfurt am Main: Akademische Verlagsgesellschaft, 1980.

Hagemeijer, W. J., and M. J. Blair, eds. *The EBCC Atlas of European Breeding Birds.* London: Poyser, 1997.

Harrison, J. A., D. G. Allan, L. G. Underhill, M. Herremans, A. J. Tree, V. Parker, and C. J. Brown, eds. *The Atlas of Southern African Birds.* Vol. 1, *Non-passerines.* Johannesburg: BirdLife South Africa, 1997.

Stattersfield, A. J., and D. R. Capper, eds. *Threatened Birds of World: The Official Source for Birds on the IUCN Red List.* Cambridge: BirdLife International, 2000.

Periodicals

Löhrl, H. "Zum Brutverhalten des Wiedehopfs *Upupa epops.*" *Vogelwelt* 98 (1977): 41–58.

Skead, C. J. "A Study of the African Hoopoe." *Ibis* 92 (1950): 434–463.

Barry Taylor, PhD

Woodhoopoes

(Phoeniculidae)

Class Aves
Order Coraciiformes
Suborder Bucerotes
Family Phoeniculidae

Thumbnail description
Medium-sized to small birds with long, slender, decurved bill; plumage mainly black, with green or purple gloss; broad, rounded wings and long, graduated tail

Size
8–15 in (20–38 cm); 0.6–3.5 oz (18–99 g)

Number of genera, species
2 genera; 6 species

Habitat
Forest, woodland, savanna

Conservation status
Not threatened

Distribution
Sub-Saharan Africa

Evolution and systematics

The fossilized remains of woodhoopoe-like birds have been found from Eocene and more recent Miocene deposits in Europe, but modern woodhoopoes are now confined to Africa. Woodhoopoes occur in two main forms, based on studies of their anatomy, DNA-DNA hybridization, and nuclear DNA. The two groups are usually separated as subfamilies, sometimes even as families, the true woodhoopoes, Phoeniculinae (genus *Phoeniculus*), which live in groups and breed cooperatively, and the smaller scimitarbills, Rhinopomastinae (genus *Rhinopomastus*), which live and breed only as pairs. The separation occurred about 10.2 million years ago, based on estimation from a molecular clock.

Woodhoopoes are most closely related to the more widespread hoopoe (*Upupa epops*), family Upupidae, with which they share a unique stirrup-like structure of the inner ear bone. The hoopoe feeds mainly while walking on the ground, whereas woodhoopoes feed mainly by hopping about on tree branches. Molecular and anatomical evidence suggests that woodhoopoes and hoopoes are close relatives of hornbills (family Bucerotidae), which lack the stirrup-like stapes but have their own unique neck vertebrae, and also have terres-

trial and arboreal members. Still more distant relatives are other members of the order Coraciiformes.

Three species of woodhoopoe are recognized in each subfamily. All species can be divided further into clearly recognizable subspecies that, in the widespread and variable green woodhoopoe (*P. purpureus*), are sometimes even treated as separate species, the violet woodhoopoe (*P. damarensis*) and the black-billed woodhoopoe (*P. somaliensis*).

Physical characteristics

Medium-sized to small birds, mainly black, with a green or purple gloss to the plumage and a dark brown iris. In four species there are patches of white across the wing and on the tips of tail feathers. The broad, rounded wings and long, graduated tail allow buoyant, dexterous and, at times, rapid flight. The bill and feet are black in juveniles of all species, but the bill becomes orange-red in adults of three species, as do the feet in two species. The bill is long, thin, and decurved, especially in two species of scimitarbill, as the name implies. The long toes, short legs, and hooked claws are thick and strong.

A male green woodhoopoe (*Phoeniculus purpureus*) offering food to a female. (Photo by Kerry T. Givens. Bruce Coleman Inc. Reproduced by permission.)

Distribution

Throughout sub-Saharan Africa, with at least one and in some areas up to three species present as in northeast Africa. Absent from treeless desert and steppes, and only two species occur in the more northern areas of tropical forest. The green woodhoopoe is the most widespread member of the subfamily Phoeniculinae, while the common scimitarbill (*R. cyanomelas*) and black woodhoopoe (*R. aterrimus*) are the most widespread members of the subfamily Rhinopomastinae.

Habitat

Woodhoopoes require sufficient tree holes to roost and nest, and adequate areas of bark and twig foraging surfaces to conceal their invertebrate food. Suitable habitat ranges from scattered trees within arid steppes, to the rainforest canopy. Two species, the forest woodhoopoe (*P. castaneiceps*) and the white-headed hoopoe (*P. bollei*), are confined to forest. The widespread green woodhoopoe, black woodhoopoe, and the common scimitarbill occupy a range of woodland and savanna types, while the Abyssinian scimitarbill (*R. minor*) is confined to more arid savanna.

Behavior

All woodhoopoes appear to be sedentary and territorial. True woodhoopoes are gregarious, usually living in groups of six to eight birds, while scimitarbills live only as pairs. All species appear to roost in a tree hole, a behavior that conserves energy in the group-roosting green woodhoopoe. True woodhoopoes are noisy and each group defends its territory with loud cackling calls and exaggerated bowing and flagging displays. Scimitarbills are much quieter

and usually advertise their occupancy only with melodious hooting.

Feeding ecology and diet

The woodhoopoe diet is mainly insects, arachnids, and their larvae, with a few fruits, other invertebrates, or small vertebrates when available. Prey is located mainly by probing into crevices and cracks, or levering off bark, on the trunks and limbs of trees. Strong feet allow woodhoopoes to hunt at all angles, including hanging upside down. Larger species tend to search on larger branches, species with thicker, straighter bills dig and lever more often, while the small scimitarbills probe into the finest holes on the smallest twigs. Some species will also feed on the ground or hawk flying insects, but no species needs to drink regularly since they obtain moisture from their prey.

Reproductive biology

All woodhoopoes nest in a tree hole. Most cavities are natural, but old nest holes excavated by barbets and woodpeckers are also occupied, and they will rarely use a hole in the ground or a building. Where known, pairs are monogamous, and the male and adult group members courtship-feed the breeding female before egg-laying and continue to provision her and later the chicks during nesting. The female lays and incubates the clutch of two to five gray or blue-green eggs with their oval, pitted shells. She broods the young chicks after hatching, but later leaves them and joins the male or group to help deliver food. The nestlings initially retain the growing feathers within their sheaths, giving them a prickly appearance. Juveniles stay with the parents for several months after fledging, especially in gregarious species where they become helpers. The breeding biology of forest species and those in remote areas is virtually unknown. Woodhoopoes, especially chicks, have a musty odor derived from the oil of the preen gland. When combined with a hissing, head-waving, threat display and excretion of copious smelly feces, this deters predators at the nest or roost. In the green woodhoopoe, the odor is the chemical product of a symbiotic *Enterococcus* bacterium, unique to woodhoopoes and only discovered in 2001. In addition to repelling potential predators, it also has a hygienic, antibacterial effect.

Conservation status

No species of woodhoopoe is threatened. Locally, in western Kenya, Uganda, and South Africa, numbers have been reduced by collection of timber for fuel or building material.

Significance to humans

Woodhoopoes have no particular significance to humans. Onomatopoeic local names recognize the noisy *Phoeniculus* species and the smelly and aggressive chicks.

1. Green woodhoopoe (*Phoeniculus purpureus*); 2. Common scimitarbill (*Rhinopomastus cyanomelas*). (Illustration by Joseph E. Trumpey)

Species accounts

Green woodhoopoe
Phoeniculus purpureus

SUBFAMILY
Phoeniculinae

TAXONOMY
Promperops purpureus J. F. Miller, 1784, Cape Province, South Africa. Eleven subspecies.

OTHER COMMON NAMES
English: Red-billed woodhoopoe; French: Irrisor moqueur; German: Baumhopf; Spanish: Abubilla-arbórea Verde.

PHYSICAL CHARACTERISTICS
13–15 in (32–37 cm), 2–3.5 oz (54–99 g). Largest, most widespread woodhoopoe. Black plumage with variable green and purple gloss, white spots on flight feathers and tip of tail, red bill and feet (bill black in some populations), male bill length and mass 18–20% more than female.

DISTRIBUTION
Sub-Saharan Africa. *P. p. guineensis*: northern Senegal, Mail, eastern to northern Ghana; *P. p. senegalensis*: southern Senegal east to Ghana; *P. p. niloticus*: northeast Zaire, Sudan to western Ethiopia; *P. p. abyssinicus*: northern Ethiopia and Eritrea; *P. p. neglectus*: central Ethiopia; *P. p. somaliensis*: southeast Ethiopia, Somalia to northeast Kenya; *P. p. marwitzi*: southern Somalia, Kenya, eastern Uganda, south to eastern South Africa; *P. p. granti*: southern Ethiopia and Kenya; *P. p. damarensis*: southwest Angola and northwest Namibia; *P. p. angolensis*: eastern Angola and eastern Namibia, east to western Zambia and western Zimbabwe; *P. p. purpureus*: southeastern South Africa.

HABITAT
Open woodland, savanna, and dry mixed scrub with a few larger trees, to over 6,560 ft (2,000 m) above sea level.

BEHAVIOR
Group-living, advertise territory with loud cackling calls and bowing displays. Often allopreen and exchange food in social behavior.

FEEDING ECOLOGY AND DIET
Probe cracks or bark on tree trunks and limbs for invertebrate food such as caterpillars, beetle larvae, and spiders. Sometimes dig in animal dung on the ground, hawk insects in flight, or pirate food from nestlings of other species.

REPRODUCTIVE BIOLOGY
Breed cooperatively, alpha pair assisted by adult and juvenile helpers. Nest in tree hole, or rarely in ground or building, usually during late summer wet season. Lay two to five eggs, incubation 17–18 days, nestling period about 30 days, female and chicks fed by group. Sometimes parasitized by greater honeyguide (*Indicator indicator*).

CONSERVATION STATUS
Not threatened. Widespread and common throughout its range, including in a number of large national parks.

SIGNIFICANCE TO HUMANS
None known, but often found in gardens and parks. ◆

Phoeniculus purpureus
▨ Resident

Common scimitarbill
Rhinopomastus cyanomelas

SUBFAMILY
Rhinopomastinae

TAXONOMY
Falcinellus cyanomelas Vieillot, 1819, Orange River, northwestern Cape, South Africa. Two subspecies.

OTHER COMMON NAMES
English: Scimitar-billed woodhoopoe, greater/black scimitarbill; French: Irrisor namaquois; German: Sichelhopf; Spanish: Abubilla-arbórea Cimitarra.

PHYSICAL CHARACTERISTICS
10–12 in (26–30 cm), 0.8–1.3 oz (24–38 g). Small, plumage black with purple gloss and white spots on primaries and tips of outer tail feathers. Black bill and feet, strongly decurved bill slightly larger in male.

DISTRIBUTION
Southeast and southern Africa. *R. c. schalowi*: southern Uganda, southwest and central Kenya and southern Somalia south to Zambia, and northeast South Africa; *R. c. cyanomelas*: southwest Angola and Namibia, east to northern South Africa.

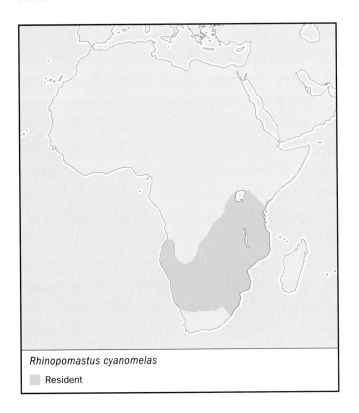

Rhinopomastus cyanomelas

Resident

HABITAT

Open woodland, savanna, and dry thorn-scrub with a few larger trees, to over 6,560 ft (2,000 m) above sea level.

BEHAVIOR

Solitary, in pairs, or in family groups after breeding,. Roosts in tree hole or, if unavailable, clings to trunk. Territorial call a mournful three to five hooting notes.

FEEDING ECOLOGY AND DIET

Mainly insect larvae and spiders, plus some other small insects, fruit, buds, and nectar, taken among finer branches and twigs. Probe at all angles, often inserting only the slender lower mandible. Join mixed species foraging flocks during dry winter months.

REPRODUCTIVE BIOLOGY

Breed as territorial pair. Nest in tree hole, usually during early summer wet season. Lay two to four eggs, white or slightly-tinted in color, incubation 17–18 days, female and chicks fed at nest by male, nestling period about 24 days during which female emerges to help male. Chicks produce smelly preen oils and feces in defense.

CONSERVATION STATUS

Not threatened. Widespread and common throughout its range, including in a number of large national parks.

SIGNIFICANCE TO HUMANS

None known. ◆

Resources

Books

Law-Brown, Janette. *Chemical Defense in the Red-billed Woodhoopoe Phoneiculus purpureus.* Unpublished M.Sc. thesis. Cape Town: University of Cape Town, 2001.

Ligon, J.D. "Family Phoeniculidae (Woodhoopoes)." In *Handbook of Birds of the World.* Vol. 6, *Mousebirds to Hornbills,* edited by J. del Hoyo, A. Elliott, and J. Sargatal. Barcelona: Lynx Edicions, 2001.

Sibley, C.G., and J.E. Ahlquist. *Phylogeny and Classification of Birds: A Study in Molecular Evolution.* New Haven and London: Yale University Press, 1990.

Stuart, Chris, and Tilde Stuart. "Wood-hoopoes." In *Birds of Africa: From Seabirds to Seed-eaters.* Cambridge, MA: MIT Press, 1999.

Periodicals

Burton, P.K.J. "Anatomy and Evolution of the Feeding Apparatus in the Avian Order Coraciiformes and Piciformes." *Bulletin of the British Museum (Natural History)* 47, no. 6 (1984): 1–113.

Johansson, U.S., T.J. Parsons, M. Irestedt, and P.G.P. Ericson. "Clades Within the 'Higher Land Birds,' Evaluated by Nuclear DNA Sequences." *Journal of Zoological Systematics and Evolutionary Research* 39 (2001): 37–51.

Ligon, J.D., C. Carey, and S.H. Ligon. "Cavity Roosting, Philopatry, and Cooperative Breeding in the Green Woodhoopoe May Reflect a Physiological Trait." *Auk* 105 (1988): 123–7.

Plessis, M.A. du, and J.B. Williams. "Communal Cavity Roosting in Green Woodhoopoes: Consequences for Energy Expenditure and the Seasonal Pattern of Mortality." *Auk* 111 (1994): 292–9.

Steyn, P. "The Breeding Biology of the Scimitarbilled Woodhoopoe." *Ostrich* 70, no. 3&4 (1999): 173–8.

Wanless, Ross. "Red-billed Woodhoopoes Go on the Defensive. Preening Power." *Africa Birds & Birding* 6, no. 1 (2001): 55–59.

Organizations

Woodhoopoe Research Project, FitzPatrick Institute of African Ornithology, University of Cape Town. P.O. Rondebosch, Cape Town, Western Cape 7700 South Africa. Phone: +27 (0)21 650-3290. Fax: +27-21-650-3295. E-mail: fitztitute@botzoo.uct.ac.za Web site: <http://www.fitztitute.uct.ac.za>

Other

Coraciiformes Taxon Advisory Group. <http://www.coraciiformestag.com>.

Alan C. Kemp, PhD

Hornbills
(Bucerotidae)

Class Aves

Order Coraciiformes

Suborder Bucerotes

Family Hornbills (Bucerotidae)

Thumbnail description
Medium to large-sized, stocky, highly vocal birds with long, slightly decurved bills topped by casques of various shapes, sizes, and colors

Size
11.8–47.3 in (30–120 cm); 0.22–13.2 lb (100g–6 kg)

Number of genera, species
14 genera; 54 species

Habitat
Forest, woodlands, and savanna

Conservation status
Endangered: 2 species; Critically Endangered: 2 species; Vulnerable: 5 species; Near Threatened: 12 species

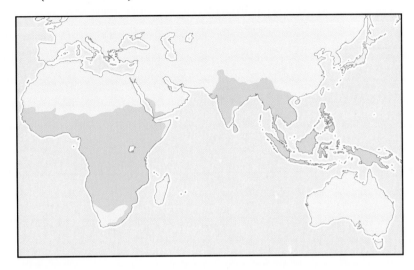

Distribution
Sub-Saharan Africa, south and Southeast Asia, New Guinea, and the Solomon Islands

Evolution and systematics

In his classic work *The Hornbills* published in 1995, Alan Kemp wrote "Trying to decide what other groups of birds are most closely related to hornbills is not quite so easy." Time has not made that decision any easier; the classification of these bizarre, large-billed birds is still debated. Most modern taxonomic treatments place hornbills within the order Coraciiformes together with their closest relatives: the hoopoe (Upupidae) and the woodhoopoes and scimitarbills (Phoeniculidae). This classification is based primarily on similarity in foot and jaw morphology, and a prolonged retention of quills by nestlings, which gives them a prickly "pin-cushion" look. All birds of these families nest in tree holes, but only hornbills seal the entrance to their cavities.

As of 2001, science recognizes 54 species of hornbills grouped within 14 genera and two subfamilies. All but two species are classified within the subfamily Bucerotinae. The exceptions are the terrestrial ground-hornbills, which fall within the subfamily Bucorvinae. The distinction between Bucerotinae and Bucorvinae is based on unique feather lice and anatomical and behavioral differences such as a greater number of neck vertebrae and the lack of nest-sealing behavior in the Bucorvinae. In 2001, S. Huebner and colleagues conducted detailed molecular studies of the two groups. They found that ground-hornbills were probably the earliest form.

All 54 hornbill species display unique anatomical features that clearly identify their affinities. These include being blessed with long, sweeping eyelashes on their upper lids and a fusion of the first two cervical vertebrae to provide support

for large bills. All hornbills lack carotid arteries as well as the short feathers under the wings that cover the primary and secondary flight feathers of other birds. Finally, hornbills have unusual kidneys in that they are two-lobed instead of three, and the Z chromosome, one of a pair of sex chromosomes, is oversized.

Physical characteristics

Hornbills are among the most flamboyant birds of their habitat. The oversized, slightly decurved bills topped by sometimes outlandish casques shaped as bumps, ridges, or horns make hornbills an unforgettable component of any landscape. Hornbills vary tremendously in size and shape, starting with the large, long-legged southern ground-hornbill (*Bucorvus leadbeateri*) weighing up to 13.2 lb (6 kg), and going down to the 0.26 lb (120 g) red-billed dwarf hornbill (*Tockus camurus*). Males are always larger and stouter than females but the greatest dimorphism often occurs in bill length with males having up to 30% longer bills. Hornbill plumage is described as "drab," lacking the brilliant colors of relatives such as the kingfishers (Alcedinidae) and rollers (Coraciidae). However, the bold black-and-white patterns of many forest hornbills and the delicate gray pied patterns of many *Tockus* species are far from dull. Add in bills and casques of brilliant orange, yellow-gold, deep crimson, or shiny black, and patches of bare skin around the eyes and throat in a kaleidoscope of garish hues, and you have a colorful group of birds.

Plumage color and size and shape of the casque identify the age and sex of an individual. Newly fledged hornbills have

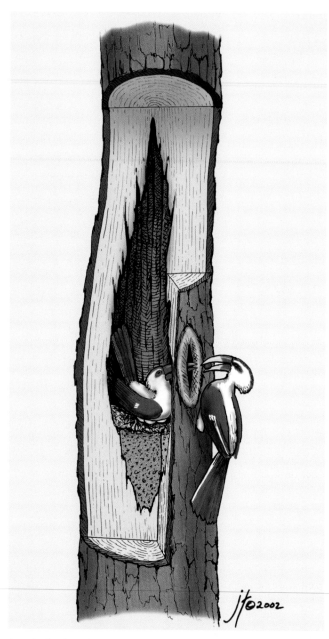

A male feeds a female Von der Decken's hornbill (*Tockus deckeni*) while she is mudded up in the nest with their eggs. (Illustration by Joseph E. Trumpey)

Numerous authors have described the noise produced by flying hornbills as that of an approaching train. This incredible "whooshing," produced in different pitches depending on the species' size, is a result of wing structure. Because hornbills lack the small feathers that normally cover the shafts of the primary and second flight feathers, each powerful stroke of the wing allows air to pass through and vibrate the large feathers.

The most outstanding feature, and the one from which hornbills acquire their common name, is the casque on the top of the bill. Casques vary from the mere ridge of the red-billed hornbill (*Tockus erythrorhynchus*) to the wash-board bumps of the wreathed hornbill (*Rhyticeros undulatus*) and the elaborate banana of the rhinoceros hornbill (*Buceros rhinoceros*). The function of casques, which may take up to six years to develop, is the topic of many debates. It is possible that casques provide structural support for a long bill. Casques may also serve an acoustic function by helping amplify a hornbill's call. Additionally, casques may be attractive to the opposite sex. The helmeted hornbill uses its casque in bizarre, aerial displays where individuals of either sex collide in mid-air, casque-to-casque. The head-butting competitions always occur near fruiting fig trees (*Ficus* spp.). Although Gustav Schneider once reported that helmeted hornbills perform this comical ritual when they are intoxicated on fermented figs, observations from Sumatra indicate that this acrobatic act may be in defense of clumped food resources.

Distribution

Hornbills occur across sub-Saharan Africa, through India and southern Asia, across the Indonesian and Philippine archipelagos, and east to the Solomon Islands. There are no hornbills in the New World. Within the hornbill family, 23 species inhabit Africa, while the remaining 31 are found in Asia. The largest and most widespread genus in Africa, *Tockus*, is represented by 13 species. The most ubiquitous of the *Tockus* species is the African gray hornbill (*Tockus nasutus*), a medium-sized gray and white bird with populations occurring from the shores of Mauritania in the west, east to the Red Sea, and as far south as Namibia and South Africa. In Asia, *Aceros* and *Rhyticeros* hornbills dominate with five genera each, occurring from Bhutan and northern India in the west to the Solomon Islands in the east. Two species of the genus *Ocyceros*, the Malabar gray hornbill (*Ocyceros griseus*) and Indian gray hornbill (*Ocyceros birostris*), occur exclusively in India while the third species, the Sri Lankan gray hornbill (*Ocyceros gingalensis*), is restricted, as its name implies, to the island of Sri Lanka. Several hornbill species, especially those occupying oceanic islands, have restricted distributions. The taritic hornbills are a prime example. The Luzon (*Penelopides manillae*), Visayan (*Penelopides panini*), and Mindanao (*Penelopides affinis*) taritic hornbills occur only on a few neighboring islands within the Philippine archipelago. A fourth Philippine species, the Mindoro taritic hornbill (*Penelopides mindorensis*) is endemic to the small island of Mindoro while the fifth species of the genus, the Sulawesi taritic hornbill (*Penelopides exarhatus*) is found only on the Indonesian island of Sulawesi. Several other species are endemic to single islands, including the Sulawesi red-knobbed hornbill (*Aceros cassidix*), the Sumba

underdeveloped casques and small bills, but after the first year of life, appearances converge on that of their adult counterparts. In species where sexes differ in color as adults, determining the gender of the young can be difficult. For example, in almost all *Aceros*, *Rhyticeros*, *Penelopides*, and *Tockus* species, the young, regardless of their sex, resemble their fathers for the first year of life. The opposite is true for the *Bycanistes* and *Ceratogymna* who resemble the adult female. Young of the northern ground-hornbill (*Bucorvus abyssinicus*) and a few *Tockus* species show plumage true to their sex while chicks of the rufous hornbill (*Buceros hydrocorax*) are radically different from both parents.

hornbill (*Rhyticeros everetti*), and the Narcondam hornbill (*Rhyticeros narcondami*).

Habitat

Arid deserts, scrubby woodlands, cool mountains, and steamy rainforests all constitute hornbill habitat. In general, however, hornbills are birds of the forest. Of the 30 species found in India and Southeast Asia, only the Indian gray hornbill lives in open savanna. In Africa, where forests are less extensive, the proportion of savanna-dwelling species increases accordingly; 13 of 23 species reside in savannas and woodlands while the remaining 10 inhabit forests. Species occupying savannas tend to have more extensive ranges but, like the red-billed hornbill, may be separated into many distinct populations by imposing bands of woodland. Endemic species are, by default, limited to habitats available within their restricted range. This is particularly true of insular species like the Sumba hornbill, which occupies all forest types on its native island.

There are key features that must be present in all hornbill habitats—an ample number of large trees for nesting, an adequate year-round supply of food, and enough habitat area to support a viable population. Each species has a particular set of requirements, which may help explain why several species can simultaneously occupy the same habitat. In the forests of Thailand, where nine hornbill species may occur together, a small Tickell's brown hornbill (*Anorrhinus tickelli*) is able to use nest holes of smaller dimensions than the larger great hornbill (*Buceros bicornis*). On Sumatra, where a similar number of species coexist, they generally forage on different diet items; when diet overlap occurs, as with rhinoceros and helmeted hornbills, they partition their habitat by feeding at different heights in the canopy. Habitat quality will influence the number of hornbills an area can support. Habitat size also limits hornbill populations. On the island of Sumba, hornbills are rare or absent from forest patches less than 3.6 mi^2 (10 km^2) in size.

Behavior

Hornbills generally wake at dawn, preen their feathers, then begin their search for food. Normally, hornbills move about in pairs, but some species are found in family groups of three to 20 individuals. Some hornbills gather in large flocks around clumped food resources. The Sulawesi red-knobbed hornbill is occasionally seen in groups of more than 100 individuals at large fruiting figs. In Thailand, wreathed hornbills roost in flocks of over 1,000 individuals. The plain-pouched hornbill (*Aceros subruficollis*) takes the record for the largest aggregations; over 2,400 individuals were counted in Malaysia in 1998 traveling to roost. Roosts may serve as "information centers" where individual birds can reduce foraging time by following a knowledgeable, long-term resident. As Alan Kemp summarizes, these massive gatherings are "wonderfully noisy and visually stunning, and must surely rate among the foremost spectacles of the bird world."

Hornbills are believed to be monogamous. The only research on the faithfulness of hornbills failed to find evidence

This male Sulawesi red-knobbed hornbill (*Aceros cassidix*) holds a fig—an important food source for many hornbills—in his bill. (Photo by Margaret F. Kinnaird. Reproduced by permission.)

of extra-pair paternity in Monteiro's hornbill (*Tockus monteiri*), boosting confidence in their monogamous behavior. Monogamy may have many variations on the theme. Among cooperative social groups, there is generally one monogamous breeding pair and a number of offspring who become "helpers" during the nesting season, delivering morsels to their mother and siblings and defending a mutual territory. Cooperative breeding occurs more often in hornbills than any other bird family, and may characterize up to one-third of all hornbill species.

Many hornbills range widely but none of these movements is considered migratory. Most hornbills are sedentary and many are territorial. The majority of *Tockus* and small-bodied forest hornbills are territorial throughout the year. Larger hornbills such as the *Aceros* and *Rhyticeros* that rely on scattered fruit resources, may range over 21 mi^2 (58 km^2) and only defend temporary territories around nest sites.

Hornbills communicate through a wide range of spectacular calls and each species can be identified by its vocalizations. Loud calls announce territories, or in the non-territorial species, aid in maintaining contact. Territorial ground-hornbills "boom" when their boundaries are invaded and non-territorial wreathed hornbills bark like dogs while coordinating flocks. While calls are important in dense forest habitats, visual displays are more prevalent in open grasslands. For example, the Hemprich's hornbill (*Tockus hemprichii*) has an elaborate territorial display that resembles the mechanical movements of a wind-up toy; the bill is pointed skyward, while the bird whistles, and lifts and fans its tail over its back.

Feeding ecology and diet

Hornbill diets span the spectrum from animals to fruits and seeds but most are omnivorous, mixing meat and fruit in their meals. Among *Tockus*, diets tend more toward insects, scorpions, lizards, snakes, and small mammals, while *Ocyceros*

A male wreathed hornbill (*Aceros undulatus*) preens. (Photo by Terry Whittaker. Photo Researchers, Inc. Reproduced by permission.)

and taritic diets include more fruit. Omnivory is the rule among the territorial, group-living hornbills. Because animal prey often occurs at low density and is available year-round, hornbills may develop defendable territories in which dietary needs for the pair or group are satisfied. Additionally, these species maximize exploitation of their territories by using abundant but ephemeral fruit resources as they become available. The availability of fruit resources within a habitat may determine the degree of omnivory observed.

Heavy reliance on fruits requires that hornbills have large home ranges, and may affect reproductive rates. Fruit diets combined with large home ranges have important consequences for forest ecology. As hornbills travel, they disperse seeds of the fruits they relish, playing a role in regenerating the forests in which they live.

Reproductive biology

Hornbill reproduction tends to coincide with rainfall and increased food supply. In seasonal African savannas, *Tockus* species begin courtship and reproduction with the rains, when invertebrates and fruits are plentiful. The opposite occurs on Sulawesi where lack of rainfall stimulates reproduction in the Sulawesi red-knobbed hornbill, so the burst in fruit supply

occurs immediately after fledging. In aseasonal Bornean rainforests, reproduction appears to be supra annual, tied to highly cyclical peaks in food supply. Breeding in these populations may be controlled by the rate at which pairs regain condition between reproductive cycles. In fig-rich forests of North Sulawesi, hornbills breed every year, usually returning to the same nest tree.

The hornbill's unique nesting behavior is the feature that has most fascinated students of nature. All hornbills are holenesters, preferring natural cavities in trees or rock crevices. Unlike any other group of birds, the female hornbill seals the entrance to her nest cavity, leaving only a narrow slit through which she, and later her chicks, receive food from her mate. In most species, the male ferries mud to the female who then works for several days to seal the cavity entrance. Where mud is a rare commodity, the female uses her own feces as building material.

Nest sealing is believed to have evolved as a form of predator defense, for protection against other intruding hornbills, and to enforce male fidelity. Nest sealing has been described as an example of male chauvinism in which the male cloisters his female, forcing her to depend on him for survival. In reality, the female incarcerates herself and later frees herself, forcing the male to provide for her and their offspring. Because the male is busy provisioning his family, he is incapable of maintaining two nests, and the female can be sure of his complete attention.

The onset of breeding begins with courtship. When in flight, courting pairs act as though they are attached by an invisible rubber band, reacting swiftly to each other's movements. They perch in cozy proximity, engage in mutual preening, and exchange food gifts as a demonstration of their ardor. Other clues of the onset of breeding include the intensification in color of the exposed fleshy areas around the face and throat, reflecting hormonal changes. Nest inspection increases in frequency until copulation occurs and the female enters the nest cavity.

The number of eggs, their size, and the length of incubation are all correlated with body size. Clutch size ranges from two to three eggs in large hornbills and up to eight for smaller hornbills. Incubation runs from 23-49 days in small and large species, respectively. Eggs hatch in intervals and the emerging chicks are naked and translucent pink with closed eyes. Feather growth begins within a few days and as chicks develop, the skin blackens and begging calls change from feeble cheeps to loud, insistent calls.

The timing of female emergence varies tremendously; some females accompany their chicks from the nests and others leave well before chicks fledge. Research on Monteiro's hornbill suggests that females emerge to ensure survival when their body condition reaches its lowest point.

Male hornbills can be impressive providers. Although many *Tockus* species carry items to the nest one-by-one, most hornbills collect multiple food items, stuffed into a bulging gullet before delivering a load to the nest. A Sulawesi redknobbed hornbill once delivered 162 fruits in one trip, a load equivalent to nearly 20% of his body weight.

Nesting success is high for those species studied. In southern Africa, chicks fledged from 90–92% of the nests of four *Tockus* species and in Thailand, 80% of great hornbill nests monitored fledged young. Sulawesi red-knobbed hornbills averaged 80% nesting success over three years, but this figure plummeted to 62% during the 1997 El Niño/ENSO fires. Smaller hornbills fledge up to four chicks, but large hornbills rarely fledge more than one chick per year.

Conservation status

Only 16% of all hornbill species are classified as being under some level of threat, ranging in increasing degree from Vulnerable to Critical and Endangered, according to the IUCN. An additional 12 species, however, are considered Near Threatened and will probably experience a decline in status within the twenty-first century. Africa presently has no hornbills in danger of extinction; only two West African forest inhabitants, the yellow-casqued hornbill (*Ceratogymna elata*) and the brown-cheeked hornbill (*Bycanistes cylindricus*), are classified as Near Threatened. All nine species suffering endangerment reside in Asia, and most (77%) occur on small oceanic islands. The Sumba and Narcondam hornbills, both single-island endemics, are classified as Vulnerable, with total populations hovering around 4,000 and 300, respectively. The situation in the Philippines is especially urgent. Rapidly dwindling forests contain two species ranked as the most endangered hornbills in the world, the Visayan and Mindoro taritic hornbills, as well as two species classified as Critical, the Sulu (*Anthracoceros montani*) and rufous-headed (*Aceros waldeni*) hornbills, and one Vulnerable species, the Palawan hornbill (*Anthracoceros marchei*). There are no rigorous population estimates for these species, but we assume populations are extremely small and may vanish within decades unless conservation measures are adopted.

The underlying threat to hornbill populations is habitat alteration resulting in forest loss and fragmentation. As forests become smaller and more isolated, hornbill populations decline, resulting in increased vulnerability to extinction from natural disasters such as disease. Protection of hornbill populations and their habitats within conservation areas of adequate size offer some hope for their long-term persistence. In the late 1990s, two parks were established on Sumba to aid in the conservation of the Sumba hornbill, and the Philippines have proposed to establish the Central Panay Mountains National Park (NP) for the Visayan taritic hornbill. In India and Africa, vast tracts of savanna and forest have been protected as parks for decades. The long-term success of hornbill conservation in these parks, however, depends on active management to ensure that they are more than "parks on paper."

Unsustainable hunting for food, pets, and body parts is also a problem. Although illegal, trade in helmeted hornbill ivory continues. Great and oriental pied hornbill casques are common souvenirs in Thai and Laotian markets. Traditions that require feathers or skulls take a toll on living birds. Female *Kenyalang* dancers of Malaysia carry up to 10 hornbill tail feathers in each hand, thus supplying a full complement of 20 dancers can cost up to 80 hornbills.

The female great (or concave-casqued) hornbill (*Buceros bicornis*) is sealed in the nest with the eggs and chicks, and the male must bring food to her. (Photo by Aaron Ferster. Photo Researchers, Inc. Reproduced by permission.)

The 1990s have seen a dramatic increase in awareness of hornbill ecology and conservation needs. The number of hornbill studies, especially those by range-country biologists, escalated during this time and continues to increase. Developments in hornbill research and conservation are quickly communicated to the global community through the IUCN Species Survival Commission's Hornbill Specialist Group and facilitated by Internet communication. Only with such global attention are we able to finance local initiatives and put pressure on a range of state governments to conserve these unusual birds.

Significance to humans

Like many other groups of birds, hornbills are hunted for food and consumed for medicine. In Africa, parts of the ground-hornbill are eaten to improve health and sagacity, whereas in India, the great hornbill, the Indian pied hornbill, and the Indian gray hornbill are rendered into oils that supposedly aid in childbirth and relieve gout and joint pains. In Indonesia, the meat of the Sumba hornbill is roasted and eaten to relieve rheumatism and asthma. Because they are easily tamed, hornbills are captured and traded for pets or exhibition. Unlike any other group of birds, however, hornbills play special roles in the folklore and ceremonies of the countries where they occur. Long, elegant tail feathers are the most sought-after hornbill part, but heads and casques are also coveted. The Nishis people of Arunachal Pradesh, India, attach the upper beak of the great hornbill to rattan *bopiah* caps as traditional male headgear. Neighboring Wanchos of eastern Arunachal use the warm, chestnut-colored neck feathers of rufous-necked hornbills to cover caps. On Borneo, the helmeted and rhinoceros hornbills reach mythical proportions in the eyes of the local inhabitants. The helmeted hornbill, in particular, is strongly associated with headhunting. C. Hose, an early twentieth century naturalist and explorer, reported that only someone who has taken a human head is allowed to wear the intricately carved earrings created from the "ivory" of the helmeted hornbill casque, or to adorn themselves with the bird's long, central tail feathers. Helmeted hornbills are also believed to judge souls leaving their mortal existence.

Today, hornbills are increasingly highlighted as local mascots or state birds. This is especially true in Asia. The great hornbill is the state bird of Arunachal Pradesh, northern India. The rhinoceros hornbill has been adopted as the state bird of Sarawak, Malaysia, where it appears on tourism advertisements, T-shirts, and even the state coat-of-arms. In Indonesia, the helmeted hornbill, the Sulawesi red-knobbed hornbill, and the Sumba hornbill proudly serve as official mascots for three provinces.

1. Sulawesi red-knobbed hornbill (*Aceros cassidix*); 2. Sumba hornbill (*Rhyticeros everetti*); 3. Silvery-cheeked hornbill (*Bycanistes brevis*); 4. Helmeted hornbill (*Rhinoplax vigil*); 5. Great hornbill (*Buceros bicornis*); 6. Monteiro's hornbill (*Tockus monteiri*); 7. Red-billed hornbill (*Tockus erythrorhynchus*); 8. Visayan tarictic hornbill (*Penelopides panini*); 9. Plain-pouched hornbill (*Rhyticeros subruficollis*); 10. Southern ground-hornbill (*Bucorvus leadbeateri*). (Illustration by Joseph E. Trumpey)

Species accounts

Southern ground-hornbill
Bucorvus leadbeateri

SUBFAMILY
Bucorvinae

TAXONOMY
Bucorvus leadbeateri Vigors, 1825, Lower Bushman River, South Africa. Monotypic.

OTHER COMMON NAMES
English: Ground hornbill; African ground hornbill; French: Bucorve du Sud; German: Kaffernhornrabe; Spanish: Cálao Terrestre Sureño.

PHYSICAL CHARACTERISTICS
35.4–39.4 in (90–100 cm); female 4.9–10.1 lb (2.23–4.58 kg), male 7.6–13.6 lb (3.46–6.18 kg). Largest hornbill species; black with white primaries and red throat skin.

DISTRIBUTION
Eastern South Africa, Botswana, northern Namibia, Angola, and southern Burundi and Kenya.

HABITAT
Woodland and savanna.

BEHAVIOR
Territorial, defending areas as large as 36 mi² (100 km²) in S. Africa. Hunts on the ground in cooperative groups.

FEEDING ECOLOGY AND DIET
Most carnivorous of all hornbills, consuming arthropods, snakes, lizards, small mammals, and carrion using powerful dagger-like bill.

REPRODUCTIVE BIOLOGY
Cooperative breeder with juvenile offspring assisting dominant breeding pair. Generally lays two eggs September through December. Nests in unsealed tree or rock face lined with dry leaves and grass. Incubation 37–43 days; fledging c. 86 days; younger hatchling starves to death within one week to a month.

CONSERVATION STATUS
Not threatened. Widespread and common but at low densities and declining in some areas of South Africa and Zimbabwe.

SIGNIFICANCE TO HUMANS
Revered, but eaten for food and medicinal purposes. ◆

Red-billed hornbill
Tockus erythrorhynchus

SUBFAMILY
Bucerotinae

TAXONOMY
Buceros erythrorhynchus Temminck, 1823, Podor, Senegal. Four subspecies.

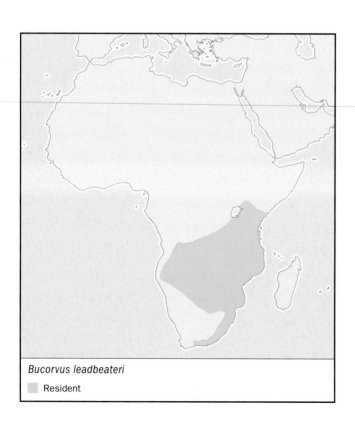

Bucorvus leadbeateri
▨ Resident

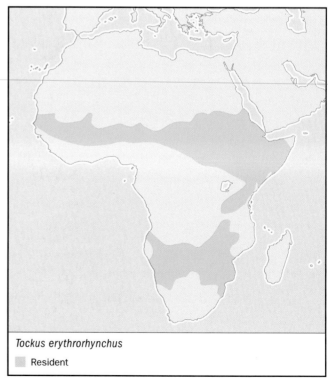

Tockus erythrorhynchus
▨ Resident

OTHER COMMON NAMES
English: African red-billed hornbill, South African red-billed hornbill, Damaraland red-billed hornbill; French: Calao à bec rouge; German: Rotschnabeltoko; Spanish: Toco Piquirorojo.

PHYSICAL CHARACTERISTICS
13.8 in (35 cm); female 0.2–0.44 lb (90–200 g), male 0.27–0.48 lb (124–220 g). Small, black and white with spotted wing coverts; long slender red bill with small casque.

DISTRIBUTION
T. e. erythrorhynchus: Niger Delta to Ethiopia and Somalia, south to Tanzania; *T. e. kempi*: Southern Mauritania, Senegal to Niger Delta; *T. e. rufirostris*: Southern Angola, northern Namibia, Zambia, southern Malawi and northeastern South Africa; *T. e. damarensis*: Northwestern and central Namibia.

HABITAT
Open savannas, woodland, and dry thorn-scrub.

BEHAVIOR
Territorial, maintaining boundaries by calling and displaying on conspicuous perch early morning.

FEEDING ECOLOGY AND DIET
Active forager. Commonly joins other birds when feeding.

REPRODUCTIVE BIOLOGY
Lays two to seven eggs one to two months after start of rains. Incubation 23–25 days; fledging 39–50 days and female emerges with first fledgling. Remaining chicks reseal the nest cavity using their own droppings and undigested food. Nesting success ranged from 90% to 94% in Transvaal and Kenya, respectively but overall productivity is around 1.5 chicks.

CONSERVATION STATUS
Not threatened. Widespread and common, mixing well with domestic stock on open range as long as sufficient nest trees available.

SIGNIFICANCE TO HUMANS
None known. ◆

Monteiro's hornbill
Tockus monteiri

SUBFAMILY
Bucerotinae

TAXONOMY
Tockus monteiri Hartlaub, 1865, Benjuela, Angola. Monotypic.

OTHER COMMON NAMES
French: Calao de Monteiro; German: Monteirotoko; Spanish: Toco Angoleño.

PHYSICAL CHARACTERISTICS
19.7 in (50 cm); female 0.59–0.93 lb (269–423 g), male averages 0.81 lb (370 g). Small, brown-and-white with white spotted wings and large, dark red bill.

DISTRIBUTION
Southwestern Angola and northwestern and central Namibia.

HABITAT
Dry scrub and thornbush, generally the driest habitat of any hornbill.

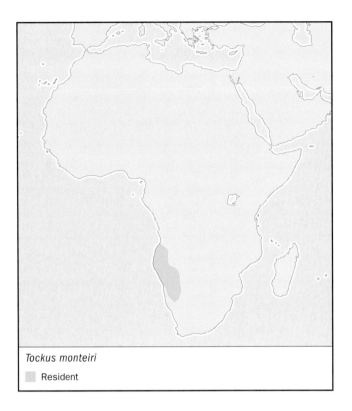

Tockus monteiri

▨ Resident

BEHAVIOR
Territorial, semi-terrestrial, highly vocal birds. Sometimes bob up and down when calling; during territorial displays bow the head, hunch the wings, and cluck.

FEEDING ECOLOGY AND DIET
Forages mainly on the ground, consuming primarily insects.

REPRODUCTIVE BIOLOGY
Usually nests in rock faces near water. Lays two to eight eggs generally after a period of rain. Incubation 24–27 days; fledging c. 45 days.

CONSERVATION STATUS
Not threatened. Common and widespread within its limited range.

SIGNIFICANCE TO HUMANS
Excellent species for research because of its open habitat and willingness to nest in artificial nest boxes. ◆

Great hornbill
Buceros bicornis

SUBFAMILY
Bucerotinae

TAXONOMY
Buceros bicornis Linnaeus, 1758, Bengkulu, Sumatra. Monotypic.

OTHER COMMON NAMES
English: Great Indian hornbill, great pied hornbill, giant hornbill, concave-casqued hornbill; French: Calao bicorne; German: Doppelhornvogel; Spanish: Cálao bicorne.

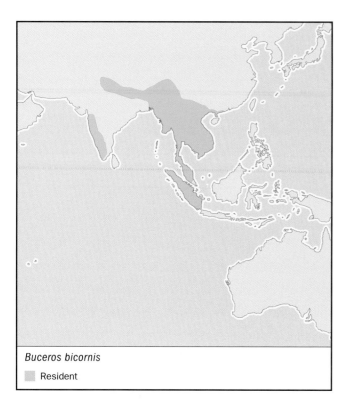

Buceros bicornis
◼ Resident

SIGNIFICANCE TO HUMANS
Hunted for trophies, food, and medicine. Considered "superior meat" in India. Heads are used to decorate ceremonial hats in India and are commonly traded as souvenirs in Thailand. ◆

Helmeted hornbill
Rhinoplax vigil

SUBFAMILY
Bucerotinae

TAXONOMY
Rhinoplax vigil Forster, 1781, Sumatra. Monotypic.

OTHER COMMON NAMES
English: Great-helmeted hornbill, solid-billed hornbill; French: Calao à casque rond; German: Schildschnabel; Spanish: Cálao de Yelmo.

PHYSICAL CHARACTERISTICS
43.3–47.3 in (110–120 cm); female 5.742–6.25 lb (2.61–2.84 kg), male 6.73 lb (3.06 kg). Very large, dark brown and white with short red bill colored with preen oil; high casque and long, white tail feathers.

DISTRIBUTION
South Myanmar and south Thailand, Malaysia, Sumatra, and Borneo.

HABITAT
Rainforest bird preferring primary habitat below altitude of 4,900 ft (1,500 m) but capable of using selectively logged forest.

PHYSICAL CHARACTERISTICS
37.4–41.4 in (95–105 cm); female 4.74–7.37 lb (2.15–3.35 kg), male 5.72–7.48 lb (2.6–3.4 kg). Large, pied hornbill with long yellowish bill (usually dyed with preen oil) and a large, flat, double-pointed casque. Black with white plumage at head, wing coverts, and tail.

DISTRIBUTION
Southwestern India, southern Himalayas, east to Myanmar, south China and Vietnam, south down the Malaysian peninsula and Sumatra.

HABITAT
Prefers large blocks of primary rainforest but in Indonesia found on forest edge and in disturbed habitats.

BEHAVIOR
Generally found as resident pairs and believed to be territorial. Makes a variety of loud calls audible from more than 873 yd (800 m). When aggressive, bounces up and down on perch and flicks the bill.

FEEDING ECOLOGY AND DIET
Feeds primarily in the canopy of fruiting trees, but also takes animal prey. Especially fond of figs.

REPRODUCTIVE BIOLOGY
Nests high in large, forest trees, generally in January through April. Lays one to four eggs; incubation 38–40 days with fledging occurring 72–90 days after hatching. Female molts wing and tail feathers after sealing and emerges well before the oldest chick fledges.

CONSERVATION STATUS
Considered Near Threatened and listed on the Convention for International Trade in Endangered Species' (CITES) Appendix II, prohibiting unauthorized possession, sale, or importation of living birds or their parts.

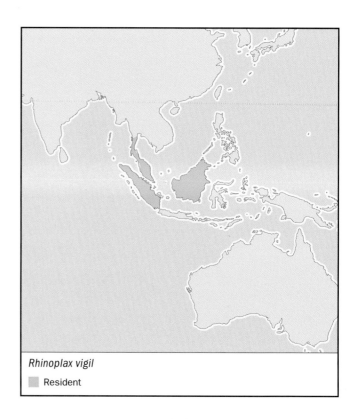

Rhinoplax vigil
◼ Resident

BEHAVIOR

Believed to be territorial. Distinctive loud call has a series of orientation "Tok" followed by a cascading laughter. Individuals of either sex occasionally engage in strange, aerial head-butting behaviors, especially near fruiting figs.

FEEDING ECOLOGY AND DIET

Appears to be a fig specialist. Studies in Malaysia, Borneo, and Sumatra confirm the diet to be 98–100% figs with up to 12 species of figs eaten regularly.

REPRODUCTIVE BIOLOGY

Little known. Reported to lay aseasonally but in southern Sumatra tends to fledge young in May/June.

CONSERVATION STATUS

Considered Near Threatened and listed on Appendix I of CITES. Locally common where habitat is intact but probably declining through most of its range due to hunting and forest destruction. Extinct in Singapore in 1950.

SIGNIFICANCE TO HUMANS

One of the most significant species in traditional Southeast Asian cultures; strongly associated with head-hunting. Feathers and "ivory" are highly coveted for traditional dances and ceremonial decorations and although illegal, carved casques are still traded internationally. ◆

Visayan tarictic hornbill
Penelopides panini

SUBFAMILY
Bucerotinae

TAXONOMY
Buceros panini Boddaert, 1783, Panay, Philippines. Two subspecies.

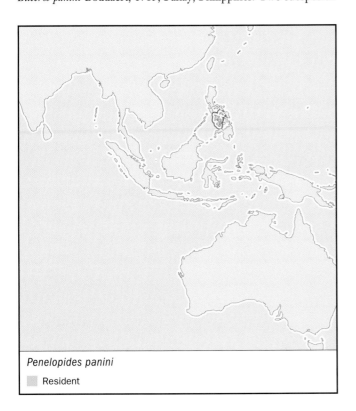

Penelopides panini

▪ Resident

OTHER COMMON NAMES

English: Visayan Hornbill, Panay tarictic hornbill, rufous-tailed hornbill; French: Calao tarictic; German: Visayan-Tariktikhornvogel; Spanish: Cálao Chico de Panay.

PHYSICAL CHARACTERISTICS

17.7 in (45 cm); weights unrecorded. Small in size with prominently ridged bill. Black with yellowish head and rufous tail and underparts; female all black with rufous tail.

DISTRIBUTION

Philippine islands of Masbate, Panay, Sicogon, Pan de Azucar, Guimaras, Negros, and Ticao.

HABITAT

Prefers primary rainforest but will visit fruiting trees in secondary habitat up to 4,900 ft (1,500 m) altitude.

BEHAVIOR

Territorial, living in family groups of two to three and, rarely, 12 birds.

FEEDING ECOLOGY AND DIET

Omnivorous. Forages mid-canopy and at forest edge.

REPRODUCTIVE BIOLOGY

Cooperative breeder; lays two to three eggs March through April; incubation not recorded but nesting cycle c. 95 days. Female molts while breeding and exits cavity with eldest fledgling.

CONSERVATION STATUS

One of the most Endangered hornbills due to habitat loss and excessive hunting.

SIGNIFICANCE TO HUMANS

Hunted for food. ◆

Sulawesi red-knobbed hornbill
Aceros cassidix

SUBFAMILY
Bucerotinae

TAXONOMY

Buceros cassidix Temminck, 1823, Sulawesi. Sometimes classified as *Rhyticeros cassidix*. Monotypic.

OTHER COMMON NAMES

English: Red-knobbed hornbill, knobbed hornbill, island hornbill, Celebes hornbill, Sulawesi wrinkled hornbill; French: Calao à cimier; German: Helmhornvogel; Spanish: Cálao grande de Célebes.

PHYSICAL CHARACTERISTICS

27.6–31.5 in (70–80 cm); female weight unknown, male 5.2–5.5 lb (2.36–2.5 kg). Black with white tail; high wrinkled casque is red; neck is rufous (male) or black (female); ridged yellow beak with blue throat skin.

DISTRIBUTION

Indonesian island of Sulawesi and neighboring islands of Lembeh, Togian, Muna, and Buton.

HABITAT

Prefers primary lowland rainforest below 3,600 ft (1,100 m) altitude.

Aceros cassidix

 Resident

BEHAVIOR
Wide-ranging, non-territorial bird. Usually seen in pairs and observed in large numbers at fruiting figs, rarely up to 120 individuals. Emits loud barking calls that can be heard more than 1.2 mi (2 km) away. High mobility and reliance on fruit makes these hornbills critical agents of seed dispersal and forest regeneration.

FEEDING ECOLOGY AND DIET
One of the most frugivorous hornbills with a diet of up to 90% fruit comprised of more than 60 species during a year. Forage primarily in the top of the canopy.

REPRODUCTIVE BIOLOGY
Nests in high densities of up to 10 pairs/km² beginning June/July at end of rainy season, so fledging is timed with fruiting peak. Lays two to three eggs; incubation 32–35 days and nestling period c. 100 days resulting in nesting cycle of c. 139 days. Female emergence highly variable, ranging from 58 to 140 days but generally before chick. Only one chick fledges but nesting success high at an average of 80%.

CONSERVATION STATUS
Not threatened. Locally common attaining densities of 130 birds/mi² (51 birds/km²) in Tangkoko Nature Reserve, North Sulawesi—the highest density ever recorded for a forest hornbill. Distribution, however, becoming restricted and patchy due to massive deforestation.

SIGNIFICANCE TO HUMANS
Feathers and casques, believed to impart power and insure invincibility, are used to decorate headdresses and drums for the *Cakalele*, a traditional warrior's dance. Less frequently, heads are hung in rafters to bring power to the homes and protect against evil spirits. Meat also eaten. ◆

Plain-pouched hornbill
Rhyticeros subruficollis

SUBFAMILY
Bucerotinae

TAXONOMY
Rhyticeros subruficollis Blyth, 1843, Tenasserim. Monotypic.

OTHER COMMON NAMES
English: Plain-pouched wreathed hornbill, Burmese hornbill, Tenasserim hornbill; French: Calao à gorge claird; German: Blythhornvogel; Spanish: Cálao Gorgiclaro.

PHYSICAL CHARACTERISTICS
25.6–27.6 in (65–70 cm); female weight unknown, male 4.0–5.0 lb (1.81–2.27 kg). Black with short white tail, yellowish bill, and low ridged casque. Male has rufous crown with yellow throat skin; female all black with blue throat skin.

DISTRIBUTION
Southern Myanmar, western and southern Thailand, and northern Malaysia.

HABITAT
Lowland rainforest and mixed deciduous hill forest.

BEHAVIOR
Reports of enormous flocks of more than 2,000 individuals traveling long distances near the Malay-Thai border suggest this species to be wide-ranging and makes extensive local movements.

FEEDING ECOLOGY AND DIET
Little known but believed to feed primarily on fruits.

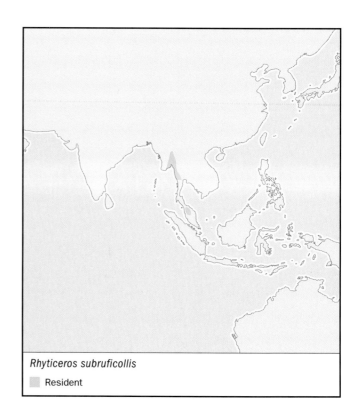

Rhyticeros subruficollis

 Resident

REPRODUCTIVE BIOLOGY
Little information but believed to lay one to three eggs around January to March.

CONSERVATION STATUS
Classified as Vulnerable and listed on CITES Appendix I. Little known regarding status due to limited research and repeated confusion with wreathed hornbill.

SIGNIFICANCE TO HUMANS
Hunted for food and kept as pets. ◆

Sumba hornbill
Rhyticeros everetti

SUBFAMILY
Bucerotinae

TAXONOMY
Rhyticeros everetti Rothschild, 1897, Sumba. Monotypic.

OTHER COMMON NAMES
English: Sumba Island hornbill, Sumba wreathed hornbill, Everett's hornbill; French: Calao de Sumba; German: Sumba-hornvogel; Spanish: Cálao de la Sumba.

PHYSICAL CHARACTERISTICS
21.7 in (55 cm); No weights available. Small and black with long, all-black tail. Head and neck rufous with blue throat skin, ridged casque, and pale yellow bill.

DISTRIBUTION
Restricted to the Indonesian island of Sumba.

HABITAT
Patches of evergreen, monsoon, and gallery forests.

BEHAVIOR
Non-territorial but may be resident within certain forest patches.

FEEDING ECOLOGY AND DIET
Highly frugivorous but no detailed studies available. Seeds of 16 fruit species found under one nest tree; most observations of feeding birds in strangling figs.

REPRODUCTIVE BIOLOGY
Little known regarding nesting cycle.

CONSERVATION STATUS
Classified as Vulnerable and listed on CITES Appendix II. Population estimated at 4,000 individuals.

SIGNIFICANCE TO HUMANS
Meat is eaten to relieve rheumatism and asthma. Species becoming increasingly popular in pet trade. ◆

Silvery-cheeked hornbill
Bycanistes brevis

SUBFAMILY
Bucerotinae

TAXONOMY
Bycanistes brevis Friedmann, 1929, Usambara Mountains, Tanzania. Monotypic.

OTHER COMMON NAMES
French: Calao à joues argent; German: Silberwangen-Hornvogel; Spanish: Cálao Cariplateado.

PHYSICAL CHARACTERISTICS
23.6–27.6 in (60–70 cm); female 2.3–3.2 lb (1.05–1.45 kg), male 2.78–3.1 lb (1.26–1.4 kg). Medium in size with distinctive silvery gray feathering on face. Black with white rump and tail coverts. High yellowish casque with dark brown bill.

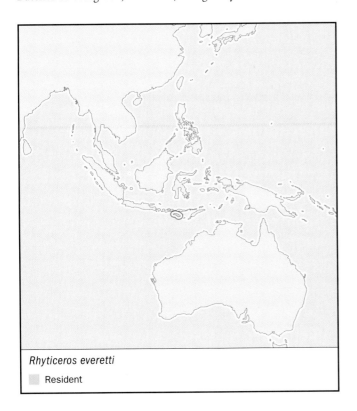

Rhyticeros everetti

◻ Resident

Bycanistes brevis

◻ Resident

DISTRIBUTION

Disjunct distribution in three major areas: Ethiopian highlands and southeastern Sudan; central Kenya through eastern and southern Tanzania and Malawi; and central Mozambique to southeastern Zimbabwe.

HABITAT

Occupies a wide range of habitats including montane and coastal evergreen forest, gallery forest and deciduous forest, and woodland.

BEHAVIOR

Non-territorial, usually found in pairs but roosts communally in groups of up to 200 birds and will fly long distances, especially during dry-season in search of rare, fruiting trees.

FEEDING ECOLOGY AND DIET

Feeds on a wide variety of fruit, especially figs and small, single seeded fruits and hard nuts.

REPRODUCTIVE BIOLOGY

Lays one to two eggs at different times of the year depending on geographic location. Incubation 40 days; fledging 77–80 days, totaling 107–138 days for the entire nesting period.

CONSERVATION STATUS

Not threatened. Locally common although patchily distributed. Catholic habitat requirements and high mobility make the species less vulnerable to deforestation. Important seed disperser.

SIGNIFICANCE TO HUMANS

Probably hunted for meat but otherwise no known significance to humans. ◆

Resources

Books

BirdLife International. *Threatened Birds of Asia: BirdLife International Red Data Book*. Cambridge, United Kingdom: BirdLife International, 2001.

del Hoyo, J., A. Elliott, and J. Sargatal, eds. *Handbook of the Birds of the World*. Vol. 6, *Mousebirds to Hornbills*. Barcelona: Lynx Edicions, 2001.

Kemp, A. *The Hornbills*. Oxford: Oxford University Press, 1995.

Poonswad, P., ed. *The Asian Hornbills: Ecology and Conservation*. Bangkok: Thai Studies in Biodiversity, 1998.

Periodicals

Anggraini, K., M. Kinnaird, and T. O'Brien. "The Effects of Fruit Availability and Habitat Disturbance on an Assemblage of Sumatran Hornbills." *Bird Conservation International* 10 (2000): 189–202.

Hadiprakarsa, Y., and M. Kinnaird. "Foraging Characteristics of an Assemblage of Four Sumatran Hornbill Species." *The Third International Hornbill Workshop Abstracts*. (2001). Hornbill Research Foundation, Bangkok, Thailand.

Huebner, S., R. Prinzinger, and M. Wink. "Phylogenetic Relationships in Hornbills (Aves, Bucerotiformes): Inferences from Nucleotide Sequences from the Mitochondrial Cytochrome *b* gene." *Journal für Ornitologie* (in press).

Kemp, A. "The Role of Species Limits and Biology in the Conservation of African Hornbills." *Ostrich* (in press)

Klop, E., T. Hahn, M. Kauth, S. Engel, L. Lastimoza, and E. Curio. "Diet Composition and Food Provisioning of the Visayan Tarictic Hornbill (*Penelopides panini panini*) During the Breeding Season." *Ecololgy of Birds* 22 (2000).

Organizations

Hornbill Research Foundation. c/o Department of Microbiology, Faculty of Science, Mahidol University, Rama 6 Rd, Bangkok, 10400 Thailand. Phone: +66 22 460 063, ext. 4006. E-mail: scpps@mucc.mahidol.ac.th

Other

Coraciiformes Taxon Advisory Group. "Hornbills." <http://www.coraciiformestag.com> (1 June 2001).

Threatened Birds of Asia. The BirdLife International Red Data Book. "Threatened Bird Species Account." <http://www.adb.or.id/> (10 June 2001).

Margaret Field Kinnaird, PhD

Piciformes
(Woodpeckers and relatives)

Class Aves

Order Piciformes

Number of families 6

Number of genera, species 62 genera; 383 species

Photo: Groove-billed barbets (*Lybius dubius*). (Photo by Daniel Zupanc. Bruce Coleman Inc. Reproduced by permission.)

Evolution and systematics

The six families in the Piciformes order are: honeyguides (Indicatoridae), woodpeckers, wrynecks, and piculets (Picidae), barbets (Capitonidae), toucans (Ramphastidae), jacamars (Galbulidae), and puffbirds (Bucconidae)

The order Piciformes takes its name from the Roman forest god Picus who, according to myth, was turned into a woodpecker by the sorceress Circe for spurning her amorous advances. Like the eponymous Picus, most piciform birds are forest dwellers, and most share a particular adaptation to life in the trees: zygodactylous or "yoke-toed" feet, with two toes pointing forward and two toes pointing backward. This arrangement of digits helps piciform birds get a grip on rough bark while hopping along branches and up and down tree trunks.

Zygodactyly is not unique to piciforms, and in the eighteenth and early nineteenth century, taxonomists grouped various bird families together on the basis of this common foot structure plus other traits. Linneaus, for example, used the trait to group parrots and cuckoos with woodpeckers and toucans in the order Picae. Illiger (1811) also used foot structure as a factor when placing these four groups plus trogons, puffbirds, and jacamars in an order he called Scansores (from the Latin *scansum*, "to climb"). Marshall and Marshall (1871) similarly recognized an order Scansores, but placed toucans, barbets, cuckoos, and turacos in this category.

In 1953, Beecher (1953) proposed that barbets, jacamars, puffbirds, toucans, woodpeckers and honeyguides form a "natural unit," noting similarities in the jaw muscles, tongue, and other traits; many modern classification schemes retain this grouping of the six families.

Subsequent workers, however, have questioned the close grouping of jacamars and puffbirds with the other piciforms. In 1972, Sibley and Ahlquist presented evidence from electrophoretic protein analyses to suggest that jacamars and puffbirds were more closely allied to kingfishers (Coraciiformes) than to woodpeckers and their allies. After completing more sophisticated DNA hybridization studies, the same researchers reported in 1990 that the evidence supports grouping of woodpeckers with honeyguides and toucans with barbets (family Capitonidae) in the order Piciformes, but that jacamars (family Galbulidae) and puffbirds (family Bucconidae) should be placed in a separate order, the Galbuliformes. In 2001, Hofling et al. also suggested removing these two families from the Piciformes, but, based on the structure of the shoulder girdle, said that Galbulidae and Bucconidae more closely resemble the Coraciiformes than the Piciformes. This volume follows the convention of many contemporary taxonomies by placing all six families in the Piciformes.

On the evidence of their DNA hybridization studies, Sibley and Ahlquist also conclude that piciform birds diverged

Green-barred woodpecker (*Chrysoptilus melanochloros*) in Goias, Brazil. (Photo by John S. Dunning. Bruce Coleman Inc. Reproduced by permission.)

from coraciiform birds in the Upper Cretaceous, 98 to 65 million years ago (mya). The oldest known fossil woodpecker bones, however, have been dated only to the Miocene, about 25 mya. A specimen of petrified wood collected in Arizona, dated to the Eocene (40–50 mya), includes a well-preserved woodpecker cavity and entrance hole.

Anatomical as well as genetic evidence supports the idea that jacamars and puffbirds differ from the other piciform families. For example, birds in former two groups have two carotid arteries, whereas members of the other four families have a single (left) carotid artery. Furthermore, jacamars and puffbirds have bare skin over the preen gland; most other piciform species have a feather-covered preen gland. Jacamars and puffbirds have an appendix; other piciformes lack an appendix. And in both jacamars and puffbirds, the syrinx is expanded and drumlike; however, this is not the case for the other families. Finally, jacamars and puffbirds are distinct from other piciforms in their nesting habits; whereas other piciforms make their nests or lay their eggs in tree cavities, these two groups most often breed in burrows that they excavate in soil.

Physical characteristics

Piciformes are small to medium-sized birds of forests and woodlands. In addition to having distinctive "X"–shaped feet, members of this group share other skeletal features: they have 14 cervical vertebrae; all of the thoracic vertebrae are unattached; the sternum has four (two pairs of) notches; and they have five complete ribs.

Piciformes share other common features in their musculature, digestive system, and plumage. For example, many members of this group have stout, sturdy beaks, which they use for gouging wood or other substrates to obtain food. Also, in most species the adult birds lack down feathers (jacamars are an exception to this pattern). And though plumage patterns and colors are quite varied for the group as a whole, combinations of black and white with accents of red and yellow are common, and males and females of a species often look alike. Woodpeckers (one of three groups in the family Picidae) are unusual in this family and indeed among all birds in having extra-stiff tail feathers, which they use to brace their bodies against tree trunks while climbing vertically or hammering with their beaks. Barbets do use their tails as a brace, but only while excavating the nest cavity.

Distribution

Most piciforms are year-round residents in their home ranges and do not migrate. Just the same, this is one of the most widespread avian orders, mostly thanks to the woodpeckers, which are represented on five continents (they are absent only from Australia and Antarctica). Members of the five other piciform families are less widely distributed; toucans, jacamars, and puffbirds are restricted to the New World tropics, and honeyguides occur only in Africa and southeast Asia. Barbets are found both in both the New World and Old World tropics.

Habitat

Almost all birds in this group are tree-dwellers, and for many species, the preferred habitat is mature forest with a closed canopy. Most piciform species are rarely, if ever, seen walking on the ground or flying across open space; typically these birds search for food in trees, nest in trees, raise young in trees, and roost at night in trees (usually in cavities). Of course there are exceptions to the rule of an arboreal life; for example toco toucans (*Ramphastos toco*) live in open, fragmented forests and woodland savannas. And many kinds of puffbirds and jacamars prefer forest edges and streambanks over forest interiors.

Behavior

Most piciform birds are not particularly social. Breeding is usually solitary, not colonial. In many species, however, male and female maintain the pair-bond and defend a shared territory year-round; this behavior is especially common in woodpeckers, puffbirds, and barbets. Some species of toucans do form small, loose flocks when foraging, and barbets and honeyguides may congregate temporarily where food is plentiful.

Though they are good climbers, many species of piciform birds are described as "weak" flyers. (Acrobatic honeyguides are one noteworthy exception.) Most members of this taxon do not migrate, although a few species (notably the yellow-bellied sapsucker [*Sphyrapicus varius*]) migrate long distances between breeding and wintering grounds.

Two behaviors exhibited by some birds in this taxon are extremely unusual in the avian world. One is drumming; woodpeckers routinely communicate in this distinctive way, hammering rhythmically, in species-specific patterns, on resonating structures such as hollow trees. A few species of barbet also communicate by drumming. Another is guiding, the behavior for which the honeyguides are named. Two honeyguide species in Africa "guide" other animals—including honey-badgers, baboons, and humans—to bees' nests. The birds alert their foraging partners to the presence of a honey-loaded hive with their calls and make short flights to indicate the direction of travel. If one of these larger animals does locate and break open the hive, the birds dart in to feast on beeswax.

Feeding ecology and diet

Feeding habits vary within this group; woodpeckers, jacamars, and puffbirds eat mainly insects, whereas toucans and barbets feed mainly on fruit (although these birds do feed their nestlings insects and similar protein-rich prey). Toucans are often described as active predators on eggs and nestlings of other birds but Remsen et al. (1993) contest this characterization.

Both jacamars and puffbirds take their insect prey in flight, behaving like oversized flycatchers. Butterflies are the preferred prey for jacamars; puffbirds most often capture flying beetles. Some woodpeckers also "hawk" after insects, but most often members of this group drill holes in tree bark to extract soft insect larvae (a few barbet species also excavate for insects). Some woodpeckers do eat fruit or nuts, and the appropriately named sapsuckers drill "sap wells" in trees and drink the sticky exudate.

Honeyguides are unique among birds for their habit of eating mainly beeswax from honeycombs; a symbiotic microorganism living in the gut helps these birds to extract nutrition from a substance that most other animals find indigestible. Honeyguides are also unusual in having an excellent sense of smell; many reports exist relating how birds are attracted to burning beeswax candles in rural churches.

Reproductive behavior

Sexual dichromatism is uncommon in piciforms. Most often, males and females look alike, probably because birds that maintain a year-round monogamous pair bond do not require elaborate courtship displays. In woodpeckers, though males and females often have different plumages, the differences between the sexes tend to be subtle, involving the color of nape patches

Chestnut-mandibled toucans (*Ramphastos swainsonii*) in Panama. (Photo by Art Wolfe. Photo Researchers, Inc. Reproduced by permission.)

or the presence or absence of "moustaches." Neotropical barbets are the exception to the rule of uniform plumage for this group; all barbet species show marked differences between males and females with regard to plumage color and/or pattern.

Most piciform birds are cavity-nesters; even the honeyguides, all of which are nest parasites, lay their eggs in the nests of other hole-nesting species such as barbets and woodpeckers. The type of cavity used varies among families. Some species of jacamars and puffbirds dig out nest sites in rotten trees where termites have nested. Other species in these two families excavate their nesting burrows in soil, often along riverbanks. Barbets and woodpeckers use their strong, sharp beaks to hammer out nest cavities in rotting trees, and the largest toucan species occupy natural tree cavities. The smaller toucan species often drive woodpeckers away from just-excavated holes, then use their powerful beaks to enlarge the nest opening.

Almost all members of this group lay white eggs. Unpigmented eggs are typical of cavity-nesting birds—with the nest hidden from predators, there is no need for the eggs to be camouflaged.

Pileated woodpecker (*Dryocopus pileatus*) young in the nest. (Photo by Joe McDonald. Bruce Coleman Inc. Reproduced by permission.)

Helping at the nest, an uncommon bird behavior, is often seen in woodpeckers, and is also known in some species of toucans.

Conservation status

Of the 383 piciform species, a total of 15 species are classified as Critically Endangered, Endangered, or Vulnerable. An additional 28 species are classified as Near Threatened. All of the most-threatened species show downward population trends. All three species listed as Critically Endangered are large woodpeckers: the imperial woodpecker (*Campephilus imperialis*), the Okinawa woodpecker (*Sapheopipo noguchii*), and the ivory-billed woodpecker (*Campephilus principalis*). (Classification as Critically Endangered means experts believe these species have no more than an estimated 50% chance of surviving over the next 10 years or three generations.) Indeed, the U.S. Fish and Wildlife Service declared the ivory-bill extinct in 1997 because the last confirmed U.S. sighting occurred almost 50 years earlier. In 1999, however, a credible report of an ivory-bill sighting in a Louisiana swamp raised hopes that this species may yet persist. A 2002 expedition to search for the species was inconclusive; no birds were spotted, but experts believe they heard the ivory-bill's distinctive "double-knock" drumming.

Though at present a comparatively small proportion of species are threatened, experts caution against complacency because, almost everywhere these birds are found, habitat loss and habitat fragmentation are occurring at a rapid rate. When forests are clear-cut for lumber, not only is habitat destroyed

in the short term, but subsequent commercial reforestation and lumber management practices produce young, even-aged stands that lack the standing dead trees many species require for nest sites and to provide insect food. Meanwhile, forest fragmentation resulting from agriculture or development activities is a problem for species that require large tracts of unbroken forest. Piciform species that prefer edge habitat, however, may be increasing in numbers; data is lacking.

Habitat loss has already been identified as a factor in the declines of some toucan species—in South America, the saffron toucanet (*Baillonius bailloni*) is threatened both by hunting and by capture for the cage-bird trade. Forest loss in the tropics is probably a problem for many puffbird species as well, although data is limited. In Indonesia, both logging and fires threaten barbet habitat. Experts believe that the collection of specimens for museums contributed to the extinction of some species.

Significance to humans

The group of birds that scientists have named for a mythical deity are prominently featured in the myths and folklore of many native cultures. Historically, native people have used toucan and woodpecker feathers and beaks as ceremonial ornaments; these large birds were also hunted for food.

With regard to popular culture in the developed world in the twenty-first century, toucans are probably best known as the mascot for a popular breakfast cereal that is (appropriately) fruit-flavored. Woody Woodpecker, the impudent cartoon character with the trademarked laugh, was created in the 1940s by Walter Lantz for Universal Studios and is still popular with children decades later. Meanwhile, their parents may consider woodpeckers to be pests because the birds sometimes damage homes when they drum on roofs or siding—either to signal possession of a territory, or to get at concealed insects infesting the home. Perhaps the most famous case of woodpecker damage occurred in 1995, when northern flickers (*Colaptes auratus*) in Florida pecked four-inch-diameter holes in the foam insulation covering the fuel tanks of the space shuttle *Discovery*.

Scientists, however, generally regard the piciform birds as beneficial to ecosystems. For example, fruit-eating species such as toucans and barbets play a key role in maintaining tropical forests because they disperse tree seeds into areas favorable for germination. Woodpeckers help suppress populations of pest insects in forests, and their abandoned nesting cavities provided crucial nest sites for such hole-nesting birds as bluebirds as well as mammals such as flying squirrels. Migratory hummingbirds are often sustained in spring by the insects attracted to the sweet maple sap dripping from holes drilled by sapsuckers.

Resources

Books

Haffer, J. *Avian Speciation in Tropical South America.* Cambridge, MA: Nuttall Ornithology Club, 1974.

Short, L. *Woodpeckers of the World.* Delaware Museum of Natural History, 1982.

Sibley, C. G., and B. L. Monroe, Jr. *Distribution and Taxonomy of Birds of the World.* New Haven: Yale University Press, 1990.

Sibley, C. G., and J. E. Ahlquist. *Phylogeny and Classification of Birds: A Study in Molecular Evolution.* New Haven: Yale University Press, 1990.

Skutch, A. F. *Life of the Woodpecker.* Santa Monica: Ibis Publishing Company, 1985.

Periodicals

Hofling, E., and M. F. Alvarenga-Herculano. "A Comparative Study of the Bones of the Shoulder Girdle in the Piciformes, Passeriformes and Caraciiformes, and also in Related Orders of Birds such as the Trogoniformes, Coliiformes, Apodiformes, Strigiformes and Carpimulgiformes." *Zoologischer–Anzeiger* 240 (2001): 196–208.

Remsen, J. V., Jr., M. A. Hyde, and A. Chapman. "The Diets of Neotropical Trogons, Motmots, Barbets, and Toucans." *Condor* 95 (1993): 178–192.

Cynthia Ann Berger, MS

Jacamars
(Galbulidae)

Class Aves
Order Piciformes
Suborder Galbulae
Family Galbulidae

Thumbnail description
Slim tree birds, similar to an oversized hummingbird, with iridescent green plumage and long pointed bills

Size
5–12 in (13–31 cm)

Number of genera, species
5 genera; 17 species

Habitat
Neotropical forests

Conservation status
Endangered: 1 species; Vulnerable: 1 species

Distribution
Central and South America, from Mexico to Argentina

Evolution and systematics

With their brilliant colors and energetic ways, jacamars resemble hummingbirds but are actually related to puffbirds, toucans, and woodpeckers. Like all members of the order Piciformes, jacamars and their relatives have zygodactyl feet, with two toes pointing forward and two facing back. Jacamars evolved with this toe arrangement, which helps them grasp branches while hunting in trees. Jacamars, like woodpeckers and other piciform birds, are cavity nesters: they tunnel into the ground to build nests. Scientists believe jacamars are closely related to Old World bee-eaters, which also prey on flying insects, have similar plumage, and raise their young in the same manner.

Jacamars tend to live near lush tropical rainforests, which have a dazzling variety of large, colorful butterfly species. Jacamars have become highly selective predators. They often make their homes near streams, drilling nest cavities into steep banks and upturned tree roots.

Because 13 of the 17 Galbulidae species belong to superspecies complexes, researcher J. Haffer concludes that jacamars had a relatively recent Pleistocene radiation of the family. Jacamars are believed to have originated in the Amazon region where they are most common, and spread to other parts of Central and South America. Unique anatomical features of this family include a long appendix, no gall bladder, a bare preen gland, and a long, thin tongue.

Physical characteristics

Wildlife enthusiasts treasure jacamars for their jewel-like colors. Their most distinctive characteristic is the long, sharp bill they use to snatch prey out of the air. In some species, the bill can be three times as long as the bird's head. Jacamars vary in size, from the brown jacamar (*Brachygalba lugubris*), at 7 in (18 cm) long, to the 1-ft-long (30 cm) great jacamar (*Jacamerops aurea*).

All jacamars, except one species, have short legs with four toes: two facing forward and two facing back. The three-toed jacamar (*Jacamaralcyon tridaetyla*), however, lacks the rear (first) toe. A jacamar has a long tail, with 10–12 graduated tail feathers. The short wings have 10 primaries, and contour feathers have a short secondary shaft (except in the genus *Malacoptila*).

Males and females have similar plumage, although females of some species may have less striking colors on the head and

A rufous-tailed jacamar (*Galbula ruficauda*) with its insect prey. (Photo by Doug Wechsler/VIREO. Reproduced by permission.)

neck. Jacamars, known for their brilliant plumage, typically have a metallic green head, reddish underparts, and a light patch on the throat or breast. Some species have color variations ranging from purple to red or chestnut brown. The paradise jacamar (*Galbula dea*) has much darker bluish black plumage with a contrasting white patch on the throat and a long, elegant tail.

While most newly hatched piciform birds are born naked, jacamars are covered with white down. By the time they leave the nest, plumage resembles that of the parents.

Jacamars are not songbirds, but A. Skutch has noted that they have loud calls, trills, and short songs that could be considered melodious.

Distribution

Jacamars occur mainly at low altitudes, ranging from southern Mexico to northern Argentina. The most widespread species, the rufous-tailed jacamar (*Galbula ruficauda*), occurs in Costa Rica, Trinidad, Mexico, Ecuador, Colombia, Brazil, and Argentina. The distinctive great jacamar ranges from Costa Rica to the Amazon basin. The green-tailed jacamar (*Galbula galbula*) is found north and south of the Amazon. More rare is the coppery-chested jacamar (*Galbula pastazae*), which has a low population. It has been observed at a small number of sites in southern Colombia and along the slopes of the Ecuadorian Andes.

Habitat

Jacamar habitat includes Neotropical rainforests, streams or riverine forest, and savanna woodland. Generally, jacamars live at the edge of forests. Jacamars prefer habitat that supports their favorite prey—large, showy butterflies. Suitable nesting sites, such as dirt masses associated with fallen trees and sandy or clay stream banks, also are key. The rufous-tailed jacamar prefers low-lying thickets that border rivers, while the paradise jacamar seeks a higher perch. The coppery-chested jacamar lives at the highest elevations of all jacamars. It has been observed at altitudes of 5,100 ft (1,550 m) in the Andes. It is a resident of the eastern slope of the Ecuadorian Andes, although its distribution is patchy, and its population is dwindling due to habitat loss.

Behavior

Jacamars are exciting to watch because they are dramatic acrobats, swooping down from perches to capture colorful prey in mid-air. They spend most of their time on a branch, alert and scanning the air for flying insects.

Typically, jacamars live in pairs that perch and hunt in the same area. Certain species will occasionally congregate in small family groups. Unlike other jacamar species, the yellow-billed jacamar (*Galbula albirostris*) often joins mixed flocks of birds. Jacamars use a variety of calls to communicate. For example, rufous-tailed jacamars signal danger or agitation with a sharp trill.

Feeding ecology and diet

As an entirely insectivorous family, jacamars prefer large, showy, flying insects. Their diet consists of butterflies, moths, wasps, dragonflies, and flying beetles. A long, forceps-like bill allows the jacamar to grasp its prey while keeping it at enough of a distance to avoid becoming blinded by fluttering wings or injured by a stinging insect. Once they catch an insect, jacamars batter it against a tree branch to kill it, and remove the wings before swallowing.

Favorite food sources include beautiful blue morpho butterflies, hawk moths, and venomous Hymenoptera such as bees and wasps. Skutch also observed a preference for butterflies and dragonflies. In his work, Chai described how young rufous-tailed jacamars in Costa Rica learned to discriminate between butterfly species by color, markings, and flight patterns. Jacamars tend to reject butterflies with chemical defenses that make them less palatable. With their specialized hunting skills, jacamars may have played a major role in the evolution of butterflies that live in jacamar habitats.

Reproductive biology

During breeding season, male jacamars engage in lively vocal performances, with a series of explosive, sharp calls. Two rival males use this display of courtship and verbal bravado to impress a potential mate. Jacamars form monogamous pairs.

Jacamars dig holes for nests in steep river banks. They use the bill to break up the soil, then remove it by kicking backwards with their feet as they burrow. These tunnels also can be found some distance from the water, on soil banks or roots of fallen trees. The nest sits at the end of the tunnel in a horizontal, oval-shaped terminal chamber. Some jacamars, including the rufous-tailed jacamar, may use termite nests for breeding if no appropriate site to dig a ground tunnel can be found. Tunnels are 12–36 in (30–91 cm) long and about 2 in (5 cm) in diameter. The nest chamber is used repeatedly and does not contain nest material, although eggs often are covered with a layer of regurgitated insect parts. In some species, male and female participate in building the nest hole; in other species only the female does this work.

Jacamars lay one to four round, glossy, white eggs. Both parents incubate the eggs during the day for one to three hours at a time. At night, the female incubates alone while the male stays nearby to defend the nest. Jacamars rarely leave eggs unattended. During incubation, the male feeds his partner several times each day. The incubation period is 20–23 days.

Both parents feed the young with insects. Chicks remain in the nest 21–26 days. Unlike other species, young pale-headed jacamars (*Brachygalba goeringi*) may return to the burrow to sleep with the parents for several months after they fledge.

Conservation status

Habitat loss is a continuing threat to jacamars. In Brazil, intensive clearance of understory vegetation in forest fragments caused a decline of jacamar populations, along with an overall lack of bird species diversity in the area.

Only one jacamar species, the three-toed jacamar, is classified as Endangered. This species prefers lowland tropical rainforest, plantations, tropical monsoon, and dry forest. The three-toed jacamar used to thrive in southeast Brazil, including eastern Minas Gerais, Espírito Santo, Rio de Janeiro, São Paulo, and Paraná. Development and agricultural land use have destroyed much of its habitat. The remaining population, restricted to Rio Paraíba valley in Rio de Janeiro and the dry regions of Minas Geraís, is small and fragmented. The three-toed jacamar is protected under national law in Brazil. The Caratinga Reserve in Brazil offers a safe haven for this endangered species.

The coppery-chested jacamar—found in montane tropical rainforests of Colombia, Ecuador, and Amazonian Brazil—is classified as Vulnerable. Its total population is believed to be small and declining due to destruction of its forest habitat.

Significance to humans

The Tupi, an indigenous population in Brazil, gave the jacamar its name. It is derived from a Tupi word, *jacama-ciri*. Over the years, people in the region have given this popular bird nicknames such as "bico-de-agulha" (needle bill) and "beija-flor grande" (big hummingbird).

1. Paradise jacamar (*Galbula dea*); 2. Great jacamar (*Jacamerops aurea*); 3. Rufous-tailed jacamar (*Galbula ruficauda*); 4. Coppery-chested jacamar (*Galbula pastazae*); 5. Green-tailed jacamar (*Galbula galbula*); 6. Yellow-billed jacamar (*Galbula albirostris*); 7. Three-toed jacamar (*Jacamaralcyon tridactyla*). (Illustration by Wendy Baker)

Species accounts

Rufous-tailed jacamar

Galbula ruficauda

TAXONOMY

Galbula ruficauda Baron Cuvier, 1816.

OTHER COMMON NAMES

French: Jacamar à queue rousse; German: Rotschwanz-Glanzvogel; Spanish: Jacamar Común.

PHYSICAL CHARACTERISTICS

9 in (23 cm); 2 in (51 mm) slender bill. Metallic green upper parts, white or buff patch on throat, rufous or reddish underside.

DISTRIBUTION

Very common from southern Mexico to northern Argentina, including Brazil, Colombia, and Ecuador. Also found in Trinidad and Tobago.

HABITAT

Forest edge, woodland, thickets, and near streams and rivers.

BEHAVIOR

Live in pairs, prefer to hunt from low shrubbery.

FEEDING ECOLOGY AND DIET

Prefers flying insects, like most jacamars. Catches prey in mid-air and batters it against a branch before consuming it.

REPRODUCTIVE BIOLOGY

Lays one to four white eggs in ground-hole nest cavity. Incubation is 20–23 days. Chicks emerge from nest after 21–26 days. Both sexes incubate, and care for chicks.

CONSERVATION STATUS

Not threatened; widespread and common, adapts to many different habitats.

SIGNIFICANCE TO HUMANS

None known. ◆

Green-tailed jacamar

Galbula galbula

TAXONOMY

Galbula galbula Linnaeus, 1766.

Galbula ruficauda

▨ Resident

Galbula galbula

▨ Resident

OTHER COMMON NAMES
French: Jacamar vert; German: Grünschwanz-Glanzvogel;
Spanish: Jacamar de Cola Verde.

PHYSICAL CHARACTERISTICS
8 in (20 cm); 2 in (51 mm) slender bill. Metallic green upperparts, white or buff patch on throat, tail shorter and more
rounded than other species, with green on top and dusky blue
underneath.

DISTRIBUTION
Brazil, Colombia, the Guianas, and Venezuela.

HABITAT
Forest edge, woodland, usually close to water.

BEHAVIOR
Like rufous-tailed jacamar, they prefer lower shrub perches for
hunting.

FEEDING ECOLOGY AND DIET
Prefer butterflies and dragonflies. Catches prey in mid-air and
batters it against a branch before consuming it.

REPRODUCTIVE BIOLOGY
Lays one white eggs in ground-hole nest cavity. Incubation is
20–23 days. Chicks emerge from nest after 21–26 days, covered
in white down. Both sexes incubate and care for chicks.

CONSERVATION STATUS
Not threatened.

SIGNIFICANCE TO HUMANS
None known. ◆

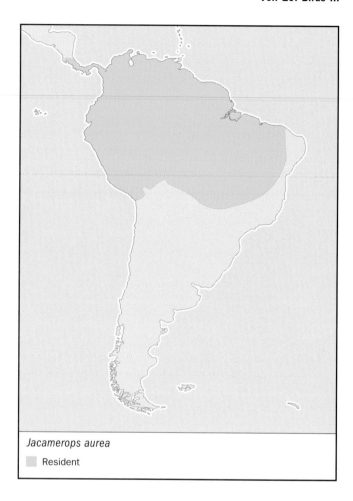

Jacamerops aurea
▨ Resident

Great jacamar
Jacamerops aurea

TAXONOMY
Jacamerops aurea Müller, 1776.

OTHER COMMON NAMES
French: Grand jacamar; German: Breitmaul-Glanzvogel; Spanish: Jacamar Grande.

PHYSICAL CHARACTERISTICS
The largest jacamar: 12 in (30 cm) long, with thick, slightly
curved bill. Metallic green upperparts, white narrow band on
throat, rufous underside, with bluish black underside of tail.

DISTRIBUTION
Venezuela, Ecuador, Costa Rica, Bolivia, and Amazonian Brazil.

HABITAT
Riverine forest, lowlands.

BEHAVIOR
Quieter and slower moving than other jacamar species. Known
for its mournful-sounding call.

FEEDING ECOLOGY AND DIET
Preys on flying insects. Catches prey in mid-air and batters it
against a branch before consuming it.

REPRODUCTIVE BIOLOGY
Lays one to four white eggs in ground-hole nest cavity. Incubation is 20–23 days. Chicks emerge from nest after 21–26
days, covered in white down. Both sexes incubate, and care for
chicks.

CONSERVATION STATUS
Not threatened.

SIGNIFICANCE TO HUMANS
None known. ◆

Three-toed jacamar
Jacamaralcyon tridactyla

TAXONOMY
Jacamaralcyon tridactyla Vieillot, 1817.

OTHER COMMON NAMES
French: Jacamar tridactyle; German: Dreizehen-Glanzvogel;
Spanish: Jacamar Tridáctilo.

PHYSICAL CHARACTERISTICS
7 in (18 cm); 0.7 oz (20 g). Has three toes, two facing forward
and one back; dark grayish green plumage and a chestnut-
brown head.

DISTRIBUTION
Southeastern Brazil.

HABITAT
Open woodland, lowlands.

Jacamaralcyon tridactyla
▨ Resident

Galbula pastazae
▨ Resident

BEHAVIOR
Similar to kingfishers in hunting behavior, color, and beak shape.

FEEDING ECOLOGY AND DIET
Prefers flying insects. Perches on tall grasses at the edge of forest to watch for prey, then darts out and catches it in mid-air.

REPRODUCTIVE BIOLOGY
Lays one to four white eggs in ground-hole nest cavity. Sometimes nests colonially. Incubation is 20–23 days. Chicks emerge from nest after 21–26 days, covered with white down. Both sexes incubate, and care for chicks.

CONSERVATION STATUS
The only Endangered jacamar species, due to habitat loss from agriculture and development. The current population is not known, but has declined dramatically and is believed to exist in very small numbers. Three-toed jacamars are protected in the Caratinga Reserve in Brazil.

SIGNIFICANCE TO HUMANS
None known. ◆

Coppery-chested jacamar
Galbula pastazae

TAXONOMY
Galbula pastazae Taczanowski and Berlepsch, 1885.

OTHER COMMON NAMES
French: Jacamar des Andes; German: Kupferglanzvogel; Spanish: Jacamar Cobrizo.

PHYSICAL CHARACTERISTICS
9 in (23 cm) long. Heavier 2 in (51 mm) bill. Metallic green upperparts, dark rufous throat, copper tail, distinctive yellowish orange eye ring.

DISTRIBUTION
Colombia, Ecuador, and Amazonian Brazil.

HABITAT
Lives in the highest forest elevation of all jacamar species.

BEHAVIOR
Alert hunter, similar to other jacamars. Gives a series of three to five loud calls.

FEEDING ECOLOGY AND DIET
Diverse variety of flying insects. Prefers to hunt from one favorite perch, capturing insects as they fly through the air.

REPRODUCTIVE BIOLOGY
Lays one to four white eggs in curved ground-hole nest cavity, so eggs are out of view. Incubation is 20–23 days. Chicks emerge from nest after 21–26 days, covered with white down. Both sexes incubate and care for chicks.

CONSERVATION STATUS
Vulnerable; thin distribution, low population limited to a few locations, primarily in Colombia and the east slope of the Andes. Threatened by deforestation.

SIGNIFICANCE TO HUMANS
None known. ◆

Paradise jacamar
Galbula dea

TAXONOMY
Galbula dea Linnaeus, 1758.

OTHER COMMON NAMES
French: Jacamar à longue queue; German: Paradeisglanzvogel; Spanish: Jacamar de Cola Larga.

PHYSICAL CHARACTERISTICS
12 in (30 cm) long; 2 in (51 cm) slender bill. Metallic bluish black color on upper and lower body, contrasting white patch on throat, long elegant tail.

DISTRIBUTION
Amazonian Brazil, the Guianas, Peru, Bolivia, Venezuela, and Ecuador.

HABITAT
Forest and forest edge or upland woodland.

BEHAVIOR
Hunts alone, in pairs, or in groups of three, may join canopy flocks.

FEEDING ECOLOGY AND DIET
Prefers butterflies and dragonflies. Perches on a branch, then darts out to capture prey in mid-air.

REPRODUCTIVE BIOLOGY
Lays one to four white eggs in ground-hole nest cavity. Incubation is 20–23 days. Chicks emerge from nest after 21–26 days, covered in white down. Both sexes incubate, and care for chicks.

CONSERVATION STATUS
Not threatened.

SIGNIFICANCE TO HUMANS
None known. ◆

Yellow-billed jacamar
Galbula albirostris

TAXONOMY
Galbula albirostris Latham, 1790.

OTHER COMMON NAMES
English: Blue-cheeked jacamar; blue-necked jacamar; French: Jacamar à bec jaune; German: Gelbschnabel-Glanzvogel; Spanish: Jacamar de Pico Amarillo.

Galbula dea
▢ Resident

Galbula albirostris
▢ Resident

PHYSICAL CHARACTERISTICS
7.5 in (19 cm) long. The only jacamar species with a yellow bill. Metallic green on upperparts, purplish brown head, white patch on throat, rufous on underparts and tail. Feet and eye ring are yellow.

DISTRIBUTION
Amazonian Brazil, the Guianas, Peru, Venezuela, Colombia, and Ecuador.

HABITAT
Prefers forest interior more than most jacamars.

BEHAVIOR
Often joins mixed flocks of other bird species.

FEEDING ECOLOGY AND DIET
Prefers butterflies and dragonflies. Captures flying insects while hunting from a perch.

REPRODUCTIVE BIOLOGY
Lays one to four white eggs in ground-hole nest cavity. Incubation is 20–23 days. Chicks emerge from nest after 21–26 days, covered with white down. Both sexes incubate, and care for chicks.

CONSERVATION STATUS
Not threatened.

SIGNIFICANCE TO HUMANS
None known. ◆

Resources

Books

Hilty, Steven L., and William L. Brown. *A Guide to the Birds of Colombia.* New Jersey: Princeton University Press, 1986.

Janzen, Daniel H., ed. *Costa Rican Natural History.* Chicago: University of Chicago Press, 1983.

Periodicals

Chai, P. "Butterfly Visual Characteristics and Ontogeny of Responses to Butterflies by a Specialized Tropical Bird." *Biological Journal of the Linnean Society* 59, no. 1 (1996): 37–67.

Chai, P. "Field Observations and Feeding Experiments on the Responses of Rufous-tailed Jacamars to Free-flying Butterflies in a Tropical Rainforest." *Biological Journal of the Linnean Society* 29, no. 3 (1986): 161–189.

Marsden, Stuart, J., Mark Whiffin, and Mauro Galetti. "Bird Diversity and Abundance in Forest Fragments and Eucalyptus Plantations Around an Atlantic Forest Reserve, Brazil." *Biodiversity and Conservation* 10, no. 5 (2001): 737–751.

Organizations

Neotropical Bird Club. c/o The Lodge, Sandy, Bedfordshire SG19 2DL United Kingdom. E-mail: secretary @neotropicalbirdclub.org Web site: <www.neotropicalbirdclub .org>

Other

Caratinga: Soundscapes from Brazil's Atlantic Rainforest. Compact disc, Earth Ear Catalog, 2001. Available online at <http://www.earthear.com/catalog/caratinga.html>

The IUCN Red List of Threatened Species. Species Information: Three-toed Jacamar. <http://www.redlist.org>

Melissa Knopper, MS

Puffbirds

(Bucconidae)

Class Aves
Order Piciformes
Suborder Galbulae
Family Bucconidae

Thumbnail description
Small to medium-sized birds, usually with short tails, rounded wings, and fairly robust bills; some are more streamlined

Size
5.12–11.42 in (13–29 cm); 0.49–3.74 oz (14–106 g)

Number of genera, species
7 genera, 32 species

Habitat
Dry and humid forest, and wooded savanna

Conservation status
Near Threatened: 1 species

Distribution
Central America and South America south to northern Argentina

Evolution and systematics

Palaeontological evidence suggests that puffbirds may have arisen from a rather ancient lineage. Certainly fossils of a similar family of birds, named Primobucconidae because they seem closest to modern puffbirds, have been found in widespread Eocene deposits, and are even tentatively identified from Europe. It seems clear that small non-passerines similar to puffbirds were dominant perching birds of the Eocene in both hemispheres.

Despite vacillations regarding the overall taxonomic placement of puffbirds, their closest relatives have never been in doubt: a convincing array of morphological features indicates a link with jacamars (Galbulidae). The two families are traditionally combined to form the Galbulae, a suborder within Piciformes. They differ in bill shape, general comportment, the form of the spinal cord (puffbirds exhibit Piciforme design, jacamars that of Coraciiformes), and number of ribs.

The relationship between these two families and other Piciformes and Coraciiformes is controversial. The association between puffbirds and other Piciforme taxa is apparently weak, supported by few characters, and discounted by an equal

number of features that suggest a Coraciiforme origin. It seems safer to treat the Galbulae in an order of its own (Galbuliformes). The generic structure of Bucconidae is contested and needs review as of 2002.

Physical characteristics

The name puffbird is applied to the family because of an unusual propensity of its members to puff up their feathers when alarmed. Even when not alarmed they seem large-headed and large-eyed, with robust, slightly curved or hook-tipped bills. Puffbird wings tend to be short and rounded (there are 10 primaries and 12 rectrices), and their tails tend to be short and narrow, though broader and longer in some nunbirds (*Monasa*). Plumage is soft and loose. Feet are small and zygodactylous (two toes before and two behind), with the first and fourth digits permanently reversed. The most divergent species is the swallow-winged puffbird (*Chelidoptera tenebrosa*), which has more tapered wings and a shorter bill than other family members.

Unlike jacamars with their gaudy garb, puffbird plumage lacks colorful tones or iridescence. Nevertheless, most have

Puffbird nest. (Illustration by Dan Erickson)

very striking plumage patterns, such as sharply demarcated breast bands, or streaked and spotted underparts. Puffbirds are generally sexually monomorphic: only two species of *Malacoptila* exhibit sex-related plumage differences, and in one of these the variation is minor.

Distribution

The puffbird family is essentially Neotropical, its northernmost limit tallying with the extent of humid forest in southern Mexico. Likewise, it extends no further south than Paraguay and northern Argentina, the southern outposts of tall forest and dry woodland. Between these extremes, puffbirds are absent from all islands beyond the continental shelf, suggesting that they do not disperse well over water. They reach their greatest diversity in northern South America, particularly Amazonia; no fewer than 70% of puffbirds occur in Brazil. Their arboreal nature ties their distribution roughly to that of evergreen or deciduous woodland, although no species has adapted to the southern beech (*Nothofagus*) forests of Chile and Argentina.

Habitat

Puffbirds are exclusively arboreal when foraging. While the majority inhabit lowland humid forest, they are not deep-forest species. Most are apparently birds of forest edge, tree-fall gaps, streamsides, lakesides, and clearings, where horizontal perches are abundant and tangled vines tumble from the canopy to the lower strata. However, this conclusion might sometimes result from observational bias: it is possible that species, particularly those generally confined to the canopy, are more difficult to find in continuous forest than at gaps or streams, where a better view is afforded of the forest's upper levels. Some of these species might range largely undetected across forest canopies.

Several species of puffbird have adapted to semi-arid habitat or open woodland. A few more occupy foothill and submontane forest in the Andes. Most of these reach only the 3,280-ft (1,000-m) contour, but the black-streaked puffbird (*Malacoptila fulvogularis*) has been recorded up to 7,550 ft (2,300 m), and the white-faced nunbird (*Hapaloptila castanea*) to 9,500 ft (2,900 m). Habitat choice depends on a complex interaction of environmental factors, including competition with related and unrelated birds. For example, the rufous-capped nunlet (*Nonnula ruficapilla*) is confined to bamboo or riverine regrowth in some areas, but is much more general in its habitat selection in others.

Behavior

Puffbirds, to varying degrees, are sit-and-wait predators that perch motionless for prolonged periods, searching their surroundings for movement. Their stillness renders them difficult to locate, and this presumably benefits them in terms of concealment from predators and prey alike. Nunbirds are more active and noisy than most, and groups often draw attention to themselves by bickering, chorusing, and more frequently switching perching position.

The flight of most puffbirds is quick and direct on whirring wings. Nunbirds behave slightly differently, often interspersing series of wingbeats with glides, and sometimes wheeling briefly under the canopy. Flight pattern and fuller tail shape derive from the more aerial foraging niche. The most aerial of all puffbirds, however, is the swallow-winged puffbird. This species perches conspicuously at the pinnacle of trees, or on high bare branches, and frequently launches itself to catch flying insects with great skill and maneuverability. It finds all its food on the wing.

Most puffbirds appear to defend year-round territories, using vocalizations to signal their presence, attract mates, and deter rivals. Although they are generally solitary birds, some occur in small groups, presumably extended family parties. Group-living nunbirds give cacophonous semi-coordinated choruses, while some pair-living species, such as the white-eared puffbird (*Nystalus chacuru*), sing loud coordinated duets. Very few reports of seasonal movements have emerged. At the upper edge of altitudinal limits, or the southernmost extent of some species ranges, minor migrations may occur in response to seasonal conditions.

Feeding ecology and diet

So reclusive are their lifestyles that little has been reported about the diet of most puffbirds. It seems that all are predominantly insectivorous, with at least the swallow-winged puffbird being entirely so. Most species take other arthropods, and several have been reported to eat lizards, small snakes, and frogs. Some puffbirds take advantage of marauding troops of primates or oropendolas, following them through the forest and pursuing prey that is forced to flee, drop, or otherwise abandon camouflage. For the same reason, some species also attend ant swarms, and nunbirds are often noted accompanying mixed-species flocks of birds passing through the canopy or mid-levels, although this latter tactic is probably adopted as much to reduce predation pressure as to increase foraging efficiency.

The importance of vegetable matter in puffbird diets is low, but some species have been reported taking fruit, berries, and buds. Fruit consumption seems greatest in higher-altitude species such as the lanceolated monklet (*Micromonacha lanceolata*) and white-faced nunbird.

Reproductive biology

Most puffbirds are monogamous, defending pair territories year round, and breeding at varying seasons throughout their ranges depending on rainfall patterns. Nunbirds breed cooperatively in groups of up to five, and the aberrant swallow-winged puffbird appears to breed in non-territorial groupings.

From the available data it seems that some genera (*Notharchus*, *Bucco*, and *Hypnelus*) usually occupy cavities excavated in arboreal termitaria (ant beds), while others (*Nystalus*, *Monasa*, and *Chelidoptera*) habitually dig terrestrial burrows into sand or soil. In intermediate cases (*Malacoptila* and *Nonnula*), both types of site have been reported. Nesting in tree holes is inconclusively documented, but at least one species is known to use the abandoned mud nests of pale-legged horneros (*Furnarius leucopus*).

Nests are usually excavated by both members of the pair, cavities being of variable length and terminating in a rounded chamber without lining. Ground nests are dug at a shallow angle into flat or gently sloping terrain. Smaller species may complete burrows only 20 in (50 cm) long; larger species may dig for 60 in (150 cm). In some species, a frame of leaves around the entrance is thought to function as camouflage. In black-fronted nunbirds (*Monasa nigrifrons*) at least, a pile of leaves is regularly tossed over the nest hole to conceal it; there are reports from the closely related black nunbird (*M. atra*) of an antechamber—a tunnel running under leaves to the entrance.

Puffbird eggs are relatively small, dull or glossy white, and normally produced in clutches of 2–3. The incubation period is about 15 days in swallow-winged puffbirds. A period of two days is thought to elapse between the laying of each egg. Both sexes incubate, and no puffbird species raises more than one brood per year.

A white-eared puffbird (*Nystalus chacuru*) with its insect prey. (Photo by Doug Wechsler/VIREO. Reproduced by permission.)

Chicks hatch blind and naked, but are remarkably mobile. Within the first day they crawl to the cavity entrance, thus allowing the adults to deliver food without entering. The male apparently contributes entirely to brooding in the first few days after hatching, the female carrying out all provisioning at this time. After about nine days, brooding is no longer necessary, but the female continues to feed the chicks far more than the male. The fledging period is 20–30 days. Young puffbirds probably remain on the natal territory for almost a year in some cases. Groups of up to five nunbirds have been counted attending an active nest, but it is not clear whether this entails a dominant pair accompanied by offspring from previous broods.

Conservation status

In general, puffbird populations must be declining throughout South and Central America simply because the area of standing forest is shrinking, although any species with a real predilection for edge habitats might be faring rather better. The swallow-winged puffbird has probably increased in numbers with the opening up of Amazonian forests by man. No species is currently considered threatened, and only the sooty-capped puffbird (*Nystactes noanamae*) is classified as Near Threatened. This species is confined to the Chocó Endemic Bird Area in western Colombia, where it is generally uncommon. Recent

A pair of white-whiskered puffbirds (*Malacoptila panamensis*) perch near each other in Ecuador. (Photo by Steven Holt/VIREO. Reproduced by permission.)

records come from very few sites, although much of its potential range awaits thorough exploration.

Significance to humans

Their insectivory, apparent lethargy, infrequent singing, and dull coloration make puffbirds unsuitable pets and they rarely appear as cage birds. Their somber aspect gives rise to common names such as "nunbirds" and "monklets." They thus have almost no significance to humans apart from being the frequent brunt of insults in the literature. Many observers, from the distant past to the present, have described puffbirds as either stupid or lazy because they perch motionless for such long periods and allow men with guns or slingshots to approach closely. These derogatory terms are rather unfairly leveled at puffbirds because stillness is an integral part of their foraging niche and antipredator response. Moreover, the lightning dashes they make across fairly long distances to snatch mobile insects are indication that they are alert and effective hunters.

1. Black-fronted nunbird (*Monasa nigrifrons*); 2. Lanceolated monklet (*Micromonacha lanceolata*); 3. Collared puffbird (*Bucco capensis*); 4. Rufous-capped nunlet (*Nonnula ruficapilla*); 5. Swallow-winged puffbird (*Chelidoptera tenebrosa*); 6. White-eared puffbird (*Nystalus chacuru*); 7. White-necked puffbird (*Notharchus macrorhynchos*); 8. White-whiskered puffbird (*Malacoptila panamensis*). (Illustration by Dan Erickson)

Species accounts

White-necked puffbird
Notharchus macrorhynchos

TAXONOMY
Bucco macrorhynchos J. F. Gmelin, 1788, Cayenne. Forms a superspecies with *N. swainsoni* of the Brazilian Atlantic forest. Two subspecies.

OTHER COMMON NAMES
French: Tamatia à gros bec; German: Weißhals-faulvogel; Spanish: Buco Picogordo.

PHYSICAL CHARACTERISTICS
11 in (25 cm); 2.9–3.7 oz (81–106 g). Black upperparts and broad chest band, white forehead, collar, throat and belly. Variable dark barring on flanks. Bill and feet black.

DISTRIBUTION
N. m. hyperrhynchus: Mexico south to Venezuela, Colombia, Ecuador, eastern Peru, northern Bolivia, and western Brazil; *N. m. macrorhynchos*: eastern Venezuela, the Guianas, and northern Brazil south to the Amazon.

HABITAT
Humid to semiarid forest, open woodland, clearings, and plantations (0–3,940 ft [0–1,200 m]).

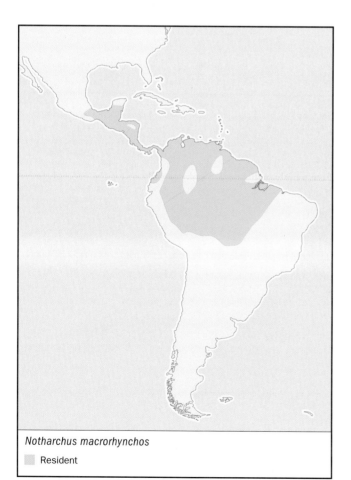

Notharchus macrorhynchos
▇ Resident

BEHAVIOR
Pair territorial and sedentary; generally found perching stolidly on high open branches, but otherwise inconspicuous.

FEEDING ECOLOGY AND DIET
Hunts at all levels, from ant swarms to upper canopy, preying on large insects and small vertebrates; some vegetable matter.

REPRODUCTIVE BIOLOGY
Nests excavated by both pair members in arboreal termitaries, usually 40–50 ft (12–15 m) up (occasionally 10–60 ft [3–18 m]). Holes in the ground or in trees are also reported.

CONSERVATION STATUS
Not threatened: scarce in Central America, widespread and often fairly numerous in South America.

SIGNIFICANCE TO HUMANS
None known. ◆

Collared puffbird
Bucco capensis

TAXONOMY
Bucco capensis Linnaeus, 1766, Cape of Good Hope; error = the Guianas. Monotypic.

OTHER COMMON NAMES
French: Tamatia à collier; German: Halsband-faulvogel; Spanish: Buco Musiú.

PHYSICAL CHARACTERISTICS
7.5 in (19 cm), 1.6–2.2 oz (46–62 g). Dumpy, with large head and short tail. Sides of face bright orange; rest of upperparts rufous, finely barred darker. Complete black collar forms band across chest, bordered buffy on nape. Underparts white, grading to rich buffy on lower flanks. Bill robust and orange; eye orange.

DISTRIBUTION
Widespread in Amazon basin from eastern Ecuador and southeastern Colombia; east to southern Venezuela, the Guianas and northern Brazil; south to southeastern Peru and northern Mato Grosso.

HABITAT
Lower to mid strata of tall humid forest, also dry hilly country, bamboo bordering rivers, and varzea; 0–5,580 ft (0–1,700 m).

BEHAVIOR
Still-hunts from horizontal perches; inconspicuous and tame. Distinctive vocalization given mainly before dawn.

FEEDING ECOLOGY AND DIET
Large insects (beetles, orthopterans, cicadas) and small vertebrates (lizards, snakes, frogs) taken from ground or foliage.

REPRODUCTIVE BIOLOGY
Excavates nests in arboreal termitaria.

Bucco capensis
☐ Resident

Nystalus chacuru
☐ Resident

CONSERVATION STATUS

Not threatened. Uncommon (or at least infrequently observed) but widespread, occurring in several protected areas.

SIGNIFICANCE TO HUMANS

None known. ◆

White-eared puffbird

Nystalus chacuru

TAXONOMY

Bucco chacuru Vieillot, 1816, Paraguay. Two subspecies recognized.

OTHER COMMON NAMES

French: Tamatia chacuru; German: Weißohr-faulvogel; Spanish: Buco Chacurú.

PHYSICAL CHARACTERISTICS

8.3–8.7 in (21–22 cm); 1.7–2.3 oz (48–64 g). Upperparts brown, finely spotted and barred paler. Large pale ear-covert spot bordered below by blackish patch extending toward nape; white on forehead extending to narrow pale coronal stripe. Underparts ochraceous. Large bill reddish.

DISTRIBUTION

N. c. uncirostris: eastern Peru, eastern Bolivia, and extreme western Brazil; *N. c. chacuru*: northeastern, eastern, and southern Brazil; eastern Paraguay; and northeastern Argentina.

HABITAT

Tropical dry forest, clearings and pastures, open woodland, savanna with scattered trees, even suburban areas where suitably wooded.

BEHAVIOR

Still-hunts from perches (including posts and wires) between ground level and upper canopy.

FEEDING ECOLOGY AND DIET

Eats mainly arthropods caught aerially or on the ground, but occasionally takes small vertebrates (lizards).

REPRODUCTIVE BIOLOGY

Nests at the ends of cavities excavated in level ground or banks; 2–4 eggs are laid.

CONSERVATION STATUS

Not threatened. A common bird in much of southeastern and central Brazil. Although less common elsewhere, the species is nevertheless secure, especially because it is favored by deforestation.

SIGNIFICANCE TO HUMANS

None known. ◆

White-whiskered puffbird

Malacoptila panamensis

TAXONOMY

Malacoptila panamensis Lafresnaye, 1847, Panama. Four subspecies recognized.

Malacoptila panamensis
■ Resident

REPRODUCTIVE BIOLOGY
Nest burrows are dug in level or sloping ground, or in arboreal termitaries; occasionally uses tree holes. Same hole can be used in successive breeding attempts. Two to three eggs are laid; incubation period unknown but fledging period is about 20 days.

CONSERVATION STATUS
Not threatened. Although rare in the northern portion of its range, it is common in southern Costa Rica and parts of western Colombia and Ecuador.

SIGNIFICANCE TO HUMANS
None known. ◆

Lanceolated monklet
Micromonacha lanceolata

TAXONOMY
Bucco lanceolata Deville, 1849, Pampa del Sacramento, upper Ucayali River, Peru. Monotypic.

OTHER COMMON NAMES
French: Barbacou lanceolé; German: Streifen-faulvogel; Spanish: Monjita Lanceolata.

PHYSICAL CHARACTERISTICS
5.1–5.9 in (13–15 cm); 0.67–0.78 oz (19–22 g). Warm brown upperparts, scaled buffy. Whitish nasal tufts and chin feathers,

OTHER COMMON NAMES
French: Tamatia de Lafresnaye; German: Weißzügel-faulvogel; Spanish: Buco Barbón.

PHYSICAL CHARACTERISTICS
7.1–8.3 in (18–21 cm); 1.2–1.6 oz (33–46 g). Male mainly rufous, with white 'whiskers'; belly and flanks streaked blackish and white. Female similar, but duller brown on head and upperparts. Bill dark with basal two-thirds of lower mandible yellow; eyes red.

DISTRIBUTION
M. p. inornata: southeastern Mexico to western Panama; *M. p. panamensis*: southwestern Costa Rica to northwestern Colombia; *M. p. magdalenae*: west-central Colombia. *M. p. poliopis*: southwestern Colombia and western Ecuador.

HABITAT
Occurs in lower strata of primary and secondary humid forest and adjacent shady pastures; often perches by small openings such as trails.

BEHAVIOR
Sedentary on year-round pair territories. Perches immobile and inconspicuous for long periods.

FEEDING ECOLOGY AND DIET
Still-hunts from horizontal perches, consuming large arthropods, especially grasshoppers; small reptiles and amphibians are also taken. Sometimes attends ant swarms or mixed-species flocks.

Micromonacha lanceolata
■ Resident

white loral patch (bordered black) extending across forehead. White underparts heavily streaked black, except for central belly; undertail coverts buffy. Bill black and iris brown.

DISTRIBUTION
Western Costa Rica, west-central Panama; also from southwestern Colombia to western Ecuador and west-central Colombia to northern Bolivia.

HABITAT
At all strata (but usually low down) and most often at borders of primary and secondary humid forest at 980–6,890 ft (300–2,100 m).

BEHAVIOR
Principally solitary, although pairs are probably sedentary and territorial. Usually found sitting unobtrusively at forest edges or sometimes accompanying mixed-species flocks.

FEEDING ECOLOGY AND DIET
Hunts insects from perches and is known to eat berries, at least seasonally.

REPRODUCTIVE BIOLOGY
Nest is placed at end of a 16-in (40-cm) tunnel into a bank. Clutch contains two eggs; estimated incubation period 15 days.

CONSERVATION STATUS
Not threatened. Nowhere common, but widespread and thought to be secure.

SIGNIFICANCE TO HUMANS
None known. ◆

Nonnula ruficapilla
▨ Resident

Rufous-capped nunlet
Nonnula ruficapilla

TAXONOMY
Lypornix ruficapilla Tschudi, 1844, Peru. Four subspecies currently recognized.

OTHER COMMON NAMES
French: Barbacou à couronne rousse; German: Grauwangenfaulvogel; Spanish: Monjita coronada.

PHYSICAL CHARACTERISTICS
5.3–5.5 in (13.5–14 cm); 0.49–0.78 oz (14–22 g). Small with fine bill. Chestnut crown; face, nape, and sides of breast gray. Upperparts brown, more rufous on underparts with whitish belly.

DISTRIBUTION
N. r. rufipectus: northeastern Peru; *N. r. ruficapilla*: eastern Peru and western Brazil south of the Amazon; *N. r. nattereri*: northern Brazil, Mato Grosso, and northern Bolivia; *N. r. inundata*: eastern Pará, Brazil.

HABITAT
Mid-levels and undergrowth in humid forest edges, secondary forest, streamside forest, and igapó. In most of range associated with either bamboo or riverine regrowth.

BEHAVIOR
Singles (or quite often pairs together) usually found sitting quietly in low vegetation where they sally for food.

FEEDING ECOLOGY AND DIET
Apparently exclusively insectivorous.

REPRODUCTIVE BIOLOGY
No information. Other members of genus reported to nest in holes in banks or trees.

CONSERVATION STATUS
Not threatened. The species seems fairly common in suitable habitat.

SIGNIFICANCE TO HUMANS
None known. ◆

Black-fronted nunbird
Monasa nigrifrons

TAXONOMY
Bucco nigrifrons Spix, 1824, Rio Solimões, Brazil. Two subspecies recognized.

OTHER COMMON NAMES
French: Barbacou unicolore; German: Schwarzstirntrappist; Spanish: Manja Unicolor.

PHYSICAL CHARACTERISTICS
10.2–11.4 in (26–29 cm); 2.4–3.5 oz (68–98 g). Relatively slender and long tailed for a puffbird, entirely sooty black, darker around bill and paler ventrally. Bill bright red and legs black.

DISTRIBUTION
M. n. nigrifrons: southeastern Colombia, eastern Ecuador, eastern Peru, much of Amazonian Brazil; *M. n. canescens*: eastern Bolivia.

Monasa nigrifrons

■ Resident

Swallow-winged puffbird
Chelidoptera tenebrosa

TAXONOMY
Cuculus tenebrosus Pallas, 1782, Surinam. Three subspecies.

OTHER COMMON NAMES
English: Swallow wing; French: Barbacou à croupion blanc; German: Schwalben-faulvogel; Spanish: Buco Golondrina.

PHYSICAL CHARACTERISTICS
5.5–5.9 in (14–15 cm); 1.1–1.5 oz (30–41.5 g). Quite small with relatively pointed wings. Mostly sooty black, with large white area on lower back, and chestnut on lower underparts. Bill short and black.

DISTRIBUTION
C. t. pallida: northwest Venezuela; *C. t. tenebrosa*: Venezuela (except northwest), eastern Colombia and the Guianas south to eastern Ecuador, eastern Peru, Amazonian Brazil, and northern Bolivia; *C. t. brasiliensis*: coastal southeastern Brazil.

HABITAT
Recorded in wide variety of mostly open habitats 0–5,740 ft [0–1,750 m]), but prefers open sandy areas with scattered trees and bushes near forest.

HABITAT
Strong affinity to tall forest or regrowth on lakesides, riversides, and floodplains; also igapó and varzea, but generally absent from terra-firme forest.

BEHAVIOR
Group territorial, up to six individuals constantly foraging together noisily, often conspicuous in mixed-species foraging flocks.

FEEDING ECOLOGY AND DIET
Groups perch from lower strata to subcanopy, primarily taking insect prey (Lepidoptera, Orthoptera, Hymenoptera) in flight, but regularly from ground or foliage. Also reported catching small lizards, and following army ant swarms and primate troops to feed on flushed prey.

REPRODUCTIVE BIOLOGY
Lays about 3 eggs in nest burrows dug into level or slightly sloping ground. Incubation and fledging periods unknown. Presumed to be a cooperative breeder as groups visit nest area.

CONSERVATION STATUS
Not threatened. A common bird throughout most of its range.

SIGNIFICANCE TO HUMANS
None known. ◆

Chelidoptera tenebrosa

■ Resident

BEHAVIOR

Spends most time in pairs or small groups, perched high up in bare branches of emergent trees or other vantage points, from which it flies to catch prey. It can stay aloft for prolonged periods, leisurely circling, stalling, then swooping after passing prey before gliding to a perch. In flight it vaguely resembles a martin (Hirundinidae) or, more closely, a wood swallow (Artamidae).

FEEDING ECOLOGY AND DIET

Diet is entirely composed of bees, wasps, flying ants (Hymenoptera), alate termites (Isoptera), and other insects caught and consumed in flight.

REPRODUCTIVE BIOLOGY

One to two eggs laid in cavity at end of burrow dug into level sand or earth banks, frequently beside rivers. Incubation and nestling periods undocumented.

CONSERVATION STATUS

Not threatened. Common and conspicuous throughout much of range, and favored by moderate deforestation.

SIGNIFICANCE TO HUMANS

None known. ◆

Resources

Books

del Hoyo, J., A. Elliott, and J. Sargatal, eds. *Handbook of the Birds of the World.* Vol. 7, *Jacamars to Woodpeckers.* Barcelona: Lynx Edicions, 2002.

Sclater, P. L. *A Monograph of the Jacamars and Puffbirds,* London: R. H. Porter, 1882.

Joseph Andrew Tobias, PhD

Barbets
(Capitonidae)

Class Aves

Order Piciformes

Family Capitonidae

Thumbnail description
Small to medium-sized, colorful, thick-billed, large-headed and short-tailed birds; noticeable bristles around bill; tongue sometimes forked or brush-tipped; foot zygodactyl, two toes pointing backwards

Size
3.2–13.8 in (8–35 cm); 0.3–7.2 oz (8.5–203 g)

Number of genera, species
13 genera, 92 species

Habitat
Mostly tropical forest and forest edge, with some species (including most African ones) thriving in secondary forest, parkland, and even suburbs with many ornamental trees; a few live in drier, thornbush habitats with large termite mounds.

Conservation status
Endangered: 1 species; Near Threatened: 9 species.

Distribution
Northern South America and southern Central America; sub-Saharan Africa; south and Southeast Asia

Evolution and systematics

The order Piciformes includes barbets, toucans, honeyguides, and woodpeckers—on the face of it a varied group but, on closer examination, a range of birds with common characteristics. The toucans and barbets are particularly closely related and share a number of physical features: the tooth-edged bills of toucans are rather like elongated, exaggerated forms of the heavy barbet bills, although the smallest barbets, such as tinkerbirds, have much smaller and simpler bills. Toucans and some barbets are essentially fruit-eaters, with a pivotal role in the dispersal of tree seeds in forests. Almost all barbets can excavate nest holes in trees, as do woodpeckers. Of these various families the barbets are the most generalized. Perhaps small species long ago (12–20 million years ago) gave rise to woodpeckers and honeyguides. American toucans and barbets more certainly arose from a common ancestor some 10 million years ago, but the toucans have become more widespread in range (both to the north and the south of the barbets) and habitat choice. It has been said that, in evolutionary terms, the toucans "left the barbets behind" as they developed while the barbets stayed put in the old, tried-and-tested form.

Physical characteristics

All barbet species perch and climb well and have the two toes forward, two toes back foot structure that allows for an especially good grip on wide branches. This is a grasping or perching rather than climbing foot, and in climbing a near-vertical branch the outer toe may be swung forwards or sideways to give a better purchase. Barbets do not support themselves on their tail feathers, as do woodpeckers, except during brief spells when they are excavating their nests. Unlike woodpeckers, barbets do not have a long, thin, sticky tongue that can be extended and inserted into ant nests, although a few woodpecker species share the brush-tipped tongues of many barbets.

The bill is stout and strong, quite stubby but solid in the smaller species and rather elongated and pointed in the bigger ones. It usually has several tufts of bristles at the base. These can be flattened against the bill and opened up to avoid damage when the bird is pecking or excavating wood, but their precise function is unclear. Some barbet bills are extremely like toucanet bills, the cutting edge of the prominently grooved upper mandible being saw-toothed, but barbet bills are not flattened nor usually keeled in the same way as the larger toucans' extravagant beaks. In many species they are nevertheless used in displays, being brightly colored or contrastingly pale against a dark face.

Neotropical barbets always show marked sexual differences in color or pattern, but males and females are usually alike in Africa and Asia. Many African species are largely black and white with patches of red, yellow, or both in dramatic patterns. African crested and ground barbets are dark above with bold white spots in a rather random, untidy, spangled pattern. Their tails are also black with white or yellowish spots, with a touch of red at the base. The brown barbets of Africa and Asia are, by contrast, mostly dull and weakly

A red-crowned barbet (*Megalaima rafflesii*) with its insect prey. (Photo by T. Laman/VIREO. Reproduced by permission.)

marked, uniformly dark brown for the most part. The tinkerbirds of Africa, small and secretive birds of thick foliage, are often strongly patterned, striped black and white or with patches of red or yellow on the head, and vivid orange or yellow rumps: these color marks are often reflected in their names. Asian barbets include many lovely green species, with patterns of red, yellow, purple, brown, and blue in complex bibs, caps, cheek marks, spots, and bars. South American species are often gaudy and differ in appearance from the rest, despite sharing the usual color palette of black, white, red, and yellow. A number have orange breast shields or red breast bands. One genus, *Eubucco*, contains several small species that are predominantly bright green above, bright red on the head, and red or yellow beneath, topped off by pale yellow bills.

Distribution

Barbets are found in greatest numbers and diversity in tropical Africa; in the Americas they are fewer in species and have a surprisingly restricted range. Asian barbets are also less varied than the African ones. The 15 American barbets are mostly lowland birds of the Amazon basin; only three extend into Central America, in marked contrast with the related toucans. Few spread south of the Amazon and, unlike toucans, there are no southern barbets, and few that live in montane forest.

Barbets occupy most of Africa south of the Sahara, from the Atlantic to Ethiopia and south to South Africa. In Africa, barbets appear to have evolved during a long period of changing forest cover. As vast forests became split into discrete areas, so populations of barbets were isolated. These groups, over time, developed into separate species as forest areas re-

mained apart, each beyond the reach of weak-flying barbets from other forest refuges. This characteristic of ever-changing habitat also forced some barbets to occupy drier, bushy places outside proper forest cover, not seen in barbets elsewhere except in recent habitats such as orchards and towns where conditions have quite incidentally been made acceptable to fruit-eating forest birds.

Tropical Asian barbets are less well studied than the others. Five species live in peninsular India, four in Sri Lanka, and five or six in northern India and Nepal, but Southeast Asia has a greater variety. Sumatra has eight species, Java six, and Borneo nine; in all, 21 of 26 Asian barbets are in Southeast Asia. This variety within a spectacular family is wonderful, but its future is surely bleak as forest cover is being destroyed at an unprecedented (and still accelerating) rate. No barbets are found east of Wallace's Line or in Sulawesi: no islands east of Bali have barbets.

Habitat

Lowland tropical forest is typical barbet habitat, especially in Asia and Central and South America. They are often associated with fruiting trees, and many species of ornamental tree that produce copious fruits attract barbets into gardens, towns, and even city parks, particularly in Asia and Africa, where cities such as Nairobi and Harare host barbets all year round. They live in a variety of artificial woodlands, too, from pine to coconut plantations. Dead wood is important for barbets to excavate nesting holes, which they also use for roosting all year round. Over-managed woodland is not really suitable for them.

In East Africa several closely related species live in thorn thickets and bushy places on the fringe of the great wide-open plains, often associated with big termite mounds. In such places D'Arnaud's barbets (*Trachyphonus darnaudii*) are frequently seen foraging around thickets, ditches, and outbuildings alongside safari lodges, becoming quite bold and confiding. Professor Grzimek must have often seen and heard these lively birds in his camps in and around the Serengeti in Tanzania, unaware of the subsequent debate over their precise classification, as this is a species that has been "split" into two or three in more recent studies. Crested barbets (*Trachyphonus vaillantii*) are particularly common around termite mounds in open savanna woodland and thornbush, foraging about the tall earthy piles and perching openly upon them in pairs. Where the two occur together, D'Arnaud's prefers the flatter, open spaces between the little cliffs, streamside bluffs, and dry watercourses that the crested occupies. Tinkerbirds prefer tall forest with dense undergrowth, forest edge, and riverside woodland with tangled growth around the base of tall, old trees.

Sometimes two similar species remain separate within a large geographical region through habitat choice. For example, the coppersmith barbet (*Megalaima haemocephala*) of India favors forest edge, plantations of figs, village and city trees, gardens, orchards, and mangroves—wherever there is fruit to eat and broken branches and stumps to nest in. It does not, however, live in cleared areas where rainforest has been felled, nor the wet rainforest of Sri Lanka, in which areas it is replaced by the crimson-throated barbet (*M. rubricapilla*). The scarlet-crowned barbet (*Capito aurovirens*) of Colombia and Ecuador is found close to water, in the *varzea* or swampy or floodplain forests, and in tall, secondary growth, but it is rarely seen close to the related black-spotted barbet (*C. niger*), which tends to replace it in the "terra firma" forest of drier, more solid ground. Many species, such as the red-headed barbet (*Eubucco bourcierii*) of Costa Rica, are found along trails and cut rides in forests, living at the forest edge, but are quickly lost once the original forest is felled. Some, such as the prong-billed barbet (*Semnornis frantzii*) of Costa Rica and Panama, prefer the high canopy of rich, damp, mountain forest where the trees are draped in long beards of moss and epiphytes.

A number of closely similar species pairs occur in Africa, separated by range rather than habitat. These are examples of allopatric species, sometimes grouped as "superspecies": their ranges are complementary and do not overlap. The bristle-nosed barbet (*Gymnobucco peli*) lives in lowland forest in West Africa while the nearly identical Sladen's barbet (*G. sladeni*) replaces it in the forests of Zaire. Whyte's barbet (*Stactolaema whytii*) occupies woodland from southern Tanzania to eastern Zimbabwe, while the similar Anchieta's barbet (*S. anchietae*) is found in similar habitats to the west of this range. The range of the red-fronted tinkerbird (*Pogoniulus pusillus*) of eastern African bush is almost exactly surrounded by the more widespread yellow-fronted tinkerbird (*P. chrysoconus*). The red-fronted prefers wetter areas than the yellow-fronted in southern Africa, but drier areas in the northeast. Both overlap with the yellow-rumped tinkerbird (*P. bilineatus*), an example of sympatric species (found living side by side in the same area, the opposite of allopatric), although where the latter bird is most common the red-fronted is usually absent. Nevertheless, red-fronted and yellow-rumped may even sing from the same tree, but at different times. Where yellow-fronts and yellow-rumps occur together, the yellow-rumped prefers wetter, lusher woodlands; if the yellow-rumped is absent, such woods are, however, occupied by the yellow-fronted.

Two more species with almost exactly complementary ranges are the bearded barbet (*Lybius dubius*) of West Africa, found in a band of woodland and forest south of the Sahara, and the black-breasted barbet (*L. rolleti*) just to the east of it, in Chad and Sudan. There is no known overlap but the two may perhaps meet in Chad. The southern edge of the combined range of these two is irregular and "fingered," precisely abutting the similarly fingered range of the double-toothed barbet (*L. bidentatus*) of more open woodland immediately to the south. These make a trio of species, two with similar habitat but different range, the third replacing them in a slightly different but immediately adjacent habitat.

Behavior

Barbets hop and clamber about trees, move rather heavily through low bushes or on the ground, and often perch quite still for long periods. The bigger species tend to be more sluggish than the tiny ones, which are quite acrobatic when feed-

A coppersmith barbet (*Megalaima haemacephala*) feeding at its nest. (Photo by V. Sinha/VIREO. Reproduced by permission.)

ing. They fly well, but look a bit heavy and ungainly in the air and generally fly only for short stretches. The large, colorful species may suddenly appear, flying into a tree with a blurry splash of color, while smaller ones are usually heard but not so easily seen. A number of African barbets perch prominently and call repeatedly from one spot, making them easy to find and watch. The tinkerbirds call repeatedly, with a monotonous and sometimes infuriating repetition of simple notes, keeping out of sight as they do so. Some of these, as well as the coppersmith barbet (*Megalaima haemacephala*) in India, were among several birds referred to as "brain fever" birds by early European colonists, because of their nonstop calls. Together with the heat, the mosquitoes, and fever, the nonstop repetition of 1,000 or more calls by an invisible tinkerbird was sometimes just too much to bear.

Pairs may "duet" at times, calling in a neat pattern of coordinated notes with remarkable synchrony; this may be taken up and expanded by others within small social groups. It is often impossible to make out which, or even how many, birds are calling in the most perfected duets or choruses, which may produce a rhythmic phrase that is repeated several times in identical form.

They are aggressive birds, but several species (especially the larger ones) are social, with "helpers" at the nest and complicated social lives. Others are strictly territorial, each pair keeping others well away from the nest. They commonly roost in nest holes all year round and data on breeding seasons are poorly known for most species because of this source of potential confusion. A few species "drum" with the bill against a branch in the manner of a woodpecker, while others have aerial displays and noisy wing-rustling displays.

Feeding ecology and diet

Barbets are fruit eaters, but their growing chicks require high protein diets and are fed almost exclusively insects. Barbets have an important role in seed dispersal (although some species regurgitate pellets of undigested seeds from fruit, and then eat them again). In dense forest, there would be little chance of a seed being carried far by wind, or rolling more than several feet along the ground. To get far beyond the parent tree, it is best eaten by a barbet and ejected in a neat package of fertilizer from some far-off perch. Rich, evergreen, tropical forests have many species of fruiting trees, shrubs, and vines that flower and fruit at different times of the year; hence, barbets have a good food source all year round in sizeable forests (but this may become limited or disrupted in areas where forests are cleared or remain only in small patches). In Thailand as many as 100 blue-eared barbets (*Megalaima australis*) may gather at a single large, fruiting tree, breaking down their normal territories to feed on a temporary abundance of food that is sufficient for all. African barbets have been known to feed on at least 50 different genera of plants, mostly eating fruit but in some cases also taking nectar or blossoms. Figs are especially important food trees in both Africa and Asia and also supply fruits to South American barbets. Cultivated papayas, mangos, bananas, peppers, and avocados are also eaten, probably where natural foods have been reduced by human activity. Usually it is the larger species of barbets that dominate a fruiting tree where several species gather together.

Large fruits are naturally taken only by larger species that are big enough to swallow them. These take longer to digest than smaller food, and a barbet that has eaten some large fruits may then rest for an hour or more while its meal is digested. The red-headed barbet (*Eubucco bourcierii*) eats flower spikes along with the many insect larvae that find temporary refuge within them, while black-spotted barbets (*Capito niger*) and several other species, including the scarlet-hooded barbet (*E. tucinkae*) take both flowers and nectar. African barbets are the most persistently insectivorous species within the family: red-and-yellow barbets (*Trachyphonus erythrocephalus*) will eat almost anything that they can find, including scraps of food put out for them (eagerly taking milk, cereals, meat, bread, and fruit). Snails, worms, lizards, spiders, centipedes, even birds' eggs and young birds are eaten by some African species, as well as the many large insects that abound at certain times. Some even have special "anvils" where they remove the wings of large prey such as locusts. Asian barbets likewise take insects when breeding, but fewer at other times; some also eat birds' eggs, lizards, and centipedes. Several species dig into bark to find beetle larvae and into termite mounds to reach termites. Flying ants and termites are also caught on the wing in quite proficient flycatching sallies from a perch.

At least one South American barbet, the spot-crowned barbet (*Capito maculicoronatus*), follows swarms of army ants and eats insects that these fearsome columns flush out of hiding places, although it is not a constant companion of the ants in the same way that some woodcreepers and antbirds are, the "professional" followers of ant swarms. Others in this genus tap on bark to stir up insects while birds of the genus *Eubucco* specialize in foraging among clusters of dead leaves, searching out insects and other small invertebrates that seek food and safety in the brittle bunches. Several of these small barbets join mixed flocks of birds that roam around the forest, eating insects as well as fruit as they move through the treetops.

Reproductive biology

Most barbets have breeding territories and are monogamous, and some pair for life. They proclaim and defend a territory by singing, but their songs may be quite varied: some have 10 or 12 particular calls that are used as song, often in complex male-female duets. Often other birds apart from the breeding pair join in the chorus of songs, apparently serving to strengthen the effect of the song by communicating the size and therefore strength of the group.

Various displays include exposure of color patches on the head, wings, rump, tail, and bill, with feathers being erected to maximize the effect. If there is a difference between the sexes, the male's special color patches are used in such displays. Pairs of barbets focus their courtship activity around the hole that will serve as their nest, and often preen one another. They are long-lived birds with strong pair bonds, supported and cemented by activities such as mutual grooming, which are maintained all year. The pair typically mates for life, so breeding season courtship displays are rather limited, being required only to synchronize breeding condition between male and female.

The nest is in a hole in a tree, usually freshly excavated. Smaller species nest in dead or dying branches, typically digging in from beneath a horizontal bough. Barbets tend to make larger entrance holes than woodpeckers of a similar size. The hole enters a vertical shaft ending in a slightly widened chamber in which the eggs are laid. In more social species several "helpers" may be involved in digging the nest hole over a number of weeks, although in some the helpers at the nest are not involved in the excavation. Wood chips are usually carried away, sometimes being swallowed and then regurgitated elsewhere, presumably to prevent telltale signs from accumulating beneath the tree, giving clues of the nest's whereabouts to predators. The eggs are still not described for many species and incubation and fledging periods are generally little known. Like those of woodpeckers, barbet eggs are, however, always pure white and the clutch size up to six (typically three) in African barbets, two to five (again, most often three) in South American species. Incubation averages around 15 days—12 days for the smallest and up to 19 days for larger species. Once the chicks are hatched, adult barbets keep the nest exceptionally clean in most cases, probably swallowing rather than carrying away the chicks' droppings. Nestlings of small species leave the nest after 17 to 23 days, larger ones remaining as long as 30 or 40 days, notably long periods for such modestly sized birds. Helpers (sometimes young of the pair from previous years) assist the pair in feeding the chicks in some species, while many are strictly territorial and only the breeding pair is involved.

Conservation status

The white-mantled barbet (*Capito hypoleucus*) of Colombia is classified as Endangered. This species has a small, very fragmented range some parts of which are subject to rapid habitat loss. The population of this species is probably declining as a result.

An additional nine species of barbets—from South America, Africa, and Southeast Asia—are considered Near Threatened. Threats to barbets are all too apparent in all forested habitats. Uncontrolled and frequently illegal logging is increasing, especially in Southeast Asia, where the loss of habitat for barbets and other wildlife is almost incalculable; certainly millions of acres disappear each year.

Good habitat is becoming both reduced in total area and also more patchily distributed, separated by unsuitable areas that barbets are unable to cross. Small populations will inevitably become threatened and less viable, so conservation requires strong protection for forests (including dead trees within them) and also maintenance of corridors that link remaining forest patches. In many areas this appears to be an unachievable aim.

Fires have had severe effects on barbet habitat, too, especially in Indonesia in recent years. In Africa, the barbets that live in open woodland, bush, and savanna will presumably do much better in the long term, but forest-living species face an uncertain future at best.

Significance to humans

Barbets have little or no commercial impact on people, taking some fruits but not usually being sufficiently numerous to be classified as pests. They are, however, very obvious components of a forest's sounds and many species, if not well known visually, are easily recognized by local people by their calls and repetitive songs.

1. Black-spotted barbet (*Capito niger*); 2. Black-collared barbet (*Lybius torquatus*); 3. Toucan barbet (*Semnornis ramphastinus*); 4. Yellow-fronted tinkerbird (*Pogoniulus chrysoconus*); 5. D'Arnaud's barbet (*Trachyphonus darnaudii*); 6. Coppersmith barbet (*Megalaima haemacephala*). (Illustration by Joseph E. Trumpey)

Species accounts

Coppersmith barbet
Megalaima haemacephala

TAXONOMY
Megalaima h. haemocephala Muller, 1776. Five subspecies.

OTHER COMMON NAMES
English: Crimson-breasted barbet; French: Barbu à plastron rouge; German: Kupferschmeid; Spanish: Barbudo de Pecho Rojo.

PHYSICAL CHARACTERISTICS
6.7 in (17 cm); 1.4–1.8 oz (39–52 g). Adult birds are unmistakable. Upperparts are dark green; underparts are pale greenish with broad, dark green streaks and a red band across the upper breast. Forecrown is red; sides of head and throat are yellow; eyestripe and submoustachial stripe are black.

DISTRIBUTION
Peninsular and northern India, northeastern Pakistan, Nepal, Bangladesh, and Sri Lanka to southwestern China, Malaysia, Sumatra, Philippines.

HABITAT
Forest edge, dry deciduous woodland, teak forest, irrigated orchards and plantations with figs and other fruiting trees; also town and city parks, gardens, ornamental trees, and edges of mangroves.

BEHAVIOR
Sings frequently with synchronized jerk of body, bob of head, and flick of tail; throat puffed out and jerk of head gives song a ventriloquial quality: a monotonous "pohp-pohp" or "tonk-tonk" all day and even on moonlit nights. Sound recalls metallic hammering.

FEEDING ECOLOGY AND DIET
Forages in tree canopy and on fruiting trees, eating figs, pipals, guavas, mangos, and custard apples as well as smaller berries and many insects; taps and chips away bark to reach invertebrates.

REPRODUCTIVE BIOLOGY
Lays 2–4 eggs (usually 3) in hole excavated in tree; both parents incubate for 14 days; chicks fledge after five weeks. Both parents feed chicks but they are abandoned as soon as they fledge, when female begins second brood.

CONSERVATION STATUS
Not threatened and very common in most of range.

SIGNIFICANCE TO HUMANS
Frequently heard and well known for its "hammering" song, even in urban areas. ◆

Yellow-fronted tinkerbird
Pogoniulus chrysoconus

TAXONOMY
Pogoniulus chrysoconus Temminck, 1832. Three subspecies.

OTHER COMMON NAMES
English: Yellow-fronted tinker barbet; French: Barbion à front jaune; German: Gelbstim-Bartvogel; Spanish: Gitano de Frente Amarillo.

Megalaima haemacephala
■ Resident

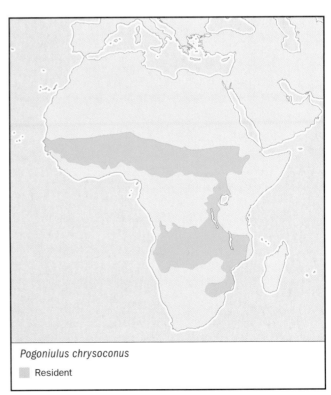

Pogoniulus chrysoconus
■ Resident

PHYSICAL CHARACTERISTICS

4.3 in (11 cm); 1.9–2.2 oz (53–63 g). Upperparts are black with white to yellow-white streaks shading to a mostly pale yellow rump; underparts are gray washed with lemon yellow. Tail is black with yellow-white edges; wings are blackish brown edged in white or yellow-white. Forecrown and center of crown are yellow to deep orange bordered in black; hindcrown is black with white streaks.

DISTRIBUTION

Sub-Saharan Africa, from Atlantic to southern Sudan but not reaching Red Sea coast; absent from central West Africa; south from Sudan to Lake Victoria, most of Central Africa south to Mozambique.

HABITAT

Many kinds of forest and riverside woodland habitats and dry, bushy vegetation, from small patches of forest to tall clumps and isolated trees in grassland.

BEHAVIOR

Solitary, occasionally joining mixed flocks briefly; flies rapidly from place to place, excavating roosting cavities in a variety of habitats. Very aggressive and alert to other barbets, approaching them when they call and even visiting nests of other species. Erects gold crown feathers in display.

FEEDING ECOLOGY AND DIET

Moves steadily upwards through foliage, pecking at insects and other invertebrates or finding fruits; takes smaller berries from bushes, often bright red, orange, or purple fruits; investigates clumps of dead leaves for insects.

REPRODUCTIVE BIOLOGY

Territorial; swings head, flicks tail, and erects bright crown and rump feathers in time with monotonously repeated popping notes: 8–120 notes per minute for up to 109 minutes. Territories defended by patrols and singing along borders. Nest excavated in dead stump or branch; 2–3 white eggs incubated for 12 days, nestlings fledge in three weeks.

CONSERVATION STATUS

Not threatened; widespread and generally quite common.

SIGNIFICANCE TO HUMANS

Constant repetition of song notes well known. ◆

Black-collared barbet

Lybius torquatus

TAXONOMY

Lybius torquatus Dumont, 1816. Seven subspecies.

OTHER COMMON NAMES

French: Barbican à collier; German: Halsband-Bartvogel; Spanish: Barbudo de Collar Negro.

PHYSICAL CHARACTERISTICS

7.8 in (20 cm); 1.6–2.8 oz (45–80 g). Forecrown to upper breast is red. A broad black band separates the red breast from the pale yellow belly. Hindcrown to upper back is black; lower back and rump are brown with fine, dark and yellowish lines. Wing feathers are dark brown edged with pale yellow.

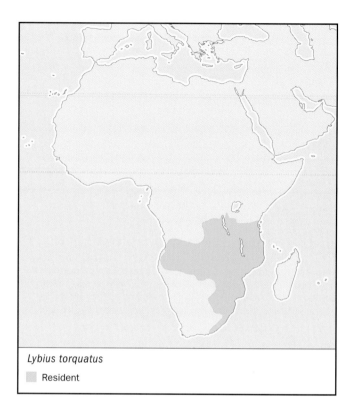

Lybius torquatus
Resident

DISTRIBUTION

Central and southern Africa from the east coast of Kenya west to Angola, south through Zimbabwe and Mozambique to eastern South Africa.

HABITAT

Open woodland, including vicinity of villages and camps, open wooded grassland, gardens, and farmland.

BEHAVIOR

Perches in pairs or groups of up to six adults on conspicuous treetop branches. Interacts with other, larger barbet species, calling frequently unless chased away. Group lives together, feeding and roosting in close association.

FEEDING ECOLOGY AND DIET

Eats fruit such as figs, guava, grapes, and many brightly colored berries; also eats many insects including termites and beetles. Catches flies in flight, sometimes drops to ground to pick up fruit.

REPRODUCTIVE BIOLOGY

Breeding requires dead trees or branches; territory of 50–125 acres (20–50 ha). Group maintains territory with frequent synchronized duets, with two calling birds using different notes, given in rapid succession, sounding like "pududut," "tay-pudit-tay-pudit" and many other variations, male calling at a lower pitch than female. Duetting increases near nesting time; primary male of group gives aerial display and pair has intricate greeting ceremonies with cocked tails, swinging bodies, bowing, and short leaps. Paired birds touch bills and male feeds mate. Nest excavated in dead stump; 1–5 eggs (typically 3–4) incubated for 18–19 days; chicks fledge at 33–35 days. Up to four broods per year. Breeding pair excavate the nest, but all members of group help with incubation and feeding young.

CONSERVATION STATUS

Not threatened; generally quite common and secure.

SIGNIFICANCE TO HUMANS
Familiar, noisy bird around human habitation. ◆

D'Arnaud's barbet
Trachyphonus darnaudii

TAXONOMY
Trachyphonus darnaudii Cretzschmar, 1826. Four subspecies.

OTHER COMMON NAMES
English: Usambiro barbet; French: Barbican d'Arnaud; German: Ohrfleck-Bartvogel; Spanish: Barbudo de D'Arnaud.

PHYSICAL CHARACTERISTICS
6.7–7.5 in (17–19 cm); 0.9–1.8 oz (25–50 g). A boldly spotted bird. Upperparts are blackish brown to brownish with white speckles; underparts are black speckled and there is a black patch on the lower throat and a black-and-white spotted band across the breast. Forehead and crown are yellow or orange and yellow speckled with black (one subspecies has black forehead and crown).

DISTRIBUTION
East Africa, including southeastern Sudan, northeastern Uganda, Kenya, northern Tanzania.

HABITAT
Dry thornbush and bushy savanna, abandoned Masai camps, open woodland.

BEHAVIOR
Found mostly in pairs, sometimes in groups of 4–5; often perches inconspicuously low down on a bush or stump. Erects crest feathers, bobs head, and flirts or sways tail during duetting calls.

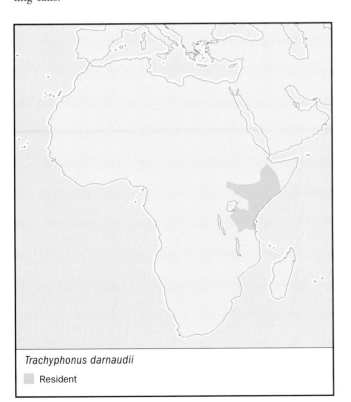

Trachyphonus darnaudii

▨ Resident

FEEDING ECOLOGY AND DIET
Eats many ants, termites and their eggs, grasshoppers and other insects, some caught in flight, others picked from leaves or the ground. Also takes many berries and small fruits and seeds.

REPRODUCTIVE BIOLOGY
Song in duet from male and female, notes differing but not consistent within either sex; typically a series of up-down notes followed by two or three high notes, such as "witch-ee-tee-ta-ta-ta" or "ker-ka-tee-tootle," in synchronized pattern. Pairs usually use existing cavity in tree, not excavating their own; 2–4 eggs incubated probably by the parent birds, but perhaps assisted by helpers. The breeding pair feeds the chicks almost or entirely unassisted by helpers within a small group. Young beg for food from parents for some days after fledging; they may become helpers in subsequent breeding attempts in the next breeding season.

CONSERVATION STATUS
Widespread and locally common; not threatened.

SIGNIFICANCE TO HUMANS
Little known to most people but frequent around old bush camps, Masai camps, and some tourist lodges. ◆

Black-spotted barbet
Capito niger

TAXONOMY
Capito niger Muller, 1776. Fifteen subspecies.

OTHER COMMON NAMES
French: Cabézon tacheté; German: Tupfenbartvogel; Spanish: Chaboclo Turero.

PHYSICAL CHARACTERISTICS
7.1–7.5 in (18–19 cm); 0.8–0.9 oz (22–25 g)

DISTRIBUTION
West Colombia to Venezuela, the Guianas, Ecuador, Peru, Bolivia, and east through Brazil north of the Amazon, but restricted to a small area south of that river.

HABITAT
Mostly in mature, lowland forest, both dry and wet floodplain forests; also upland forest, forest edge, gardens, orchards, plantations, and elfin mossy forest at high altitudes in Peru; forest patches in savanna and coastal forest in the Guianas.

BEHAVIOR
Usually solitary or in pairs, foraging through canopy, sometimes descending to lower levels of forest; also joins roving bands of various flycatchers, woodcreepers, manakins, and tanagers. Typically acrobatic when feeding, searching leaf clusters, lichen, and old bark, often at tips of tiny branches and twigs.

FEEDING ECOLOGY AND DIET
Eats insects and fruit of many kinds; picks clusters of leaves to shreds and breaks into bunches of dead leaves to find insects, but 80% of food is fruit and oily seeds. Sometimes holds large fruit and tough insects with its feet and pecks them into pieces.

REPRODUCTIVE BIOLOGY
Song is commonly heard in tropical forest, a low-pitched, double note, "hoop-oop" repeated for 6–20 or even 60 seconds

Capito niger
▢ Resident

Semnornis ramphastinus
▢ Resident

without a break, sometimes fading away or continued at a lower volume. Otherwise, courtship, display, and breeding cycle are little known. Both sexes excavate a cavity in a tree stump and 3–4 white eggs are laid; incubation by both parents, period unknown. Chicks fly when 34 days old and fed by parents for an additional 23 days or so.

CONSERVATION STATUS
Not threatened; generally common throughout its large range.

SIGNIFICANCE TO HUMANS
None known. ◆

Toucan barbet
Semnornis ramphastinus

TAXONOMY
Semnornis ramphastinus Jardine, 1855. Two subspecies.

OTHER COMMON NAMES
French: Cabézon toucan; German: Tukanbartvogel; Spanish: Capitán Tucán.

PHYSICAL CHARACTERISTICS
9.8 in (25 cm); 3.0–3.9 oz (85–110 g).

DISTRIBUTION
Southwestern Colombia and western Ecuador.

HABITAT
Wet subtropical forest and montane tropical forest, secondary growth, and more open pastures with scattered fruiting trees.

BEHAVIOR
Usually in or around fruiting trees and bushes in groups of up to six birds, typically a territorial pair and their young. Relatively heavy, hopping about on branches or climbing up through low, tangled, bushy growth, from almost ground level to the high canopy. Sometimes associates with other species in small mixed groups if they pass through the territory. Sometimes perches still for long periods, remaining very inconspicuous.

FEEDING ECOLOGY AND DIET
Mostly eats fruit, but takes many insects as well. Eats fruits of 62 species of plants, many of them typical of disturbed areas (only 11 of pristine forest); many of these are seasonal and insects and other invertebrates must be relied upon at times. May trespass into neighboring territories to reach fruiting trees if not challenged.

REPRODUCTIVE BIOLOGY
Song is series of low, short, foghorn-like notes repeated for several minutes, often in duets, but these are simple simultaneous songs and not properly synchronized. Tail often cocked; this and many short calls help birds keep in touch in dense foliage. Territory established around roosting holes in dead tree; breeding pair then drives away older offspring and other adults from group, but some retain one or two helpers. Primary female remains near nest; number of eggs unknown; incubation about 15 days, mostly by male, but also by primary female and helpers. Young fed by pair and helpers, if present, for 43–46 days. Young beg for food for some time afterwards; if second brood is reared, young from first remain as helpers while still begging for food themselves at times. In next breeding season, these older young disperse to establish new groups while younger offspring remain.

CONSERVATION STATUS

Near Threatened; occupies only some 7,700 square miles (20,000 square kilometers), but still common in parts of that small range; suffers from trapping and loss of habitat.

SIGNIFICANCE TO HUMANS

Trapped as cage bird. ◆

Resources

Books

Fry, C. H., S. Keith, and E. K. Urban., eds. *The Birds of Africa*. Vol. 3. New York: Academic Press, 1988.

Grimmett, Richard, Carol Inskipp, and Tim Inskipp. *A Guide to the Birds of India, Pakistan, Nepal, Bangladesh, Bhutan, Sri Lanka, and the Maldives.* Princeton: Princeton University Press, 1998.

Short, Lester, and Jennifer Horne. *Toucans, Barbets, and Honeyguides.* Oxford: Oxford University Press, 2001.

Skutch, A. F. *Birds of Tropical America*. Austin: University of Texas Press, 1983.

Zimmerman, Dale A., Donald A. Turner, and David J. Pearson. *Birds of Kenya and Northern Tanzania.* Princeton: Princeton University Press, 1996.

Robert Arthur Hume, BA

Toucans

(Ramphastidae)

Class Aves

Order Piciformes

Family Ramphastidae

Thumbnail description
Medium-sized to large birds, instantly recognizable by their strikingly large and colorful bills; often associate in small flocks when foraging

Size
13–24 in (33–60 cm); 4–30 oz (113–850 g)

Number of genera, species
6 genera; 41 species

Habitat
Predominately tropical and montane rainforest

Conservation status
Endangered: 1; Near Threatened: 3

Distribution
South America south to northern Argentina

Evolution and systematics

The toucans (Ramphastidae) are a very striking group. No fossilized remains of this family are known; recent remains of a toco toucan (*Ramphastos toco*), from the Pleistocene (20,000 years old) have been found, in Minas Gerais, Brazil.

The toucans' closest relatives are barbets (Capitonidae), and the two groups are thought to have evolved from a common ancestor; in their landmark 1993 study, Sibley and Ahlquist describe toucans as "New World barbets with big bills." Woodpeckers (Picidae) are closely allied to the toucans, which are also similar to hornbills (Bucerotidae), albeit more distantly related.

J. Haffer proposed the most widely accepted theory of speciation in toucans, and his model has since been applied to other Amazonian bird groups. Toucan speciation apparently occurred during Pleistocene glaciations (characterized by dry forest expansions) and interglaciations (characterized by tropical forest expansions). The rapid formation and disappearance of new Pleistocene biomes enhanced speciation rates in this region. Species derived from drier biomes later became sympatric (inhabiting the same range but not interbreeding) when tropical forest expanded in Amazonia, forcing the dry forest specialists to adapt to tropical forest or perish. There

are 6 genera with 41 species. The green toucanets (*Aulacorhynchus*) are represented by seven species; the aracaris (*Pteroglossus*) by 12 species; the black toucanets (*Selenidera*) by six species; the saffron toucanet (*Baillonius*) is monotypic and in Peters is lumped into genus *Andigena*; the mountain toucans (*Andigena*) are represented by four species; and the true toucans (*Ramphastos*) by 11 species.

Social mimicry, a relatively rare evolutionary event in nature, is seen in this group. Social mimicry is the situation in which two or more distantly related species are more similar to one another than they are to other, more closely related species. Among toucans, it is apparent in the yellow-ridged toucan (*Ramphastos vitellinus culminatus*), which mimics the color and appearance of the larger white-throated toucan (*Ramphastos tucanus cuvieri*). The mimic benefits because it avoids being attacked by the larger model, while its appearance deters smaller species from trying to share feeding sites.

Physical characteristics

Toucans are characterized by brightly colored plumage and an unusually long and bulky, but very lightweight, bill

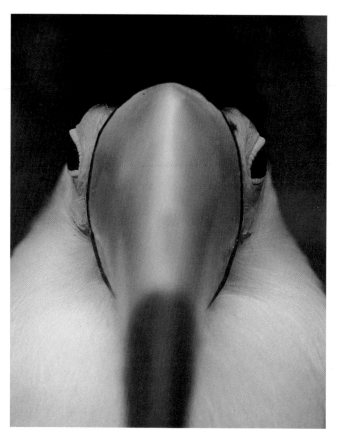

An unusual view of a keel-billed toucan's (*Ramphastos sulfuratus*) colorful bill. (Photo by T.J. Ulrich/VIREO. Reproduced by permission.)

with a downcurved tip and serrated edges. Another trait characteristic of toucans is a tongue with a bristly or brushy tip. Toucans range in length from 13–24 in (33–60 cm). In almost all species, male and female look alike; only the black toucanets (*Selenidera*) and green aracari (*Pteroglossus viridis*) are sexually dimorphic.

The toe arrangement is zygodactylous, with two toes projecting towards the front and two to the back.

Naturalists have long puzzled over the significance of the toucan's large bill. Originally, observers suggested that the bill was a weapon used to defend the nest cavity. This is not so; when toucans sense danger, they come out of the cavity entrance in a hurry, threatening the enemy only out in the open, if at all. Instead, a long bill enables these rather heavy birds to pluck berries from the tips of branches without leaving a stable perch. A thin, dark-colored bill would, however, be just as useful for this purpose. Possibly the toucan's bill plays a role in pair formation and in the social life of the group. According to E. Thomas Gilliard, it acts as a signal. However, toucans can also use their bills to threaten those birds whose nests they plunder. Tyrant flycatchers and even small raptors are frightened by the giant bill, which is even more effective because of its lively colors, and they fly about helplessly while the toucans devour their young or eggs. Other birds will attack a toucan only while it is in flight, because it is then unable to defend itself with its bill.

Distribution

Toucans are restricted to the Neotropics, where they are distributed through most of tropical South America as far as northern Argentina, with some individuals found as far north as Mexico. Although most species are lowland residents, there are some exceptions: most of the toucanets (*Aulacorhynchus*) have home ranges at 8,200 ft (2,500 m) or higher in Central America, and the mountain toucans (*Andigena*) range 3,900–11,000 ft (1,200–3,350 m) in the South American Andes.

The nation with by far the highest diversity of toucans is Colombia, with 21 species; Venezuela, Ecuador, and Brazil are home to 17 species each. The country with the lowest diversity of toucans is Trinidad, with a single species, the channel-billed toucan (*Ramphastos vitellinus*).

Habitat

Most toucans are canopy specialists in tropical or montane rainforest. Secondary vegetation may be inhabited by true toucanets, aracaris, and some of the true toucans. Some of the aracaris and the toco toucan are riverine specialists, and species such as red-breasted (*Ramphastos dicolorus*) and toco toucans may inhabit palm savanna.

Behavior

Toucans wander through the forests and adjacent clearings in small family groups and flocks; such flocks rarely consist of more than twelve birds. Toucans are not intensely sociable; they never take flight in a tight group, but instead wander about in loose groups. The agile aracaris fly swiftly and in a straight line; large toucans are poor fliers. After beating their wings a few times, they hold them out and glide downward, as if pulled down by the weight of their large bills. Then they begin to beat their wings again. The flight, as a result, is both undulating and brief.

An emerald toucanet (*Aulacorhynchus prasinus*) feeds on a berry. (Photo by B. Miller/VIREO. Reproduced by permission.)

Toucans preen one another, particularly on the head and nape, with the tip of the bill. When sleeping, toucans lay the bill up over the back, and tip up the tail, forming a roof over the back and bill. One can see how advantageous such a "feather ball" is to the aracaris, which sometimes sleep five or six together in an old woodpecker tree hole or in rotted hollow tree trunks; the last bird to squeeze in enters the hole backwards, with its tail laid over its back.

Toucans utter melodious calls many times in the late afternoon, continuing when most other birds have gone to rest, but are inactive at night. They also vocalize more during morning hours, and after rains. They prefer to remain high in trees, even bathing in the rainwater pools that form in the fork of a tree or on a thick horizontal branch.

Although many species have ranges that overlap, D. Brooks found that species living in the same region have bills of different lengths and take different prey, thus avoiding competition. The only case of competitive exclusion where two similar-sized species coexist involves pale-billed aracaris (*Pteroglossus flavirostris*), a species that is restricted to the forest canopy in the Peruvian Amazon by chestnut-eared aracaris (*Pteroglossus castanotis*), which occupy forest edge almost exclusively. Competitive exclusion is assumed because pale-billed aracaris in Venezuela will use edge forest, not just forest canopy, in the absence of chestnut-eared aracaris; meanwhile chestnut-eared aracaris in Paraguay, where there are no pale-billed aracaris, remain in forest-edge habitat.

Most toucan species are year-round residents in their home range but montane species may undergo seasonal altitudinal migrations, moving downslope in fall and upslope in spring. For some lowland species there are records of huge flocks invading areas with fruiting trees after the breeding season, when fruits become scarce.

Feeding ecology and diet

All toucans forage in the forest canopy (as opposed to the understory or on the ground). Fruits (and some flowers) are a toucan's primary food. Fruits eaten include those of palms (such as *Mauritia*, *Euterpe*, *Oenocarpus*), nutmeg (*Virola*), figs (*Ficus*), guava (*Psidium*), red pepper (*Capsicum fructescens*), and other fruits such as *Casearia corymbosa*, *Cecropia*, *Didymopanax*, *Phytolacca*, *Rapanea*, and *Ehretia tinifolia*. Although most of the toucan diet is composed of fruit, these birds occasionally eat small mammals such as bats, plus baby birds and bird eggs. They also eat insects such as crickets, cicadas, spiders, and termites, other small invertebrates, and lizards and snakes. Toucans often drink water from epiphytic bromeliads rather than descend to the ground to drink from a pool or stream. Toucans that plunder nests to eat eggs and chicks are often mobbed by the other birds.

Though specialization on fruit may decrease with increasing body size, toucans as a group are primarily frugivorous. Remsen and his colleagues found fruit in 96.5% of all toucan stomachs they examined. Arthropod prey was found in only 5.5% of stomachs, and vertebrate prey was found in fewer than 1% of samples.

A toucan's colorful bill is its most distinctive characteristic. Pictured here is a toco toucan (*Ramphastos toco*). (Photo by K. Schafer/VIREO. Reproduced by permission.)

Toucans pluck fruit while perched. Getting a small berry from the tip of the huge bill into the throat is quite a task; toucans perform it by jerking the head back while the bill is open. Large lumps of food are held under the foot and reduced to smaller lumps with the bill. When the soft part of a berry has been digested, large seeds are regurgitated on the spot; smaller seeds pass through the digestive tract and are dispersed. This seed dispersal helps to regenerate the tropical forests where toucans live.

Reproductive biology

Typically the male feeds berries to the female as part of courtship. Mutual preening is also an important aspect of courtship. Toucans form monogamous pairs.

Eggs are elliptical and white. The clutch typically ranges from two to four eggs, though as many as six may be laid in some species. The egg may be as heavy as 5% of the weight of the female in some species. Incubation lasts approximately 18 days, but this depends on the species. Both parents incubate.

Toucans nest in tree holes, usually high above the ground; the larger species prefer natural cavities in rotted trees, while the smaller ones often use woodpecker holes. Chestnut-eared aracaris will nest in abandoned tree-termite nests. Toucans

A pair of collared aracaris (*Pteroglossus torquatus*) share a branch. (Photo by T.J. Ulrich/VIREO. Reproduced by permission.)

sometimes drive woodpeckers away from newly excavated nest holes, and then enlarge the entrance if it is too narrow for them. The floor of the hole is typically 3–8 in (8–20 cm) below the opening. Toucans sometimes lay their eggs in the same hole year after year. The eggs lie on wood debris, or often on a lining of seed pellets that the birds have regurgitated before egg laying.

Newly hatched toucans have bare red skin that is not covered with down. Their bills are short, with the lower bill longer and wider than the top, as is the case with young woodpeckers. Thick, horny swellings on the heels protect these joints from friction. The heel pads have sharp outgrowths that may form a ring; these "tarsal calluses" fall off after the young leave the nest.

Both parents brood the naked nestlings, and parents also share the work of feeding their young. They bring some of the food in the tips of the bill, but most food is carried in the

throat or esophagus and regurgitated at the nest. Parents and older nestlings clean the nest hole and carry refuse away in their bills. Young toucans open their eyes after they are three weeks old. Their feathers grow so slowly that young still show much bare skin even after they are a month old. Young of the largest species probably remain in the nest for about 50 days. Fledglings typically begin feeding themselves 8–10 days after leaving the nest.

Conservation status

The main enemies of adult toucans seem to be raptors, which sometimes catch toucans almost as large as themselves. Of the 41 toucan species, only the yellow-browed toucanet (*Aulacorhynchus huallagae*) is considered Endangered. The main threat to this Peruvian species is loss of habitat.

Three additional species are considered Near Threatened, however. The saffron toucanet (*Baillonius bailloni*) of central South America is threatened by habitat destruction, capture for the cage bird industry, and hunting to some degree. Two of the mountain toucans—the plate-billed (*Andigena laminirostris*) and gray-breasted (*Andigena hypoglauca*)—are threatened by habitat destruction in the Andes, where they are endemic.

Significance to humans

A Brazilian folk tale relates that the toucan wanted to be ruler of all birds, so it hid inside a tree cavity with only its large bill visible through the hole. Seeing the massive bill, the other birds accepted the toucan as king—until it emerged from the cavity. Then the thrush noted that despite its large bill, the toucan had a small body, and the toucan was humiliated by all birds.

Amazonian Indians use colorful toucan feathers and bills for decoration, especially those of true toucans (*Ramphastos*) and aracaris (*Pteroglossus*). In some cultures toucan flesh is relished, and the birds are regarded as trophies of sorts.

Native South Americans sometimes take young toucans from their nests and keep them as free-flying pets. The first European settlers in South America often kept pet toucans as well. Such tame toucans have been known to dominate chickens in villages and farmyards.

1. Toco toucan (*Ramphastos toco*); 2. Yellow-browed toucanet (*Aulacorhynchus huallagae*); 3. Chestnut-eared aracari (*Pteroglossus castanotis*); 4. Gray-breasted mountain toucan (*Andigena hypoglauca*); 5. Saffron toucanet (*Baillonius bailloni*); 6. Plate-billed mountain toucan (*Andigena laminirostris*); 7. White-throated toucan (*Ramphastos tucanus*). (Illustration by Joseph E Trumpey)

Species accounts

Chestnut-eared aracari
Pteroglossus castanotis

TAXONOMY
Pteroglossus castanotis Linnaeus, 1758. This genus falls between green toucanets (*Aulacorhynchus*) and black toucanets (*Selenidera*).

OTHER COMMON NAMES
French: Aracari à oreillons roux; German: Braunohr-Arassari; Spanish: Arasari Orejicastano.

PHYSICAL CHARACTERISTICS
The smallest and most slender of the toucans; compared to others in the genus, this species is comparatively large and heavy bodied. Length about 18 in (46 cm); average weight 8.7–9.7 oz (247–275 g) males; 8.5–9.5 oz (240–271 g) females. This species is named for the chestnut "ear" patches on its black head. Plumage is typical of this genus, with a dark-green back and yellow underparts; the species is distinguished from close relatives by the single red band across its yellow belly. Bill is dark-brown to black with a yellow streak extending to the tip of the upper mandible and distinct, ivory-colored "teeth." Eye is usually white but sometimes straw yellow; skin around the eyes is blue.

DISTRIBUTION
The most widely distributed aracari, found throughout the tropics of South America, from the Colombian Amazon through eastern Ecuador, Peru, and Boliva, western Brazil, and eastern Paraguay to northeastern Argentina.

HABITAT
This species occurs in moist lowland tropical rainforest, up to 2,950 ft (900 m) elevation, often in swamp forests or along lake and river edges; it is the most common aracari on river islands. Unlike most other toucan species, often found in open forest habitat and even gardens around homes; commonly seen at forest edges and roadsides.

BEHAVIOR
Hilty and Brown describe the call as "a sharp, inflected *skeez-up.*" Often seen in pairs but birds also forage in small family groups of three to five. Individuals roost nightly in woodpecker holes, sometimes evicting the official occupant.

FEEDING ECOLOGY AND DIET
Mainly eat fruit of *Cecropia*, *Ocotea*, *Ficus*, and *Coussapoa*; will hang upside down to reach fruit. Occasionally take insects and have been documented raiding nests of yellow-rumped caciques (*Cacicus cela*) among others. Sometimes forage in mixed flocks with other toucan species.

REPRODUCTIVE BIOLOGY
Not well known. This species nests both in tree holes and in abandoned arboreal termitaria.

CONSERVATION STATUS
Not threatened.

SIGNIFICANCE TO HUMANS
Sometimes hunted for meat. ◆

Pteroglossus castanotis
Resident

Yellow-browed toucanet
Aulacorhynchus huallagae

TAXONOMY
Aulacorhynchus huallagae Carriker, 1933. This species shares the genus *Aulacorhynchus* with six other species of toucanets.

OTHER COMMON NAMES
French: Toucanet à sourcils jaunes; German: Gelbbrauen-Arassari; Spanish: Tucancito de Cuello Dorado.

PHYSICAL CHARACTERISTICS
Like other members of this genus, has a green back; underparts are red with golden-yellow undertail coverts and white band around base of tail. Length 14–16 in (37–41 cm); weight 9.3 oz (264 g) male; 9.8 oz (278 g) female.

DISTRIBUTION
This species is restricted to two small and isolated highland forest fragments in central Peru.

HABITAT
This species occupies the canopy of cool, humid, cloud-forests from 6,970–8,240 ft (2,125–2,510 m) in elevation, where trees are thickly covered with epiphytes and under-growth is dense.

Aulacorhynchus huallagae
▨ Resident

Baillonius bailloni
▨ Resident

BEHAVIOR
Not well known. Calls are low-pitched and frog-like.

FEEDING ECOLOGY AND DIET
Not well known; has been seen feeding on flowers and melastome (family Melastomataceae) fruits. Remsen et al. examined stomach contents of three specimens and found only fruit; no arthropods or small vertebrates.

REPRODUCTIVE BIOLOGY
Not known.

CONSERVATION STATUS
Endangered. Habitat loss due to widespread deforestation is the principal threat. The geographic range is estimated at 175 mi² (450 km²). Although this species occurs within the Rio Abiseo National Park, the population may be small. Precise numbers are unknown, but populations appear to be declining.

SIGNIFICANCE TO HUMANS
None known. ◆

Saffron toucanet
Baillonius bailloni

OTHER COMMON NAMES
French: Toucan de baillon; German: Goldtukan; Spanish: Tucán Amarillo.

TAXONOMY
Baillonius bailloni Vieillot, 1819. The saffron toucanet is the

sole representative of the genus *Baillonius*, which is the only monotypic ramphastid genus. Peters (1934–1986) groups this species with the genus *Andigena*.

PHYSICAL CHARACTERISTICS
A medium-sized toucanet, about 10 in (35 cm) long. Average weight 5.7 oz (163 g). In the field, the only toucanet to appear mostly yellow; bill is red and green. Eye is yellow and surrounded by red skin.

DISTRIBUTION
This species is limited to the south Atlantic forests of eastern Paraguay, southeastern Brazil, and northeastern Argentina.

HABITAT
Not well known. This species ranges from lowland tropical evergreen forest up to 5,100 ft (1,550 m) in lower montane forest.

BEHAVIOR
Calls are similar to those of aracaris *Pteroglossus*. Described as silent and secretive in its habits.

FEEDING ECOLOGY AND DIET
Has been seen eating figs, palmito fruits, and *Cecropia* fruits, but feeding ecology not well known. Captive birds hold fruits with feet while breaking into smaller pieces. Remsen et al. examined stomach contents of five specimens and found only fruit, no arthropods or small vertebrates.

REPRODUCTIVE BIOLOGY
Little is known. Courtship feeding and allopreening occur. Nests in tree cavities; male and female sometimes drum like woodpeckers near the nest entrance.

CONSERVATION STATUS

Near Threatened due to habitat loss, capture for the cage bird industry, and to some degree hunting. Habitat destruction is mostly a consequence of agrarian conversion.

SIGNIFICANCE TO HUMANS

Sometimes hunted for meat. ◆

Plate-billed mountain toucan
Andigena laminirostris

TAXONOMY

Andigena laminirostris Gould, 1850. This species shares the genus *Andigena* with three other species of mountain toucans. The group appears to be a "link" between the saffron toucanet (*Baillonius*) and true toucans (*Ramphastos*).

OTHER COMMON NAMES

French: Toucan montagnard; German: Leistenschnabel-tukan; Spanish: Tucán Andino Piquilaminado.

PHYSICAL CHARACTERISTICS

Length about 20 in (51 cm). Weight 11.1 oz (316 g) male; 10.7 oz (305 g) female. A dark olive-brown bird with slate-blue underparts; the dark bill has a raised, rectangular, cream-colored plate near the base; skin surrounding the eye is blue above and yellow below.

DISTRIBUTION

Comparatively small range, restricted to a band on the west slope of the Andes from southwestern Colombia through west-

Andigena laminirostris
▨ Resident

ern Ecuador. Thought to undergo seasonal altitudinal migration in Colombia.

HABITAT

This species occupies humid montane rainforest where bromeliads and mosses are abundant. Although usually found at 3,900–10,500 ft (1,200–3,200 m), seasonal altitudinal migration occurs and birds may be found as low as 990 ft (300 m).

BEHAVIOR

Call (equivalent of song) is nasal, whining sound that rises with each note. Often seen foraging in pairs or small groups, and sometimes accompanies mixed-species flocks. Larger flocks of a dozen or more birds form in the fall, after nestlings fledge; at this time birds in Colombia may migrate downslope in search of food.

FEEDING ECOLOGY AND DIET

Adult birds forage in small groups of three to six, usually up in the forest canopy; daily movements of small flocks vary depending on location of ripe fruit. At least 49 plant species produce fruits eaten by this toucan; most are in the *Cecropia* family and most are swallowed whole. Remsen et al. examined stomach contents of eight specimens and found only fruit; no arthropods or small vertebrates; however nestlings are fed beetles as well as birds' eggs, snails, and mice along with fruit.

REPRODUCTIVE BIOLOGY

Breeding season peaks in June and July, when the weather is driest and fruit is most abundant. When a pair of toucans is ready to begin nesting, they drive away the other members of the small group with which they usually forage. Nesting pairs also sometimes evict toucan-barbets from active nests to get control of a tree cavity. The cavities selected are often in trees in the familiy Laureaceae; birds are thought to do some excavation to modify the cavity and entrance hole. Courtship-feeding occurs before copulation. Both male and female incubate the eggs and feed the young. The nestlings fledge 46–60 days after hatching but the parents continue to feed them for two or three weeks; a second clutch sometimes follows.

CONSERVATION STATUS

Near Threatened due to habit loss from intensive logging, human settlement, cattle grazing, mining, and palm cultivation. Large tracts of suitable habitat do remain, however with some areas already protected. The species may be threatened to a lesser extent by the international cage bird trade.

SIGNIFICANCE TO HUMANS

A desired sighting for bird watchers. ◆

Gray-breasted mountain toucan
Andigena hypoglauca

TAXONOMY

Andigena hypoglauca Gould, 1833. This species shares the genus *Andigena* with three other species of mountain toucans.

OTHER COMMON NAMES

French: Toucan bleu; German: Blautukan; Spanish: Tucán Andino Pechigris.

PHYSICAL CHARACTERISTICS

Length about 18 in (46 cm). Weight 12 oz (341 g) male; 10 oz (287 g) female. Head is black, contrasting with pale gray collar

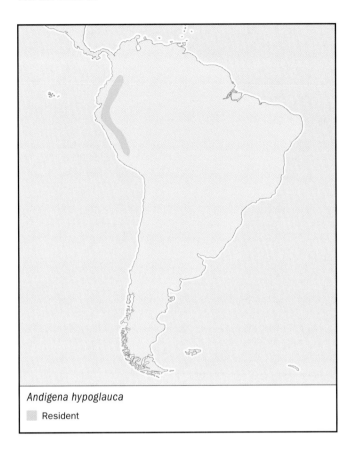

Andigena hypoglauca
■ Resident

on nape and blue skin around eyes. Chest slaty blue; back and wings bronze-green; tail is black with red undertail-coverts. Distinctive multi-colored bill is red and black at the tip and yellow-green at the base, with two black, oval "thumbprints," one on each side of the lower bill.

DISTRIBUTION
This species is restricted to the west slope of the central Andes, from central Colombia through Ecuador to southeast Peru. Usually occurs at higher elevations than all other mountain toucans.

HABITAT
Occupies montane rainforest and elfin forest from 5,900–11,300 ft (1,800–3,450 m), but typically not lower than 7,800 ft (2,400 m), where trees are draped with moss and other epiphytes.

BEHAVIOR
An uncommon and secretive bird; habits are not well known. Typically stays high in the canopy. The call is a series of nasal rising pitches, repeated over and over at four- to five-second intervals.

FEEDING ECOLOGY AND DIET
Not well known; Remsen et al. examined stomach contents of 14 specimens and found only fruit, particularly those of *Cecropia* and blackberries; no arthropods or small vertebrates. Feeds quietly, sometimes hanging upside down to reach fruits, usually foraging alone or in pairs but sometimes in small, possibly family, groups.

REPRODUCTIVE BIOLOGY
Not much known; adults with offspring have not been reported by scientific observers. Males in breeding condition have been collected in late January and early February.

CONSERVATION STATUS
Near Threatened due to rapid deforestation resulting from intensive agrarian expansion, logging, and mining. The species may also be threatened to a lesser extent by the international cage bird trade.

SIGNIFICANCE TO HUMANS
A desired sighting for bird watchers. ◆

White-throated toucan
Ramphastos tucanus

TAXONOMY
Ramphastos tucanus Linnaeus, 1758. The genus *Ramphastos* is made up of 11 species, the largest members of the toucan family. The clade is most closely allied to the mountain toucans (*Andigena*).

OTHER COMMON NAMES
English: Red-billed toucan, Cuvier's toucan; French: Toucan de Cuvier; German: Weissbrusttukan; Spanish: Tucan Goliblanco.

PHYSICAL CHARACTERISTICS
Length about 24 in (55 cm); weight about 21.8–24.8 oz (618–702 g) male; 19.4–24.2 oz (551–687 g) female. A large bird, jet-black from the top of the head down the nape and back to the rump. White at the throat, as the name indicates, with narrow red breast band; below that, black belly and thighs. Yellow uppertail-coverts and red undertail-coverts set off the black tail. The ring of bare skin around the eye is blue; the bill is mostly red (subspecies *tucanus*) or black (subspecies

Ramphastos tucanus
■ Resident

cuvieri) with a narrow yellow stripe along keel of upper bill; base of bill is yellow on top and blue on bottom.

DISTRIBUTION
Primarily Amazonian in distribution, ranging from Colombia and eastern Venezuela through Bolivia and Brazil. Often occurs in the same areas as the similar-looking (though smaller) channel-billed toucan.

HABITAT
Occupies tropical lowland Amazonian forest almost exclusively, ranging up to 3,600 ft (1,100 m). Also seen in second-growth forest, edge habitat and clearings, and around towns and cities.

BEHAVIOR
A conspicuous, large toucan that typically perches on dead branches above the forest canopy. The yelping call—considered the characteristic bird call of the Amazon—is said to resemble a puppy's barking; it begins with a sharp, descending, whining note that is followed by two to four additional monosyllabic pitches. Birds usually travel in pairs or small family groups of up to six birds, although larger flocks are sometimes seen. Aggressive and dominant in interactions with other species at fruit trees. Flockmates seem to engage in play, including bill-fencing, tag, and pass-the-fruit.

FEEDING ECOLOGY AND DIET
Not well known but eats a variety of fruits plus insects and lizards and takes eggs and nestlings from a variety of passerines. An active bird when feeding, often plucks fruit with tip of bill, then tosses head so fruit falls backward into mouth. Wipes bill on tree branch afterward, or scrapes bill with foot. To drink, turns bill sideways and swipes across surface of water, then lifts bill up and tilts head back.

REPRODUCTIVE BIOLOGY
Courtship behaviors include bill-nibbling, allopreening, and courtship feeding. The breeding season varies with location. Nests are constructed in natural tree cavities, 3–60 ft (1–18 m) off the ground; birds often re-use cavities where they have nested successfully in the past. The two to three white eggs are elliptical in shape; the parents share incubation duties and the eggs hatch after about 15 days. Both parents feed the young, which open their eyes after 29 days and fledge after about 49 days. Capuchin monkeys are a threat to toucan eggs and young.

CONSERVATION STATUS
Not threatened.

SIGNIFICANCE TO HUMANS
Hunted by Native Americans for food. ◆

Toco toucan
Ramphastos toco

TAXONOMY
Ramphastos toco Mueller, P.L.S., 1776. The genus *Ramphastos* includes the largest members of the toucan family. The clade is closest allied to the mountain toucans (*Andigena*).

OTHER COMMON NAMES
French: Toucan toco; German: Riesentukan; Spanish: Tucán Toco.

Ramphastos toco

▨ Resident

PHYSICAL CHARACTERISTICS
The largest of all toucans. Length about 24 in (60 cm). Weight 26.8 oz (760 g) male; 20.7 oz (587 g) female. The only *Ramphastos* species with a predominately black-and-white body (other species in this group may have orange, orange-red, or yellow on the breast or throat). Enormous orange bill with black spot at the tip is unmistakable.

DISTRIBUTION
This species ranges from the mouth of the Amazon River southward to southern Brazil and northern Argentina.

HABITAT
This species inhabits more macrohabitats than any other toucan and is the species most likely to be seen in open and semi-open habitat. For example, it may be found in tropical forest, Chiquitano forest, pantanal, seasonal inundated savanna, and secondary vegetation. A lowland species, the toco toucan ranges up to 3,900 ft (1,200 m) in altitude.

BEHAVIOR
Of all the toucans, this is the species most likely to be seen in flight over rivers and across open areas; often perches high in trees, on dead branches. Flap-glide flight is slow and undulating. The call is perhaps the lowest in pitch of all toucans; has been compared to snoring or the croaking of a toad. Often rattles bill or raps tongue against closed bill.

FEEDING ECOLOGY AND DIET
May feed alone, with a mate, or in a small flock of up to nine birds. Very agile and often hangs head down to reach a fruit or probe a crevice. Typically feeds in the canopy, but unusual among toucans in sometimes alighting on the ground to take fallen fruits. Eats a variety of fruits plus arthropods and has

been seen hunting cooperatively for eggs and nestlings in nesting colonies of yellow-rumped caciques.

REPRODUCTIVE BIOLOGY
Thought not to breed until they are two years old. Pairs engage in mutual preening and bill-fencing. Often nest in palm tree cavities which they modify by excavating with their sturdy bills; also nest in burrows in earthen banks, and in tree-termite nests (after nests have been opened by woodpeckers). The male and female incubate the two to four white eggs for about 18

days; nestlings are feed insects initially, then diet becomes mostly fruit. Young birds fledge at 43 to 52 days.

CONSERVATION STATUS
Not threatened. May be expanding its range into newly cleared areas in Amazonia.

SIGNIFICANCE TO HUMANS
A flagship species of the Neotropics, this toucan is often depicted in South American art. Still hunted for food throughout most of its range; young birds often taken for pets. ◆

Resources

Books

Birdlife International. *Threatened Birds of the World.* Barcelona: Lynx Edicions, 2000.

Brooks, D.M. "¿Son la competencia, el tamaño y la superposición de dietas pronosticadores de la composición de Ramphastidae?" In *Manejo de Fauna Silvestre en la Amazonia*, edited by T.G. Fang et al., pp. 283–288. Bolivia: United Nations Development Program, 1997.

Hilty, Steven L., and William L. Brown. *A Guide to the Birds of Colombia.* New Jersey: Princeton University Press, 1986.

Ridgely, R.S., and P.J. Greenfield.*The Birds of Ecuador.* Vol. 1, *Status, Distribution, and Taxonomy.* Ithaca: Cornell University Press, 2001.

Short, L.L., and J.F.M. Horner. *Toucans, Barbets, and Honeyguides.* New York: Oxford University Press, 2001.

Sibley, C.G., and J.E. Ahlquist. *Phylogeny and Classification of Birds: A Study in Molecular Evolution.* New Haven, CT: Yale University Press, 1993.

Sick, Helmut. *Birds in Brazil: A Natural History.* New Jersey: Princeton University Press, 1993.

Stotz, Douglas F., et al. *Neotropical Birds: Ecology and Conservation.* Chicago: University of Chicago Press, 1996.

Periodicals

Buhler, P. "Size, Form and Coloration of the Ramphastid Bill as Basis of the Evolutionary Success of the Toucans?" *Journal of Ornithology* 136 (1995): 187–193.

Haffer, J. "Avian Speciation in Tropical South America." *Publication of the Nuttall Ornithology Club* 14 (1974).

Remsen, James V., et al. "The Diets of Neotropical Togons, Motmots, Barbets, and Toucans." *Condor* 95 (1993): 178–192.

Organizations

The Neotropical Bird Club. c/o The Lodge, Sandy, Bedfordshire SG19 2DL United Kingdom. E-mail: secretary@neotropicalbirdclub.org Web site: <http://www .neotropicalbirdclub.org>

Cynthia Ann Berger, MS
Daniel M. Brooks, PhD

▲
Honeyguides
(Indicatoridae)

Class Aves

Order Piciformes

Family Indicatoridae

Thumbnail description
Small, generally drably colored tropical birds resembling barbets and finches

Size
4–8 in (10–20 cm)

Number of genera, species
4 genera; 17 species

Habitat
Forest, scrub

Conservation status
Near Threatened: 3 species

Distribution
Tropical and temperate Africa south of the Sahara. There are two species outside Africa, the yellow-rumped honeyguide (*Indicator xanthonotus*), along the southern foothills of the Himalayas, and the Malaysian honeyguide (*I. archipelagicus*), in Southeast Asia.

Evolution and systematics

The family Indicatoridae is within the order Piciformes, which also includes the barbets (Capitonidae), woodpeckers (Picidae), jacamars (Gabulidae), puffbirds (Bucconidae), and toucans (Ramphastidae). DNA-DNA hybridization studies seem to place honeyguides nearer the woodpeckers, but morphology and behavior strongly suggest them to be nearer the barbets.

The Indicatoridae are subdivided into four genera, *Prodotiscus*, with three species; *Indicator*, with 11 species; *Melignomon*, with two species; and *Melichneutes*, which is monotypic, with only the lyre-tailed honeyguide (*M. robustus*).

African honeyguides eat the wax of only one bee species, the common bee (*Apis mellifera*), despite the presence of other bee types throughout Africa, suggesting that their adaptations for cercophagy (wax-eating) are relatively recent, perhaps in the last one to two million years.

Physical characteristics

Body size of honeyguide species runs from 4 to 8 in (10–20 cm). The plumage is nondescript and cryptic, mostly olive-greens, grays, browns, black, and white, with occasional touches of yellow, depending on species. Only three species, including the greater (*Indicator indicator*) and lyre-tailed honeyguides, show sexual dichromatism in their plumage.

The bill is short and sturdy, well fitted to gouging wax and probing for insects in tree bark, and in most species the nostrils have raised rims to protect them from influxes of beeswax, honey, and other sorts of comb contents. The tail is long, often marked with white bars that the bird displays in flight, as guides for juveniles or in mating displays. The legs and toes are strong for clinging to tree bark, and the claws are long and hooked. As in all Piciformes, two toes are directed forward, the other two backward.

The wings are long, narrow, and pointed, allowing for vigorous flight, complicated maneuvering, and aerial acrobatics, which some species use to advantage in mating and territorial displays. The wings of the greater scaly-throated (*I. variegatus*), and lyre-tailed honeyguides make a distinctive noise in flight.

Honeyguide vision and hearing are acute. The olfactory lobe of the brain is well developed, although there are no studies showing to what extent it uses olfaction to track down bee nests. There are many accounts of the birds flying into missionary churches and attacking the beeswax candles, probably alerted by the odor, which had been intensified and spread by the candle flames.

Distribution

Honeyguides are found across nearly all Africa south of the Sahara, and there are two species in Asia. Distribution of species necessarily parallels that of their brood hosts.

A scaly-throated honeyguide (*Indicator variegatus*) feeds on honeycomb. (Photo by P. Davey/VIREO. Reproduced by permission.)

Habitat

Species live variously in dense primary forest, secondary forest, gallery forest in semiarid country, and a mix of trees, shrubs, and grassland. Altitude of habitation runs from sea level to near the treeline in mountainous territory.

Behavior

The honeyguides would surely be on any list of the most interesting birds on Earth, not because of their physical appearance, which would scarcely attract notice, but because of their behavior. Two species deliberately lead animals, including humans, to honey sources. All species are cercophagous, or wax-eating, and have digestive systems able to handle the substance. All are obligate brood parasites, the females laying fertile eggs in the nests of other birds.

Honeyguides are tough, aggressive birds, whether they are harassing brood host birds, mobbing wax sources, or insistently leading humans to bee nests. Although solitary most of the time, dozens of individuals of up to four species may converge on a wax source. Scaly-throated honeyguides will shoulder aside greater and lesser (*I. minor*) honeyguides, while immature greater honeyguides out-bully all other species. Several African honeyguide species, notably the greater and scaly-throated honeyguides, routinely follow and watch human activities, flying about camping sites and inspecting everything, even the tents and vehicles.

All honeyguides (except the yellow-rumped [*I. xanthonotus*]) sing, and add to their singing a wide and various repertoire of sounds fitted to situations. Singing honeyguides add posture to their songs, as if using their entire bodies to belt out every note, arching their necks, fluffing out rump feathers, or all feathers, and quivering the tail in time with the notes. Several species enhance aggressive or mating voicings with rustling wing sounds.

Feeding ecology and diet

No honeyguide species depends entirely on beeswax as a food source, but all partake of it as a major food item, while a few species, among them Zenker's honeyguide (*Melignomon zenkeri*) and the yellow-footed honeyguide (*Melignomon eisentrauti*), get some of their wax requirement by eating scale insects (*Coccidae*) that produce a waxy exudate that covers their dorsal surfaces. The digestive system of honeyguides produces specialized enzymes for digesting wax. Similar enzymes have been found in seabirds and other landbirds, although not in other Piciformes.

All honeyguides supplement wax with insects and spiders, including bee larvae, ants, waxworms (*Galleria* spp.), termites, flies, and caterpillars. Honeyguides will occasionally eat fruits and other plant matter. All honeyguides do some insect feeding by flycatching, snagging airborne insects in mid-flight.

Honeyguides feed on beeswax directly, if the nest is accessible, by flying up to a bee nest, gripping the tree's surface alongside the outer part of the comb, and biting off and swallowing pieces of wax. The birds' hides are tough and parchmentlike, offering some but not complete protection against stings, and a ferocious onslaught of defensive bees can kill a honeyguide.

Only two species of honeyguide partake in the behavior that has made the family famous, and given it its name *Indicatoridae*—from the Latin for "to point toward"—deliberately alerting, then leading animals and humans to honey sources. These species are the greater, or black-throated, honeyguide and the variegated or scaly-throated honeyguide.

The greater honeyguide alerts an animal or human to a honey source by flying close to them and calling "churr-churr-churr-churr" or "tirr-tirr-tirr-tirr." Once it has attracted the attention of the partner, it flies ahead, stopping at intervals and calling. When both parties reach the nest, the bird circles near the bee nest, gives a short, toned-down "indication call," then perches silently in the vicinity of the nest and waits for the chunks of comb to fall. A honeyguide seems to know in advance where a nest is, most likely through regularly surveying its territory.

H. A. Isack of the National Museum of Kenya, and H. U. Reyer of the University of Zurich conducted a three-year study to ascertain the truth of the honeyguide-human guiding relationship. In the mid-1980s, the two ornithologists followed Boran tribesmen through dry bush country in northern Kenya, searching for honey. The Boran claimed to have worked out an elaborate communication system with the greater honeyguide, with certain signals understood by both parties. The birds would indicate direction and decreasing distance of the nest, as the parties approached it, through flight patterns and calls. Isack and Reyer confirmed all this in the field. The birds signaled by calls and movements and perchings, the tribesmen by attracting the bird with special calls, then making noise as they followed the bird to keep communication going. Nearly all the nests the Boran reached with honeyguide help were inaccessible to the honeyguide, necessitating help from other quarters. In spite of the research of Isack and Rider, and others, some researchers have disputed accounts of honeyguides leading animals, leading humans, or leading at all, asserting that such accounts are not reliable.

The favored animal accomplice is the ratel, or honey badger (*Mellivora capensis*), a medium-sized, powerfully built, omnivorous viverrid. The relationship between bird and follower is mutualistic; neither species needs the other to survive, but both benefit from the relationship. The honey badger, which is well equipped to deal with hive-raiding, gouges out hunks of comb, and the honeyguide helps itself to fallen pieces of comb with little risk of injury from bee stings. The ratel-honeyguide relationship has not been proven, but abundant testimony leans toward it being true.

Humans who follow the honeyguides and gorge on the honey habitually leave the birds pieces of comb, as a reward and to keep the relationship intact. Most interesting, in relation to this custom, is that in suburban areas where people prefer sugar over honey, honeyguides have been reported to have lost the ability or the desire for guiding behavior.

Reproductive biology

Honeyguides besmirch their rather charming image as guides for honey-lovers with a less charming aspect of their behavior. All honeyguide species are brood parasites. There is little or no pair bonding between sexes. Adult female honeyguides lay fertile eggs among those of other bird species, usually one egg per nest, occasionally two. A female will lay six eggs or more, but divide them up among host nests. All host nests are cavity types—in trees, in the ground, in termite mounds, or in carton ant nests.

The honeyguide female will invade a host nest when the host female is out foraging, or perhaps sit and wait if the host female is home. Although the host female must leave to feed sooner or later, the honeyguide cannot wait too long at one egg-laying spot because the host female or a bonded male and female will sometimes chase off the intruder.

When the female honeyguide does lay an egg in a host nest, she enters the cavity nest, deposits the egg and leaves in 10 to 15 seconds (perhaps in as few as three seconds). Females of some honeyguide species will puncture or remove a host's egg before replacing it with her own. Most honeyguide eggs are white, except those of the green-backed honeyguide (*Prodotiscus zambesiae*), which are blue, matching the color of the eggs of white-eyes, their most frequent hosts.

The list of host species for honeyguides is long, but all are cavity-nesters and all are mainly insectivorous. Host choices of a honeyguide species may be as specific as one or two species, or a wide selection. The most frequently chosen hosts are barbets, but honeycreepers choose hosts from among the kingfishers (Alcedinidae), bee-eaters (Meropidae), hoopoes (Upudidae), woodpeckers (Picidae), and others.

The parasitic honeyguide eggs may hatch when the host's eggs are still unhatched, or after their hatching. The young honeyguides hatch with hooked bills and formidably sharp talons, with which they puncture the eggs or savage the young of the host. Some honeyguide hatchlings heave the host hatchlings from the nest, or the host mother, following hygiene instincts, will toss out the dead chicks or punctured eggs. The host mothers, bound and trapped by their strong instincts for parental care, raise the intruders as their own.

Honeyguides are obligate brood parasites, i.e., brood parasitism is the only option open to them for raising the next generation. No honeyguide species builds nests and none are able to raise young by themselves. Of necessity, the breeding seasons of honeyguide species closely accord with those of host species.

Conservation status

No species is listed as threatened. The Malaysian honeyguide (*Indicator archipelagicus*), yellow-rumped honeyguide (*Indicator xanthonotus*), and dwarf honeyguide (*Indicator pumilio*) are listed as Near Threatened. In the long view, most or all honeyguide species are threatened by rampant deforestation in Africa and Asia.

Significance to humans

The species that lead people to honey sources help humans secure a treasured food. Because of that famous role, honeyguides have found their way into African folklore. A typical honeyguide tale cautions thoughtfulness and fair reciprocation, based on the human custom of leaving chunks of comb for a honeyguide as reward for its help. In such a tale, a human refuses to pay the honeyguide his fee and reaps dire consequences.

1. Lyre-tailed honeyguide (*Melichneutes robustus*); 2. Malaysian honeyguide (*Indicator archipelagicus*); 3. Scaly-throated honeyguide (*Indicator variegatus*); 4. Spotted honeyguide (*Indicator maculatus*); 5. Yellow-rumped honeyguide (*Indicator xanthonotus*); 6. Greater honeyguide (*Indicator indicator*). (Illustration by Wendy Baker)

Species accounts

Greater honeyguide
Indicator indicator

TAXONOMY
Indicator indicator Sparrman, 1777.

OTHER COMMON NAMES
English: Black-throated honeyguide; French: Grand Indicateur; German: Schwarzkehl-Honiganzer; Spanish: Indicador de Garganta Negra.

PHYSICAL CHARACTERISTICS
Up to 8 in (20 cm). Sexes show only minor differences; male is olive-brown on the crown, nape, wings and dorsal surface; black throat patch extends up around the eyes in a mask. Yellowish patch on cheek area, another on the wing where it meets the shoulder at rest; underparts are whitish with some grayish touches. Bill is yellow and pink. Female is more brownish and lacks throat and cheek patches. As in most honeyguides, the outer retrices of the tail are white in both sexes and are shown to advantage in flight.

DISTRIBUTION
Nearly all tropical and temperate Africa south of the Sahara; the most widely distributed species.

HABITAT
Dense bush country, grasslands with scattered trees and bushes, and dense forest, from sea level to 8,000 ft (2,438 m) above sea level. They prefer territory with large resident aardvark (*Oryctoperus afer*) population. Ratels shelter in aardvark

burrows, while bee-eaters (*Melittophagus* spp.), ant-eating chats (*Myrmecochlica* spp.), and swallows, all parasitized by the greater (and other) honeyguides, nest in the roofs of the burrows.

BEHAVIOR
Large for a honeyguide, lively and aggressive. Individuals patrol known territories, visiting and spot-checking bee nests, and keeping their senses alert for new ones. They will also follow humans and visit human campsites, not always for beckoning and guiding. The male's most oft-heard call is "vic-tor," repeated from six to 11 times in rapid staccato, perching erect with throat feathers slightly puffed out.

FEEDING ECOLOGY AND DIET
Diet is typically wax, bee larvae, other insects, and spiders. Greater honeyguides and scaly-throated honeyguides are the only species known to guide animals, including humans, to honey sources.

REPRODUCTIVE BIOLOGY
Males put on aerial attraction displays for females, simplified versions of the lyre-tailed honeyguide's performance. The greater honeyguide male may also put on an act wherein he circles around and above a female, making drumming sounds with his wings. The male will alight on a branch near the perching female, then slowly approach her, spreading his white-bordered tail, fluttering his wings, puffing his feathers, and making a low, "chrrr" call. Females parasitize at least 40 species, including barbets, kingfishers, bee-eaters, hoopoes, and woodpeckers.

CONSERVATION STATUS
Not threatened.

SIGNIFICANCE TO HUMANS
Helps humans secure a treasured food and plays a significant role in folklore. ◆

Malaysian honeyguide
Indicator archipelagicus

TAXONOMY
Indicator archipelagicus Temminck, 1832.

OTHER COMMON NAMES
English: Sunda honeyguide; French: Indicateur archipélagique; German: Malaien-Honiganzeiger; Spanish: Indicador Malayo.

PHYSICAL CHARACTERISTICS
2.5 in (16 cm). Plumage brownish gray with small, bright yellow shoulder patch, underparts are white, and breast light gray. Eyes and bill are brown, legs and feet black. Female lacks yellow shoulder patch.

DISTRIBUTION
Malaysian Peninsula, Sumatra, and Borneo.

HABITAT
Tropical rainforest from sea level to 3,280 ft (1,000 m) above sea level.

Indicator indicator
Resident

Indicator archipelagicus
☐ Resident

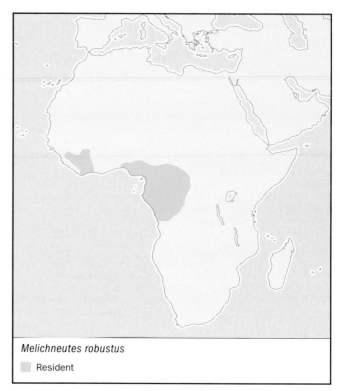

Melichneutes robustus
☐ Resident

BEHAVIOR
Calls with harsh, catlike "miaow," followed by a churring "miaow-krrruuu" or "miao-miao-krruuu," rising in pitch.

FEEDING ECOLOGY AND DIET
Often seen near nests of Asian honeybees (*Apis dorsata* and *A. florea*). Eats beeswax, bee larvae, and adults of those species and other insects. Some individuals sing in presence of humans, but none exhibits guiding behavior.

REPRODUCTIVE BIOLOGY
Little is known other than the voice of the singing male, although presumed to be brood parasites like other honeyguide species. Breeding seasons thought to be February into May in Malaya, during August in Thailand, May into June in Sumatra, and from January into March in Borneo.

CONSERVATION STATUS
Listed as Near Threatened due to deforestation.

SIGNIFICANCE TO HUMANS
None known. ◆

Lyre-tailed honeyguide
Melichneutes robustus

TAXONOMY
Melichneutes robustus Bates, 1909.

OTHER COMMON NAMES
French: Indicateur à queue-en-lyre; German: Leierschwanz-Honiganzeiger; Spanish: Indicador Cola de Lira.

PHYSICAL CHARACTERISTICS
6 in (17 cm). Long, lyre-shaped tail is most distinctive feature; two median pairs of retrices are curved outward at distal ends, outermost retrices are narrow and short.

Undersurface of tail is white; bird shows the white part conspicuously in flight. Both sexes uniformly olive-green above and whitish below; female shows some gray streaks on rear underbelly, and tail is not as large or exaggerated, but has similar shape.

DISTRIBUTION
Two separate areas of distribution in western Africa, one including Sierra Leone, Liberia, and the Ivory Coast, the other, larger area centering around Cameroon.

HABITAT
Lowland tropical rainforest.

BEHAVIOR
Only incomplete information on behavior, since species is rare and individuals are difficult to spot. The spectacular aerial displays are described below. Individuals have been seen perched in trees near bees' nests, and associated with barbets and tinkerbirds, presumably as preferred brood hosts.

FEEDING ECOLOGY AND DIET
Includes beeswax, bee larvae, and other arthropods.

REPRODUCTIVE BIOLOGY
Most spectacular mating display of all honeyguides. The male, airborne, voices several "peee" notes, then executes rapid, steep dive with tail feathers spread. Those feathers produce a signatory "kwa-ba kwa-ba" series of sounds. Male may also ascend and descend in spirals.

CONSERVATION STATUS
Not threatened.

SIGNIFICANCE TO HUMANS
None known. ◆

Scaly-throated honeyguide

Indicator variegatus

TAXONOMY

Indicator variegatus Lesson, 1830.

OTHER COMMON NAMES

French: Indicateur écaillé; German: Strichelstirn-Honiganzeiger; Spanish: Indicador de Garganta Escamosa.

PHYSICAL CHARACTERISTICS

Source of common name "scaly-throated" is network of small, alternating light and dark spots (not of scales, but feathers) adorning the throat and extending down onto the breast like a long bib. Otherwise, colored as most honeyguides, gray-olive on upperparts, off-white along breast and belly (except for spots).

DISTRIBUTION

Eastern South Africa to southern Sudan, Ethiopia, and Somalia; also in west Africa from coastal Angola to Zambia.

HABITAT

Forest patches and wooded and dense vegetation along rivers in dry areas.

BEHAVIOR

Show wide range of vocalizations fitted to situations. Males assert or threaten with a high-pitched whistling. Females in similar mode make loud chatterings. Both sexes accompany aggressive voice with rustling wing sounds.

FEEDING ECOLOGY AND DIET

An aggressive species, individuals generally dominate a beeswax source as a feeding post, but will make off with small chunks of comb and eat them elsewhere. Also eats bee larvae and adults, waxmoth larvae, beetles, and other insects, including termites, on which they gorge on the wing when the insects emerge in hordes. Occasionally takes fruits and seeds. Small groups sometimes join mixed-species flocks of foraging birds.

REPRODUCTIVE BIOLOGY

Male displays for female similarly to greater honeyguide.

CONSERVATION STATUS

Not threatened.

SIGNIFICANCE TO HUMANS

Alleged to lead humans to honey sources. ◆

Spotted honeyguide

Indicator maculatus

TAXONOMY

Indicator maculatus G. R. Gray, 1847.

OTHER COMMON NAMES

French: Indicateur tacheté German: Tropfenbrust-Honiganzeiger; Spanish: Indicador Moteado.

PHYSICAL CHARACTERISTICS

6.75 in (16.5 cm). Olive-brown with some green tinges above, brownish underparts with parallel rows of yellow-white spots running from the throat to tail. Crown has array of small spots that echo rows on the underside. No appreciable sexual dichromatism. Bill is short, stout, and conical.

DISTRIBUTION

Equatorial Africa from Senegambia to eastern Zaire.

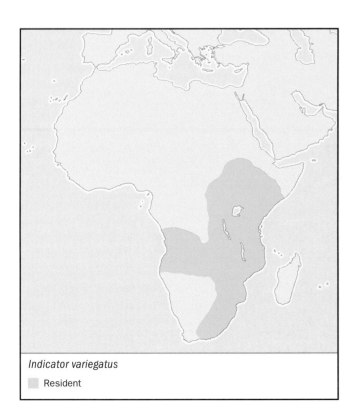

Indicator variegatus

▨ Resident

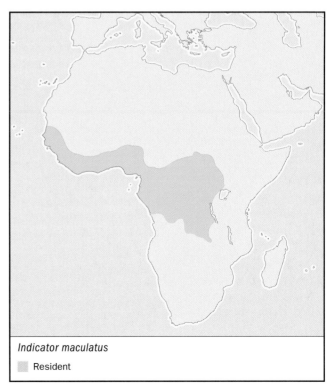

Indicator maculatus

▨ Resident

HABITAT
Dense, lowland forest along rivers.

BEHAVIOR
Similar to that of its close relative, the scaly-throated honeyguide. Individuals are solitary, roaming about the forest canopy most of the time, but regularly foraging at lower levels. Most prominent vocalization is trill like that of the scaly-throated honeyguide's, described as low "brrrr" or "prrrr." Other voicings include a loud "woe-woe-woe" similar to calls of falcons or eagles, with tail fanned and body feathers puffed.

FEEDING ECOLOGY AND DIET
Forages throughout all forest levels, but prefers the canopy. Regularly search out beeswax sites, and vary that diet with caterpillars, beetles, other insects, and spiders.

REPRODUCTIVE BIOLOGY
Preferred host species is buff-spotted woodpecker (*Campethera nivosa*). Plumage approximates that of woodpecker, so they enhance their brood parasitism by imitating the woodpecker's voice.

CONSERVATION STATUS
Not threatened.

SIGNIFICANCE TO HUMANS
None known. ◆

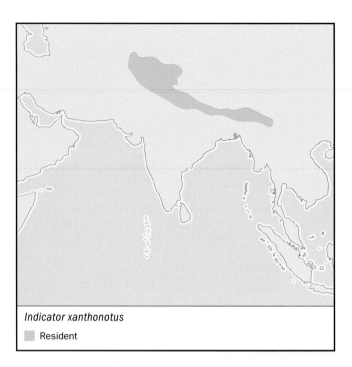

Indicator xanthonotus
■ Resident

HABITAT
Forests up to 7,000 ft (2,134 m) above sea level, in coniferous, dry deciduous, and lowland tropical rainforests.

BEHAVIOR
No song has been noted. Instead, males stake out territory in forests near rock cliffs where the preferred honey producers, rock bees (*Apis dorsata*), build their nests in fissures well off the ground. Males feed off wax of chosen nest, while guarding and defending the valuable food source against rival males, and enticing females with the promise of food.

FEEDING ECOLOGY AND DIET
Wax and other comb ingredients in rock bee nests; also flycatch winged insects. Males perch alone in a tree near a honeycomb, erect or hunched, for hours.

REPRODUCTIVE BIOLOGY
Females parasitize local species of woodpeckers and barbets.

CONSERVATION STATUS
Listed as Near Threatened. Threats include deforestation and increased human consumption of rock bee honey.

SIGNIFICANCE TO HUMANS
None known. ◆

Yellow-rumped honeyguide
Indicator xanthonotus

TAXONOMY
Blyth 1842

OTHER COMMON NAMES
English: Indian honeyguide, orange-rumped honeyguide; French: Indicateur à dos jaune; German: Gelbbürzel-Honiganzeiger; Spanish: Indicador Hindú.

PHYSICAL CHARACTERISTICS
5 in (15 cm). Most brightly colored honeyguides. Both sexes adorned with golden caps and more gold on the throat and lower back; female's gold patches not as extensive as male's. Remainder of livery for both sexes is gray and olive-green.

DISTRIBUTION
Foothills of the southern Himalayas, from Afghanistan to Burma.

Resources

Books

Ali, Salim, Dillon Ripley, and John Henry Dick. *A Pictorial Guide to the Birds of the Indian Subcontinent.* Oxford: Oxford University Press, 1996.

Barlow, Clive, Tim Wacher, and Tony Disley. *A Field Guide to the Birds of the Gambia and Senegal.* New Haven: Yale University Press, 1998.

Bernard, Robin, and Nneka Bennett. *Juma and the Honey Guide: An African Story.* Parsippany, NJ: Silver Press, 1996.

Friedmann, Herbert. *The Honey-Guides.* United States National Museum Bulletin 208. Washington, DC: Smithsonian Institution, 1955.

Keith, Stuart, Emil Urban, C. Hilary Fry, eds. *Birds of Africa.* London: Academic Press, 1997.

Resources

Short, Lester, and Jennifer Horne. *Toucans, Barbets and Honeyguides: Rhamphastidae, Capitonidae and Indicatoridae* Oxford: Oxford University Press, 2001.

Species Survival Commission (SSC) Red List Programme. *2000 Red List of Threatened Species.* Cambridge: IUCN Publications Unit, 2000.

Strange, Morten. *A Photographic Guide to the Birds of Southeast Asia, Including the Philippines and Borneo.* Boston: Charles E. Tuttle Co., 2001.

Periodicals

Isack, H. A., and H. U. Reyer. "Honeyguides and Honey Gatherers: Interspecific Communication in a Symbiotic Relationship." *Science* 243, (1989): 1343–1346.

May, Robert M. "Honeyguides and Humans." *Nature* 338, (1989): 707–709.

Organizations

African Bird Club, c/o Birdlife International. Wellbrook Court, Girton Road, Cambridge, Cambridgeshire CB3 0NA United Kingdom. Phone: +44 1 223 277 318. Fax: +44-1-223-277-200. E-mail: birdlife@birdlife.org.uk Web site: <http://www.africanbirdclub.org>

BirdLife International. Wellbrook Court, Girton Road, Cambridge, Cambridgeshire CB3 0NA United Kingdom. Phone: +44 1 223 277 318. Fax: +44-1-223-277-200. E-mail: birdlife@birdlife.org.uk Web site: <http://www.birdlife.net>

IUCN Species Survival Commission. 219c Huntingdon Road, Cambridge, Cambridgeshire CB3 0DL United Kingdom. E-mail: ssc@iucn.org Web site: <http://www.iucn.org/themes/ssc>

Other

The IUCN Species Survival Commission. "2000 IUCN Red List of Threatened Species" <http://www.redlist.org/> (1 April 2002).

Kevin F. Fitzgerald, BS

Woodpeckers, wrynecks, and piculets
(Picidae)

Class Aves

Order Piciformes

Family Picidae

Thumbnail description
Small to medium, primarily arboreal; often with patterns of brown, green, or black-and-white; woodpeckers (Picinae) and piculets (Picumninae) often sexually dichromic, males with red or yellow on the head and females lacking it or with less color; wrynecks (Jynginae) sexes similar; woodpeckers have stiff rectrices used for support while climbing on tree surfaces; wrynecks and piculets do not; most have four toes, arranged two forward and two back (zygodactyl)

Size
Wrynecks: 6.3–7.5 in (16–19 cm); 0.78–2.1 oz (22–59 g); piculets: 3–6.3 in (7.5–16 cm); 0.24–1.2 oz (6.8–33 g); woodpeckers: 4.7–24 in (12–60 cm); 0.6–21+ oz (17–600+ g)

Number of genera, species
26 genera, 213 species; wrynecks, 1 genus, 2 species; piculets, 2 genera, 29 species; woodpeckers, 23 genera, 182 species

Habitat
Forests, woodlands, parks, and savannas; a few species found in grasslands and deserts

Conservation status
Critically Endangered: 3 species; Endangered: 1 species; Vulnerable: 7 species; Near Threatened: 12 species

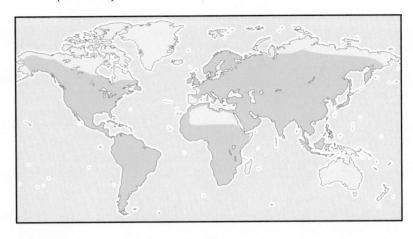

Distribution
Worldwide except absent from Australia, New Zealand, New Guinea, Madagascar, Ireland, many oceanic islands, and treeless polar regions; wrynecks limited to Eurasia and Africa; piculets to Asia, Central and South America, and Hispaniola

Evolution and systematics

The picids are an ancient, very distinctive, and thus easily recognized group. Early picids were present at least by the Eocene in both Northern and Southern Hemispheres, but the fossil record is limited and sheds little light on picid relationships. Within the Piciformes they seem to be most closely related to the barbets (Capitonidae), toucans (Ramphastidae), and honeyguides (Indicatoridae). Wrynecks are considered the most primitive picids and they lack many adaptations of the family for tree-climbing and excavation. Woodpeckers may have originated in the New World and there are incredible parallels between Neotropical and African woodpecker groups.

Physical characteristics

Wrynecks are cryptically colored above in brown, gray, and black, and lighter below. They have a slender, pointed bill, rounded wings, and a relatively long tail with rounded tail feathers that lack the stiffness found in woodpecker rec-

trices. They have short legs and four toes in a zygodactyl (two toes forward, two back) arrangement. Sexes are alike.

Piculets are like miniature woodpeckers, but tail feathers, though pointed, are not stiff and are not used for support. Piculet plumage tends to be soft, and brown and black dominate their color patterns. As with woodpeckers, the sexes are often distinguished by the presence of red on the head of the male. Also like woodpeckers, mechanical tapping on wood is sometimes used for communication.

Woodpeckers have a relatively large head, a straight, sharply pointed to chisel-tipped bill, and a long cylindrical tongue that is often barbed or brushlike at the tip for extracting insect prey from tunnels and crevices. Short legs and three or four toes in a zygodactyl arrangement, and strongly curved claws facilitate climbing. Stiff rectrices are used as a prop for climbing on vertical surfaces and probably also as a "spring" to maximize efficiency of pecking motions. The major tail feathers are mostly black, the melanin adding strength that is needed to counter wear resulting from contact with

Different tongue structure and uses: Lineated woodpecker (*Dryocopus lineatus*) (top) rakes out insects with its stiff, barbed tongue after chiseling wood away with its powerful bill; red-breasted sapsucker (*Sphyrapicus ruber*) (middle) drills shallow holes and uses its bristled tongue to obtain sap; Eurasian green woodpecker (*Picus viridis*) (bottom) probes the ground for ants with its long, sticky tongue. (Illustration by Gillian Harris)

tree surfaces. Many species are crested, such as the pileated woodpecker (*Dryocopus pileatus*).

Distribution

Woodpeckers are found in forested regions throughout the world except for Madagascar, the Australian region, and some oceanic islands. Some woodpeckers (e.g., flickers, *Colaptes*) ex-

tend beyond the limits of forest, making nest and roost cavities in utility poles, fence posts, or dirt banks. Wrynecks are confined to the Old World, and piculets to the tropics.

Habitat

Picidae habitats include virtually any environment with woody vegetation, and some without. A major component of woodpecker habitats that has been given little attention but is worthy of consideration is water. High relative humidity, frequent precipitation, and the local presence of standing or running water contribute to abundance and diversity of picid species within regions. The link between water and picids is the requisite moist wood for fungal decay, which facilitates cavity excavation and provides suitable habitats for the wood-boring arthropods that so many picids depend on. Larger woodpeckers, of course, need larger trees in which to excavate their nest and roost cavities. They also often feed on larger prey and need more extensive habitat in order to find adequate food resources. Some smaller woodpeckers that have become specialized for unique habitat conditions also have ex-

An acorn woodpecker (*Melanerpes formicivorus*) stores acorns for winter food. (Photo by Tim Davis. Photo Researchers, Inc. Reproduced by permission.)

European wryneck (*Jynx torquilla*) at its nest. (Photo by Hans Reinhard. Bruce Coleman Inc. Reproduced by permission.)

tensive habitat needs. The red-cockaded woodpecker (*Picoides borealis*) can require 200–1,000 or more acres (80–400 ha) of pine forest per pair depending on habitat quality.

Behavior

Flight of picids is often undulating, but the largest woodpeckers tend toward less undulation and more level flight. Picid wings tend to be relatively short and rounded, providing better control for maneuvering in forest habitats. Some species, such as the yellow-bellied sapsucker (*Sphyrapicus varius*) and northern flicker (*Colaptes auratus*) in North America are migrant (at least in northern populations). The yellow-bellied sapsucker shows a distinct pattern of differential migration by the sexes, females going farther south. Others such as the three-toed woodpecker (*Picoides tridactylus*) are somewhat irruptive, departing areas when food supplies are low and moving to areas of food concentration such as epidemic beetle outbreaks. Most occupy similar habitats year round, but some can make drastic seasonal shifts. The red-headed woodpecker (*Melanerpes erythrocephalus*), for example, is generally a bird of very open habitats where it feeds on beetles, grasshoppers, other arthropods, and some fruit during summer, but in winter it often moves to bottomland forest and focuses its foraging on acorns and other mast (nuts found on the forest floor).

Vocalizations are often simple, with single notes often used as contact calls between mates and "whinny" or "rattle" calls found with some variation across the family. In social species such as the California woodpecker (*Melanerpes formicivorus*) and red-cockaded woodpecker, the vocal repertoire can be more complex. Communication by production of mechanical sounds through tapping on resonant wood is common in picids and the loud, rolling tattoo of such large species as the black woodpecker (*Dryocopus martius*) of Eurasia and the pileated woodpecker of North America are truly magnificent.

Feeding ecology and diet

The diet of picids is heavily biased toward forest insects and other arthropods, but also includes varying, and sometimes substantial, proportions of fruit, nuts, and tree sap. Many of the physical adaptations of picids are specializations for obtaining food from tree surfaces and subsurfaces. A chisel-like bill is used for excavating to retrieve wood-boring beetle larvae, ants, termites, and other invertebrates from within wood or other substrates. It is also used to reach mast and produce sap wells from which the birds obtain liquid nourishment. The exceptionally long vermiform tongue with a barbed tip is used like a rake to retrieve prey from tunnels and crevices, its efficiency enhanced by a coating of sticky saliva produced by the large salivary glands that characterize the group. The barbs at the tip of a sapsucker's tongue are short and abundant, making the tongue more "brushlike," aiding in obtaining liquid nourishment provided by sap.

Great spotted woodpecker (*Dendrocopos major*) feeding young at nest in Lincolnshire, England. (Photo by Holt Studios/Peter Wilson. Photo Researchers, Inc. Reproduced by permission.)

An ocellated piculet (*Picumnus cirratus*). (Photo by J. Dunning/VIREO. Reproduced by permission.)

Reproductive biology

All members of the Picidae are cavity nesters. Most excavate their own nest and roost cavities, a process that takes about two weeks and is shared by monogamous pair members. The wrynecks do not excavate cavities, but may enlarge one. They also differ from typical woodpeckers by sometimes adding grass or moss as a nest lining. In the red-cockaded woodpecker, cavity excavation, which is characteristically in a living pine, can take several years. A woodpecker nest cavity is usually the roost of the male. No nest material is brought in, though woodpeckers generally leave a layer of fresh chips on the bottom of the cavity and may add more chips, excavated from cavity walls during laying, incubation, and brooding of small nestlings. All picids lay shiny white eggs. At first these are somewhat translucent and may even appear pinkish; with development they become opaque. Clutch size varies within and among species, but usually averages three to five eggs. Incubation begins with laying of the last egg and is shared by both parents. Incubation periods are very short, usually 10–12 days. Young are very altricial and remain naked and with closed eyes for four to seven days. Nestlings fledge at three to six weeks and may be dependent on parents for weeks to months. Nesting success is generally high, although brood reduction through starvation of the last-hatched chick is common.

Conservation status

Seventeen woodpecker and five piculet species or subspecies were included on the 2000 IUCN Red List of Threatened Species. The ivory-billed (*Campephilus principalis*), imperial (*Campephilus imperialis*), and the Okinawa woodpecker (*Sapheopipo noguchii*) are Critically Endangered by loss of old-growth forest. The red-cockaded, Arabian (*Dendrocopos dorae*), helmeted (*Dryocopus galeatus*), and Sulu (*Picoides ramsayi*) woodpeckers are all classified as Vulnerable as a re-

sult of habitat losses. The red-cockaded is classified as Endangered under the U.S. Endangered Species Act. Ten other woodpecker species are listed by IUCN in their Lower Risk category. The tawny (*Picumnus fulvescens*), ochraceous (*Picumnus limae*), and speckle-chested (*Picumnus steindachneri*) piculets are all listed as Vulnerable, and the rusty-necked (*Picumnus fuscus*) and mottled (*Picumnus nebulosus*) piculets are included as Lower Risk.

The greatest threat to picids is habitat destruction and modification. Clearing of forests for non-forest uses reduces and fragments populations and allows invasion of forest-edge species that compete with woodpeckers for cavities or that prey on woodpeckers. Clearcutting followed by harvesting of trees before they reach natural maturity reduces habitat quality, availability of nest sites, and abundance, diversity, and stability of food supplies. All endangered and threatened woodpecker species are suffering impacts of habitat losses. Introduction of exotic secondary cavity-nesting species, such as the European starling (*Sturnus vulgaris*), has increased competition for woodpecker cavities and contributed to population declines and possibly shifts in woodpecker nesting phenology that further upset woodpecker roles within ecosystems. For example, starling competition with early-nesting red-bellied woodpeckers (*Melanerpes carolinus*) seems to have resulted in later renesting of that species, placing it in greater competition with the later nesting red-headed woodpeckers.

Significance to humans

The brilliant red feathers on the head of many male woodpeckers have been sought by indigenous peoples in many areas of the world. In North America, the scalps and bills of ivory-billed woodpeckers were sought and traded far outside the range of the species to be used to adorn war pipes and ceremonial dress. Red-headed woodpecker feathers were similarly used by the Ojibway Indians of Canada. In California, scalps of woodpeckers became essentially the basis of a monetary system among indigenous peoples. Woodpecker tongues and other body parts have been used in folk medicine and woodpeckers have been eaten in many cultures. In Italy, however, the tapping of woodpeckers is considered unlucky, a belief perhaps handed down from the Romans. At the end of the nineteenth century, skins of rare species such as the ivory-billed and imperial woodpeckers had a high market value and were the subject of intense collecting pressure. In the late twentieth century the endangered red-cockaded woodpecker of the southeastern United States became a symbol of conflict between the forest industry and environmental action groups. Several species of woodpeckers have been eaten by local cultures, including flickers, pileated woodpeckers, and ivory-billed woodpeckers in North America. The latter two were both known to early settlers in North America as "Indian hens," perhaps a reference to their edibility. One early writer suggests that ivory-billed woodpecker tasted as good as "pintail duck."

Woodpeckers are very important components of forest ecosystems because of their role in providing nest and roost sites for many secondary cavity-nesting species, their control of forest insect pests, and to some extent dispersal of seeds. Woodpeckers are also blamed for considerable damage to buildings, some damage to crops (including sugar cane, cacao, corn, oranges, and other fruit), to commercially valuable trees, and sometimes to eggs of poultry. Often, however, the perceived damage is a perception only and the "services" provided by the birds far outweigh any real damage. Wood-

peckers have a lot of popular appeal and have contributed to human culture in such diverse ways as through the cartoon "Woody Woodpecker" (patterned after the pileated woodpecker), door-knockers shaped like woodpeckers, and toothpick dispensers that include a miniature woodpecker that picks up a toothpick for the user. In Brazil, muzzle-loading shotguns are called "woodpeckers."

1. Black woodpecker (*Dryocopus martius*); 2 Three-toed woodpecker (*Picoides tridactylus*); 3. Smoky-brown woodpecker (*Veniliornis fumigatus*); 4. Northern flicker (*Colaptes auratus*); 5. Gray-faced woodpecker (*Picus canus*); 6. Great slaty woodpecker (*Mulleripicus pulverulentus*); 7. Okinawa woodpecker (*Sapheopipo noguchii*); 8. Lesser flame-backed woodpecker (*Dinopium benghalense*); 9. Rufous woodpecker (*Celeus brachyurus*); 10. Ivory-billed woodpecker (*Campephilus principalis*). (Illustration by Gillian Harris)

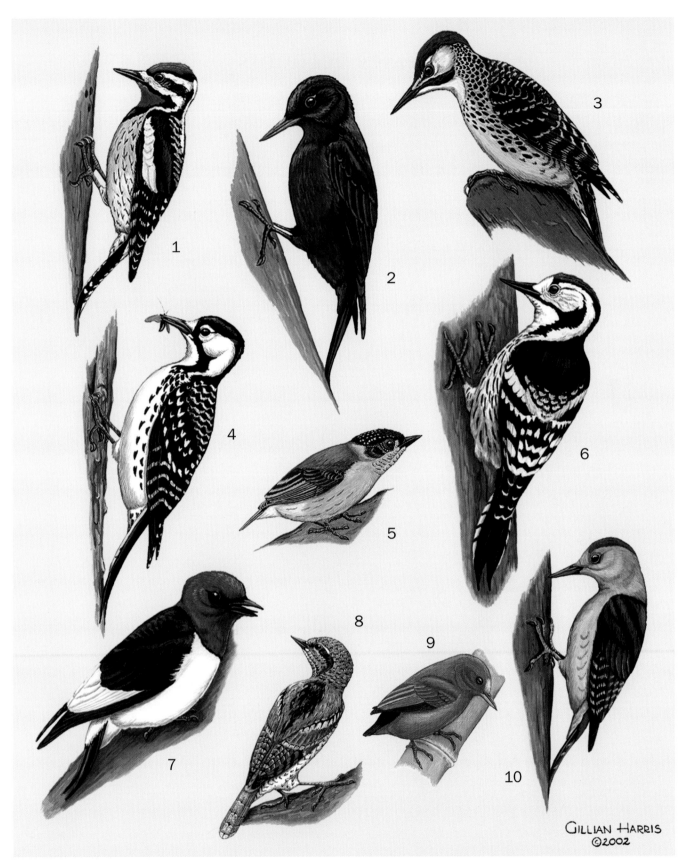

1. Yellow-bellied sapsucker (*Sphyrapicus varius*); 2. Guadeloupe woodpecker (*Melanerpes herminieri*); 3. Bennett's woodpecker (*Campethera bennettii*); 4. Red-cockaded woodpecker (*Picoides borealis*); 5. Olivaceous piculet (*Picumnus olivaceus*); 6. White-backed woodpecker (*Dendrocopos leucotos*); 7. Red-headed woodpecker (*Melanerpes erythrocephalus*); 8. Northern wryneck (*Jynx torquilla*); 9. Rufous piculet (*Sasia abnormis*); 10. Gray woodpecker (*Dendropicos goertae*). (Illustration by Gillian Harris)

Species accounts

Northern wryneck

Jynx torquilla

SUBFAMILY
Jynginae

TAXONOMY
Jynx torquilla Linnaeus, 1758, Sweden. Four subspecies.

OTHER COMMON NAMES
English: European wryneck, Eurasian wryneck; French: Torcol fourmilier; German: Wendehals; Spanish: Torcecuello de África Tropical.

PHYSICAL CHARACTERISTICS
6.3–6.7 in (16–17 cm); 0.8–1.9 oz (22–54 g), weight extremes associated with migratory preparation and losses resulting from migration; a small, aberrant woodpecker with an overall gray appearance and lacking stiff tail feathers of typical woodpeckers; upperparts are gray mottled with brown and buff in a pattern much like some nightjars (Caprimulgiformes); diamond-shaped dark patch on back extends to the nape; breast is lighter gray; sexes alike and juveniles similar to adults.

DISTRIBUTION
Breeds from northern Eurasia south through temperate Eurasia to Japan; disjunct breeding population in western Asia and northwestern Africa; winters central Eurasia south to drier areas of central and West Africa, India, Southeast Asia, southern China, southern Japan. *J. t. torquilla* , most of Eurasia; *J. t. tschusii*, Corsica, Sardinia, Italy, eastern Adriatic coast; *J. t. mauretanica* , northwestern Africa; *J. t. himalayana*, Kashmir.

HABITAT
Open forests, clearings, edge habitats with sparse ground cover.

BEHAVIOR
Takes its name from head movements produced when cornered in the nest; described as mimicry of a snake. Migrant, travels mainly at night. Moves about home range alone, as pairs during breeding season, or as post-breeding family groups. Rarely climbs a vertical surface.

FEEDING ECOLOGY AND DIET
Forages for arthropods, especially ants, their larvae and pupae, mainly on ground; captures prey with its sticky tongue.

REPRODUCTIVE BIOLOGY
Nests in old woodpecker holes, nest boxes, and other natural and humanmade cavities; may slightly enlarge a well-rotted cavity. Typically nest is 3–49 ft (1–15 m) up. Nesting is May–June. Unlike woodpeckers, nest bottom is sometimes lined with sparse grass or moss. Typical clutch size is 7–12 eggs, but fewer eggs at some nests and as many as 18–23 at others where more than one female is laying. Incubation period is about 11 days and young usually fledge in 20–22 days. Both parents care for young for 10–14 days post-fledging. Second nest attempt may quickly follow first.

CONSERVATION STATUS
Not threatened, but declines in Europe associated with loss of unimproved pasture and orchard habitats and increases in conifer forests.

SIGNIFICANCE TO HUMANS
In Greek mythology, the king of the Greek gods, Zeus, was bewitched by Inyx, daughter of Echo and Peitha, and in revenge, Hera, Zeus's wife, turned Inyx into a wryneck. As a result, the wryneck is considered a "love charm." No doubt this tale is linked to the sinuous, somewhat sensual movements of the disturbed wryneck in its nest. ◆

Olivaceous piculet

Picumnus olivaceus

SUBFAMILY
Picumninae

TAXONOMY
Picumnus olivaceus Lafresnaye, 1845.

OTHER COMMON NAMES
French: Picumne olivâtre; German: Olivrücken-Zwergspecht; Spanish: Carpinterito Olicáceo, Telegrafista.

PHYSICAL CHARACTERISTICS
3.5–3.7 in (9–10 cm), 0.39–0.53 oz (11–15 g); tiny, short, pointed bill; olive above, black cap with white spots, dusky cheeks with white streaks; pale olive to dusky below with light flank streaking. Male with yellow-orange streaked crown; female with no yellow-orange.

DISTRIBUTION
Atlantic slope of Central America from northeast Guatemala south into northern South America to Colombia, northwest Venezuela, western Ecuador to northwest Peru.

HABITAT
Humid tropical evergreen forest and forest edge, including plantations; often in cutover areas; seems absent from mature forest; lowlands to about 7,000 feet (2,100 m).

Jynx torquilla

☐ Breeding ☐ Nonbreeding

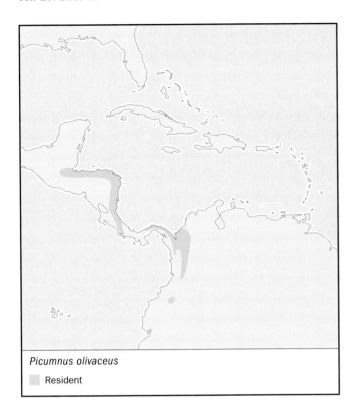

Picumnus olivaceus
  Resident

Sasia abnormis
  Resident

BEHAVIOR
Constantly moving, almost nuthatch-like, moving over small branches both high and low within the forest, but favoring thickets and vines and avoiding large trunks and limbs. The Spanish common name *telegrafista* comes from the resemblance of its feeding percussion blows to the sound of Morse code being tapped out by telegraph.

FEEDING ECOLOGY AND DIET
Feeds largely on ants, especially those that tunnel in dead twigs; also takes other insects and their eggs and larvae.

REPRODUCTIVE BIOLOGY
Nest cavity excavated in soft wood, in a low stub, by both members of a pair. Pair roosts together in the cavity prior to nesting. Clutch of 1–3 white eggs incubated for about 14 days by both parents; young fed by both parents; fledge at about age 24–26 days.

CONSERVATION STATUS
Not threatened.

SIGNIFICANCE TO HUMANS
None known. ◆

PHYSICAL CHARACTERISTICS
3.5 in (9 cm), 0.25–0.42 oz (7.2–12 g); tiny, green above, rust below; male with yellow forehead; female with rufous forehead. Very short tail.

DISTRIBUTION
Malay Peninsula, Sumatra, Java, Billiton, Borneo, Nias.

HABITAT
Secondary forest with much decaying wood; bamboo stands; low dense vegetation.

BEHAVIOR
Somewhat social, seen in groups of 4–5; fast moving through understory.

FEEDING ECOLOGY AND DIET
Gleans insects from surface and excavates small holes to retrieve insects. Diet includes beetles, ants, other small insects, their eggs, and larvae, and small spiders.

REPRODUCTIVE BIOLOGY
Little known; nests found in bamboo; nestlings found May–June; brood size 2.

CONSERVATION STATUS
Not threatened.

SIGNIFICANCE TO HUMANS
None known. ◆

Rufous piculet
Sasia abnormis

SUBFAMILY
Picumninae

TAXONOMY
Picumnus abnormis Temminck, 1825.

OTHER COMMON NAMES
French: Picumne roux; German: Malaienmausspecht; Spanish: Carpinterito Rufo.

Northern flicker
Colaptes auratus

SUBFAMILY
Picinae

TAXONOMY
Colaptes auratus Linnaeus, 1758, based on the drawing of the "golden-winged wood-pecker" by Mark Catesby. Has at times

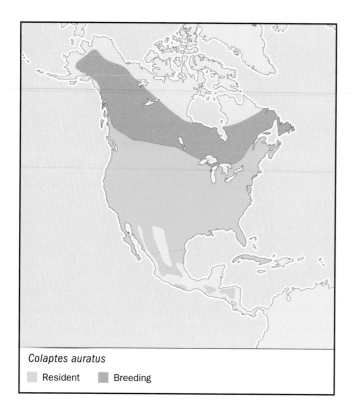

Colaptes auratus

☐ Resident ■ Breeding

HABITAT
Very open forest to savanna with sparse understory, urban and suburban parks and landscapes, less commonly in treeless grasslands.

BEHAVIOR
Northern populations migratory, many others resident. The Cuban form seems more arboreal than continental forms. Solitary much of the time, but migrants found in small flocks. May roost in or on buildings.

FEEDING ECOLOGY AND DIET
Generally feeds on the ground on ants and other arthropods; also feeds on arthropods in well-rotted wood; fruits, and seeds in season.

REPRODUCTIVE BIOLOGY
Monogamous; often nests in near treeless areas, excavating cavities in utility poles, occasionally dirt banks; sometimes uses nest boxes. Not a strong excavator and often uses available cavities. Nesting occurs February–August (earlier in warmer latitudes, later in colder areas). Clutch size 3–12 eggs, 4–9 common; incubation 11–12 days by both parents; young fledge at 25–28 days. Young are fed by regurgitation. Two broods possible. Often suffers from competition for cavities with the introduced European starling (*Sturnus vulgaris*).

CONSERVATION STATUS
Not threatened. Guadeloupe flicker (*Colaptes auratus rufipileus*) of Guadeloupe Island off Baja California is extinct.

SIGNIFICANCE TO HUMANS
Often mentioned in folklore, feathers were used decoratively by Native Americans; sometimes eaten and considered a game bird; occasionally causes problems by excavating into the siding of homes and other buildings. ◆

been separated into red-shafted (*C. cafer*) and yellow-shafted species (*C. auratus*), but considerable hybridization occurs in the North American Great Plains. Four major racial groups including nine races are recognized. The gilded flicker (*C. chrysoides*) is sometimes considered as another race of the northern flicker.

OTHER COMMON NAMES
English: Common flicker, yellow-shafted flicker, red-shafted flicker, Guatemalan flicker, Cuban flicker; French: Pic flamboyant; German: Goldspecht; Spanish: Carpintero Escapulario.

PHYSICAL CHARACTERISTICS
11.8–13.8 in (30–35 cm); 3.1–5.8 oz (88–164 g). Primarily a ground-feeding woodpecker that is camouflaged with earth-toned colors and black spotting on back and wings, a disruptive black "V" on its breast, and heavy black spotting on belly and flanks; western forms have red-vaned flight feathers, eastern and southern forms yellow-vaned flight feathers, hybrids have orange; male red-shafted flickers have a red "moustache" stripe, yellow-shafted have a black moustache, hybrids intermediate.

DISTRIBUTION
Throughout North America from west central Alaska to the northern regions of the Yukon, Manitoba, Ontario, and Quebec, to southern Labrador and Newfoundland, south throughout North America and adjacent islands to northern Baja California, southern Mexico, north central Nicaragua, Cuba, Isle of Pines, Cayman Islands. Distribution of the four major racial groups: *C. a. auratus*, northern and eastern North America; *C. a. cafer*, western North America and Mexico; *C. a. mexicanoides*, highlands from Chiapas south to Nicaragua; *C. a. chrysocaulosus*, Cuba and Cayman Islands.

Rufous woodpecker
Celeus brachyurus

SUBFAMILY
Picinae

TAXONOMY
Picus brachyurus Vieillot, 1818. Nine subspecies.

OTHER COMMON NAMES
French: Pic brun; German: Rostspecht; Spanish: Carpintero Rufo.

PHYSICAL CHARACTERISTICS
8.3–9.8 in (21–25 cm), 1.9–4.0 oz (55–114 g); a reddish brown woodpecker with black barring on rufous back, wings, tail, and flanks; reddish eye; male with a red cheek, female without; juveniles like adults but sometimes more, sometimes less barring.

DISTRIBUTION
Southeast Asia from southwest India and Sri Lanka to Nepal, southern China, Hainan, Borneo, Sumatra, and Java; several disjunct and island populations resulting in considerable variation recognized as subspecies.

HABITAT
Lowland forest to 5,580 ft (1,700 m) in some areas; occupies a diversity of forest types, including bamboo, mangroves, and

Celeus brachyurus
☐ Resident

Melanerpes erythrocephalus
☐ Resident ☐ Breeding

scrub, as well as both primary and secondary evergreen and deciduous forest.

BEHAVIOR
A shy bird usually found away from people; found in pairs; seems to prefer open forest, but seeks shaded areas; very vocal.

FEEDING ECOLOGY AND DIET
Forages throughout the forest, sometimes in mixed species flocks. Feeds extensively on tree-dwelling ants and other insects; also takes fruits, nectar, and sap.

REPRODUCTIVE BIOLOGY
Male and female share excavation of nest, often in a nest of tree ants; clutch usually of 2–3 shiny white eggs incubated by both sexes for 12–14 days; young fed by regurgitation.

CONSERVATION STATUS
Not threatened.

SIGNIFICANCE TO HUMANS
None known. ◆

Red-headed woodpecker
Melanerpes erythrocephalus

SUBFAMILY
Picinae

TAXONOMY
Picus erythrocephalus Linnaeus, 1758, based on Mark Catesby's drawing of the "Red-headed Wood-Pecker."

OTHER COMMON NAMES
French: Pic à tête rouge; German: Rotkopfspecht; Spanish: Carpintero de Cabeza Roja.

PHYSICAL CHARACTERISTICS
9–10 in (24–26 cm); 2–3.4 oz (56–97 g). Adults have a completely red head, black back and tail, white breast, black wings with white secondaries that appear as a white shield when the wings are folded over the back. Sexes are alike. Juveniles have a gray-black head and some black markings on white secondaries.

DISTRIBUTION
Breeds east of Rocky Mountains in North America from southern regions of Saskatchewan, Manitoba, Ontario, and Quebec, south to the Gulf of Mexico and south central Florida; winters mostly in southern two-thirds of its breeding range, but farther north during mild winters.

HABITAT
Open woodland, especially with oaks and beech; also roadsides and open areas with scattered trees or utility poles. Often seasonally moves from open areas in summer to bottomland forest in fall and winter.

BEHAVIOR
Solitary or as pairs in summer, often found in loose social groups in winter. Often very vocal in social groups. Interspecifically territorial with red-bellied woodpeckers, but normally more of an open-country bird than the red-bellied.

FEEDING ECOLOGY AND DIET
Often sallies from a perch to capture flying insects or to seize arthropods and other small animals from the ground. It then often takes its captured prey to the top of a stub or utility pole that it regularly uses as a "chopping block" to remove legs and other hard parts before eating. Takes considerable fruit, acorns, beechnuts, and other nuts in season and will sometimes cache these in cavities.

REPRODUCTIVE BIOLOGY

Poor cavity excavator and frequently uses old cavities or usurps a fresh cavity from other species such as the red-bellied woodpecker. When it excavates a cavity of its own, it is usually in a well-rotted stub with a pre-existing crack that often forms a flat side to the cavity entrance. Both pair members share cavity excavation, incubation, and care of young. Nesting normally occurs between April and August. Clutch size varies from 3 to 10 eggs, but is usually 4–5 eggs; incubation lasts 12–13 days; young fledge at about 37–29 days. The red-headed woodpecker often loses cavities to European starlings.

CONSERVATION STATUS

Not threatened, but suffers high mortality due to being hit by vehicles when it comes to roads for insects; also suffers in competition with European starlings.

SIGNIFICANCE TO HUMANS

Among the Chitimacha Indians of Louisiana, there is a folktale about a great flood from which the red-headed woodpecker escaped drowning by clinging to a cloud, but its tail hung down into the dark water and to this day it has a black tail as a result. The red feathers of red-headed woodpeckers have been highly valued by Native Americans and used decoratively. ◆

Bennett's woodpecker

Campethera bennettii

SUBFAMILY

Picinae

TAXONOMY

Chrysoptilus bennettii A. Smith, 1836, western Transvaal, South Africa. Two races recognized.

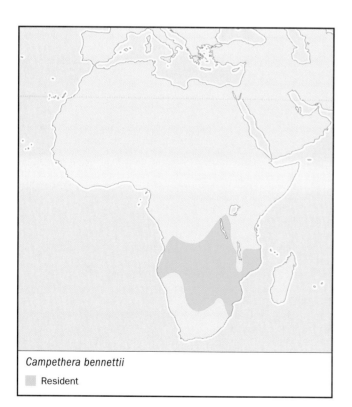

Campethera bennettii

▨ Resident

OTHER COMMON NAMES

English: Specklethroated woodpecker, Reichenow's woodpecker; French: Pic de Bennett; German: Bennettsspecht; Spanish: Pico de Bennett.

PHYSICAL CHARACTERISTICS

About 9.5 in (24 cm); 2.2–3 oz (61–84 g). A small, brownish yellow woodpecker with heavy spotting on underparts; male has a red forehead to the nape, a red "moustache," and white ear coverts; female has a red nape, black-and-white mottled forehead, brown throat and ear coverts, a buff wash on the breast, and less intense spotting on the underparts; juveniles are darker above and more spotted below, with a white-spotted black crown.

DISTRIBUTION

Lake Victoria region, western Tanganyika and southeastern Congo to Angola, central Kalahari Desert, Damaraland, southern Zimbabwe, and Transvaal. *C. b. bennettii*, most of range except for southern Angola, southwestern Zambia, northern Namibia, and northern Botswana; *C. b. capricorni*, southwest parts of range including southern Angola, southwestern Zambia, northern Namibia, and northern Botswana. Absent from large areas.

HABITAT

Uses a wide range of open forest and bush habitats, especially acacia, miombo, and *Brachystegia*.

BEHAVIOR

A territorial, but social species found in pairs or family groups. Much of its time is spent on bare ground or in short-grass areas, including lawns. May show some migration in drier parts of its range.

FEEDING ECOLOGY AND DIET

Highly terrestrial in its search for food, but also forages on low trunks and larger limbs of trees. Often accompanies glossy starlings (*Lamprotornis*) when foraging. Diet includes mainly ants, termites, and their larvae and pupae, but also other arthropods.

REPRODUCTIVE BIOLOGY

Monogamous. Breeds from August to February, with nesting peaking in October and November in Zimbabwe and Transvaal. Nests are often in open areas and often in cavities excavated by other species. Clutch size 2–5 eggs; incubation lasts 15–18 days; parental duties carried out by both parents. Nest cavities may be reused.

CONSERVATION STATUS

Not threatened.

SIGNIFICANCE TO HUMANS

None known. ◆

Gray woodpecker

Dendropicos goertae

SUBFAMILY

Picinae

TAXONOMY

Picus goertae P. L. S. Müller, 1776, Senegal. At least eight races described, four currently recognized.

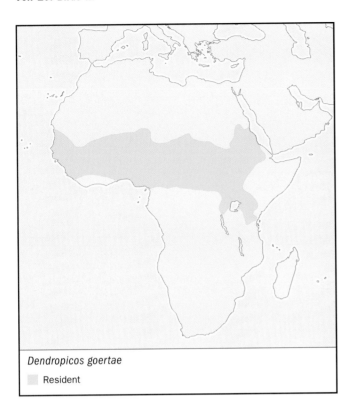

Dendropicos goertae

■ Resident

September to November in eastern Africa; sexes share nest cavity excavation at 1–60 ft (0.3–18.3 m) above ground in a dead tree or dead stub of a live tree. Clutch size 2–4 eggs. No data on incubation, parental care, or fledging.

CONSERVATION STATUS
Not threatened.

SIGNIFICANCE TO HUMANS
None known. ◆

Red-cockaded woodpecker
Picoides borealis

SUBFAMILY
Picinae

TAXONOMY
Picus borealis Vieillot, 1807, North America (specific locality unknown, but arbitrarily decided to be Mount Pleasant, South Carolina). Two subspecies described, *Picoides borealis borealis* from most of the species range, and *P. b. hylonomus* from central and southern Florida. Researchers have discounted the latter race in the late twentieth century.

OTHER COMMON NAMES
French: Pic boreal; German: Kokardenspecht; Spanish: Carpintero de Cresta Roja.

PHYSICAL CHARACTERISTICS
About 8.7 in (22 cm); 1.4–1.9 oz (40–55 g). Medium black-and-white woodpecker; primary distinguishing characteristics include large white cheek patches with ladder back; males have several tiny red feathers between white cheek patch and black

OTHER COMMON NAMES
English: African gray woodpecker; French: Pic gris; German: Graubrustspecht; Spanish: Pico Gris.

PHYSICAL CHARACTERISTICS
About 8 in (20 cm); 1.4–1.9 oz (40.5–52.5 g). A small woodpecker with a long, straight, rather broad bill; upperparts are unbarred green or brownish green with a red rump and barred brown tail; underparts are gray with an orange to yellow belly patch and some barring on flanks. Male has a pale, striped, gray head with a red hindcrown and nape; female lacks red on head.

DISTRIBUTION
Found in a broad swath of forest and savanna habitats in central and West Africa; from sea level to 9,800 ft (3,000 m). *D. g. goerte*, West Africa to Sudan, south to northeastern Zaïre; *D. g. koenigi*, the Sahel zone across central eastern Mali, Niger, Chad, to western Sudan; *D. g. abessinicus*, eastern and northern Sudan and northern and western Ethiopia; *D. g. meridionalis*, south central Zaire, probably to southern Gabon and northwestern Angola.

HABITAT
Wooded and savanna areas, thickets with large trees, riverine forest, gardens, mangroves in some areas.

BEHAVIOR
Found in pairs and family groups, moves rapidly through its habitat. Often at forest edge.

FEEDING ECOLOGY AND DIET
Forages on the ground and in live and dead trees, especially trunk and larger limbs. Occasionally takes insects in flight. Diet includes ants, termites, beetle larvae, and other arthropods.

REPRODUCTIVE BIOLOGY
Nesting is from December to June in west, December to February and July to September in Zaire, and February to July and

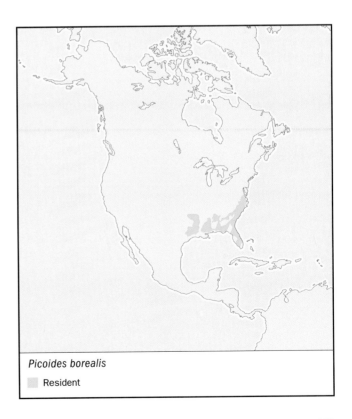

Picoides borealis

■ Resident

crown, but the red is usually concealed; females lack the red; immature males may have irregular red patch on forehead, immature females tend to have white flecks on lower forehead.

DISTRIBUTION
Southeastern Oklahoma and eastern Texas, southern Missouri, south central Kentucky, central Tennessee, to southeastern Maryland, south to southern Florida and across the Gulf coast. Now extirpated from Missouri, Kentucky, Tennessee, and Maryland. Vagrants have shown up as far north as Illinois and New Jersey.

HABITAT
Extensive, open, old-growth pine forest, naturally maintained by lightning-started fires.

BEHAVIOR
The red-cockaded woodpecker is a very social species that lives in extended family groups including one breeding pair, their offspring from recent nesting efforts, and males from earlier nesting efforts. The group forages over an area averaging about 200 acres (80 ha) in good habitat and more than 1,000 acres (400 ha) in poor habitat. Cavities are in living pines, usually below the lowest branch, in trees infected with the red-heart fungus. Cavities persist sometimes for decades and are used multiple years as nest and roost sites, inherited by males who remain with the group. Birds peck tiny holes, called resin wells, above and below each cavity. These are continually worked so they provide a steady flow of sticky resin, which is an effective barrier against tree-climbing rat snakes (*Elaphe*).

FEEDING ECOLOGY AND DIET
Feeds primarily on tree-surface arthropods obtained from the surface and by scaling loose bark from the tree. Males tend to forage mostly on limbs and trunk of pines above the lowest branch, females on the trunk below the lowest branch.

REPRODUCTIVE BIOLOGY
Monogamous, but a cooperative breeder; nest is in the roost cavity of the breeding male. Clutch size typically 2–5 eggs; incubation period 10–11 days; young fledge at 26–29 days; offspring cared for by both parents and helpers.

CONSERVATION STATUS
Vulnerable, and listed as Endangered under the U.S. Endangered Species Act; populations have suffered from habitat fragmentation and deforestation, cutting of old-growth pine forests, and control of natural fire.

SIGNIFICANCE TO HUMANS
Needs of this species conflict with desires of forest industry to cut young pines; birders travel to the southeastern United States specifically to see this bird. ◆

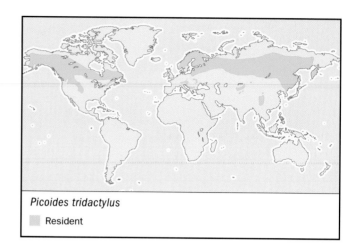

Picoides tridactylus
▨ Resident

Three-toed woodpecker
Picoides tridactylus

SUBFAMILY
Picinae

TAXONOMY
Picus tridactylus Linnaeus, 1758, Sweden. Although North American and Eurasian populations appear similar, recent studies suggest significant genetic differences. Three races are recognized in the New World, five in Eurasia.

OTHER COMMON NAMES
English: Northern three-toed woodpecker; French: Pic tridactyl; German: Dreizehenspecht; Spanish: Pico Tridáctilo.

PHYSICAL CHARACTERISTICS
8–9.5 in (20–24 cm); 1.6–2.7 oz (46–76 g) a medium-sized woodpecker with black above, white below, varying extent of barring on flanks and white on back; male has a yellow forehead and crown, female has a whitish crown with fine black streaks; immature has duller plumage that is somewhat brown; only three toes on each foot.

DISTRIBUTION
Resident from near the tree line in northern Alaska, northern Canada, northern Eurasia, south to northern tier of United States, mountains of southern Europe, western China, northern Mongolia, northern Korea, and Japan. Wanders south in winter to New England, north central United States. *P. t. tridactylus*, northern Eurasia, Scandinavia, Latvia, to Mongolia, southeastern Siberia, and Sakhalin Island; *P. t. crissoleucus*, northern taiga from Urals to Sea of Okhotsk; *P. t. albidor*, Kamchatka; *P. t. alpinus*, mountains of central, southern, and southeastern Europe, northeastern Korea, Hokkaido, Japan; *P. t. funebris*, southwestern China to Tibet; *P. t. dorsalis*, Rocky Mountains, Montana to Arizona and New Mexico; *P. t. fasciatus*, western North America, Alaska, and Yukon south to Oregon; *P. t. bacatus*, eastern North America, Alberta east to Labrador and Newfoundland, south to Minnesota and New York.

HABITAT
Coniferous forest; less often mixed coniferous-deciduous forest.

BEHAVIOR
Nonmigratory, often very quiet; very arboreal; populations respond to forest insect outbreaks.

FEEDING ECOLOGY AND DIET
Feeds primarily on wood-boring insects, their larvae and pupae; forages lower in winter than at other times; when sexes forage together, females forage higher than males.

REPRODUCTIVE BIOLOGY
Monogamous; breeding activities generally extend from mid-March through June; nest cavities usually excavated in dead stub; pair share excavation, incubation, and care of young; clutch size 3–6 eggs; incubation period 11–14 days; young fledge at 22–26 days.

CONSERVATION STATUS
Fairly common, not threatened.

SIGNIFICANCE TO HUMANS
None known. ◆

Smoky-brown woodpecker
Veniliornis fumigatus

SUBFAMILY
Picinae

TAXONOMY
Picus fumigatus d'Orbigny, 1840, Corrientes Province, Argentina, and Yungas, Bolivia. Five races recognized.

OTHER COMMON NAMES
English: Brown woodpecker; French: Pic enfumé German: Russpecht; Spanish: Carpintero Pardo.

PHYSICAL CHARACTERISTICS
5.9–6.7 in (15–17 cm); 1.1–1.8 oz (31–50 g). A very plain woodpecker with no color pattern evident in its plumage and no crest; smoky brown overall; adult male has dark gray nape feathers tipped with red, back and scapular feathers are tawny-olivaceous with a golden, sometimes orange-red wash; adult female similar to male, but with nape feathers tipped with brown; juveniles similar to adult, but plumage duller.

DISTRIBUTION
Found from Nayarit and southeastern San Luis Potosi in Mexico, through Central America to Colombia and northern Venezuela, south along the west slope of the Andes to north central Peru and along the east slope of the Andes to northwestern Argentina. *V. f. oleaginous*, eastern Mexico; *V. f. sanguinolentus*, central Mexico to western Panama; *V. f. reichenbachi*, eastern Panama, northern Venezuela, Colombia to eastern Ecuador; *V. f. fumigatus*, upper Amazonia; *V. f. obscuratus*, northwestern Peru to northwestern Argentina.

HABITAT
Evergreen forests of mountains and lowlands, including secondary forests of tropical and subtropical areas; seems to prefer smaller tree trunks to larger ones.

BEHAVIOR
A resident species that often travels in pairs and, after nesting, in family groups, often in mixed species flocks; a rather inconspicuous species that moves about the forest canopy as well as tangled vines of the understory.

FEEDING ECOLOGY AND DIET
Forages high in broken canopy and lower at edges on small branches or vines; seems to prefer edges in lowland forest where it may forage low, but as with many species, this "preference" could be a function of where birders can most easily see them and careful study is needed. Often uses second growth. Diet seems to favor small wood-boring beetles and their larvae.

REPRODUCTIVE BIOLOGY
Nesting occurs February–May. Nest excavated in a fence post, utility pole, or tree trunk, 5–25 feet (1.5–7.6 m) up. Reported clutch size of 4 eggs; no information on young or parental care.

CONSERVATION STATUS
Not threatened.

SIGNIFICANCE TO HUMANS
None known. ◆

Guadeloupe woodpecker
Melanerpes herminieri

SUBFAMILY
Picinae

TAXONOMY
Picus herminieri Lesson, 1830.

OTHER COMMON NAMES
French: Pic de la Guadeloupe; German: Guadeloupespecht; Spanish: Carpintero de Guadeloupe.

PHYSICAL CHARACTERISTICS
9.4 in (24 cm), 3.1–3.5 oz (87–100 g); glossy black above and black with dull red overtones below. Sexes alike. Juveniles less glossy and with dull red-orange tinge below.

DISTRIBUTION
Found only in Guadeloupe in the West Indies; 75% of the population of 10,000+ birds are on the more forested island of Basse-Terre.

HABITAT
Semi-deciduous to evergreen forest, including upland, mangrove, and swamp forest.

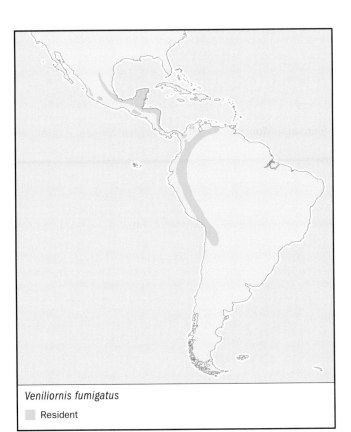

Veniliornis fumigatus

☐ Resident

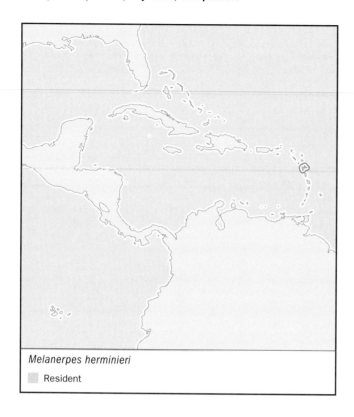

Melanerpes herminieri
  Resident

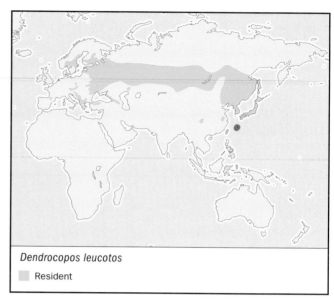

Dendrocopos leucotos
  Resident

BEHAVIOR
Moves deliberately through the forest; returns frequently to fruit-bearing trees.

FEEDING ECOLOGY AND DIET
Forages mainly on trunks and larger branches; diet includes insects, fruit, and seeds.

REPRODUCTIVE BIOLOGY
Typically excavates nest cavity in the trunk of a dead stub; nesting occurs February–August; both parents share incubation of 3–5 eggs; incubation period 14–16 days; young fledge at 33–37 days.

CONSERVATION STATUS
The Guadeloupe woodpecker is considered a Near Threatened species; clearcutting, conversion of forest habitats to other uses, and removal of dead trees are major threats; introduced rats may prey on eggs and nestlings. ◆

White-backed woodpecker
Dendrocopos leucotos

SUBFAMILY
Picinae

TAXONOMY
Picus leucotos Bechstein, 1802.

OTHER COMMON NAMES
English: Owston's woodpecker; French: Pic à dos blanc; German: Weissrückenspecht; Spanish: Carpintero de Lomo Blanco.

PHYSICAL CHARACTERISTICS
9.8–11.0 in (25–28 cm), 3.2–5.6 oz (92–158 g); A pied woodpecker with white cheeks, white forehead and lower back, white breast shading to pale pink and deeper pink vent area; male with prominent red cap; female with black cap. Birds in southeast Europe have vermiculated white back.

DISTRIBUTION
Found in a broad band across forested areas of northern Eurasia from Fennoscandia to Kamchatka and Japan; many isolated populations in montane and island areas

HABITAT
Wet mixed forest, often near rivers or lakes

BEHAVIOR
Has a large home range, moving great distances to areas with many dead and dying trees in order to find preferred foods. Drumming has been likened to a bouncing ping-pong ball— a strong beginning accelerating to a weaker end "bouncing to a halt."

FEEDING ECOLOGY AND DIET
Feeds primarily on insects, especially wood–boring beetles; spends considerable time excavating beetle larvae from near the base of willows and alders.

REPRODUCTIVE BIOLOGY
Courtship often begins in February; nests often high in rotted stub or utility pole; clutch of 3–5 eggs incubated by both parents for 14–16 days; both adults tend nestlings which fledge at 27–28 days.

CONSERVATION STATUS
Not threatened globally, but considered regionally threatened by forest clearing and disturbance.

SIGNIFICANCE TO HUMANS
None known. ◆

Black woodpecker
Dryocopus martius

SUBFAMILY
Picinae

TAXONOMY
Picus martius Linnaeus, 1758, Sweden. Two races recognized.

OTHER COMMON NAMES
French: Pic noir; German: Schwarzspecht; Spanish: Pito Negro.

PHYSICAL CHARACTERISTICS
17.7–22.4 in (45–57 cm); 9.2–13 oz (260–370 g). Crow-sized; black, pale bill, and whitish eyes; male has raspberry red crown; female a red nape; juveniles similar to adults, but duller, looser-textured plumage.

DISTRIBUTION
Cool-temperate Eurasia; from western Europe north to the Arctic Circle in Scandinavia and east to Japan and Kamchatka. *D. m. martius*, most of range except for southwestern China and Tibet; *D. m. khamensis*, southwestern China and Tibet.

HABITAT
Mature coniferous, mixed, or deciduous forest. A pair normally needs 750–1,000 acres (300–400 ha) of forest.

BEHAVIOR
Resident; solitary or in pairs; spring drumming on resonant limb or stub is very loud and low in tone; climbs by "hopping" up a trunk or limb; flies with direct "crowlike" flight.

FEEDING ECOLOGY AND DIET
Feeds mostly on ants and their larvae and pupae, but also takes larvae of wood-boring beetles, occasionally other insects, nuts, and seeds, and rarely fruit.

REPRODUCTIVE BIOLOGY
Courtship may begin in January, but nesting is primarily late March to May; nest usually placed high in a large stub. About 75% of nests are in decayed but living trees. Clutch size 3–5 eggs; incubation lasts 12–14 days; young usually fledge at 27–28 days. Both parents incubate and care for young; young are fed by regurgitation.

CONSERVATION STATUS
Not threatened; many populations increased during the late twentieth century, though there were local declines associated with habitat fragmentation and loss.

SIGNIFICANCE TO HUMANS
None known, except as a symbol of "wildness." ◆

Gray-faced woodpecker
Picus canus

SUBFAMILY
Picinae

TAXONOMY
Picus canus Gmelin, 1788. Hybridization with Eurasian green woodpecker known. Eleven subspecies recognized.

OTHER COMMON NAMES
English: Gray-headed woodpecker, gray-headed green woodpecker, ashy woodpecker, black-naped woodpecker; French: Pic cendré; German: Grauspecht; Spanish: Pito Cano.

PHYSICAL CHARACTERISTICS
10.2–13.0 in (26–33 cm); 3.9–7.3 oz (110–206 g); back olive green, yellow-green rump, gray head, narrow black "moustache," breast light gray-green; male with small red patch on lower forehead; female without red.

DISTRIBUTION
Central and eastern Europe through central Asia to Himalayas; Southeast Asia through China, Manchuria, Korea, Hokkaido, Hainan, Taiwan, Sumatra. Subspecies can be divided into two major groups: 2 subspecies found in northern Eurasia; 9 found in southeast and east Asia.

HABITAT
Found in a great diversity of forest habitats: moist bottomland forest to open park-like, to uplands with many conifers.

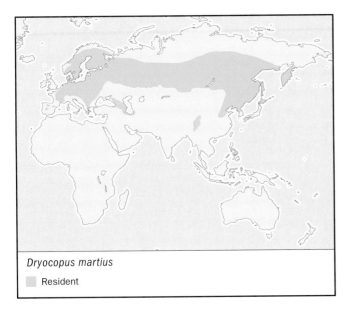

Dryocopus martius

■ Resident

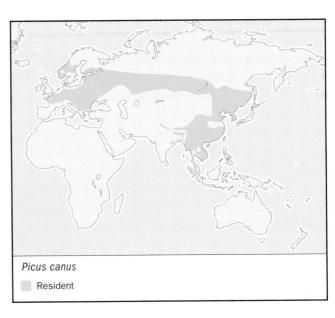

Picus canus

■ Resident

BEHAVIOR

Monogamous, normally solitary; territorial during breeding. Often winters in riparian areas and closer to humans; some nomadic winter movements.

FEEDING ECOLOGY AND DIET

Diet is mostly ants and termites and their brood, but also includes other arthropods, fruit, nuts, nectar, and eggs of other birds; in many areas it especially frequents old aspen trees.

REPRODUCTIVE BIOLOGY

Nest cavity in decayed wood excavated by both sexes; clutch of 4–9 white eggs incubated by both parents for 14–17 days; young cared for by both parents (rarely by a helper); fledge at 23–27 days.

CONSERVATION STATUS

Not threatened.

SIGNIFICANCE TO HUMANS

None known. ◆

Lesser flame-backed woodpecker

Dinopium benghalense

SUBFAMILY

Picinae

TAXONOMY

Dinopium benghalense Linnaeus, 1758.

OTHER COMMON NAMES

English: Black-rumped flameback, black-rumped golden-backed woodpecker, lesser golden-backed woodpecker; French: Pic du Bengale; German: Orangespecht; Spanish: Pico Lomo en Llamas.

Dinopium benghalense

Resident

PHYSICAL CHARACTERISTICS

10.2–11.4 in (26–29 cm), 3.0–4.7 oz (86–133 g); a medium red-crested woodpecker; black mantle, lower back, and rump; yellow to yellow-green mid-back and wings; black tail; breast white with feathers edged in black. Male's forehead is red to bill; female's forehead is black with white spots. Race from Sri Lanka has back and wings deep red, more black on head.

DISTRIBUTION

Indian subcontinent.

HABITAT

Diverse forest and cultivated areas.

BEHAVIOR

Seen in pairs and in mixed species flocks. Pair members keep in contact with one another using frequent loud rattling calls. Breaks into leaf nests of ants. Frequents coconut plantations and wooded gardens.

FEEDING ECOLOGY AND DIET

Primary food is ants, but takes other arthropods, fruit, and nectar.

REPRODUCTIVE BIOLOGY

Nests in March to April in most areas, again in July and August in south; from December to September in Sri Lanka. Clutch of 2–3 white eggs is incubated by both parents for 17–19 days; both adults feed young by regurgitation; young fledge at 21–23 days; sometimes a second brood is raised.

CONSERVATION STATUS

Not threatened.

SIGNIFICANCE TO HUMANS

None known. ◆

Yellow-bellied sapsucker

Sphyrapicus varius

SUBFAMILY

Picinae

TAXONOMY

Picus varius Linnaeus, 1766, based on a drawing by Mark Catesby from South Carolina.

OTHER COMMON NAMES

English: Common sapsucker; French: Pic maculé; German: Feuerkopf-Saftlecker; Spanish: Carpintero de Paso.

PHYSICAL CHARACTERISTICS

7.5–8.7 in (19–22 cm); 1.4–2.2 oz (40–62 g). Small black-and-white woodpecker with short, chisel-tipped bill; easily distinguished by white stripe that extends down the wing of perched birds. Adult male has a red throat, forehead, and crown; female a white throat and a somewhat paler red forehead and crown. Juveniles have considerable brown and buff and initially upperparts are somewhat barred; they also have much less white and much less red in crown.

DISTRIBUTION

Breeds in northern North America east of the Rocky Mountains across Canada from northeastern British Columbia to southern Labrador and Newfoundland, south to North Dakota and Connecticut, with some disjunct populations in the Appalachians of eastern Tennessee and northern Georgia.

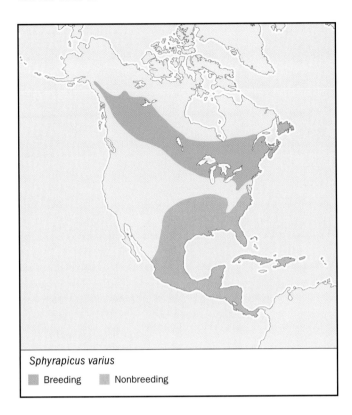

Sphyrapicus varius

▨ Breeding   ▨ Nonbreeding

sapsuckers and the trees they feed on suggests that they select injured and diseased trees because these trees produce a sweeter sap. Many other animals take advantage of the sapsucker's sap wells. ◆

Okinawa woodpecker
Sapheopipo noguchii

SUBFAMILY
Picinae

TAXONOMY
Picus noguchii Seebohm, 1887, Okinawa.

OTHER COMMON NAMES
English: Pryer's woodpecker; French: Pic d'Okinawa; German: Okinawaspecht; Spanish: Pico de Okinawa.

PHYSICAL CHARACTERISTICS
12.2–13.8 in (31–35 cm). An earth-toned bird; the male has a rusty red cap from the forehead to the nape; female has a black cap from forehead to nape; both have a gray throat and belly with deep red tones on the back and wings; prominent white spotting on primary feathers; black at edge of cap accents a lighter gray-brown face; rump red, tail black; immatures are duller and grayer.

DISTRIBUTION
Found only in the central mountain range of Yambaru, the northern part of the island Okinawa, Japan.

HABITAT
Restricted to old-growth subtropical evergreen broadleaf forest; breeding range seems limited by a need for large dead limbs for nest and roost cavity excavation.

Winters in eastern United States through eastern and southern Mexico and Central America, Bahamas, and West Indies.

HABITAT
Within its breeding range it is found in deciduous and mixed forest and is especially associated with aspen (*Populus*), in which it often excavates nest cavities, and birch (*Betula*) and hickory (*Carya*), which also provide sap resources. Winters in many wooded habitats, including urban parks.

BEHAVIOR
Typically solitary and often inconspicuous outside the breeding season. Sapsuckers maintain an "orchard" of trees with sapwells from which they obtain food. Males migrate shorter distances than females and return earlier than females to breeding areas.

FEEDING ECOLOGY AND DIET
Feeds on beetles and their larvae, ants, other arthropods, and extensively on sweet sap from diseased trees, which it obtains by pecking small holes (sap wells) into the cambium. Also takes insects attracted to the sap wells. Berries are also taken and sometimes fed to nestlings.

REPRODUCTIVE BIOLOGY
Most nests are in living trees that are infected with a heartrot fungus. Cavity entrances are very small, such that a sapsucker often has to squeeze to get in. Clutch size averages 4–5 eggs, but varies geographically, increasing from south to north. Incubation lasts 12–13 days and is shared by parents; young fledge at 25–29 days of age and become independent about two weeks later.

CONSERVATION STATUS
Not threatened.

SIGNIFICANCE TO HUMANS
At times considered a pest and damaging to shade and fruit trees. More detailed knowledge of interrelationships between

Sapheopipo noguchii

▨ Resident

BEHAVIOR
A highly vocal species that spends most of its time foraging at lower levels in the forest.

FEEDING ECOLOGY AND DIET
In spite of its rarity, the Okinawa woodpecker seems to have a broad foraging niche, searching for arthropods on larger branches and trunks, among canopy leaves, on downed wood, and in leaf litter on the ground; also opportunistically feeds on other small animals and on fruit.

REPRODUCTIVE BIOLOGY
Nesting activity begins as early as February, but typically in March and continues through mid-June. It excavates nest cavities primarily in old, partially dead *Castonpsis cuspidate* and *Machilus thunbergii* trees. Typically one or two nestlings are raised. No other details available.

CONSERVATION STATUS
Critically Endangered due to habitat destruction and population fragmentation. Population estimates since 1950 have ranged from 40 to about 200 birds. In 1977, undisturbed forest was limited to about 1,100 acres (450 ha) and has since declined. The Okinawa woodpecker has been declared a "Natural Monument" and "Special Bird for Protection" by the Japanese government. The population remains highest in a military training area that is off-limits to civilians. Some expansion into secondary forest was noted in the late twentieth century.

SIGNIFICANCE TO HUMANS
None known. ◆

Mulleripicus pulverulentus

▨ Resident

BEHAVIOR
Nonmigratory; often seen in pairs or small (family?) groups; voice described as "almost honking" to a distinctive whinny; flight is less undulating than smaller woodpeckers. Displays include head-swinging with both wings and tail extended; drums loudly.

FEEDING ECOLOGY AND DIET
Forages mostly in tall trees, where it excavates larvae of wood-boring beetles and other arthropods, but also feeds on ants on the ground and occasionally hawks flying ants and other insects.

REPRODUCTIVE BIOLOGY
Nests from March through August; both sexes excavate nest cavity, but male dominates; nest is generally high (27–135 ft; 8.2–41 m), dug into very large stubs or branches. Clutch includes 2–4 eggs; no data on incubation period or age at fledging; both sexes incubate and care for young; young may remain with parents until next nesting season.

CONSERVATION STATUS
Not threatened, but uncommon to rare (e.g., Java and Sumatra), and threatened locally by deforestation.

SIGNIFICANCE TO HUMANS
None known, but probably eaten when accessible. ◆

Great slaty woodpecker
Mulleripicus pulverulentus

SUBFAMILY
Picinae

TAXONOMY
Picus pulverulentus Temminck, 1826, Java and Sumatra. Two subspecies.

OTHER COMMON NAMES
French: Pic meunier; German: Puderspecht; Spanish: Pico Pizarro.

PHYSICAL CHARACTERISTICS
19–20 in (48–50 cm); 12.7–20 oz (360–563 g). The largest Old World woodpecker; "lanky" in appearance; male is gray on top of the head and hind neck, with a slight crest, a pale red "moustache," yellow-white throat with red-tipped feathers, and the rest of the body dark gray, darkest on the wings and tail; female is similar, but lacks the red; immature is dark gray tinged with brown.

DISTRIBUTION
Northern India to southwest China, Southeast Asia to Sumatra, Java, Borneo, and Philippines. *M. p. pulverulentus*, Malaysia, Sumatra, Riouw Archipelago, Java, Borneo, North Natuna Islands, east to Palawan; *M. p. harterti*, India, Nepal, east to southwestern China and Indochina.

HABITAT
Extensive forested areas, including second growth, up to about 1,000 feet (300 m).

Ivory-billed woodpecker
Campephilus principalis

SUBFAMILY
Picinae

TAXONOMY
Picus principalis Linnaeus, 1758, based on Mark Catesby's drawing of the "Largest White-bill Woodpecker" from South Carolina. Two subspecies recognized.

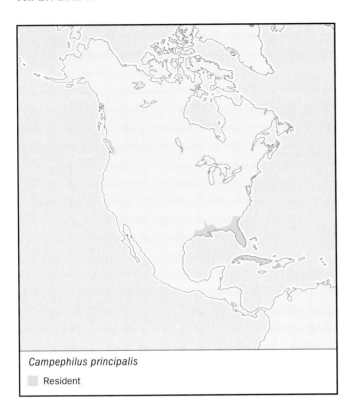

Campephilus principalis

Resident

OTHER COMMON NAMES
French: Pic à bec ivoire; German: Elfelbeinspecht; Spanish: Carpintero Real.

PHYSICAL CHARACTERISTICS
18.5–21 in (47–54 cm); 15.5–18.3 oz (440–570 g). A very large, black woodpecker with white lines extending down the neck on each side to the upper base of the wing, white secondaries and inner primaries, a very robust, chisel-tipped, ivory-colored bill; male has a pointed crest that is black in front and scarlet behind; female has a longer, more pointed, somewhat recurved solid black crest.

DISTRIBUTION
C. p. principalis formerly found in southeastern United States from eastern Texas to North Carolina and north to southern Illinois and southern Ohio; *C. p. bairdi* formerly in forested areas throughout Cuba. Most recent known populations are from northeastern Louisiana, Florida, and northeastern Cuba. May now be extinct, though continued unverified reports in southeastern Cuba, southeastern Louisiana, and Florida provide hope.

HABITAT
Extensive old-growth forest, especially bottomland forest, but also pine uplands in both the United States and Cuba; habitat losses resulted in last North American populations being in bottomlands and last Cuban populations being in upland pines.

BEHAVIOR
Wanders over a home range of 6 sq mi (15.5 sq km) or more; perhaps somewhat social, often seen in family groups; characteristic call is a plaintive, single- or double-note nasal tooting that has been likened to a child's "tin horn" and that can be mimicked by blowing on a clarinet mouthpiece; mechanical sound produced is a hard single pound on a resonant surface followed immediately by another such that the second sounds like an echo of the first. This mechanical sound is characteristic of *Campephilus* woodpeckers and is called the "double rap."

FEEDING ECOLOGY AND DIET
Visits recently dead trees and with its heavy, chisel-like bill, knocks large slabs of bark from the tree to reveal subsurface arthropods. Feeds extensively on the larvae of large wood-boring beetles, especially Cerambycidae; also takes other arthropods and fruit in season.

REPRODUCTIVE BIOLOGY
Monogamous; known to breed from January through April in North America and March through June in Cuba, but few data are available. Nest cavity is in a large dead tree or in a live tree with extensive heartrot. Recorded nests have been 24–50 ft (7.3–15.2 m) up; cavity entrance typically taller than wide, but shape varies. Clutch 2–4 eggs; incubation by both parents; incubation period and age at fledging not known; young may remain with parents until next breeding season.

CONSERVATION STATUS
Critically Endangered by all criteria; may be extinct. The major factor leading to current status has been loss and fragmentation of old-growth forest, but other factors have been nineteenth century killing of birds by scientists, amateur collectors, Native Americans, and hunters, and probably more recent limitation of natural fire. In North America, confusion with the similar-sized and similar-appearing pileated woodpecker (*Dryocopus pileatus*) leads to many false sightings.

SIGNIFICANCE TO HUMANS
Bills and scalps of males were culturally important to Native Americans, apparently symbolic of successful warfare. They were often used to decorate war pipes and medicine bundles and were widely traded outside the range of the species. Early Europeans in North America also killed the birds for their bills and used them for such things as watch fobs. In the late 1800s, there was a brisk trade in skins and eggs among private and professional collectors. In both the United States and Cuba, ivory-bills were occasionally eaten. The ivory-bill has become symbolic of rarity. Collector prints, ceramic ivory-bills, trade cards with ivory-bills on them, and use of ivory-bills in advertisements have drawn much attention to the species. ◆

Resources

Books

Frugis, S., G. Malaguzzi, G. Vicini, and P. Cristina. *Guida ai Picchi del Mondo.* Torino, Italy: Museo Regionale di Scienze Naturali, Monografia VII, 1988.

Fry, C. H., S. Keith, and E. K. Urban., eds. *The Birds of Africa.* Vol. 3. New York: Academic Press, 1988.

Garrido, O. H., and A. Kirkconnell. *Field Guide to the Birds of Cuba.* Ithaca: Cornell University Press, 2000.

Resources

Jackson, J. A. "Ivory-billed Woodpecker Campephilus principalis principalis." In *Rare and Endangered Biota of Florida: Birds*, edited by J. A. Rodgers and H. W. Kale, II. Gainesville: University of Florida Press, 1996.

MacKinnon, J., and K. Phillipps. *A Field Guide to the Birds of Borneo, Sumatra, Java, and Bali.* Oxford: Oxford University Press, 1993.

Moore, W. S., and V. R. DeFilippis. "The Window of Taxonomic Resolution for Phylogenies Based on Mitochondrial Cytochrome b." In *Avian Molecular Evolution and Systematics*, edited by D. P. Mindell. San Diego: Academic Press, 1997.

Sick, H. (translated by W. Belton). *Birds in Brazil.* Princeton: Princeton University Press, 1993.

Snow, D. W., and C. M. Perrins et al. *The Birds of the Western Palearctic.* Vol. 1. Concise edition. Oxford: Oxford University Press, 1998.

Winkler, H., D. A. Christie, and D. Nurnie. *Woodpeckers: A Guide to the Woodpeckers of the World.* Boston: Houghton Mifflin Company, 1995.

Periodicals

Jackson, J. A. "Red-cockaded Woodpecker *Picoides borealis.*" *Birds of North America* 85 (1994).

Other

Bent, A. C. *Life Histories of North American Woodpeckers.* United States National Museum Bulletin 174, 1939.

Lawrence, L. D. K. *A Comparative Life-History Study of Four Species of Woodpeckers.* AOU Ornithological Monographs No. 5, 1967.

Short, L. L. *Woodpeckers of the World.* Greenville, Delaware: Delaware Museum of Natural History, Monograph Series Number 4, 1982.

Tanner, J. T. *The Ivory-billed Woodpecker.* Research Report Number 1, New York: National Audubon Society, 1942.

Jerome A. Jackson, PhD

Passeriformes

(Perching birds)

Class Aves

Order Passeriformes

Number of families Approximately 74

Number of genera, species Approximately 1,161 genera and 5,700 species

Photo: Long-tailed titmice (*Aegithalos caudatus*). (Photo by Hans Reinhard. Bruce Coleman Inc. Reproduced by permission.)

Evolution and systematics

With more than 5,700 species—approximately 59% of the total number of bird species—passerines make up the single largest order of birds in the world. Indeed, many researchers roughly divide all birds into two major categories: passerines, and everything else.

Other than the large size of the order, however, there is very little that can be generally said about passerines. Members of this species-rich order are usually small, morphologically uniform, terrestrial birds that eat mainly seeds, fruit, nectar, and/or insects. Australian lyrebirds (*Menura*) and ravens (*Corvus corax*; weight approximately 60 oz [1,700 g]) are the largest members of the order, and bushtits (*Psaltriparus*) and pygmy tits (*Psaltria exilis*) are the smallest.

Nearly all taxonomists agree that Passeriformes are monophyletic, meaning they share a common ancestor, but beyond that there is little agreement about the evolutionary history and genetic relationships of this order. Robert J. Raikow defines the group with five derived characters: a wing tendon architecture that is unique; except in one genus (*Conopophaga*), a distinctive palate called "aegithognathous"; unique, bundled sperm structure; and a highly specialized foot and leg that facilitate perching, with a large hallux (rear toe) that is specially arranged, deep plantar tendons, and simplified foot muscles. Researchers agree that these characters are unique to the order, but some also cite other, traditional morpho-

logical features as defining the group. These features include, among others, the arrangement of the toes (anisodactyl, or three toes forward, one toe pointing rearward), an incumbent (non-elevated), independently acting hallux, and distinctive syringeal architecture. Raikow, however, claims that all these features are more general within birds as a whole, and therefore not useful for establishing passerine monophyly.

DNA analysis may prove to be a critical tool for defining the phylogeny, or genealogical relationships, among birds, especially in the case of subdivision of the order. Except for the architecture of the syrinx and feet, passerines are remarkably similar in morphology. Differences in the syrinx have allowed for two generally recognized suborders, Tyranni (suboscines) and Passeri (oscines). However, beyond these large suborders, classification has been extremely problematic. Convergent and parallel evolution has produced structural features and behaviors that are remarkably similar in birds that are, in fact, not closely related. However, divergent features have evolved in some species that are closely related. As a consequence, many species and genera have been traditionally, but possibly incorrectly, grouped together in families on the basis of similar morphologic features. Charles Sibley, Jon Ahlquist, and Burt Monroe have proposed a reorganization of passerines, particularly oscines, using DNA hybridization techniques. However, though these techniques promise to provide additional insight into these relationships,

An Altamira oriole (*Icterus gularis*) at its nest in south Texas. (Photo by Larry Ditto. Bruce Coleman Inc. Reproduced by permission.)

it may still take some time before passerine systematics are fully understood.

Fossil record

Although Passeriformes fossils are well represented in the Miocene, there is little in the way of a fossil record for the first passerine species. The oldest known fossils of passerine origin were two bones found in Upper and Uppermost Oligocene strata in France.

There are two competing hypotheses for the time and place of origin of the passerines. The paucity of fossil evidence for passerines prior to the Oligocene and the presence of numerous early fossils in the Northern Hemisphere are used to support the hypothesis that passerines arose in Laurasia sometime after the Cretaceous extinction. However, techniques using DNA molecular clock interpretations suggest that passerines may have had a southern, Gondwanan origin in the Early Cretaceous. According to some researchers, the primitive suboscines originated from early passerines in Gondwana when South America separated from Africa. This is a hypothesis that is strongly supported by the overwhelming num-

ber of passerines found in South America today, where there are more than 3,000 species.

Suborders Tyranni and Passeri

In spite of the many confusing relationships within the order, there is still a great deal of information that is known about the phylogeny of passerines. The two broadly defined suborders, oscine and suboscines, are rather easily separated, primarily on the basis of differences in the architecture of the syrinx, the special structure that birds use to produce sound.

In oscines, the so-called songbirds, there are more than three pairs of intrinsic syringeal muscles, while the more primitive suboscines have much less elaborate musculature. In spite of the complexity of the oscine syrinx, however, the musculature is very uniform throughout the suborder, which strongly suggests a monphyletic relationship among the species. On the other hand, the syringes (plural of syrinx) of suboscines are much more variable in design, and monophyly is suggested by a unique middle ear ossicle design that is highly uniform among suboscine species.

The complexity of the syrinx, as well as the fossil record, suggests that oscines are more evolved than primitive suboscines. In fact, oscines are considered to be among the most advanced of all bird species. As a result, they are traditionally placed at the very end of taxonomic lists of avian orders.

Three groups of birds, New Zealand wrens (Acanthisittidae), Australian lyrebirds (Menuridae), and scrub-birds (Atrichornithidae), are highly problematic when it comes to classification. Prior to 1975, these were considered to be suboscine, however, some later researchers placed them in the oscine suborder. Given the primitive nature of these birds, however, this has not been entirely satisfactory, with some researchers suggesting that they should be placed in separate suborders.

Tyranni (suboscines)

OLD WORLD SUBOSCINES The most primitive suboscines (excluding the controversial New Zealand wrens) belong to three families found in the Old World: broadbills (Eurylaimidae), pittas (Pittidae), and asities (Philepittidae). Broadbills are brightly colored, arboreal, insectivors, and frugivorous birds that occur in Africa and Asia. Pittas are brightly colored, chubby, ground-dwelling carnivores found in Asia, Maylaysia, and parts of Africa and Australia. Asities resemble pittas in size, shape, and coloration, and feed on nectar and insects.

NEW WORLD SUBOSCINES New World suboscines are often subdivided into two superfamilies, the Furnarioidea (ovenbirds, antbirds, tapaculos, and woodcreepers) and the Tyrannoidea (flycatchers, sharpbills, plantcutters, cotingas, and manakins). This subdivision as a formal classification is controversial, though it is not in dispute that these two groups, informally at least, represent two major radiations of South American passerines.

Tyrant flycatchers, the largest family in the suborder with more than 300 species, have a number of different body shapes owing to the diversity of their feeding strategies. Cotingas are medium-sized frugivores with broad bills. Manakins, also broad-billed frugivores, are small, with stubby tails and wings.

Passeri (oscines)

Passeri, or songbirds, make up about four-fifths of the passerine order with anywhere from 4,100 to 4,500 species. Although extensive adaptive radiation of oscines has produced a wide variety of ecotypes, the morphological features that are used for classification in this order are very uniform, making it difficult to assign them to families and genera with certainty. Many researchers have adopted Sibley's and his colleagues' classification of oscines, though more work will undoubtedly refine the systematics even further in the future.

Sibley and Ahlquist, using DNA hybridization, have subdivided Passeri into two broad groupings, the parvorders Corvida and Passerida, each further subdivided into three superfamilies. The parvorder Corvida consists of crows and allies, further subdivided into in the superfamilies Menuroidea, Meliphagoidea, and Corvoidea. The sister group Passerida consists of Muscicapoidea (thrushes and allies), Sylvioidea (sylvoid oscines and allies), and Passeroidea (oscine weavers and allies).

A savannah sparrow (*Passerculus sandwichensis*) singing. (Photo by Dwight R. Khun. Bruce Coleman Inc. Reproduced by permission.)

Physical characteristics

In spite of the very large size of the order Passeriformes, the exceptionally large distribution, and the diversity of life histories, there are some physical characteristics that most or all of the species share. Passerines are generally small in size (exceptions being the corvids and lyrebirds), with large wings relative to their total body mass, and all possess unique sperm. Two of the most notable physical features of the group, and the reasons they are commonly called songbirds or perching birds, are a distinctive syrinx and highly specialized feet and legs.

The syrinx is found at the junction between the trachea and the two primary bronchii and is responsible for the vast range of vocalizations in birds. When air passes across two thin, clear membranes—the internal tympaniform membranes, located on either side of the junction—the membranes vibrate and produce sound. Some birds are restricted to very simple sounds like grunts or hisses because they lack muscles to control the action of the syrinx. In birds that produce more complex sounds, such as the melodic sounds we call songs, there are special muscles to control the action of the syrinx. In passerines, especially oscines, these are highly developed and include up to six pairs of intrinsic syringeal muscles as well as two extrinsic muscles.

Although the anisodactyl foot is the most common design among birds and is not unique to passerines, the design is most commonly associated with the so-called perching birds. Anisodactyl feet evolved in arboreal species for the purpose of gripping tree branches, but modern passerines use them to

exploit all kinds of environments in addition to trees, such as grasses, telephone and fence wires, feeders, or anything that offers a place to perch. Structure of the anisodactyl foot is non-webbed and consists of toes two to four pointing forward, or anterior, and the large toe, the hallux, pointing rearward, or posterior. This arrangement differs from some other non-passerine perching birds such as woodpeckers, cuckoos, trogons, and owls that have zygodactyl feet (two toes forward and two toes rearward).

The hallux of the passerine foot is incumbent, or non-elevated. (The elevated hallux is commonly found in ground-dwelling birds rather than tree dwellers, and is usually reduced in size.) When the bird sits, deep tendons in the leg and foot flex and the large, opposable (independently moving) hallux of passerines enables the foot to automatically grasp the perch, effectively fastening the foot to the perch. Birds are thus able to sleep while perched. When the tendons are relaxed, for instance when a bird rises up from the sitting position to fly, the hallux releases and the grip is opened.

Although the specialized foot is uniform among Passeriformes species, the rich diversity of bills of passerines truly demonstrates how extraordinarily successful these birds have been at exploiting virtually any available ecological niche on land. For birds, the size and shape of the bill are functional morphologies that reflect the diet of individual species. Bill morphology is tremendously variable within Passeriformes. Their widely diverse diet has produced a variety of different

Black-throated finches (*Poephila cincta*) in eastern Australia. (Photo by Erwin & Peggy Bauer. Bruce Coleman Inc. Reproduced by permission.)

most researchers believe the order originated in the Southern Hemisphere, possibly in the Cretaceous. Fossil evidence for the first, primitive suboscines exists only from the early Tertiary. Sometime during the Middle Miocene, an explosive adaptive radiation occurred in which passerines quickly established their modern distribution patterns.

Old World suboscine birds (Pittidae, Eurylaimidae, and Philepittidae) occur in tropical Africa and Asia, including the Philippines and Sumatra. New World suboscines in the suborder Tyranni are found in Central, South, and North America, Africa, Asia, Australia, New Zealand, Madagascar. Oscines are found throughout the world.

Habitat

Given the large numbers and diversity of Passeriformes, it is not at all surprising that they have managed to exploit a variety of habitats and ecological niches, far more than any other order of birds. They are found in grasslands, woodlands, scrublands, forests, deserts, mountains, and urban environments, and in arid to wet, temperate to tropical climates. In short, almost anywhere there is habitat without a permanent snow cover, one or more established passerine species can be found.

It is difficult to describe a typical habitat for passerines. Old World warblers, which consist of more than 350 species in the family Muscicapidae, can be found in virtually all terrestrial habitats. In general, however, passerines are arboreal birds that can be found in any woodland and forest setting. Also, there are numerous species that are found in what could be called atypical habitat, though they are certainly common enough within the order. These include species like the savannah sparrow (*Passerculus sandwichensis*), a bird of the grasslands, or the rock wren (*Salpinctes obsoletus*), a musical songbird that prefers to make a nest in the crevices provided by a rocky landscape.

bill types that ranges from tiny, needle-like bills of insect-eating warblers and vireos, to the generally massive, vise-like bills of finches, designed to crack the hard shells of seeds.

Isolated islands provide a wonderful breeding ground for speciation, or variability that arises from a single ancestor species. Another example of the rapid speciation that occurs under these conditions is found in the case of the Hawaiian honeycreepers, family Drepanididae. In the case of the honeycreepers, the many small mountains and valleys of the Hawaiian Islands have created small pocket environments that have allowed the finches to evolve from a single, primitive ancestral finch into several divergent species, a process called allopatric speciation. These birds also provide us with an excellent example of convergent evolution as some species of these finches evolved bills that mimic bills of the true honeycreepers of Central and South America.

Distribution

Passerines have a phenomenally widespread distribution and can be found on all continents except Antarctica. The historical distribution of passerines is somewhat unclear, though

Behavior

If a general behavior exists that can be associated with Passeriformes, it would probably be their complex, even melodic vocalizations. The physical differences in suboscine and oscine syrinxes have been discussed above, and primarily center on the architecture of the musculature. However, the differences between the two suborders extend beyond the physical characteristics of the syrinx. Oscines, in general, have more complex vocalizations than suboscines, and tend to learn their song repertoires through mimicry. There are three ways in which avian song can be acquired: through learning, inheritance, and invention. All birds have an innate ability to vocalize, although non-passerines, except for parrots and their relatives (Psittaciformes) and hummingbirds (Apodiformes), do not possess the ability to learn song. But even within passerines, learning of song repertoires tends to be more of an oscine specialty.

Passerines also make great mimics: 15–20% of the order practice some form of vocal copying. Many songbirds will listen to the songs of competing males within their own species and then incorporate parts of those repertoires into their own. Several species are famous for their accomplished mimicry, including northern mockingbirds, bowerbirds, European starlings, Australian lyrebirds, and scrub-birds. These birds are also capable of copying the sounds of insects, frogs, and mechanical sounds.

Feeding ecology and diet

Passerines, being generally small to medium-sized, have a high basal metabolic rate. Consequently, passerines need to feed often, and on a high-energy diet in order to maintain their energy reserves. Their diet, as a whole, is as diverse as the order.

There is a wide variety of feeding strategies among passerines, and these can range from the gleaning of insects from bark by creepers, to the hawking of insects by flycatchers, and to the specialized seed eating by the finches. Many species are highly opportunistic, like the truly omnivorous Corvidae, that have been known to feed on anything from carrion to potato chips.

Most species eat their food as they find it; some, however, will store their food for later consumption. Shrikes have a particularly unusual strategy for storing food. They impale prey, which are usually insects, but also small birds or lizards, on a thorn or barbed wire. This unusual practice has earned them the rather gruesome nickname of "butcher bird."

Dippers, so named for their habit of bobbing up and down while perched on a rock, are truly odd passerines in both their feeding ecology and habitat. Though they are undisputedly oscines, dippers qualify as water birds rather than land birds because they forage on aquatic invertebrates. With their stubby wings, thick down, strong legs and toes, and specialized eyes that can see both above and below the water, dippers are uniquely designed to walk or swim along the bottoms of fast-flowing streams and rivers.

Reproductive biology

All passerines are altricial, which means that they are born naked, blind, and completely helpless, making them totally dependent on the parents for food. Chicks grow very rapidly, however, and fledge between eight and 45 days.

Altricial eggs are small, and usually colored or otherwise marked. There are anywhere from one to 16 eggs in passerine clutches and incubation usually lasts around 14 days, but can last up to 28 days in some large species, and up to 50 days in lyrebirds. Some passerines are indeterminate layers, that is, they are capable of replacing eggs that are lost or destroyed.

Passerines as architects and builders

One of the great mysteries in ornithology has been the question of why there are so many passerines in the world. One interesting hypothesis looks to another example of rich passerine diversity, the nest, as a possible explanation. Some

researchers believe that a remarkable ability to build elaborate nests out of a wide variety of materials and, in many cases, the skill to camouflage them coupled with the ability to utilize a diversity of nest sites, has provided passerines with a decided evolutionary advantage over non-passerines.

Passerines, more than any other group of birds, are notable for the diversity and architectural complexity of their nests. Nicholas Collias has suggested that the incredibly rich diversity of modern passerine nests stems from only three nest types utilized by the earliest members of the order: hole nests, nests open above, and domed nests with constructed roofs. Modern passerine nests are all variations on these three nest types, but they range from relatively simple to highly elaborate. Although other bird species may also utilize these types of nest structures, species of Passeriformes have done so with unparalleled frequency. The domed nest, in particular, is found with much more frequency among passerine birds than among non-passerines.

There is such an incredible diversity of nest structures among the Passeriformes that it is impossible to discuss any of one type as being passerine in nature. A few, non-open nests, however, are worth noting simply for their ingenious architecture. Oscines, especially, are known for their remarkable nest structures. In particular, weavers (Ploceidae) are certainly among the champions of building nests. Most passerine nests are constructed from grass and twigs and have some sort of coherence because the nest material interlocks. But the weavers take this construction technique to a level that is unmatched by any other family of birds by literally weaving or stitching grass blades into a grass "fabric" that becomes the walls and floor of the nest.

Though there are more than 100 species of weavers, there are essentially only four nest shapes that are utilized: globular, kidney-shaped, retort-shaped with a funnel-shaped entrance, and retort-shaped with a much longer, hanging tunnel entrance. Construction of a weaver nest is a highly ritualized, six-stage event. It begins with knotting grass to form clumps on two separate, vertical stems or twigs, and then proceeds through constructing a ring between the two clumps, building up the roof and walls, adding a floor, and reinforcing the walls and floor. Finally, a platform is added.

Ovenbirds (Furnariidae) are also well known for their elaborate nests. With more than 200 species, there are numerous variations on the nest structure. Although they are all domed, nests range in design from excavated burrows in stream banks, to adobe-like ovens, to virtual avian mansions with many rooms. The most studied ovenbird, the rufous hornero (*Furnarius rufus*) of southern and central South America, builds the type of nest for which the family is named. A heavy, domed structure built of clay, the nest closely resembles the primitive clay ovens that are common throughout Latin America. The oven-like nest consists of an antechamber, or vestibule, that is separated by a wall from another room in which the eggs reside. The nest is around 9 in (23 cm) tall and a little bit wider at the base, with walls about 1 in (2.5 cm) thick. Horsehair, fibrous rootlets, and cow dung are common materials used as binders for the mud used to shape the walls. The nest is usually built on top of some other structure such as a fence post or roof, and away from the ground. In spite of being nearly indestructible when construction is complete, birds build new nests each year and abandon the old ones. However, Skutch reports that if a pair of birds particularly likes a site, it will build a second, or sometimes even a third, nest on top of the first.

Swallows and martins (Hirundinidae) also build nests of mud, often in association with humanmade structures. These nests are generally small, and may be cup-shaped or retort-shaped. Construction of the nests is sturdy enough that they may be used for many years.

Though the architectural diversity of the passerine nest certainly may have contributed to the enormous radiation of the order, nest site selection has played an equally important role. That many species are opportunistic in site selection probably has increased species survival. The wren family (Troglodytidae) illustrates both the diversity of nest types as well as the variety of sites. Cactus wrens (*Campylorhynchus brunneicapillus*) bury their football-shaped globular nests deep inside the thorns of cacti to discourage predators. Rock wrens (*Salpinctes obsoletus*) build their nest in holes in the ground, the crevices of rocky outcrops, or similar small shelters. Nests are usually cup-shaped structures, made of grasses, rootlets, or bark, and lined with finer materials like animal hair or spider webs. House wrens (*Troglodytes aedon*) also make small, cup-shaped nests, but evidence of their creativity in site selection appears in photographs showing a house wren nest tucked inside an unusual humanmade item such as an old shoe or flowerpot. Nests are sometimes constructed for purposes other than brooding. Male wrens will occasionally build several decoy nests in their territory, believed to discourage predators searching for eggs.

Verdins (*Auriparus flaviceps*, family Remizidae) are scrappy little birds of the North American desert Southwest, also known to build several nests each year, some of which are used for hibernating during winter months. A separate nest is used for brooding.

Other structures built by birds are unrelated to either brooding or sleeping, and these might be the oddest of all. Bowerbirds (Ptilonorhynchidae) of Australia and New Guinea are, as a group, uniformly plain-looking birds. This has not, however, been an insurmountable problem for male bowerbirds that have turned to constructing elaborately decorated bowers, or secluded retreats, in order to attract and court females through visual stimulation. Gill reports that some researchers seem to have identified an apparent relationship between the absence of showy plumage and the ornate, somewhat fussy constructions fashioned by bower males. It seems that the plainer the plumage of the species, the more exquisitely bedecked the bower. Bowers come in two types: simple to complex mats of sticks built around a slender sapling, called maypole bowers, and bizarre tower structures that line the south side of a display court, called avenue bowers. The bowers, and the paths leading up to them, might be embellished with bright objects like fresh flowers (live orchids are popular), feathers, blue berries, yellow leaves or straw, or even, if the opportunity arises, shiny humanmade objects like coins or car keys, stolen from nearby camps. The importance of the decorations to some species cannot be underestimated. Some adornments are so highly prized that competing males often steal them from other bowerbirds in order to use them for their own structures.

Towers and mats are also often painted by the bowerbirds with shredded grass or charcoal dust and saliva or pulp from fruits. Bowerbirds are very particular about the arrangements of these decorations and will quickly tidy up anything that has been disturbed. Wilted flowers and leaves also are discarded and replaced on a daily basis.

Resources

Books

Brooke, M. "Order: Passeriformes." In *The Cambridge Encyclopedia of Ornithology*, edited by M. Brooke, and T. Birkhead. Cambridge: Cambridge University Press, 1991.

Catchpole, C. K., and P. J. B. Slater. *Bird Song: Biological Themes and Variations.* Cambridge, Cambridge University Press, 1995.

Ehrlich, P. R., D. S. Dobkin, and D. Wheye. *The Birder's Handbook.* Cambridge: Cambridge University Press, 1995.

Feduccia, A. *The Origin and Evolution of Birds.* 2nd ed. New Haven: Yale University Press, 1999.

Gill, F. B. *Ornithology.* 2nd ed. New York: W. H. Freeman and Company, 1995.

Short, L. L. *The Lives of Birds.* New York: Henry Holt and Company, 1993.

Sibley, C. G., and J. E. Ahlquist. *Phylogeny and Classification of Birds.* New Haven: Yale University Press, 1990.

Sibley, C. G., and B. L. Monroe Jr. *Distribution and Taxonomy of Birds of the World.* New Haven: Yale University Press, 1990.

Skutch, A. *Antbirds and Ovenbirds.* Austin: University of Texas Press, 1996.

Periodicals

Bledsoe, A. H. "Nuclear DNA Evolution and Phylogeny of the New World Nine-primaried Oscines." *Auk* 105 (1988): 559–571.

Collias, N. "On the Origin and Evolution of Nest Building by Passerine Birds." *Condor* 99 (1997): 253–270.

Conway, C. J., and T. E. Martin. "Evolution of Passerine Incubation Behavior: Influence of Food, Temperature, and Nest Predation." *Evolution* 54, no. 2 (2000): 670–685.

Cracraft, J. "Avian Evolution, Gondwana Biogeography and the Cretaceous-Tertiary Mass Extinction Event." *Proc. Royal Society London, Series B* 268 (2001): 459–469.

Resources

Olson, S. L., R. J. Raikow, and A. H. Bledsoe. "Why So Many Kinds of Passerine Birds?" *BioScience* 51, no. 4 (2001): 268–70.

Raikow, R. J., and A. H. Bledsoe. "Phylogeny and Evolution of the Passerine Birds." *BioScience* 50, no. 6 (2000): 487–499.

Sheldon, F. H., and D. Winkler. "Nest Architecture and Avian Systematics." *Auk* 116, no. 4 (1999): 875–877.

Susan L. Tomlinson, PhD

▲
Broadbills
(Eurylaimidae)

Class Aves

Order Passeriformes

Suborder Tyranni (Suboscines)

Family Eurylaimidae

Thumbnail description
Small to medium-sized birds, some very colorful, most with a broad head, broad bill wide gape, and large eyes

Size
4.5–10.8 in (11.5–27.5 cm); 0.35–6.0 oz (10–171 g)

Number of genera, species
8 genera; 15 species

Habitat
Mostly humid tropical lowland or montane forest

Conservation status
Vulnerable: 3 species; Near Threatened: 3 species

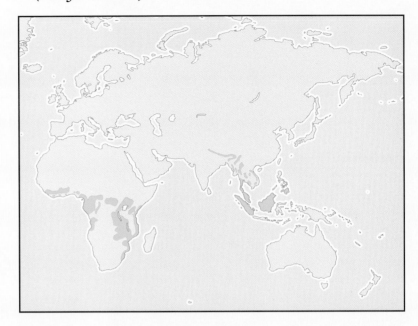

Distribution
Sub-Saharan Africa and Southeast Asia including Hainan Island, Borneo, Sumatra, and Java

Evolution and systematics

The broadbills are a small family with only 15 species and eight genera. Their taxonomic position was somewhat of a mystery for most of the twentieth century, but recent phylogenetic studies using morphological and molecular (DNA sequencing) data have shown that the broadbills form a monophyletic group with the pittas (Pittidae) and asities (Philepittidae). Together, these three families comprise the Old World suboscines, a group that shares a tracheobronchial syrinx, or vocal apparatus. Richard Prum suggests that the asities are the sister group of the African green broadbill (*Pseudocalyptomena graueri*). His Eurylaimidae thus consists of five subfamilies: Smithornithinae, Calyptomenanae, Eurylaiminae, Pseudocalyptomenanae, and the Philepittinae.

The eight genera in this family are quite distinct and do not appear to be closely related. They may be the last survivors of a once much more diverse group that was slowly replaced by a radiation of oscine songbirds. The origin of the broadbills is still a matter of speculation. Suboscines probably evolved on Gondwana, so it seems reasonable to hypothesize an African origin (giving rise to the ancestors of *Smithornis*) following the break up with South America. Broadbills then diversified by spreading into Asia (*Calyptomena, Eurylaimus, Corydon, Serilophus*) and then back to Africa (*Pseudocalyptomena* and asities).

Physical characteristics

Broadbills are small to medium-sized birds ranging from 4.5 to 10.8 in (11.5–27.5 cm) in length and weighing 0.35–6.0 oz (10–171 g). Broadbills share a host of characters in their syrinx and hindlimb musculature. Except for *Calyptomena* they are unique among the passerines in having 11 primaries. Most species have an exceptionally wide bill and gape. The dusky broadbill (*Corydon sumatranus*) beats all records among the passerines with a pink, hooked bill that is as wide as it is long. It even surpasses the skull in width. The more frugivorous *Calyptomena* and *Pseudocalyptomena* have a much narrower bill, but have retained a wide gape.

In terms of plumage coloration, the broadbills are a diverse group. Apart from a few dull-colored species, most broadbills are quite colorful, spanning the range from blue to red. Some species also have an area of bare skin around the eyes that is blue in the two Philippine *Eurylaimus* and pink in the dusky broadbill. Whereas the two sexes look alike in some species, more or less subtle difference can be found in others. Sexual dimorphism is most obvious in *Calyptomena*. Here the males are more intensely colored and the loral feathers form a forward facing tuft that covers the base of the bill, resembling a small crest.

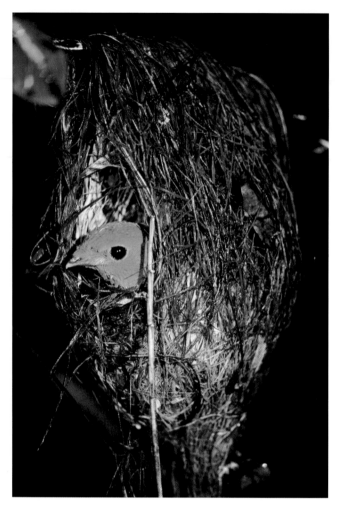

A green broadbill (*Calyptomena viridis*) in its nest. (Photo by W.K. Fletcher. Photo Researchers, Inc. Reproduced by permission.)

Distribution

The highly fragmented distribution of broadbills is limited to tropical and subtropical sub-Saharan Africa, Himalayan India, Thailand, Cambodia, Vietnam, extreme southern China (also Hainan Island), Borneo, Sumatra, Java, peninsular Malaysia, and the Philippines. *Smithornis* and *Pseudocalyptomena* are the only genera occur in Africa. Hose's broadbill (*Calyptomena hosii*) and Whitehead's broadbill (*C. whiteheadi*) are restricted to Borneo and the Mindanao and Visayan wattled broadbill (*Eurylaimus steerii* and *E. samarensis*) are restricted to a few islands in the Philippine archipelago.

Habitat

Broadbills are inhabitants of humid forests. Only the African broadbill uses somewhat drier areas. Interestingly, most broadbills inhabit mountainous terrain. Those species that primarily occur in the tropical lowlands move seasonally into montane forests as a response to changing resource levels. Many species are tolerant of some human caused disturbance, but most require primary forest for long-term persistence.

Behavior

Little is known about broadbill mating systems and general behavior. Most species appear to be monogamous. However, the frugivores (*Calyptomena*, *Pseudocalyptomena*) may be polygynous as they appear to form leks. Territoriality is another area that requires more research. Most broadbills join single or mixed species flocks, but it is still unclear if they remain in, or always return to, the same territory. There is evidence, however, that *Calyptomena* ranges over wide areas in search of fruit.

Broadbills perform a variety of displays that may be related to territory maintenance or courtship. The best-described species are the *Smithornis* broadbills, which make characteristic elliptical flights, and the green broadbill, which has a series of complex displays. The songs are rather uncomplicated, consisting of whistles, trills, dove-like cooing, and variable series of notes described with different qualities, from bubbly to screaming.

Feeding ecology and diet

Most broadbills are insectivorous. Some species also eat small vertebrates, such as lizards and small fish. The three *Calyptomena* species are frugivorous, depending to a large degree on figs. Fruit also makes up a large component of the African green broadbill's diet. Most of the food is either gleaned from leaves or branches, or caught in flight. Some broadbills (*Smithornis*, *Cymbirhynchus*) also drop to the ground to catch prey.

Reproductive biology

The reproductive season depends mostly on the local rainfall regimes. Some species tend to nest during the dry season and others during the rainy season. All broadbills make pendant nests. Common nest materials include fibers, strips of leaves from monocotyledonous plants, such as grasses, bamboo, and palms, and other leaves. *Smithornis* also interweave black fungal strands. Spider webs and co-

The colorful bill of a black-and-red broadbill (*Cymbirhynchus macrorhynchos*). (Photo by Doug Wechsler/VIREO. Reproduced by permission.)

coons, moss, and other materials camouflage the nests. Except in *Smithornis* and *Calyptomena*, nests are suspended from the tips of branches, and are hard for predators to reach. Locating the nests above water further deters predators. *Cymbirhynchus* suspends its nests 5–26 ft (1.5–8 m) above rivers or other water bodies. While protected from predators, nests in the lower range are often destroyed by rising water levels.

Two to six eggs are laid. With the possible exception of *Calyptomena*, both sexes build the nest. Male parental care appears to be common in most species. The dusky and long-tailed broadbills may even be cooperative breeders as more than two individuals of the species have been observed around nests. As an interesting side note, female green broadbills' heads protrude from the nest entrance. Why? Nobody knows. Much needs to be learned about broadbill reproductive behavior.

Conservation status

Six out of 15 species are on the IUCN Red List. Three species are considered Vulnerable and three Near Threatened. With no more than 10,000 individuals each, the vulnerable species all have tiny ranges that are threatened by deforestation, mining activities, and/or guerilla warfare (on Mindanao, Philippines). Deforestation of lowland forest threatens Hose's, Whitehead's and black-and-yellow (*Eurylaimus ochromalus*) broadbills.

Significance to humans

Most broadbills are colorful birds. However, little information is available on their significance in the pet trade. The silver-breasted broadbill (*Serilophus lunatusis*) is sold at local markets in Thailand. The African broadbill has the distinction of being the first suboscine whose entire mitochondrial genome has been sequenced.

1. Male Hose's broadbill (*Calyptomena hosii*); 2. Male silver-breasted broadbill (*Serilophus lunatus*); 3. Female black-and-yellow broadbill (*Eurylaimus ochromalus*); 4. Long-tailed broadbill (*Psarisomus dalhousiae*); 5. Male African broadbill (*Smithornis capensis*); 6. Male Visayan wattled broadbill (*Eurylaimus samarensis*); 7. Dusky broadbill (*Corydon sumatranus*); 8. African green broadbill (*Pseudocalyptomena graueri*); 9. Black-and-red broadbill (*Cymbirhynchus macrorhynchos*). (Illustration by Bruce Worden)

Species accounts

African broadbill
Smithornis capensis

SUBFAMILY
Eurylaiminae (Smithornithinae)

TAXONOMY
Platyrhynchus capensis A. Smith, 1840, Natal (coastal forests of northern Zululand). Nine subspecies are recognized.

OTHER COMMON NAMES
French: Eurylaime du cap; German: Kapbreitrachen; Spanish: Pico Ancho Africano.

PHYSICAL CHARACTERISTICS
4.7–5.5 in (12–14 cm); 0.7–1.1 oz (20–31 g). Brownish head and upperparts. Underparts buffy streaked with blackish.

DISTRIBUTION
S. c. capensis: South Africa in coastal Natal and southern Zululand. *S. c. camerunensis*: Cameroon, Gabon, Central Africa. *S. c. delacouri*: Sierra Leone, Liberia, Ivory Coast, and Ghana. *S. c. albigularis*: Central Africa, in northern Malawi, northern Zambia, Democratic Republic of Congo, Tanzania, and isolated in Angola. *S. c. meinertzhageni*: highlands of northeastern Democatic Republic of Congo and adjacent Rwanda and Uganda, western Kenya. *S. c. medianus*: highlands of central Kenya and northeastern Tanzania. *S. c. suahelicus*: from southeastern Kenya south to Mozambique, as far inland in Tanzania as the Uluguru and Nguru Mountains. *S. c. conjunctus*: from southern Angola through northeastern Namibia to northwestern Mozambique. *S. c. cryptoleucus*: from southwestern Tanzania

and southern Malawi south to South Africa (Zululand). Although more common below 2,300 ft (c. 700 m) elevation, it can be found as high as 8,000 ft (2,440 m) in the Usumbara Mountains, Tanzania.

HABITAT
Variable; generally inhabits the understory of primary and secondary forests, riparian forests, a variety of woodlands, dense thickets and brush, disturbed areas, montane forests, and open agricultural lands.

BEHAVIOR
Territorial. Both sexes commonly perform elliptical display flights. During courtship both birds face each other on a horizontal branch and flick their wings, changing between a perching and a hanging position.

FEEDING ECOLOGY AND DIET
Insectivorous. Forage by sallying and occasionally tumbling to the ground.

REPRODUCTIVE BIOLOGY
Breeding season very variable depending on locality. Lays two to three eggs.

CONSERVATION STATUS
Not threatened. Locally common, but scarce in some areas. This species has suffered from habitat destruction in some places.

SIGNIFICANCE TO HUMANS
None known. ◆

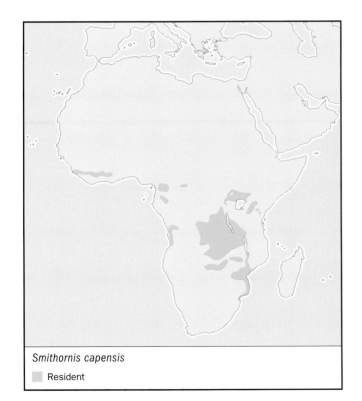

Smithornis capensis
▨ Resident

Hose's broadbill
Calyptomena hosii

SUBFAMILY
Calyptomeninae

TAXONOMY
Calyptomena hosii Sharpe, 1892, Mt. Dulit, Borneo.

OTHER COMMON NAMES
English: Blue-bellied broadbill, magnificent green broadbill; French: Eurylaime de Hose; German: Azurbreitrachen; Spanish: Pico Ancho Magnífico.

PHYSICAL CHARACTERISTICS
7.5–8.3 in (19–21 cm); female, 3.3 oz (92 g); male, 3.6–4.1 oz (102–115 g). Green head and upperparts with black spots about head and neck, and on wings. Deep blue underparts.

DISTRIBUTION
Endemic to Borneo. Most commonly found between 2,000 and 4,000 ft (610–1,220 m), but also as low as 1,000 ft (300 m) in Sabah and as high as 5,512 ft (1,680 m) in the Kelabit Uplands.

HABITAT
Understory and midstory of submontane forest; locally in lowland rainforest.

BEHAVIOR
Little known. Occurs in pairs or small groups.

Calyptomena hosii
▪ Resident

Cymbirhynchus macrorhynchos
▪ Resident

FEEDING ECOLOGY AND DIET
Feeds mostly on fig fruits, but also some insects and leaf buds.
Small groups can be seen aggregating in fruiting fig trees.

REPRODUCTIVE BIOLOGY
Breeding probably begins in March to April.

CONSERVATION STATUS
Rare to locally common in very restricted distribution. Considered Near Threatened.

SIGNIFICANCE TO HUMANS
None known. ◆

Black-and-red broadbill
Cymbirhynchus macrorhynchos

SUBFAMILY
Eurylaiminae

TAXONOMY
Todus macrorhynchus Gmelin, 1788, no locality. Four subspecies recognized.

OTHER COMMON NAMES
French: Eurylaime rouge et noir; German: Kellenschnabel;
Spanish: Pico Ancho Negro y Rojo.

PHYSICAL CHARACTERISTICS
8.3–9.4 in (21–24 cm); 1.8–2.7 oz (50–76.5 g). Black head,
back, and tail feathers. Red underparts and on rump and
throat. Black wings with white stripe. Bill is light blue above
and yellow underneath.

DISTRIBUTION
C. m. macrorhynchus: Borneo. *C. m. affinis*: western Myanmar.
C. m. malaccensis: Myanmar, southern Thailand, southern Laos,

south Vietnam, and peninsular Malaysia. *C. m. lemniscatus*:
Sumatra.

HABITAT
Evergreen forest always near water, also in degraded areas.

BEHAVIOR
Not well known. Usually found in pairs or small groups. Male
may incubate eggs.

FEEDING ECOLOGY AND DIET
Mostly insectivorous. Occasionally takes mollusks, crabs, and
small fish.

REPRODUCTIVE BIOLOGY
Reproduces mostly in dry season. Lays two to three eggs.

CONSERVATION STATUS
Not threatened; fairly common, but range contracting.

SIGNIFICANCE TO HUMANS
None known. ◆

Long-tailed broadbill
Psarisomus dalhousiae

SUBFAMILY
Eurylaiminae

TAXONOMY
Eurylaimus dalhousiae Jameson, 1835, northern India. Four subspecies recognized.

OTHER COMMON NAMES
French: Eurylaime psittacin; German: Papageibreitrachen;
Spanish: Pico Ancho de Cola Larga.

PHYSICAL CHARACTERISTICS
9.1–10.2 in (23–26 cm); 1.9–2.4 oz (53–67 g). Yellow face,
black head with blue and yellow spots. Light green underparts,

Psarisomus dalhousiae

☐ Resident

Serilophus lunatus

☐ Resident

darker green upperparts and wings. Primaries black and blue. Long blue tail.

DISTRIBUTION
P. d. dalhousiae: Himalayas to northeast India and southeastern Bangladesh, south to northern Thailand, Laos, and north Vietnam. *P. d. psittacinus*: peninsular Malaysia and Sumatra. *P. d. borneensis*: northwestern Borneo. *P. d. cyanicauda*: southern Indochine peninsula.

HABITAT
Tropical and subtropical evergreen or semi-evergreen forest up to 6,560 ft (2,000 m) elevation in Himalayas.

BEHAVIOR
Travel in flocks during nonbreeding season, but pairs tend to be secretive during the breeding season. Often sits motionless in lower canopy.

FEEDING ECOLOGY AND DIET
Insectivorous, taking prey by gleaning or sallying.

REPRODUCTIVE BIOLOGY
Start of breeding season depends on locality, but generally between March and June. Lays five to six eggs.

CONSERVATION STATUS
Not threatened; common throughout its range.

SIGNIFICANCE TO HUMANS
None known. ◆

Silver-breasted broadbill
Serilophus lunatus

SUBFAMILY
Eurylaiminae

TAXONOMY
Eurylaimus lunatus Gould, 1833 (1834), near Rangoon. Eight subspecies recognized.

OTHER COMMON NAMES
English: Gould's broadbill, collared broadbill, Hodgson's broadbill; French: Eurylaime du Gould; German: Würgerbreitrachen; Spanish: Pico Ancho de Pecho Plateado.

PHYSICAL CHARACTERISTICS
6.3–6.7 in (16–17 cm); 0.9–1.2 oz (25–35 g). Grayish head with black behind eyes. Whitish breast and wingtips. Wings alternate black and blue with some rusty color. Back is light gray, turning to dark rust colored at rump and black on the tail.

DISTRIBUTION
S. l. lunatus: Myanmar and northwestern Thailand. *S. l. rubropygius*: Nepal east to northeastern India, Myanmar. *S. l. elisabethae*: northeastern Myanmar, eastern Thailand, southern China, Vietnam, and north central Laos. *S. l. impavidus*: Bolovens Plateau, southern Laos. *S. l. rothschildi*: penisular Malaysia and southern peninsular Thailand. *S. l. intesus*: Sumatra. *S. l. polionotus*: mountains of Hainan. *S. l. stolidus*: southern Myanmar and northern peninsular Thailand. Elevational range variable, depending on location from lowlands to 7,320 ft (2,230 m) in Thailand.

HABITAT
Evergreen and semi-evergreen tropical and subtropical forests. Often associated with bamboo.

BEHAVIOR
In pairs or mixed and single species flocks. Both parents, and possibly helpers, care for brood. Some altitudinal and short distance movements.

FEEDING ECOLOGY AND DIET
Mostly insectivorous; also eats snails and small lizards. Feeds by gleaning from foliage and branches, ocassionally sallying.

REPRODUCTIVE BIOLOGY
Breeding begins in March to July depending on locality. In the north coincides with abundant rain and in the south with dry conditions. Lays four to seven eggs.

CONSERVATION STATUS
Not threatened; rare to locally common. Sold in the local pet trade in Thailand.

SIGNIFICANCE TO HUMANS
Used as cage bird. ◆

Black-and-yellow broadbill
Eurylaimus ochromalus

SUBFAMILY
Eyrylaiminae

TAXONOMY
Eurylaimus ochromalus Raffles, 1822, Singapore Island. Three subspecies recognized.

OTHER COMMON NAMES
French: Eurylaime à capuchon; German: Halsband-Breitrachen; Spanish: Pico Ancho Negro y Amarillo.

PHYSICAL CHARACTERISTICS
5.3–5.9 in (13.5–15 cm); 1.1–1.4 oz (31–39 g). Blue bill, black head, white band at throat. Whitish to rosy breast to yellow behind abdomen. Wings and tail black with yellow bands on wings, white spots on tail.

DISTRIBUTION
E. o. ochromalus: peninsular Thailand and Malaysia, Sumatra, Borneo, Riau and Lingga Archipelagoes, Belitung, Bangka, Batu, and North Natuna Islands. *E. o. mecistus*: Tuangku Island to the northwest of Sumatra. *E. o. kalamantan*: Sarawak.

HABITAT
Evergreen forests, lower montane rainforests, peat swamp and tidal swamp forests, and mixed dipterocarp forests. Also in logged forests, secondary vegetation, and overgrown plantations.

BEHAVIOR
During courtship display the male stretches its wings and wags its tail, whereas the female shivers her half-spread wings, exposing a white rump.

FEEDING ECOLOGY AND DIET
Insectivorous, eats little fruit. Catches insects by sallying.

REPRODUCTIVE BIOLOGY
Breeding between February and October depending on locality. Lays two to three eggs.

CONSERVATION STATUS
Not threatened.

SIGNIFICANCE TO HUMANS
None known. ◆

Visayan wattled broadbill
Eurylaimus samarensis

SUBFAMILY
Eurylaiminae

TAXONOMY
Sarcophanops samarensis Steere, 1890, Catbalogan, Samar.

OTHER COMMON NAMES
English: Samar broadbill, Visayan broadbill; French: Eurylaime de Steere; German: Philippinenbreitrachen; Spanish: Pico Ancho Caranculado.

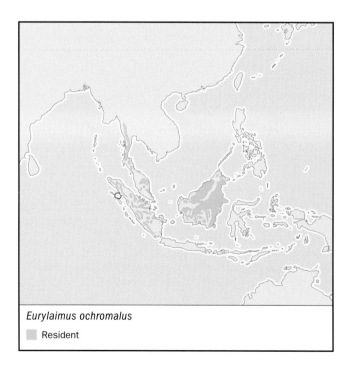

Eurylaimus ochromalus
▪ Resident

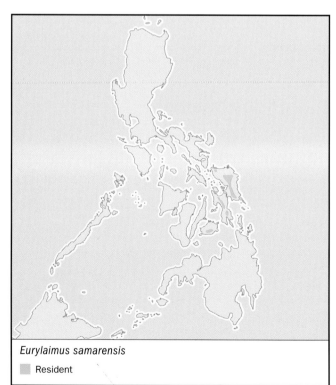

Eurylaimus samarensis
▪ Resident

PHYSICAL CHARACTERISTICS
5.7–5.9 in (14.5–15.0 cm); 1.2–1.5 oz (33.5–41.5 g). Dark red-dish head, upperparts, and tail. Black on wings and under bill. White band around throat and on wings. Upper breast is rosy, turning to white below.

DISTRIBUTION
Leyte, Samar, and Bohol in the Visayan Islands, Philippines.

HABITAT
Understory of primary forest between 330 and 1,975 ft (100–600 m).

BEHAVIOR
Little known. Usually found in pairs, small groups, or mixed-species flocks.

FEEDING ECOLOGY AND DIET
Insectivorous, may eat some fruit.

REPRODUCTIVE BIOLOGY
Probably breeds February to June.

CONSERVATION STATUS
Vulnerable. Threatened by deforestation. This species has a very small occupied range and a small population.

SIGNIFICANCE TO HUMANS
None known. ◆

Corydon sumatranus
▨ Resident

Dusky broadbill
Corydon sumatranus

SUBFAMILY
Eurylaiminae

TAXONOMY
Coracius sumatranus Raffles, 1822, interior of Sumatra. Three subspecies recognized.

OTHER COMMON NAMES
French: Eurylaime corydon; German: Reisenbreitrachen; Spanish: Pico Ancho Sombrío.

PHYSICAL CHARACTERISTICS
9.4–10.8 in (24–27.5 cm); about 4.9 oz (140 g); has exceptionally broad bill and wide gape. Black body, white at throat and white banding on tail.

DISTRIBUTION
C. s. sumatranus: Sumatra, peninsular Malaysia and Thailand, and Penang Island. *C. s. laoensis*: Patchily in Myanmar, northern Thailand, Laos, Cambodia, and Vietnam. *C. s. brunnescens*: Borneo and North Natuna Islands. Possibly up to 6,600 ft (2,000 m).

HABITAT
Canopy of rainforests and primary and logged evergreen and deciduous forests.

BEHAVIOR
Probably a cooperative breeder. Usually found in groups.

FEEDING ECOLOGY AND DIET
Feeds on large insects (up to 3.1–3.9 in [8–10 cm] in length) and lizards. Usually gleans after a short flight.

REPRODUCTIVE BIOLOGY
In northern part of the range breeding starts at the end of the dry season, in the southern part during the rainy season. Lays four to six eggs.

CONSERVATION STATUS
Not threatened, though habitat loss due to logging and deforestation may have led to a range contraction.

SIGNIFICANCE TO HUMANS
None known. ◆

African green broadbill
Pseudocalyptomena graueri

SUBFAMILY
Eurylaiminae (Pseudocalyptomeninae)

TAXONOMY
Pseudocalyptomena graueri Rothschild, 1909.

OTHER COMMON NAMES
English: Grauer's broadbill; French: Eurylaime de Grauer; German: Blaukehl-Breitrachen; Spanish: Pico Ancho Verde Africano.

PHYSICAL CHARACTERISTICS
5.35–6.14 in (13.6–15.6 cm); 1.0–1.2 oz (29.0–32.5 g). Grayish head, whitish below bill onto breast. Upper- and underparts green. Dark tail.

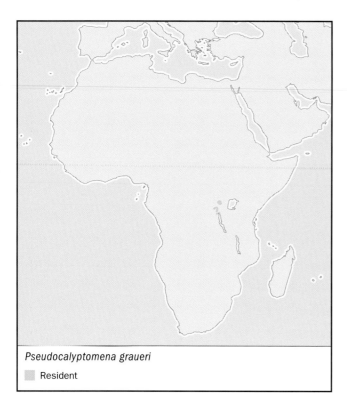

Pseudocalyptomena graueri

☐ Resident

DISTRIBUTION
Known only from a few localities in extreme eastern Democratic Republic of Congo and western Uganda.

HABITAT
Primary montane forest with dense bamboo, forest edge, and cultivated areas between 5,770 and 8,140 ft (1,760–2,480 m).

BEHAVIOR
Little known; it can be found singly, in small groups, or mixed-species flocks. May perform display flights.

FEEDING ECOLOGY AND DIET
Omnivorous, including fruit, seeds, flowers, insects, and small snails in its diet. Probably feeds by flycatching.

REPRODUCTIVE BIOLOGY
Breeding reported between April and July.

CONSERVATION STATUS
Vulnerable. Threatened in its tiny range by deforestation, commercial logging, and mining.

SIGNIFICANCE TO HUMANS
None known. ◆

Resources

Books
BirdLife International. *Threatened Birds of Asia: The BirdLife International Red Data Book*. Cambridge, UK: BirdLife International, 2001.

Lambert, Frank, and Martin Woodcock. *Pittas, Broadbills and Asities*. Sussex, UK: Pica Press, 1996.

Periodicals
Irstedt, Martin, Ulf S. Johansson, Thomas J. Parsons, and Per G. P. Ericson. "Phylogeny of Major Lineages of Suboscines (Passeriformes) Analysed by Nuclear DNA Sequence Data." *Journal of Avian Biology* 32 (2001): 15–25.

Prum, Richard O. "Phylogeny, Biogeography, and Evolution of the Broadbills (Eurylaimidae) and Asities (Philepittidae) Based on Morphology." *Auk* 110 (1993): 304–324.

Organizations
BirdLife International Indonesia Programme. P. O. Box 310/Boo, Bogor, Indonesia. Phone: +62 251 357222. Fax: +62 251 357961. E-mail: prue@burung.org Web site: <http://www.birdlife-indonesia.org>

Other
World Conservation Monitoring Center. "Threatened Animals of the World. UNEP-WCMC Animal Database." <http://www.unep-wcmc.org/>

Markus Patricio Tellkamp, MS

False sunbirds and asities
(Philepittidae)

Class Aves
Order Passeriformes
Suborder Tyranni (Suboscines)
Family Philepittidae

Thumbnail description
Medium or small birds; two species frugivorous with short bills, two species nectarivorous with long, curved bills; males have brightly colored wattles during the breeding season; plumage usually dark blue, black, or yellow

Size
3.5–6.5 in (9–16.5 cm); 0.2–1.3 oz (6–37 g)

Number of genera, species
2 genera; 4 species

Habitat
Canopy and understory of tropical rainforest

Conservation status
Endangered: 1 species; Near Threatened: 1 species

Distribution
Endemic to Madagascar

Evolution and systematics

The ecological and morphological diversity of these Madagascan endemics has caused considerable confusion over their proper classification. As currently recognized, the philepittids are a family of suboscine birds including two genera, the asities (*Philepitta*) and the sunbird-asities (*Neodrepanis*). With their frugivorous habits and short bills, the asities were originally associated with several of the oscine passerines, including the starlings, sunbirds, and birds of paradise. However, by the late 1800s, it was recognized that these birds were in fact suboscine passerines. Sunbird-asities, with their long, decurved bills and nectarivorous foraging habits, were also originally assumed to be oscine passerines and classified as sunbirds (Nectariniidae). However, it was not until 1951 that investigations of syrinx morphology demonstrated that these unique birds were also suboscines.

The placement of Philepittidae within the Old World suboscines remains problematic. The most recent morphological analysis, published in 1993 by Richard Prum, suggests that philepittids are not a true family, but should be considered a subfamily (Philepittinae) of the Eurylaimidae. Under this hypothesis, the Philepittidae represent a radiation from a broad-bill ancestor that probably originated from Africa. In contrast, a phylogeny based on nuclear DNA gene sequences published by Martin Irestedt and coworkers in 2001 found strong support for placing the Philepittidae outside the Eurylaimidae, thus justifying its classification as a true family.

Physical characteristics

Of the philepittids, the asities are the larger of the two genera, measuring 4.7–6.3 in (12–16 cm) in length. These birds are primarily frugivorous and have short bills, relatively large heads, round bodies, and short tails. In contrast, sunbird-asities, like many nectarivorous passerines, are relatively small, measuring only 3.5–4.3 in (9–11 cm) in length. They have long, curved bills, and specialized tongues for extracting nectar from forest flowers. Sunbird-asities have plump bodies and short tails.

The plumage patterns of the philepittids are generally characterized by black, dark blue, olive-green and bright yellow. All species are sexually dimorphic; females are generally olive-green and cryptic, while males have much brighter and contrasting plumage. Perhaps the most unique feature of these birds is the bright blue, yellow, and green caruncles (wattles) of bare facial skin developed by males during the breeding season. Richard Prum has performed an extensive investigation of these structures from velvet asities (*Philepitta castena*) and common sunbird-asities (*Neodrepanis coruscans*) and has demonstrated that the bright colors are produced by light reflecting from unique alignments of collagen fibers. This form of structural coloration is not known to occur in any other animals.

Distribution

Asities and sunbird-asities are endemic to the forests of Madagascar.

Habitat

These birds are found in the understory, lower levels of canopy, and occasionally in the canopy of primary rainforest, dry deciduous forest, degraded or logged forest, and in secondary forest. The genus *Philepitta* occurs in many forest habitats of Madagascar, with one species (*P. schlegeli*) found on the drier eastern side of the island and the other (*P. castanea*) found in the moister rainforest on the west side of the island. In contrast, *Neodrepanis* is found only in the rainforests on the west side of the island. These birds occur from sea level up to 5,900 ft (1,800 m), but some species have strong affinities for relatively narrow altitudinal zones.

Behavior

Philepittids may be found either alone, in pairs, or in small groups. Several individuals may occur together at food resources, such as flowering or fruiting trees. Such aggregations may lead to intraspecific displays and defense of resources, especially among *Neodrepanis*. Philepittids will also join mixed-species foraging flocks when feeding on fruit or insects.

Males call frequently, especially during the breeding season, and a wide variety of male displays has been recorded. Much of this behavior appears to be the basis of a lek mating system. There is no evidence that philepittids migrate, but the possibility that some species are altitudinal migrants cannot be ruled out.

Feeding ecology and diet

The asities in the genus *Philepitta* are primarily frugivorous, but also feed on nectar and insects. Fruits are consumed while the birds are perched. Although *Philepitta* does not have the specialized bill morphology of *Neodrepanis*, nectar is probably an important component of the diet. Initial investigations of tongue morphology have suggested that a brush-like tip and the ability to roll it into a tube-like form may increase the efficiency with which nectar can be consumed. Pollen may also be an important food item.

Sunbird-asities (*Neodrepanis*) are specialized for consuming nectar. These birds have long, decurved bills for entering flower corollas and long, tube-like tongues for the efficient extraction of nectar. They feed on the nectar of a variety of flowers including Balanophoraceae, Balsaminaceae, Loranthaceae, Rubiaceae, Clusiaceae, Melastomataceae, and Zingiberaceae. Insects and spiders are also an important component of their diet. These species have been observed searching in moss and along branches for invertebrates and fly-catching small insects.

Reproductive biology

Although there are few breeding records for any of the philepittids, it appears that most nesting attempts are made between July and January. Local breeding seasonality may be timed so that young fledge at the onset of the rainy season.

Most observations suggest that the philepittids are polygynous. The most convincing evidence comes from detailed observations of the velvet asity. The males of this species defend territories no more than 66–100 ft (20–30 m) in diameter, within which males perform three displays, including wing-flapping, an open-mouth display, and an open-mouth display coupled with hanging inverted from their perch. Other observations, such as groups of male Schlegel's asities (*Philepitta schlegeli*) calling simultaneously and an upside-down open-mouth display by the yellow false-sunbird, suggest that a polygynous mating system may be widespread in this family.

All philepittids build ragged, pendulant nests hung from the low branches of trees. Nests are built of moss, palm fibers, dead leaves, and fine twigs. Based on observations of velvet asities and common sunbird-asities, nest construction is by the female alone and proceeds in an unique fashion: a complete orb is first constructed, then, once finished, the female pokes a hole through the side to form the entrance.

Clutch size varies from two to three eggs. There is no information on incubation or nestling periods. For most species that have been observed, the female incubates and feeds the young.

Conservation status

The philepittids are especially vulnerable to the widespread destruction of forested habitats occurring on Madagascar. The most seriously threatened species is the yellow-bellied sunbird-asity (*Neodrepanis hypoxantha*), which is considered Endangered. It was also added to Appendix I of CITES in 1995. This species was recently reclassified from Critically Endangered due to studies that indicate its population may be larger than previously thought. However, it remains threatened because of its very small, fragmented range and continuing decline of its habitat.

Significance to humans

For many years, these small, inconspicuous birds have lived in a remote area inhabited by few people, and have been essentially overlooked. More recently, the growing popularity of bird-watching and ecotourism has lead to an increased interest in these species.

1. Common sunbird-asity (*Neodrepanis coruscans*); 2. Velvet asity (*Philepitta castanea*). (Illustration by Dan Erickson)

Species accounts

Velvet asity
Philepitta castanea

SUBFAMILY
Philepittinae

TAXONOMY
Turdus castanea P.L.S. Muller, 1776, Madagascar.

OTHER COMMON NAMES
French: Philépitte veloutée; German: Seidenjala; Spanish: Asitis de Terciopelo.

PHYSICAL CHARACTERISTICS
5.5–6.5 in (14–16.5 cm); 1.1–1.3 oz (31.5–37 g). Breeding male: almost completely black with greenish caruncle over each eye and blue stripe between caruncle and eye. Female: olive-green with pale yellow blotches on underparts.

DISTRIBUTION
Eastern Madagascar.

HABITAT
Understory, lower levels of canopy, and occasionally in the canopy of primary rainforest, degraded or logged forest, and in secondary forest; from sea level up to 5,900 ft (1,800 m).

BEHAVIOR
Often forages in small groups or pairs, but sometimes alone. May also join mixed-species flocks.

FEEDING ECOLOGY AND DIET
Primarily frugivorous (e.g., Melastomataceae, Pittosporaceae, and Rubiaceae), but also feeds on nectar and insects.

REPRODUCTIVE BIOLOGY
Breeds July–January; northern birds breed earlier than southern birds. Pear-shaped nest, woven of moss and palm fibers and lined with leaves, is hung from a low branch in a shaded location. Usually lays three white eggs. Only the female has been observed to incubate, but both males and females have been recorded feeding young.

CONSERVATION STATUS
Not threatened, and common in suitable habitat. However, loss of rainforest habitat in Madagascar may threaten its long-term persistence.

SIGNIFICANCE TO HUMANS
None known. ◆

Common sunbird-asity
Neodrepanis coruscans

SUBFAMILY
Neodrepanidinae

TAXONOMY
Neodrepanis coruscans Sharpe, 1875, Madagascar.

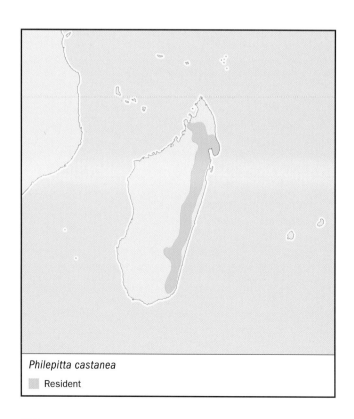

Philepitta castanea
░ Resident

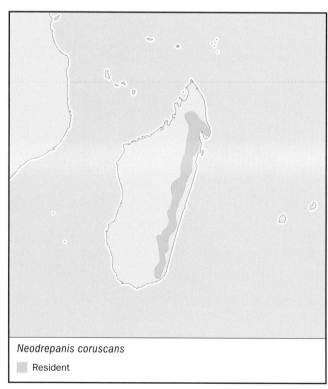

Neodrepanis coruscans
░ Resident

OTHER COMMON NAMES
French: Philépitte faux-souimanga caronculé German: Langschnabel-Nektarjala; Spanish: Asitis Caranculado.

PHYSICAL CHARACTERISTICS
3.7–4.1 in (9.5–10.5 cm); 0.2–0.3 oz (6.2–6.6 g). Small bird with long, decurved bill. Male: blue-brown upperparts, dull yellow underparts with brown streaks on breast, blue caruncle around eye. Female: brown upperparts, light brown head, dull underparts with yellow on flanks and under tail.

DISTRIBUTION
Eastern Madagascar.

HABITAT
Understory, lower levels of canopy, and occasionally in the canopy of rainforest and in secondary forest from sea level up to 4,000 ft (1,200 m).

BEHAVIOR
Forage alone in or pairs, actively moving through the understory to visit flowers; sometimes feed in association with other nectarivorous birds, such as white-eyes and sunbirds.

FEEDING ECOLOGY AND DIET
Primarily nectarivorous; observed feeding at a variety of flowers including Balanophoraceae, Balsaminaceae, Loranthaceae, Rubiaceae, Clusiaceae, Melastomataceae, and Zingiberaceae. Also eats insects and spiders, often searching in moss or on branches for these invertebrates.

REPRODUCTIVE BIOLOGY
Poorly known. Breeds August–December. Pendulant nest of moss, leaves, and twigs is hung from a low branch. One nest contained two pale green eggs. Only female has been observed incubating and feeding young.

CONSERVATION STATUS
Not threatened, and common in suitable habitat. However, loss of rainforest habitat may threaten its long-term survival.

SIGNIFICANCE TO HUMANS
None known. ◆

Resources

Books

Lambert, F., and M. Woodcock. *Pittas, Broadbills and Asities.* Sussex, UK: Pica Press, 1996.

Langrand, O. *The Birds of Madagascar.* New Haven: Yale University Press, 1990.

Periodicals

Irestedt, M., U. Johansson, T. Parsons, and P. Ericson. "Phylogeny of Major Lineages of Suboscines (Passeriformes) Analysed by Nuclear DNA Sequence data." *Journal of Avian Biology* 32 (2001): 15–25.

Prum, R. "Phylogeny, Biogeography, and Evolution of the Broadbills (Eurylaimidae) and Asities (Philepittidae) Based on Morphology." *Auk* 110 (1993): 304–324.

Prum, R., R. Torres, C. Kovach, S. Williamson, and S. Goodman. "Coherent Light Scattering by Nanostructured Collagen Arrays in the Caruncles of the Malagasy Asities (Eurylaimidae: Aves)." *Journal of Experimental Biology* 202 (1999): 3507–3522.

Prum, R., and V. Razafindratsita. "Lek Behavior and Natural History of the Velvet Asity (*Philepitta castanea:* Eurylaimidae)." *Wilson Bulletin* 109 (1997): 371–392.

Nathaniel E. Seavy, MS

Pittas
(Pittidae)

Class Aves

Order Passeriformes

Suborder Tyranni (Suboscines)

Family Pittidae

Thumbnail description
Medium-sized with round bodies, large heads, and long legs; brightly colored; terrestrial; difficult to observe

Size
5.9–11.0 in (15–28 cm); 1.6–7.1 oz (45–202 g)

Number of genera, species
1 genus; approximately 30 species

Habitat
Understory of lowland tropical forests

Conservation status
Critically Endangered: 1 species; Vulnerable: 8 species; Near Threatened: 4 species

Distribution
Primarily southeastern Asia, including Indonesia, China, Japan, and India; Australia; West and East Africa

Evolution and systematics

Original taxonomic treatments of the pittas led to their inclusion in the crow family, and subsequently in the thrush family. It was not until the early 1800s that the pittas were designated as a distinct family and correctly classified as suboscines. More recently, DNA hybridization and morphological analyses have convincingly demonstrated Pittidae is monophyletic and a sister taxon to the broadbills of Africa and Asia.

Although classification at the family level is widely accepted, there are conflicting opinions regarding the appropriate number of genera and species. Although as many as six genera have been proposed, and preliminary estimates of genetic divergence support these distinctions, most authors have chosen to recognize only the genus *Pitta*. Recent taxonomic treatments recognize 29–31 species. This number will undoubtedly change as molecular methods generate a better understanding of the evolutionary history of the Pittidae.

Physical characteristics

Secretive and rarely seen, pittas are often described as "jewels of the forest" on account of their brilliant plumage coloration. Many species are characterized by patches of red, green, purple, black, white, chestnut, and turquoise, often adjacent and sharply contrasting. In many cases these colors are on the breast, chin, or on areas of the body that can be con-

cealed by more drab-colored wing feathers, presumably in an attempt to avoid detection by predators. In most species, females and males share these brilliant colors. Cryptically colored females occur in 11 species, and in only a single species is cryptic coloration shared by males and females. In contrast, nearly all juvenile and immature birds are cryptic.

Pittas have round bodies, large heads, long legs, and short tails. These features reflect the terrestrial habits shared by all pittas. The pittas are also strikingly similar in size, with most species measuring about 8 in (20 cm) in length. Pittas have stout bills, often hooked at the tip, not unlike the bills of many thrushes (Turdidae).

Distribution

Pittas occur in Asia, Indonesia, Australia, and Africa. The greatest diversity of species is found in peninsular Malaysia, Borneo, Sumatra, and Java. Only a single species is found throughout most of India, two species occur in Africa, and only two species occur regularly in Australia.

Habitat

Most pittas are found on the ground in tropical rainforests. In many cases they appear to prefer areas that are moist, often near rivers or streams or in shaded ravines, with a rich layer of leaf litter in which to forage. Some species occur in

Pitta nest. (Illustration by Michelle Meneghini)

adjacent territories face off and perform bowing displays sometimes in conjunction with "purring" vocalizations.

When approached, a number of species give alarm calls in conjunction with distraction displays, such as flashing a conspicuous white patch of the wing, spreading the tail, or fanning out the bright feathers of the breast. In other cases threats are responded to with behaviors that may reduce conspicuousness, in which species lower their brightly colored breasts to the ground and remain motionless.

Most pittas are nonmigratory or make local movements outside the breeding season. However, Indian pittas (*Pitta brachyura*), blue-winged pittas (*P. moluccensis*), and fairy pittas (*P. nympha*), as well as a subspecies of the African pitta (*P. a. longipennis*) and populations of hooded pittas (*P. sordida*) and red-bellied pittas (*P. erythrogaster*) are migrants. Although most species migrate over land, it is believed that the fairy pitta may fly nonstop from Vietnam across the ocean to Borneo, a flight of approximately 620 mi (1,000 km)! Given the short, rounded wings of pittas, it is somewhat surprising these birds make long migratory flights.

Feeding ecology and diet

Pittas forage terrestrially, hopping along the forest floor, sometimes remaining motionless to search for exposed invertebrates, sometimes searching noisily through the leaf-litter or digging in the soil for earthworms. The primary food items are invertebrates, including spiders, a wide variety of insects, snails and slugs, and annelid worms. Some of the larger species may also take small vertebrates, including small frogs, snakes, and even mice. Seeds have also been found in the stomachs of several species, but whether fruit is regularly consumed or simply eaten from the forest floor after it is infested with insects remains unknown. Using stone "anvils" for smashing snails to remove the shells has been observed in at least six species.

Earthworms figure prominently in the diets of many pittas, especially during the nesting season. In Australia, the diet of the rainbow pitta varies seasonally; earthworms comprise most of the diet during the wet season, while other invertebrates are more important during the dry season.

Reproductive biology

Almost all pittas breed seasonally, with breeding timed to coincide with the onset of the rainy season. An exception to this pattern is the superb pitta (*Pitta superba*), which apparently nests throughout the year on the island of Manus. In most species there are relatively few unique displays prior to copulation, and most pittas probably are monogamous. However, the African pitta performs a unique display prior to the breeding season. During display bouts, this species repeatedly jumps about 10 in (25 cm) into the air, parachuting back to the perch with several shallow wing-beats. During this display the red belly is prominently displayed and the birds often give a "prrt-wheet" vocalization.

The domed nest typical of the pitta family is the size and shape of a "rugby football." Both the male and female par-

moist, montane forest up to elevations of 8,200 ft (2,500 m), but the majority of species are found near sea level. In Australia these birds use monsoon and eucalypt forest, and in Africa they inhabit rainforest and drier bush and woodlands. Although they avoid open habitats, a number of species are relatively tolerant of habitat modification, persisting in degraded forest, forest fragments, and secondary forest.

Behavior

Pittas are secretive birds, uncommonly encountered, and difficult to observe in the poor light and dense vegetation of the forest understory. As a result, there are few behavioral observations for the majority of species. With the publication of *Pittas, Broadbills, and Asities*, by Frank Lambert and Martin Woodcock in 1996 and *Pittas of the World*, by Johannes Erritzoe and Helga Boullet Erritzoe in 1998, there is now a solid foundation synthesizing the information from what few species have been studied and highlighting the large gaps in our knowledge that remain.

Pittas are found alone or in pairs and are territorial. Territories may vary widely in size depending on the species and habitat; African pitta (*Pitta angolensis*) territories may be as small as 0.75 acre (0.3 ha), rainbow pittas (*Pitta iris*) defend areas larger than 2.5 acres (1 ha), and for some species only a single pair may be found in an area as large as 50 acres (20 ha).

Pittas give short calls, usually one, two, or occasionally three syllables, which can be either whistled or buzzy. The role that these calls play in territorial defense is evidenced by the fact that many species can be drawn out of dense vegetation by playing a recording of their call. In a natural setting, such a response may lead to encounters between males on adjacent territories. For rainbow pittas and elegant pittas (*Pitta elegans*) biologists have described displays in which males from

A blue-winged pitta (*Pitta moluccensis*) suns itself. (Photo by R. & N. Bowers/VIREO. Reproduced by permission.)

ticipate in the construction of this bulky nest, which is loosely built leaves and twigs, often on top of a platform constructed of larger sticks. Although this "sloppy" construction may decrease the durability of the nests, it has been hypothesized that it may also decrease the ease with which they are detected by predators. The entrance to the nest is through a hole in the side, often facing out onto a path or other opening in the vegetation. The interior of the nest is lined with fibers or finer leaves. The nest may be located on the ground or 3–50 ft (1–15 m) above the ground in a tree or small bush. Ground nesting species often build a "door mat" of fine twigs, but the door mat built by the rainbow pitta is often constructed of mammal dung.

Clutch size varies from two to six eggs; most species lay three to four eggs. The incubation period lasts 14–16 days. For most species that have been observed, both the male and female share the task of incubation. The eggs apparently hatch asynchronously. The altricial young hatch naked, blind, and with limited mobility. The male and female share the tasks of brooding and feeding the young.

The young fledge from the nest after only 11–17 days, at which time they are usually already able to fly. They continue to be fed by the adults, usually for a week to ten days, but this period may last up to a month.

Conservation status

Many pittas have restricted ranges and depend on forested habitats that are rapidly being cleared. Additionally, the bright colors of these birds have made them popular cage birds and they are also popular targets of hunters in many parts of the world. As a result, there is considerable concern about the population viability of many pitta species. The most seriously threatened species is Gurney's pitta (*Pitta gurneyi*), which is considered Critically Endangered. It was added to Appendix I of CITES in 1995. Additionally, eight other species have been recognized by the IUCN and BirdLife International as Vulnerable: Schneider's pitta (*Pitta schneideri*), superb pitta, azure-breasted pitta (*P. steerii*), whiskered pitta (*P. kochi*), fairy pitta, black-faced pitta (*P. anerythra*), graceful pitta (*P. venusta*), and blue-headed pitta (*P. baudii*). Of these, whiskered pittas are listed in Appendix I of CITES and fairy pittas and banded pittas (*P. guajana*) are listed in Appendix II. Four species are designated as Near Threatened—the giant pitta (*P. caerulea*), Sula pitta (*P. dohertyi*), garnet pitta (*P. granatina*), and mangrove pitta (*P. megarhyncha*). Effective conservation of these species depends on habitat preservation and protection from hunting and trapping. Although this may seem an impossible task, an increasing awareness of the shrinking population of Gurney's pitta has begun to shift the economic value of this species away from illegal trade toward conservation-based ecotourism that relies on habitat preservation. Perhaps this widely publicized project can serve as a model for the protection of other threatened pittas.

Significance to humans

Their bright colors have made pittas popular birds in the wild bird trade. Pittas have also been hunted for food. This has probably been most extensive along migratory routes where pittas can be captured in large numbers, often with snares that are set in the vegetation. More recently, the growing popularity of bird-watching and ecotourism has lead to an increased interest in these species.

1. Hooded pitta (*Pitta sordida*); 2. Indian pitta (*Pitta brachyura*); 3. Superb pitta (*Pitta superba*); 4. Graceful pitta (*Pitta venusta*); 5. Gurney's pitta (*Pitta gurneyi*); 6. African pitta (*Pitta angolensis*); 7. Rainbow pitta (*Pitta iris*). (Illustration by Michelle Meneghini)

Species accounts

Gurney's pitta
Pitta gurneyi

TAXONOMY
Pitta gurneyi Hume, 1875, Tenasserim, Burma.

OTHER COMMON NAMES
English: Black-breasted pitta; French: Brève de Gurney; German: Goldkehlpitta; Spanish: Pita de Gurney.

PHYSICAL CHARACTERISTICS
8.3 in (21 cm); approximately 1.8–2.5 oz (50–70 g). Male has black face, blue crown; white under bill and yellow band on upper breast. Underparts are black but yellow banded with black at the sides. Back and wings brownish with blue tail. Female is buffy to brownish from crown to nape; buffy underparts with light brownish banding.

DISTRIBUTION
Peninsular Thailand and Tenasserim, Myanmar.

HABITAT
Semi-evergreen rainforest, secondary forest, and degraded forest fragments; often near streams and gullies; from sea level up to 515 ft (160 m).

BEHAVIOR
Primarily terrestrial; occurs alone or in pairs. Males defend territories by calling, which may be accompanied with a "wing-flicking" display.

FEEDING ECOLOGY AND DIET
Forages for invertebrates in the leaf-litter of the forest floor and digs for earthworms with its bill. Food items include small spiders, insects, insect larvae, slugs, snails, earthworms, and frogs.

REPRODUCTIVE BIOLOGY
Breeds May to August. Domed nest is located 3–10 ft (1–3 m) above the ground, often in palm trees. Nest constructed from dead leaves and twigs on a base of larger sticks; lined with fine rootlets. Clutch size usually three to four. Eggs similar to those of the hooded pitta; white with dark purple or brownish spots over gray markings, most numerous on widest end. Female and male share incubation, brooding, and provisioning of young.

CONSERVATION STATUS
This species is considered Critically Endangered and only just survives. It has a single, very small, declining population and a similarly small, declining range. Threats to the species from habitat destruction are still compounded by trapping for the cage bird trade.

SIGNIFICANCE TO HUMANS
Historically, this species was trapped for the cage bird trade. However, with increasing awareness of the dwindling population size, the economic value of Gurney's pitta has begun to shift from illegal trade to conservation-based ecotourism. ◆

Hooded pitta
Pitta sordida

TAXONOMY
Turdus sordida P.L.S. Müller, 1776, Philippines. Twelve subspecies recognized.

Pitta gurneyi
■ Resident

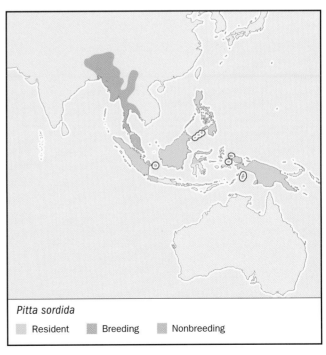

Pitta sordida
■ Resident ■ Breeding ■ Nonbreeding

OTHER COMMON NAMES
English: Green-breasted pitta, black-headed pitta; French: Brève a capuchon; German: Kappenpitta; Spanish: Pita Encapuchada.

PHYSICAL CHARACTERISTICS
7.5 in (19 cm); 1.6–2.5 oz (42–70 g). Black head and bill. Underparts greenish, with red under tail. Upperparts and wings are darker greenish; light bands on wings.

DISTRIBUTION
Widespread throughout Southeast Asia, from the foothills of the Himalayas to Indonesia, the Philippines, and New Guinea. *P. s. cucullata* is migratory; breeds in the foothills of the Himalayas to Myanmar, Yunnan, and Thailand; moves south to winter in peninsular Malaysia, Borneo, Sumatra, and Java. Eleven other subspecies, most restricted to small groups of islands.

HABITAT
Forested and wooded habitats, including primary rainforest, degraded or logged forest, secondary forest, bamboo, scrub, plantations, and even cultivated areas adjacent to forests; from sea level up to 4,900 ft (1,500 m).

BEHAVIOR
Primarily terrestrial; occurs alone or in pairs. Hops rapidly along ground to forage and to escape if disturbed, but can also fly strongly. Territorial. Displays observed include bowing, head-bobbing, wing flicking, and wing/tail fanning, possibly serving as alarm or distraction displays.

FEEDING ECOLOGY AND DIET
Forages terrestrially for insects, worms, berries, and snails.

REPRODUCTIVE BIOLOGY
Breeds February to August; varies geographically. Dome-shaped nest, usually on ground, constructed from roots, leaves (often bamboo), and twigs. Clutch size usually three, but ranges from two to five. Eggs white with gray, brown, or dark purple spots more numerous on widest end. Female and male share in nest construction, incubation, and provisioning of young.

CONSERVATION STATUS
Not threatened; common throughout much of its range where habitat is suitable.

SIGNIFICANCE TO HUMANS
None known. ◆

Pitta superba

▨ Resident

Superb pitta
Pitta superba

TAXONOMY
Pitta superba Rothschild and Hartert, 1914, Manus Island, Admiralty Islands.

OTHER COMMON NAMES
English: Black-backed pitta; French: Brève superbe; German: Mohrenpitta; Spanish: Pita Soberbia.

PHYSICAL CHARACTERISTICS
8.5 in (21–22 cm); c. 3.5–4.6 oz (100–130 g). Black head, wings, upperparts, and underparts to chest. Red under abdomen and tail; white bands on wings.

DISTRIBUTION
Island of Manus, Admiralty Islands.

HABITAT
Forested and wooded habitats, including primary forest, secondary forest, bamboo, and scrub.

BEHAVIOR
Poorly known, but assumed to be similar to other pittas. Terrestrial and secretive, occurring alone or in pairs. Territorial, responds to playback of its call.

FEEDING ECOLOGY AND DIET
Reported to feed on snails and smash them on stone "anvils."

REPRODUCTIVE BIOLOGY
Probably breeds year round, as do other forest birds of Manus. Single nest was dome-shaped; constructed from roots, leaves (including bamboo), and twigs; and contained two eggs. Eggs white with purplish gray and purplish brown spots, more numerous on the widest end.

CONSERVATION STATUS
Because this species is restricted to the island of Manus, the total population size is very small and the species is considered Vulnerable.

SIGNIFICANCE TO HUMANS
None known. ◆

Graceful pitta
Pitta venusta

TAXONOMY
Pitta venusta S. Müller, 1835, Sumatra.

OTHER COMMON NAMES
English: Black-and-scarlet pitta, black-crowned pitta; French: Brève gacieuse; German: Granatpitta; Spanish: Pita de Corona Negra.

Pitta venusta
▨ Resident

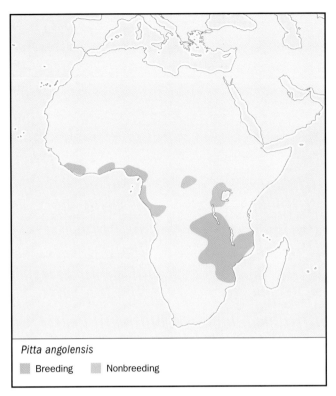

Pitta angolensis
▨ Breeding   ▨ Nonbreeding

PHYSICAL CHARACTERISTICS
5.7–7.3 in (14.5–18.5 cm); weight unrecorded. Black head with light blue feather behind eye. Black breast and red underparts. Deep red back and black wings.

DISTRIBUTION
Western Sumatra.

HABITAT
Understory and lower levels of forest, especially in most ravines; 1,300–4,600 ft (400–1,400 m).

BEHAVIOR
Terrestrial and secretive, occurring alone or in pairs.

FEEDING ECOLOGY AND DIET
Diet includes insects, snails, seeds, and worms.

REPRODUCTIVE BIOLOGY
Probably breeds February to June. A single nest was dome-shaped; constructed from roots, leaves (including bamboo), and soft, rotting material. Clutch size two to three. Eggs described as pure white or dull cream with buffish and brown spots and lines over gray-lilac markings. Markings are evenly distributed over the egg.

CONSERVATION STATUS
Considered Vulnerable as of 2000. Its small population may be decreasing due to deforestation and habitat loss.

SIGNIFICANCE TO HUMANS
None known. ◆

African pitta
Pitta angolensis

TAXONOMY
Pitta angolensis Vieillot, 1816, Angola. Three subspecies recognized.

OTHER COMMON NAMES
English: Angolan pitta; French: Brève d'Angola; German: Angolapitta; Spanish: Pita Africana.

PHYSICAL CHARACTERISTICS
6.7–8.7 in (17–22 cm); 1.6–3.5 oz (45–98 g). Black head with yellow stripe on side. Whitish under bill to yellow at breast and red under tail. Back and wings are green, with blue and black banding on wings. black tail, and blue on upper tail.

DISTRIBUTION
P. a. longipennis: migratory; breeds in central Tanzania, Malawi, southeast Democratic Republic of Congo, eastern Zambia, Zimbabwe, and possibly into northern South Africa; nonbreeding migrant in northern Tanzania, Rwanda, Burundi, Democratic Republic of Congo, Central African Republic, Uganda, and coastal Kenya. *P. a. pulih*: West Africa; resident in Sierra Leone lowlands, Ghana, Liberia, Ivory Coast, Nigeria, and coastal Cameroon. *P. a. angolensis*: West Africa; southern Cameroon, Guinea, Congo, Democratic Republic of Congo, and Angola

HABITAT
Evergreen bush, forest-like thickets along watercourses, and secondary forest; also in tall semi-deciduous and evergreen rainforest; from sea level up to 4,100 ft (1,250 m).

BEHAVIOR
Primarily terrestrial, occurring alone or, especially on breeding grounds, in pairs. Hops rapidly along ground to forage and often flies only a short distance if disturbed before dropping back to the forest floor. Territorial, often singing from the ground or low perch.

FEEDING ECOLOGY AND DIET
Foraging birds stand motionless watching for prey, then hop to a new spot to continue scanning. Periods of scanning are often followed by the pursuit of insects or other invertebrates among

the leaf-litter of the forest floor. Food items include ants, termites, beetles, insect larvae, slugs, snails, millipedes, caterpillars, and earthworms.

REPRODUCTIVE BIOLOGY
Appears to breed during the wet season. Nest is an untidy dome, placed 7–26 ft (2–8 m) above the ground, often protected by thorns. Constructed from roots, sticks, twigs, dried leaves, rootlets, and fine fibers. Clutch size usually three, but ranges from one to four. Eggs creamy-white, sometimes greenish or pinkish, with reddish brown and purplish spots and lines over gray-lilac markings, most numerous on widest end.

CONSERVATION STATUS
Not threatened. Common, especially in breeding range of East Africa, but deforestation is probably contributing to habitat loss and reductions in populations.

SIGNIFICANCE TO HUMANS
None known. ◆

Indian pitta
Pitta brachyura

TAXONOMY
Corvus brachyurus Linneaus, 1776, Sri Lanka.

OTHER COMMON NAMES
English: Bengal pitta; French: Brève du Bengale; German: Neunfarbenpitta; Spanish: Pita de Alas Azules.

PHYSICAL CHARACTERISTICS
5.9–7.5 in (15–19 cm); 1.7–2.3 oz (47–66 g). Black mask through eyes with white and brown stripe above. Buffy under bill to yellow at breast and red under tail. Back and wings green; black tail with blue tip; blue, black and white bands on wings.

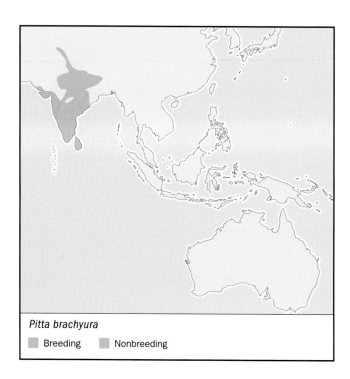

Pitta brachyura
- Breeding
- Nonbreeding

DISTRIBUTION
Breeds in Pakistan, Himachal Pradesh, southern Nepal, southern Sikkim, wet areas of Rjasthan, Kanara, and Bangladesh; and central India; nonbreeding migrant in southern India and Sri Lanka.

HABITAT
Breeds in understory of evergreen and deciduous forest, often near ravines with dense brush or bamboo; nonbreeding migrants use forested areas, including small fragments and wooded gardens; from sea level up to 5,600 ft (1,700 m).

BEHAVIOR
Primarily terrestrial, occurring alone or in pairs. During breeding, males defend territories by calling, often in conjunction with tail bobbing and moving the head forward and backward. When rival males are encountered, territorial males often extend their wings, flashing the white patches at the base of the primaries. Males also defend territories during nonbreeding season, and chase out intruding males.

FEEDING ECOLOGY AND DIET
Feeds by foraging for invertebrates in the leaf-litter of the forest floor and digging for earthworms with their bill. Food items include ants, termites, insect larvae, slugs, snails, millipedes, and earthworms.

REPRODUCTIVE BIOLOGY
Breeds May to August. Nest located on the ground or low in a small tree. Constructed from leaves, grass, twigs, and moss, and lined with grass and bamboo or tamarisk leaves. Clutch size usually four to five. Eggs glossy white, sometimes pinkish, with purplish or black spots and specks over dull lavender and purple markings, most numerous on widest end. Nest construction probably primarily by female.

CONSERVATION STATUS
Not threatened; still common throughout much of range, despite habitat loss and trapping during migration.

SIGNIFICANCE TO HUMANS
Trapped during migration for use as food, especially along the southern coast of India. ◆

Rainbow pitta
Pitta iris

TAXONOMY
Pitta iris Gould, 1842, N. Australia.

OTHER COMMON NAMES
English: Black-breasted pitta; French: Brève iris; German: Rogenboenpitta; Spanish: Pita Arco Iris.

PHYSICAL CHARACTERISTICS
5.9–6.9 in (15–17.5 cm); 1.9–2.5 oz (54–72 g) Black head and breast with brownish band above eye to back of head. Red under tail. Back, wings, and tail green with lighter green on upper wing.

DISTRIBUTION
Northern Territory, Australia.

HABITAT
Found in a variety of forest-like habitats, primarily monsoon forest, but also gallery forest along rivers, mangrove edges, eucalypt forest, and scrub; at low elevations along coast.

Pitta iris
▨ Resident

BEHAVIOR

Primarily terrestrial, occurring alone or in pairs. Birds defend territories and remain on these territories throughout the year. Calling primarily by males, and may be accompanied with a

"bowing" display, in which males from adjacent territories assume an upright position then lower their breasts nearly to the ground in slow motion. These displays are believed to maintain territorial boundaries. The species also performs a "wing-flicking" display, perhaps used to alarm or distract predators, and a "ducking posture" that may be used to avoid detection by predators.

FEEDING ECOLOGY AND DIET

Feeds by foraging in leaf-litter of the forest floor. Food items include spiders, insects, insect larvae, centipedes, snails, earthworms, small frogs, skinks, and Carpentaria palm fruits. Diets vary seasonally, with earthworms dominating during the wet season.

REPRODUCTIVE BIOLOGY

Breeds October to March. Domed nest is either on the ground or up to a height of 10 ft (3 m). Nest constructed primarily from small twigs and sticks, but also leaves, bark, ferns, vines, and palm fronds. Clutch size usually three to four. Eggs glossy white or creamy-white with dark purple or brownish spots over purple-gray markings, most numerous on widest end. Female and male share nest construction, incubation, and brooding and provisioning of young. Incubation period is 14–15 days, nestling period 14 days.

CONSERVATION STATUS

Not threatened; common throughout much of its range where habitat is suitable.

SIGNIFICANCE TO HUMANS

None known. ◆

Resources

Books

BirdLife International. *Threatened Birds of the World*. Barcelona, Spain and Cambridge, UK: Lynx Edicions and BirdLife International, 2000.

Erritzoe, J., and H. Erritzoe. *Pittas of the World*. Cambridge: Lutterworth Press, 1998.

Keith, S., E. Urban, and C. Fry. *The Birds of Africa*, Vol. 4. London: Academic Press, 1992.

Lambert, F., and M. Woodcock. *Pittas, Broadbills and Asities*. Sussex, UK: Pica Press, 1996.

Periodicals

Dutson, G., and J. Newman. "Observations on the Superb Pitta (*Pitta superba*) and Other Manus Endemics." *Bird Conservation International* 1 (1991): 215–222.

Gretton, A., M. Kohler, R. Lansdown, T. Pankhurst, J. Parr, and C. Robson. "The Status of Gurney's Pitta (*Pitta gurneyi*) 1987–1989." *Bird Conservation International* 3 (1993): 351–367.

Rozendaal, F. "Species Limits Within the Garnet Pitta-complex." *Dutch Birding* 16 (1994): 239–245.

Nathaniel E. Seavy, MS

New Zealand wrens

(Acanthisittidae)

Class Aves
Order Passeriformes
Suborder Tyranni (Suboscines)
Family Acanthisittidae

Thumbnail description
Small, compact, superficially wren-like birds occupying the ecological niches of small rodents and insectivores

Size
3–4 in (8–10) cm

Number of genera, species
2 genera, 4 species

Habitat
Forest, scrubland, alpine

Conservation status
Of the four species known in historic times, two are extinct, and two are still fairly common and fully protected by law

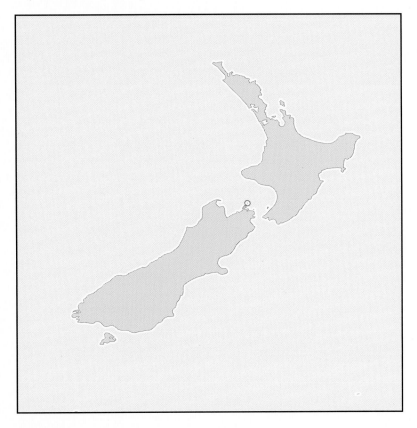

Distribution
Endemic to North and South Islands, and some satellite islands, of New Zealand

Evolution and systematics

New Zealand wrens have nothing in common with the more familiar wrens, Northern Hemisphere birds of the family Troglodytidae. Visible similarities between the families are superficial. DNA comparison and morphological studies strongly suggest that the Acanthisittidae are living relatives of the earliest passeriform birds, which date back at least 85 million years, when New Zealand severed from West Antarctica during the later stages of the Gondwana supercontinent breakup. Nevertheless, placing them in the oscine (Passeres) or suboscine (Tyranni) passerine suborders, or perhaps in a suborder all their own, is an unsettled issue.

The Acanthisittidae show a mix of oscine and suboscine traits. The syrinx (a vocal organ in the throat) differs in anatomy and position in the body from that of a typical oscine syrinx. On the other hand, the Acanthisittidae lack typical suboscine inflated stapes (a bone in the inner ear). The stapes are unique and more like oscine stapes. DNA hybridization studies among New Zealand wrens and 10 other passerine bird species by Sibley and Ahlquist (1990) showed the greatest similarity of New Zealand wrens with an oscine (bowerbird [Ptilonorhynchidae]), and the least with one Old-

and one New-World suboscine, respectively (pitta [Pittidae] and tyrannt flycatcher [Tyrannidae]).

Acanthisittidae is divided into two genera that are living or recently extinct (*Acanthisitta, Xenicus*), and another two that were extinct before historical times (*Dendroscansor, Pachyplichas*).

Genus *Acanthisitta* is monotypic with only one species, the rifleman (*Acanthisitta chloris*).

Genus *Xenicus* includes three species: one living (rock wren [*Xenicus gilviventris*]) and two recently extinct (bush wren [*Xenicus longipes*], Stephens Island wren [*Xenicus lyalli*]).

The rock wren survives in the highlands of South Island. The bush wren is probably extinct; none have been seen since 1972. The extinct Stephens Island wren is probably the most generally well known of the New Zealand wrens, due to ironic circumstances of its habitat, discovery, and demise.

Three fossil species, known from the Holocene of New Zealand, have been found: *Pachyplichas yaldwyni* (Millener, 1988), *Pachyplichas jagmi* (Millener, 1988), and *Dendroscansor decurvirostris* (Millener and Worthy, 1991).

Physical characteristics

New Zealand wrens are among the smallest of birds, at 3–4 in (8–10 cm), so compactly built and short-tailed that at rest they may look truncated or almost spherical. They have short wings and stout legs with strong, gripping feet. The toes are long and slender, and the third and fourth toes are joined at their bases. The bill is straight or slightly upturned, slender, and pointed. Coloring among the Acanthisittidae runs to greens, browns, and white.

Distribution

As a family, New Zealand wrens are—or were in historical times—common throughout the two main islands of New Zealand and several satellite islands. Today, the rifleman (*Acanthisitta chloris*) is still fairly common in forest and scrub on both main islands. The rock wren (*Xenicus gilviventris*) is somewhat less common on South Island, inhabiting rocky areas and alpine scrub at or above timberline in mountainous areas from Nelson to Fiordland.

Habitat

The living Acanthisittidae are well adapted to forest, scrubland, and alpine environments, all of which are likely to harbor abundant larders of insects, the wrens' main food source.

Behavior

Behavior among Acanthisittidae species has often been likened to that of small rodents and insectivores like mice or shrews, and they may fill niches of small feeders on New Zealand that, until historic times, had no such indigenous mammal types.

Feeding ecology and diet

Typically, an individual, a bonded male and female couple, or a family group forages on the ground or crawls over the bark and within the leafy parts of trees to search for, snag, and eat small arthropods.

Reproductive biology

Breeding season is November–March. Males and females form strong, long-lasting monogamous pair bonds. Both construct the elaborate nest—riflemen in tree crevices, rock wrens in rock crevices. Males feed nesting females and both parents feed chicks.

Conservation status

Of four species known in historic times, two have been exterminated by introduced domestic cats, stoats, ferrets, weasels, and rats. The Pacific Island rat was brought to New Zealand by colonizing Maori many centuries ago, the black rat and Norway rat were brought later by colonizing Europeans.

The bush wren or matuhi, a 4-in (9-cm) forest insectivore, was formerly widespread throughout North, South, and Stewart Islands. The last population lived on rat-free Big South Cape Island (near Stewart Island, off the southeast coast of South Island) until Norway rats jumped ship onto the island in 1961 and subsequently exterminated the bush wren. None have been seen anywhere since 1972.

The Stephens Island wren was endemic to the tiny islet between North and South Islands. The entire population was exterminated by a single cat.

The rifleman and rock wren are still fairly common, protected by New Zealand law and a vigorous conservation program. Neither are included in the 2000 IUCN Red List of Threatened Species, but the rock wren is listed as threatened by the New Zealand Department of Conservation.

Significance to humans

None known.

1. Female rifleman (*Acanthisitta chloris*); 2. Male rifleman; 3. Male Stephens Island wren (*Xenicus lyalli*). (Illustration by Barbara Duperron)

Species accounts

Rifleman
Acanthisitta chloris

SUBFAMILY

TAXONOMY
Acanthisitta chloris Sparrman, 1787.

OTHER COMMON NAMES
French: Xénique grimpeur; German: Grenadier; Spanish: Reyezuelo de Nueva Zelanda Fusil.

PHYSICAL CHARACTERISTICS
The riflemen, the smallest living bird species in New Zealand, averages about 3 in (8 cm). There is considerable sexual dichromatism and dimorphism. The female is larger than the male, an odd reversal of the normal state of affairs in bird life. Male dorsal parts are bright yellow-green above; female dorsal parts are striped darker and lighter brown and riddled with red-brown flecks. Both sexes have white ventral parts, white superciliary streaks, and yellowish rumps and flanks. The wings each sport a yellow bar and a white spot posterior to the bar. Bills of both sexes are slightly upturned, the female's a little more emphatically.

DISTRIBUTION
The rifleman is the most cosmopolitan of Acanthisittidae, fairly common and at home throughout most of lowland New Zealand, including the lower two-thirds of North Island, all of South Island, Stewart Island (off the southeast coast of South Island), and the Great Barrier and Little Barrier Islands.

Acanthisitta chloris

◼ Resident

Some ornithologists recognize two subspecies—South Island rifleman *(Acanthisitta chloris chloris)* and North Island rifleman *(Acanthisitta chloris granti)*—although the two differ only slightly in color, *granti* trading chloris's yellow rump for a greenish one.

HABITAT
The rifleman thrives easily in various habitats, including forests, farmlands, disturbed and regenerating habitats, and scrublands. It has even adapted well to landscapes partly composed of non-native plant species.

BEHAVIOR
Riflemen are lively, diurnal birds. The call is a sharp, high-pitched, cricket-like *zipt*, single or in a rapid staccato. Birds spend their days foraging in trees, winging from one to another, usually over an accustomed route, and only rarely on the ground. A rifleman sometimes displays an odd behavior that Acanthisittidae alone may claim as theirs: an individual will perch on a branch and energetically flick its wings, as if showing off.

FEEDING ECOLOGY AND DIET
Sexual dichromatism relates to feeding methods. Both sexes feed on insects, spiders, and other small invertebrates, but they split up feeding strategies. The male gleans from the leaves of a tree while the female works the bark, both going about their work meticulously and minutely. Thus, either sex has proper camouflage for its particular gleaning grounds. The female's slightly more upcurved bill may give her an advantage in poking into and prying at loose bark.

REPRODUCTIVE BIOLOGY
Male and female form strong, long-lasting pair bonds. Pairs breed August–January; females lay 2–4 white eggs. A typical pair builds a rather elaborate nest in a tree crevice, sometimes with a dome-like roof, floored and wallpapered inside with spider webs and mosses. The male feeds the brooding female and both parents feed chicks. A bonded pair typically fledge two broods in one season, fledged chicks of the first brood often pitching in to help feed chicks of the later brood.

CONSERVATION STATUS
The species is widespread, fairly common, and protected by law. It is not threatened.

SIGNIFICANCE TO HUMANS
None known. ◆

Stephens Island wren
Xenicus lyalli

TAXONOMY
Xenicus lyalli Rothschild, 1894.

OTHER COMMON NAMES
English: Stephens wren; French: Xénique de Stephen; German: Stephenschlüpfer; Spanish: Reyezuelo de Stephen.

PHYSICAL CHARACTERISTICS
A typical individual was 4 in (10 cm). Both sexes were colored similarly, the female being merely duller. Both had small but

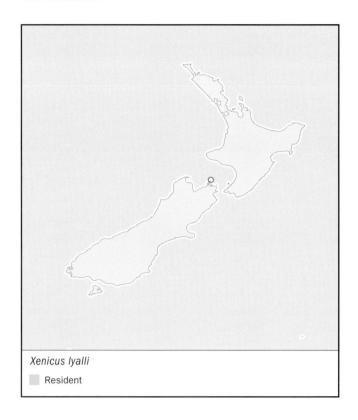

Xenicus lyalli

Resident

HABITAT
Steep, rocky outcroppings; a small forest, grass, and scrub.

BEHAVIOR
All that is known about this species, including its behavior, was recorded by a single person, George Lyell. The birds ran and skittered about on the ground, similar to mice, whose niche the birds likely filled. The species could not fly, or flew very little and ineffectively—a handy adaptation to life on a very small island, but marking them for certain death from introduced predators. The short, rounded wings and soft plumage attest as well to diminished or lost powers of flight. The voice was never described.

FEEDING ECOLOGY AND DIET
The wrens were apparently most active during twilight hours and may have been nocturnal. They would emerge from holes in rocks and spend some time poking about, alternately running about and hiding, most likely hunting for small arthropods.

REPRODUCTIVE BIOLOGY
Unknown.

CONSERVATION STATUS
The Stephens Island wren is emphatically extinct. Its discovery and extirpation are a masterpiece of cruel irony. The birds went unnoticed and were safe until the New Zealand government built a lighthouse on the islet and in 1894 staffed it with George Lyell, who brought his cat, Tibbet, to the island with him. The consequences are predictable. Within a few months, Tibbet killed, then ate or brought home as show-off gifts for his master, the entire population of Stephens Island wrens. Lyell sent nine corpora to prominent ornithologists Walter Lawry Buller and Walter Rothschild, who declared them a previously unknown species of New Zealand wren. By the time the glad news reached Lyall, the wrens were extinct. As if in a final petulant jest, the birds were scientifically dubbed *Traversia lyalli*, later changed to *Xenicus lyalli*, after the owner of the cat who wiped them out. Ten specimens still exist, distributed throughout five museums.

SIGNIFICANCE TO HUMANS
Among biologists and conservationists, the Stephens Island wren has become a poignant symbol of the fragility of isolated island species with limited space and populations. ◆

stout, strong bills. The lower mandible was light brown, as were legs and feet, the upper mandible dark brown with a horn-colored tip. The tail was little more than a stub. Although the overall body color was brown, the superciliary streak, chin, and throat were greenish yellow. Light-brown feather margins on partly overlapping body feathers decorated male and female with rows of roundish, fuzzy-edged spots on a darker brown backround. Rows, parallel to one another while following body contours, ran head to tail and covered the entire body, lending the birds a passing resemblance to pinecones. The female's spots were more softly applied.

DISTRIBUTION
The species inhabited only this small island, a mere 100 ft (30.5 m) square, but steep-sided, with an elaborate ecology.

Resources

Books

Flannery, Tim, and Peter Schouten. *A Gap in Nature: Discovering the World's Extinct Animals.* New York: Atlantic Monthly Press, 2001.

Moon, Geoff. *The Hand Guide to the Birds of New Zealand.* Mechanicsburg, PA: Stackpole Books, 1999.

Robertson, H. A., B. D. Heather, and D. J. Onley. *The Reed Field Guide to New Zealand Birds.* Oxford: Oxford University Press, 2001.

Sibley, C. E., and J. E. Ahlquist. *The Phylogeny and Classification of Birds: A Study in Molecular Evolution.* New Haven: Yale University Press, 1991.

Worthy, Trevor H., R. N. Holdaway, and Rod Morris. *The Lost World of the Moa: Prehistoric Life of New Zealand.* Bloomington: Indiana University Press, 2002.

Periodicals

Cracraft, Joel. "Gondwana Genesis." *Natural History.* Dec 2001–Jan 2002

Feduccia, A. "Morphology of the Bony Stapes in the Menuridae and Acanthisittidae: Evidence for Oscine Affinities." *Wilson Bulletin.* 87 (1975): 418–420.

Feduccia, A., and S. L. Olson. "Morphological Similarities between the Menurae and Rhinocryptidae, Relict Passerine

Resources

Birds of the Southern Hemisphere." *Smithsonian Contributions to Zoology*. 366, iii (1982).

Hunt, G. R., and I. G. McLean. "The Ecomorphology of Sexual Dimorphism in the New Zealand Rifleman, *Acanthisitta chloris*." *EMU: Austral Ornithology*. Vol. 93 (1993): 71–78.

Sibley, C. G., Williams, G. R., and J. E. Ahlquist. "The Relationships of New Zealand Wrens (Acanthisitiidae) as Indicated by DNA-DNA Hybridization." *EMU: Austral Ornithology*. 84 (1982): 236–241.

Organizations

The Ornithological Society of New Zealand. P.O. Box 12397, Wellington, North Island New Zealand. E-mail: OSNZ@xtra.co.nz Web site: <http://osnz.org.nz>

Other

New Zealand Birds <http//www.nzbirds.com>

Payne, Robert B. *Bird Families of the World: A Resource of the University of Michigan Museum of Zoology, Bird Division* <http://www.ummz.lsa.umich.edu/birds/birddivresources/families.html>.

Kevin F. Fitzgerald, BS

Ovenbirds
(*Furnariidae*)

Class Aves

Order Passeriformes

Suborder Tyranni (Suboscines)

Family Furnariidae

Thumbnail description
Small to medium-sized, brownish colored, insectivorous songbirds

Size
Body length 5–11 in (13–28 cm)

Number of genera, species
34 genera; about 218 species

Habitat
Occur in forests of various types, brushlands, pampas (grasslands), alpine habitats, and semidesert

Conservation status
Critically Endangered: 3 species; Endangered: 9 species; Lower Risk: 18 species, Vulnerable: 15 species.

Distribution
Range from central Mexico to southern South America

Evolution and systematics

The ovenbirds (Furnariidae) are a family of songbirds within the extremely diverse order of perching birds (Passeriformes). They are most closely related to the woodcreepers (Dendrocolaptidae), ant thrushes (Formicariidae), cotingas (Cotingidae), manakins (Pipridae), and tyrant flycatchers (Tyrannidae). Some avian taxonomists consider the woodcreepers to be a subfamily of the Furnariidae, naming them as Dendrocolapinae. The family is divided into three subfamilies. The true ovenbirds (Furnariinae) are about 40 species of long-legged songbirds found predominantly in southern South America that nest on the ground or use moist soil for nest-building. The bushcreepers (Synallaxeinae) are 95 species of small birds found mainly in tropical South America, often with a fringed or long tail; they build a ball-like nest. The leafcreepers (Philydorinae) are 84 species found predominantly in tropical America that forage on tree-trunks or in foliage and usually build their nest in an excavated tunnel.

Physical characteristics

Species in the ovenbird family have a range of body length of 5–11 in (13–28 cm). Their wings are relatively short, and may be rounded or pointed at the tips. The legs and feet are of medium length, and the front toes are joined at the base. The bill is slender, short to long, and pointed. They usually have a brownish, relatively inconspicuous coloration on the back, and range from light to brown-and-white speckled or streaked on the belly. Many species have a white throat. They have a light stripe over the eye, known as a superciliary line. The wing bands are often brownish red or white. The sexes are usually similarly colored.

Distribution

Species in the ovenbird family range from central Mexico to Patagonia in southern South America.

Wing-banded hornero (ovenbird) (*Furnarius figulus*) at its nest in Brazil. (Photo by Anthony Mercieca. Photo Researchers, Inc. Reproduced by permission.)

Habitat

Ovenbirds are non-migratory birds that inhabit forests of various kinds as well as brushlands, pampas (grasslands), alpine habitats, and semi-desert.

Behavior

Ovenbirds may occur as solitary individuals or as a breeding pair, or sometimes in small groups. Some ovenbirds occur with other birds in mixed-species foraging flocks. They occur on the ground and in trees; the ground-foraging species tend to walk and hop, while some of the arboreal species forage acrobatically within foliage and finer branches, and others on tree-trunks. The flight of some species is rather weak, but it is strong in others, although not over a long distance. The calls are harsh and scolding, and the song consists of series of whistles and trills.

Feeding ecology and diet

Ovenbirds feed mostly on insects, spiders, and other invertebrates. Some species also eat small seeds. They forage among litter on the ground or in foliage and on bark and epiphytes of shrubs and trees.

Reproductive biology

The nests of ovenbird species are extremely variable in their shape and mode of construction. Many species build a loose nest of plant fibers inside of a natural cavity in a tree or among rocks. The birds that are actually called "ovenbirds" are species in the genus *Furnarius*; these species are also called horneros, Spanish for "a baker of bread." Their nest is built by both members of a breeding pair and is made of thousands of lumps of moist clay, each about 0.1 oz (3 g) in weight and carried to the nest-site in the bill. Initially, a nest-base of varying thickness is built, depending on the nature of the supporting structure beneath, which is often a stout tree branch, but may also be on the ground. Next, the pair of birds builds the outer walls, which are then joined to form a roof, thus creating a structure that superficially resembles an oven. An entrance hole is left on one of the sides to permit access to the nest cavity, which is lined with fine fibers of grass and other plant tissues. Nests typically weigh about 10 lb (4 kg), but they can weigh as much as 15 lb (6.8 kg). One pair of ovenbirds may work on constructing as many as four nests at a time, either working together on all of them or each bird making only one. At the beginning of the egg-laying season, however, usually only two nests are completed and ready for use for breeding or roosting. Often other species of birds, such as swallows, use abandoned ovenbird nests for their own breeding.

The miners (*Geositta*) dig tunnels 3–10 ft long (1–3 m) into an earthen bank or cliff. Spine-tails (*Synallaxis*) build small, spherical, hanging nests in trees, which are entered through a hole from below. Canasteros (*Asthenes*) build a huge, roughly spherical nest about 14 ft (4 m) high in a tree. The nest is about 15–17 in in diameter (40 cm) and is entered by a hole on the side. Thornbirds (*Phacellodomus*) build the largest nests, which can be 3 ft (1 m) high, spherical, and made of twigs. It often contains several chambers, all of which are entered from below.

Species in the ovenbird family lay two to six eggs that are usually colored white, or sometimes blue or greenish. Both parents share in the incubation of the eggs and in the care of the nestlings and fledglings.

Conservation status

The IUCN lists 45 species of ovenbirds as being at risk. Critically Endangered species are the royal cinclodes (*Cinclodes*

Rufous hornero (ovenbird) (*Furnarius rufus*) at its nest in Brazil. (Photo by Erwin & Peggy Bauer. Bruce Coleman Inc. Reproduced by permission.)

aricomae) of Bolivia and Peru, the Alagoas foliage-gleaner (*Philydor novaesi*) of Brazil, and the plain spinetail (*Synallaxis infuscata*) of Brazil. Endangered species are the Cipo castanero (*Asthenes luizae*) of Brazil, the Bolivian spinetail (*Cranioleuca henricae*) of Bolivia, the white-browed tit-spinetail (*Leptasthenura xenothorax*) of Peru, the hoary-throated spinetail (*S. kollari*) of Brazil and Guyana, the blackish-headed spinetail (*S. tithys*) of Ecuador and Peru, the Bahia spinetail (*S. whitneyi*) of Brazil, the russet-bellied spinetail (*S. zimmeri*) of Peru, the russet-mantled softtail (*Thripophaga berlepschi*) of Peru, and the striated softtail (*T. macroura*) of Brazil. Most of the designated species at-risk have declined in range and abundance because of the conversion of their habitat into agricultural or residential land-uses, or habitat degradation associated with timber harvesting or other disturbances. These same sorts of stres-
sors are also affecting many other species in the family and are causing them to decline in range and abundance, but not yet to the degree that they are considered to be at-risk.

Significance to humans

The rufous hornero (*Furnarius rufus*) is the national bird of Argentina, in popular recognition of its bold and jaunty demeanor; thus, it is of some cultural significance. Other than this species, members of the ovenbird family are not of much direct importance to humans. They are an interesting and diverse group of birds, however, and viewings of them are widely sought by birdwatchers and other naturalists, resulting in local economic benefits through ecotourism.

1. Bar-winged cinclodes (*Cinclodes fuscus*); 2. Coastal miner (*Geositta peruviana*); 3. Scale-throated earthcreeper (*Upucerthia dumetaria*); 4. Bolivian earthcreeper (*Ochetorhynchus harterti*); 5. Thorn-tailed rayadito (*Aphrastura spinicauda*); 6. Des Murs's wiretail (*Sylviorthorhynchus desmursii*); 7. Striolated tit-spinetail (*Leptasthenura striolata*); 8. Campo miner (*Geobates poecilopterus*); 9. Rufous hornero (*Furnarius rufus*); 10. Band-tailed earthcreeper (*Eremobius phoeincurus*). (Illustration by Jonathan Higgins)

1. Wren-like rushbird (*Phleocryptes melanops*); 2. Rufous-tailed xenops (*Xenops milleri*); 3. Mouse-colored thistletail (*Schizoeaca griseomurina*); 4. Pale-breasted spinetail (*Synallaxis albescens*); 5. Streak-capped spinetail (*Cranioleuca hellmayri*); 6. Great spinetail (*Siptornopsis hypochondriacus*); 7. Greater thornbird (*Phacellodomus ruber*); 8. White-throated treerunner (*Pygarrhichas albogularis*); 9. Rufous-necked foliage-gleaner (*Syndactyla ruficollis*); 10. Cinnamon-rumped foliage-gleaner (*Philydor pyrrhodes*); 11. Short-billed leaftosser (*Sclerurus rufigularis*). (Illustration by Jonathan Higgins)

Species accounts

Campo miner
Geobates poecilopterus

SUBFAMILY
Furnariinae

TAXONOMY
Geobates poecilopterus Wied, 1830.

OTHER COMMON NAMES
French: Géositte des campos; German: Camposerdhacker; Spanish: Caminera de Campo.

PHYSICAL CHARACTERISTICS
Body length is about 5 in (12.5 cm). Bill is short, slightly downcurved, and pointed. The tail is short. The sexes are similar. The overall coloration is light brown, with a lighter buff-brown belly, a whitish throat, and a light stripe over the eye.

DISTRIBUTION
Occurs in interior regions of south-central Brazil and northeastern Bolivia.

HABITAT
Inhabits grassy glades within tropical forest and open grassland, but needs at least a few scattered trees. Appears to favor

areas that have recently been burned. Generally occurs at 1,600–3,950 ft (500–1,200 m).

BEHAVIOR
Non-migratory. Pairs of breeding birds defend a territory. The song is usually given during a hovering display flight, and is a simple, repeated series of buzzy notes.

FEEDING ECOLOGY AND DIET
Forages for insects on the ground. Sometimes perches in shrubs or trees.

REPRODUCTIVE BIOLOGY
Builds an oven-shaped nest of clay. Both the male and female incubate the eggs and rear the nestlings.

CONSERVATION STATUS
Not threatened. A locally abundant species.

SIGNIFICANCE TO HUMANS
None known. ◆

Coastal miner
Geositta peruviana

SUBFAMILY
Furnariinae

TAXONOMY
Geositta peruviana Lafresnaye, 1847.

OTHER COMMON NAMES
French: Géositte du Péerou; German: Küstenerdhacker; Spanish: Caminera de la Costa.

PHYSICAL CHARACTERISTICS
Body length is about 5.5 in (14 cm). Bill is short, slightly downcurved, and pointed. The sexes are similar. The tail is moderately short, and overall coloration is light gray-brown, with a whitish belly and throat, and a light stripe over the eye.

DISTRIBUTION
Occurs in coastal regions of western Peru.

HABITAT
Occurs in open, arid, often-sandy, desert-like barrens of the Pacific coast. Habitat ranges from almost non-vegetated to having scattered shrubs. Occurs as high as about 1,300 ft (400 m).

BEHAVIOR
A non-migratory species. Usually occurs singly or in pairs. Defends a breeding territory. The song is given by the male during a hovering display flight and is a lengthy, musical twittering.

FEEDING ECOLOGY AND DIET
Forages actively by running and hopping on the ground, seeking its food of insects and other small invertebrates. Sometimes perches in low shrubs or on walls of buildings.

Geobates poecilopterus

　Resident

Geositta peruviana

Resident

Upucerthia dumetaria

Resident

REPRODUCTIVE BIOLOGY
Builds an oven-shaped nest out of clay. Both the male and female incubate the eggs and rear the nestlings.

CONSERVATION STATUS
Not threatened. A locally abundant species.

SIGNIFICANCE TO HUMANS
None known. ◆

Scale-throated earthcreeper
Upucerthia dumetaria

SUBFAMILY
Furnariinae

TAXONOMY
Upucerthia dumetaria Geoffroy Saint-Hilarie, 1832.

OTHER COMMON NAMES
French: Upucerthie des buissons; German: Schuppenkehl-Erdhacker; Spanish: Bandurrita Común.

PHYSICAL CHARACTERISTICS
Body length is about 8.5 in (21.5 cm). Bill is long, strongly downcurved, and pointed. The tail is long. The sexes are similar. Overall coloration is dull gray-brown, with a whitish belly, pale tips of the tail-feathers, a scaly white-on-brown pattern on the throat, and a light stripe over the eye.

DISTRIBUTION
Occurs in the Andean region of western Bolivia, extreme southern Peru, Chile, and southern and western Argentina through southern Patagonia.

HABITAT
Occurs in montane and alpine slopes and plains, with cover ranging from shrubby to more-open grasslands. Occurs as high as about 12,800 ft (3,900 m).

BEHAVIOR
A non-migratory species. Usually occurs singly or in pairs. Defends a breeding territory. Tends to skulk among cover on the ground or in dense near-ground cover. Often cocks its long tail erect. The song is a musical trilling.

FEEDING ECOLOGY AND DIET
Forages actively by running and hopping on the ground, seeking its prey of insects and other small invertebrates.

REPRODUCTIVE BIOLOGY
Builds a nest in a tunnel dug into an earthen bank. Both the male and female incubate the eggs and rear the nestlings.

CONSERVATION STATUS
Not threatened. A locally abundant species, particularly in southern parts of its range.

SIGNIFICANCE TO HUMANS
None known. ◆

Bolivian earthcreeper
Ochetorhynchus harterti

SUBFAMILY
Furnariinae

TAXONOMY
Ochetorhynchus harterti Berlepsch, 1892.

OTHER COMMON NAMES
French: Upucerthie de Bolivie; German: Braunkappen-Erd-hacker; Spanish: Bandurrita Boliviana.

PHYSICAL CHARACTERISTICS
Body length is about 6 in (17 cm). Bill is rather long, some-what downcurved, and pointed. The tail is long. The sexes are similar. Overall coloration is dull brown on the back, with a lighter belly, white throat, and a tan stripe over the eye.

DISTRIBUTION
A local (or endemic) species of the Andean region of southern Bolivia.

HABITAT
Occurs in foothills and lower slopes of Andean valleys within its limited range. Occurs near edges of deciduous woods and in dry shrubby habitats. Often occurs in microhabitats with a high density of terrestrial bromeliads. Occurs within an altitudinal range of 4,700–9,700 ft (1,430–2,960 m).

Ochetorhynchus harterti

■ Resident

BEHAVIOR
A non-migratory species. Usually occurs singly or in pairs. Defends a breeding territory. Tends to skulk among cover on the ground or in dense near-ground cover. Often cocks its long tail erect. The song is a series of loud, piercing, steady or descending notes.

FEEDING ECOLOGY AND DIET
Forages in low shrubs and trees and on the ground for insects and other small invertebrates.

REPRODUCTIVE BIOLOGY
The nest has not yet been observed, but a closely related species builds a nest of twigs within a natural tree-hollow or in a clump of rocks, or sometimes in an 'oven' abandoned by another species of ovenbird. Both the male and female incubate the eggs and rear the nestlings.

CONSERVATION STATUS
An endemic and rather uncommon species, but not considered at risk.

SIGNIFICANCE TO HUMANS
None known. ◆

Band-tailed earthcreeper
Eremobius phoenicurus

SUBFAMILY
Furnariinae

TAXONOMY
Eremobius phoenicurus Gould, 1839.

OTHER COMMON NAMES
French: Annumbi rougequeue; German: Dornschlüpfer; Spanish: Bandurrita Turca.

PHYSICAL CHARACTERISTICS
Body length is about 7 in (18 cm). Bill is rather long, slightly downcurved, and pointed. The tail is long. The sexes are similar. Overall coloration is dull olive-brown on the back, with a lighter brown-streaked belly, white throat, rufous on the margins of an otherwise blackish tail, and a whitish stripe over the eye.

DISTRIBUTION
Occurs in southeastern Argentina and barely into extreme southern Chile.

HABITAT
Inhabits cool, sparsely shrubby, level grasslands of the prairie (steppe) of Patagonia. Occurs as high as about 3,900 ft (1,200 m).

BEHAVIOR
Non-migratory. Usually occurs singly or in pairs. Defends a breeding territory. A largely terrestrial bird that runs over the ground, and only sometimes perches in shrubs. Often cocks its long tail erect. The song is a short, rapid trill.

FEEDING ECOLOGY AND DIET
Forages for insects and other small invertebrates on the ground, often by probing into soft earth with its bill.

REPRODUCTIVE BIOLOGY
Builds a nest of twigs in a low shrub. Both the male and female incubate the eggs and rear the nestlings.

Eremobius phoenicurus

Resident

Cinclodes fuscus

Resident

CONSERVATION STATUS
An uncommon species, but not considered at risk.

SIGNIFICANCE TO HUMANS
None known. ◆

Bar-winged cinclodes
Cinclodes fuscus

SUBFAMILY
Furnariinae

TAXONOMY
Cinclodes fuscus Vieillot, 1818.

OTHER COMMON NAMES
French: Cinclode brun; German: Binden-Uferwipper; Spanish: Ticotico de Cuello Blanco.

PHYSICAL CHARACTERISTICS
Body length is about 7 in (17–17.5 cm). Bill is rather short, almost straight, and pointed. The tail is of medium length. The sexes are similar. The overall coloration is dull brown on the back, with a tan belly, white throat finely barred with brown, conspicuous whitish or tan wing-stripes visible in flight, and a white stripe over the eye. There is significant geographic variation in the plumage coloration of this widespread species.

DISTRIBUTION
Occurs in isolated pockets of the Andean region from southern Venezuela through Colombia, and more continuously through Ecuador, Peru, western Bolivia, and Chile and western Argentina throughout Patagonia.

HABITAT
Inhabits open grasslands at higher altitudes of the mountains and at lower levels in Patagonia. Usually occurs in the vicinity of surface water, such as streams, rivers, ponds, or lakes. Occurs as high as about 16,400 ft (5,000 m).

BEHAVIOR
Mostly a non-migratory species, although Patagonian populations may migrate northward to spend their winter in a lower latitude. Usually occurs singly or in pairs. Defends a breeding territory. A largely terrestrial bird that runs and hops over the ground, and also perches in shrubs. The song is a short, rapid trill, often given in flight.

FEEDING ECOLOGY AND DIET
Forages for insects and other small invertebrates on the ground, often by probing into soft earth with its bill.

REPRODUCTIVE BIOLOGY
Builds a nest in a burrow that it excavates itself, or in a natural cavity in an earthen bank or rock pile. Both the male and female incubate the eggs and rear the nestlings.

CONSERVATION STATUS
Not threatened. A widespread and abundant species within its habitat.

SIGNIFICANCE TO HUMANS
None known. ◆

Rufous hornero

Furnarius rufus

SUBFAMILY
Furnariinae

TAXONOMY
Furnarius rufus J.F. Gmelin, 1788.

OTHER COMMON NAMES
English: Rufous ovenbird; French: Fournier roux; German:
Rosttöpfer; Spanish: Hornero Común.

PHYSICAL CHARACTERISTICS
Body length is 7–8 in (18–20 cm). Bill is rather short, almost
straight, and pointed. The tail is of medium length. The sexes
are similar. The overall coloration is brown on the back, with a
tan belly, white throat, somewhat rufous tail, and a tan stripe
over the eye.

DISTRIBUTION
A widespread species, occurring in Bolivia, much of southern
Brazil, Paraguay, Uruguay, and northern and central Ar-
gentina.

Furnarius rufus
▢ Resident

HABITAT
Inhabits a wide variety of arid and other open habitats. Often
occurs in the vicinity of surface water, such as streams,
rivers, ponds, or lakes. Commonly occurs in the vicinity of
human habitation and along roads. Mostly occurs as high as
about 8,200 ft (2,500 m), but can be as high as 12,150 ft
(3,500 m).

BEHAVIOR
Non-migratory. Usually occurs singly or in pairs. Defends a
breeding territory. A largely terrestrial bird that boldly runs
and hops over the ground, and also perches in exposed shrubs.
The song is a loud, fast, raucous series of notes, often performed as a duet by a mated pair of birds.

FEEDING ECOLOGY AND DIET
Forages for insects and other small invertebrates on the
ground, among leaf litter, and by probing into soft earth with
its bill.

REPRODUCTIVE BIOLOGY
Constructs a large nesting structure of thousands of billfuls of
moist mud, used to make a spherical, oven-like structure
perched on a natural stump, fencepost, or telephone pole. The
internal nest cavity is accessed through a side-hole entrance.
The nesting structure is used once and then abandoned, although disused nests may persist for several years and are often
used by other species. If posts are of limited supply, a new nest
may be constructed on top of an old one. Both the male and
female incubate the eggs and rear the nestlings.

CONSERVATION STATUS
Not threatened. A widespread and abundant species within its
habitat.

SIGNIFICANCE TO HUMANS
National bird of Argentina, largely in popular recognition of its
bold and jaunty demeanor, and so it is of cultural significance. ◆

Des Murs's wiretail

Sylviorthorhynchus desmursii

SUBFAMILY
Synallaxeinae

TAXONOMY
Sylviorthorhynchus desmursii Des Murs, 1847.

OTHER COMMON NAMES
French: Synallaxe de Des Murs; German: Sechsfedern-
schlüpfer; Spanish: Colilarga Común.

PHYSICAL CHARACTERISTICS
Body length is 7.5–9 in (19–23 cm). The rather long bill is
about the same length as the head, straight, and sharply
pointed. The body is small and the tail is extremely long and
thin; it is about twice the length of the main part of the body.
Most of this lengthy tail is formed of the elongated central pair
of tail-feathers (or retrices), with the lateral pair of retrices being about half as long as the central ones. The sexes are similar. The overall coloration is rufous-brown on the back, with a
tan belly, a reddish crown of the head, and a light-tan stripe
over the eye.

Sylviorthorhynchus desmursii

☐ Resident

Thorn-tailed rayadito
Aphrastura spinicauda

SUBFAMILY
Synallaxeinae

TAXONOMY
Aphrastura spinicauda J.F. Gmelin, 1789.

OTHER COMMON NAMES
French: Synallaxe rayadito; German: Stachelschwanzschlüpfer;
Spanish: Rayadito Común.

PHYSICAL CHARACTERISTICS
Body length is about 5.5 in (14–14.5 cm). Bill is short, straight,
and sharply pointed. The body is small and the tail is long and
tipped with sharp spines emanating from the tips of the feathers. The sexes are similar. The back is colored with lengthwise
stripes of alternating dark-brown and tan, the tail is bright rufous, the throat and belly are white, the wings have two buff-
colored bands, and there is a tan stripe over the eye.

DISTRIBUTION
Occurs in southern and central Chile and adjacent western Ar-
gentina. Occurs on many coastal islands, and rarely on the
Falkland Islands.

HABITAT
Inhabits a variety of forests and wooded habitats, including primary temperate forest dominated by southern beech (*Nothofagus*
species), mature and younger secondary woodland, and low-

DISTRIBUTION
Occurs in southern and central Chile and adjacent western
Argentina.

HABITAT
Inhabits the dense undergrowth vegetation of primary temperate forest dominated by southern beech (*Nothofagus*
species), as well as mature secondary woodland containing
dense stands of the bamboo *Chusquea*. Occurs as high as
about 3,300 ft (1,000 m).

BEHAVIOR
Non-migratory. Usually occurs singly or in pairs. Defends a
breeding territory. It is a skulking, largely terrestrial bird. The
song is a loud series of notes.

FEEDING ECOLOGY AND DIET
Forages for insects and other small invertebrates, mostly within
foliage.

REPRODUCTIVE BIOLOGY
Constructs a ball-shaped, enclosed nest of grasses and other
fibers, with a side-hole entrance. The nest is placed close to
the ground surface. Both the male and female incubate the
eggs and rear the nestlings.

CONSERVATION STATUS
Not threatened. A locally abundant species within its habitat.

SIGNIFICANCE TO HUMANS
None known. ◆

Aphrastura spinicauda

☐ Resident

shrub and tussock-grass scrub. Mostly occurs as high as about 3,950 ft (1,200 m), and sometimes up to 6,550 ft (2,000 m).

BEHAVIOR
A non-migratory species. Occurs as pairs during the breeding season, and in small groups of up to 15 individuals during the winter. May also occur in mixed-species foraging flocks during the non-breeding season. An active and bold species, often holding the tail cocked erect. The song is an extended, buzzy trill.

FEEDING ECOLOGY AND DIET
Forages energetically for insects and other small invertebrates within foliage at all levels of the forest canopy.

REPRODUCTIVE BIOLOGY
Constructs a nest of grasses and other fibers within a tree-hole or in a space behind loose bark. Both the male and female incubate the eggs and rear the nestlings.

CONSERVATION STATUS
Not threatened. An abundant species within its habitat.

SIGNIFICANCE TO HUMANS
None known. ◆

Striolated tit-spinetail
Leptasthenura striolata

SUBFAMILY
Synallaxeinae

TAXONOMY
Leptasthenura striolata Pelzeln, 1856.

OTHER COMMON NAMES
French: Synallaxe striolé; German: Strichelschlüpfer; Spanish: Coludito Estriado.

PHYSICAL CHARACTERISTICS
Body length is about 6.5 in (16–16.5 cm). Bill is short, straight, and sharply pointed. The body is small and slender, and the tail is long and tipped with two sharp spines emerging from the tips of the central pair of tail-feathers. The sexes are similar. The back is colored brown with buffy streaks, the tail is brown with rufous outer feathers, the wings are uniformly brown, the throat and belly are reddish brown with brown speckles, the crown of the head is black with rufous streaks, and there is a buffy-white stripe over the eye.

DISTRIBUTION
A locally distributed species of southeastern Brazil.

HABITAT
Inhabits a variety of forested and shrubby habitats, and well-vegetated gardens. Mostly occurs at 1,650–3,600 ft (500–1,100 m).

BEHAVIOR
Non-migratory. Occurs as pairs during the breeding season, or in small groups. Sometimes joins mixed-species foraging flocks. The song is a high-pitched, descending series of notes and trills.

FEEDING ECOLOGY AND DIET
Forages energetically for insects and other small invertebrates within the shrub and tree canopy, often hanging upside-down while inspecting foliage, twigs, and flowers for prey.

Leptasthenura striolata

☐ Resident

REPRODUCTIVE BIOLOGY
Constructs a nest in a cavity in a tree, rock pile, wall, or earthen bank, or in an abandoned oven-nest of another species of ovenbird. Both the male and female incubate the eggs and rear the nestlings.

CONSERVATION STATUS
Not threatened. An endemic species, abundant within its local habitat.

SIGNIFICANCE TO HUMANS
None known. ◆

Wren-like rushbird
Phleocryptes melanops

SUBFAMILY
Synallaxeinae

TAXONOMY
Phleocryptes melanops Vieillot, 1817.

OTHER COMMON NAMES
French: Synallaxe troglodyte; German: Rohrschlüpfer; Spanish: Junquero Trabajador.

PHYSICAL CHARACTERISTICS
Body length is about 5.5 in (13.5–14.5 cm). Bill is short, slightly downcurved, and pointed. The body is small and chunky, and the

Phleocryptes melanops

☐ Resident

tail is moderately short and tipped with two short spines emerging from the tips of the central pair of tail-feathers. The sexes are similar. The back is colored brown with whitish streaks, the tail is brown, the wings are brown with rufous patches, the throat and belly are whitish bordered with buff, the crown of the head is dark brown, and there is a buff-white stripe over the eye.

DISTRIBUTION
Occurs widely in southern South America, including western Peru, western Bolivia, Chile, southern Brazil, Paraguay, Uruguay, and Argentina.

HABITAT
Inhabits reedbeds of marshes and lake margins, including both fresh and brackish waterbodies. Occurs as high as about 14,100 ft (4,300 m).

BEHAVIOR
Northern populations are non-migratory, but southern ones may migrate to spend the winter in northern parts of the species range. Occurs as pairs during the breeding season. The song is a quiet series of ticking notes.

FEEDING ECOLOGY AND DIET
Forages for insects and other small invertebrates on muddy ground, among reedy vegetation, among floating plants, and even in shallow water.

REPRODUCTIVE BIOLOGY
Constructs a spherical nest attached to reeds, with a side entrance near the top. Both the male and female incubate the eggs and rear the nestlings.

CONSERVATION STATUS
Not threatened. A widespread and locally abundant species within its reedy habitat.

SIGNIFICANCE TO HUMANS
None known. ◆

Pale-breasted spinetail
Synallaxis albescens

SUBFAMILY
Synallaxeinae

TAXONOMY
Synallaxis albescens Temminck, 1823.

OTHER COMMON NAMES
French: Synallaxe albane; German: Temminckschlüpfer; Spanish: Pijuí de Cola Parda.

PHYSICAL CHARACTERISTICS
Body length is about 6.5 in (16–16.5 cm). Bill is short, straight, and pointed. The body is small and slender, and the tail is long and indistinctly tipped with two short spines. The sexes are similar. The back and tail are colored olive-brown, the wings are olive-brown with bright rufous patches, the cheeks, throat, and belly are whitish, the crown of the head is rufous, and there is a whitish stripe over the eye.

Synallaxis albescens

☐ Resident

DISTRIBUTION
Occurs widely from southern Central America through much of South America. It occurs in Costa Rica, Panama, northern Colombia, Venezuela, the Guianas, most of Brazil, Paraguay, Bolivia, eastern Peru, Uruguay, and northern and central Argentina.

HABITAT
Occurs in open savannahs and grassy meadows with scattered trees and shrubs. Occurs as high as about 4,900 ft (1,500 m).

BEHAVIOR
Mostly occurs as skulking, inconspicuous pairs. The song is a nasal, two-noted vocalization.

FEEDING ECOLOGY AND DIET
Forages for insects and other small invertebrates on the ground or in dense vegetation.

REPRODUCTIVE BIOLOGY
Constructs a bulky globular nest of sticks and grassy fibers attached to a shrub, with a side entrance. Both the male and female incubate the eggs and rear the nestlings.

CONSERVATION STATUS
Not threatened. A widespread and abundant species.

SIGNIFICANCE TO HUMANS
None known. ◆

Siptornopsis hypochondriacus

▨ Resident

Great spinetail
Siptornopsis hypochondriacus

SUBFAMILY
Synallaxeinae

TAXONOMY
Siptornopsis hypochondriacus Salvin, 1895.

OTHER COMMON NAMES
French: Synallaxe à poitrine rayée; German: Salvinschlüpfer; Spanish: Canastero Grande.

PHYSICAL CHARACTERISTICS
Body length is about 7.5 in (18.5 cm). Bill is short, slightly downcurved, and pointed. The body is relatively robust, and the tail is long and indistinctly forked. The sexes are similar. The back and tail are colored olive-brown, the wings are olive-brown with rufous patches, the throat and belly are whitish streaked with brown on the flanks, the crown of the head is olive, and there is a white stripe over the eye.

DISTRIBUTION
An endemic, little-known species that only occurs in a small area, in the Rio Maranon Valley of northern Peru.

HABITAT
Occurs on slopes in humid, dense montane shrubs and forest. Mostly occurs at elevations of 6,550–9,850 ft (2,000–3,000 m).

BEHAVIOR
Not well known. Occurs as pairs. The song is a loud chatter.

FEEDING ECOLOGY AND DIET
Forages for insects and other small invertebrates.

REPRODUCTIVE BIOLOGY
Constructs a large, bulky, roofed nest of sticks. Both the male and female incubate the eggs and rear the nestlings.

CONSERVATION STATUS
An endemic species, and listed as Vulnerable because of its small range and few known populations. Its habitat is thought to be declining in area because of conversion into agricultural land-use and other disturbances. This little-known species should be better-studied, and its critical habitat conserved.

SIGNIFICANCE TO HUMANS
None known. ◆

Streak-capped spinetail
Cranioleuca hellmayri

SUBFAMILY
Synallaxeinae

TAXONOMY
Cranioleuca hellmayri Bangs, 1907.

OTHER COMMON NAMES
French: Synallaxe des broméliades; German: Strichelkopfschlüpfer; Spanish: Pijuí de Reiser.

Cranioleuca hellmayri

Resident

PHYSICAL CHARACTERISTICS
Body length is about 5.5 in (14 cm). Bill is short, almost straight, and pointed. The body is slender, and the tail long and tipped with small spines. The sexes are similar. The back is colored dark olive-brown, the tail is rufous, the wings are olive-brown with large rufous patches, the throat is white, the belly is light olive, the crown of the head is rufous streaked with black, and there is a whitish stripe over the eye.

DISTRIBUTION
An endemic species that only occurs in a small area, in the Santa Marta Mountains of northern Colombia.

HABITAT
Inhabits slopes with montane humid forest, mature second growth woodland, and forest edges. Mostly occurs at elevations of 5,250–9,850 ft (1,600–3,000 m).

BEHAVIOR
Occurs as pairs. The song is a series of shrill notes, falling in intensity and pitch.

FEEDING ECOLOGY AND DIET
Forages acrobatically in the forest canopy, even hanging up-side-down, for insects and other small invertebrates.

REPRODUCTIVE BIOLOGY
Constructs a large, bulky, roughly spherical nest of mosses and other fibers, with a side entrance, and attached to a drooping outer limb of a tree. Both the male and female incubate the eggs and rear the nestlings.

CONSERVATION STATUS
Not threatened. A very local species, but abundant within its highly restricted range.

SIGNIFICANCE TO HUMANS
None known. ◆

Mouse-colored thistletail
Schizoeaca griseomurina

SUBFAMILY
Synallaxeinae

TAXONOMY
Schizoeaca griseomurina P.L. Sclater, 1882.

OTHER COMMON NAMES
French: Synallaxe souris; German: Grau-Distelschwanzschlüpfer; Spanish: Piscuiz Gris.

PHYSICAL CHARACTERISTICS
Body length is about 7.5 in (18.5–19 cm). Bill is short, straight, and pointed. The body is slender, and the tail is very long, tipped with spines, and has a frayed appearance. The sexes are similar. The back, tail, and wings are colored dull olive-brown, the belly and throat are light grayish, there is a white eye-ring, and there is a whitish stripe over the eye.

Schizoeaca griseomurina

Resident

DISTRIBUTION
An endemic species that only occurs in a small area in the Andes of southern Ecuador and extreme northern Peru.

HABITAT
Inhabits slopes with humid montane forest and woodland and their edges, just below or near the altitudinal tree-line and in woody clumps above it. Occurs at elevations of 9,200–10,800 ft (2,800–3,300 m).

BEHAVIOR
Occurs singly or as pairs. The song is a high-pitched trill.

FEEDING ECOLOGY AND DIET
Forages in the dense forest canopy, often quite acrobatically, for insects and other small invertebrates among leaves and twigs.

REPRODUCTIVE BIOLOGY
Constructs a large, bulky, roughly spherical nest, with a side entrance, and attached to a limb of a tree. Both the male and female incubate the eggs and rear the nestlings.

CONSERVATION STATUS
Not threatened. A very local species, but abundant within its highly restricted range.

SIGNIFICANCE TO HUMANS
None known. ◆

Phacellodomus ruber

▨ Resident

Greater thornbird
Phacellodomus ruber

SUBFAMILY
Synallaxeinae

TAXONOMY
Phacellodomus ruber Vieillot, 1817.

OTHER COMMON NAMES
French: Synallaxe rouge; German: Rotschwingen-Bündelnister; Spanish: Espinero Grande.

PHYSICAL CHARACTERISTICS
Body length is about 8 in (20.5 cm). Bill is short, slightly down-curved, and pointed. The body is stout, and the tail is long. The sexes are similar. The back is colored brown, the tail and wings are rufous, the belly and throat are whitish, the cap of the head is rufous, and there is a light-brown stripe over the eye.

DISTRIBUTION
A widespread species occurring in Bolivia, central Brazil, Paraguay, northern Argentina, and likely extreme northern Uruguay.

HABITAT
Inhabits the undergrowth of humid tropical forest near ponds and other surface water. Occurs at elevations up to 4,600 ft (1,400 m).

BEHAVIOR
A skulking bird that occurs singly or as pairs. The song is a long series of loud, abrupt, accelerating notes.

FEEDING ECOLOGY AND DIET
Forages on the forest floor and at nearby edges of waterbodies for insects and other small invertebrates.

REPRODUCTIVE BIOLOGY
Constructs a large, conspicuous, bulky, roughly cylindrical nest of sticks and twigs, often containing several chambers, and attached to an outer, drooping branch of a tree. Both the male and female incubate the eggs and rear the nestlings.

CONSERVATION STATUS
Not threatened. A widespread and locally abundant species.

SIGNIFICANCE TO HUMANS
None known. ◆

White-throated treerunner
Pygarrhichas albogularis

SUBFAMILY
Philydorinae

TAXONOMY
Pygarrhichas albogularis King, 1831.

OTHER COMMON NAMES
French: Picotelle à gorge blanche; German: Spechttöpfer; Spanish: Picolezna Patagónico.

PHYSICAL CHARACTERISTICS
Body length is about 6 in (14.5–15 cm). Bill is of medium length, slightly upturned, and sharply pointed. The body is

Pygarrhichas albogularis

■ Resident

chunky, and the tail is short and tipped with short spines emanating from the tips of the tail-feathers. The sexes are similar. The back and the top of the head are colored brown, the tail is rufous, the wings have rufous patches, the belly is brown spotted with white, and the throat and chest are bright white.

DISTRIBUTION
Occurs in central and southern Chile and adjacent western Argentina through most of Tierra del Fuego. Occurs on many coastal islands.

HABITAT
Inhabits deciduous temperate forest dominated by southern beech (*Nothofagus* species), as well as clearings having some mature trees present. Occurs as high as about 3,950 ft (1,200 m).

BEHAVIOR
Non-migratory. Occurs singly or as pairs. Often occurs with other birds in mixed-species flocks. Can be quite tame with humans. The song is a loud, repeated, metallic, one- or two-syllable note.

FEEDING ECOLOGY AND DIET
Forages nuthatch-like for insects and other invertebrates on the trunks of trees and branches, sometimes moving downwards head-first.

REPRODUCTIVE BIOLOGY
Constructs a nest within a cavity dug into rotten wood of a branch. Both the male and female incubate the eggs and rear the nestlings.

CONSERVATION STATUS
Not threatened. A locally abundant species within its habitat.

SIGNIFICANCE TO HUMANS
None known. ◆

Rufous-tailed xenops
Xenops milleri

SUBFAMILY
Philydorinae

TAXONOMY
Xenops milleri Chapman, 1914.

OTHER COMMON NAMES
French: Sittine à queue rousse; German: Rotschwanz-Baum-späher; Spanish: Picolezna de Cola Rufa.

PHYSICAL CHARACTERISTICS
Body length is about 4 in (11 cm). Bill is short, straight, and pointed. The body is slender, and the tail is moderately long. The sexes are similar. The back, top of head, and underparts are brown heavily streaked with buffy-white, the wing and tail are rufous, the wing shows a conspicuous rufous wing-band in flight, and there is a white stripe over the eye.

Xenops milleri

■ Resident

DISTRIBUTION
Occurs in northern South America, in southern Venezuela, French Guiana, Surinam, southeastern Colombia, eastern Ecuador, eastern Peru, Amazonian Brazil, and likely northern Bolivia.

HABITAT
Inhabits humid, lowland tropical forest, occurring in both the tree canopy and at forest-edges. Occurs as high as about 2,000 ft (600 m).

BEHAVIOR
A non-migratory species. Occurs singly or as a breeding pair. Often occurs with other birds in mixed-species foraging flocks. The song is not known, but is likely a series of shrill notes, similar to other xenops.

FEEDING ECOLOGY AND DIET
Forages on tree branches and in dense vine-tangles for insects and other invertebrates.

REPRODUCTIVE BIOLOGY
Constructs a nest within a cavity dug into rotten wood of a branch or tree trunk, or uses a natural cavity or one excavated and abandoned by another species of bird. Both the male and female incubate the eggs and rear the nestlings.

CONSERVATION STATUS
Not threatened. A locally abundant species.

SIGNIFICANCE TO HUMANS
None known. ◆

Syndactyla ruficollis
▢ Resident

Rufous-necked foliage-gleaner
Syndactyla ruficollis

SUBFAMILY
Philydorinae

TAXONOMY
Syndactyla ruficollis Taczanowski, 1884.

OTHER COMMON NAMES
English: Red-necked foliage-gleaner; French: Anabate á cou roux; German: Rothals-Baumspäher; Spanish: Trepamusgo de Cuello Rufo.

PHYSICAL CHARACTERISTICS
Body length is about 7 in (18–18.5 cm). Bill is short, straight, rather stout, and pointed. The body is slender, and the tail is long. The sexes are similar. The back, neck, and top of the head are rufous-brown, the underparts are brown streaked with buff, the tail is rufous, and there is a buffy stripe over the eye.

DISTRIBUTION
An endemic species that occurs only in a small Andean region of extreme southern Ecuador and northern Peru.

HABITAT
Inhabits humid, lowland and montane forest, secondary woodland, and forest edges. Mostly inhabits evergreen humid forest, but also occurs in somewhat drier, deciduous forest. Occurs mostly at 4,250–9,400 ft (1,300–2,700 m), but as low as 1,950 ft (600 m) in undisturbed primary forest.

BEHAVIOR
A non-migratory species. Occurs singly, as a breeding pair, or in a small group. Often occurs with other birds in mixed-

species foraging flocks. The song is an accelerating series of harsh, nasal notes.

FEEDING ECOLOGY AND DIET
Forages on tree branches and trunks for insects and other invertebrates hidden among bark or in epiphytic mosses and bromeliads.

REPRODUCTIVE BIOLOGY
Constructs a nest within a burrow dug into an earthen bank. Both the male and female incubate the eggs and rear the nestlings.

CONSERVATION STATUS
This endemic species is listed as Vulnerable, largely because its highly restricted habitat is being fragmented by conversion into agricultural land-use and further reduced by other disturbances.

SIGNIFICANCE TO HUMANS
None known. ◆

Cinnamon-rumped foliage-gleaner
Philydor pyrrhodes

SUBFAMILY
Philydorinae

TAXONOMY
Philydor pyrrhodes Cabanis, 1848.

Philydor pyrrhodes

Resident

OTHER COMMON NAMES
French: Anabate flamboyant; German: Zimtbürzel-Blattspäher; Spanish: Ticotico Acanelado.

PHYSICAL CHARACTERISTICS
Body length is about 6.5 in (16.5–17 cm). Bill is moderately long, slightly downcurved, and pointed. The body is stout, and the tail is long. The sexes are similar. The back, wings, and top of the head are brown, the underparts, tail, rump, and throat are bright cinnamon-brown, and there is a cinnamon stripe over the eye.

DISTRIBUTION
A widespread species that occurs in the Guianas, southern Venezuela, southeastern Colombia, eastern Ecuador, eastern Peru, northern Bolivia, and Amazonian Brazil.

HABITAT
Inhabits humid, lowland, tropical forest, including terra firme (or non-flooded) and wetter stands. Tends to occur where palms are abundant. Occurs as high as about 2,300 ft (700 m).

BEHAVIOR
Non-migratory. Occurs singly, or as a breeding pair. Sometimes associates with mixed-species foraging flocks. The song is not known, but is likely a long chatter of notes, similar to other species in its genus.

FEEDING ECOLOGY AND DIET
A furtive species that forages in dense foliage and thick cover for insects and other invertebrates.

REPRODUCTIVE BIOLOGY
Constructs a nest within a cavity in a tree or snag, but may also dig a nesting burrow in an earthen bank. Both the male and female incubate the eggs and rear the nestlings.

CONSERVATION STATUS
Not threatened. A widespread but uncommon species.

SIGNIFICANCE TO HUMANS
None known. ◆

Short-billed leaftosser

Sclerurus rufigularis

SUBFAMILY
Philydorinae

TAXONOMY
Sclerurus rufigularis Pelzeln, 1868.

OTHER COMMON NAMES
English: Short-billed leafscraper; French: Sclérure à bec court; German: Kurzschnabel-Laubwender; Spanish: Raspahojas de Pico Corto.

PHYSICAL CHARACTERISTICS
Body length is about 6 in (16 cm). Bill is short, straight, and pointed. The body is stout, and the tail is moderately long. The sexes are similar. The upperparts are dark-brown, the tail

Sclerurus rufigularis

Resident

is blackish brown, the underparts are cinnamon-brown, and there is a tan stripe over the eye.

DISTRIBUTION
A widespread species that occurs in the Guianas, southern Venezuela, southeastern Colombia, eastern Ecuador, eastern Peru, northern Bolivia, and Amazonian Brazil.

HABITAT
Inhabits humid, lowland, tropical forest and humid montane forest. Mostly occurs up to about 1,650 ft (500 m), and rarely as high as 5,900 ft (1,800 m).

BEHAVIOR
A non-migratory species. Occurs singly, or as a breeding pair. Sometimes associates with mixed-species foraging flocks. The song is a long trill or chatter.

FEEDING ECOLOGY AND DIET
A furtive species that forages in dense foliage on or near the ground for insects and other invertebrates. It searches among leaf litter, often tossing debris with its bill to search beneath for prey.

REPRODUCTIVE BIOLOGY
Constructs a nest within a long burrow dug into an earthen bank. Both the male and female incubate the eggs and rear the nestlings.

CONSERVATION STATUS
Not threatened. A widespread but uncommon species.

SIGNIFICANCE TO HUMANS
None known. ◆

Resources

Books

BirdLife International. *Threatened Birds of the World.* Barcelona, Spain and Cambridge, UK: Lynx Edicions and BirdLife International, 2000.

Ridgely, R.S., and G. Tudor. *The Birds of South America.* Volume 2, *The Suboscine Passerines.* Austin, Texas: University of Texas Press, 1994.

Skutch, A.F. *Antbirds and Ovenbirds: Their Lives and Homes.* Austin, Texas: University of Texas Press, 1996.

Organizations

BirdLife International. Wellbrook Court, Girton Road, Cambridge, Cambridgeshire CB3 0NA United Kingdom. Phone: +44 1 223 277 318. Fax: +44-1-223-277-200. E-mail: birdlife@birdlife.org.uk Web site: <http://www.birdlife.net>

IUCN–The World Conservation Union. Rue Mauverney 28, Gland, 1196 Switzerland. Phone: +41-22-999-0001. Fax: +41-22-999-0025. E-mail: mail@hq.iucn.org Web site: <http://www.iucn.org>

Bill Freedman, PhD

Woodcreepers
(Dendrocolaptidae)

Class Aves

Order Passeriformes

Suborder Tyranni (Suboscines)

Family Dendrocolaptidae

Thumbnail description
Non-migratory, plain-colored songbirds that forage for invertebrates on the bark surface of trees

Size
Body length ranges from 5.5–14 in (14–36 cm)

Number of genera, species
13 genera; 52 species

Habitat
Occurs in many types of humid tropical and montane forests and brushlands

Conservation status
Vulnerable: 1 species

Distribution
Occurs throughout the tropics of Central and South America, ranging from southern Mexico to northern Argentina

Evolution and systematics

The woodcreepers (Dendrocolaptidae) are a family of songbirds within the extremely diverse order of perching birds (Passeriformes). They are believed to be most closely related to the ovenbirds (Furnariidae), ant thrushes (Formicariidae), cotingas (Cotingidae), manakins (Pipridae), and tyrant fly-catchers (Tyrannidae). Some avian systematists, however, consider the Dendrocolaptidae to be a subfamily of the Furnariidae (naming them as Dendrocolapinae). The wood-creepers resemble the creepers (Certhiidae), a Northern Hemisphere family, but this resemblance is because of con-vergent evolution, as the families are not closely related. The woodcreepers are divided into 13 genera and 52 species (de-pending on the taxonomic treatment).

Physical characteristics

Woodcreepers have a body length that can range from 5.5 to 14 in (14–36 cm). They have a rather slim body, long rounded wings, and a long graduated tail with 12 outer feath-ers (or retrices). The shafts of the tail feathers project beyond

the broad vanes as sharp, downward-curving spines. The legs of woodcreepers are short and the feet are strong, with long toes and strong claws. The bill varies greatly in size and shape, ranging from a 3-in (7 cm) long, strongly down-curved, sickle-shaped bill at one extreme, to a short, laterally compressed bill at the other. The plumage is typically dull brown, brown-ish olive, brownish red, or brownish yellow, and is usually striped, banded, or spotted. The sexes have a similar external anatomy and coloration.

Distribution

Woodcreepers occur throughout the tropics of Central and South America, ranging from southern Mexico to northern Argentina. The greatest richness of species occurs in tropical Amazonia.

Habitat

Woodcreepers occur in many types of humid tropical and montane forests and brushlands.

A red-billed scythebill (*Campylorhamphus trochilirostris*) forages for insects on a tree trunk. (Photo by Doug Wechsler/VIREO. Reproduced by permission.)

Behavior

Woodcreepers are non-migratory birds, remaining all year within their general breeding area. They do not form flocks, but some species live in pairs all year. Some species associate with mixed-species foraging flocks of other species. Woodcreepers roost at night, always singly, in natural tree-cavities or old woodpecker holes. Their songs are pure, simple, clear melodies. Often the song is a soft trill or a long sequence of loud, ringing, similar tones. They sing mostly at dusk, often while feeding.

Feeding ecology and diet

Woodcreepers feed on insects, spiders, and other invertebrates, for which they search in clefts in bark and among mosses, lichens, and epiphytic plants growing on tree-limbs. While looking for food, they clamber up tree trunks, often vertically, using the stiff tail as a support. When they reach the top of a tree, they fly directly to a low point on another one, where they again start foraging upwards. Some species, notably those of the genus *Dendrocincla*, may follow foraging columns of army ants, seeking insects flushed out of their hiding places by the ants. Some woodcreepers also catch flying insects on the wing, and they may take small lizards. Rarely, they may eat small fruits.

Reproductive biology

Woodcreepers build their nests in old woodpecker excavations or in natural cavities in trees. The nest is made of fine rootlets, pieces of bark, lichens, and strands of plant fibers. They lay two or three white eggs. In almost all species, both

parents share in building the nest, incubating the eggs, and rearing the young. In the tawny-winged woodcreeper (*Dendrocincla anabatina*), however, only the female incubates the eggs. The incubation period is 15–21 days, and the time to fledging is 19–23 days. In most species, the older nestlings and fledglings roost in different holes from each of the parents.

Conservation status

IUCN lists one species, the moustached woodcreeper (*Xiphocolaptes falcirostris*), as Vulnerable. This species inhabits the dry tropical forests of the interior of eastern Brazil. It has declined greatly in range and abundance because almost all of this habitat type has been converted to agricultural land-uses or is being harvested to provide raw material to manufacture charcoal or paper pulp. The greater scythebill (*Campylorhamphus pucherani*) of the tropical forests of Colombia, Ecuador, and Peru is listed by IUCN as Near Threatened. Other species have also declined significantly in range and abundance, mostly because of extensive habitat loss or disturbance, but are not yet considered to be at risk.

Significance to humans

Woodcreepers are of no direct importance to humans, other than the indirect economic benefits of ecotourism associated with tropical birdwatching.

A ruddy woodcreeper (*Dendrocincla homochroa*) uses its tail for support while climbing a tree trunk and foraging for insects. (Photo by R. & N. Bowers/VIREO. Reproduced by permission.)

1. Long-tailed woodcreeper (*Deconychura longicauda*); 2. Scimitar-billed woodcreeper (*Drymornis bridgesii*); 3. Uniform woodcreeper (*Hylexetastes uniformis*); 4. Lesser woodcreeper (*Lepidocolaptes fuscus*); 5. Plain-brown woodcreeper (*Dendrocincla fuliginosa*); 6. Red-billed scythebill (*Campylorhamphus trochilirostris*); 7. Long-billed woodcreeper (*Nasica longirostris*); 8. Great rufous woodcreeper (*Xiphocolaptes major*). (Illustration by Amanda Humphrey)

Species accounts

Plain-brown woodcreeper
Dendrocincla fuliginosa

TAXONOMY
Dendrocincla fuliginosa Vieillot, 1818.

OTHER COMMON NAMES
English: Thrush-like woodcreeper; French: Grimpar enfumá; German: Grauwangenbaumsteiger; Spanish: Trepatronco Pardo.

PHYSICAL CHARACTERISTICS
Body length 8–8.5 in (19.5–21.5 cm). Has a stout, chisel-shaped bill. Overall coloration is rufous-brown, redder on the rump and tail and lighter on the belly. There is geographic variation in coloration among races of this widespread species.

DISTRIBUTION
Occurs throughout much of tropical Central and South America, from Honduras in the north through to Amazonian Brazil, Peru, and Bolivia. Also occurs along the Pacific coast of Ecuador and Colombia.

HABITAT
Occurs in a range of types of humid tropical rainforest and in mature secondary forest. Inhabits the lower part of the canopy. Occurs as high as about 4,300 ft (1,300 m).

BEHAVIOR
Usually occurs singly or in pairs, or sometimes in small groups in the vicinity of a swarm of army ants. The song is a prolonged series of high-pitched notes.

FEEDING ECOLOGY AND DIET
Often attends swarms of army ants along with other species in a mixed foraging flock. Forages from a perch on a tree trunk, making sallies to catch insects disturbed by the ants. Also forages for arthropods on bark surfaces.

REPRODUCTIVE BIOLOGY
Lays two or three eggs in a nest in a tree-cavity or abandoned woodpecker hole. The sexes share incubation and care of the nestlings.

CONSERVATION STATUS
Not threatened. A widespread and abundant species.

SIGNIFICANCE TO HUMANS
None known. ◆

Long-tailed woodcreeper
Deconychura longicauda

TAXONOMY
Deconychura longicauda Pelzeln, 1868.

OTHER COMMON NAMES
French: Grimpar à longue queue; German: Langschwanz-Baumsteiger; Spanish: Trepatronco de Cola Larga.

PHYSICAL CHARACTERISTICS
7.5–8.5 in (19–21.5 cm). Has a relatively long tail and a stout, chisel-shaped bill. Overall coloration is rufous-brown, redder on the rump and tail, with a buff-colored throat.

DISTRIBUTION
Occurs throughout much of tropical Central and South America, from Honduras in the north, through Costa Rica, Panama, and parts of Venezuela, the Guianas, Colombia, Ecuador, and Amazonian Brazil, Peru, and Bolivia. It has disjunct populations in the northern parts of the range, which could represent separate species.

HABITAT
Occurs in humid tropical and montane forest, especially in terra firme (or unflooded) forest. Occurs in the lower and middle levels of the canopy. Occurs as high as about 4,300 ft (1,300 m).

BEHAVIOR
Usually occurs singly or in pairs, or sometimes in mixed-species foraging flocks. The song is a series of high-pitched whistled notes, but it varies among geographic races (which may actually be separate species).

Dendrocincla fuliginosa
  Resident

Deconychura longicauda

☐ Resident

Nasica longirostris

☐ Resident

makes up about one-third of the body length. The back and tail are colored rufous-brown, the neck and back of head are brown speckled with white, and the throat and chest are white.

DISTRIBUTION
Occurs throughout much of tropical South America, including southwestern Venezuela, eastern parts of Colombia, Ecuador, Peru, and Bolivia, and most of Amazonian Brazil.

HABITAT
Inhabits humid, lowland, non-flooded tropical forest, usually close to surface water, as high as about 1,000 ft (300 m). Occurs in the middle and higher levels of the canopy.

BEHAVIOR
Usually occurs singly or in pairs. The song is a series of three or four long, eerie, whistled notes.

FEEDING ECOLOGY AND DIET
Forages for arthropods on tree-trunks and stout branches, often near forest-edges in the vicinity of a body of water.

FEEDING ECOLOGY AND DIET
Forages on tree-trunks and stout branches.

REPRODUCTIVE BIOLOGY
Lays two or three eggs in a nest in a tree-cavity or abandoned woodpecker hole. The sexes share incubation and care of the nestlings.

CONSERVATION STATUS
Not threatened. A widespread and locally abundant species.

REPRODUCTIVE BIOLOGY
Lays two to three eggs in a nest in a tree-cavity or abandoned woodpecker hole. The sexes share incubation and care of the nestlings.

SIGNIFICANCE TO HUMANS
None known. ◆

CONSERVATION STATUS
Not threatened. A widespread and locally abundant species.

Long-billed woodcreeper
Nasica longirostris

SIGNIFICANCE TO HUMANS
None known. ◆

TAXONOMY
Nasica longirostris Vieillot, 1818.

OTHER COMMON NAMES
French: Grimpar nasican; German: Langschnabel-Baumsteiger; Spanish: Trepatronco de Pico Largo.

PHYSICAL CHARACTERISTICS
Body length 14–14.5 in (35–36 cm). A large woodcreeper with a long tail and a stout, slightly downcurved, white-colored bill that

Scimitar-billed woodcreeper
Drymornis bridgesii

TAXONOMY

Drymornis bridgesii Eyton, 1849.

OTHER COMMON NAMES

French: Grimpar porte-sabre; German: Degenschnabel-Baumsteiger; Spanish: Chinchero Grande.

PHYSICAL CHARACTERISTICS

Body length 12 in (30–31 cm). A large woodcreeper with a long tail and a stout, strongly downcurved, blackish bill that makes up about one-third of the body length. The back and tail are colored olive-brown, with white stripes along the side of the face, a white throat, and a brown-and-white striped belly.

DISTRIBUTION

Occurs in southern Bolivia, southern Brazil, western Paraguay, and northern and central Argentina.

HABITAT

Inhabits relatively open, lowland tropical forest and scrub, as high as about 1,650 ft (500 m).

BEHAVIOR

Usually occurs singly or in pairs. The song is a series of loud, fast, shrieks.

FEEDING ECOLOGY AND DIET

Forages for arthropods on tree-trunks and stout branches, and sometimes on the ground.

REPRODUCTIVE BIOLOGY

Lays two or three eggs in a nest in a tree-cavity or abandoned woodpecker hole. The sexes share incubation and care of the nestlings.

CONSERVATION STATUS

Not threatened. A widespread and locally abundant species.

SIGNIFICANCE TO HUMANS

None known. ◆

Uniform woodcreeper
Hylexetastes uniformis

TAXONOMY

Hylexetastes uniformis Hellmayr, 1909.

OTHER COMMON NAMES

French: Grimpar uniforme; German: Wellenbauch-Baumsteiger; Spanish: Trepatronco de Pico Rayado.

PHYSICAL CHARACTERISTICS

Body length 10.5 in (27 cm). A large woodcreeper with a long tail and a stout, short, reddish bill. The back and tail are uni-

Drymornis bridgesii
  ▨ Resident

Hylexetastes uniformis
  ▨ Resident

formly colored reddish brown, with a somewhat lighter belly, and few distinct markings.

DISTRIBUTION
Occurs in southeastern Bolivia and central Amazonian Brazil.

HABITAT
Inhabits lowland, humid, tropical forest, as high as about 1,650 ft (500 m).

BEHAVIOR
Usually occurs singly or in pairs. The song is a series of four to six loud, piercing whistles.

FEEDING ECOLOGY AND DIET
Forages for arthropods on tree-trunks and stout branches.

REPRODUCTIVE BIOLOGY
Lays two or three eggs in a nest in a tree-cavity or abandoned woodpecker hole. The sexes share incubation and care of the nestlings.

CONSERVATION STATUS
Not threatened. A widespread but not abundant species.

SIGNIFICANCE TO HUMANS
None known. ◆

Great rufous woodcreeper
Xiphocolaptes major

TAXONOMY
Xiphocolaptes major Vieillot, 1818.

OTHER COMMON NAMES
French: Grand Grimpar; German: Riesenbaumsteiger; Spanish: Trepatronco Castaño.

PHYSICAL CHARACTERISTICS
Body length 11–12 in (28–31 cm). A large woodcreeper with a long tail and a stout, rather long, slightly downcurved bill. The back and tail are uniformly colored rufous-brown, with a somewhat lighter cinnamon-brown head and underparts.

DISTRIBUTION
Occurs in north and central Bolivia, southwestern Brazil, Paraguay, and northern Argentina.

HABITAT
Inhabits lowland subtropical forest and open woodland, as high as about 4,900 ft (1,500 m).

BEHAVIOR
Usually occurs singly or in pairs. The song is a series of loud, piercing whistles.

FEEDING ECOLOGY AND DIET
Forages for arthropods on tree-trunks and stout branches, and sometimes on the ground.

REPRODUCTIVE BIOLOGY
Lays two or three eggs in a nest in a tree-cavity or abandoned woodpecker hole. The sexes share incubation and care of the nestlings.

CONSERVATION STATUS
Not threatened. A widespread but not abundant species.

SIGNIFICANCE TO HUMANS
None known. ◆

Xiphocolaptes major

▨ Resident

Lesser woodcreeper
Lepidocolaptes fuscus

TAXONOMY
Lepidocolaptes fuscus Vieillot, 1818.

OTHER COMMON NAMES
French: Grimpar brun; German: Schlankschnabel-Baumsteiger; Spanish: Chinchero Enano.

PHYSICAL CHARACTERISTICS
Body length about 7 in (17–18 cm). A medium-sized, rather slender woodcreeper with a long tail and a slim, short, down-curved bill. The back and tail are colored rufous-brown, the throat is whitish, and the underparts are brown-and-white streaked.

DISTRIBUTION
Occurs widely in northeastern South America, in eastern Brazil, Paraguay, and northeastern Argentina.

HABITAT
Inhabits humid lowland tropical forest, mature secondary forest, and montane forest as high as about 4,300 ft (1,300 m).

BEHAVIOR
Usually occurs singly or in pairs, but may also accompany mixed-species foraging flocks. The song is a trill-like series of notes.

FEEDING ECOLOGY AND DIET
Forages for arthropods on tree-trunks and stout branches.

Lepidocolaptes fuscus
▨ Resident

Campylorhamphus trochilirostris
▨ Resident

REPRODUCTIVE BIOLOGY
Lays two or three eggs in a nest in a tree-cavity or abandoned woodpecker hole. The sexes share incubation and care of the nestlings.

CONSERVATION STATUS
Not threatened. A widespread and locally abundant species.

SIGNIFICANCE TO HUMANS
None known. ◆

Red-billed scythebill
Campylorhamphus trochilirostris

TAXONOMY
Campylorhamphus trochilirostris M.H.K. Lichtenstein, 1820.

OTHER COMMON NAMES
English: Black-billed scythebill; French: Grimpar à bec rouge; German: Rotrücken-Sensenschnabel; Trauersensenschnabel; Spanish: Picapalo Rojizo.

PHYSICAL CHARACTERISTICS
Body length 9.5–11 in (24–28 cm). A large woodcreeper with a long tail and a slender, long, strongly downcurved, reddish bill (length 2.5–3.5 in; 6.5–9 cm). The back and tail are colored ru-

fous-brown, with lighter cinnamon-brown underparts, and brown-and-white streaked head and throat.

DISTRIBUTION
Occurs widely in three disjunct regions, including areas in Panama, Venezuela, Colombia, Ecuador, Peru, Bolivia, Brazil, Paraguay, and northern Argentina.

HABITAT
Inhabits lowland humid tropical forest, mature secondary forest, open woodland, and montane forest as high as about 6,600 ft (2,000 m).

BEHAVIOR
Usually occurs singly or in pairs, but may accompany mixed-species foraging flocks. The song is a series of ascending or descending musical notes.

FEEDING ECOLOGY AND DIET
Forages for arthropods on tree-trunks and stout branches.

REPRODUCTIVE BIOLOGY
Lays two or three eggs in a nest in a tree-cavity or abandoned woodpecker hole. The sexes share incubation and care of the nestlings.

CONSERVATION STATUS
Not threatened. A widespread and locally abundant species.

SIGNIFICANCE TO HUMANS
None known. ◆

Resources

Books

BirdLife International. *Threatened Birds of the World.* Barcelona, Spain, and Cambridge, UK: Lynx Edicions and BirdLife International, 2000.

Ridgely, R.S., and G. Tudor. *The Birds of South America.* Vol. II, *The Suboscine Passerines.* Austin, Texas: University of Texas Press, 1994.

Organizations

BirdLife International. Wellbrook Court, Girton Road, Cambridge, Cambridgeshire CB3 0NA United Kingdom. Phone: +44 1 223 277 318. Fax: +44-1-223-277-200. E-mail: birdlife@birdlife.org.uk Web site: <http://www.birdlife.net>

IUCN–The World Conservation Union. Rue Mauverney 28, Gland, 1196 Switzerland. Phone: +41-22-999-0001. Fax: +41-22-999-0025. E-mail: mail@hq.iucn.org Web site: <http://www.iucn.org>

Bill Freedman, PhD

Ant thrushes

(Formicariidae)

Class Aves
Order Passeriformes
Suborder Tyranni (Suboscines)
Family Formicariidae

Thumbnail description
Small to medium-sized songbirds with short, rounded wings, a short or long tail, and a stout or slender bill slightly hooked at the tip, feed on insects on the ground or in trees or thickets; prey is usually gleaned from foliage, although some species also catch flying insects. Some species participate in mixed-species foraging flocks that follow columns of army ants to catch insects and other small animals

Size
4–14 inches (10–36 cm)

Number of genera, species
52 genera; 244 species

Habitat
Shrublands and forests in subtropics and tropics

Conservation status
Critically Endangered: 4 species; Endangered: 16 species; Vulnerable: 16 species; Near Threatened: 18 species; Data Deficient: 1 species

Distribution
Southern Mexico to northern Argentina, with most species in the Amazonian region of tropical South America

Evolution and systematics

As treated here, the Formicariidae includes two closely related groups of perching birds (Passeriformes), which are sometimes considered as separate families. These are the ground antbirds and the typical antbirds (separated by some taxonomists into the family Thamnophilidae). These are large and highly varied groups of birds, consisting of about 56 species of ground antbirds and 188 species of typical antbirds. Their greatest species radiation and diversity occur in the Amazonian basin of tropical South America, where some locations may have as many as 30–40 species of Formicariidae present. Many species have only recently been described, and little is known of the life history, behavior, or ecology of any but the most abundant species in the family.

Physical characteristics

The family Formicariidae contains about 244 species of birds variously known as ant thrushes, antbirds, antcatchers, antpittas, antshrikes, or antwrens. They are among the most widespread and abundant birds of tropical and subtropical re-

gions of Central and South America. Because there are so many species, the typical characteristics of the group are not readily described. Moreover, the more widespread species may exhibit considerably geographic variation in plumage patterns and coloration, and sometimes in foraging ecology, song, and other qualities as well.

The wings are generally short and rounded. Some species have a short tail, which they typically hold erect, while in others it is longer than the body. The bill of the larger species is relatively stout and has a hooked tip. The bill may also have a single serration on the side, known as a tooth, similar to that of shrikes (Laniidae). Smaller species in the antbird family have a finer, smooth bill and lack the tooth. The legs of species that live on or close to the ground are muscular and long, particularly in the antpittas, although the toes are relatively short. The plumage is typically full and soft, especially on the back and sides of the body of the antshrikes. The sexes of the Thamnophilidae group are usually strongly dimorphic in the coloration of their plumage, while those of the Formicariidae are mostly similar, or monomorphic. In sexually dimorphic species, the coloration is generally dark in

A female plain antvireo (*Dysithamnus mentalis*) brooding nestlings in Costa Rica. (Photo by Michael Fogden/Animals Animals. Reproduced by permission.)

restrial in their foraging preference, while those in the Formicariidae group are mostly arboreal. Only a few species, however, forage in the uppermost tree-crown part of the canopy or inhabit open shrubby areas exposed to the sun. Occasionally, they may be seen bathing in shallows of quiet forest streams or in rain puddles. None of the ant thrushes or antbirds flies far; rather, they make short-distance flights within a territory or local foraging habitat. When they sense an invasion of their territory by a competitor, males tend to fly directly through the undergrowth towards the source of a specific song; some species respond strongly to playbacks of their own songs, making it easier to see them in dense vegetation.

When foraging, or if agitated, some species move about in noisy jumps, while others slink through the dense foliage. Often, however, they will fly up to a branch to briefly perch for a better view of an intruder. Most species have bright, white signal spots, particularly on the back, that are hidden deep in the plumage when the bird is at rest. When they feel threatened, the birds display these prominent spots in an alternating on-and-off-again manner, alerting nearby individuals to the presence of possible danger. Almost all species vocalize frequently and loudly, and are much more often heard than seen. Songs generally consist of short rhythmical phrases. These are rather non-melodious and quacking in many of the ant-shrikes, and other species have pure whistling sounds. However, the loud, flute-like, long-lasting scales of the ant thrushes are among the most beautiful, harmonious, and characteristic avian sounds of forests of the tropical Americas. Females also sing, and sometimes a pair will vocalize as a coordinated duet. Fledged young

males, and often black, gray, with some white, while females are generally brown and more strikingly patterned with bright or paler spots on the body, wings, and tail. Females of some widespread species show considerably more geographic variation in their coloration than do the males, an unusual pattern referred to as heterogynism.

Distribution

Distribution of species in the family Formicariidae ranges from warm, humid regions of southern Mexico to Paraguay and northern Argentina in South America. The largest number of species is found in the Amazon basin and other regions of tropical rainforest.

Habitat

Formicariids occur in a variety of shrubby and forested habitats in subtropical and tropical regions. Species occur in a wide range of lowland forests and woodlands, including secondary forest, and range as high as montane cloud-forest.

Behavior

Species of ant thrushes and antbirds tend to forage on the ground or into the medium levels of the forest canopy. In general, those in the Thamnophilidae group are mainly ter-

A male slaty antwren (*Myrmotherula schisticolor*) at its nest in Costa Rica. (Photo by Michael Fogden/Animals Animals. Reproduced by permission.)

sometimes make themselves noticeable by making conspicuous location calls.

Feeding ecology and diet

The tooth on the bill is used effectively in killing prey, including such arthropods as crickets, bugs, beetles, spiders, centipedes, and woodlice, and sometimes other kinds of invertebrates such as land snails. Larger species of these birds may also eat small frogs, lizards, snakes, mice, and young nestlings. Some species supplement their diet with seeds and small fruits.

The "ant-" prefix of the name of many species derives from the habit of seeking and following foraging swarms of army ants. They do this to snatch up the many arthropods and other small animals flushed by the foraging ants. Some ant-following species are so closely adapted to army ants that they are rarely found far from swarms of these insects. The behavior of antbirds and other species near a foraging swarm of army ants is a thrilling spectacle. Antbirds themselves are a useful guide to finding ant swarms, because their loud calls and songs betray the presence of the insects. The birds are particularly attracted to swarms of the red army ant (*Eciton burchelli*) and the smaller black rain ant (*Labidus praedator*). During their periods of mass foraging, huge numbers of heavily armed army ants are on the move, sometimes in fronts several feet wide, but often in narrower columns. The ground appears to come alive at the front of an advancing column as many small animals and insects run to escape the aggressively foraging ants. Crickets, in particular, may rise up in astonishing numbers. Even larger animals such as lizards, mice, and bird nestlings are potential prey for the swarms of army ants.

When found and killed, the dead prey are cut into pieces by the ants and hauled to the central, staging location of the swarm. Meanwhile, antbirds and other birds pick off some of the smaller prey as they flush into the open. The antbirds commonly hang off a woody shoot close to the ground or from a low vine, or perch upon a stump, waiting for the flushing of prey. For some species, the local presence of one or more army-ant swarms is a crucial attribute of habitat quality, which may be vigorously defended against intruders of the same or other species. Outside the breeding season, ant-dependent species may be nomadic to some degree, seeking active swarms of army ants. Dominant individuals, particularly adult males and owners of nearby territories, often drive off juvenile birds of their species. During the breeding season, antbird pairs usually restrict themselves to ant swarms that pass near their nest. A behavior known as anting, or the rubbing of live ants into the plumage, has been observed in antbirds as well as other birds, likely serving to kill skin parasites by releasing the formic acid of the ants.

Reproductive biology

Most formicariids appear to be sedentary, staying within their breeding territory. Most species, possibly all, appear to be monogamous, mating for life. Many species construct a deep, open-cup nest lightly fitted into a narrow branch-fork

A rusty-backed antwren (*Formicivora rufa*) at its nest in Brazil. (Photo by Fabio Colombini/Animals Animals. Reproduced by permission.)

of thinner branches of a shrub or low tree, often hanging over water. Other species build well-closed, spherical, oven-shaped nests with a side entrance on the forest floor. Still others build a woven pouch-nest. Some species breed in natural cavities in rotted trees, logs, or stumps. The typical clutch is two eggs, colored white or yellowish with fine spots, or sometimes an unspotted white or uniform blue-green. For most species, incubation is 14–17 days. Although most species have not been studied well, it appears that the general pattern is for parents to share in the incubation of eggs (though females usually brood at night), feeding of the young, and tending of the fledglings. Young leave the nest soon after hatching and follow their parents about, seeking food and shelter. Many species appear to remain paired all year, and most remain in or close to their territory. Some species, however, become sociable after the breeding season and wander about in mixed foraging flocks with ovenbirds, wood-creepers, tanagers, and other small birds.

Conservation status

As of 2001, the IUCN listed 36 species as being threatened, plus an additional 19 species that are considered Near Threatened or Data Deficient. Little is known, however, about the conservation status of many other species in the family Formicariidae. As additional research is done, further species will certainly be listed as threatened. Many of these and other tropical and subtropical birds are declining rapidly in abundance because of destruction of natural forest habitats. Threatened species listed by the IUCN include:

- The white-bearded antshrike (*Biatas nigropectus*) is a rare species in bamboo-containing forest of southeastern Brazil and nearby Argentina. It is considered Vulnerable because of destruction of most of its original forest habitat in montane and lowland zones.

- The recurve-billed bushbird (*Clytoctantes alixii*) occurs only in a few isolated localities in westernmost Venezuela and nearby Colombia and is considered Endangered. The forest habitat of this species in lowlands and foothills has mostly been destroyed to develop agricultural lands.

- The speckled antshrike (*Xenornis setifrons*) inhabits steep, damp slopes and ravines in eastern Panama and adjacent northwestern Colombia and is considered Vulnerable. Its habitat of humid lowland and foothill forest has mostly been cleared for agricultural development and highway construction.

- The Alagoas antwren (*Myrmotherula snowi*) is a Critically Endangered species that persists only in a tiny area of upland forest in northeastern Brazil, the rest of its original habitat having been lost to deforestation.

- The ash-throated antwren (*Herpsilochmus parkeri*) is an Endangered species occurring in a tiny range in northern Peru. Its humid montane forest is being lost to agricultural deforestation.

- The black-hooded antwren (*Formicivora erythronotos*) is an Endangered species whose only known surviving habitat near Rio de Janeiro, Brazil, is being degraded by tourism and recreational development.

- The rufous-fronted ant thrush (*Formicarius rufifrons*) is a Near Threatened species of southeastern Peru. Its habitat of riverine floodplain thickets is at risk from agricultural development.

- The giant antpitta (*Grallaria gigantea*) is Endangered and occurs in moist cloud-forest habitat in southwestern Colombia and nearby Ecuador. Its habitat is being destroyed by agricultural deforestation.

- The bicolored antpitta (*Grallaria rufocinerea*) is a Vulnerable species of the Central Andean region of Colombia. Its habitat of cloud-forest and humid montane forest has mostly been cleared for agricultural land-use.

Significance to humans

Formicariids are rarely hunted as food. As such, they are not of much direct importance to humans. However, views of these and other tropical and subtropical birds are widely sought by birdwatchers and other ecotourists, and this can bring significant economic benefits to accessible areas that retain natural forest habitat.

1. Warbling antbird (*Hypocnemis cantator*); 2. Black-throated antbird (*Myrmeciza atrothorax*); 3. Black-faced antbird (*Myrmoborus myotherinus*); 4. Spot-backed antbird (*Hylophylax naevia*); 5. Gray antbird (*Cercomacra cinerascens*); 6. Giant ant-pitta (*Grallaria gigantea*); 7. Black-faced antthrush (*Formicarius analis*); 8. Fulvous-bellied ant-pitta (*Hylopezus dives*); 9. Thrush-like ant-pitta (*Myrmothera campanisona*). (Illustration by Dan Erickson)

1. Spot-crowned antvireo (*Dysithamnus puncticeps*); 2. Gray antwren (*Myrmotherula menetriesii*); 3. Ash-winged antwren (*Terenura spodioptila*); 4. Black-capped antwren (*Herpsilochmus atricapillus*); 5. Scaled antbird (*Drymophila squamata*); 6. Undulated antshrike (*Frederickena unduligera*); 7. Barred antshrike (*Thamnophilus doliatus*); 8. Fasciated antshrike (*Cymbilaimus lineatus*); 9. Giant antshrike (*Batara cinerea*); 10. Cinereous antshrike (*Thamnomanes caesius*). (Illustration by Dan Erickson)

Species accounts

Fasciated antshrike
Cymbilaimus lineatus

TAXONOMY
Cymbilaimus lineatus Leach, 1814.

OTHER COMMON NAMES
English: Bamboo antshrike; French: Batara fascié German: Zebra-Ameisenwürger; Spanish: Batará Franjeado.

PHYSICAL CHARACTERISTICS
7 in (17–18 cm); heavy hooked bill and red iris.

DISTRIBUTION
Southern Central America and north-central South America; from Honduras to Panama, and in Venezuela, Colombia, Ecuador, Guyana, eastern Peru, northern Bolivia, and western Amazonian Brazil.

HABITAT
Typically below 3,300 ft (1,000 m) in humid tropical forest; vine-tangled and shrubby borders of streams and rivers and tree-fall openings in intact forest; also, mature secondary forest.

BEHAVIOR
Nonmigratory, territory-defending pairs forage widely at various levels of a dense forest canopy. Song is a series of 6–8 soft, repeated whistles.

FEEDING ECOLOGY AND DIET
Feed in dense foliage on insects and other arthropods.

REPRODUCTIVE BIOLOGY
Monogamous pairs bond for life, typically lay two eggs, and share incubation and care of nestlings and fledglings.

CONSERVATION STATUS
Not threatened. Locally widespread and abundant.

SIGNIFICANCE TO HUMANS
No direct significance, except for the indirect economic benefits of bird-watching and ecotourism. ◆

Undulated antshrike
Frederickena unduligera

TAXONOMY
Frederickena unduliger Pelzeln, 1868.

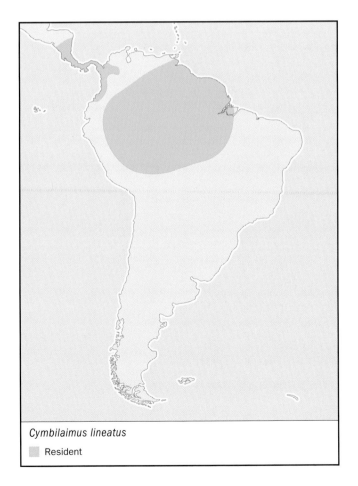

Cymbilaimus lineatus

■ Resident

Frederickena unduligera

■ Resident

OTHER COMMON NAMES
French: Batara ondé German: Mormor-Ameisenwürger; Spanish: Batará Ondulado.

PHYSICAL CHARACTERISTICS
9 in (23 cm); relatively short tail, a massive hooked bill, and brown to pale-orange iris.

DISTRIBUTION
Northwestern South America, including Colombia, Ecuador, Peru, Bolivia, and western Amazonian Brazil.

HABITAT
Typically below 2,300 ft (700 m) in humid, lowland tropical forest; dense undergrowth vegetation and vine-laden tree-falls within terra firme (non-flooded) forest.

BEHAVIOR
Nonmigratory, territory-defending pairs that forage close to or on the ground. Song is a series of 11–16, high-pitched, repeated notes. The head crest may be raised when calling, and the tail may wag as well.

FEEDING ECOLOGY AND DIET
Feed in dense foliage on insects and other arthropods.

REPRODUCTIVE BIOLOGY
Monogamous pairs bond for life, typically lay two eggs, and share incubation and care of nestlings and fledglings.

CONSERVATION STATUS
Uncommon to rare species, but not formally threatened.

SIGNIFICANCE TO HUMANS
No direct significance, except for the indirect economic benefits of bird-watching and ecotourism. ◆

Batara cinerea
☐ Resident

Giant antshrike
Batara cinerea

TAXONOMY
Batara cinereus Vieillot, 1819.

OTHER COMMON NAMES
French: Batara géant; German: Batará; Spanish: Batará Grande.

PHYSICAL CHARACTERISTICS
Largest species of the formicariids: 12.5–14 in (30.5–35.5 cm), with a relatively long tail and a massive bill.

DISTRIBUTION
East-central South America, including the eastern slope of the Andes Mountains in Bolivia, northern Argentina, and in a separate range in southeastern Brazil and northeastern Argentina.

HABITAT
Up to 9,800 ft (3,000 m) in humid tropical and montane forest, forest-edges, and dense thickets in higher woodland; dense vegetation of the lower or middle parts of the forest canopy.

BEHAVIOR
Nonmigratory pairs defend a relatively large territory. They forage close to the ground or in the middle canopy. They are a rarely seen shy and skulking bird. Song is a loud, rather fast series of repeated, ringing, musical notes.

FEEDING ECOLOGY AND DIET
Feed in dense foliage on insects and other arthropods.

REPRODUCTIVE BIOLOGY
Monogamous pairs bond for life, typically lay two eggs, and share incubation and care of nestlings and fledglings.

CONSERVATION STATUS
Uncommon species but can be locally abundant, and not threatened.

SIGNIFICANCE TO HUMANS
No direct significance, except for the indirect economic benefits of bird-watching and ecotourism. ◆

Barred antshrike
Thamnophilus doliatus

TAXONOMY
Thamnophilus doliatus Linnaeus, 1764.

OTHER COMMON NAMES
English: Chapman's antshrike; French: Batara rayé; German: Bindenwollrücken; Spanish: Choca Barreada.

PHYSICAL CHARACTERISTICS
6.5 in (16 cm), with a yellow iris, long tail, and a large hooked bill.

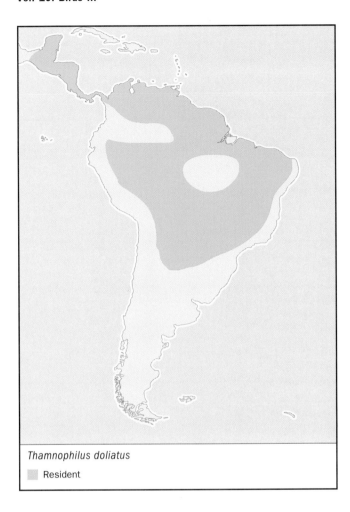

Thamnophilus doliatus

Resident

DISTRIBUTION
Much of tropical South America and Central America; east of the Andes as far south as Bolivia, Paraguay, and northern Argentina; widespread farther north except for the heart of Brazilian Amazon; as far north as southern Mexico.

HABITAT
Up to 6,600 ft (2,000 m) in tropical forest-edges, thickets, open woodland, and in vegetated clearings and gardens, ranging from humid to more arid habitats.

BEHAVIOR
Nonmigratory pairs defend a breeding territory. Both sexes sing a fast series of nasal notes; there are also several other calls.

FEEDING ECOLOGY AND DIET
Feed in dense foliage on insects and other arthropods.

REPRODUCTIVE BIOLOGY
Monogamous pairs bond for life, typically lay two eggs, and share incubation and care of nestlings and fledglings.

CONSERVATION STATUS
Not threatened. Widespread and relatively abundant.

SIGNIFICANCE TO HUMANS
No direct significance, except for the indirect economic benefits of bird-watching and ecotourism. ◆

Cinereous antshrike
Thamnomanes caesius

TAXONOMY
Thamnomanes caesia Temminck, 1820.

OTHER COMMON NAMES
English: Bluish-slate antshrike; French: Batara cendré German: Buschwürgerling; Spanish: Choca Guayanesa.

PHYSICAL CHARACTERISTICS
6 in (14.5 cm), with a long tail.

DISTRIBUTION
Much of northern tropical South America; east of the Andes in Venezuela, Guyana, Colombia, Ecuador, northeastern Peru, Amazonian Brazil, and northeastern Bolivia; coastal eastern Brazil.

HABITAT
Up to 2,650 ft (800 m) in humid, lowland tropical forest, especially in terra firme (or non-flooded) habitats and mature secondary forest; lower-canopy habitats.

BEHAVIOR
Nonmigratory pairs defend a breeding territory, but may also associate with mixed-species flocks. The vocalizations are loud and distinct, and help to organize local, mixed-species flocks. Song is a series of notes and trills.

Thamnomanes caesius

Resident

FEEDING ECOLOGY AND DIET
Feed in dense foliage on insects and other arthropods. Prey is often caught in the air.

REPRODUCTIVE BIOLOGY
Monogamous pairs bond for life, typically lay two eggs, and share incubation and care of nestlings and fledglings.

CONSERVATION STATUS
Not threatened. Widespread and relatively abundant.

SIGNIFICANCE TO HUMANS
No direct significance, except for the indirect economic benefits of bird-watching and ecotourism. ◆

Spot-crowned antvireo
Dysithamnus puncticeps

TAXONOMY
Dysithamnus puncticeps Salvin, 1866.

OTHER COMMON NAMES
French: Batara ponctué; German: Perlkappenwürgerling; Spanish: Choquita de Corona Moteada.

PHYSICAL CHARACTERISTICS
4.5 in (11.5 cm), with a whitish iris and moderate-length tail.

Dysithamnus puncticeps

■ Resident

DISTRIBUTION
Pacific slope of the Andes Mountains in western Colombia and northwestern Ecuador.

HABITAT
Up to 3,300 ft (1,000 m) in humid tropical forest, especially in lower-growth parts of the canopy.

BEHAVIOR
Nonmigratory pairs defend a breeding territory. Song is a series of soft whistled notes.

FEEDING ECOLOGY AND DIET
Feed in dense foliage on insects and other arthropods.

REPRODUCTIVE BIOLOGY
Monogamous pairs bond for life, typically lay two eggs, and share incubation and care of nestlings and fledglings.

CONSERVATION STATUS
Not threatened. Locally abundant.

SIGNIFICANCE TO HUMANS
No direct significance, except for the indirect economic benefits of bird-watching and ecotourism. ◆

Black-capped antwren
Herpsilochmus atricapillus

TAXONOMY
Herpsilochmus atricapillus Pelzeln, 1868.

OTHER COMMON NAMES
English: Bahia antwren, creamy-bellied antwren, pileated antwren; French: Grisin mitré; German: Schwarzkopf-Ameisenfänger; Spanish: Tiluchí de Cabeza Negra.

PHYSICAL CHARACTERISTICS
5 in (12 cm), with a rather long tail.

DISTRIBUTION
East-central South America in Brazil, eastern Bolivia, Paraguay, and northwestern Argentina.

HABITAT
Up to 3,600 ft (1,100 m) in humid tropical forest and woodlands.

BEHAVIOR
Nonmigratory pairs defend a breeding territory. Both sexes sing an accelerating trilled song; males often echoed by females. Tail rapidly vibrates while singing.

FEEDING ECOLOGY AND DIET
Gleans insects and other arthropods from dense foliage throughout the canopy.

REPRODUCTIVE BIOLOGY
Monogamous pairs bond for life, typically lay two eggs, and share incubation and care of nestlings and fledglings.

CONSERVATION STATUS
Not threatened. Widespread and locally abundant.

Herpsilochmus atricapillus

■ Resident

Myrmotherula menetriesii

■ Resident

SIGNIFICANCE TO HUMANS
No direct significance, except for the indirect economic benefits of bird-watching and ecotourism. ◆

Gray antwren
Myrmotherula menetriesii

TAXONOMY
Myrmotherula menetriesii d'Orbigny, 1837.

OTHER COMMON NAMES
French: Myrmidon gris; German: Buntflügel-Ameisenschlüpfer; Spanish: Hormiguero de Garganta Gris.

PHYSICAL CHARACTERISTICS
4 in (9.5-10 cm), with a rather short tail.

DISTRIBUTION
Northern South America, including southern Venezuela, Guyana, eastern Colombia, eastern Ecuador, eastern Peru, northern Bolivia, and much of Amazonian Brazil.

HABITAT
Up to 2,950 ft (900 m) in humid tropical forest, forest-edges, and openings within forest. Mostly restricted to terra firme (or non-flooded) forest.

BEHAVIOR
Nonmigratory pairs defend a breeding territory. Often join mixed-species foraging flocks, but tend to feed higher in the

canopy than other species. Song is a weak series of about 12 repeated notes.

FEEDING ECOLOGY AND DIET
Glean insects and other arthropods from foliage in the middle and upper canopy.

REPRODUCTIVE BIOLOGY
Monogamous pairs bond for life, typically lay two eggs, and share incubation and care of nestlings and fledglings.

CONSERVATION STATUS
Not threatened. Widespread and locally abundant.

SIGNIFICANCE TO HUMANS
No direct significance, except for the indirect economic benefits of bird-watching and ecotourism. ◆

Ash-winged antwren
Terenura spodioptila

TAXONOMY
Terenura spodioptila P.L. Sclater & Salvin, 1881.

OTHER COMMON NAMES
French: Grisin spodioptile; German: Grauschwingen-Ameisenfänger; Spanish: Tiluchí Piojito.

Terenura spodioptila

Resident

Scaled antbird

Drymophila squamata

TAXONOMY

Drymophila squamata M.H.K. Lichtenstein, 1823.

OTHER COMMON NAMES

French: Grisin écaillé; German: Schuppenameisenfänger; Spanish: Tiluchí Escamado.

PHYSICAL CHARACTERISTICS

5 in (11.5 cm), with a long tail.

DISTRIBUTION

Eastern coastal Brazil.

HABITAT

Below 2,000 ft (600 m) in understory vegetation of humid tropical forest, forest-edges, and mature secondary forest.

BEHAVIOR

Nonmigratory pairs defend a breeding territory. Song is a raspy series of descending notes, sometimes echoed by the female.

FEEDING ECOLOGY AND DIET

Glean insects and other arthropods from foliage in dense vegetation near ground level.

REPRODUCTIVE BIOLOGY

Monogamous pairs bond for life, typically lay two eggs, and share incubation and care of nestlings and fledglings.

PHYSICAL CHARACTERISTICS

4 in (10 cm), with a long tail.

DISTRIBUTION

Northern South America, including southern Venezuela, Guyana, southeastern Colombia, northeastern Ecuador, eastern Peru, and the northern Amazonian Brazil.

HABITAT

Up to 3,600 ft (1,100 m) in humid tropical forest and forest-edges, mostly in terra firme (or non-flooded) forest. Utilize higher parts of the canopy.

BEHAVIOR

Nonmigratory pairs defend a breeding territory. Often in mixed-species foraging flocks. Song is an accelerating trill.

FEEDING ECOLOGY AND DIET

Glean insects and other arthropods from foliage in the upper parts of the forest canopy.

REPRODUCTIVE BIOLOGY

Monogamous pairs bond for life, typically lay two eggs, and share incubation and care of nestlings and fledglings.

CONSERVATION STATUS

Not threatened. Locally abundant.

SIGNIFICANCE TO HUMANS

No direct significance, except for the indirect economic benefits of bird-watching and ecotourism. ◆

Drymophila squamata

Resident

CONSERVATION STATUS
Not threatened. Locally abundant.

SIGNIFICANCE TO HUMANS
No direct significance, except for the indirect economic benefits of bird-watching and ecotourism. ◆

Gray antbird
Cercomacra cinerascens

TAXONOMY
Cercomacra cinerascens P.L. Sclater, 1857.

OTHER COMMON NAMES
French: Grisin ardoisé; German: Aschkopf- Ameisenfänger; Spanish: Hormiguerito Gris.

PHYSICAL CHARACTERISTICS
6 in (16 cm), with a long tail.

DISTRIBUTION
Northern South America, including southern Venezuela, Guyana, eastern Colombia, eastern Ecuador, eastern Peru, northern Bolivia, and widely in Amazonian Brazil.

HABITAT
Below 2,300 ft (700 m) in the mid- and upper-canopy of humid tropical forest and mature secondary forest, particularly in terra-firme (or non-flooded) forest.

BEHAVIOR
Nonmigratory pairs defend a breeding territory. Sometimes associated with mixed-species foraging flocks. Song is a rough series of notes, sometimes echoed by the female.

FEEDING ECOLOGY AND DIET
Glean insects and other arthropods from foliage in dense vegetation in mid- and upper-canopy habitats.

REPRODUCTIVE BIOLOGY
Monogamous pairs bond for life, typically lay two eggs, and share incubation and care of nestlings and fledglings.

CONSERVATION STATUS
Not threatened. Widespread and locally abundant.

SIGNIFICANCE TO HUMANS
No direct significance, except for the indirect economic benefits of bird-watching and ecotourism. ◆

Black-faced antbird
Myrmoborus myotherinus

TAXONOMY
Myrmoborus myotherinus Spiz, 1825.

OTHER COMMON NAMES
French: Alapi masqué; German: Schuppenflügel- Ameisenschnäpper; Spanish: Hormiguero Ratonero.

Cercomacra cinerascens

▦ Resident

Myrmoborus myotherinus

▦ Resident

PHYSICAL CHARACTERISTICS
5 in (13 cm), with a bright red iris, and short tail.

DISTRIBUTION
Amazonian region of northern South America, including southern Venezuela, southeastern Colombia, eastern Ecuador, eastern Peru, northern Bolivia, and widely in Amazonian Brazil.

HABITAT
Below 2,300 ft (700 m) in the mid- and upper-canopy of humid terra-firme (non-flooded) tropical forest and mature secondary forest.

BEHAVIOR
Nonmigratory pairs defend a breeding territory. Sometimes associated with mixed-species foraging flocks. Song is a loud, fast series of notes.

FEEDING ECOLOGY AND DIET
Glean insects and other arthropods from foliage in dense lower-canopy vegetation.

REPRODUCTIVE BIOLOGY
Monogamous pairs bond for life, typically lay two eggs, and share incubation and care of nestlings and fledglings.

CONSERVATION STATUS
Not threatened. Widespread and relatively abundant.

SIGNIFICANCE TO HUMANS
No direct significance, except for the indirect economic benefits of bird-watching and ecotourism. ◆

Hylophylax naevia
▨ Resident

Spot-backed antbird
Hylophylax naevia

TAXONOMY
Hylophylax naevia J.F. Gmelin, 1789.

OTHER COMMON NAMES
French: Alapi paludicole; German: Braunflecken-Waldwächter; Spanish: Hormiguero de Espalda Punteada.

PHYSICAL CHARACTERISTICS
4.5 in (11.5 cm), with a gray iris, and short tail.

DISTRIBUTION
Amazonian region of northern South America, including southern Venezuela, Guyana, southeastern Colombia, eastern Ecuador, eastern Peru, northern Bolivia, and widely in Amazonian Brazil.

HABITAT
Below 3,600 ft (1,100 m) in undergrowth vegetation of terra-firme (non-flooded) and flooded tropical forest.

BEHAVIOR
Nonmigratory pairs defend a breeding territory. Song is a fast, high-pitched series of wheezy notes.

FEEDING ECOLOGY AND DIET
Glean insects and other arthropods from foliage in dense lower-canopy vegetation.

REPRODUCTIVE BIOLOGY
Monogamous pairs bond for life, typically lay two eggs, and share incubation and care of nestlings and fledglings.

CONSERVATION STATUS
Not threatened. Widespread and relatively abundant.

SIGNIFICANCE TO HUMANS
No direct significance, except for the indirect economic benefits of bird-watching and ecotourism. ◆

Warbling antbird
Hypocnemis cantator

TAXONOMY
Hypocnemis cantator Boddaert, 1783.

OTHER COMMON NAMES
French: Alapi carillonneur; German: Singameisenschnäpper; Spanish: Hormiguero Cantarín.

PHYSICAL CHARACTERISTICS
5 in (12 cm), with a black iris and short tail.

DISTRIBUTION
Amazonian region of northern South America, including southern Venezuela, Guyana, southeastern Colombia, eastern Ecuador, eastern Peru, northern Bolivia, and widely in Amazonian Brazil.

HABITAT
Below 3,300 ft (1,000 m) in dense undergrowth vegetation of the borders of humid tropical forest and secondary forest, often in the vicinity of wet areas.

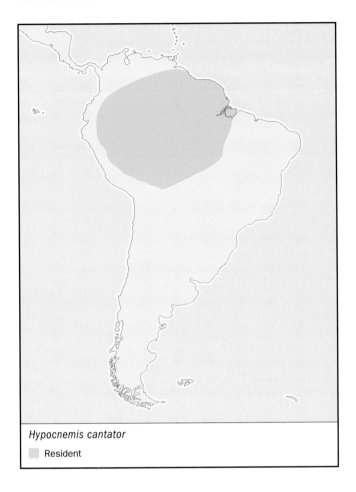

Hypocnemis cantator

Resident

Myrmeciza atrothorax

Resident

BEHAVIOR
Nonmigratory pairs defend a breeding territory. Song of males is a rapid series of notes, sometimes echoed by the female.

FEEDING ECOLOGY AND DIET
Glean insects and other arthropods from foliage in dense lower-canopy vegetation.

REPRODUCTIVE BIOLOGY
Monogamous pairs bond for life, typically lay two eggs, and share incubation and care of nestlings and fledglings.

CONSERVATION STATUS
Not threatened. Widespread and relatively abundant.

SIGNIFICANCE TO HUMANS
No direct significance, except for the indirect economic benefits of bird-watching and ecotourism. ◆

Black-throated antbird
Myrmeciza atrothorax

TAXONOMY
Myrmeciza atrothorax Boddaert, 1783.

OTHER COMMON NAMES
French: Alapi de Buffon; German: Pechbrust-Ameisenvogel; Spanish: Hormiguero de Garganta Negra.

PHYSICAL CHARACTERISTICS
5.5 in (14 cm), with a black iris and moderately long tail.

DISTRIBUTION
Amazonian region of northern South America, including southern Venezuela, Guyana, southeastern Colombia, northeastern Ecuador, eastern Peru, northern Bolivia, and widely in Amazonian Brazil.

HABITAT
Below 1,600 ft (500 m) in dense undergrowth vegetation of the borders of humid tropical forest, secondary forest, and savanna woodland, usually in the vicinity of wet areas.

BEHAVIOR
Nonmigratory pairs defend a breeding territory. May forage in larger groups. Song of males is a rapid, high-pitched series of notes.

FEEDING ECOLOGY AND DIET
Glean insects and other arthropods from foliage in dense vegetation close to the ground.

REPRODUCTIVE BIOLOGY
Monogamous pairs bond for life, typically lay two eggs, and share incubation and care of nestlings and fledglings.

CONSERVATION STATUS
Not threatened. Widespread and relatively abundant.

SIGNIFICANCE TO HUMANS
No direct significance, except for the indirect economic benefits of bird-watching and ecotourism. ◆

Black-faced antthrush
Formicarius analis

TAXONOMY
Formicarius analis d'Orbigny & Lafresnaye, 1837.

OTHER COMMON NAMES
French: Tétéma coq-de-bois; German: Schwarzkehl-Ameisendrossel; Spanish: Chululú Enmascarado.

PHYSICAL CHARACTERISTICS
7 in (17–18 cm), with a black iris and white eye-ring, and short tail held erect.

DISTRIBUTION
Amazonian region of northern South America and in tropical Central America; from tropical southern Mexico, through appropriate habitats in Guatemala, Honduras, Nicaragua, Costa Rica, Panama, coastal Venezuela, the Guyanas, northern and central Colombia, eastern Ecuador, eastern Peru, northern Bolivia, and widely in Amazonian Brazil.

HABITAT
Below 3,300 ft (1,000 m) in extremely dense undergrowth vegetation of humid tropical forest and mature secondary woodland.

BEHAVIOR
Nonmigratory pairs defend a breeding territory. May forage near swarms of army ants. Song of males is a series of up to 10 fading notes.

FEEDING ECOLOGY AND DIET
Glean insects and other arthropods from foliage in dense vegetation close to the ground.

REPRODUCTIVE BIOLOGY
Monogamous pairs bond for life, typically lay two eggs, and share incubation and care of nestlings and fledglings.

CONSERVATION STATUS
Not threatened. Widespread and relatively abundant.

SIGNIFICANCE TO HUMANS
No direct significance, except for the indirect economic benefits of bird-watching and ecotourism. ◆

Giant ant-pitta
Grallaria gigantea

TAXONOMY
Grallaria gigantea Lawrence, 1866.

Formicarius analis
    Resident

Grallaria gigantea
    Resident

OTHER COMMON NAMES
French: Grallaire géante; German: Riesenameisenpitta; Spanish: Chululú Gigante.

PHYSICAL CHARACTERISTICS
One of the largest birds in family; 9.5 in (24 cm), with a black iris and tan eye-ring, heavy bill, and very short tail.

DISTRIBUTION
Sporadic, local distribution on the western slopes of the Andes Mountains in southwestern Colombia and western Ecuador.

HABITAT
Between 7,200 to 9,850 ft (2,200–3,000 m) in montane primary and mature secondary forest, and sometimes in rough pasture near forest.

BEHAVIOR
Nonmigratory pairs defend a breeding territory. Song of males is a series of quavering notes lasting about five seconds.

FEEDING ECOLOGY AND DIET
Forages for insects and other arthropods on or very close to the ground.

REPRODUCTIVE BIOLOGY
Monogamous pairs bond for life, typically lay two eggs, and share incubation and care of nestlings and fledglings.

CONSERVATION STATUS
Endangered. Very rare species surviving in only a few, isolated populations. Its surviving critical habitats must be protected against damages caused by economic development.

SIGNIFICANCE TO HUMANS
No direct significance, except for the indirect economic benefits of bird-watching and ecotourism. ◆

Myrmothera campanisona
▨ Resident

Thrush-like ant-pitta
Myrmothera campanisona

TAXONOMY
Myrmothera campanisonam Hermann, 1783.

OTHER COMMON NAMES
French: Grallaire grand-beffroi; German: Fleckenbrust-Ameisenjäger; Spanish: Chululú Campanero.

PHYSICAL CHARACTERISTICS
6 in (15 cm), with a black iris, stout bill, and very short tail.

DISTRIBUTION
Amazonian region of northern South America, including southern Venezuela, Guyana, eastern Colombia, eastern Ecuador, eastern Peru, northwestern Bolivia, and widely in Amazonian Brazil.

HABITAT
Below 3,950 ft (1,200 m) in humid tropical forest, especially where there is dense undergrowth.

BEHAVIOR
Nonmigratory pairs defend a breeding territory. Song of males is a series of 5–6 whistled notes.

FEEDING ECOLOGY AND DIET
Forage for insects and other arthropods on or very close to the ground.

REPRODUCTIVE BIOLOGY
Monogamous pairs bond for life, typically lay two eggs, and share incubation and care of nestlings and fledglings.

CONSERVATION STATUS
Not threatened. Widespread and relatively abundant.

SIGNIFICANCE TO HUMANS
No direct significance, except for the indirect economic benefits of bird-watching and ecotourism. ◆

Fulvous-bellied ant-pitta
Hylopezus dives

TAXONOMY
Hylopezus dives Salvin, 1865.

OTHER COMMON NAMES
English: White-lored antpitta; French: Grallaire à ventre fauve; German: Schwarzkappen-Ameisenpitta; Spanish: Chululú de Buche Canela.

Hylopezus dives

▪ Resident

PHYSICAL CHARACTERISTICS
6 in (14.5 cm), with a black iris and very short tail.

DISTRIBUTION
From Honduras, Nicaragua, and Panama in Central America to the Pacific lowlands of western Colombia and adjacent northwestern Ecuador.

HABITAT
Below 2,950 ft (900 m) in dense vegetation along the edges of humid tropical forest and regenerating clearings.

BEHAVIOR
Nonmigratory pairs defend a breeding territory. Song of males is a series of 6–8 whistled notes.

FEEDING ECOLOGY AND DIET
Forage for insects and other arthropods on or very close to the ground.

REPRODUCTIVE BIOLOGY
Monogamous pairs bond for life, typically lay two eggs, and share incubation and care of nestlings and fledglings.

CONSERVATION STATUS
Not threatened. Relatively abundant.

SIGNIFICANCE TO HUMANS
No direct significance, except for the indirect economic benefits of bird-watching and ecotourism. ◆

Resources

Books

BirdLife International. *Threatened Birds of the World.* Barcelona, Spain, and Cambridge, UK: Lynx Edicions and BirdLife International, 2000.

Ridgely, R. S., and G. Tudor. *The Birds of South America.* Vol. II, *The Suboscine Passerines.* Austin: University of Texas Press. 1994.

Organizations

BirdLife International. Wellbrook Court, Girton Road, Cambridge, Cambridgeshire CB3 0NA United Kingdom. Phone: +44-1-223-277-318. Fax: +44-1-223-277-200. E-mail: birdlife@birdlife.org.uk Web site: <http://www.birdlife.net>

IUCN–The World Conservation Union. Rue Mauverney 28, Gland, 1196 Switzerland. Phone: +41-22-999-0001. Fax: 41-22-999-0025. E-mail: mail@hq.iucn.org Web site: <http://www.iucn.org>

Bill Freedman, PhD

Tapaculos

(Rhinocryptidae)

Class Aves

Order Passeriformes

Suborder Tyranni (Suboscines)

Family Rhinocryptidae

Thumbnail description
Wren- to thrush-sized birds, terrestrial or skulking in dense undergrowth; bill straight, in some species elevated at base; tail short to medium and carried half-cocked; large feet; most species dull-colored

Size
3.9–9.1 in (10–23 cm); 0.37–6.53 oz (10.4–185 g)

Number of genera, species
12 genera; 54 species

Habitat
Undergrowth and ground of forest and scrub, often bamboo thickets; a few species prefer bunch grass, rocks, or tall grass

Conservation status
Critically Endangered: 2 species; Endangered: 1 species; Vulnerable: 1 species; Near Threatened: 5 species

Distribution
Central and South America

Evolution and systematics

Tapaculos are among the most primitive of passerine birds. Based on both DNA and morphological comparisons, their closest relatives are believed to be the ground antbirds (Formicariidae, *sensu stricto*), particularly the genera *Formicarius* and *Chamaeza*, and the gnateaters (Conopophagidae). Together, these three form the superfamily Formicarioidea.

Twelve genera with 54 species are currently recognized, but one form of the genus *Melanopareia* that is currently treated as a subspecies may deserve full species rank, and the taxonomy of the genus *Scytalopus* is far from resolved. Several species are so alike that they cannot be told apart by external morphology, not even in the hand, rendering *Scytalopus* the most problematic of all bird genera. The species differ genetically, in vocalizations, elevational distribution, and frequently in body mass (but often not in other measurements), but plumage differences have not evolved as an important part of species recognition, probably owing to the dark haunts of these birds. Voices of several yet undescribed forms are known, and it seems possible that as many as ten or more species will be recognized through further taxonomic revision and description of new taxa.

Three genera are so aberrant that they may not belong in the same family as the others. *Melanopareia* and possibly with it the poorly studied *Teledromas* differ in so many characteristics that they are probably best placed in a family of their own. The monotypic *Psilorhamphus* has been variously placed with gnatwrens, wrens, and true antbirds, but an anatomical study showed that it shares some characteristics with the tapaculos, with which it was then placed. However, its affinities may yet be shown to lie elsewhere.

Physical characteristics

Tapaculos comprise a rather diverse array of forms with only a few common characteristics. They all have nostrils covered by a tactile flap and possess a sternum with four posterior notches. A similar sternum is only found in the gnateaters and some ground antbirds, both families being close relatives of the tapaculos. Presumably as a consequence of their limited use of flight, tapaculos have no keel on the sternum. The bill is straight and fairly weak, in some species with an elevated base.

The bones are exceptionally soft, and in the genus *Scytalopus* the brain case never ossifies at all. Except for

An elegant crescentchest (*Melanopareia elegans*) perches on a branch. (Photo by J. Dunning/VIREO. Reproduced by permission.)

rather bright chestnut, black and white colors in striking patterns.

Distribution

Tapaculos are distributed from Costa Rica through South America to Tierra del Fuego. The large majority of species are montane or live in temperate regions. Only one tapaculo, *Liosceles*, inhabits the Amazonian lowlands. Six species of *Scytalopus* and *Merulaxis*, and the peculiar *Psilorhamphus*, are confined to southeastern Brazil, some of them also occurring in adjacent parts of Argentina and Paraguay. Two of the four species of *Melanopareia* are found in the arid lowlands of northwestern Peru and southwestern Ecuador, the other two in the arid parts of Brazil, Paraguay, Bolivia and northern Argentina. *Teledromas* occurs in desert scrub in western Argentina; *Rhinocrypta* in the chaco and drier part of the pampas.

In the Andes the most widespread genus, *Scytalopus*, reaches its greatest diversity. Up to five species may occur on a single slope, some showing strikingly sharp altitudinal replacements. In Central America and north-east of the "Táchira Gap" in the Andes near the border of Colombia and Venezuela, the diversity is smaller. In the coastal mountains of Venezuela only a single *Scytalopus* species is found, none on the Tepuis or in the Guianas. Only two genera, *Myornis* and *Acropternis* are endemic to the northern Andes. Three genera, *Pteroptochos*, *Scelorchilus*, and *Eugralla* are endemic to the southern Andes.

Melanopareia and *Teledromas* the humerus is distinctly curved. *Melanopareia* (and possibly *Teledromas*) also differ by the shape of the stapes, a small bone in the inner ear. Most tapaculos have disproportionally strong feet and large claws. *Acropternis* has a very long hind claw, the function of which has been disputed.

The feather tract of the flank is fused with that of the back except in *Melanopareia* and *Teledromas*. The body is densely feathered, particularly on the rump, and the feathers fall off easily, probably to confuse predators. The feathers of the lores are stiff and erect in many tapaculos, protecting the eyes from dirt and ants. These feathers are most evolved in *Merulaxis*, where they are greatly elongated and allow the bird to see while boring its head into litter.

Tapaculos are nearly flightless. Their wings are short and rounded with ten primary flight feathers. The tail, usually carried half-cocked, is composed of a variable number of feathers ranging from eight to 14, the number varying even within the same population. The more or less graduated tail is decidedly short in many species, but medium long in some forms. The sexes are fairly similar in most species, with females appearing somewhat smaller and duller. In *Merulaxis*, however, the sexes are distinctly differently colored. Plumage colors are generally dull, grayish or brown, and without marked pattern, but *Melanopareia* exhibits some

Habitat

Tapaculos occur in a wide range of habitats. Most frequent dense undergrowth in humid or wet montane forest, and a few species have even adapted to tussock grassland and scree adjacent to forest in Patagonia and above treeline in the Andes. One inhabits the lowland forest of Amazonia, another is confined to uniform stands of tall grass and bulrush in south Brazilian marshes. A few inhabit semi-humid forest and scrub, and five species prefer arid scrub.

Behavior

Tapaculos live such secluded lives that little is known about their behavior. They try to avoid the open so much that in most cases they can be observed only briefly. Except for *Acropternis*, which hops slowly on the forest floor, most of the large tapaculos walk or run, making sudden halts. The fairly small *Teledromas* also runs, taking long strides as it speeds over the bare ground. Most of the small species are more constantly active and tend to hop rather than run while on the ground, and all tapaculos hop when moving in the vegetation above ground. Most tapaculos can be very swift when surprised and have been observed attacking a tape-recorder after playback of their song, but both *Acropternis* and *Liosceles* appear to be always slow-moving. *Acropternis* is perhaps best described as lethargic.

Feeding ecology and diet

Tapaculos take a variety of food, but mainly feed on insects and spiders. It has recently been found that some add a substantial amount of berries to their diet, at least seasonally.

The foraging mode varies between species. *Pteroptochos* and *Scelorchilus* walk or run quickly, then stop to scrape the ground with one foot, or with both feet simultaneously, throwing earth and leaves backwards in an awkward jump. Such "jump-scratching" has also been seen in *Eugralla* and is frequently seen in *Acropternis*. The grotesquely long hind claw of the latter may be an adaptation to this behavior. *Scytalopus* tapaculos glean prey from moss, litter, earth and rotting vegetation as they move along quickly in the undergrowth or on the ground with a mouse-like appearance. *Myornis* perch gleans clumps of bamboo. *Merulaxis* has been seen digging its head into leaf litter on the ground. *Liosceles* picks prey from the ground as it walks slowly, occasionally scratching with one foot, and appears to feed entirely on bugs. *Rhinocrypta* and *Teledromas* feed only on the ground, running quickly to cross open ground, the latter sometimes scratching the ground with one foot. *Melanopareia* runs on the ground like *Teledromas*, but also perch gleans insects while working through low branches like a *Synallaxis* spinetail. *Psilorhamphus* mainly hops through viny thickets of bamboo.

Reproductive biology

Some, perhaps all, tapaculos might form permanent pair bonds. Most small birds quickly replace a lost mate, and tapaculos are no exception. In *Scytalopus* tapaculos, a new male appears almost immediately after the old one has been removed.

Although nests of 18 of the 54 species of tapaculos are known, details of nesting have been studied in only a single species, *Rhinocrypta lanceolata*.

A few build cup nests, but most construct closed nests with a side entrance or place their nest in a tunnel. The nest is fairly soft, made of root fibers, grass, moss, and a few small twigs. Nests of *Eugralla* and *Rhinocrypta* are bulky. *Rhinocrypta*, *Eugralla*, and sometimes *Melanopareia* place their nest above ground, but most tapaculos nest at the end of a tunnel or hollow trunk. Tunnels may be dug by the bird or an abandoned rodent burrow may be used.

Most tapaculos lay two to three eggs that are white, large for the size of the bird, rounded and lacking in sheen. *Melanopareia* is an exception, as it lays ovoid and spotted eggs.

The incubation period is 15–17 days and the gestation period is 14–15 days for *Rhinocrypta*. Males take part in the incubation in some species, but apparently not in *Scytalopus*, where brood patches have been found only in females.

The young hatch naked. Both parents care for nestlings, but apparently fledglings are sometimes fed by the female alone.

Conservation status

Two species of tapaculo are listed as Critically Endangered and may have gone extinct over the last two decades. They

The oscellated tapaculo (*Acropternis orthonyx*) spends much of its time in the forest undergrowth. (Photo by Doug Wechsler/VIREO. Reproduced by permission.)

are *Merulaxis stresemanni* and *Scytalopus psychopompus*, both restricted to small areas in coastal Bahía in eastern Brazil, where deforestation continues at an alarming rate. The former has not been seen since 1995 and the latter not since the 1980s. Hope for their survival is slight.

The recently discovered *Scytalopus iraiensis* inhabits the few remaining rushy marshes in eastern Paraná in southern Brazil, a habitat under constant pressure from human development. It is considered Endangered and needs human support to insure its survival.

Scytalopus panamensis is confined to the Tacarcuna Massif in the Darién gap on the border of Panama and Colombia. Owing to forest clearance within its very restricted range, the species is considered Vulnerable. It is still common, but if the Panamerican Highway is completed as planned, the pressure on its habitat will accelerate drastically.

Five species are considered Near Threatened owing to small ranges and continuing loss of habitat. These are *Scytalopus novacapitalis* of swamp gallery forest in a small area in central Brazil; *Scytalopus indigoticus*, *Merulaxis ater*, and *Psilorhamphus guttatus*, all confined to the rapidly dwindling Atlantic forests of east and southeast Brazil; and *Melanopareia maranonica* of dense arid scrub in a small area in the Río Marañón drainage in northern Peru, a habitat increasingly disturbed and giving way to agriculture.

Additionally, *Scytalopus robbinsi*, which is confined to an area in southern Ecuador that will be nearly completely deforested over the next few decades unless protective measures are taken, should be added to the species of concern. Several other species of *Scytalopus* have very small ranges and are therefore vulnerable but not presently considered to be at risk.

Significance to humans

Tapaculos are small and do not occur in concentrations, so although they taste good, they have never been hunted.

Their secluded existence is revealed only through their loud calls, giving rise to onomatopoetic local names, the name tapaculo itself being apparently derived from the distinctive "tá-pa-koo" call of *Scelorchilus albicollis*. The loud voice of *Scelorchilus rubecula* figures in folklore in southern Chile, and the distinctive voice of *Liosceles* heard through large parts of the Amazon basin is so loud that it must be part of some folklore. No tapaculo held in captivity has ever been reported except for a single *Acropternis orthonyx*, which survived for only a few months in New York Zoo.

1. Chucao tapaculo (*Scelorchilus rubecula*); 2. Ocellated tapaculo (*Acropternis orthonyx*); 3. Collared crescentchest (*Melanopareia torquata*); 4. Ochre-flanked tapaculo (*Eugralla paradoxa*); 5. Moustached turca (*Pteroptochos megapodius*); 6. Crested gallito (*Rhinocrypta lanceolata*); 7. Black-throated huet-huet (*Pteroptochos tarnii*); 8. Spotted bamboowren (*Psilorhamphus guttatus*); 9. Slaty bristlefront (*Merulaxis ater*); 10. Rusty-belted tapaculo (*Liosceles thoracicus*). (Illustration by Brian Cressman)

Species accounts

Black-throated huet-huet
Pteroptochos tarnii

TAXONOMY
Hylactes tarnii King, 1831, Chiloé Island, and Port Otway, Gulf of Penas, Chile.

OTHER COMMON NAMES
English: Huet huet; French: Tourco huet-huet; German: Schwarzkehl-Huëthuët; Spanish: Huet-huet del Sur.

PHYSICAL CHARACTERISTICS
9 in (23 cm); 5.3–6.3 oz (150–180 g). Crown, forehead, and breast dark chestnut. The rest of body is black.

DISTRIBUTION
Southern Chile from Río Bío Bío to Brunswick Peninsula, and adjacent Argentina from Neuquén to northwestern Santa Cruz.

HABITAT
Humid forest.

BEHAVIOR
Terrestrial.

FEEDING ECOLOGY AND DIET
Walks slowly. Flips debris with bill and sometimes scrapes the ground with one foot. Feeds on invertebrates and berries.

REPRODUCTIVE BIOLOGY
Nest placed in excavated tunnel or a hollow trunk. Two or three eggs.

CONSERVATION STATUS
Not threatened.

SIGNIFICANCE TO HUMANS
None known. ◆

Moustached turca
Pteroptochos megapodius

TAXONOMY
Pteroptochos megapodius Kittlitz, 1830, Valparaiso, Chile.

OTHER COMMON NAMES
English: Turco; French: Tourco à moustaches; German: Turco; Spanish: Huet-huet Turco.

PHYSICAL CHARACTERISTICS
8.9 in (22.5 cm). Brown. Brow and broad moustache are whitish. Belly barred dusky.

DISTRIBUTION
P. m. atacamae: Northern Chile in Atacama. *P. m. megapodius*: Central Chile from Coquimbo to Concepción.

Pteroptochos tarnii
☐ Resident

Pteroptochos megapodius
☐ Resident

HABITAT
Semi-arid scrub.

BEHAVIOR
Terrestrial.

FEEDING ECOLOGY AND DIET
Walks and runs, then stops and scrapes the ground with one foot.

REPRODUCTIVE BIOLOGY
Nest placed in excavated tunnel. Two, rarely three eggs.

CONSERVATION STATUS
Not threatened.

SIGNIFICANCE TO HUMANS
None known. ◆

Chucao tapaculo
Scelorchilus rubecula

TAXONOMY
Pteroptochos rubecula Kittlitz, 1830, Concepción, Chile.

OTHER COMMON NAMES
French: Tourco rougegorge; German: Rotkehl-Tapaculo; Spanish: Tapacola Chucao.

PHYSICAL CHARACTERISTICS
7.3–7.5 in (18.5–19 cm); 1.5–1.6 oz (43–45 g). Brown. Throat and breast orange. Belly gray with irregular black and white bars.

DISTRIBUTION
S. r. rubecula: Southern Chile from Colchagua to Aysén, and adjacent Argentina from Neuquén to Chubut. *S. r. mochae*: Southern Chile on Mocha Island.

HABITAT
Bamboo thickets in humid forest and woodland.

BEHAVIOR
Mainly terrestrial.

FEEDING ECOLOGY AND DIET
Walks on the ground. Food includes invertebrates and berries.

REPRODUCTIVE BIOLOGY
Nest placed in excavated tunnel. Two or three eggs.

CONSERVATION STATUS
Not threatened.

SIGNIFICANCE TO HUMANS
None known. ◆

Crested gallito
Rhinocrypta lanceolata

TAXONOMY
Rhinomya lanceolata Geoffroy Saint-Hilaire, 1832, Patagonia, on the banks of the Río Negro.

OTHER COMMON NAMES
French: Tourco huppé; German: Schopfgallito; Spanish: Gallito Copetón.

Scelorchilus rubecula
▨ Resident

Rhinocrypta lanceolata
▨ Resident

PHYSICAL CHARACTERISTICS
8.3 in (21 cm); 1.8–2.3 oz (52–64 g). Gray. Crest rufous with thin white streaks. Sides and flanks chestnut. Lower belly is white.

DISTRIBUTION
R. l. saturata: Southeastern Bolivia, and Paraguay. *R. l. lanceolata*: Argentina from Catamarca and Buenos Aires to Río Negro.

HABITAT
Semi-arid thorny woodland.

BEHAVIOR
Mainly terrestrial.

FEEDING ECOLOGY AND DIET
Walks on the ground or hops through low bushes.

REPRODUCTIVE BIOLOGY
Nest globular and bulky with a side entrance, placed in bush. Two eggs. Incubation takes 16–17 days, gestation is 14–15 days.

CONSERVATION STATUS
Not threatened.

SIGNIFICANCE TO HUMANS
None known. ◆

Rusty-belted tapaculo
Liosceles thoracicus

TAXONOMY
Pteroptochus thoracicus Sclater, 1865, Salto do Girao, Rio Madeira, Brazil.

OTHER COMMON NAMES
French: Tourco ceinturé German: Brustband-Tapaculo; Spanish: Gallito Pardo.

PHYSICAL CHARACTERISTICS
7.7 in (19.5 cm); 1.4–1.5 oz (39–42 g). Upper coloring is dark gray-brown. Has narrow white brow. Sides of neck are gray. Throat and breast are white, bordered on the sides by a black line and with a more or less complete rufous breast band. Belly is barred black, rufous and white.

DISTRIBUTION
L. t. dugandi: Southeast Colombia and adjacent western Brazil on the Rio Solimoes. *L. t. erithacus*: From east Ecuador south to the mouth of Río Urubamba, eastern Peru. *L. t. thoracicus*: Southeast Peru and southwest Amazonian Brazil, east to Rio Tapajóz.

HABITAT
Humid forest.

BEHAVIOR
Terrestrial.

FEEDING ECOLOGY AND DIET
Walks and hops slowly. Feeds almost entirely on bugs.

REPRODUCTIVE BIOLOGY
Nest is globular with top entrance and placed underground among roots.

CONSERVATION STATUS
Not threatened.

Liosceles thoracicus
▢ Resident

SIGNIFICANCE TO HUMANS
None known. ◆

Slaty bristlefront
Merulaxis ater

TAXONOMY
Merulaxis ater Lesson end of 1830 or beginning of 1831, Mexico; error, Rio de Janeiro substituted by Hellmayr (1924).

OTHER COMMON NAMES
French: Mérulaxe noir; German: Bürstentapaculo; Spanish: Galltio Gris.

PHYSICAL CHARACTERISTICS
7.3 in (18.5 cm); 1.2–1.3 oz (33–37 g). Loral feathers are stiff and elongated. Tail is fairly long, black. Male is uniform dark gray. Female is brown, palest below.

DISTRIBUTION
Southeast Brazil from south Bahía to east Paraná.

HABITAT
Humid forest.

BEHAVIOR
Terrestrial, usually in pairs, occasionally ascending a bush or log to look out.

FEEDING ECOLOGY AND DIET
Walks, hops and runs. Bores head into litter.

Merulaxis ater
Resident

Melanopareia torquata
Resident

REPRODUCTIVE BIOLOGY
Unknown.

CONSERVATION STATUS
Near Threatened.

SIGNIFICANCE TO HUMANS
None known. ◆

Collared crescentchest
Melanopareia torquata

TAXONOMY
Synallaxis torquatus Wied, 1831, Campo Geral of inner Brazil = campos on the Bahia-Minas Gerais border.

OTHER COMMON NAMES
French: Cordon-noir à col roux; German: Zimtbandvogel; Spanish: Gallito de Collar.

PHYSICAL CHARACTERISTICS
5.7 in (14.5 cm); 0.56–0.81 oz (16–23 g). Tail is fairly long. Above gray-brown (crown black in *bitorquata*), nuchal collar rufous, concealed interscapular patch white. Brow white, sides of head black. Throat buff, breast band black, rest of underparts dark reddish brown, palest on lower belly.

DISTRIBUTION
M. t. torquata: Eastern Brazil. *M. t. rufescens*: Central Brazil and northeastern-most Paraguay. *M. t. bitorquata*: Eastern Bolivia.

HABITAT
Semi-arid scrub.

BEHAVIOR
Alone or in pairs. Runs on the ground and hops in low parts of the vegetation.

FEEDING ECOLOGY AND DIET
Unknown.

REPRODUCTIVE BIOLOGY
Unknown.

CONSERVATION STATUS
Not threatened.

SIGNIFICANCE TO HUMANS
None known. ◆

Spotted bamboowren
Psilorhamphus guttatus

TAXONOMY
Leptorhynchus guttatus Ménétriès, 1835, Cuiabá-Sabará, Minas Gerais, Brazil.

OTHER COMMON NAMES
French: Mérulaxe des bambous; German: Trugzaunkönig; Spanish: Gallito Overo.

PHYSICAL CHARACTERISTICS
5.3 in (13.5 cm); 0.39–0.46 oz (11–13 g). Tail is fairly long, graduated. Male is gray above and dotted white. Rump rufous. Tail is brown, barred buff and tipped white. Throat and breast

Psilorhamphus guttatus
Resident

Eugralla paradoxa
Resident

are white, belly is buff, all underparts are dotted black. Female is similar, but gray above replaced by brown and white below by buff.

DISTRIBUTION
Southeast Brazil and adjacent northeast Argentina.

HABITAT
Viny bamboo thickets in humid forest.

BEHAVIOR
Alone or in pairs. Within dense vegetation, occasionally on the ground.

FEEDING ECOLOGY AND DIET
Hops tirelessly through the branches a few meters above the ground. Feeds on insects and caterpillars.

REPRODUCTIVE BIOLOGY
Unknown.

CONSERVATION STATUS
Near Threatened.

SIGNIFICANCE TO HUMANS
None known. ◆

Ochre-flanked tapaculo
Eugralla paradoxa

TAXONOMY
Troglodytes paradoxus Kittlitz, 1830, Concepción, Chile.

OTHER COMMON NAMES
French: Mérulaxe à flancs ocre; German: Rostflanken-Tapaculo; Spanish: Churrín de la Mocha.

PHYSICAL CHARACTERISTICS
5.7 in (14.5 cm); 0.92–1.02 oz (26–29 g). Bill elevated at base. Dark gray, rump and flanks rufous. Legs bright yellow.

DISTRIBUTION
Chile from Santiago to Chiloé and adjacent Argentina in Río Negro.

HABITAT
Undergrowth of humid *Nothophagus* forest, mainly bamboo thickets.

BEHAVIOR
Moves in close pairs.

FEEDING ECOLOGY AND DIET
Forages near or on the ground within dense thickets, sometimes digging with both feet simultaneously in an awkward jump.

REPRODUCTIVE BIOLOGY
Nest is bulky and globular with a side entrance, placed hidden 3–6 ft (1–2 m) above ground, occasionally higher. Two, rarely three eggs. Both parents tend the nestlings.

CONSERVATION STATUS
Not threatened.

SIGNIFICANCE TO HUMANS
None known. ◆

Ocellated tapaculo
Acropternis orthonyx

TAXONOMY
Merulaxis orthonyx Lafresnaye, 1843, Colombia.

OTHER COMMON NAMES
French: Mérulaxe ocellé; German: Perlenmanteltapaculo;
Spanish: Churrín Ocelado.

PHYSICAL CHARACTERISTICS
8.5 in (21.5 cm); 2.9–3.5 oz (81–100 g). Hindclaw very long
and straight. Bill elevated at base. Mainly black with large
white spots. Rump and flanks chestnut. Forehead, face, throat
and upper breast rufous.

DISTRIBUTION
A. o. orthonyx: Andes of northwest Venezuela, and eastern and
central chains of Colombia. *A. o. infuscata*: Andes of Ecuador
and northern Peru.

HABITAT
Mainly bamboo thickets in humid forest.

BEHAVIOR
Almost invariably in close pairs rummaging slowly on the forest floor or hopping lethargically through tangles of bamboo,
occasionally onto mossy trunks and thick branches several meters above the ground.

FEEDING ECOLOGY AND DIET
Digs with both feet simultaneously in an awkward jump. Food
includes both arthropods and plant material.

REPRODUCTIVE BIOLOGY
Unknown.

CONSERVATION STATUS
Not threatened.

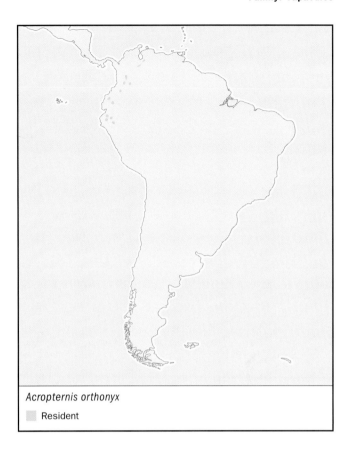

Acropternis orthonyx

▨ Resident

SIGNIFICANCE TO HUMANS
None known. A single individual in New York Zoo survived a
few months in captivity. ◆

Resources

Books

Fjeldså, J., and N. Krabbe. *Birds of the High Andes.* Copenhagen
and Svendborg, Denmark: Zoological Museum, University
of Copenhagen and Apollo Books, 1990.

Fraga, R., and S. Narosky. *Nidificación de las Aves Argentinas
(Formicariidae a Cinclidae).* Buenos Aires: Asociación
Ornitológica del Plata, 1985.

Hilty, S. L., and W. L. Brown. *A Guide to the Birds of Colombia.*
Princeton, New Jersey: Princeton University Press, 1986.

Johnson, A. W. *The Birds of Chile and Adjacent Regions of
Argentina, Bolivia, and Peru.* Vol. 2. Buenos Aires: Platt
Establecimientos Gráficos, 1967.

Meyer de Schauensee, R. *A Guide to the Birds of South America.*
Edinburgh: Oliver and Boyd, 1970.

de la Peña, M. R. *Nidos y Huevos de Aves Argentinas.* Santa Fe,
Argentina: Fundación Habitat & Desarrollo, 1999.

Ridgely, R. S., and P. J. Greenfield. *The Birds of Ecuador.* Vols.
1 and 2. London: Christopher Helms, 2001.

Ridgely, R. S., and G. Tudor. *The Birds of South America.* Vol.
2. Austin, Texas: University of Texas Press, 1994.

Schönwetter, M. *Handbuch der Oologie.* Vol. 2. Berlin:
Akademie-Verlag, 1967.

Skutch, A. F. *Studies of Tropical American Birds.* Cambridge,
Massachusetts: Publications of the Nuttall Ornithological
Club 10, 1972.

Periodicals

Ames, P. L., M. A. Heimerdinger, and S. L. Warter. "The
anatomy and systematic position of the antpipits
Conopophaga and *Corythopis.*" *Peabody Museum of Natural
History Bulletin* 114 (1968).

Ames, P. L. "The morphology of the syrinx of passerine
birds." *Peabody Museum of Natural History Bulletin* 37 (1971).

Baudet, G. "Primeira observaçao do entufado-baiano
(*Merulaxis stresemanni*) na natureza." *Tangara* 1 (2001):
51–56.

Behn, K. F. "Contribución al estudio del *Pteroptochos castaneus*
Phillippi et Landbeck." *Hornero* 8 (1944): 464–470.

Belton, W. "Birds of Rio Grande do Sul, Brazil. Part 2.
Formicariidae through Corvidae." *Bulletin of the American
Museum of Natural History* 180 (1985): 1–241.

Resources

Bornschein, M. R., B. L. Reinert, and M. Pichorim. "Descriçao, ecologia e conservaçao de um novo *Scytalopus* (Rhinocryptidae) do sul do Brasil, com comentários sobre a morfologia da família." *Ararajuba* 6 (1998): 3–36.

Bornschein, M. R., M. Pichorim, and B. L. Reinert. "Novos registros de *Scytalopus iraiensis*." *Nattereria* 2 (2001): 29–33.

Brooks, T. M., et al. "New information on nine birds from Paraguay." *Ornitologia Neotropical* 6 (1995): 129–134.

Buckley, P. A, et al, eds. "Neotropical ornithology." *Orn. Monogr.* 36 (1985).

Bullock, D. S. "Las aves de la Isla de la Mocha." *Revista Chilena de Historia Natural* 39 (1935): 232–253.

Chesser, R. T. "Molecular systematics of the Rhinocryptid genus *Pteroptochos*." *Condor* 101 (1999): 439–446.

Correa, A., et al. "La dieta del chucao (*Scelorchilus rubecula*), un Passeriforme terrícola endémico del bosque templado húmedo de Sudamérica austral." *Rev Chilena Hist Nat* 63 (1990): 197–202.

Heimerdinger, M. A., and P. L. Ames. "" *Peabody Museum of Natural History Bulletin* 105 (1967).

Howell, S. N. G., and S. Webb. "Species status of the chestnut-throated huet-huet *Pteroptochos castaneus*." *Bulletin of the British Ornithology Club* 115 (1995): 175–177.

Mezquida, E. T. "Aspects of the breeding biology of the crested gallito." *Wilson Bulletin* 104 (2001): 104–108.

Navas, J.R., and N.A. Bó. "Aportes al conocimiento de la distribución, la cría y el peso de aves de las provincias de Mendoza Y San Juan, Rep. Argentina. Segunda parte (Aves: Falconidae, Scolopacidae, Thinocoridae, Columbidae, Psittacidae, Strigidae, Caprimulgidae, Apodidae, Furnaridae, Rhinocriptidae y Tyrannidae)." *Hornero* 16 (2001): 31–37.

Olalla, A. M. "Notas de campo. Observaciones biologicas." *Revista do Museu Paulista* 223 (1937): 281–297.

Pearman, M. "Some range extensions and five species new to Colombia, with notes on some scarce or little known species." *Bulletin of the British Ornithology Club* 113 (1993): 66–75.

Pearman, M. "Notes on a population of chestnut-throated huet-huet *Pteroptochos castaneus* in Neuquén Province: A new Rhinocryptid for Argentina." *Hornero* 15 (2000): 145–150.

Rosenberg, G. H. "The nest of the rusty-belted tapaculo (*Liosceles thoracicus*)." *Condor* 88 (1986): 98.

Short, L. L. "Observations on three sympatric species of tapaculos (Rhinocryptidae) in Argentina." *Ibis* 111 (1969): 239–240.

Sick, H. "Zur Kenntnis von *Ramphocaenus* (Sylviidae) und *Psilorhamphus* (Formicariidae)." *Bonn Zool Beitr* 5 (1954): 179–190.

Sick, H. "Zur Systematik und Biologie der Bürzelstelzer (Rhinocryptidae), speziell Brasiliens." *Journ. Orn.* 101 (1960): 141–174.

Sieving, K. E., M. F. Willson, and T. L. De Santo. "Habitat barriers to movement of understory birds in fragmented south-temperate rainforest." *Auk* 113 (1996): 944–949.

Sieving, K. E., M. F. Willson, and T. L. De Santo. "Defining corridor functions for endemic birds in fragmented south-temperate rainforest." *Conservation Biology* 14 (2000): 1120–1132.

Stiles, E. W. "Nest and eggs of the white-browed tapaculo (*Scytalopus superciliaris*)." *Condor* 81 (1979): 208.

Vielliard, J. M. E. "Estudo bioacústico das aves do Brasil: o gênero *Scytalopus*." *Ararajuba* 1 (1990): 5–18.

Wege, D. C. "Threatened birds of the Darién highlands, Panama: A reassessment." *Bird Conservation International* 6 (1996): 175–179.

Whitney, B. M., J. L. Rowlett, and R. A. Rowlett. "Distributional and other noteworthy records for some Bolivian birds." *Bulletin of the British Ornithology Club* 114 (1994): 149–162.

Willson, M. F., T. L. De Santo, C. Sabag, and J. J. Armesto. "Avian communities in fragmented south temperate rainforest in Chile." *Conservation Biology* 8 (1994): 508–520.

Zimmer, J. T. "Studies of Peruvian birds. 32." *Amer Mus Novit* 1044 (1939): 18 pp.

Niels K. Krabbe, PhD

Tyrant flycatchers
(Tyrannidae)

Class Aves
Order Passeriformes
Suborder Tyranni (Suboscines)
Family Tyrannidae

Thumbnail description
Small to medium-sized perching birds with simple coloration (most are olive-green, gray, brown, or pale yellow); broad, flat bills; and specific vocalizations that are used to differentiate species

Size
3.5–11 in (9–28 cm); 0.2–2.4 oz (5.7–68 g)

Number of genera, species
110 genera; 375 species

Habitat
Riparian woodlands

Conservation status
Critically Endangered: 2 species; Endangered: 9 species; Vulnerable: 14 species; Near Threatened: 23 species

Distribution
North, Central, and South America

Evolution and systematics

Passeriformes, the largest order of birds, is divided by taxonomists into two groups: the suboscines and the oscines. Tyrannidae belong to the suboscines and is the only member of this primarily Central and South American group whose distribution extends into North America. Suboscines and oscines differ in the following ways: suboscines have a simpler syrinx (the respiratory-tract structure that produces sound); the small bone that transmits sound through the middle ear is differently shaped in the two groups; the mitochondrial DNA is differently organized; and oscines learn their songs while suboscines do not.

Taxonomists recognize four Tyrannid subfamilies based on skull and syrinx characteristics. The subfamilies are loosely enough defined that future classification modifications seem likely. The subfamily Elaeniinae (tyrannulets and elaenias) includes more than 180 species, and all but one—the northern beardless tyrannulet (*Camptostoma imberbe*)—breed in Central and South America. The subfamily Platyrinchinae (tody flycatchers and flatbills) includes genera exclusive to Central and South America. The subfamilies Fluvicolinae (fluvicoline

flycatchers) and Tyranninae (tyrannine flycatchers) include genera from across the Tyrannid distribution. Two newly recognized Tyrannid genera, the becards (*Pachyramphus*) and the tityras (*Tityra*), had not been placed in subfamilies as of 2001. These two genera were formerly classified in the family Cotingidae.

The Tyrannidae, with 110 recognized genera and 375 species, form one of the largest bird families; indeed, the family has the largest number of species among Western Hemispheric birds. Many genera contain species that are nearly indistinguishable by sight and can be identified only by their distinct vocalizations. Divisions of what was formerly recognized as one species into two have occurred in numerous cases, as differences in vocalizations have allowed the identification of two non-interbreeding groups. For instance, the former Traill's flycatcher now consists of the alder flycatcher (*Empidonax alnorum*) and the willow flycatcher (*E. traillii*). These two species differ only in vocalization. Previously unknown tyrant flycatchers continue to be identified, sometimes several per year, and often due to their distinctive vocalizations. Most newly identified tyrant flycatchers are resident in South America.

The length and number of rictal bristles differ among birds. Shown here are the gray kingbird (*Tyrannus dominicensis*) and the western wood-pewee (*Contopus sordidulus*). (Illustration by Wendy Baker)

Nest structure is also a useful indicator of phylogenetic relationships among Tyrannidae. Pewees (*Contopus*), phoebes (*Sayornis*), kingbirds (*Tyrannus*), and flycatchers (*Empidonax*) build cup-shaped nests (the most common shape among tyrannids). *Myiarchus* and the sulphur-bellied flycatcher (*Myiodynastes luteiventris*) nest in tree cavities. An enclosed dome-shaped nest with a side entrance is built by the northern beardless tyrannulet, the great kiskadee (*Pitangus sulphuratus*), and the rose-throated becard (*Pachyramphus aglaiae*).

Physical characteristics

The Tyrannidae family includes many species that look quite different from one another. However, certain physical characteristics are shared by the family as a whole. Tyrannidae are small to medium-sized, ranging from 3.5 to 11 inches (9 to 28 cm) in length—excluding tail streamers. They are usually simply colored, with shades of olive-green, gray, and brown on top and lighter colors (pale yellow, beige, and whitish) on the underparts. A few tyrannids are more brightly colored; the male vermilion flycatcher (*Pyrocephalus rubinus*) has a bright red crown and underparts. Several species of kingbird have bright yellow breasts. The kiskadees have a bright red, orange, yellow, or white spot on the crown that is visible only when the feathers are erected or spread out in excitement. The royal flycatcher (*Onychorhynchus coronatus*) performs perhaps the most spectacular visual display among tyrannids. This species has a crest that is hardly visible when not erected. When courting, however, the male's crest becomes erect, and the forehead appears surrounded by a widespread crown embedded with brownish-purple and velvety-black dots. The female's crest is almost as wide but paler.

The Tyrannid bill is generally short, wide, and slightly hooked at the tip; this characteristic distinguishes the family from most other passerines. The size of the bill varies with a species' food preference. Species that capture small insects like gnats and midges have a short bill, and those that eat larger insects like dragonflies, bees, and beetles are endowed with a longer, sturdier bill. In most species, bills are equipped with stiff rictal bristles (modified feathers), presumably to help direct flying insects into the open bill. Studies in the 1990s challenged this assumption. In experiments, flycatchers whose rictal bristles were either clipped off or taped back were just as adept at catching insects as their counterparts with intact bristles. A new hypothesis is that the bristles may help prevent insects from entering the eyes on collision with the bird.

Tyrannids' third and fourth toes are joined along the most basal segment, and there are horny plates on the outer side of the tarsus.

The tail consists usually of 12 feathers but sometimes 10, and varies in shape from square to graduated and forked. *Tyrannus* has greatly elongated central tail feathers. The fork-tailed flycatcher (*Tyrannus savana*) and the scissor-tailed flycatcher (*Tyrannus forficata*) measure up to about 6 in (16 cm) in length, but when the tail is included these birds measure an impressive 14 in (36 cm) from the head to the tip of the tail feathers.

The sexes are visually similar, although the female is paler in many species. Young resemble adults, although in species that sport brighter colors, adults are brighter and more colorful.

Ash-throated flycatcher (*Myiarchus cinerascens*) at its nest in a saguaro cactus in southeast Arizona. (Photo by G.C. Kelley. Photo Researchers, Inc. Reproduced by permission.)

The willow flycatcher (*Empidonax traillii*) builds cup-shaped nests, the great kiskadee (*Pitangus sulphuratus*) builds globular-shaped twig nests, and the rose-throated becard (*Pachyramphus aglaiae*) builds a nest that hangs from a branch. (Illustration by Wendy Baker)

Distribution

Tyrant flycatchers are distributed throughout the New World, from Tierra del Fuego to beyond the Arctic Circle in Canada and Alaska. Warm tropical lowlands are the areas of greatest species abundance and diversity. Most species that spend the summer in North America migrate to Central and South America for the winter.

Habitat

Tyrant flycatchers are found in a habitats ranging from hot, wet tropical forests to dry deserts and inhospitable mountains at heights that insects still can live. They are absent only from the coldest alpine tundra and Arctic regions. Most tyrannids require trees in combination with open areas, so they can sight prey from a perch and fly out to catch insects in midair. Some species stay mainly below the lower canopy, in shrublike vegetation; others perch within the higher canopy where tree vegetation is sparse and affords room to maneuver.

In many cases, two closely related species occupy the same general location but avoid competition by remaining in habitats separated by height, density of vegetation, wetness, or type of tree. Such distinctions can help identify a species. For instance, Hammond's flycatcher (*Empidonax hammondii*) frequents the dense, high canopy of tall, mature conifers; while the dusky flycatcher (*Empidonax oberholseri*), which has a like appearance, lives in a similar habitat and region but stays lower to the ground in more open areas.

Most migrant tyrant flycatchers winter in a habitat similar to their breeding habitat.

Behavior

Many tyrant flycatchers have loud vocalizations. The largest are usually quite vocal; their voices may be loud and rough or soft and melancholic, depending on the species. Continuous calls are generally heard only at dawn. A few species repeat the twilight song at the end of the day. The birds rarely call or sing in full daylight except during courtship or territorial dispute. The twilight song is generally given for several minutes almost without interruption. The large sulphur-bellied flycatcher (*Myiodynastes luteiventris*) and related species sing at dawn in sweet, melodious tones, which contrast with the shrill calls they utter later in the day. Studies

The feathers of this royal flycatcher (*Onychorhynchus coronatus*) make a distinctive display. (Photo by Doug Wechsler/VIREO. Reproduced by permission.)

Flight for catching prey differs between the willow flycatcher (*Empidonax traillii*) (top) and the kingbird (*Tyrannus tyrannus*) (bottom). (Illustration by Wendy Baker)

with a few North American tyrannids suggest that their calls are innate rather than learned.

A few species perform display flight. At dusk, the lesser elaenia (*Elaenia chiriquensis*) flies up steeply from thickets where it spends the day and sings a short, rough-sounding song until it is above the crowns of the trees. It then makes a steep dive into the bushes and becomes silent for the night. Spectacular courtship displays also take place in flight in some species. The western kingbird (*Tyrannus verticalis*) male flies straight upward for up to 50 ft (15 m) and then flings himself back downward, tumbling wildly.

The name tyrant is derived from the kingbirds, which boldly attack raptors and other enemies, such as snakes or

squirrels, in defense of their nesting territory. Only rarely do kingbirds molest their smaller neighbors. Male tyrannids also protect their territory against birds of the same species, and two species may enter into disputes when their ranges overlap. During an aggressive chase, tyrannids typically fly above an intruder, pecking and sometimes clawing at its back.

Tyrant flycatchers that nest in the tropics are generally resident, while most North American flycatchers are migrants. Most tyrant flycatchers migrate at night, but a few, including the western kingbird, the eastern kingbird, and the scissor-tailed flycatcher, sometimes fly southward during the day. The eastern kingbird travels thousands of miles during migration, sometimes in flocks of dozens to hundreds. When over-water crossings are delayed by rough weather, huge flocks gather along coastlines to await more favorable conditions. In late August 1964, a flock of an estimated one million eastern kingbirds was reported off Florida.

Feeding ecology and diet

Though predominantly insectivorous, tyrant flycatchers supplement their diet with all sorts of additional foods. Many species take spiders, caterpillars, berries, and fruit. The largest tyrannids often catch small vertebrates like fish, frogs, lizards, and even mice; occasionally some, like kiskadees and ground tyrants, become nest robbers.

Flycatchers' methods of obtaining food are variable. Many species rest on perches from which they seem to dive into the air to catch flying insects, returning to the perch after each flight. Whenever these birds make a catch or even just snap their bills in vain, the movement often makes an audible noise. A few species beat large insects forcefully against a branch until twitching stops. Many small species flit unobtrusively from twig to twig through thickets or in crowns of trees while catching insects. They may pick up insects and spiders from foliage or bark. Some tyrannids snatch larvae, insects, or small fish from shallow, fast-flowing water.

A number of tyrant flycatchers pick up food from the ground. The black phoebe (*Sayornis nigricans*), yellow-bellied flycatcher (*Empidonax flaviventris*), and scissor-tailed flycatcher, often sit on a fairly low perch and watch the ground beneath; they then fly down to catch crawling insects or worms. On the open steppes of southeastern South America and in the high Andes, ground tyrants walk or run on the ground and pick up worms, insects, and small vertebrates. They also make short flights in pursuit of flying insects.

The curved bill of some species is adapted to taking insects from the undersurfaces of leaves during flight or while hovering.

Reproductive biology

Most tyrannid species are solitary or occur in pairs. In many species, pairs remain together year round. Most species have been assumed to mate monogamously, with a few exceptions, such as the eastern phoebe (*Sayornis phoebe*), in which the male occasionally mates with more than one female. However, DNA studies suggest that extra-pair copulations might not be as uncommon as thought. In the eastern kingbird (*Tyrannus tyrannus*), for example, these studies show that a male other than the mate regularly fertilizes the eggs. Such promiscuity is less likely to occur in species like the great crested flycatcher (*Myiarchus crinitus*), in which the male guards the female closely after mating.

In many species both partners participate in nest building. In some species, the female does the bulk of the task while the male accompanies her as she gathers nest material, or at least greets her when she arrives with a billful of nest material. A few species do not form pairs, and the female builds the nest alone.

A black-and-white becard (*Pachyramphus albogriseus*) perches on a branch. (Photo by J. Dunning/VIREO. Reproduced by permission.)

The diversity of nests in this family is quite rare. With the exception of the ovenbirds, no group of birds in America and possibly anywhere builds such a diversity of nest types in such a variety of locations. Tyrant flycatchers build open nests in bushes or trees and, rarely, on the ground. These open structures range from the orderly, firmly knitted, lichen-covered cup of the yellow-bellied elaenia (*Elaenia flavogaster*); to the wide, flat, shallow, disorderly cup of the tropical kingbird (*Tyrannus melancholicus*). Some species nest in holes in trees, posts, or cliffs, or in buildings or nest boxes. In places where trees and shrubs are rare, ground tyrants often build their nests in clefts in banks or stone walls.

Sometimes, existing holes are in short supply due to a large number of cavity-nesting birds in an area. In one remarkable case, a pair of ash-throated flycatchers (*Myiarchus cinerascens*) was reported to have attempted to build a nest in a pair of overalls hanging on a clothesline. The nest material they brought to the site kept falling out the bottom end of the pant leg, but then the owner tied the leg closed. The pair filled the leg and raised a brood of eggs.

Phoebes build nests that are large hemispheres of mud and plant material; then they place soft lining materials in the hollow upper part of the nest. They attach these nests to cliffs, bridge pylons, or sometimes onto the wall of a deserted house—always in a spot where the nest is protected from rain, because rain would detach the nest from the wall. There are also large nests with roofs and with a side entrance placed in the fork of two branches. The vermilion flycatcher saves itself much trouble by simply building its nest over an abandoned nest.

Tyrant flycatcher chicks on Valdés Peninsula, Patagonia, Argentina. (Photo by Des & Jen Bartlett. Bruce Coleman Inc. Reproduced by permission.)

Many tyrant flycatchers build hanging nests that are suspended from thin twigs or hang from a single strand instead of being supported from below. These structures vary greatly in form, but always consist of interlocking nest materials. First, the flycatcher fixes a loosely connected base of fibers onto the selected spot. Then it forces the fibers apart and forms a nest chamber; next it lines the interior with additional material.

Many tyrant flycatchers make it difficult to enter their nest. For example, the blackish nest of the yellow-olive flycatcher (*Tolmomyias sulphurescens*) is shaped like a chemist's retort, suspended so that the opening faces straight down; the bird must enter the nest from below. The royal flycatcher's nest almost always hangs over running water. It is a loose, elongated mass of fibers sometimes 5 ft (1.5 m) long. The eggs are laid and the young are reared in a flat, hidden niche in the center of the nest.

The noisy, cantankerous piratic flycatcher (*Legatus leucophaius*) does not build its own nest. It takes over the nest of some other species by simply throwing out its victim's eggs or young. After bringing in a few dry leaves, the pirates begin their egg-laying.

Many tropical tyrant flycatchers lay two eggs; some lay three or occasionally four. Two to six eggs are laid in the higher latitudes of the northern and southern hemispheres. Eggs are white, pale gray, yellowish brown, or cream-colored. They are

sometimes unspotted, but in many species have brownish red, brown, or pale lavender spots or blotches. Only the female incubates in all species about which reproductive information has been gathered; females of many smaller species sit rather restlessly on the eggs. If weather is good, she flies off the nest for a short while every few minutes. In phoebes, pewees, and a few other species, the male brings food to the incubating partner, but this is exceptional in the tyrant flycatcher family. Incubation varies from 12 to 23 days, depending on the species; it is longest in some of the smallest forms.

The nestlings are blind and helpless after hatching. They are sparsely covered with down or, in a few small tropical species, naked. The interior of a nestling's mouth is yellow or orange-yellow. The young are brooded for warmth by the mother but are fed by both parents in species that live in pairs. Most species bring insects or berries in their bills to the nest. Fledging occurs 14 to 28 days. In general, young of species with open nests leave the nest much sooner than those reared in hanging nests. The plumage of fledging is much like that of the adults. Young do not return to the nest to roost; however, in some species, the female continues to use the nest as a sleeping place.

Conservation status

Human activities have extended the distribution of a few tyrannid species. Newly built structures and freshly planted trees provide nesting sites for these species in areas where

Great crested flycatchers (*Myiarchus crinitus*) at their nest. (Photo by B. Henry/VIREO. Reproduced by permission.)

Tyrannidae rarely live, such as open plains. Both the eastern and western kingbirds extended their ranges to the Great Plains of North America during the twentieth century, nesting on utility poles, towers, and planted trees.

Most cases of distribution change among Tyrannidae during the twentieth century, however, have been characterized by restriction rather than expansion. The range of the alder flycatcher has shifted northward, possibly due to climatic warming or the natural regrowth of woodlands. The latter possible cause is feasible because this species' habitat is swamp and meadow thicket.

Declines in population were reported during the 1990s for numerous species of tyrant flycatcher. The 2000 IUCN Red List of Threatened Species lists the following species as facing the most critical conservation challenges among tyrannids.

Critically Endangered:

- Alagoas tyrannulet (*Phylloscartes ceciliae*) (Sibley and Monroe classification)

- Minas Gerais tyrannulet (*Phylloscartes roquettei*)

Endangered:

- Ash-breasted tit tyrant (*Anairetes alpinus*)

- Fork-tailed pygmy tyrant (*Hemitriccus furcatus*)

- Kaempfer's tody tyrant (*Hemitriccus kaempferi*)

- Santa Marta bush tyrant (*Myiotheretes pernix*)

- Atlantic royal flycatcher (*Onychorhynchus coronatus swainsoni*)

- Urich's tyrannulet (*Phyllomyias virescens urichi*)

- Bahia tyrannulet (*Phylloscartes beckeri*) (Clements classification)

- Antioquia bristle tyrant (*Phylloscartes lanyoni*) (Sibley and Monroe classification)

- Giant kingbird, also called Cuban flycatcher (*Tyrannus cubensis*)

The southwestern subspecies of the willow flycatcher was on the United States Endangered Species List in 2001. Its population was estimated to be only a few hundred individuals in southern California, Arizona, and New Mexico. This subspecies breeds in dense vegetation along moving water, and its decline may be due to destruction of this habitat by cutting, fire, and cattle grazing. Large declines were also reported for the olive-sided flycatcher (*Contopus borealis*), the western wood-pewee (*Contopus sordidulus*), and the eastern wood-pewee (*Contopus virens*). Declines in the willow subspecies and the western wood-pewee are thought to be caused by deforestation of the birds' winter habitat in the South American Andes. The eastern wood-pewee may owe its decline to increases in the population of white-tailed deer in eastern North America. The deer's grazing behavior clears the vegetation structure of the forest floor, making the habitat less hospitable to the bird.

Significance to humans

Tyrant flycatchers do not pose any danger or particular usefulness to humans, nor have they been significant in art or myth. As insect foragers, they may inconspicuously play a role in keeping in check populations of insects that humans consider to be pests. At one time species that prey on bees were thought to be among the causes of a dangerous decline in bee populations. However, it has been shown that these species do not eat enough bees to be considered a serious threat, particularly when compared to the damage that parasitic mites wreak on bee colonies.

Since many tyrannid genera include species that are visually indistinguishable, this family provides challenging identification tasks to birders. Many species require that vocalization and nest-building behavior are considered before identification can be confirmed.

1. Hammond's flycatcher (*Empidonax hammondii*); 2. Sulphur-bellied flycatcher (*Myiodynastes luteiventris*); 3. Nutting's flycatcher (*Myiarchus nuttingi*); 4. Great crested flycatcher (*Myiarchus crinitus*); 5. Greater pewee (*Contopus pertinax*); 6. Western wood-pewee (*Contopus sordidulus*); 7. Eastern phoebe (*Sayornis phoebe*); 8. Say's phoebe (*Sayornis saya*). (Illustration by Wendy Baker)

1. Willow flycatcher (*Empidonax traillii*); 2. Northern beardless-tyrannulet (*Camptostoma imberbe*); 3. Scissor-tailed flycatcher (*Tyrannus forficata*); 4. Olive-sided flycatcher (*Contopus borealis*); 5. Western kingbird (*Tyrannus verticalis*); 6. Vermillion flycatcher (*Pyrocephalus rubinus*); 7. Rose-throated becard (*Pachyramphus aglaiae*); 8. Great kiskadee (*Pitangus sulphuratus*). (Illustration by Wendy Baker)

Species accounts

Rose-throated becard

Pachyramphus aglaiae

TAXONOMY

Platyrhynchus aglaiae Lafresnaye, 1839, Jalapa, Veracruz, Mexico. Monotypic.

OTHER COMMON NAMES

French: Bécarde à gorge rose; German: Dickkopfbekarde; Spanish: Bacaco de Garganta Rosada.

PHYSICAL CHARACTERISTICS

(6.5–7.25 in (16.5–18.5 cm). The male has dark gray upperparts, pale gray underparts, a blackish cap and nape, and a bright pink patch at the throat. The female has a gray crown, grayish brown or cinnamon upperparts, buff underparts, and a whitish throat. Body shape is stocky with a relatively big head. Juveniles are similar in color to adult females.

DISTRIBUTION

Central America and Mexico. Also occurs in parts of southeast Arizona and southwest Texas during breeding season.

HABITAT

Open forests, forest edges, wooded canyons, and mountains. As the nest hangs from a tree branch high above the ground, the species requires areas with tall trees.

BEHAVIOR

Lives singly or in pairs, sometimes joins foraging flocks. Vocalizations include a soft, down-slurred whistle "tseeoou!" sometimes preceded by some reedy chatter. At dawn, its song is a reedy, complaining, long "wheeuu-whyeeeuuur, wheeuu-whyeeeuuur!"

FEEDING ECOLOGY AND DIET

Sits nearly motionless on a branch, hidden among leaves, watching for insects from the middle levels of clearings or forest edges. Sallies forth to snag insects from foliage or in flight and returns to same perch. Diet consists of insects, their larvae, and sometimes wild fruits and berries.

REPRODUCTIVE BIOLOGY

Breeds in monogamous pairs, once per year, and male and female share nest-building duties, although the female carries a larger burden. Nests are spherical and hang from a branch of a deciduous tree. Clutches include two to six eggs, which the female incubates for 15 to 17 days. Juveniles fledge at 19 to 21 days and are fed by both parents.

CONSERVATION STATUS

Not threatened. Rarely hosts cowbird parasitism.

SIGNIFICANCE TO HUMANS

None known. ◆

Scissor-tailed flycatcher

Tyrannus forficata

SUBFAMILY

Tyranninae

TAXONOMY

Muscicapa forficatus Gmelin, 1789, Mexico. Monotypic.

Pachyramphus aglaiae
■ Resident ■ Breeding

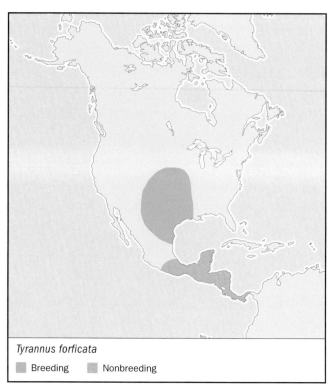

Tyrannus forficata
■ Breeding ■ Nonbreeding

OTHER COMMON NAMES
French: Tyran à longe queue; German: Scherentyrann; Spanish: Pitirre Tijereta.

PHYSICAL CHARACTERISTICS
11.5–15 in (29–38 cm); half of which is tail. Characterized by a long tail that opens and closes like a pair of scissors. Plumage includes pale gray upperparts, a pale gray head, white throat and underparts, pale salmon-pink sides and flanks, and dark brown wings with white edges. Bill, legs, and feet are black. Sexes are similar; juveniles are paler overall. Weight is 1.5 oz (42 g).

DISTRIBUTION
Oklahoma, Texas, limited surrounding areas in neighboring U.S. states, and far northern Mexico. Accidental across much of North America. Winters in southern Mexico and Central America.

HABITAT
Inhabits open land with scattered trees, prairies, scrublands, and farmlands.

BEHAVIOR
Lives singly or in pairs during the day, roosts at night in groups of up to 200. Capable of acrobatic flight. During courtship, the male makes a sudden plunge from a hundred feet (30 m) above ground, flies downward and diagonally back and forth singing with a cackle, and proceeds to an upward flight followed by several backward somersaults. This display persists through courtship and nesting until the eggs hatch. Main vocalization is a sharp "bik!" or "kew!"; other calls include a chattering "ka-quee-ka-quee!" and a repeated string of "ka-lup!".

FEEDING ECOLOGY AND DIET
From a main perch on branches, utility wires, and fences, watches for bees, wasps, and other flying insects, and then sallies forth, hovering momentarily over prey and dipping to catch it. Returns to same perch. Also hunts near the ground for crickets and grasshoppers.

REPRODUCTIVE BIOLOGY
Breeds monogamously once per year; female builds nest and incubates a clutch of three to six eggs for 14 to 17 days. Nest is cup-shaped and built on either deciduous or coniferous branches, on shrubs, and in human-made structures. Young are fed by both parents and remain in nest for 14 to 16 days.

CONSERVATION STATUS
Not threatened. Rarely hosts cowbird parasitism.

SIGNIFICANCE TO HUMANS
None known, other than interest in viewing the spectacular flight display of the courting male. ◆

Western kingbird
Tyrannus verticalis

SUBFAMILY
Tyranninae

TAXONOMY
Tyrannus verticalis T. Say, 1823, La Junta, Colorado. Monotypic.

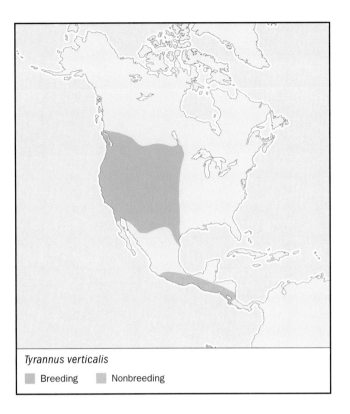

Tyrannus verticalis
■ Breeding   ■ Nonbreeding

OTHER COMMON NAMES
English: Arkansas kingbird; French: Tyran de l'Ouest; German: Arkansastyrann; Spanish: Pitirre Occidental.

PHYSICAL CHARACTERISTICS
8.75 in (22 cm); 1.4 oz (40 g). Plumage includes pale ashy gray head, neck, and breast, olive-green tinted back, bright lemon-yellow underparts, and dark brown wings. Tail is squared and black with white outer edges. Feet and legs are black. Bill is small, flat, and black. The crown, rarely erect, hides a red-orange patch. Sexes are similar.

DISTRIBUTION
Occurs throughout the western half of the continental United States, with limited extensions into western Canada and northern Mexico. Winters in southwestern Mexico and Central America. Vagrants are common during migration in the southeastern United States.

HABITAT
Semiarid open areas and grasslands with scattered trees. During the twentieth century, range expanded with the spread of agriculture; buildings, utility structures, and fences provide new foraging perches and nest sites. Lives gregariously in urban areas; up to three pairs can nest in the same tree.

BEHAVIOR
Lives singly, in pairs, or in small groups. Male performs courtship flight display, involving upward darting flight, fluttering and vibrating of feathers, and trilling vocalizations. Regular call is a quiet, quick "bek!" Also chatters abrasively: "ker-er-ip, ker-er-ip, pree preee pr-prrr." Known for being aggressively territorial, often chasing large birds such as hawks, crows, and ravens away from its nesting area.

FEEDING ECOLOGY AND DIET
Sallies from low, middle, and high perches in open areas to catch insects in midair, returning to the same perch. Hovers

momentarily over prey before dipping to catch. Also takes fruits and berries.

REPRODUCTIVE BIOLOGY
Monogamous breeders, in solitary pairs or in small colonies. Nests are built by both sexes, near the trunk on a horizontal limb or on a cross-arm of a human-made structure. Nest is cup-shaped. Clutches of three to seven eggs are incubated 18 to 19 days by the female; and young are fed by both parents and fledge after 16 to 17 days.

CONSERVATION STATUS
Not threatened.

SIGNIFICANCE TO HUMANS
Well-known locally. ◆

Great kiskadee
Pitangus sulphuratus

SUBFAMILY
Tyranninae

TAXONOMY
Lanius sulphuratus Linnaeus, 1766, Cayenne. Monotypic.

OTHER COMMON NAMES
English: Kiskadee flycatcher; French: Tyran quiquivi; German: Bentevi; Spanish: Benteveo Común.

Pitangus sulphuratus
■ Resident

PHYSICAL CHARACTERISTICS
9.8 in (25 cm). Plumage includes white forehead and eyebrows, black crown, wide black eye line, brown back and rump, reddish brown wings and tail, bright yellow crissum, yellow underparts, and white cheeks, chin, and throat. Bill is black and stout; legs and feet are also black. Underwings are yellow, and in flight, the yellow of the underwings and belly contrast with the reddish brown of the overwings and tail. A yellow crown patch is usually concealed. Sexes are similar.

DISTRIBUTION
Common in the tropics of Central and South America. Common in southwest Texas; casual in coastal Louisiana, southeast Arizona, southeast New Mexico, southeast Texas, western Oklahoma. Introduced and established on Bermuda.

HABITAT
Wet woodlands, open areas with scattered trees, forest edges, scrub vegetation, bushes, and lakes and rivers.

BEHAVIOR
The great kiskadee gets its name from its call, a loud, slow, screaming "kiss-ka-dee!" or "k-reah!" It is energetic, noisy, and aggressively territorial, chasing away much larger birds from its nesting area. Lives solitary or in pairs. It is easily spotted, drying its feathers on an open, conspicuous perch after diving for aquatic prey. Nonmigratory.

FEEDING ECOLOGY AND DIET
Sits on perch to watch for prey that includes aquatic insects, small fish, frogs, tadpoles, baby birds, lizards, mice, and both crawling and flying insects. Returns to perch to eat. It often beats larger prey against a branch before swallowing. Takes fruits and berries when other food is unavailable.

REPRODUCTIVE BIOLOGY
Breeds monogamously, two to three clutches of two to five eggs per year. Nests are spherical, built by both sexes in thorny trees, palm trees, or on braces of utility poles. Female incubates eggs an estimated 13 to 15 days, and young fledge at 12 to 21 days. Young are fed by both parents.

CONSERVATION STATUS
Some in the United States have declined due to habitat loss caused by deforestation and development. The species is still common in the Central and South American tropics.

SIGNIFICANCE TO HUMANS
None known. ◆

Vermilion flycatcher
Pyrocephalus rubinus

SUBFAMILY
Fluvicolinae

TAXONOMY
Pyrocephalus rubinus Boddaert, 1783. Monotypic.

OTHER COMMON NAMES
French: Moucherolle vermillion; German: Purpurtyrann; Spanish: Sangre de toro.

PHYSICAL CHARACTERISTICS
6 in (15.25 cm). Among the most colorful tyrannids, and certainly the most brightly colored tyrannid in North America.

Pyrocephalus rubinus

■ Resident

Plumage of males includes bright red crown and underparts, blackish brown tail and upperparts, and dark brown lores and mask joining at nape. Bill is short, broad, flat, and black; legs and feet are brownish black. Females have a white chin, throat, and chest, a pale pinkish belly and crissum, grayish brown upperparts, and a thin white supercilliary stripe. The juvenile male is similar to the adult male but more pale; the juvenile female is similar to the adult female but with a yellow wash to the belly and crissum.

DISTRIBUTION
Occurs from the southernmost regions of California, Arizona, New Mexico, and Texas south to Argentina.

HABITAT
Desert, semidesert, scrub vegetation, forest edge, open forest, and grassland with scattered trees. Often found near water bodies, such as lakes, ponds, irrigation ditches, and cattle tanks.

BEHAVIOR
Lives singly or in pairs. Has a quiet and tame disposition; very approachable. Often chooses a low perch. Wags and pumps the tail. Vocalization follows a pattern during display flight. While in direct flight, the call is a piercing, metallic "pseeup!"; this is followed by a hovering phase with tail spread out and crest erected, which is accompanied by a rapid "pi-pi-li-li-li-sing!"

FEEDING ECOLOGY AND DIET
Perches on a low branch, sights prey and sallies forth, hovering to catch it. Returns to the same perch. Sometimes forages on the ground. Feeds on insects, particularly bees.

REPRODUCTIVE BIOLOGY
Monogamously breeds twice a year. Cup-shaped nest is constructed by the female in the fork of a horizontal branch or on top of an abandoned nest. Eggs typically number two to four. Female incubates them for 14 to 15 days, and the young, fed by both parents, fledge after 14 to 16 days.

CONSERVATION STATUS
Not threatened, though populations in southeast California and Texas declining; cause unknown.

SIGNIFICANCE TO HUMANS
Attractive to birders for its bright red color. ◆

Northern beardless-tyrannulet
Camptostoma imberbe

SUBFAMILY
Elaeniinae

TAXONOMY
Camptostoma imberbe Slater, 1839, San Andres Tuxtla, Veracruz, Mexico. Monotypic.

OTHER COMMON NAMES
English: Northern beardless flycatcher; French: Tyranneau imberbe; German: Chaparral-Fliegenstecher; Spanish: Piojito Norteño.

PHYSICAL CHARACTERISTICS
Length is 4.5 in (11.5 cm); perches in very upright posture. Plumage includes a gray crown with a bushy crest, gray-olive upperparts, grayish brown wings, and white or pale yellow underparts. Bill is small and slightly curved, with brown tip and creamy pink base.

DISTRIBUTION
From southern Arizona and Texas to Costa Rica. More common in the southern half of its range.

Camptostoma imberbe

■ Resident   ■ Breeding

HABITAT
Low woods, mesquite, stream thickets, brush, lower canyons.

BEHAVIOR
Lives singly or in pairs. Often wags tail while perching. Song is a high, thin whistle "peert!" or "pee-yerp!" Also sings three or more down-slurred notes "dee, dee, dee, dee."

FEEDING ECOLOGY AND DIET
Hawks insects in midair. Also gleans insects form twigs and leaves and takes berries.

REPRODUCTIVE BIOLOGY
Breeds monogamously once or twice a year. Nest is spherical, built by female, and located on the outer branches of a deciduous tree. Clutch is one to three eggs, incubated by female for undetermined time; age of young at first flight also unknown.

CONSERVATION STATUS
Not threatened, though some populations declining with loss of streamside habitat, possibly due to cattle grazing.

SIGNIFICANCE TO HUMANS
None known. ◆

Olive-sided flycatcher
Contopus borealis

SUBFAMILY
Fluvicolinae

TAXONOMY
Contopus borealis Swainson, 1832. Monotypic.

OTHER COMMON NAMES
English: Boreal peewee; French: Moucherolle à côtés olive; German: Fichtentyrann; Spanish: Pibí Boreal.

PHYSICAL CHARACTERISTICS
7.5 in (19 cm). Stout flycatcher with a large head, short neck, and short tail. Plumage includes brownish olive upperparts, head, crest, and wings; and dull white throat, center breast strip, belly, and undertail coverts. Bill is large and mostly black, with a dull orange lower mandible.

DISTRIBUTION
Breeding regions include Alaska, most of Canada, the northwest United States, California, and the Rocky Mountains. Winters from Southern Central American to Peru.

HABITAT
Mountainous terrain and coniferous forest. Also frequents burns, bogs, and swamps.

BEHAVIOR
Solitary and reclusive; often perches on a high, exposed limb or the top of a dead or living tree. Vocalization is often described as "quick, THREE beers!" with the second note higher. Also trebles a "pip!" Aggressively defends nesting territory against predators and humans.

FEEDING ECOLOGY AND DIET
Often from a dead branch, hawks large insects (up to the size of cicadas, beetles, and honeybees) in mid-flight and returns to the same perch.

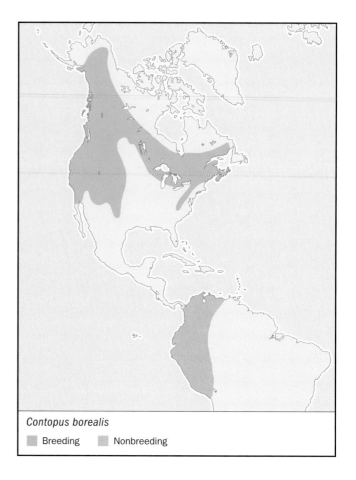

Contopus borealis
■ Breeding ■ Nonbreeding

REPRODUCTIVE BIOLOGY
Breeds monogamously once per year. Female builds cup-shaped nest, usually on horizontal branches of coniferous trees. Clutch consists of three to four eggs, incubated 14 to 17 days by the female; young fledge at 21 to 23 days.

CONSERVATION STATUS
Not threatened, but many areas host declining populations; a loss of wintering habitat is the suspected cause.

SIGNIFICANCE TO HUMANS
None known. ◆

Willow flycatcher
Empidonax traillii

SUBFAMILY
Fluvicolinae

TAXONOMY
Empidonax traillii Audubon, 1828. Five subspecies.

OTHER COMMON NAMES
French: Moucherolle des saules; German: Weidentyrann; Spanish: Mosqueta Saucera.

PHYSICAL CHARACTERISTICS
5.75 in (14.5 cm). Plumage includes a brownish to brownish green head, brownish green upperparts, dark wings with buff

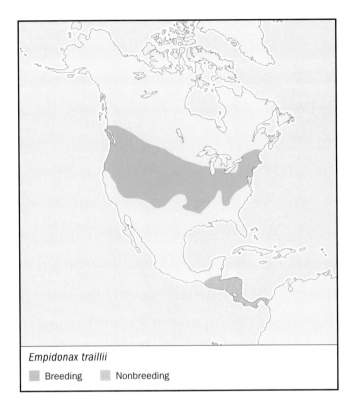

Empidonax traillii

▪ Breeding ▪ Nonbreeding

grazing animals. Nests are parasitized by the brown-headed cowbird (*Molothrus ater*). May also be imperiled by loss of tropical wintering habitat due to deforestation.

SIGNIFICANCE TO HUMANS
None known. ◆

Nutting's flycatcher
Myiarchus nuttingi

SUBFAMILY
Tyranninae

TAXONOMY
Myiarchus nuttingi Ridgway, 1882. Monotypic.

OTHER COMMON NAMES
English: Pale-throated flycatcher; French: Tyran de Nutting; German: Blasskehltyrann; Spanish: Atrapamoscas de Nutting.

PHYSICAL CHARACTERISTICS
7.25 in (18.5 cm). Plumage includes a dark gray crown (sometimes with a tinge of cinnamon), olive-brown upperparts, two white wingbars, wing coverts and secondaries edged with white, cinnamon-edged primaries, dusky central tail feathers, yellow belly and undertail coverts, and pale gray throat and breast. Interior of mouth is orange. Sexes are similar. Almost identical in appearance to ash-throated flycatcher (*Myiarchus cinerascens*); distinguishable in the field only by song.

DISTRIBUTION
Western Mexico to western Costa Rica; accidental to southeastern Arizona.

HABITAT
Prefers semiarid deciduous slopes and thorny thickets.

to yellow wing bars, pale yellow trim on tertials and secondaries, a dark tail, a thin, pale eye ring, pale lores, whitish underparts, and dusky side flanks tinged with yellow. Feet and legs are blackish, and bill is blackish with a yellowish pink lower mandible. Plumage color varies somewhat with region; for example, northwestern races have a dark head, while southwestern races have a pale head. Sexes are similar.

DISTRIBUTION
Breeding is mostly restricted to the continental United States, including the northwestern states, Rocky Mountains, Great Plains, Midwest, and northeastern regions.

HABITAT
Prefers shrubs and undergrowth, willow thickets, fresh water marshes, ponds, rivers, and lakes.

BEHAVIOR
Silent in migration, but otherwise sings a sharp "fitz-bew!" or "fitz-be-yew!" Also releases a loud "whit!" Perches low, below top of vegetative layer; chooses an exposed perch to sing. During courtship, males chase females in flight.

FEEDING ECOLOGY AND DIET
Perches to spot prey, sallies forth to catch prey in midair, and returns to the same perch. Feeds on flying insects, insects gleaned from foliage, spiders, and occasional berries.

REPRODUCTIVE BIOLOGY
Monogamous breeders. Nest is cup-shaped and compact, often with hanging streamers, built by female in the fork of a deciduous tree. One clutch per year of two to four eggs, incubated by female for 12 to 15 days. Juveniles remain in the nest for 12 to 14 days, fed by both sexes.

CONSERVATION STATUS
Not threatened, though populations on the west coast are declining due to loss of streamside habitat, particularly caused by

Myiarchus nuttingi

▪ Resident

BEHAVIOR
Lives singly or in pairs. Nonmigratory. Song is a quick, loud, chattering "wheep! wheep!", in addition to a repeated "ki di-di-dir!"

FEEDING ECOLOGY AND DIET
Eats insects and some berries. Most often snatches prey from foliage while hovering; also hawks prey in midair and returns to perch.

REPRODUCTIVE BIOLOGY
Monogamous. Nest is built by both sexes in a preformed burrow and lined with grasses, rootlets, weeds, and feathers. Female incubates one to two clutches of three to five eggs per year for 14 days. Young are fed by both parents and fledge at 14 to 16 days.

CONSERVATION STATUS
Not threatened.

SIGNIFICANCE TO HUMANS
None known. ◆

Say's phoebe
Sayornis saya

SUBFAMILY
Fluvicolinae.

TAXONOMY
Sayornis saya Bonaparte, 1825. Monotypic.

OTHER COMMON NAMES
French: Moucherolle à ventre roux; German: Sayphoebe; Spanish: Mosquero Llanero.

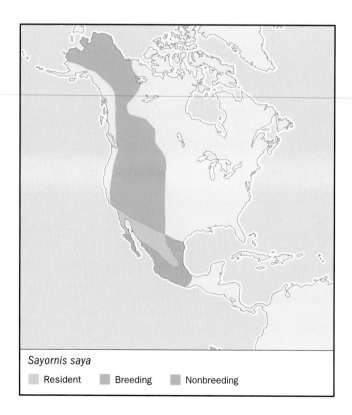

Sayornis saya

■ Resident   ■ Breeding   ■ Nonbreeding

PHYSICAL CHARACTERISTICS
7.5 in (19 cm). Plumage includes brownish gray upperparts, pale grayish brown throat and breast, tawny buff belly and undertail coverts, and blackish brown tail feathers. Bill is small and black; legs and feet are also black. Sexes are similar.

DISTRIBUTION
Alaska to Texas along the western half of North America (excluding the coast). Winters throughout Mexico.

HABITAT
Savannas, farmlands, and open brushlands. Not as tied to watercourses as other phoebes.

BEHAVIOR
While perched, song is a whistled, down-slurred "phee-eur!" or "chu-weer!" In flight, utters a quick "pit-se-ar!" Frequently sings at dawn. Lives singly or in pairs. Conspicuously perches on exposed branches, wires, posts, buildings, and other structures. Migratory.

FEEDING ECOLOGY AND DIET
Eats insects and rarely berries; sometimes regurgitates insect exoskeletons. Eyes prey from perch or while hovering, and sallies forth to capture in midair (often with a loud snap of the mandibles).

REPRODUCTIVE BIOLOGY
Nest, built by the female, is cup-shaped and adheres to the vertical wall of a cave, cliff, bridge, or building. Monogamously breeds once to twice per year. Female incubates clutch of three to seven eggs for 12 to 14 days. Juveniles' fledge at 14 to 16 days.

CONSERVATION STATUS
Not threatened. Rarely hosts cowbird parasitism.

SIGNIFICANCE TO HUMANS
None known. ◆

Eastern phoebe
Sayornis phoebe

SUBFAMILY
Fluvicolinae

TAXONOMY
Sayornis phoebe Latham, 1879. Monotypic.

OTHER COMMON NAMES
French: Moucherolle phébi; German: Phoebe; Spanish: Mosquero Fibí.

PHYSICAL CHARACTERISTICS
7 in (18 cm). Plumage includes dark brownish gray head and upperparts, dark brownish wings and tail, white underparts (touched with yellow in fall in first-year birds), and olive-tinted sides and breast. Bill is small and black, as are feet and legs.

DISTRIBUTION
Eastern North America from central Canada to the Midwest and northeastern states; winters in the southern states and eastern Mexico.

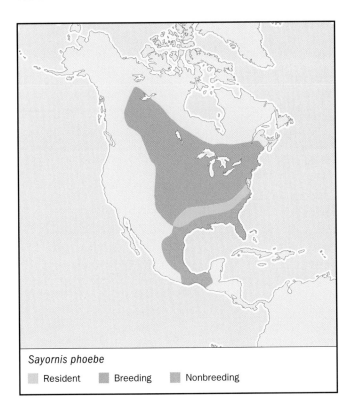

Sayornis phoebe

▢ Resident ▢ Breeding ▢ Nonbreeding

HABITAT
Lives near lakes and rivers in forest edges, open areas with scattered trees, and rocky areas.

BEHAVIOR
Often wags tail on perch. Lives singly or in pairs. Songs include a sharp, repeated "chip!" and a "FEE-be!" with an accent on the first syllable.

FEEDING ECOLOGY AND DIET
Perches to watch for insects, catches prey in midair, and returns to perch. Also takes insects from foliage and from the ground. Sometimes takes fruit, berries, and small fish.

REPRODUCTIVE BIOLOGY
Breeds two to three times per year. Mostly monogamous, but sometimes a male breeds with more than one female. Female builds a cup-shaped nest attached to a vertical wall or on a shelf. Nest may be located on a cliff, building, or bridge.

CONSERVATION STATUS
Not threatened by IUCN standards. Blue-listed by the National Audubon Society in 1980 and listed as Special Concern in 1986, due to decreases in several areas across the Midwest, south Atlantic, and Great Lakes regions. Some populations in the 1990s were reported to be stable or increasing, and ranges of some populations were expanding, possibly due to the species' tolerance for human-made structures as nesting sites.

SIGNIFICANCE TO HUMANS
Migrates early and indicates the coming of spring to the southern states. The first bird-banding experiment in North America, carried out by John James Audubon in 1840, used the eastern phoebe to gather information about longevity, dispersal, migratory movements, and site fidelity. ◆

Hammond's flycatcher
Empidonax hammondii

SUBFAMILY
Fluvicolinae

TAXONOMY
Empidonax hammondii Xantus, 1858. Monotypic.

OTHER COMMON NAMES
French: Moucherolle de Hammond; German: Tannentyrann; Spanish: Mosqueta de Hammond.

PHYSICAL CHARACTERISTICS
5.5 in (14 cm). Small bird with large head and short tail. Plumage includes gray head, white eye ring, grayish olive back, dark gray wings and tail, whitish wing bars, gray or olive tint on the breast and sides, and belly washed with pale yellow. Bill is narrow and short. Base of lower mandible is pale orange. Keeps a fairly horizontal stance while perching.

DISTRIBUTION
Summer resident in southeastern Alaska, western Canada, northwestern United States, and Rocky Mountains. Winters throughout Latin America.

HABITAT
Inhabits wide range of forest types, but prefers coniferous forest at higher elevations than other *Empidonax* flycatchers.

BEHAVIOR
Active bird, frequently flicking tail and wings while perched. Can be silent for long periods; when vocal, call is a low, rapid "sill-it!" or "chip-it!", also a low, rough "greep!" or "pweet!" Migratory.

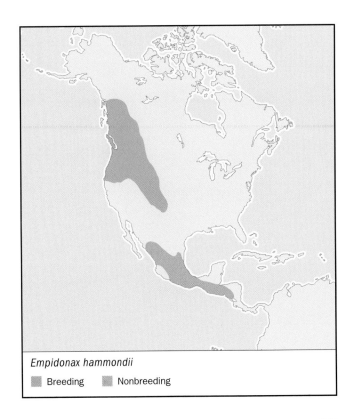

Empidonax hammondii

▢ Breeding ▢ Nonbreeding

FEEDING ECOLOGY AND DIET
Eats flying insects. Perches high to spot prey, hawks in midair, and returns to same perch. Also gleans insects from foliage.

REPRODUCTIVE BIOLOGY
A shallow, cup-shaped nest is built by the female, who incubates one clutch of three to four eggs once per year. Breeding is monogamous.

CONSERVATION STATUS
Not threatened. Habitat vulnerable to deforestation of high-elevation conifers.

SIGNIFICANCE TO HUMANS
None known. ◆

Greater pewee
Contopus pertinax

SUBFAMILY
Fluvicolinae

TAXONOMY
Contopus pertinax Cabanis and Heine, 1859.

OTHER COMMON NAMES
English: Coues' flycatcher, smoke-colored peewee; French: Moucherolle bistré German: Couestyrann; Spanish: Pibí Ahumado.

PHYSICAL CHARACTERISTICS
8 in (20 cm). Plumage includes grayish olive head and upperparts, whitish throat and chin, pale gray breast and underparts, a yellow wash on the belly, and a long tail appearing notched when folded. The slender, tufted crest is a distinctive identifier. Bill has a black upper mandible and an orange lower mandible.

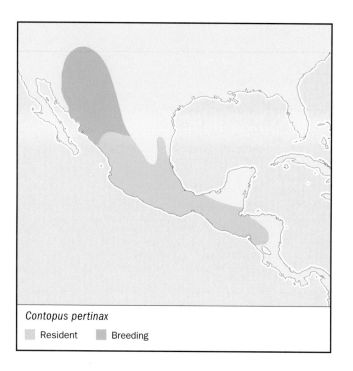

Contopus pertinax
- Resident
- Breeding

DISTRIBUTION
Southern Arizona and New Mexico, south through Mexico and into Nicaragua.

HABITAT
Montane pine-oak woodlands and wooded canyons.

BEHAVIOR
Lives singly or in pairs. Frequently perches in dead pines. Whistles "ho-sa, ma-re-ah!" and chirps a steadily repeated "pip-pip-pip!" Defends nesting territory aggressively against larger birds, snakes, and squirrels. Migratory.

FEEDING ECOLOGY AND DIET
From middle-level perch, hawks insects in midair.

REPRODUCTIVE BIOLOGY
Breeds monogamously once per year. Female builds a cup-shaped nest in the fork of a conifer or sycamore. Clutch consists of three to four eggs. Young are fed by both sexes.

CONSERVATION STATUS
Not threatened. Habitat vulnerable to logging of coniferous forest.

SIGNIFICANCE TO HUMANS
None known. ◆

Western wood-pewee
Contopus sordidulus

SUBFAMILY
Fluvicolinae

TAXONOMY
Contopus sordidulus Slater, P.L., 1859.

OTHER COMMON NAMES
French: Pioui de l'Ouest; German: Forst-Piwih; Spanish: Pibí Occidental.

PHYSICAL CHARACTERISTICS
6.25 in (16 cm). Dark grayish brown plumage overall, with paler underparts and two thin white bars on wings. Bill is dark with yellow-orange lower mandible base.

DISTRIBUTION
Central Alaska south across most of the western half of North America, through western Mexico and Central America. Winters from Panama to Peru.

HABITAT
Inhabits riparian woodlands, and open, mountainous, mixed conifer and hardwood forests.

BEHAVIOR
Solitary dweller, remains mostly quiet and hidden. Sings "tsee-tee-teet!" on breeding grounds. Also uses soft, nasal whistle "peeer!" Frequently sings until after dark and before daylight. Shakes its wings when landing on perch. Migratory.

FEEDING ECOLOGY AND DIET
Perches to watch for food; hawks prey in midair. Eats variety of flying insects; occasionally feeds on spiders and berries.

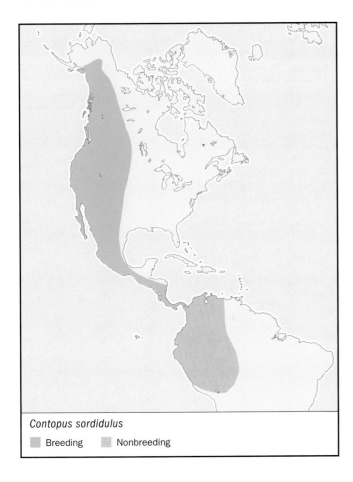

Contopus sordidulus

▪ Breeding ▪ Nonbreeding

Myiarchus crinitus

▪ Breeding ▪ Nonbreeding

REPRODUCTIVE BIOLOGY
Breeds monogamously once per year. Nest is cup-shaped, built by the female, and sits on a horizontal branch of a (usually coniferous) tree, bound to the branch by spider web. Brood is two to four eggs, incubated by the female for 12 to 13 days.

CONSERVATION STATUS
Not threatened, though some populations in California are declining for unknown reasons. Vulnerable to deforestation in wintering areas.

SIGNIFICANCE TO HUMANS
None known. ◆

Great crested flycatcher

Myiarchus crinitus

SUBFAMILY
Tyranninae

TAXONOMY
Myiarchus crinitus Linnaeus, 1758. Monotypic.

OTHER COMMON NAMES
French: Tyran huppé; German: Schnäppertyrann; Spanish: Atrapamoscas Copetón.

PHYSICAL CHARACTERISTICS
8.5 in (21.5 cm). Plumage includes a dark gray crown, olive-green upper parts, gray throat and upper breast, yellow belly and undertail coverts, two white wing bars, and reddish inner webs on tail feathers. Bill is heavy and black.

DISTRIBUTION
Eastern half of the United States, extending into southeastern Canada. Winters from eastern Mexico to Columbia.

HABITAT
Prefers thickly wooded areas and forest edges.

BEHAVIOR
Lives singly or in pairs. Aggressively territorial; males will battle in the air with other males, clawing and pulling out feathers. Songs include a strong whistle of "wheeeep!" and a rolling "prrrrrrreeet!" Often perches high in the canopy on exposed or dead limbs. Migratory.

FEEDING ECOLOGY AND DIET
Hawks large insects in midair, higher in the air than most flycatchers; also gleans prey from foliage. Sallies from and back to a single perch. Takes beetles, crickets, katydids, caterpillars, moths, butterflies, and some fruits and berries.

REPRODUCTIVE BIOLOGY
Breeds monogamously once per year. Male chases female in flight during courtship. Nest, built by both sexes, is located in a preformed cavity such as the abandoned hole of another bird or a bird box. Nest lining is often covered with a shed snakeskin or piece of discarded plastic. Female incubates four to eight eggs for 13 to 15 days.

CONSERVATION STATUS
Not threatened. Habitat is vulnerable to deforestation.

SIGNIFICANCE TO HUMANS
None known. ◆

Sulphur-bellied flycatcher

Myiodynastes luteiventris

SUBFAMILY
Tyranninae

TAXONOMY
Myiodynastes luteiventris Slater, P.L., 1859. Monotypic.

OTHER COMMON NAMES
French: Tyran tigré German: Weisstirntyrann; Spanish: Benteveo de Buche Amarillo.

Myiodynastes luteiventris

■ Breeding ■ Nonbreeding

PHYSICAL CHARACTERISTICS
8.5 in (21.5 cm). Like only one other tyrannid (the other is the streaked flycatcher *Myiodynastes maculatus*), is streaked both above and below. Plumage includes olive-green upperparts with heavy streaking, pale yellow belly with dark brown streaking, reddish rump and tail, whitish secondaries and wing coverts, a blackish malar mark, and white stripes on face above and below dark eye patch. Bill is thick and black. Yellow patch in center of crown is visible only when crown is erect, during passion or aggression while courting.

DISTRIBUTION
Southeastern Arizona to Costa Rica; winters from eastern Ecuador to northern Bolivia.

HABITAT
Sycamore canyons, open woods, forest edges, and plantations.

BEHAVIOR
Lives singly or in pairs. Often perches high in canopy, remaining hidden. Early-morning song is a soft, repeated "tree-le-reere!" During courtship, both sexes sing a loud "kee-ZEE-ik!" Migratory.

FEEDING ECOLOGY AND DIET
Spots prey from perch, hawks in midair, and typically returns to perch to eat. Also gleans prey from foliage while hovering. Takes large insects, caterpillars, and spiders, but will also eat fruits and berries.

REPRODUCTIVE BIOLOGY
Male and female chase each other in flight during courtship. Breeds monogamously, once per year, later in year than most other flycatchers. Clutch of two to four eggs are incubated by the female in a preformed cavity nest located in a tree knot, abandoned nest, or bird box.

CONSERVATION STATUS
Not threatened.

SIGNIFICANCE TO HUMANS
None known. ◆

Resources

Books

American Ornithologists' Union. *Check-list of North American Birds: the Species of Birds of North America from the Arctic through Panama, including the West Indies and Hawaiian Islands.* 7th edition. Washington, DC: American Ornithologists' Union, 1998.

Finch, Deborah M., and Scott H. Stoleson, eds. *Status, Ecology, and Conservation of the Southwestern Willow Flycatcher.* Ogden, UT: United States Department of Agriculture, Forest Service, Rocky Mountain Research Station, 2000.

McCabe, Robert A. *The Little Green Bird: Ecology of the Willow Flycatcher.* Madison, WI: Rusty Rock Press, 1991.

Poole, Alan F., P. Stettenheim, and Frank B. Gill, eds. *The Birds of North America.* Philadelphia: The Academy of Natural Science; Washington, DC: American Ornithologists' Union, 1992--.

Skutch, Alexander Frank. *Life of the Flycatcher.* Norman, OK: University of Oklahoma Press, 1997.

Periodicals

Brosseau, La Ree. "Southwestern Willow Flycatcher." *Endangered Species Technical Bulletin* 25 (2000): 32.

Busch, Joseph D. et al. "Genetic Variation in the Endangered Southwestern Willow Flycatcher." *Auk* 117 (2000): 586.

Mezquida, Edouardo T. and Luis Marone. "Breeding Biology of Gray-Crowned Tyrannulet in the Monte Dessert." *Condor* 102 (2000): 205.

Resources

Taroff, Scott A. and Laurene Ratcliffe. "Pair Formation and Copulation Behavior in Least Flycatcher Clusters." *Condor* 104 (2000): 832.

Organizations

Center for Biological Diversity. P.O. Box 710, Tucson, Arizona 85702-0701 USA. Phone: (520) 623-5252. Fax: (520) 623-9797. E-mail: center@biologicaldiversity.org Web site: <http://www.biologicaldiversity.org/swcbd/index.html>

Other

Mangoverde World Bird Guide. "Tyrant Flycatchers." <http://www.mangoverde.com/birdsound/tyrflycatch/>. (11 December 2001).

Tyrant Flycatchers. <http://www.montereybay.com/creagrus/flycatchers.html>. (11 December 2001).

Tamara Schuyler, MA

Sharpbills
(Oxyruncidae)

Class Aves

Order Passeriformes

Suborder Tyranni (Suboscines)

Family Oxyruncidae

Thumbnail description
Smallish tyrannid with spotted plumage and sharp bill

Size
6–7 in (15.2–17.8 cm); 1.6 oz (44 g)

Number of genera, species
1 genus; 1 species

Habitat
Humid montane forests

Conservation status
Unknown

Distribution
Tropical Central and South America

Evolution and systematics

The exact affinities of the sharpbill (*Oxyruncus cristatus*) have been in dispute since the genus *Oxyruncus* was first described in 1820. Since the late nineteenth century most authors have given the sharpbill family status, despite its widely scattered distribution. Sharpbills are obviously related to the tyrannid passerines, particularly the tyrant flycatchers, cotingas, and manakins. However, the sharpbills' exact relations with these groups remain unclear.

In the 1980s, genetic comparisons seemed to indicate that sharpbills are cotingas and also are closely related to tityras and becards. However, for the purposes of this discussion the sharpbills are treated as a separate family (Oxyruncidae) within the Passeriformes as per Peters checklist.

This monotypic family has been divided into five races, differentiated slightly by color and size, but mainly by their distribution. *Oxyruncus cristatus frater*, the Costa Rican sharpbill, ranges from northeastern Costa Rica to western Panama, where it is known by its Spanish name, Pico Agudo. *O. c. brooksi* is found in the Darien region of Panama. *O. c. hypoglaucus* inhabits the highlands of southern Venezuela, Guyana, and Suriname. *O. c. cristatus* and *O. c. tocantinsi* are

both found in Brazil (the latter may be synonymous with *O. c. hypoglaucus*).

Physical characteristics

Sharpbills are small, stocky birds with a muted but distinctive plumage. The back, scapulars, and rump are olive green. The sides and flanks shade from dull white to greenish yellow. Tear-shape spots fade from buffy on the breast (except in the center of the abdomen, where they are absent) to smaller, darker, and denser on the head. The wings are blackish, with two yellowish bars. The tail is blackish, and the short, stout feet are a dull gray.

The sharpbill's head is marked by a red eye and the straight, gray bill that gives the genus its name. This instrument tapers from a broad base to an unusually pointed tip. Short rictal bristles encircle its conical base. A median crest ranges between races from bright crimson to orange. It is raised only when the bird is excited.

Adult female plumage is more muted, and the crest is less conspicuous. In general, though, write Stiles and Whitney, "the sexes are too similar in appearance to be safely distinguished

Male sharpbill (*Oxyruncus cristatus*). (Illustration by Bruce Worden)

in the field under any but the most favorable circumstances." Immature sharpbills resemble mature females, but with an even smaller scarlet crown.

Distribution

Sharpbills are scattered in discontinuous patches in Central America and South America. This curious dispersal may mark the remnants of a more widespread and continuous historical distribution.

In Central America, sharpbills inhabit northwest and central Costa Rica, including the Dota Mountains, and a majority of Panama. In South America, the species ranges from the Pantepui of southern Venezuela, Guyana, and Suriname, to northeastern and southeastern Brazil, including the Amazon. Sharpbills are also found in southern Paraguay and in central and eastern Peru.

Habitat

Sharpbills inhabit rain and cloud forests at 1,300–5,900 ft (400–1,800 m). In these humid montane regions, the birds can be found in dense primary forest, as well as along forest edges and in secondary growth. In some locations they have been observed descending to lowlands during nonbreeding seasons.

Behavior

The few accounts of sharpbills in the wild describe them as "stolid," even "sluggish" birds prone to brief bouts of abrupt movement. In between lengthy periods of branch sitting, sharpbills will launch into fast, direct flights between trees, or make rapid sorties for food.

They are most often solitary birds, and usually hang back even when part of mixed foraging flocks. Owing to their generally inconspicuous coloration and demeanor, sharpbills can be difficult to observe.

Stiles and Whitney observed male Costa Rican sharpbills (*O. c. cristatus*) alternate between vocalizing from conspicuous perches and active, but silent, defense of overlapping territories from other males. Noting that the defenders seemed happy to welcome females into those same territories, the authors concluded that the cluster of ranges functioned as an exploded lek (mating ground).

The sharpbill call is a high, rough, slightly descending trill, transcribed by one observer as "eeeeuuuurrrr." It has been likened to the call of cicadas, certain cotingas, and the three-toed sloth.

Feeding ecology and diet

Sharpbills mainly eat fruit and invertebrates. They either hop about the densely leafed canopy, or make short sallies among the outer twigs and leaves. In South America, sharpbills have been observed feeding alone, in pairs, and in mixed flocks alongside tanagers, cotingas, woodpeckers, woodcreepers, and other small birds. Spiders, ants, berries, and seeds have been found in sharpbill stomachs.

To find its food, a sharpbill will often hang upside down from a branch, like a tit. It will also probe with its bill into tufts of moss, epiphytes, fruit pods, and tightly rolled leaves. Often the bird will then open its bill to reveal arillate seeds, or insect egg cases concealed inside the leaves.

This foraging behavior, known as "pry-and-gape," has been observed in other members of the Icteridae family. Still, it is considered a unique specialization among the Neotropical tyrannids, and it may provide an evolutionary explanation for the sharpbill's namesake appendage.

Reproductive biology

As late as the 1980s, the sharpbill was one of the last avian families whose mating and nesting habits remained a mystery. The first nest was not found until 1980, and courtship behavior has been rarely, if ever, observed. In their observations of the Costa Rican sharpbill, Stiles and Whitney concluded that some of the male activity, including perching hopefully next to females and following them into the foliage, qualified as courtship behavior.

The sharpbill breeding season probably occurs at the same time as its singing season. This extends from late February or early May, to late May, or early June.

The nest described by Brooke, Scott, and Teixeira was built by a female near the top of a 100-ft (30-m) tree in southeastern Brazil, 30 mi (50 km) from Rio de Janeiro. It consisted of a simple, shallow cup 3 in (7.8 cm) in diameter, slung underneath a slim horizontal branch. A thin outer surface of mosses, spider's webs, liverworts, and leaves may have been held together and suspended by dried saliva. Two eggs were incubated for 14–24 days, and the female fed the young by regurgitation. Observations suggested a nestling period of 25–30 days.

Conservation status

Since the species has yet to be studied in any depth, sharpbills' conservation status and population size remain un-

known. They have been described as uncommon or rare over much of their range. However, sharpbills are regularly seen in primary forests near Rio de Janeiro.

Significance to humans

Sharpbills have no known significance to humans. ◆

Resources

Books

Perrins, C. M., ed. *The Illustrated Encyclopedia of Birds*. New York: Prentice Hall, 1990.

Sibley, C. G., and B. L. Monroe Jr. *Distribution and Taxonomy of Birds of the World*. New Haven: Yale University Press, 1990.

Wetmore, A. *The Birds of the Republic of Panama*. Vol. 3. Washington, DC: Smithsonian Institution, 2000.

Periodicals

Brooke, M., D. Scott, and D. Teixeira. "Some Observations Made at the First Recorded Nest of the Sharpbill *Oxyruncus cristatus*." *Ibis* 125, 2 (1983): 259–261.

Da Silva, J. M. C. "The Sharpbills in the Sierra dos Carajas, Para, Brazil, With Comments on Altitudinal Migration in the Amazon Region." *Journal of Field Ornithology* 64, 3 (1993): 310–315.

Lanyon, S. M. "Molecular Perspective on Higher-Level Relationships in the Tyrannoidea Aves." *Systematic Zoology* 34, 4 (1985): 404–418.

Sibley, C. G. "The Relationships of the Sharpbill." *Condor* 86 (1984): 48–52.

Stiles, F. G., and B. Whitney. "Notes on the Behavior of the Costa Rican Sharpbill." *Auk* 100, 1 (1983): 117–125.

Julian Smith, MS

Manakins

(Pipridae)

Class Aves
Order Passeriformes
Suborder Tyranni (Suboscines)
Family Tyrannidae

Thumbnail description
Small, stocky birds of Neotropical woodland and rainforest with short bills, short tails, and big eyes; males are marked with boldly patterned or exceptionally colorful plumage; females have dull, olive-brown plumage; the colorful males attract the drab females by performing elaborate displays, often on special display grounds called leks; the females alone build nests and raise the young; small fruits and some insects are plucked on the wing during sallying flight

Size
Length 3–5.9 in (7.5–15 cm); weight approximately 0.35–0.70 oz (10–20 g)

Number of genera, species
17 genera, 54 species

Habitat
Understory of subtropical woodlands to lush tropical rainforests

Conservation status
Critically Endangered: 1 species; Endangered: 1 species; Vulnerable: 2 species; Near Threatened: 1 species

Distribution
Neotropics from Mexico to Argentina, and the islands of Trinidad and Tobago

Evolution and systematics

No fossil manakins have been reported. Peters Checklist considers the manakins to be a distinct family (Pipridae) with 17 genera and about 54 species. Based upon DNA-DNA studies and other characteristics, many experts now consider the true manakins to be a subfamily (Piprinae) of the suboscine passerine family Tyrannidae (tyrant flycatchers). This subfamily is comprised of 11 genera and 41 species, including the genera *Heterocercus, Chloropipo, Xenopipo, Chiroxiphia, Antilophia, Manacus, Ilicura, Corapipo, Masius, Machaeropterus,* and *Pipra.* The following six genera, comprising 13 species, are no longer grouped with true manakins under this taxonomic system and will not be covered further in this discussion: *Sapayoa, Schiffornis, Tyranneutes, Neopelma, Neopipo,* and *Piprites.*

Physical characteristics

The manakins are beautiful, stocky little passerines, most less than 4.9 in (12.5 cm) long. They have short, somewhat broadened and very slightly curved bills; rounded, short wings, sometimes with feathers modified in the males to produce sound effects. The legs are short. They have three toes in front

and one in back of the foot, but the front middle toe is fused at its base with one of the adjoining toes. The eyes are large. Sexes are different. Female and juvenile plumage is typically drab olive-green. Male coloration is stunning—basic black and olive wings are contrasted with patches of intense white, blue, red, or yellow on areas such as the crown, neck, and mantle. Juvenile males may go through several intermediate subadult molts before acquiring full adult male coloration.

Distribution

Neotropics from Mexico to Argentina, and the islands of Trinidad and Tobago. These fascinating perching birds are found only in the Neotropics. They are widely distributed in the understory of subtropical woodlands and tropical forests of Central and South America, and a few nearby islands. All manakins are resident, non-migratory species within their range.

Habitat

Understory of subtropical woodlands to lush tropical rainforests.

A crimson-hooded manakin (*Pipra aureola*) spreads its wings. (Photo by C.H. Greenewalt/VIREO. Reproduced by permission.)

Behavior

Birdwatchers, eco-tourists, and professional ornithologists alike find manakins to be among the most beautiful and enchanting of all the world's birds. The displays may include distinctive songs and calls, and mechanical sounds (such as "whirring" and "wing snapping" noises) made with modified flight feathers. The ritualized displays may be conducted at leks by a single male, or, in some species, multiple males (from 2–3) cooperate during courtship, with copulation usually going to the most dominant male of the duo or trio.

Feeding ecology and diet

Small, berry-sized fruits and insects are taken during quick, sallying flights. Some manakins seem to be particularly fond of fruits which are bluish or purplish in color. The bluish feces often contain seeds.

Reproductive biology

Manakins do not form a lasting pair-bond, but are polygamous, using a leh (courtship area) where females choose and mate with a male. After copulation, the females fly off alone to build the nests, incubate the eggs, and raise the young. The nest is constructed using woven fibers and grasses to form a tiny hammock in small trees or ferns usually over water. Incubation lasts from 17–21 days, with short fledging times of 13–15 days for one to two young.

Manakin courtship display. (Illustration by Brian Cressman)

A male long-tailed manakin (*Chiroxiphia linearis*) sings in Costa Rica. (Photo by Kenneth W. Fink. Bruce Coleman Inc. Reproduced by permission.)

Many manakins have colorful plumage, such as this wire-tailed manakin (*Pipra filicauda*). (Photo by J. Alvarez A./VIREO. Reproduced by permission.)

Conservation status

According to IUCN, the newly discovered Araripe manakin (*Antilophia bokermanni*) is Critically Endangered due to its extremely small Brazilian range and population, coupled with pressure on its habitat due to development. The golden-crowned manakin (*Lepidothrix (Pipra) vilasboasi*) is Vulnerable on the basis of its very small range in Brazil. The yellow-headed manakin (*Chloropipo flavicapilla*), Wied's tyrant-manakin (*Neopelma aurifrons*), and the black-capped manakin (*Piprites pileatus*) are considered Near Threatened, Endangered, and

Vulnerable, respectively. No manakin is currently listed by the Convention on International Trade in Endangered Species (CITES).

Significance to humans

Manakins may be of indirect economic importance to countries which cater to birdwatchers and eco-tourists. Their images are popular on postage stamps of their range countries, as well as t-shirts and local artwork.

1. Red-capped manakin (*Pipra mentalis*); 2. Striped manakin (*Machaeropterus regulus*); 3. Long-tailed manakin (*Chiroxiphia linearis*); 4. Golden-headed manakin (*Pipra erythrocephala*); 5. Scarlet-horned manakin (*Pipra cornuta*); 6. Wire-tailed manakin (*Pipra filicauda*); 7. Araripe manakin (*Antilophia bokermanni*). (Illustration by Brian Cressman)

Species accounts

Long-tailed manakin
Chiroxiphia linearis

SUBFAMILY
Piprinae

TAXONOMY
Pipra linearis Bonaparte, 1838, Mexico = Santa Efigenia, Oaxaca.

OTHER COMMON NAMES
French: Manakin fastueux; German: Langschwanzpipra; Spanish: Saltarín Toledo, Saltarín de Cola Larga.

PHYSICAL CHARACTERISTICS
Sexes differ. Female's length is 5.5 in (14 cm), including 1 in (2.5 cm), elongated, central tail feathers. Male's length is 8.5–10.5 in (21.5–26.5 cm), including 3.9–5.9 in (10–15 cm), elongated, central tail feathers. Weight is 0.7 oz (19 g). The male is mostly black with an azure blue back, a red crown with a rear projecting crest, and long, central tail feathers. Females are olive-green. Distinctive orange legs and feet.

DISTRIBUTION
Southern Mexico to Costa Rica.

HABITAT
Open vine tangles and thick undergrowth of dry or humid forest, secondary forest and plantation borders, and borders of mangroves swamps.

BEHAVIOR
Mercedes S. Foster conducted classic observations of the long-tailed manakin. In the advertising call, male pairs or trios synchronously repeat, "To-lay-do," from which their Spanish name has been derived. In the Up-Down Dance, males alternately make fluttering jumps straight upward into the air. In the Cartwheel Dance, each male in turn flutters up and backward in a vertical circle to land on the spot previously occupied by his dance partner. As many as 100 jumps may be completed per cartwheel sequence. The dominant male, who gets all copulations, finally ends the cooperative display and dismisses his dance partner with a single piercing note *pweet!*.

FEEDING ECOLOGY AND DIET
Uses sallying flight to pluck small fruits from tropical, evergreen understory trees like *Ardisia revoluta* (Myrsinaceae), as well as from shade-intolerant secondary growth trees such as *Cecropia peltata*.

REPRODUCTIVE BIOLOGY
Following copulation, the female leaves to build the nest, incubate eggs, and raise the young on her own. The nest is a shallow cup of fibers, mosses, grasses, and dry leaves, attached by its rim and suspended from horizontal forks in small trees, approximately 27 ft (8 m) above the ground. The nest is not placed with any obvious connection to the lek. One, or usually two, buffy eggs with heavy brown spotting are laid. Fruit is included in the diet of the offspring.

CONSERVATION STATUS
Not threatened. Common in its preferred habitat; abundant in some areas.

SIGNIFICANCE TO HUMANS
Eco-tourists and birdwatchers enjoy seeing the males. ◆

Chiroxiphia linearis
▨ Resident

Araripe manakin
Antilophia bokermanni

SUBFAMILY
Piprinae

TAXONOMY
Antilophia bokermanni Coelho and Silva, 1998, Brazil.

OTHER COMMON NAMES
English: Araripe's soldier; French: Manakin de Araripe; German: Helmpipra; Spanish: Bailarín de Araripe.

PHYSICAL CHARACTERISTICS
Sexes differ. Length is approximately 5.7 in (14.5 cm) in both sexes. The male is mostly white with a red upstanding frontal crest, crown, nape, and mid-back. The female is olive, with a visibly upstanding frontal crest.

DISTRIBUTION
Rare. Central Brazil, extremely limited range near Chapada do Araripe. Wet forest, altitude around 2,625 ft (800 m).

HABITAT
Wet forest at low elevations.

BEHAVIOR
Unknown to date.

Antilophia bokermanni

◻ Resident

Machaeropterus regulus

◻ Resident

FEEDING ECOLOGY AND DIET
Probably takes small fruits and insects during quick, sallying flights as in other manakins.

REPRODUCTIVE BIOLOGY
Unknown.

CONSERVATION STATUS
This newly discovered (1998) species is categorized as Critically Endangered. Its known population and home range are both extremely small, and its habitat is under pressure from land developers.

SIGNIFICANCE TO HUMANS
None known. ◆

Striped manakin
Machaeropterus regulus

SUBFAMILY
Piprinae

TAXONOMY
Machaeropterus regulus Hahn, 1819.

OTHER COMMON NAMES
French: Manakin rubis; German: Streifenpipra; Spanish: Manaquin Franjeado.

PHYSICAL CHARACTERISTICS
Sexes differ, but not as markedly as in other manakins. Length is 3.5–3.7 in (9–9.5 cm). The male is olive above with a shiny

red crown and nape, and the underparts are streaked reddish brown and white. The female is plain olive above with some reddish brown streaking below. In both sexes the iris is dark red, and the legs purplish flesh.

DISTRIBUTION
Colombia, Venezuela, Ecuador, Peru, Brazil.

HABITAT
Understory of the interior of humid forest.

BEHAVIOR
Difficult to see, they travel alone or in pairs. Communal displays and noticeable lekking have not been recorded. Males have been recorded clinging to a horizontal perch with their feet, while hanging upside down and rocking to and fro while making buzzing sounds with their wings. Vocalizations are weak, hummingbird-like "chips."

FEEDING ECOLOGY AND DIET
Small fruits and insects are taken during quick, sallying flights.

REPRODUCTIVE BIOLOGY
Unknown.

CONSERVATION STATUS
Not threatened. Uncommon to fairly common in localized areas of humid forest.

SIGNIFICANCE TO HUMANS
None known. ◆

Red-capped manakin

Pipra mentalis

SUBFAMILY
Piprinae

TAXONOMY
Pipra mentalis Sclater, 1857, Cordova = Córdoba, Vera Cruz, Mexico.

OTHER COMMON NAMES
French: Manakin à cuisses jaunes; German: Gelbhosenpipra; Spanish: Saltarín de Capa Roja.

PHYSICAL CHARACTERISTICS
Sexes differ. Length is 3.9 in (10 cm). Males are velvety black except for bright yellow thighs and pale yellow underwing coverts, with a distinctive bright scarlet head. The shafts of the flight feathers are thickened, and both thickened and curved in the secondary feathers. The female is dull olive above. Males have white eyes, females have brown eyes. The legs are dull brown.

DISTRIBUTION
Western Colombia and western Ecuador.

HABITAT
Lower and middle understory of humid and wet forest.

BEHAVIOR
Lekking males gather in loose groups in low to middle forest understory. The modified shafts of the rectrices and secondaries produce mechanical wing snaps, and wing whirring and rustling buzzes.

FEEDING ECOLOGY AND DIET
Small fruits and insects are taken during quick, sallying flights.

REPRODUCTIVE BIOLOGY
The female alone makes a shallow cup-shaped nest attached to a horizontal branch fork from 5–10 ft (1.5–3 m) above the forest floor. The clutch consists of two grayish buff eggs, with a wreath of mottled brown around the large end.

CONSERVATION STATUS
Not threatened. Common in its preferred habitat.

SIGNIFICANCE TO HUMANS
Eco-tourists and birdwatchers enjoy seeing the males. Its image has been used on postage stamps. ◆

Scarlet-horned manakin

Pipra cornuta

SUBFAMILY
Piprinae

TAXONOMY
Pipra cornuta Spix, 1825, Brazil.

OTHER COMMON NAMES
French: Manakin à cornes rouges; German: Schopfpipra; Spanish: Saltarín Encopetado.

PHYSICAL CHARACTERISTICS
Sexes differ. Length is 4.6 in (11.7 cm). The male is mostly glossy blue-black. The whole head is brilliant scarlet, including a long bilobed crest on the hindcrown projecting back and

Pipra mentalis
◻ Resident

Pipra cornuta
◻ Resident

slightly upward; the thighs are scarlet. The female is dull olive. Bills are pale flesh color in both sexes.

DISTRIBUTION
Venezuela (tepuis), Guyana, and possibly extreme north Brazil.

HABITAT
Low and middle understory of humid forest and mature secondary woodland.

BEHAVIOR
Curious and confiding. Males display in traditional leks, vocalizing as they fly between perches, and making some mechanical whirring sounds with their wings.

FEEDING ECOLOGY AND DIET
Small fruits and insects are taken during quick, sallying flights.

REPRODUCTIVE BIOLOGY
Unknown.

CONSERVATION STATUS
Not threatened. Uncommon to locally fairly common in preferred habitat.

SIGNIFICANCE TO HUMANS
None known. ◆

Pipra erythrocephala
▨ Resident

Golden-headed manakin
Pipra erythrocephala

SUBFAMILY
Piprinae

TAXONOMY
Pipra erythrocephala Linnaeus, 1758.

OTHER COMMON NAMES
French: Manakin à tête d'or; German: Goldkopfpipra; Spanish: Saltarín Cabecidorado.

PHYSICAL CHARACTERISTICS
Sexes differ. Length is 3.6 in (9.1 cm). The male is glossy black with a brilliant golden-yellow head; thighs are red and white. Adult males have white eyes. The female is dull olive with gray eyes. Bills are yellowish white and legs are pale or flesh-toned in both sexes.

DISTRIBUTION
Eastern Panama southward to northeast Peru, Brazil north of the Amazon, the Guianas, and quite numerous on Trinidad.

HABITAT
Upper understory and middle growth of both humid and relatively deciduous forest and mature secondary woodland.

BEHAVIOR
Lek displays are noisy and conspicuous. Pairs of males often seem to be competing with each other. Established males may maintain their residency at the lek for up to eight years. In typical manakin fashion, the males reach a fevered pitch of display when a female approaches, and in full display expose their red and white thigh feathers.

FEEDING ECOLOGY AND DIET
Small fruits and insects are taken during quick, sallying flights. Feed at fruiting trees up to the height of the forest canopy.

REPRODUCTIVE BIOLOGY
Following copulation, the female alone builds the nest, incubates the eggs, and raises the young. The nest is a thinly woven cup of fibers attached to a horizontal fork in branches located 3.3–33 feet (1–10 m) off the ground. The eggs, usually two, are pale greenish yellow with many brown streaks around the large end of the shell.

CONSERVATION STATUS
Not threatened. Rather common in preferred habitats.

SIGNIFICANCE TO HUMANS
Eco-tourists and birdwatchers enjoy seeing the males. ◆

Wire-tailed manakin
Pipra filicauda

SUBFAMILY
Piprinae

TAXONOMY
Pipra filicauda Spix, 1825, Brazil.

OTHER COMMON NAMES
French: Manakin filifère; German: Fadenpipra; Spanish: Saltarín Cola De Hilo.

PHYSICAL CHARACTERISTICS
Sexes differ. Length is 4.2 in (10.7 cm). The gaudy males are black above, with scarlet red crown, nape, and upper back, and intense golden yellow undersides, forehead and sides of head.

Pipra filicauda

Resident

Both sexes have tail feather shafts that project like long (2 in/5 cm) wire filaments; slightly shorter in the female. Irises are milk white in both sexes.

DISTRIBUTION
Venezuela, Colombia, Ecuador, Peru, Amazonian Brazil.

HABITAT
Near streams in secondary forest, gallery and seasonally flooded forest.

BEHAVIOR
Males form widely scattered leks in forest 3.3–26 ft (1–8 m) above the forest floor. The courtship displays include swooping and slow butterfly-like short flights, lateral side-jumps, side-to-side twisting with head lowered, crouching and erecting their body feathers, and raising the tail. The wings make mechanical sounds. When a female approaches closely, the male brushes his raised tail filaments against her face and throat. According to Ridgely and Tudor, "This is believed to be the only instance among birds in which modified tail feathers are used primarily in a tactile, as opposed to a visual, manner."

FEEDING ECOLOGY AND DIET
Small fruits and insects are taken during quick, sallying flights.

REPRODUCTIVE BIOLOGY
Following copulation, the female alone builds a cup nest in small trees besides streams to lay and incubate eggs and raise the young.

CONSERVATION STATUS
Not threatened. Locally common in preferred habitats.

SIGNIFICANCE TO HUMANS
Eco-tourists and birdwatchers enjoy seeing the males. ◆

Resources

Books

Bateman, G., ed. *All The World's Animals: Songbirds.* New York: Torstar Books Inc., 1985.

Dunning, J. S. *South American Land Birds: A Photographic Guide to Identification.* Newtown Square, PA: Harrowood Books, 1982.

Foster, M. S. "*Chiroxiphia linearis* (Saltanix Colilargo, Toledo, Long-tailed Manakin)." In *Costa Rican Natural History,* edited by D. H. Janzen. Chicago: University of Chicago Press, 1983.

Hilton-Taylor, C., comp. *2000 IUCN Red List of Threatened Species.* Gland, Switzerland and Cambridge, UK.: IUCN, 2000.

Hilty, S. L., and W. L. Brown. *A Guide to the Birds of Colombia.* Princeton, NJ: Princeton University Press, 1986.

Johnsgard, P. A. *Arena Birds: Sexual Selection and Behavior.* Washington, DC: Smithsonian Institution, 1994.

Meyer de Schauensee, R., and W. H. Phelps. *A Guide to the Birds of Venezuela.* Newtown Square, PA: Harrowood Books, 1982.

Ridgley, Robert S., and G. Tudor. *The Birds of South America.* Vol. II: *The Suboscine Passerines.* Austin: University of Texas Press, 1994.

Sibley, C. G., and B. L. Monroe, Jr. *Distribution and Taxonomy of Birds of the World.* New Haven, CT: Yale University Press, 1990.

Sibley, C. G., and B. L. Monroe, Jr. *A Supplement to Distribution and Taxonomy of Birds of the World.* New Haven, CT: Yale University Press, 1993.

Stiles, F. G., and A. F. Skutch. *A Guide to the Birds of Costa Rica.* Utica, NY: Cornell University Press, 1989.

Periodicals

Coelho, G., and W. Silva. "A New Species of *Antilophia* (Passeriformes: Pipridae) from Chapada do Araripe, Ceará, Brazil." *Ararajuba* 6 (1998): 81–84.

Foster, M. S. "Odd Couples in Manakins: A Study of Social Organization and Cooperative Breeding in *Chiroxiphia linearis.*" *American Naturalist* 111 (1977): 845–853.

Prum, R. O. "Phylogenetic Analysis of the Evolution of Display Behavior in the Neotropical Manakins (Aves: Pipridae)." *Ethology* 84 (1990): 202–231.

Prum, R. O. "Sexual Selection and the Evolution of Mechanical Sound Production in Manakins (Aves: Pipridae)." *Animal Behaviour* 55 (1998): 977–994.

Resources

Organizations

Association for BioDiversity Information. 1101 Wilson Blvd., 15th Floor, Arlington, VA 22209 USA. Web site: <http://www.infonatura.org/>

University of Michigan. 3019 Museum of Zoology, 1109 Geddes Ave, Ann Arbor, MI 48109-1079 USA. Phone: (734) 647-2208. Fax: (734) 763-4080. E-mail: rbpayne @umich.edu Web site: <http://www.ummz.lsa.umich .edu/birds/index.html>

Other

Attenborough, D. *The Life of Birds, Episode 7: Finding Partners.* BBC Video: British Broadcasting Corporation. 1998.

Charles E. Siegel, MS

Cotingas
(Cotingidae)

Class Aves

Order Passeriformes

Suborder Tyranni

Family Cotingidae

Thumbnail description
A spectacular group of birds that range in size from a canary to a crow, and are characterized by extreme colors, vocalizations, and/or body ornamentation

Size
3.25–20 in (8–51 cm); 0.04–1.25 lb (18–571 g)

Number of genera, species
25 genera; 61 species

Habitat
Predominately tropical forest, including seasonally inundated and dry tropical forest

Conservation status
Critically Endangered: 1 species; Endangered: 4 species; Vulnerable: 10 species; Near Threatened: 5 species

Distribution
Southern Mexico to northern Argentina and eastern Brazil

Evolution and systematics

The cotingas (Cotingidae) are a striking family of birds. Not widely investigated in the past, cotingas are becoming more widely researched, and the number of studies is increasing. There are no known fossils of this family at the present time.

The common ancestor in cotinga lineage was perhaps similar to the Old World family Eurylaimidae. In addition to giving rise to cotingas, this ancestor also gave rise to manakins and tyrant flycatchers. Indeed, Tityras and Becards, were, until recently, placed within the family Cotingidae, but now are considered tyrant flycatchers (family Tyrannidae). Some species of cotingas such as red-cotingas (*Phoenicircus*) are considered by some to be the link between the cotinga and manakin families, based upon morphological and behavioral characteristics. More than one-half of the genera are represented as "superspecies", where there are several closely related "sister taxa" with non-overlapping geographic ranges.

Peters Checklist recognizes 25 genera with 61 species, including: 1. The true cotinga (*Cotinga*). The males are brilliant blue to purple in different patterns, while the females are dull brown. These birds are found in the tropical zone of the Ama-

zon forests. There are seven species, including the banded cotinga (*Cotinga maculata*) which is found in Brazil; 2. The fruiteaters (*Pipreola*) comprise at least eight species, including the barred fruiteater (*Pipreola arcuata*); 3. The red-ruffed fruit-crow (*Pyroderus scutatus*); 4. The capuchin bird (*Perissocephalus tricolor*); 5. The Amazonian umbrella bird (*Cephalopterus ornatus*), one of the largest species, with a length of up to 22 in (51 cm); 6. The cocks-of-the-rock (*Rupicola*) are plump, short-tailed, broad-footed birds. There are two species, including the andean cock-of-the-rock (*Rupicola peruviana*); and 7. The bellbirds (*Procnias*), of which there are four species: a. The white bellbird (*Procnias alba*); b. The bare-throated bellbird (*Procnias nudicollis*); c. The three-wattled bellbird (*Procnias tricarunculata*); and d. The bearded (or mossy-throated) bellbird (*Procnias averano*).

Physical characteristics

Cotingas are characterized by compact bodies, large heads and wide mouths, often with a hook-tipped bill. The tarsi (feet) are surrounded only by band-like plates in front, but covered at the rear with very small platelets which are not all contiguous. Although the legs are short relative to the size of

This capuchinbird (*Perissocephalus tricolor*) has bare skin around its eyes and forehead, which is common in some cotingas. (Photo by C.H. Greenewalt/VIREO. Reproduced by permission.)

the bird, the feet are of sufficient to perch comfortably; this is perhaps enhanced by long, sharp claw in some species (e.g., *Rupicola, Cephalopterus*).

Cotingas show more variation in size than any other group of passerines, ranging from the size of a canary to a crow. The length is 3–20 in (8–50 cm). There is a significant amount of sexual dimorphism, with mass being greater in females of the smaller species (e.g., *Iodopleura, Porphyrolaema, Cotinga, Lipaugus,* and *Phoenicircus*), but the reverse situation (i.e., greater in males) in the larger species (e.g., *Gymnodoerus, Cephalopterus*). Dimorphism also extends to plumage, with males being the more colorful sex.

Many species are quite beautiful; they have striking colors, decorative plumes, crests, inflatable throat sacs, strands of skin or bare leppets on the forehead or at the angle of the beak. These ornaments are more strongly accentuated in males. Many of the larger cotingas are distinguished not only by the gloss and brightness of their plumage and their quite unusual appendages, but also by their tuneful, far-reaching calls. The vocal muscles are strong.

Distribution

Cotingas are restricted to the Neotropics, distributed from southern Mexico through most of tropical South America as

far as northern Argentina. Although most species are found at sea level, there are several Andean forms, with species such as the white-cheeked cotinga (*Ampelion stresemanni*) ranging up to 14,000 ft (4,300 m).

Countries harboring the most different species of cotingas include Brazil (approximately 33 species), and the northwest Andean countries (Colombia 35, Peru 31, Ecuador 30, Venezuela 27). In contrast, the countries furthest south (Argentina) and north (Middle American countries) of the equator contain only two or three species.

Most cotingas are regionally restricted. While several species are found throughout most of the Amazon basin, the species with the widest distribution is perhaps the purple-throated fruitcrow (*Querula purpurata*), which ranges from Costa Rica through Bolivia. In contrast, the rarest species is the kinglet calyptura (*Calyptura cristata*) that is restricted to a 0.4 mi² (1 km²) patch of forest north of Rio de Janeiro, Brazil. This species went unreported for most of the 1900s, though some recent reports from 1996 suggest that it is still present.

Habitat

Most cotingas are shy, unobtrusive avoiders of civilization, and as such they inhabit the upper and middle tree levels of continuous forest areas, as residents. Only a few species are also found in open landscapes or secondary forest. Many of the larger species (e.g., *Gymnodoerus* and *Cephalopterus ornatus*) are riverine specialists, but some of the smaller species (e.g., *Cotinga maynana*) will inhabit riverine habitat or swamp edges as well. They are often visually inconspicuous and at the same time widely distributed.

Behavior

While many of the smaller species (e.g., *Porphyrolaema*) are solitary, the larger species (e.g., *Gymnodoerus*) will often travel in small flocks.

There are general "tradeoffs" in adaptations used by male cotingas to attract females. In general terms, males of the smaller species (e.g., *Cotinga*) tend to be brighter colored and less vocal, whereas the medium-sized species tend to be more vocal and less brightly colored (*Querula*); the largest species (e.g., *Cephalopterus*) tend to have more apparent body ornamentation, such as throat wattles or lappets.

While many of the cotingas have a very subtle or soft call, some of the more "drab" species compensate what they lack in plumage with a series of resonating, and sometimes far-reaching, whistles (e.g., *Lipaugus, Tijuca, Querula*). Some of the larger species are able to expand parts of the trachea and pharynx, to release with the exhale a sound similar to the "mooing" of a cow, thus the name "Calfbird" (*Perissocephalus*); another species displaying this "mooing" vocalization is the Umbrellabird (*Cephalopterus*). Other species (e.g., *Rupicola, Procnias nudicollis*) are quite vocal as well. Some species are able to produce noises of sorts with their wings (e.g., *Cotinga, Xipholena, Phoenicircus, Rupicola*).

Many cotinga species form "leks" (loose to tight associations of several males vying for females through elaborate display), although this trend appears to be strongest in the medium-sized species (e.g., *Lipaugus, Phoenicircus*).

Cotingas do not exhibit any great degree of territoriality. For example, different species of cotingas (e.g., *Cotinga, Querula, Gymnodoerus*) may be perched in the same tree with no agonistic behavior. Similarly, these same species have been observed occupying trees with other families of birds (e.g., kites [*Ictinia plumbea*], parrots [*Graydidascalus brachyurus, Brotogeris cyanoptera*], flycatchers [*Empidonomus aurantioatrocristatus, Tyrannopsis luteiventris*], and other passerines [*Laniocerca hypopyrrha, Scaphidura oryzivora, Thraupis episcopus*]) present without incident. Helmut Sick notes that species such as *Piprites, Querula purpurata*, and *Oxyruncus cristatus* join mixed species flocks of birds regularly. During more than 10 weeks of observation in the northern Peruvian Amazon, only a single incident of agonistic behavior was observed. This involved a female spangled cotinga *Cotinga cayana* and female Cinereous mourner *Laniocerca hypopyrrha* (a non-Cotingid) simultaneously mobbing a female bare-necked fruitcrow *Gymnodoerus* through habitat that was atypical for the latter species. The mobbing behavior led to the *Gymnodoerus* flying out of the area. In Brazil, Sick noted a *Procnias nudicollis* displacing a female *Xipholena atropurpurea* from a tree.

Like most Passerines, cotingas are inactive at night, and most active during the early morning light. The secondary peak of activity tends to occur in the late afternoon, just before dusk.

Although some short-distance migration patterns (or altitudinal migration) may characterize some species of cotingas, the family is for the most part non-migratory.

Feeding ecology and diet

Cotingas have wide gaping mouths, adapted to eating berries and other fruits. The larger species and those which inhabit open country also like to take insects as well. Fruits eaten include those of palms (*Euterpes, Livistonia*) and Cecropias, as well as fruits of the plant families Lauraceae, Burseraceae, Loranthaceae, Melostomataceae, and Myrsinaceae (e.g., *Rapanea ferruginea*).

There appears to be increased dietary specialization in larger species versus smaller species, which are more generalized in their diets. Additionally, many of the smaller species tend to be "gorgers", settling in the lower parts of bushes to feed on masting fruits. Feeding is done while flying, perching, or hopping through branches. As a relatively passive group, cotingas display little intraspecific competition or aggression at fruiting trees, with several individuals (even males) foraging without incident in at least some species (*Cotinga*).

Smaller seeds of the fruits they consume are passed through and dispersed without being digested, whereas larger seeds are regurgitated on the spot. Seed dispersal helps regenerate the tropical forests where cotingas live, as seeds of their preferred food plants are distributed throughout the forests.

A male black-necked red cotinga (*Phoenicircus nigricollis*) on its display perch in the Amazon rainforest, Peru. (Photo by Michael Fogden. Bruce Coleman Inc. Reproduced by permission.)

Reproductive biology

Cotingas are polygynous birds. Several species of cotingas (e.g., *Pyroderus scutatus, Perissocephalus tricolor*, and *Phoenicircus*), form "tight" leks where the males compete for the attention of a female through elaborate displays. Other more drab species such as *Lipaugus* and *Querula purpurata*, will compete for females in "loosely-attended" leks through their loud calls that carry far in the tropical forest. In yet other species, a single male will court a female, but not without the presence of other males. "Flags" (signals designed to attract attention) during flight serve as courtship signals in species such as *Xipholena*, whereas an elaborate flight entails the courtship for species such as *Gymnodoerus* and *Haemotoderus*.

Most species lay a single egg, concordant with Rensch's rule of clutch sizes decreasing in more equatorial species; however, clutch size may reach three eggs in species such as *Phibalura flavirostris*. While egg color varies from yellow to brown among species, most taxa have flecking at the blunt end. Females incubate alone in some genera, such as *Cotinga* and *Querula*, whereas in others males assist during nest building (e.g., *Phibalura*) or incubation (e.g., *Iodopleura* and *Phibalura*). Aggressive nest defense has been observed in certain species as well. Incubation is generally 25 to 28 days for the larger species, but may be of shorter duration for the smaller species.

Nests vary considerably from species to species. Species of most genera (e.g., *Cotinga, Querula, Xipholena, Perissocephalus, Lipaugus*, and *Cephalopterus*) build a small platform of sticks in the fork of a tree. Other species build a smaller nest with a shallow cup (e.g., *Phibalura, Gymnodoerus*, and *Iodopleura*).

The chicks hatch blind and featherless, quite dependent upon the parents. Females alone care for the brood in genera such as *Cotinga* and *Procnias*. In other species (e.g., *Querula purpurata*) multiple helpers care for the brood. The young leave the nest at, or slightly more than, one month of age. Some cotingas have more than one clutch per year.

"Lekking" in the rainforest understory—a female Guianan cock-of-the-rock (*Rupicola rupicola*) (center) visits a "lek," a male courtship arena, to select a mate. (Illustration by Emily Damstra)

Conservation status

Of the 61 species, the kinglet calyptura (*Calyptura cristata*) is considered Critically Endangered, 4 species are considered Endangered (*Iodopleura pipra, Cotinga maculata, Xipholena atropurpurea,* and *Carpodectes antoniae*), 10 are considered Vulnerable (*Laniisoma elegans, Tijuca condita, Carpornis melanocephalus, Doliornis remseni, Lipaugus uropygialis, L. lanioides, Cotinga ridgwayi, Cephalopterus glabricollis, C. penduliger,* and *Procnias tricarunculata*), and 5 are considered Near Threatened. This makes nearly one-third of the species of real or potential conservation concern.

Habitat destruction is the main threat to cotingas. Of the 20 species that are of potential conservation concern, 12 are from the Brazilian coastal Atlantic forests, which suffer extensively from forest fragmentation. Of the remaining species, four are from the Andes, and four are from Middle America, both areas which also suffer forest fragmentation.

Once thought to be extinct, the kinglet calyptura caused great excitement among birdwatchers when it was spotted by Brazilian bird expert Ricardo Parrini in Rio de Janeiro on October 27, 1996. The first sighting of this tiny creature in over 100 years, the find was documented in a 2001 edition of *Cotinga* magazine.

Significance to humans

Several indigenous tribes use cotinga feathers in their ornamentation. One of the most frequently seen groups is *Cotinga,* which is commonly represented in costumes of certain Amazonian tribes. Perhaps as many as 10–15% of artifacts have *Cotinga* feathers, although the most commonly used feathers are those of Psittacids (*Ara* and *Amazona*). During the late 1990s, cocks-of-the-rock were threatened due to demand of their feathers to make fishing flies. Additionally some species may be hunted incidentally as a protein source. The head and beard ornamentation of species such as *Cephalopterus ornatus* are sometimes seen in Amazonian riverboats, but the associated belief, whether aphrodisiac or mere folklore, is unknown.

Amazonian umbrellabird (*Cephalopterus ornatus*). (Illustration by Emily Damstra)

1. Turquoise cotinga (*Cotinga ridgwayi*); 2. Plum-throated cotinga (*Cotinga maynana*); 3. Spangled cotinga (*Cotinga cayana*); 4. Purple-breasted cotinga (*Cotinga cotinga*); 5. Banded cotinga (*Cotinga maculata*); 6. Andean cock-of-the-rock (*Rupicola peruviana*); 7. Guianan cock-of-the-rock (*Rupicola rupicola*). (Illustration by Emily Damstra)

1.Three-wattled bellbird (*Procnias tricarunculata*); 2. Bare-throated bellbird (*Procnias nudicollis*); 3. Bearded bellbird (*Procnias averano*); 4. Long-wattled umbrellabird (*Cephalopterus penduliger*); 5. Bare-necked umbrellabird (*Cephalopterus glabricollis*); 6. White bellbird (*Procnias alba*). (Illustration by Emily Damstra)

Species accounts

Spangled cotinga
Cotinga cayana

TAXONOMY
Cotinga cayana Linnaeus, 1766.

OTHER COMMON NAMES
French: Cotinga de Cayenne; German: Halsbandkotinga; Spanish: Cotinga Grande.

PHYSICAL CHARACTERISTICS
The average weight is 2.7 oz (76 g). Sexually dimorphic. Females are darkish brown with a light brown, spotted breast. Males are a stunning turquoise color with shimmering iridescent feathers and a band of blue across the chest.

DISTRIBUTION
This species is found throughout the Amazon. It is the only species within the genus that overlaps the geographic distribution of other congeners.

HABITAT
This species is a canopy specialist in lowland tropical evergreen forest. While principally a lowland species, it may range up to 0.5 mi (800 m).

BEHAVIOR
The quiet behavior of the members of this genus is in contrast with their vivid colors. During courtship male spangled cotin-

gas will flatten themselves horizontally, moving the wings and spreading the tail while emitting a soft and mournful "hooooo."

Various congeners will forage in the same tree with spangled cotingas, such as plum-throated cotinga (*Cotinga maynana*). The spangled cotinga will also forage with other species of cotingas including the purple-throated (*Querula purpurata*) and bare-necked fruitcrows (*Gymnodoerus foetidus*).

While conspecifics are often found in close association, there is at least one record of a male spangled cotinga chasing another male from the area. A female spangled cotinga and female cinereous mourner simultaneously mobbed a female bare-necked fruitcrow through habitat that was atypical for the latter species. On another occasion a male spangled cotinga was observed chasing a black-headed parrot (*Pionites melanocephala*).

FEEDING ECOLOGY AND DIET
Fruit and berries are consumed, often "gorging" at a masting tree or bush such as mistletoe. The fruits are often plucked on the wing. Although the seeds of larger species (e.g., mistletoe) might be regurgitated, smaller seeds are often swallowed. Insects are also taken.

REPRODUCTIVE BIOLOGY
The mating system is not completely known within this group. However, there is some evidence that loose lek associations may be in place.

The nest is platform type, often high in a tree fork, or next to an epiphyte. The female incubates and cares for the young alone.

CONSERVATION STATUS
Not threatened.

SIGNIFICANCE TO HUMANS
Several indigenous tribes use cotinga feathers in their ornamentation. ◆

Cotinga cayana

    ▨ Resident

Purple-breasted cotinga
Cotinga cotinga

TAXONOMY
Cotinga cotinga Linnaeus, 1766.

OTHER COMMON NAMES
French: Cotinga de Daubenton; German: Purpurbrust Kotinga; Spanish: Continga de Pecho Morado.

PHYSICAL CHARACTERISTICS
Average weight is 2.5 oz (70 g). Males are predominantly navy blue in color, with black wings and tail, and violet on the throat and breast. Their subcutaneous and perivisceral fat often takes on the blue color of the berries they prefer.

DISTRIBUTION
This species is found in northern Amazonia, in eastern Colombia, the Guinan Shield, and northern Brazil. The only species

Cotinga cotinga
◻ Resident

within the genus that overlaps its geographic distribution is the spangled cotinga (*Cotinga cayana*).

HABITAT
This species is a canopy specialist in lowland tropical evergreen forest. While principally a lowland species, it may range up to 0.5 mi (800 m).

BEHAVIOR
The quiet behavior of the members of this genus is in contrast with their vivid colors. However, the male emits a sharp, loud "whirr" with his wings when in flight.

FEEDING ECOLOGY AND DIET
Fruit and berries are consumed, often "gorging" at a masting tree or bush such as mistletoe. The fruits are often plucked on the wing. Although the seeds of larger species (e.g., mistletoe) might be regurgitated, smaller seeds are often swallowed. Insects are also taken.

REPRODUCTIVE BIOLOGY
The mating system is not completely known within this group, although for the most part it appears that males display solitarily.

The nest is platform type, often high in a tree fork, or next to an epiphyte. The female incubates and cares for the young alone.

CONSERVATION STATUS
Not threatened.

SIGNIFICANCE TO HUMANS
Several indigenous tribes use cotinga feathers in their ornamentation. ◆

Banded cotinga
Cotinga maculata

TAXONOMY
Cotinga maculatus Mueller, 1776.

OTHER COMMON NAMES
French: Cotinga cordonbleu; German: Pracht Kotinga; Spanish: Continga Franjeada.

PHYSICAL CHARACTERISTICS
Weight for this genus is around 2.5–2.8 oz (70–80 g). This species is starling-sized. Males are predominantly ultramarine-blue coloration, with black on the wings and tail, and separate patches of violet on the throat and breast. Their subcutaneous and perivisceral fat often takes on the blue color of the berries they prefer.

DISTRIBUTION
This species is restricted to a small area of the coastal forests of Brazil. The only species within the genus that overlaps its geographic distribution is the spangled cotinga (*Cotinga cayana*).

HABITAT
This species is a canopy specialist in lowland tropical evergreen forest. One of the most lowland-dwelling forms in this genus, they rarely exceed 660 ft (200 m) in elevation.

BEHAVIOR
The quiet behavior of the members of this genus is in contrast with their vivid colors.

Cotinga maculata
◻ Resident

FEEDING ECOLOGY AND DIET

Fruit and berries are consumed, often "gorging" at a masting tree or bush such as mistletoe. The fruits are often plucked on the wing. Although the seeds of larger species (e.g., mistletoe) might be regurgitated, smaller seeds are often swallowed. Insects are also taken.

REPRODUCTIVE BIOLOGY

The mating system is not completely known within this group, although for the most part it appears that males display solitarily.

The nest is typically platform type, often high in a tree fork, or next to an epiphyte. Although there is a report of one nesting inside an arboreal termite nest. The female incubates and cares for the young alone.

CONSERVATION STATUS

Endangered, with habitat fragmentation being the principal threat. Additionally, populations were reduced in the past from over-collecting for the live-bird industry, as well as to provide feathers for "feather flowers" made by Indians and Bahian nuns. Today however the bird is on CITES Appendix I and is protected by Brazilian law. Its geographic range is estimated at 780 km^2. Its numbers are estimated at less than 1,000, with populations declining.

SIGNIFICANCE TO HUMANS

Several indigenous tribes use cotinga feathers in their ornamentation. One of the most frequently seen groups is *Cotinga*, which is commonly represented in costumes of certain Amazonian tribes. Perhaps as many as 10–15% of artifacts have *Cotinga* feathers, although the most commonly used feathers are those of Psittacids (*Ara* and *Amazona*). ◆

Cotinga maynana
◼ Resident

Plum-throated cotinga

Cotinga maynana

TAXONOMY

Cotinga maynana Linnaeus, 1766.

OTHER COMMON NAMES

French: Cotinga de Maynas; German: Veilchenkehl Kotinga; Spanish: Continga de Garganta Morada.

PHYSICAL CHARACTERISTICS

The average weight is 2.5 oz (70 g). This species is starling-sized, and the males are predominantly blue in color, with a violet colored throat. Their subcutaneous and perivisceral fat often takes on the blue color of the berries they prefer.

DISTRIBUTION

This species is found in western Amazonia, from southeastern Colombia to northern Bolivia and western Brazil. The only species within the genus that overlaps its geographic distribution is the spangled cotinga (*Cotinga cayana*).

HABITAT

This species, like other members of this genus, can be found in canopies of lowland tropical evergreen forest. In stark contrast to other members of this genus however, the plum-throated cotinga tends to inhabit more aqueous environs, such as flooded forest, blackwater swamps, and river edge. Additionally, it may be found in secondary forest. It may range up to 3,900 ft (1,200 m) in Ecuador.

BEHAVIOR

The quiet behavior of the members of this genus is in contrast with their vivid colors.

Various congeners will forage in the same tree with Plum-throated cotingas, such as the spangled cotinga (*Cotinga cayana*). Additionally, the plum-throated cotingas has been observed foraging in the same tree with parrots (short-tailed parrots [*Graydidascalus brachyurus*] and cobalt-winged parakeet [*Brotogeris cyanoptera*]).

FEEDING ECOLOGY AND DIET

Fruit and berries are consumed, often "gorging" at a masting tree or bush such as mistletoe. The fruits are often plucked on the wing. Although the seeds of larger species (e.g., mistletoe) might be regurgitated, smaller seeds are often swallowed. Insects are also taken.

REPRODUCTIVE BIOLOGY

The mating system is not completely known within this group, although for the most part it appears that males display solitarily.

The nest is platform type, often high in a tree fork, or next to an epiphyte. The female incubates and cares for the young alone.

CONSERVATION STATUS

Not threatened.

SIGNIFICANCE TO HUMANS

Several indigenous tribes use cotinga feathers in their ornamentation. One of the most frequently seen groups is *Cotinga*, which is commonly represented in costumes of certain Amazonian tribes. Perhaps as many as 10–15% of artifacts have

Cotinga feathers, although the most commonly used feathers are those of Psittacids (*Ara* and *Amazona*). ◆

Turquoise cotinga
Cotinga ridgwayi

TAXONOMY
Cotinga ridgwayi Ridgway, 1887.

OTHER COMMON NAMES
English: Ridgway's cotinga; French: Cotinga turquoise; German: Ridgway-Kotinga; Spanish: Continga de Ridgway.

PHYSICAL CHARACTERISTICS
Weight for this genus is around 2.5–2.8 oz (70–80 g). This species is starling-sized. Males are predominantly ultramarine-blue in color, with black on the wings and tail, and separate patches of violet on the throat and breast. Their subcutaneous and perivisceral fat often takes on the blue color of the berries they prefer.

DISTRIBUTION
This species is restricted to southwest Costa Rica, barely ranging into western Panama.

HABITAT
This species, like other members of this genus, can be found in canopies of lowland tropical evergreen forest. Additionally, it may be found in secondary forest. It may range up to 5,550 ft (1,850 m).

BEHAVIOR
The quiet behavior of the members of this genus is in contrast with their vivid colors.

FEEDING ECOLOGY AND DIET
Fruit and berries are consumed, often "gorging" at a masting tree or bush such as mistletoe. The fruits are often plucked on

the wing. Although the seeds of larger species (e.g., mistletoe) might be regurgitated, smaller seeds are often swallowed. Insects are also taken.

REPRODUCTIVE BIOLOGY
The mating system is not completely known within this group, although for the most part it appears that males display solitarily.

The nest is platform type, often high in a tree fork, or next to an epiphyte. The female incubates and cares for the young alone.

CONSERVATION STATUS
Vulnerable, with habitat alteration due to agrarian encroachment being the principal threat. Its geographic range is estimated at 3,200 mi² (8,400 km²). Its numbers are estimated at less than 10,000, with populations declining.

SIGNIFICANCE TO HUMANS
Several indigenous tribes use cotinga feathers in their ornamentation. One of the most frequently seen groups is *Cotinga*, which is commonly represented in costumes of certain Amazonian tribes. Perhaps as many as 10–15% of artifacts have *Cotinga* feathers, although the most commonly used feathers are those of Psittacids (*Ara* and *Amazona*). ◆

Bare-necked umbrellabird
Cephalopterus glabricollis

TAXONOMY
Cephalopterus glabricollis Gould, 1861.

OTHER COMMON NAMES
English: Bullbird; French: Coracine ombrelle; German: Nacktkehl-Schirmvogel; Spanish: Pájaro Paraguas de Cuello Desnudo.

PHYSICAL CHARACTERISTICS
Umbrellabirds have sharp and powerful claws to secure good grips on branches during calling. This group comprises the largest of the cotingas, being about the size of a crow. As is the case with most cotingas, the females are smaller and less dramatic than the males in terms of ornamentation. The males in this group are entirely black, except for a red throat pouch in the male.

Additionally, their ornamentation and calls make umbrellabirds among the most unique of the cotingas. The head carries a canopy-like metallic glistening crest along its entire length; this crest projects over the tip of the heavy beak and is reminiscent of an umbrella, providing the name "umbrellabird." In addition, an apron-like feathered wattle hangs down from the breast. The much-widened trachea enables umbrella birds to utter "terrible roaring" sounds which have earned them the name of "bullbirds."

DISTRIBUTION
This species is restricted to the Caribbean slope and central highlands of Costa Rica and northeastern Panama. It ranges in the foothills at 330–6,600 ft (100–2,000 m).

HABITAT
Umbrellabirds usually inhabit the mid-level to upper story of tall trees.

Cotinga ridgwayi

■ Resident

Cephalopterus glabricollis

■ Resident

SIGNIFICANCE TO HUMANS
Various tribes may use the wattles for ornamentation in their
artifacts. ◆

Amazonian umbrellabird
Cephalopterus ornatus

TAXONOMY
Cephalopterus ornatus Geoffroy Saint-Hilaire, 1809.

OTHER COMMON NAMES
English: Fifebird, bullbird; French: Coracine ornée; German:
Kurzlappen-Schirmvogel; Spanish: Pájaro Paraguas.

PHYSICAL CHARACTERISTICS
Umbrellabirds have sharp and powerful claws to secure good
grips on branches during calling. This group comprises the
largest of the cotingas, being about the size of a crow. As is
the case with most cotingas, the females are smaller and less
dramatic than the males in terms of ornamentation. For example, the male Amazonian umbrellabird is 1.65 times the
weight of females, with male weights ranging 1.5–1.6 lb
(680–745 g). Both sexes are entirely black, and the male has a
whitish eye.

DISTRIBUTION
This species is found in the western and central Amazonian
basin, at lower elevations typically not exceeding 4,300 ft
(1,300 m).

BEHAVIOR
The bare-necked umbrellabird leaves the breeding grounds in
the highlands (2,600–6,600 ft [800–2,000 m]) in late July or
August, returning there from the lowlands in March. The sexes
are segregated between altitudes to some degree during the
nonbreeding season, with males often found at 330–1,600 ft
(100–500 m), and females found below 660 ft (200 m).

The call is a plaintive combination between a "roar" and
bleating calf, often occurring in the morning or afternoon.
Umbrellabirds have a very characteristic slow-flapping during
flight with the crest laying flat. Once perched they will often
hop clumsily from branch to branch. Animal prey is often
beaten against a tree branch before swallowing

FEEDING ECOLOGY AND DIET
The umbrellabirds consume fruits such as berries and palms.
They also eat nuts. Larger seeds of the fruits they consume are
regurgitated. This helps regenerate the tropical forests they
live in, as seeds of their preferred food plants are dispersed
throughout the forests. Insects, larvae and some spiders are
taken as well. Animal matter is consumed especially during the
rainy season when fruits are more scarce.

REPRODUCTIVE BIOLOGY
Nest is built above ground, often in fork of a tree, and constructed very roughly of loose twigs such that the single egg or
chick can be seen from underneath.

CONSERVATION STATUS
The bare-necked umbrellabird is considered Vulnerable.
Global numbers are estimated at fewer than 10,000 individuals
for the species, with populations declining.

The principal threat is habitat fragmentation. In Costa Rica
this is manifested through conversion to banana plantations,
cattle ranches and non-sustainable logging. Agrarian conversion is the main factor driving habitat destruction in northeastern Panama. The birds' geographic range is estimated at
4,600–5,800 mi² (12,000–15,000 km²).

Cephalopterus ornatus

■ Resident

HABITAT

This species tends to be a riverine island specialist in the Amazonian lowlands, often associated with riverine vegetation (e.g., *Cecropia*). However, along the eastern fringe of the Andes, this species ranges up into montane evergreen forest, more similar to the primary habitat of the other species of umbrellabirds.

BEHAVIOR

Unlike the other two species of umbrellabirds, the Amazonian species appears to be sedentary. The call is a plaintive combination between a "roar" and bleating calf, often occurring in the morning or afternoon. Umbrellabirds have a very characteristic slow-flapping during flight with the crest laying down flat. Once perched they will often hop clumsily from branch to branch. Animal prey is often beaten against a tree branch before swallowing

FEEDING ECOLOGY AND DIET

The umbrellabirds consume fruits such as berries and palm fruits and nuts. Larger seeds of the fruits they consume are regurgitated. This helps regenerate the tropical forests they live in, as seeds of their preferred food plants are dispersed throughout the forests. Insects, larvae and some spiders are taken as well. Animal matter is consumed especially during the rainy season when fruits are more scarce.

REPRODUCTIVE BIOLOGY

Males are organized into widely spaced, exploded leks and may displace other males from calling perches.

The nest is platform type and built very roughly of loose twigs such that the single egg or chick can be seen from underneath. The nest is often located high in a tree fork. The single egg is 2.2 by 1.4 in (56 by 36 mm), oblong and rather pointed at one end, with khaki coloring with brownish spotting and stippling.

CONSERVATION STATUS

Not threatened.

SIGNIFICANCE TO HUMANS

The head and beard ornamention are sometimes seen in Amazonian riverboats, but the associated belief is unknown. However, various tribes use the wattles for ornamentation in their artifacts. ◆

Long-wattled umbrellabird

Cephalopterus penduliger

TAXONOMY

Cephalopterus penduliger Sclater, 1859.

OTHER COMMON NAMES

English: Bullbird; French: Coracine casquée; German: Langlappen-Schirmvogel; Spanish: Pájaro Paraguas Caranculado.

PHYSICAL CHARACTERISTICS

Umbrellabirds have sharp and powerful claws to secure good grips on branches during calling. This group comprises the largest of the cotingas, being about the size of a crow. As is the case with most cotingas, the females are smaller and less dramatic than the males in terms of ornamentation. The males are entirely black.

Wilhelm Meise describes them as follows: "The inflated throat sac, which looks somewhat like a pine cone with spread

Cephalopterus penduliger
▢ Resident

scales, is moved to and fro like a pendulum; soft sounds are heard with this movement. With the utterance of the loud, low-pitched rumbling courtship call, the head is thrown back and the wattle swings forward.". The much-widened trachea enables umbrella birds to utter "terrible roaring" sounds which have earned them the name of "bullbirds."

DISTRIBUTION

This species is restricted to the Pacific slope from southwestern Colombia through Ecuador. They are found in the foothills between 460 and 5,900 ft (140–1,800 m).

HABITAT

Umbrellabirds usually inhabit the mid-level to upper story of tall trees.

BEHAVIOR

This species may be an altitudinal migrant, but there are both highland and lowland populations known to be sedentary. The call is a plaintive combination between a "roar" and bleating calf, often occurring in the morning or afternoon. Males may displace other males from calling perches. Umbrellabirds have a very characteristic slow-flapping during flight with the crest laying down flat. Once perched they will often hop clumsily from branch to branch. Animal prey is often beaten against a tree branch before swallowing.

FEEDING ECOLOGY AND DIET

The umbrellabirds consume fruits such as berries and palm fruits and nuts. Larger seeds of the fruits they consume are regurgitated. This helps regenerate the tropical forests they live in, as seeds of their preferred food plants are dispersed throughout the forests. Insects, larvae and some spiders are taken as well. Animal matter is consumed especially during the rainy season when fruits are more scarce.

REPRODUCTIVE BIOLOGY
The nest is platform type and built very roughly of loose twigs such that the single egg or chick can be seen from underneath. The nest is often located high in a tree fork.

CONSERVATION STATUS
Vulnerable. Global numbers are estimated at less than 10,000 individuals for the species, with populations declining.

The long-wattled umbrellabird is threatened with deforestation and consequent habitat fragmentation. The habitat fragmentation is due to logging and agrarian development, such as livestock ranching, and oil palm and banana plantations. The geographic range of this species is estimated at 21,000 mi² (54,000 km²).

SIGNIFICANCE TO HUMANS
Various tribes may use the wattles for ornamentation in their artifacts. ◆

Andean cock-of-the-rock
Rupicola peruviana

TAXONOMY
Rupicola peruviana Latham, 1790.

OTHER COMMON NAMES
French: Coq-de-roche péruvien; German: Andenklippenvogel; Spanish: Gallito de Rocas Peruano.

PHYSICAL CHARACTERISTICS
Cocks-of-the-rock have sharp and powerful claws to secure good grips on branches during courtship. The pigeon-sized cocks-of-the-rock, with their teased-out feathers on the forehead, back and wings, have a particularly striking coloration. While the female is a drab brown color, the male's plumage is scarlet; the head is decorated with a helmet-like erect crest.

DISTRIBUTION
This species is distributed through the Andes from extreme western Venezuela through Colombia, Ecuador, and Peru to western Bolivia, ranging between 3,000 and 7,900 ft (900–2,400 m).

HABITAT
This species inhabits the lower to mid strata of tropical montane forest.

BEHAVIOR
When flying, a loud "hissing" sound is produced from the modified remige of the wing tip. A spectacular array of vocalizations are produced, including different "popping" noised produced by snapping the bill. Unlike the Guiana cock-of-the-rock (*Rupicola rupicola*), the Andean cock-of-the-rock will only dance in trees, rather than on the ground as well.

FEEDING ECOLOGY AND DIET
Like most cotingas, cocks-of-the-rock consume fruits primarily, but will consume more animal matter as fruits become scarce. Captive individuals are known to eat small lizards (*Anolis* sp.) and baby laboratory mice (*Mus* sp.).

REPRODUCTIVE BIOLOGY
Nests, built by the female, are typically located near the male lekking grounds, and sometimes several females build nests

Rupicola peruviana
▢ Resident

close to each other. The cup shaped nests are typically plastered to a damp rock face within crevices of cliffs or ravines, often over a stream. Unusual nesting sites have been discovered, such as under a well-trafficked bridge. The nest may weigh nearly 2.2 lb (1 kg), and is made of clay mixed with vegetable fibers and is often covered with lichens.

CONSERVATION STATUS
While Andean cocks-of-the-rock are not listed as Threatened or Endangered, the Andes Mountains are the subject of significant deforestation. Additionally, quite a few individuals were taken during the 1900s for the live bird trade. However trade is much more restricted today.

SIGNIFICANCE TO HUMANS
Andean cock-of-the-rock is the national bird of Peru. Natives may eat these birds for food. ◆

Guianan cock-of-the-rock
Rupicola rupicola

TAXONOMY
Rupicola rupicola Linnaeus, 1766.

OTHER COMMON NAMES
French: Coq-de-roche orange; German: Cayenne Klippenvogel; Spanish: Gallito de Rocas Guayanés.

PHYSICAL CHARACTERISTICS
While the female is a drab brown color, the male's plumage is a bright orange; the head is decorated with a helmet-like erect crest. The bright coloration is derived from zeaxanthin, the

Rupicola rupicola

■ Resident

demonstrated its grace and readiness to dance, gave way to a third male." Up to 50 males have been observed at a lek.

FEEDING ECOLOGY AND DIET
Like most cotingas, the cocks-of-the-rock consume fruits primarily, but will consume more animal matter as fruits become scarce.

REPRODUCTIVE BIOLOGY
Polygynous. Nests are typically located near the male lekking grounds, and sometimes several females build nests close to each other. The cup shaped nests are typically plastered to a damp rock face within crevices of cliffs or ravines, often over a stream. The nest may weigh nearly 2.2 lb (1 kg), and is made of clay mixed with vegetable fibers and is often covered with lichens. The female lays two spotted brownish eggs, and the incubation period is 27–28 days. When the chicks hatch the males can be distinguished from females as their feet and bills are yellow and black, respectively.

CONSERVATION STATUS
While neither species is listed as Threatened or Endangered, quite a few individuals were taken during the 1900s for the live bird trade. However, trade is much more restricted today.

SIGNIFICANCE TO HUMANS
Natives may eat cock-of-the-rock flesh. Because of their bright plumage, cocks-of-the-rock are hunted by men of numerous Indian tribes. The Emperor of Brazil had a mantle made of cock-of-the-rock feathers. ◆

same pigment found in corn (*Zea mays*), which it is named for. This pigmentation often fades rapidly in taxidermied specimens.

DISTRIBUTION
This species is found in northern Amazonia and the Guianan shield, from southeastern Colombia through southern Venezuela and northern Brazil, eastward through French Guiana. It is found at lower elevations, typically not exceeding 4,900 ft (1,500 m).

HABITAT
Lowland forest.

BEHAVIOR
When flying, a loud "hissing" sound is produced from the modified remige of the wing tip. A spectacular array of vocalizations are produced, including different "popping" noised produced by snapping the bill. In the courtship season, males gather on rocks amid the foam of river rapids to display their colors in most unusual dances. Robert Schomburgk (1804–1865), the well-known South American traveler, described these dances as follows: "A whole troop of these wonderful birds was holding their dance on the smooth, flat upper surface of a tremendous rock. About twenty admiring observers, both males and females, sat on the bushes nearby while a male moved about over the top of the rock in every direction with some rather unusual movements. It would spread its wings, toss its head in every direction, scratch the rock with its primaries, and hop upwards at varying speeds, always from the same point; again it would fan out and erect its tail and once more walk about coquettishly with proud steps. When it seemed to be tired, it uttered a different phrase from the usual call and, flying to the nearest twig, it left its place on the rock to another male. After awhile, this second bird, having first

White bellbird
Procnias alba

TAXONOMY
Procnias alba Hermann, 1783.

OTHER COMMON NAMES
French: Araponga blanc; German: Zapfenglöckner; Spanish: Campanero Blanco.

PHYSICAL CHARACTERISTICS
The bellbirds are distinguished by compact bodies, flat beaks, short tarsi and a plumage of small feathers. Males have among the loudest calls of any birds. They are completely white, with a long inflatable, "horn-like" wattle on the head; the "horn" is covered with small white feathers which can be erected during display. Females, which are silent, are predominantly green and somewhat smaller.

DISTRIBUTION
This species is found in the Guiana Shield, ranging from 1,500 to 4,900 ft (450–1,500 m). It may be a local altitudinal migrant.

HABITAT
Bellbirds live in tropical lowland or montane evergreen rainforest.

BEHAVIOR
They prefer high perches in the canopy, often on bare tree branches, which project above the crowns of surrounding trees. The calls sound as if an anvil were being struck with a hammer.

FEEDING ECOLOGY AND DIET
These birds feed on fruit. The short bills with a wide gape are adaptations for gorging on quantities of fruit.

Procnias alba
Resident

Procnias averano
Resident

REPRODUCTIVE BIOLOGY
The nest is of sparse construction and is built on open branches. One or two eggs are laid per clutch. Female bellbirds care for the young alone, regurgitating fruit and cleaning the nest of fecal sacks and regurgitated seeds.

CONSERVATION STATUS
Not threatened.

SIGNIFICANCE TO HUMANS
None known. ◆

Bearded bellbird

Procnias averano

TAXONOMY
Procnias averano Hermann, 1783.

OTHER COMMON NAMES
French: Araponga barbu; German: Flechtenglöckner; Spanish: Campanero Herrero.

PHYSICAL CHARACTERISTICS
The bellbirds are distinguished by compact bodies, flat beaks, short tarsi and a plumage of small feathers. Male bellbirds have among the loudest calls of any birds. Males also differ from the females in their plumage coloration. The male has a bare throat with beard-like threads of skin set in bundles around the skin of the throat. Its head is brown, and the flight feathers and tail are black; the rest of the plumage is a

light pearl-gray. Females are predominantly green, and somewhat smaller.

DISTRIBUTION
This species is patchily distributed in the north-central Amazon and Guiana Shield, and is also found in northeastern Brazil. Although primarily a lowland species, it may range up to 6,200 ft (1,900 m).

HABITAT
Bellbirds live in tropical lowland or montane evergreen rainforest. They prefer high perches in the canopy, often on bare tree branches which project above the crowns of surrounding trees.

BEHAVIOR
The far-reaching bell-like calls (often described as "bockk") of the males are characteristic of their jungle home. In display this species opens up its gape like a frog's mouth so that the threads of the "beard" (which are comparable to a wreath of tuning forks) reproduce its pure bell-like tones.

FEEDING ECOLOGY AND DIET
These birds feed on fruit. The short bills with a wide gape are adaptations for gorging on quantities of fruit.

REPRODUCTIVE BIOLOGY
The nest is made of very little construction material, and is built on open branches. One or two eggs are laid per clutch. The female cares for the young alone, regurgitating fruit and cleaning the nest of fecal sacks and regurgitated seeds. The chicks leave the nest at 33 days, and take three years to come into full color.

CONSERVATION STATUS
Not threatened.

SIGNIFICANCE TO HUMANS
None known. ◆

Bare-throated bellbird
Procnias nudicollis

TAXONOMY
Procnias nudicollis Vieillot, 1817.

OTHER COMMON NAMES
English: Naked-throated bird; French: Araponga à gorge nue;
German: Nacktkehlglöckner; Spanish: Campanero de Garganta
Desnuda.

PHYSICAL CHARACTERISTICS
The bellbirds are distinguished by compact bodies, flat beaks,
short tarsi and a plumage of small feathers. Male bellbirds have
among the loudest calls of any birds. This species accomplishes
this by having a very muscular syrinx, and filling the interclav-
icular air sacs. Males also differ from the females in their
plumage coloration. The male is white, and is distinguished by
bare wattles and a bare, inflatable throat skin of greenish color.
Females are predominantly green, and somewhat smaller.

DISTRIBUTION
This species is found in the Atlantic rainforest belt of central
Brazil through eastern Paraguay and northeastern Argentina,
ranging up to 3,370 ft (1,150 m).

Procnias nudicollis

▨ Resident

HABITAT
Bellbirds live in tropical lowland or montane evergreen rainforest. They prefer high perches in the canopy, often on bare
tree branches, which project above the crowns of surrounding
trees.

BEHAVIOR
The far-reaching bell-like calls (often described as "bockk") of
the males characterize their jungle home. Male bellbirds defend perches jealously against rivals, including other species of
cotingas on occasion. For example, a bare-throated bellbird
displaced a female white-winged cotinga *Xipholena atropurpurea*
from a tree. While the species appears to be migratory in at
least some regions, this apparently varies among populations.
For example, the species appears to migrate in southeastern
Brazil, be transient in northeastern Argentina, and resident in
Paraguay.

FEEDING ECOLOGY AND DIET
These birds feed on fruit. The short bills with a wide gape are
adaptations for gorging on quantities of fruit, such as (*Rapanea
ferruginea*). This species has also been observed eating in a *Cecropia* tree.

REPRODUCTIVE BIOLOGY
This species builds a shallow nest that is approximately 6.3 in
(16 cm) across. The nest is made of very little construction
material, and is built on open branches. One or two eggs are
laid per clutch; the eggs are oval and reddish brown, with dark
spots at the rounder end. Female cellbirds care for the young
alone, regurgitating fruit and cleaning the nest of fecal sacks
and regurgitated seeds.

CONSERVATION STATUS
Near Threatened due to habitat fragmentation, development
such as road building, and exploitation for the cage bird
trade (especially in Brazil) being the main factors. The fragmentation is primarily due to agrarian conversion and deforestation for mining concessions. In all likelihood this species
is declining

SIGNIFICANCE TO HUMANS
None known. ◆

Three-wattled bellbird
Procnias tricarunculata

TAXONOMY
Procnias tricarunculata Verreaux and Verreaux, 1853.

OTHER COMMON NAMES
French: Araponga tricarunculé; German: Hämmerling; Spanish: Procnias tricarunculata.

PHYSICAL CHARACTERISTICS
Male is approximately 12 in (30 cm) long; female is approximately 10 in (25 cm). The plumage of the adult male is:
chestnut brown, except for head, neck, and chest, which are
white. The adult female is olive-green above; yellow underside striped with dark olive-green. They are famous not only
because of the truly enchanting calls of the males, but also
because of the inflatable skin appendages about the heads of
the males.

Procnias tricarunculata

▨ Resident

DISTRIBUTION

This species is found in two patches in Middle America: eastern Honduras to northern Nicaragua, and the southern tip of Nicaragua through Costa Rica to central Panama. Although this species may be found up to 9,800 ft (3,000 m), it may locally migrate to the lowlands.

HABITAT

Bellbirds live in tropical lowland or montane evergreen rainforest. They prefer high perches in the canopy, often on bare tree branches, which project above the crowns of surrounding trees.

BEHAVIOR

The far-reaching bell-like calls (often described as "bockk") of the males are characteristic of their jungle home. This species breeds in foothill and highland forest between 2,500 and 6,900 ft (750–2,100 m), though the lower elevation for breeding is typically 3,900 ft (1,200 m). The breeding season may occur from March through September, but this is variable. During the nonbreeding season extensive migrations are taken.

FEEDING ECOLOGY AND DIET

These birds feed on fruit. The short bills with a wide gape are adaptations for gorging on quantities of fruit, and has been seen regurgitating mistletoe with much effort.

REPRODUCTIVE BIOLOGY

The nest is made of very little construction material, and is built on open branches. One or two eggs are laid per clutch. Female bellbirds care for the young alone, regurgitating fruit and cleaning the nest of fecal sacks and regurgitated seeds.

CONSERVATION STATUS

Vulnerable, with habitat fragmentation due to logging, and conversion to banana plantations and cattle ranches, being the principal threats. Its geographic range is estimated at 9,000–44,000 mi^2 (23,000–114,000 km^2). Its numbers are estimated at less than 10,000, with populations declining.

SIGNIFICANCE TO HUMANS

None known. ◆

Resources

Books

BirdLife International. *Threatened Birds of the World*. Barcelona: Lynx Edicions, 2000.

Ridgely, Robert S., and Guy Tudor. *The Birds of South America*. Vol. 2, *The Suboscine Passerines*. Austin: University of Texas Press, 1994.

Sick, Helmut. *Birds in Brazil: A Natural History*. New Jersey: Princeton University Press, 1993.

Snow, David W. *The Cotingas*. Ithaca: Cornell University Press, 1982.

Snow, David W. *The Web of Adaptation: Bird Studies in the American Tropics*. New York: Quadrangle Times Book Co., 1985.

Stotz, Douglas F., et al. *Neotropical Birds: Ecology and Conservation*. Chicago: University of Chicago Press, 1996.

Periodicals

Berry, Robert J. and Rochelle Plasse. "Breeding the Scarlet Cock-of-the-Rock (*Rupicola peruviana*) at the Houston Zoological Gardens." *International Zoo Yearbook*. 22 (1982): 171–175.

Brooks, Daniel M. "Comparative Life History of Cotingas in the Peruvian Amazon." *Orn. Neotrop.* 10 (1999): 193–206.

Cuervo, Andres M., et al. "A New Species of Piha (Cotingidae: *Lipaugus*) from the Cordillera Central of Colombia." *Ibis* 143 (2001): 353–368.

Jahn, Olaf, et al. "The Life-history of the Long Wattled Umbrellabird *Cephalopterus penduliger* in the Andean Foothills of North-west Ecuador: Leks, Behaviour, Ecology and Conservation." *Bird Conservation International* 9 (1999): 81–94.

Pacheco, José Fernando, and Paulo Sérgio Moreira da Fonseca. "The Remarkable Rediscovery of the Kinglet Calyptura *Calyptura cristata*." *Cotinga* 16 (2001): 44-47.

Sick, Helmut. "An Egg of the Umbrellabird." *Wilson Bulletin* 63 (1951): 338–339.

Snow, Betty K. "Notes on the Behavior of Three Cotingidae." *Auk* 78 (1961): 150–161.

Snow, David W. "The Classification of the Cotingidae" *Breviora* 409 (1973): 1–27.

Trail, Pepper W., and Paul Donahue. "Notes on the Behavior and Ecology of the Red-cotingas (Cotingidae: *Phoenicircus*)." *Wilson Bulletin* 103 (1991): 539–551.

Resources

von Hagen, Wolfgang. "On the Capture of the Umbrellabird (*Cephalopterus penduliger* Sclater)." *Proceedings of the Zoological Society of London* (1937): 25–30.

Wallace, Alfred R. "On the Umbrellabird (*Cephalopterus ornatus*)." *Proceedings of the Zoological Society of London* 1849: 206–207.

Organizations

Neotropical Bird Club. c/o The Lodge, Sandy, Bedfordshire SG19 2DL United Kingdom. E-mail: secretary@ neotropicalbirdclub.org

Daniel M. Brooks, PhD

Plantcutters

(Phytotomidae)

Class Aves
Order Passeriformes
Suborder Tyranni (Suboscines)
Family Phytotomidae

Thumbnail description
Small, herbivorous, finchlike birds with serrated bills for cutting plant material

Size
Approximately 7 in (17–20 cm), 1.5 oz (40 g)

Number of genera, species
1 genus, 3 species

Habitat
Forest, scrubland, and desert

Conservation status
Endangered: 1 species

Distribution
Central temperate to subtropical South America, from Peru and Bolivia to Chile, Uruguay, and Argentina

Evolution and systematics

The issue has not been settled as to whether the suboscine Phytotomidae represent a distinct family or a genus within the family Cotingidae. Lanyon and Lanyon consider the Phytotomidae a genus (*Phytotoma*) within the family Cotingidae. Prum and colleagues provisionally place the plantcutters as a subfamily (Phytotominae) within family Cotingidae. Sibley and Monroe list the plantcutters as genus *Phytotoma* in their subfamily Cotinginae within the family Tyrannidae. Most ornithologists recognize a distinct family Cotingidae.

In any case, the plantcutters are most closely related to the tyrants and becards (family Tyrannidae), cotingas (Cotingidae), sharpbills (Oxyruncidae), and manakins (Pipridae). All are New World suboscine families.

There are three species of plantcutter: the Peruvian plantcutter (*Phytotoma raimondii*), the red-breasted (or white-tailed) plantcutter (*Phytotoma rutila*), and the rufous-tailed plantcutter (*Phytotoma rara*).

There is no reliable fossil evidence of Phytotomidae ancestry.

Physical characteristics

Plantcutters (in South America, also called cortarramas, cortaplantas, and raras) look at first glance like rather ordinary, finchlike birds with a few exotic splashes of red applied to mostly gray or brown plumage. Head plumage rises to a short crest in the Peruvian and red-breasted plantcutters. Individuals are about 7 in (18–20 cm) long and weigh about 1.5 oz (40 g). They are short-crested and have short, stout, conical bills. The wings are short and pointed, the tail long, and the legs short with large, strong feet.

The sexes show considerable dichromatism. Males are more brightly colored, especially with reds, while females are more grayish to brownish.

A close look at the plantcutter bill reveals a feature rarely seen among birds: rows of tiny, strong, sharp, toothlike projections run the lengths of the tomae, or bill edges, on both sides of each mandible. These are not bony teeth but, rather, outgrowths of the keratin substance of the bill. The birds use these bills to saw through and to chew leafy vegetation, their main food source. Peering at a close-up photograph of the head, with the rows of sharp, forward-leaning pseudo-teeth, accentuated by the glaring red or golden eye, is more like coming face to face with a dragon than with a small, herbivorous bird.

Distribution

The Peruvian plantcutter is endemic to the dry scrublands of northwestern Peru. The rufous-tailed and red-breasted plantcutters live from temperate southern Argentina and Chile, northward to subtropical Paraguay and Bolivia.

Habitat

The Peruvian plantcutter lives in near-desert conditions. The rufous-tailed and red-breasted plantcutters prefer a mix of open forest, scrubland, grassland, and farmland.

Behavior

Herbivory is a rare lifestyle among birds. Herbivorous birds tend to be rather passive and slow-moving. Plantcutters, however, are lively and energetic. They are also quite capable flyers, patrolling their areas throughout daylight hours for food.

Feeding ecology and diet

Very few bird species subsist mainly on leaves as a food source. Those that do usually pay a significant price because of the high fiber content and diluted energy availability of leafy foods. The trade-offs may involve larger size, flightlessness, and low activity to conserve energy. There may also be elaborate modifications of the digestive tract. The New Zealand takahe (*Porphyrio mantelli*), whose dietary mainstay is alpine tussock grasses, is flightless. It has an inefficient digestive system and must eat almost continuously during its waking hours to ensure adequate nutrition. The leaf-eating hoatzins (*Opisthocomus hoazin*) of tropical South America lodge passively in dense riverside thickets for safety and are poor fliers. Their crops are enlarged, having become extra stomachs full of symbiotic bacteria able to digest the tough, fibrous cell walls of leaf tissues.

The plantcutters, though, seem not to have paid a high price for their choice of diet. The secret lies in the species'

feeding methods and digestive system. Chilean biologists recently undertook several studies of food intake and processing in the rufous-tailed plantcutter. Data support the efficiency of the bird's feeding and digestion, and this likely applies to the other plantcutter species. Plantcutters chew their leafy food into a pulp to rupture the tough plant cell walls and free the nutritious cell interiors for digestion. Food passes rapidly through the digestive system, which has little in the way of elaboration (although the intestine is abundantly supplied with mucous cells along its length, concentrating them toward its nether end). This combination of chewing, rapid passage of food through the digestive system, and efficient digestion allow the bird to process hefty amounts of plant material over shorter times and thereby maintain a high metabolic level.

Plantcutters spice up their diets with some intake of fruits and insects, but leaves of many plant species, depending on availability and type of habitat, are the mainstay. Rufous-tailed plantcutters have developed an affinity for cereal leaves, among them wheat and oat.

Reproductive biology

Little is known about reproduction in the Phytotomidae, mostly due to a lack of field studies. The birds build loose nests, and the females lay up to four eggs.

Conservation status

One species, the Peruvian plantcutter, is Endangered due to loss of numbers and habitat. The other species are more widespread and are not threatened.

Significance to humans

Plantcutters at times make nuisances of themselves by raiding farms and vineyards to feed on young leaves of cereal crops and grapevines. The Peruvian plantcutter has a limited distribution and is relatively easy to observe. It thus attracts birdwatchers from all over the world who wish to see this unusual and now endangered species. This generates some modest local income and gives conservationists another good reason to push for protective measures for the species. The Peruvian plantcutter has become a rallying symbol for conservation.

Species accounts

Peruvian plantcutter
Phytotoma raimondii

TAXONOMY
Phytotoma raimondii Taczanowski, 1883.

OTHER COMMON NAMES
French: Rara du Pérou German: Graubrust-Pflanzenmäher;
Spanish: Cortaplantas Peruana.

PHYSICAL CHARACTERISTICS
7–8 in (18–20 cm) long, about 1.5 oz (40 g). Both sexes are mainly medium gray, with bright yellow eyes and a cardinal-like crest. The male adds red patches on the forehead and lower breast.

Phytotoma raimondii

DISTRIBUTION
A wide area around the northern town of Talara, some small forests near Chiclayo (south of Talara), and a small forest farther south from Chiclayo; total known population estimated at 500–1,000, with perhaps 80% in the habitat around Talara.

HABITAT
Dry scrubland vegetation with bushes widely dispersed; part of the Tumbesian ecosystem.

BEHAVIOR
Diurnal; the call has been described as a donkeylike braying or like the movement of a rusty hinge; little specific information, due to lack of thorough field studies.

FEEDING ECOLOGY AND DIET
Follows the general description for plantcutters, although Peruvian plantcutters consume leaves from wild plants and have no known liking for cereal leaves. The birds will eat leaves of the widespread algarrobo (*Prosopis* spp.), chilco (*Baccharis* spp.), zapote (*Maytenia* spp.), and vichayo broadleaf bush.

REPRODUCTIVE BIOLOGY
Almost nothing is known, due to lack of field observations.

CONSERVATION STATUS
Endangered. In 1992 Peruvian plantcutters could be found in 14 sites along the north Peruvian coast; in 1998 Engblom revisited these sites but found plantcutters at only three. He then found three new sites southward.

Never very widespread, the Peruvian plantcutter is adapted to the native arid scrub forest of northwestern Peru, most of which has nearly disappeared or been degraded by goat grazing, extraction of firewood and timber, and conversion of land to sugarcane fields. The long-term security of the species will depend primarily on protecting its stronghold in the Talara region. To this end the Peruvian organization ProAves Peru, partly funded

Phytotoma raimondii
▨ Resident

by the U.S. National Audubon Society, is working toward the declaration of a reserve, environmental education at the local level, and restoration of plantcutter-friendly habitat.

SIGNIFICANCE TO HUMANS
The Peruvian plantcutter has become a rallying symbol for conservation. ◆

Rufous-tailed plantcutter
Phytotoma rara

TAXONOMY
Phytotoma rara Molina, 1782.

OTHER COMMON NAMES
English: Chilean plantcutter; French: Rara à queue rousse;
German: Rotschwanz-Pflanzenmäher; Spanish: Cortaplantas Chileno.

PHYSICAL CHARACTERISTICS
Considerable sexual dichromatism; the gaudy male outshines the more muted female. Weight about 1.5 oz (48 g), body length about 7 in (17 cm). The male shows reddish on the crown, throat, breast, and underparts. The back of the neck and dorsal plumage are dark olive green, with darker stripes. The blackish wings bear a distinctive white stripe. The tail is mainly blackish, with a red stripe down its middle. The female tends to muted browns, shading to grayish. The breast and abdomen are

Phytotoma rara

Resident

HABITAT

Forest, scrublands, and farmland from sea level to 6,000 ft (2,000 m) above sea level.

BEHAVIOR

Little has been recorded of the bird's daily activities, aside from what is described in the feeding ecology.

FEEDING ECOLOGY AND DIET

Phytotoma rara

The species often lives near farms because of its special fondness for young cereal leaves, although leaves of native plants will also be eaten. During the austral summer, the bird adds fruit and insects to the leaf diet.

REPRODUCTIVE BIOLOGY

The species prefers to nest in the forks of tree branches, but will nest in higher shrubs. The nest is made with root fibers and large twigs on the outside and smaller twigs inside. Brooding takes place in the austral summer. Egg-laying begins in October and produces two to four eggs of a clear bluish green color with some blackish spots.

CONSERVATION STATUS

Not threatened, although uncommon.

SIGNIFICANCE TO HUMANS

The rufous-tailed plantcutter can be a nuisance to farmers because of its fondness for cereal leaves. ◆

whitish with dark, longitudinal streaks in the breast and flanks. The wings are blackish like the male's, but lack the white stripe. The eyes are a vivid, emphatic red in both sexes.

DISTRIBUTION

Chile and Argentina, from Vallenar in the north to Chiloe in the south, and into Chilean and Argentinean Patagonia.

Resources

Books

Best, Brinley J., and Michael Kessler. *Biodiversity and Conservation in Tumbesian Ecuador and Peru.* Cambridge: Birdlife International Publications, 1995.

de la Pena, Martin R., and Maurice Rumboll. *Guide to the Birds of Southern South America and Antarctica.* Princeton: Princeton University Press, 1998.

Feduccia, Allan. *The Origin and Evolution of Birds.* New Haven: Yale University Press, 1999.

Ridgely, Robert S., and Guy Tudor. *The Birds of South America.* vol. II, *The Suboscine Passerines.* Austin: University of Texas Press, 1994.

Sibley, C.G., and B.L. Monroe. *Distribution and Taxonomy of Birds of the World.* New Haven: Yale University Press, 1990.

Periodicals

Lanyon, S. M., and W. E. Lanyon. "The Systematic Position of the Plantcutters, *Phytotoma*." *Auk* 106 (1989): 422–432.

Lopez-Calleja, M. V., and F. Bozinovic. "Energetics and Nutritional Ecology of Small Herbivorous Birds." *Revista Chilena de Historia Natural* 73 (September 2000): 411–420.

Lopez-Calleja, M. V. and F. Bozinovic. "Feeding Behavior and Assimilation Efficiency of the Rufous-Tailed Plantcutter: A Small Avian Herbivore." *Condor* 101 (August 1999): 705–710.

Meynard, C., M. V. Lopez-Calleja, and F. Bozinovic. "Digestive Enzymes of a Small Avian Herbivore, the Rufous-Tailed Plantcutter." *Condor* 101 (November 1999): 904–907.

Prum, R. O., N. H. Rice, J. A. Mobley, and W. W. Dimmick. "A Preliminary Phylogenetic Hypothesis for the Cotingas (Cotingidae) Based on Mitochondrial DNA." *Auk* 117 (2000).

Organizations

ProAves Peru. P.O. Box 07, Piura, Peru. E-mail: proaves@ e-mail.udep.edu.pe

Kevin F. Fitzgerald, BS

Lyrebirds

(Menuridae)

Class Aves

Order Passeriformes

Suborder Passeri (Oscines)

Family Menuridae

Thumbnail description
Large, brown, ground-foraging birds with strong legs and long tails, highly ornamental in the males

Size
28–42 in (71–107 cm)

Number of genera, species
1 genus; 2 species

Habitat
Forest

Conservation status
No species threatened or extinct

Distribution
Endemic to south-eastern Australia

Evolution and systematics

Lyrebirds probably originated in the Antarctic beech (*Nothofagus brassii*) forests and subtropical rainforests that covered much of Australia at the beginning of the Tertiary period. An early Miocene fossil species, *Menura tyawanoides*, has been described from Riversleigh in northwestern Queensland about 1,000 mi (1,700 km) from the northern limit of lyrebird distribution.

The open-floored nature of the beech forests (as can still be seen in New Zealand) was conducive to the evolution of visual courtship displays. The dense undergrowth of the wet sclerophyll forests and subtropical rainforests which replaced the beech forests probably necessitated the development of elaborate vocal displays.

When first discovered, the superb lyrebird was called a "native pheasant" and regarded as Gallinaceous, but it is now accepted that lyrebirds belong in the Passeriformes, and that their nearest relatives are the scrub-birds (*Atrichornis*). As of 2001 there was no unequivocal evidence linking the lyrebirds and scrub-birds to any other passerines. Proposed relationships with the bowerbirds (Ptilonorhynchidae) or with the tapaculos (Rhinocryptidae) have not been accepted.

The *Records of the Australian Museum* provide a detailed historical account of the systematics of the lyrebirds and scrub-birds. A meticulously planned series of studies on a single noisy scrub-bird specimen filled the 143 pages of that monograph. W. J. Bock and M. H. Clench summarize the research and conclusions reached. They accept that "the scrub-birds and lyrebirds form a monophyletic group of unknown affinities within the Oscines." They do not agree that the two genera form a single family as proposed on somewhat tenuous DNA hybridization grounds, and place them in separate families within a superfamily, the Menuroidea.

As of 2001, the most recent treatise on the systematics of Australian birds is *The Directory of Australian Birds*, published in 1999. The authors, R. Schodde and I. J. Mason, retain the two families, and recognize two species of lyrebirds, one with three subspecies.

Physical characteristics

Lyrebirds are among the largest of the songbirds, with male superb lyrebirds (*Menura novaehollandiae*) attaining a length of up to 39 in (100 cm), including the tail of 27 in (70

Front view of male Albert's lyrebird (*Menura alberti*) display. The bird's tailfeathers are thrown forward over its head to form a domed veil during its courtship display. (Photo by Wayne Lawler. Photo Researchers, Inc. Reproduced by permission.)

cm) and a weight of 2.2 lb (1 kg). Albert's lyrebird (*Menura alberti*) is slightly smaller. Lyrebirds have very strong legs and feet. The bill is short, sharp, and slightly down-curved.

Both species are brown, with Albert's being more rufous. The ornamental tail of the adult male distinguishes it from the female. Young males and females—"plain-tails"—are indistinguishable.

A lyrebird's tail has 16 rectrices (main feathers) and is best described using the terminology of Len Smith in his classic work, *The Life of Lyrebird*, published in 1988. An adult male has an outer pair of broad, fully webbed feathers ("lyrates"), a central very narrow pair ("medians"), and six pairs of "filamentaries" in which the barbs are separated except near the base. A young male superb lyrebird does not acquire a filamentary feather until at least his fourth year, and he is at least six years old before he gets a full set. Albert's lyrebird tail development is unknown.

Distribution

Lyrebirds are endemic to eastern Australia south from latitude 28° south, (between Brisbane and Melbourne) and up to about 100 mi (160 km) inland in some places. Albert's lyrebird occurs only in the northernmost 100 mi (160 km) of this range. Superb lyrebirds were introduced into Tasmania. A lyrebird fossil found at Riversleigh in northwestern Queensland shows that lyrebirds once extended much further north.

With both species, much habitat has been lost through European settlement, but the overall range has changed little in historical times.

Habitat

Lyrebirds are ground dwellers, though they rest off the ground during the day (several hours a day are spent in preening), and they roost high in the trees. They occur mainly in dense subtropical and temperate rainforest and wet sclerophyll forest, but superb lyrebirds extend into dry sclerophyll forest and even woodland where there is sufficient shrub growth to provide reasonable cover, but with areas of open ground for foraging.

Lyrebird nests are mostly built on the ground or on ledges of cliffs or large boulders, rarely in trees. Lyrebirds are poor flyers and nests are usually sited so that the female can glide downwards away from the nest if danger threatens.

Habitat requirements are trees in which to roost, sufficient vegetative cover to screen them from aerial and ground predators, an accessible year-round supply of invertebrate fauna for food in the leaf litter and surface soil, and suitable nesting sites.

Behavior

Noted ornithologist John Gould wrote that, of all the birds he had ever met, the superb lyrebird was by far the shyest and most difficult to stalk. Albert's lyrebird is even more wary, and so has been little studied. Observations of a population of superb lyrebirds that have largely lost their fear of people (in Sherbrooke Forest, Victoria, Australia) have provided much of what is known of lyrebird behavior.

Lyrebirds are sedentary, though individuals may wander away from their defended territories outside the breeding season, or to drink and bathe, which they do daily. They are generally solitary, but occasionally two or more may be seen together, mainly outside the breeding season.

Female superb lyrebirds also maintain territories and defend them against other females. Their territories overlap male territories but do not coincide with them. The situation with female Albert's lyrebird territories is unknown.

The three major components of male lyrebird vocalization are: loud territorial songs; a display song consisting largely of mimicry and aimed at attracting females; and sequences of peculiar rhythmic sounds—the so-called "pilik song" of superb lyrebirds, and the "gronking song" of Albert's lyrebirds. Lyrebird vocalizations are culturally transmitted from generation to generation. All males in a local area use the same territorial song or songs, for example, although there is great regional variation.

Male Albert's lyrebirds weave mimicked sounds into a fixed sequence, forming a stereotyped song about 40–50 seconds long that may be repeated many times without a break. All males in a local area have the same sequential song, clearly demonstrating that it is culturally transmitted. Superb lyrebird mimicry appears to come in random order.

Lyrebirds are renowned for their powers of mimicry and it is widely held that they mimic mechanical sounds of human origin such as axe-blows and mill whistles. However, they rarely do so in the wild, and never as part of their breeding season song. Both species regularly produce with the voice the sounds of feathered wings. Albert's lyrebirds mimic the brush-tailed possum (*Trichosurus vulpecula*), a local nocturnal marsupial. Out of the breeding season, lyrebirds mimic a greater variety of sounds on an irregular basis.

Lyrebirds learn to sing by copying older lyrebirds. This applies to choice of mimicry as well as the lyrebirds' own sounds, though obviously hearing the mimicked species enables them to keep the mimicry accurate. When superb lyrebirds were introduced into Tasmania they retained in their mimicry calls of eastern whipbirds (*Psophodes olivaceus*) and of pilotbirds (*Pycnoptilus floccosus*), neither of which occur there. For several decades these calls remained clearly recognizable but were no longer recognizable in 2001.

During winter, the species mimicked by the lyrebirds are not themselves breeding and are mostly silent, and so potential confusion of auditory signals is avoided.

Both species have a variety of alarm and threat calls. The principal alarm call, a loud piercing shriek, is common to both species.

Lyrebirds fly poorly but can leap vertically 6 ft (2 m) or more. They prefer to gain height by leaping and climbing in the vegetation. This way, and flying as little as possible, they can gain the forest canopy for roosting.

At dawn a lyrebird waits until there is enough light on the forest floor before descending from the roost. Mature males often spend this time in intermittent song. Once on the ground in the breeding season, those males that are defending territories first spend up to two hours singing and displaying before commencing to forage.

Feeding ecology and diet

Macro-invertebrate fauna, especially earthworms, insects, and insect larvae, are obtained by scratching in the leaf litter and upper layers of soil. Lyrebird legs, toes, and claws are long and powerful and can move surprisingly large stones and fallen branches in search of prey.

Lyrebirds require fresh water to drink and in which to bathe, preferring still pools to running water.

Reproductive biology

Lyrebirds have a dispersed lek mating system, each male having a number of display arenas within his large territory that is vigorously defended. Elaborate visual and vocal displays are used to attract females. Superb lyrebird males mate with any female they can attract, and do not form pair-bonds; Albert's lyrebirds are probably similar in behavior.

The female superb lyrebird undertakes all domestic duties, building the nest, incubating the single egg (rarely two), feeding the chick in the nest, and caring for it for up to nine

A male Albert's lyrebird (*Menura alberti*) performs its seldom-witnessed courtship display deep in the McPherson Range rainforest in Australia. (Photo by Wayne Lawler. Photo Researchers, Inc. Reproduced by permission.)

months after it leaves the nest. The nest is a large domed structure with side entrance, built of sticks with moss sealing the interstices; it is usually well camouflaged, and often has a platform at the entrance on which the female can stand. The nest is lined with feathers plucked by the female from her own body.

Incubation lasts six to eight weeks and the chick remains in the nest for another six weeks, so that the female remains close to the nest for about three months. As nests are often on or near the ground, nest hygiene is important; smell could disclose the location to a predator. The droppings of a lyrebird chick in the nest are produced in a gelatinous sac that the chick excretes directly to the bill of its mother. This she disposes of by burying or placing in a stream.

Both species breed in winter, approximately May to August, though superb lyrebird nests have been found with eggs as late as October. Winter breeding appears advantageous for superb lyrebirds in that most of their range has a winter rainfall, so that the chick is present when food is normally plentiful and more readily accessible.

Albert's lyrebirds also breed during the winter even though the different climatic pattern in their range results in the nestling being present during the driest time of the year. This suggests that the species may have evolved under different climatic conditions.

Conservation status

The extent of lyrebird habitat has been greatly reduced by human activities, first, and to an unknown extent, by Aboriginal fire management of the environment, probably over some tens of thousands of years, and second, by European settlement during the last two centuries. By the end of the nineteenth century, an extensive lowland area of rainforest in northern New South Whales, within the range of the Albert's lyrebird, had been cleared for dairying—a loss of some 185,000 acres (75,000 ha).

Both species of lyrebirds, however, appear secure with much of their remaining habitat being in conservation reserves. Some concern remains, nonetheless, for Albert's lyrebird because of its small total range. Some small isolated populations of both species are regarded as being at risk. Public opinion, as well as the law, now protects lyrebirds from hunting.

Cats, domestic and feral, dogs, and the introduced European red fox (*Vulpes vulpes*) are all potential predators, and have been a serious problem in some areas including Sherbrooke Forest. That population also suffers from deterioration of habitat. Lyrebirds need patches of open ground where they can forage. In the past, periodic bushfires ensured this by reducing the ground cover. As of 2001, however, the immediate proximity of residential development required total fire exclusion and increasing ground cover reduces foraging space.

Because they are so intensely shy, it is impossible to census a lyrebird population. Singing males can be counted, giving some indication of the extent and density of a local population, but that is all. And even this can be extremely difficult because of the dense vegetation and rugged terrain of most lyrebird habitat.

Significance to humans

The superb lyrebird has been extensively photographed and is widely regarded as "the" lyrebird. Although Albert's lyrebird was described as a species in 1850, it is so shy that it was not photographed in the wild until 1968, and as of 2001, only five people had photographed the display, three of them photographing the same individual. Thus it is the superb lyrebird that is of main significance to humans.

The name "lyrebird" arose from the resemblance of the tail of a male superb lyrebird to the shape of that ancient musical instrument. However, the bird rarely holds his tail in that position and then only briefly. The specific name, *tyawanoides* of the lyrebird fossil, is based on an Aboriginal name for lyrebird.

Lyrebirds of both species were occasionally eaten by European settlers, and male lyrebirds were extensively hunted for their ornamental tails until protective measures became effective early in the twentieth century. By 1920, populations were recovering.

In a few places where residential areas abut lyrebird habitat, superb lyrebirds forage in gardens and their vigorous scratching can be very destructive. As of 2001, Albert's lyrebirds were also invading gardens in one locality, notwithstanding their normal shyness.

The English ornithologist John Gould, who produced the first major publication on Australian birds, considered the superb lyrebird as the most appropriate bird to be the emblem of Australia: "it being not only strictly peculiar to that country, but one which will always be regarded with the highest esteem both by the people of Australia and by ornithologists in Europe, from whom it has received the specific appellations of *superba*, *paradisea*, and *mirabilis*."

While it has not achieved that distinction, it is the emblem of the National Parks and Wildlife Service of New South Wales, and it has appeared on Australian stamps. Interestingly, a superb lyrebird is featured on the logo of the Beaudesert Shire Council in Queensland, surely a tribute to that species' public renown, for the lyrebirds in that Shire are in fact all Albert's.

The song of the superb lyrebird was broadcast live from Sherbrooke Forest on July 5, 1931, and relayed to all Australian states. The first sound recordings (a gramophone record and sound film), made in Australia of a wild bird in its natural surroundings, were of a superb lyrebird in Sherbrooke Forest. The sound film was broadcast from Sydney to America in 1931.

Lyrebird song features in the music of the noted French composer Olivier Messiaen. One complete section of *Eclairs sur L'Au-Dela* (a major work commissioned by the New York Philharmonic) is based on superb lyrebird song Messiaen heard in the Brindabella Ranges near Canberra. And in Messiaen's *Des Canyons Aux Etoiles Section* IX, "Le Moquer Polyglotte" includes references to both species.

It is difficult to keep lyrebirds in aviaries, and very difficult to get them to breed in captivity. This has rarely been achieved. In captivity, where a lyrebird cannot learn from other lyrebirds, he may mimic what he does hear, such as the noise of a camera shutter and a camera motor winding on a film.

One group of lyrebirds, east of Armidale in New South Wales, has a remarkable flute-like territorial song. In the 1920s a male superb lyrebird chick was taken from the nest and raised in captivity with the domestic chickens on a farm. It could not hear other lyrebirds, and so learned to sing like the flute it heard being practiced by the farmer's son. When later this bird was released back into the wild, its "flute" calls became adopted as the territorial songs of that population. The accuracy of this story has been disputed on the grounds that the same song occurs some 60 mi (100 km) away from the release site, but it could have been culturally transmitted that far in 70 years.

1. Male Albert's lyrebird (*Menura alberti*); 2. Female Albert's lyrebird; 3. Female superb lyrebird (*Menura novaehollandiae*); 4. Male superb lyrebird. (Illustration by Barbara Duperron)

Species accounts

Albert's lyrebird
Menura alberti

TAXONOMY
Menura (Harriwhitea) alberti Bonaparte, 1850, Turanga (now Terania) Creek, Richmond River, Australia. Monotypic.

OTHER COMMON NAMES
English: Prince Albert's lyrebird, Northern lyrebird; French; Ménure d'Albert; German: Braunrücken-Leierschwanz; Spanish: Ave Lira de Alberti.

PHYSICAL CHARACTERISTICS
34–37 in (86–94 cm); 2.0 lb (0.92 kg); female weight not recorded. Tail (longest feathers), male 20 in (51 cm), female 16 in (40 cm). The male's outer pair of tail feathers are plain and fully webbed, dark brown above and dark gray below, and are the shortest at about 15 in (38 cm); the next six pairs of "filamentary" feathers, dark brown above and light gray underneath, are about 20 in (51 cm). The central pair are about 21 in (53.5 cm), but only 0.5 in (1.3 cm) wide.

DISTRIBUTION
Southeastern Queensland and northeastern New South Wales: approximately Laidley to Ballina.

HABITAT
Subtropical rainforest, wet sclerophyll forest with rainforest understory, Antarctic beech (*Nothofagus*) forest.

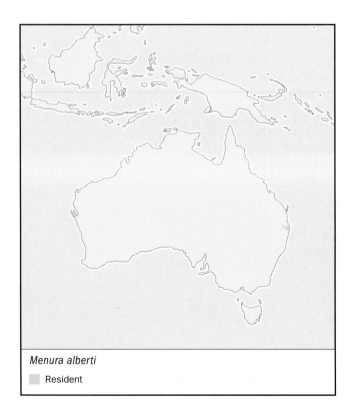

Menura alberti
Resident

BEHAVIOR
Males defend separate individual territories in dispersed leks, using vocal and visual displays to attract females for mating. Display arenas consist of crossed thin vines and sticks lying loosely on the ground and able to move with the movement of the bird's feet. During the display, the male grasps a vine and vigorously moves it up and down during his gronking song. When the vines and sticks are dry, this makes a tapping sound in perfect time with the rhythmic notes of the song. In effect he is using "rhythm sticks"—possibly the only bird in the world to accompany its song with a musical instrument.

FEEDING ECOLOGY AND DIET
Scratches in leaf litter and surface soil for invertebrate fauna.

REPRODUCTIVE BIOLOGY
Female alone builds nest and raises the chick. Single egg laid mostly in June.

CONSERVATION STATUS
Not threatened. No population estimates. Probably secure as a species, but some concern because of limited distribution. Several small isolated populations at risk.

SIGNIFICANCE TO HUMANS
Some shot for food or ornamental tails early in twentieth century. Now held in high regard by the public. Significant in ecotourism. ◆

Superb lyrebird
Menura novaehollandiae

TAXONOMY
Menura novaehollandiae Latham, 1802, Upper Nepean River, New South Wales. Three subspecies.

(The specific name *novaehollandiae* has been and remains in general use. However, its adoption was based on a mistake as to the date of publication, and on a strict interpretation of the International Code of Zoological Nomenclature, the name superba, Davies, should take priority. An application has been made to the International Commission for suppression of *superba* as an unused senior synonym.)

OTHER COMMON NAMES
English: Superb lyrebird, Edward lyrebird, Prince Edward lyrebird, Victoria lyrebird, Queen Victoria lyrebird; French; Ménure superbe; German: Graurücken-Leierschwanz; Spanish; Ave Lira Soberbia.

The lyrebird was well-known to the Aboriginal people. Names from various language groups included: balangara, bulan-bulan, beleck-beleck, golgol, and woorayl. The double names almost certainly were derived from the loud double notes of the lyrebird's "pilick" song.

PHYSICAL CHARACTERISTICS
30–39 in (76–100 cm); female 1.9 lb (0.88 kg), male 2.3 lb (1.06 kg). Male has highly ornamental tail, 28 in (71 cm). Outer pair of feathers elongated S-shape, decorated with semi-transparent "windows"; six pairs of filamentary feathers; central pair very narrow. Very strong legs and feet. Long claws span 6 in (15 cm).

Menura novaehollandiae

■ Resident

DISTRIBUTION

M. n. edwardi: Hunter River north to near Stanthorpe; *M. n. novaehollandiae*: Hunter River to Victorian border; *M. n. victoriae*: Victoria, east of Melbourne, plus Snowy Mountains to Brindabella Range in New South Wales.

HABITAT

Wet sclerophyll forest, subtropical and temperate rainforest, Antarctic beech (*Nothofagus*) forest, dry sclerophyll forest, eucalyptus woodland.

BEHAVIOR

Separate male and female territories. Males sing and display on arenas consisting of low earth mounds. Sedentary. Mostly solitary, occasionally two or more together.

FEEDING ECOLOGY AND DIET

Forage on the ground for invertebrates in soil and litter, scratching and digging to a depth of several inches (10 cm).

REPRODUCTIVE BIOLOGY

Males promiscuous. Female alone builds nest, incubates the single egg, and cares for the chick, which sometimes still begs from its mother at start of next breeding season.

CONSERVATION STATUS

All three subspecies are not threatened, though considerable reduction in habitat through European settlement. Some small isolated populations may be at risk.

SIGNIFICANCE TO HUMANS

Hunted for food and ornamental tails by early European settlers. This is the species that is now well-known and highly regarded by the public. ◆

Resources

Books

Higgins, P. J., J. M. Peter, and W. K. Steele, eds. *Tyrant-flycatchers to Chats.* Vol. 5 of *Handbook of Australian, New Zealand and Antarctic Birds.* Melbourne: Oxford University Press, 2001.

Schodde, R., and I. J. Mason. *The Directory of Australian Birds—Passerines.* Collingwood, Australia: CSIRO Publishing, 1999.

Smith, L. H. *The Life of the Lyrebird.* Richmond, Australia: William Heinemann Australia, 1988.

Periodicals

Bock, W. J., and M. H. Clench. "Morphology of the Noisy Scrub-bird, *Atrichornis clamosus,* (Passeriformes: Atrichornithidae): Systematic Relationships and Summary." *Records of the Australian Museum* 37, nos. 3 and 4 (1985).

Curtis, H. S. "The Albert Lyrebird in Display." *Emu* 72 (1972): 81–84.

Robinson, F. N., and H. S. Curtis. "The Vocal Displays of the Lyrebirds (Menuridae)." *Emu* 96 (1996): 258–275.

Smith, L. H. "Structural Changes in the Main Rectrices of the Superb Lyrebird *Menura novaehollandiae* in the Development of the Filamentary Feathers." *Emu* 99 (1999): 46–59.

H. Sydney Curtis, BSc
Darryl N. Jones, PhD

Scrub-birds

(*Atrichornithidae*)

Class Aves
Order Passeriformes
Suborder Passeri (Oscines)
Family Atrichornithidae

Thumbnail description
Small, brown, ground-dwelling birds with short wings, powerful legs, longish tail, pointed bill, and loud penetrating calls

Size
7–9 in (17–23 cm); 1–1.7 oz (30–52 g)

Number of genera, species
1 genus; 2 species

Habitat
Moist forest and thicket

Conservation status
Vulnerable: 1 species; Near Threatened: 1 species

Distribution
Limited ranges in Australia

Evolution and systematics

The two species of scrub-bird form a small monogeneric family endemic to Australia. Anatomical and molecular evidence suggest that the scrub-birds are most closely related to the lyrebirds (Menuridae), with these two families diverging from each other 30–35 million years ago. However, the broader affinities of the scrub-birds and lyrebirds are less clear and have been the subject of a long and complex debate. They exhibit a number of anatomical characteristics (particularly of the syrinx, sternum, and clavicles) that are unusual within the passerines. These characteristics have, at times, been used as an argument for placing the two families in their own suborder. It has even been argued that the anatomy of these birds indicates an affinity with the tapaculos (Rhinocryptidae) of South America. However, recent evidence from DNA analysis suggests that the scrub-birds and lyrebirds are instead related to the other endemic Australian passerine families.

Physical characteristics

Scrub-birds are small, solidly built birds, well adapted to living on the ground within extremely dense vegetation. They have stout legs, short rounded wings, and a longish graduated tail, which is often held cocked. The head has a long flat forehead tapering to a pointed triangular-shaped bill. The plumage of adults is generally brown, with fine dark barring on the upper parts. Males also have black markings on the throat and breast. Juveniles have similar plumage to adults, but slightly duller.

Distribution

Both species of scrub-bird have very restricted ranges, with the rufous scrub-bird (*Atrichornis rufescens*) confined to the central east coast of Australia and the noisy scrub-bird (*A. clamosus*) occurring only in the far southwestern corner of Western Australia. The two species are therefore on opposite sides of the Australian continent, separated by a distance of almost 2,000 mi (3,200 km) of mostly arid or semiarid lands. This current pattern suggests that scrub-birds were once distributed more widely when moist forests covered much of southern Australia during the middle Tertiary, before the continent started to dry out about 16 million years ago.

Habitat

Both scrub-bird species prefer habitat with very dense vegetation cover close to the ground, a thick layer of leaf litter, and a moist microclimate. In the case of the rufous scrub-bird, suitable habitat occurs mainly in rainforest, associated with natural or human-induced openings in the canopy, or in adjoining moist eucalyptus forest that has been well buffered from fire. The noisy scrub-bird is currently confined to coastal low forest, thicket, and heath, although historically it also occurred in moist areas within taller eucalyptus forest.

Behavior

Scrub-birds spend most of their time on, or near, the ground and are incapable of more than a few yards of sustained flight. They are, however, very adept at moving quickly through dense vegetation. Males occupy permanent territories, which they defend with remarkably loud and penetrating territorial song, particularly during the breeding season. Although territories are usually widely spaced, each individual male spends most of its time within a core area covering little more than 2.5 acres (1 ha).

Feeding ecology and diet

Both species of scrub-bird forage in leaf litter on the ground, eating a wide variety of invertebrates.

Reproductive biology

Scrub-birds are generally monogamous, with the female occupying an area on the periphery of the male's territory. The noisy scrub-bird breeds in winter, while the rufous scrub-bird breeds in spring. A domed nest is built close to the ground and is lined with a thin cardboard-like pulp of wood and grass. The clutch size is one for the noisy scrub-bird, but may be two for the rufous scrub-bird. The female incubates and feeds the young, which take three to four weeks to fledge.

Conservation status

The noisy scrub-bird was thought to have gone extinct in the late nineteenth century, until a small population was rediscovered in 1961. For many years following its rediscovery, the species had the dubious distinction of being Australia's rarest passerine. However, thanks to a highly successful program of translocation and management, the bird's status has now been reclassified from Endangered to Vulnerable. The rufous scrub-bird, while reasonably rare, has not suffered the same level of decline as that experienced by the noisy scrub-bird.

Significance to humans

None known.

Species accounts

Rufous scrub-bird
Atrichornis rufescens

TAXONOMY
Atrichia rufescens Ramsay, 1866, New South Wales, Australia.
Two subspecies recognized.

OTHER COMMON NAMES
English: Eastern scrub-bird; French: Atrichorne roux; German:
Rostbauch-Dickichtvogel; Spanish: Achaparrado Rufa.

**PHYSICAL
CHARACTERISTICS**
Male 7.1 in (18 cm),
female 6.5 in (16.5
cm). Dark rufous-
brown with fine
black barring above,
and buff belly. Male
has black mottling
on throat and breast.
Bill, eyes, and legs
dark brown.

DISTRIBUTION
Central east coast of
Australia (northeast-
ern New South
Wales and far southeastern Queensland).

Atrichornis rufescens

HABITAT
Patches of dense ground cover within rainforest or adjacent
moist eucalyptus forest. Now mainly confined to areas above
1,968 ft (600 m) altitude.

BEHAVIOR
Males defend territories using a loud "chipping" song, supplemented by accomplished mimicry of other species, particularly
when disturbed. Females much less vocal, producing only soft
"ticking" and "squeaking" calls.

FEEDING ECOLOGY AND DIET
Feeds on leaf-litter invertebrates obtained by turning leaves
with the bill and scratching with strong legs and claws.

REPRODUCTIVE BIOLOGY
Breeds September to November. Domed nest constructed
mainly of grass, inside completely lined with cardboardlike
pulp of wood and grass. Clutch size two, but one egg may be
infertile.

CONSERVATION STATUS
Listed as Near Threatened. Total population has declined
markedly since European settlement, mainly due to habitat
loss. Status currently being monitored by annual surveys.

SIGNIFICANCE TO HUMANS
None known. ◆

Noisy scrub-bird
Atrichornis clamosus

TAXONOMY
Atrichia clamosa Gould, 1844, Western Australia.

OTHER COMMON NAMES
English: Western scrub-bird; French: Atrichorne bruyant; German: Braunbach-Dickichtvogel; Spanish: Achaparrado Occidental.

**PHYSICAL
CHARACTERISTICS**
Male 9.1 in (23 cm),
female 7.7 in (19.5
cm); male 1.7 oz
(52g); female 1.2 oz
(34 g). Upperparts
brown with fine dark
barring. Rufous-
brown on lower
belly, grading to off-
white on breast.
Male also has black
patch on upper
breast and throat.
Bill, eyes, and legs
brown.

Atrichornis clamosus

DISTRIBUTION
Far southwestern corner of Western Australia.

HABITAT
Low forest, thicket, and heath with dense lower stratum of
shrubs and sedges.

Atrichornis rufescens

▢ Resident

Atrichornis clamosus
■ Resident

BEHAVIOR

Males employ a loud territorial song supplemented by various calls. Mimicry rarely used. Females generally silent, producing only soft calls.

FEEDING ECOLOGY AND DIET

Feeds on invertebrates, and occasionally small vertebrates, flushed from leaf litter or low vegetation.

REPRODUCTIVE BIOLOGY

Breeds April to October. Dome nest similar to rufous scrub-bird except only lower half of inside lined with cardboard-like substance. Clutch size is one.

CONSERVATION STATUS

Vulnerable. Recently downgraded from Endangered thanks to the success of an ongoing translocation and management program initiated over 35 years ago. Since rediscovery of the species in 1961, its total population size has been increased from less than 50 to almost 600 breeding territories.

SIGNIFICANCE TO HUMANS

None known ◆

Resources

Books

Danks, A., A. A. Burbidge, A. H. Burbidge, and G. T. Smith. *Noisy Scrub-bird Recovery Plan. Western Australian Wildlife Management Program.* No. 12. Perth: Department of Conservation and Land Management, 1996.

Ferrier, S. "Habitat Requirements of a Rare Species, the Rufous Scrub-bird." In *Birds of Eucalypt Forests and Woodlands: Ecology, Conservation, Management*, edited by A. Keast, H. F. Recher, H. Ford, and D. Saunders. Sydney: Surrey Beatty and Sons, 1985.

Garnett, S., ed. *Threatened and Extinct Birds of Australia.* RAOU Report Number 82. Melbourne: Royal Australasian Ornithologists Union, 1992.

Higgins, P. J., J. M. Peter, and W. K. Steele, eds. *Handbook of Australian, New Zealand and Antarctic Birds.* Vol. 5, *Tyrant-Flycatchers to Chats.* Melbourne: Oxford University Press, 2001.

Pizzey, G. *The Field Guide to the Birds of Australia.* Sydney: Harper Collins, 1997.

Schodde, R., and S. C. Tidemann, eds. *Reader's Digest Complete Book of Australian Birds.* Sydney: Reader's Digest, 1986.

Periodicals

Bock, W. J., and M. H. Clench. "Morphology of the Noisy Scrub-bird, *Atrichornis clamosus* (Passeriformes:

Atrichornithidae): Systematic Relationships and Summary." *Records of the Australian Museum* 37 (1985): 243–254.

Chisholm, A. H. "The Story of the Scrub-birds." *Emu* 51 (1951): 89–112, 285–297.

Danks, A. "Conservation of the Noisy Scrub-bird: A Review of 35 years of Research and Management." *Pacific Conservation Biology* 3 (1997): 341–349.

Sibley, C. G., and J. E. Ahlquist. "The Phylogeny and Classification of the Australo-Papuan Passerine Birds." *Emu* 85 (1985): 1–14.

Smith, G. T., "Habitat Use and Management for the Noisy Scrub-bird." *Bird Conservation International* 6 (1996): 33–48.

Webster, H. O. "Rediscovery of the Noisy Scrub-bird." *Western Australian Naturalist* 8 (1962): 57–59.

Other

Ferrier, S. *The Status of the Rufous Scrub-bird (Atrichornis rufescens): Habitat, Geographical Variation and Abundance.* Unpublished PhD Thesis. Armidale: University of New England, 1984.

Simon Ferrier, PhD

Larks

(Alaudidae)

Class Aves
Order Passeriformes
Suborder Passeri (Oscines)
Family Alaudidae

Thumbnail description
Small, highly terrestrial passeriforms with usually cryptic plumage; well known for continuous song and spectacular aerial song-display of males in most species

Size
3.9–9.0 in (10–23 cm); 0.4–2.6 oz (12–73 g)

Number of genera, species
15 genera; 78 species

Habitat
Open landscape, grassland, steppes, arid and semiarid habitats, wastelands, and cultivated fields

Conservation status
Endangered: 2 species; Critical: 2 species; Vulnerable: 4 species; Near Threatened: 3 species

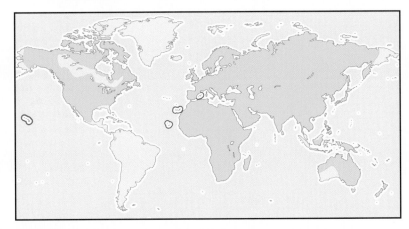

Distribution
Africa, Europe, and Asia; one species through Wallacea to Australia, one species in North America with local population in South America

Evolution and systematics

Larks (Alaudidae) are distinguished from other songbirds by several characters. The back of the tarsus is rounded and covered with individual horny scutes on its rear side; unlike in all other oscines, where the back side of the tarsus is generally covered with one single and sharply edged scute. The lark's syrinx has no pessulus, or small cartilaginous knob at the point of fusion of the bronchiae. The pessulus is present in all other passeriform birds, with the exception of Funarioidea and a few tyrannid species.

Relationships within the Passeri remain uncertain. Larks are predominantly ground-dwelling birds; they evolved, however, from more arboreal ancestors. The recapitulation of arboreal habits can be observed during the development of juvenile larks. When leaving the nest before they can fly, they first hop on the ground like thrushes, but are able to walk within a few days just like their parents. Strong indications for their long adaptation to terrestrial habits are their courtship display and nest building on the ground, egg-rolling behavior, and distraction-display. When alarmed, larks first squat onto the ground, trusting their cryptic plumage to hide them instead of flying away.

The terrestrial lifestyle of the larks is similar to that of pipits and wagtails (Motacillidae), but this is not due to common ancestry. Motacillidae share with Fringillidae, Emberizidae, Parulidae, and Icteridae a sequence of nine additional nucleotides within a specific gene, a sequence missing in larks and in all other birds investigated.

Fifteen genera of larks have been distinguished, and 78 species are listed in Peters Checklist, about a third of which are members of genus *Mirafra*. However, according to recent reinvestigations, including molecular, morphological, and behavioral studies, eight species should be added which had been thought to be subspecies.

Members of the Alaudidae should not be confused with other birds commonly called larks: meadowlarks (*Sturnella*) belong to the Icteridae, the magpie-larks (Grallinidae) of Australia and New Guinea are members of the corvid assemblage, and the Australian songlarks (*Cinclorhamphus*) are thought to be Sylviidae.

Physical characteristics

Larks range in size from the size of a finch to that of a thrush, i.e., 3.9–9.0 in (10–23 cm) and weight 0.4–2.6 oz (12–73 g). In some species males are larger and heavier than females. The plumage is usually inconspicuous, the upperparts being heavily streaked or unmarked, with a general brownish to grayish color. The plumage is often adapted to the color of local soil so that many larks are remarkably camouflaged. Nonmigrating species with large breeding ranges extending over many habitats with different soil types tend to contain several subspecies that differ in color. The underparts are usually light and without any pattern. Several species have crown feathers which they can raise to a crest; most conspicuous is the crested lark (*Galerida cristata*) and its congenerics. In most species both sexes look very similar. Even both sexes

Four-day-old wood lark (*Lullula arborea*) chicks sit in their nest. (Photo by G. Synatzschke/OKAPIA. Photo Researchers, Inc. Reproduced by permission.)

of the horned lark (*Eremophila alpestris*) and Temminck's lark (*E. bilopha*) possess a contrasting colored pattern of breast and head, and they have tiny elongated feathers above their eyes that form conspicuous horns. However, the breeding plumage of the male black lark (*Melanocorypha yeltoniensis*) is entirely black. The males of the seven known sparrow-larks (*Eremopterix*) have black underparts and a species-specific black-and-white pattern on their head. Only the male black-eared sparrow-lark (*E. australis*) is totally black-headed; females of all these species are colored as larks in general.

Juveniles often have a spotted plumage, which distinguishes them from their parents. This is most conspicuous in the juveniles of horned larks and Temminck's larks, where the feather-horns of the adult plumage are also missing. Adult larks molt completely once per year after the breeding season. With the exception of *Alaemon* and *Eremopterix*, the juvenal plumage is replaced by the adult plumage immediately after the bird becomes fully fledged. This juvenal molt is very unusual among nontropical oscines.

In general, the nostrils are concealed with small tufts of feathers. They are clearly visible, however, in the case of *Alaemon*, *Certhilauda*, *Heteromirafra*, *Mirafra*, and *Pinarocorys*.

The bills of larks are astonishingly diverse in shape, ranging from short, heavy, and conical to elongated, thin, and pointed. These differences reflect adaptations to a variety of food and feeding techniques. The thick-billed lark (*Ramphocoris clotbey*) stands out as an extreme; its bill is massive, short, and very deep, reminiscent of the bill of a grosbeak. This lark is, however, unique in having a toothlike projection on its lower mandible which fits into a notch in the upper mandible. The short bills of *Calandrella* and *Eremopterix* are less heavy, but similar to those of finches due to their conical shape. All these species feed mainly on seeds, but insects are more common in the diet of other larks. The horned lark has a short, pointed bill like some pipits; the same applies to the wood

lark (*Lullula arborea*) and fawn-colored lark (*Mirafra africanoides*). Others like the red-winged lark (*Mirafra hypermetra*) or the crested lark have long and robust bills which are also used for digging. Dupont's lark (*Chersophilus duponti*), the greater hoopoe lark (*Alaemon alaudipes*), and the Cape long-billed lark (*Certhilauda curvirostris*) are characterized by elongated, slightly decurved, and slender bills. Some species are sexually dimorphic in respect to bill length; the male has a longer bill than the female.

Larks usually have short legs, but they are fairly long in *Alaemon* and *Certhilauda*. Larks have strong feet. The hind claw is much longer than the toes, especially in species living on soft soil. However, the hind claw is shorter in larks that live on hard ground or are fast runners such as *Alaemon*.

All lark nestlings have a characteristic brightly colored gape, with one black spot inside the tip of the upper and lower mandible. There are two black spots on the base of the tongue in *Ammomanes*, some *Certhilauda*, and *Mirafra*; and an additional third spot on the tip of the tongue in *Alauda*, *Calandrella*, some *Certhilauda*, *Eremophila*, *Galerida*, and *Lullula*. After hatching, larks are covered with scanty down on their foreheads, napes, backs, shoulders, wings, and thighs.

Distribution

Most lark species occur in Africa, and many have a discontinuous area of distribution and a limited range. Such distribution patterns, in combination with a highly sedentary lifestyle that prevents gene flow between separated populations, are important preconditions for speciation. For example, southwestern Africa is inhabited by six sedentary and closely related *Certhilauda* species; the short-clawed lark (*C. chuana*) is probably the sister taxon to a group consisting of five species which were thought to be conspecific until 1999. They are allopatric species; they do not interbreed as their ranges neither overlap, nor are they in contact.

Africa is also inhabited by genera that are primitive among larks in having unconcealed nostrils (*Alaemon*, *Certhilauda*, *Heteromirafra*, *Mirafra*, and *Pinarocorys*), so it is probable that

A rufous-naped bush lark (*Mirafra africana*) at its nest in South Africa. (Photo by W. Tarboton/VIREO. Reproduced by permission.)

larks are of African origin, and that their first radiation took place on this continent. Madagascar was colonized once by the African ancestor of the Madagascar lark (*Mirafra hova*). Today, several *Mirafra* larks occur in Asia. However, it is not possible to say how many species colonized Asia independently, since it is unknown whether the group of Asiatic *Mirafra* species is monophyletic, i.e., evolved from one common ancestor.

During a second radiation, several lineages derived from an ancestor with concealed nostrils, and perhaps this ancestral species was also of African origin. Some lineages certainly evolved in Africa. One lineage led to *Spizocorys*, others (probably in Asia) led to *Alauda*, *Eremophila*, and *Lullula*.

The colonization of Australia by the Australasian bushlark (*Mirafra javanica*) and of North America by the horned lark happened in the Pleistocene at the latest. During Pleistocene glacial periods, sea level was much lower than today. Sumatra, Kalimantan, and Java (Sunda shelf) were therefore part of Asiatic mainland, and New Guinea was connected to north Australia via the Torres land bridge, forming the dispersal route of the Australasian bushlark.

At the same time in the Northern Hemisphere, the Bering land bridge, the continental shelf between Siberia and Alaska, was exposed. The horned lark spread from Asia into North America along this land bridge. Its recent breeding range in North America reaches from northern Alaska to the Gulf of Mexico, and a separated population established itself on the Andean slopes of Bogotá, Colombia.

The skylark (*Alauda arvensis*) reached Siberia from the west approximately 11,000 years ago, after the Bering Strait was formed due to rising sea-levels. This barrier slowed down its eastward dispersion. However, there are records of possible breeding in the 1970s and two well-documented, but probably unsuccessful, broods in 1995 on the Pribilof Islands, Alaska. The skylark's northward expansion toward Central Europe from southern and eastern Europe is closely connected to increasing deforestation and agriculture during the seventh to thirteenth centuries, which provided appropriate habitats for this species.

The crested lark reached Central Europe from southwestern and eastern Europe later than the skylark and was widespread in the sixteenth century, but receded during the Little Ice Age in the seventeenth and eighteenth centuries. However, it spread into Central Europe again from the middle of the nineteenth century onward, helped by global warming and man-made habitats such as roads and railway stations. Crested larks even reached Scandinavia in 1900, but became extinct in the 1990s, possibly because of climatic change.

Habitat

Larks are characteristic birds of the open landscapes like grasslands, steppes, stony plains, and heaths. Most larks prefer areas with sparse vegetation. Some species, including the wood lark and flapped lark (*Mirafra rufocinnamomea*), depend on a mixture of vegetation types within the same habitat, such as grasses for nest-building and scattered bushes and small

An ashy-crowned finch-lark (*Eremopterix grisea*) feeds its young in the nest. (Photo by V. Sinha/VIREO. Reproduced by permission.)

trees for perching. Human activities provide further suitable habitats for certain species. In North America, skylarks and horned larks regularly breed on cultivated areas such as fields and uncultivated areas such as wastelands.

Larks inhabit extreme regions such as deserts, semideserts, and arctic tundras, and areas varying in altitude from about sea level to high mountains. The horned lark, for example, breeds at 14,750 ft (4,500 m) in the Rocky Mountains, and the skylark and Tibetan lark (*Melanocorypha maxima*) breed at 14,450 ft (4,400 m) and 15,100 ft (4,600 m) in the Himalayas, respectively. Generally, such extreme habitats are left after the breeding season.

Behavior

Larks are diurnal, ground-dwelling birds, sleeping on the ground in self-made depressions. They scratch their head by indirect method, and frequently take baths in dust or sand like chickens. They may bathe in rain, but not in water. Larks move on the ground by walking and running, reaching high speeds. Their flight is typically strong and undulating, with wings closed periodically.

Larks inhabiting arid climates have evolved certain strategies to cope with these severe conditions. They avoid contact with hot soil by perching on elevated stones and shrubs, and during the hottest part of the day, they become inactive and shelter in the shade provided by vegetation or stones. In the Arabian desert, larks rest in lizard burrows. Parents shade their nestlings by standing over them with wings spread.

Larks are famous for their melodious and continuous songs, which last from minutes to up to almost an hour. Some species even sing at night. Many species enlarge their reper-

Courting postures of (top left to right): Calandra lark (*Melanocorypha calandra*), sky lark (*Alauda arvensis*); (bottom left to right): crested lark (*Galerida cristata*), black lark (*Melanocorypha yeltoniensis*), and black-crowned sparrow-lark (*Eremopterix nigriceps*). (Illustration by Emily Damstra)

toire with imitations of other bird songs and calls. Because of this behavior, the Mongolian lark (*Melanocorypha mongolica*) is called "Hundred Melodies" in China and is a favored cage bird in East Asia. The Latakoo or melodious lark (*Mirafra cheniana*) is known to imitate 57 different bird species—even ducks, guineafowl, and bee-eaters—and single males can be distinguished by the set of birds they imitate. Some species, such as the crested lark, may even imitate human whistling.

Song is performed during aerial song-displays while males circle about their territories. Some species rise almost vertically from ground or perch and ascend up to 330 ft (100 m) or more before gliding or dropping with closed wings back to the ground. Continual hovering and singing is characteristic for the skylark. Several lark species frequently utter their songs from the ground and elevated perches such as stones, tops of bushes, or trees.

During its song-flight, the male of the black lark claps its wings above its back, reminiscent of the flight-display of pigeons. Some species within the genus *Mirafra*, as well as the Dupont's lark, can create rattling sounds with their flight feathers, a sound often compared to the song of cicadas. Wing flapping is generally performed during the lark's ascending phase of song-flight. The frequency of flaps is species-specific and individually variable, and regional dialects can be distinguished. The extent of the wing-flapping behavior is also species specific. Other than its melodious song, the white-tailed lark (*Mirafra albicauda*) utters only soft instrumental sounds. Nonvocal sounds are more prominent in rufous-naped lark (*M. africana*),

and clapper lark (*M. apiata*), where clattering-flight is still followed by singing. In the flappet lark, however, sound created by the wings replaces the bird's song almost entirely.

As far as one knows, courtship behavior is displayed on the ground. The male hops and steps around the female in upright posture spreading and cocking its tail-feathers. The undertail-coverts are presented to the female (they are entirely black in the black and the black-crowned sparrow-lark). The wings are drooped and also spread to some degree quivering slightly. The crown feathers are raised even in species without elongated crest feathers. During display, the male utters song fragments. Occasionally, the male presents food items to its mate immediately before copulation (courtship-feeding).

Several desert-inhabiting larks, including sparrow-larks, are nomadic, and their movements depend on rainfall and food supply. Migratory and nomadic species have to some extent a flocking behavior, and some species form sex-specific flocks during winter months. The granivorous larks, members of the genera *Eremopterix* and *Calandrella*, are very gregarious outside their breeding period, forming large flocks.

Feeding ecology and diet

Larks feed on arthropods, as well as seeds, green plant material, buds, and fruits. Food items are taken directly from the ground or pecked from plants. Some larks even prey on venomous insects or arthropods that have chemical defense

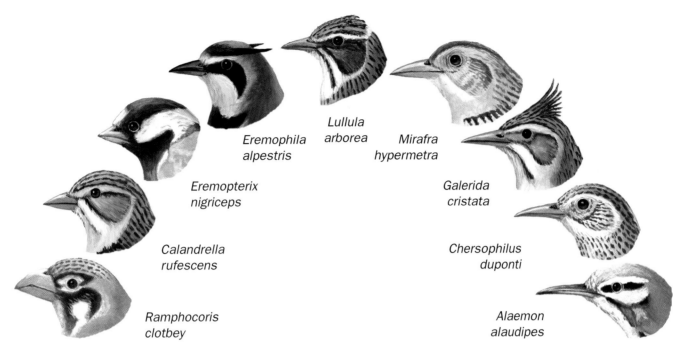

Bills of several different larks. (Illustration by Emily Damstra)

strategies, such as ants, darkling beetles, stink bugs, and millipedes. Rarely, flying insects are taken in aerial pursuit. The bill can be used for digging and probing. Depending on their diet, more or less insectivorous and granivorous species can be distinguished, but seasonal changes occur; the crested lark, wood lark, and skylark take fewer seeds during breeding season than in winter. All larks feed arthropods to their young, only Stark's lark (*Eremalauda starki*) feeds a high proportion of unripe grass seeds, even to newly hatched chicks.

In mainly insectivorous larks, the male is larger and has a longer bill than the female. This is most conspicuous in the greater hoopoe, the long-billed lark, and their relatives, which use their slender and decurved bills for digging in the ground in search of insect larvae. Sexual dimorphism in bill and body size also occurs in the bar-tailed lark (*Ammomanes cincturus*) and Gray's lark (*A. grayi*), which feed mainly on seeds. Such differences in size enable both sexes of the same species to exploit different food resources within the same habitat.

Most larks swallow whole seeds, which are crushed in their stomach using grit. Indigestible remains are ejected as small pellets. Larks in the genera *Calandrella*, *Eremopterix*, and *Melanocorypha* de-husk seeds in a finchlike manner, fixing the grain between the tongue and palatine and breaking it up. Crested larks, wood larks, and skylarks remove husks from seeds by beating them against the ground. They use the same technique for removing the legs and wings of large insects. Like the song thrush (*Turdus philomelos*), greater hoopoe larks crack the shells of snails using stones like an anvil. The same behavior was observed once in the crested lark in Morocco, but never in Central Europe. The greater hoopoe lark also frequently drops snails onto stones until their shells break.

Many larks satisfy their thirst and maintain body weight by drinking dew when water is not available. Various species, including the black, desert, Gray's, and Stark's lark, as well as the black-crowned and black-eared sparrow-lark, drink brackish or even salty water.

Reproductive biology

Larks are monogamous, and pairs stay together at least for one breeding season to raise one or two, rarely three, broods. The breeding period runs from March to July in the Palearctic, in the Tropics and South Africa it depends on the rainy season. Even erratic rainfall can induce egg-laying in some nomadic lark species. Most larks nest on the ground. The only exception is the greater hoopoe lark, which commonly builds its nest in tops of shrubs up to 24 in (60 cm) above the ground. Nests are cup-shaped depressions scratched into the ground, sheltered by grass tufts or stones and mainly lined with plant material. Some *Mirafra* and *Certhilauda* species frequently cover their nests with a dome made of grass and supported by vegetation behind the nest. If a nest is built on solid ground and it is impossible to scratch a depression, some larks add a rim or rampart of small pebbles, plant material, or pieces of dung or mud. This behavior has been observed for *Ammomanes*, *Calandrella*, *Eremophila*, *Eremopterix*, and *Ramphocoris*.

In most species the female builds the nest alone. However, in the calandra lark (*Melanocorypha calandra*) and Oriental skylark (*Alauda gulgula*), as well as several species within the *Ammomanes*, *Calandrella*, and *Galerida*, both sexes are involved in nest building. The male of the chestnut-backed sparrow-lark (*Eremopterix leucotis*) presents spider webs to its mate, which applies them to its nest. Ritualized nest building behavior as part of the courtship display is reported from two species. The male of the Rudd's lark (*Heteromirafra ruddi*) walks in front of the female with its bill full of nest material, and the male of the thick-billed lark offers the female small pebbles used for nest building.

The eggs are creamy-white or yellowish and are more or less spotted. Egg-laying occurs daily in early morning, and only one egg per day is laid. In favorable habitats the clutch size ranges from two to five, rarely up to seven, eggs. Inhabitants of arid or semiarid areas generally have smaller clutches. Gray-backed sparrow-larks (*Eremopterix verticalis*) in the Karoo, South Africa, lay significantly larger clutches in years with heavy rainfall. Females of the closely related black-crowned sparrow-lark (*Eremopterix nigriceps*) lay no more than three eggs, but only two young per brood are reared because each adult feeds only one single chick after they have left the nest. This "split brood-care" behavior, as well as the reduction of clutch size, might be an adaptation to the severe desert conditions, which make egg production and providing food to the young difficult.

Incubation starts with the last egg and, depending on the species, takes 11–16 days. Usually, only the female incubates and broods the young. However, sparrow-larks and a few other species seem to be an exception to this rule.

In all larks, both sexes feed the chicks and eat or remove their feces. After eight to 13 days, before they are able to fly, lark chicks leave the nest, still supplied with food by their parents. If the female starts a second brood, the male cares for the young alone. Steyn observed cooperative breeding in the spike-heeled lark (*Chersomanes albofasciata*); three adults fed two chicks in one nest in the Karoo.

Conservation status

According to the IUCN, several lark species are in need of special protection. Ash's lark (*Mirafra ashi*) is known only from six specimens collected in 1981 in a very small range in Somalia. Because its minute population seems to decline, this species is classified as Endangered. Botha's lark (*Spizocorys fringillaris*) is also Endangered. It is patchily distributed within South Africa, and its population is declining due to habitat destruction by agriculture.

The situation is critical for the Raso lark (*Alauda razae*) and Rudd's lark. The former species is endemic to the uninhabited Islet of Raso (Cape Verde Islands), and its population is very small (92 birds counted in 1998). Accidentally introduced predators are therefore a serious threat to eggs and nestlings. A rapid population decline of Rudd's lark could be prevented by protecting its breeding range in South Africa.

Four species are classified as Vulnerable because of their restricted and unprotected ranges and small populations. The ferruginous or red lark (*Certhilauda burra*) breeds in the Cape Province only, where habitat fragmentation continues. Archer's lark (*Heteromirafra archeri*) is known only from two

small areas in Somalia, where it was recorded last in 1955. Only a few specimens and observations of Degodi lark (*Mirafra degodiensis*) and Sidamo bushlark (*Heteromirafra sidamoensis*) are known from different localities in Ethiopia. Information about breeding range and population size is even more scanty for Friedmann's lark (*Mirafra pulpa*) from northeastern Africa, Obbia lark (*Spizocorys obbiensis*) from Somalia, and Williams's lark (*Mirafra williamsi*) from Kenya, making it almost impossible to assess the extent of the threat to their extinction.

The population of the Agulhas long-billed lark (*Certhilauda brevirostris*) is estimated to be around 1,000 mature individuals. Estimates for Sclater's lark (*Spizocorys sclateri*) are much higher, but the population is severely fragmented throughout Namibia and South Africa. Finally, the population of the Latakoo lark in southeastern Africa could decline drastically during the following decades due to habitat loss. As a result, these three species are considered to be Near Threatened.

In the countries of the European Union (EU), the calandra lark, Dupont's lark, short-toed lark (*Calandrella brachydactyla*), Thekla lark (*Galerida theklae*), and wood lark are listed in Annex I of the EU Birds Directive as species which require special conservation efforts. All these species suffer habitat destruction related to agricultural intensification and afforestation, as well as illegal hunting and predation. The European populations of crested larks and skylarks declined during the last years, probably due to climatic change, the loss of suitable habitats, and the increasing use of pesticides. The crested lark became extinct in northern Russia, Norway, Sweden, and Switzerland in the 1990s; declined in Denmark, the Netherlands (3,000–4,000 breeding pairs in 1979 vs. 400 in 1991), France, Germany, and Poland. Skylark numbers in Great Britain declined by 51% between 1968 and 1995, a loss of approximately three million breeding birds. In 1997, only about one million pairs were breeding in Great Britain.

Significance to humans

In Central Europe, larks are regarded as harbingers of spring, and they were honored by the ancient Greeks as mediators between heaven and earth, and by the Celts as being beneficial for their harvest. Their melodious and long song, especially that of the skylark, is celebrated in verse by poets, and Shakespeare called the lark "the herald of the morn" in Romeo and Juliet. Skylarks were successfully naturalized in Australia in the 1850s, in New Zealand and on the Hawaiian Islands in the mid 1860s, as well as on Vancouver Island, British Columbia, in 1903 (although declining seriously in the 1990s). Skylarks were also released at other places in North America, but they were not able to build up stable populations and thus became extinct.

Even today, larks are trapped and shot legally in France and in the Mediterranean region. Spaepen calculated that five times as many skylarks are killed by humans in France than anywhere else in Europe, and the number of skylarks killed annually in southwest France is estimated at five to 10 million birds!

1. Female Australasian bushlark (*Mirafia javanica*); 2. Greater hoopoe-lark (*Alaemou alaudipes*); 3. Black-crowned sparrow-lark (*Eremopterix nigriceps*); 4. Thick-billed lark (*Ramphocoris clotbey*); 5. Long-billed lark (*Certhilauda curvirostris*); 6. Calaudra lark (*Melanocorypha calaudra*); 7. Female greater short-toed lark (*Calaudrella brachydactyla*); 8. Sky lark (*Alauda arvensis*); 9. Female wood lark (*Lullula arborea*); 10. Crested lark (*Galerida cristata*); 11. Horned lark (*Eremophila alpestris*). (Illustration by Emily Damstra)

Species accounts

Australasian lark
Mirafra javanica

TAXONOMY
Mirafra javanica Horsfield, 1821, Java.

OTHER COMMON NAMES
English: Singing bushlark; French: Alouette de Java; German: Horsfieldlerche; Spanish: Alondra de Australiana.

PHYSICAL CHARACTERISTICS
4.7–5.9 in (12–15 cm); 0.7 oz (20 g). Plumage inconspicuous other than rufous wing patch, sexes alike.

DISTRIBUTION
Myanmar, southern and central Thailand, Indochina, Philippine Islands, Kalimantan, Java, Bali, Lesser Sunda Islands, local in New Guinea, northwestern, northern, eastern, and southeastern Australia.

HABITAT
Open grassland with scattered bushes, salty marshes, and edges of cultivated fields.

BEHAVIOR
Perches on trees and wires. Song includes imitations of other birds; song-flight takes up to 40 minutes. Forms small groups outside breeding period. Migratory in southern Australia.

FEEDING ECOLOGY AND DIET
Feeds on seeds and arthropods.

REPRODUCTIVE BIOLOGY
Monogamous. Builds domed nests in shelter of tussocks; in southern Australia, lays two to four eggs November through January.

CONSERVATION STATUS
Not threatened.

SIGNIFICANCE TO HUMANS
None known. ◆

Cape long-billed lark
Certhilauda curvirostris

TAXONOMY
Alauda curvirostris Hermann, 1783, Cape of Good Hope.

OTHER COMMON NAMES
English: Long-billed lark; French: Alouette à long bec; German: Langschnabellerche; Spanish: Alondra Picuda.

PHYSICAL CHARACTERISTICS
Male 7.5–7.9 in (19–20 cm); female 6.3–6.7 in (16–17 cm); weight unknown. Males clearly larger than females, but sexes do not differ in respect to plumage color or pattern. Plumage brown-grayish, heavily streaked, breast marked with dark spots. Species reminiscent of a thrush with a long, decurved bill. Hind claw is long and straight.

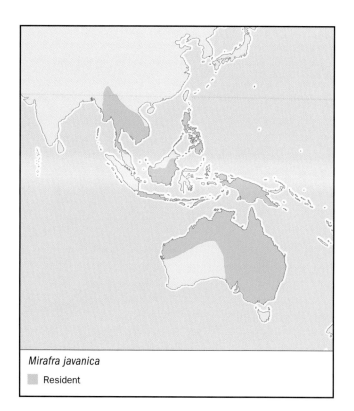

Mirafra javanica
▨ Resident

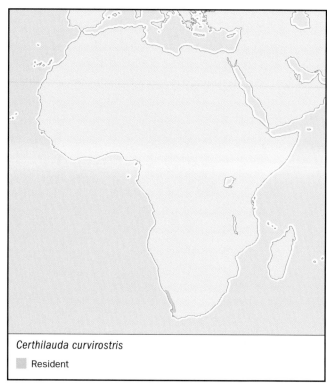

Certhilauda curvirostris
▨ Resident

DISTRIBUTION

Small strip from southern Namibia to southwestern South Africa, Cape Province.

HABITAT

Coastal dunes and arid fields, avoids more rocky areas.

BEHAVIOR

Solitary or in pairs. Male performs spectacular aerial song-display, characteristic for all *Certhilauda* species. Male first flies low over the ground, then rises nearly vertically several feet into the air and closes its wings before reaching the high point. After that it nose-dives and descends vertically, not opening the wings before coming close to the ground. Whistles while descending; same song is also uttered from prominent perches.

FEEDING ECOLOGY AND DIET

Feeds on termites, grasshoppers, beetles, and ants, as well as on seeds and berries. Uses long bill for digging. Since sexes differ in size of bill, they probably exploit different food resources.

REPRODUCTIVE BIOLOGY

Monogamous. Breeds September through December; nest often domed, clutch size ranges from two to three eggs; both parents feed the young.

CONSERVATION STATUS

Not threatened. Common in its limited southwestern African range. Range may have increased due to agriculture.

SIGNIFICANCE TO HUMANS

None known. ◆

Eremopterix nigriceps

▨ Resident

Black-crowned sparrow-lark

Eremopterix nigriceps

TAXONOMY

Pyrrhalauda nigriceps Gould, 1841, São Tiago, Cape Verde Islands.

OTHER COMMON NAMES

English: Black-crowned finch-lark, pallid finch-lark; French: Moinelette à frout-blanc; German: Weißstirnlerche; Spanish: Aloudra gottión de Corouna Negra.

PHYSICAL CHARACTERISTICS

4.5–4.9 in (11.5–12.5 cm); male 0.5–0.6 oz (14–16 g); female 0.4 oz (12 g); one of the smallest lark species. Sparrow-larks, like finches or sparrows, have a proportionally large head and a strong conical bill. Sexes are dimorphic with respect to plumage color. Female mainly pale brown and streaked, male with white forehead, cheeks, sides of neck, and collar of nape. Crown, stripe through eye to base of bill, and lower border of cheek black. Underparts entirely black, upperparts grayish brown.

DISTRIBUTION

Cape Verde Islands, southern Morocco, northern Mauritania to western Sahara, southeast Egypt. From North Africa, through most of the Arabian Peninsula, including Island of Sokrota, to Pakistan and northwestern India. Expanded its area to northern Nigeria, possibly due to desertification. Vagrant to Israel, Jordan, and Algeria.

HABITAT

Semidesert and sandy areas with sparse vegetation, also close to cultivated land.

BEHAVIOR

Territorial during breeding season, gregarious other times. In song-flight, male ascends at a 45° angle with legs dangling up to 66 ft (20 m), then flies in roughly circular paths in an undulating style to descend in stages. Song also performed from elevated perch.

FEEDING ECOLOGY AND DIET

Feeds more on seeds than arthropods; de-husks seeds; can drink salty or brackish water.

REPRODUCTIVE BIOLOGY

Monogamous. Female builds cup-shaped nest, male may also be involved in nest-building. Rim of nest frequently surrounded by small lumps of soil or stone. Breeding period irregular; two to three eggs laid, incubated by both sexes for 11–12 days. Both parents feed and brood. Young leave nest after eight days before being able to fly. Each parent cares for one single chick, so that only two young per brood can be reared (split brood-care).

CONSERVATION STATUS

Not threatened.

SIGNIFICANCE TO HUMANS

None known. ◆

Greater hoopoe-lark

Alaemon alaudipes

TAXONOMY

Upupa alaudipes Desfontaines, 1789, between Gafsa and Tozzer, Tunisia.

OTHER COMMON NAMES

English: Hoopoe lark, bifasciated lark; French: Sirli du désert; German: Wüstenläuferlerche; Spanish: Alondra Ibis.

PHYSICAL CHARACTERISTICS

7.1–7.9 in (18–20 cm); male 1.4–1.8 oz (39–51 g); female 1.1–1.6 oz (30–47 g); one of the largest lark species, females

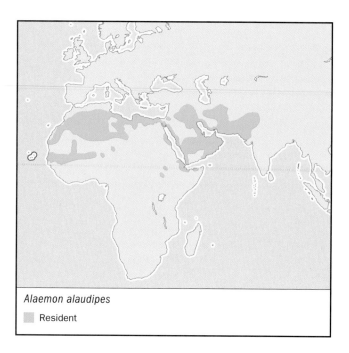

Alaemon alaudipes

Resident

Thick-billed lark

Ramphocoris clotbey

TAXONOMY
Melanocorypha clot-bey Bonaparte, 1850, Egyptian desert.

OTHER COMMON NAMES
English: Clotbey lark; French: Alouette de Clotbey; German: Knackerlerche; Spanish: Alondra de Pico Grueso.

PHYSICAL CHARACTERISTICS
6.7 in (17 cm); male 1.8–1.9 oz (52–55 g); female 1.6 oz (45 g). Strong lark with large head. Short but massive bill most conspicuous. Blunt, toothlike projection on middle of lower mandible fits into notch on upper mandible. Upperparts uniformly pink/gray-brown, chest to vent with large black spots, sides of head blackish, throat and eye-ring white; plumage of females of less contrasting color and not so heavily streaked.

DISTRIBUTION
Patchily distributed in North Africa, north of Sahara from northern Mauritania and Morocco to Algeria, Tunisia, and Libya. Further east, distribution uncertain, but reported as breeding bird from Egypt, Syria, Jordan, and Arabia.

HABITAT
Stony deserts with sparse vegetation, avoids sand dunes.

BEHAVIOR
Solitary or in small groups during breeding season, but large flocks observed in winter. Male rises from ground and starts to sing before descending in parachute style.

smaller than males. Bill very long, slender, and slightly decurved. Plumage sand-colored above, underside white, throat and breast with dark spots, similar in both sexes. Wings long and broad with conspicuous black-and-white pattern, comparable to that of African subspecies of hoopoe (*Upupa epops*). This lark was first described as a hoopoe, and its original scientific name means "hoopoe with legs of a lark."

DISTRIBUTION
Cape Verde Islands, North Africa from Mauritania to Egypt and Sudan, across the Middle East to northwestern India.

HABITAT
Semideserts and deserts.

BEHAVIOR
Tame and confiding, allowing humans to approach to within a few feet. Male defends territory against conspecific intruders and other birds by wing-spread behavior. Usually solitary or in pairs. Song ringing and far-carrying. For song-flight, male jumps up from perch and starts to sing, ascending almost vertically up to 33 ft (10 m) and performs somersaults, displaying its contrasting pattern on tail and wings. It then closes its wings, nose-dives back to perch, not opening its wings until landing. Song-flights can be repeated frequently within up to one hour.

FEEDING ECOLOGY AND DIET
Mainly arthropods and snails. Most food items obtained by digging with bill. Female bill is 30% shorter than bill of male, so sexes probably exploit different food resources. Two methods for cracking snail shells: snails dropped onto stone in flight or smashed against stones directly until shells break.

REPRODUCTIVE BIOLOGY
Monogamous. Only lark species building nests frequently on top of low shrubs, up to 24 in (60 cm) above ground. Also building cup-shaped nests on ground. Two to four eggs incubated by female for about 14 days. Both parents feed young, which leave nest after 12–13 days, before being able to fly. Young remain for at least one month with parents.

Ramphocoris clotbey

Resident

FEEDING ECOLOGY AND DIET
Feeds on seeds and green plant material as well as on insects. Uses bill to cut off plant material and to dig for food, and maybe for cracking solid cuticles of large beetles. However, even hard seeds are swallowed whole and not husked with bill, instead grit is taken with food to aid digestion.

REPRODUCTIVE BIOLOGY
Monogamous. During ground display, male presents to female pebbles that are used in nest building. Nest cup-shaped, frequently surrounded by small lumps of soil or small stones. Two to three eggs incubated by female March through May, both parents feed young.

CONSERVATION STATUS
Not threatened.

SIGNIFICANCE TO HUMANS
None known. ◆

Calandra lark
Melanocorypha calandra

TAXONOMY
Alauda calandra Linnaeus, 1766, Pyrenees.

OTHER COMMON NAMES
French: Alouette calandre; German: Kalanderlerche; Spanish: Calandria Común.

PHYSICAL CHARACTERISTICS
7.1–7.5 in (18–19 cm); male 2.0–2.6 oz (57–73 g); female 1.8–2.2 oz (50–65 g); larger and stronger than skylark. Bill conical and heavy. Upperparts brown and streaked, black patches on each side of upper breast characteristic. Sexes alike.

DISTRIBUTION
North Africa, southern Europe east to Ural steppes, from Asia Minor to central Asia, missing between Caspian Sea and Lake Aral.

Melanocorypha calandra

■ Resident

HABITAT
Open lowlands, steppe, grasslands, cultivated farmland, and meadows.

BEHAVIOR
Resident in southern Europe, Near East, and North Africa, migratory in Russia. Forms flocks of up to 2,500 individuals autumn and winter; frequently associated with other larks and corn bunting (*Miliaria calandra*). Male utters continuous song from ground or perch. Song-flight performed in circles, ascending in spirals. Several males often sing close together. Song similar to skylark, contains imitations of other birds.

FEEDING ECOLOGY AND DIET
Diet changes from insects during summer to seeds in winter. Bill used for digging.

REPRODUCTIVE BIOLOGY
Monogamous. Two broods April through June; both sexes build cup-shaped nest, clutch size ranges from three to six eggs, incubated by female, but brood patch also observed in several males. Young hatch after 16 days; fed by both parents. Leave nest after 10 days before being able to fly.

CONSERVATION STATUS
Not threatened, though population is declining in southern Europe due to agricultural intensification and possibly hunting; listed in Annex I of the European Birds Directive.

SIGNIFICANCE TO HUMANS
Hunted in the Mediterranean region. ◆

Greater short-toed lark
Calandrella brachydactyla

TAXONOMY
Alauda brachydactila Leisler, 1814, France and Italy = Montpellier, France.

OTHER COMMON NAMES
English: Short-toed lark; French: Alouette calandrelle; German: Kurzzehenlerche; Spanish: Terrera Grande.

PHYSICAL CHARACTERISTICS
5.1–5.5 in (13–14 cm); male 0.7–1.0 oz (21–28 g); female 0.6–0.9 oz (17–26 g). Small lark with dull, cryptic plumage, no streaks on chest. Bill short and finchlike. Sexes alike.

DISTRIBUTION
North Africa, southern Europe, eastward from Asia Minor through Mongolia to China.

HABITAT
Steppe with sparse vegetation, cultivated land, seashores, and saline areas. Avoids desert, moist areas, and vicinity of forests.

BEHAVIOR
Populations of Europe and Middle East migratory, wintering south to Sahel and Red Sea. Birds from central Asia winter in India. Highly gregarious outside breeding season. Song-flight performed by male in wide circles in sequence of deep and shallow undulations. Male ascends stepwise, uttering first phrase of melodious song with imitations of other birds, then stops singing and beating wings, drops down, ascends, and drops down again before next ascent, while singing starts again. Song-flight ends with descent or glide-in stages. Song sometimes performed from ground or perch.

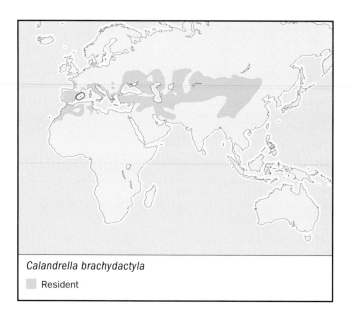

Calandrella brachydactyla

◼ Resident

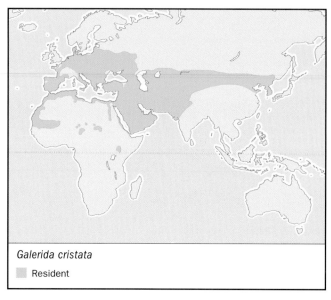

Galerida cristata

◼ Resident

FEEDING ECOLOGY AND DIET
Diet changes from insects and seeds in summer to nearly exclusively seeds during winter. Can go for months without drinking, even drinks brackish water.

REPRODUCTIVE BIOLOGY
Monogamous. Breeds April through June; cup-shaped nest often surrounded by pieces of mud, dung, and small stones. Female lays four to five eggs, incubates and broods alone, but both parents feed young.

CONSERVATION STATUS
Not threatened, though decreasing in France; listed in Annex I of the European Birds Directive.

SIGNIFICANCE TO HUMANS
None known. ◆

Crested lark
Galerida cristata

TAXONOMY
Alauda cristata Linnaeus, 1758, "in Europae viis" = Vienna.

OTHER COMMON NAMES
French: Cochevis huppé; German: Haubenlerche; Spanish: Cogujada Común.

PHYSICAL CHARACTERISTICS
6.7 in (17 cm); male 1.3–1.8 oz (37–52 g); female 1.3–1.7 oz (37–48 g); not as large as skylark, more robust, with stronger bill and longer crest. Uniformly dull-colored plumage, upperparts and breast heavily streaked, sexes alike. Very similar to Thekla lark, its sibling species.

DISTRIBUTION
From southern Europe as far as southern Scandinavia. North Africa south to Senegambia, eastward through Chad and Sudan to Ethiopia and Somalia. Does not occur in the Sahara. Breeds from Arabian Peninsula toward India, Nepal, Mongolia, China,

and Korea. In southern France and North Africa east of Libya, sympatric with Thekla lark.

HABITAT
Variety of habitats, mainly open areas with sparse vegetation, also cultivated land and other man-made semideserts such as railways, airfields, and wastelands. Where it co-occurs with the Thekla lark, the crested lark occupies the plains, the Thekla inhabits rocky and bushy slopes.

BEHAVIOR
Largely sedentary, not gregarious, low flight distance. Song-flight starts from perch, male ascends silently at angle up to 230 ft (70 m), then utters loud and melodious song ascending further up to 330 ft (100 m) and more, flying wide circles over territory. Song, uttered on ground or from perch, lasts four minutes on average, but up to 30 minutes have been observed. Known to perfectly imitate other bird songs and calls. One extraordinary example, the imitation of a shepherd's whistle reproduced so accurately that sheep dogs obey the signals as if the shepherd has given them.

FEEDING ECOLOGY AND DIET
Diet consists mainly of vegetal food; seeds and green plant material taken from ground or picked directly from plants; food remains in horse-droppings also exploited. Can husk seeds. Animal food taken to larger extent during breeding season, but proportion is negligible during rest of the year. Young fed with insects and worms.

REPRODUCTIVE BIOLOGY
Monogamous. Breeds April–June in Europe; female builds cup-shaped nest alone, lays three to five, rarely seven, eggs. Incubation by female alone, 11–17 days. Both parents feed young, which leave nest after eight to 11 days before being able to fly.

CONSERVATION STATUS
Not threatened.

SIGNIFICANCE TO HUMANS
None known. ◆

Wood lark
Lullula arborea

TAXONOMY
Alauda arborea Linnaeus, 1758, Europe-Sweden.

OTHER COMMON NAMES
French: Alouette lulu; German: Heidelerche; Spanish: Totovía.

PHYSICAL CHARACTERISTICS
5.9 in (15 cm); male 0.7–1.2 oz (21–35 g); female 1.1–1.2 oz (30–35 g); smaller and more slender than skylark and crested lark. Plumage buff brown, upperparts and chest streaked, distinguished from other larks by broad, white supercilium which continues to nape. Black-and-white pattern of alula (first digit of wing) feathers very conspicuous. Crown feathers can be raised to a small crest.

DISTRIBUTION
Northern West Africa, Europe from the Mediterranean to southern Scandinavia, Asia Minor east to Iran and Turkmeniya.

HABITAT
Requires habitat with short grass for feeding, higher vegetation for nesting, exposed trees or bushes as song-perches.

BEHAVIOR
Not gregarious, even during migration. Northern populations migratory, wintering in southern Europe and North Africa; southern populations mostly resident. Arboreal, walks on branches and perches on treetops, bushes, and wires. For song-flight, males take off from perch in tree top, ascend at an angle, spiral upward and fly circles on nearly same level 165–330 ft (50–100 m) above ground, singing all the time (unpaired males fly higher than paired ones). Descent either in stages while still singing or sometimes silently with wings closed. Song-flight takes two minutes on average, but unpaired males have been watched singing for 70 and 94 minutes. Males also sing from perch or ground, and frequently on moonlit nights. Song composed of pleasing and soft phrases, hesitant at the beginning, then increasingly forceful and louder. High site-fidelity of males has been shown by ringing experiments.

FEEDING ECOLOGY AND DIET
Feeds on small insects and spiders during breeding season, otherwise mainly granivorous.

REPRODUCTIVE BIOLOGY
Monogamous. Breeds March through June; cup-shaped nest built by female. Two, sometimes three broods annually; clutch size normally three to five eggs, female incubates for 11–15 days. Young fed by both parents, leave nest after eight days before being able to fly. If female starts incubation of second brood, male cares for young of first clutch alone.

CONSERVATION STATUS
Not threatened, though declining populations in Europe caused by habitat loss; listed in Annex I of the European Birds Directive.

SIGNIFICANCE TO HUMANS
None known. ◆

Sky lark
Alauda arvensis

TAXONOMY
Alauda arvensis Linnaeus, 1758, Europe.

OTHER COMMON NAMES
French: Alouette des champs; German: Feldlerche; Spanish: Alondra Común.

PHYSICAL CHARACTERISTICS
7.1–7.5 in (18–19 cm); male 0.9–1.9 oz (27–55 g); female 0.6–1.7 oz (17–47 g). Extensively streaked brown plumage, crown feathers can be raised to a short crest; bill stronger than bill of wood lark; sexes alike. Most common lark species within its western Palearctic part of distribution.

DISTRIBUTION
North Africa, Europe, and Asia; introduced in Australia, Canada (Vancouver Island, British Columbia), Hawaii, and New Zealand.

HABITAT
Dense grasslands, cultivated farmland, airfields and sports grounds.

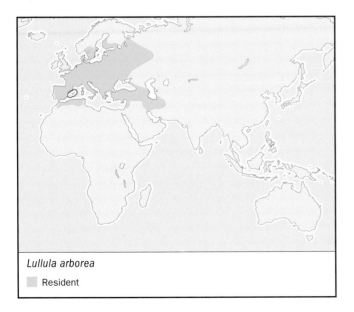

Lullula arborea
　Resident

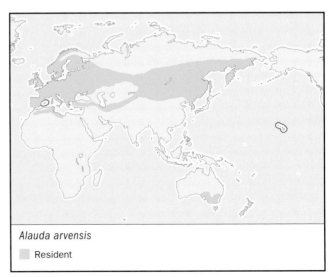

Alauda arvensis
　Resident

BEHAVIOR

Perches only rarely on wires and trees. High-level song-flight in breeding season. Male rises with rapid wing-beats, after prolonged hovering falls parachute style and drops to ground suddenly, singing all the time loudly and forcefully. Populations of the western Palearctic are migratory, hibernate in the Mediterranean region; breeding birds of the British Isles are sedentary. Eastern Asiatic populations migrate to southeastern China; populations of central Asia hibernate in northern India, Afghanistan, and Iran; and eastern Palearctic populations migrate to Turkey, Syria, and Jordan.

FEEDING ECOLOGY AND DIET

Mainly arthropods during summer, but seeds in winter.

REPRODUCTIVE BIOLOGY

Monogamous. Breeds April–July; two, rarely three, broods. Cup-shaped nest in depression on ground. Three to five, exceptionally up to seven, eggs. Female incubates for 12–14 days, chicks leave nest when eight to 10 days old. Both parents care for young.

CONSERVATION STATUS

Decreasing populations in Europe, but not considered threatened.

SIGNIFICANCE TO HUMANS

Still hunted legally in France, Greece, and Italy. ◆

Horned lark

Eremophila alpestris

TAXONOMY

Alauda alpestris Linnaeus, 1758, North America = coast of North Carolina.

OTHER COMMON NAMES

English: Shore lark; French: Alouette hausse-col; German: Ohrenlerche; Spanish: Alauda Cornuda.

PHYSICAL CHARACTERISTICS

5.9–6.7 in (15–17 cm); male 1.1–1.7 oz (30–48 g); female 0.9–1.5 oz (26–42 g). Named after elongated black feathers at sides of crown reminiscent of horns; head and breast contrastingly colored, somewhat brighter in males than in females.

DISTRIBUTION

Holarctic; widespread in North America, northern and southeastern Europe, northern and central Asia. Separated populations in Morocco, Lebanon, northern Israel, and South America (Colombia).

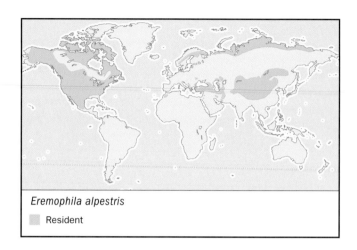

Eremophila alpestris

Resident

HABITAT

Steppes, semideserts, arctic tundra; prefers open, bare, and stony areas.

BEHAVIOR

Perches on stones, fence posts, or shrubs. Song uttered while perching and during song-flight up to 300 ft (91 m) above ground. Populations of arctic tundra are migratory, hibernate between 55° and 35° N. Present year round in most areas in North America, from southern Canada southward; some may be permanent residents. Migratory in far north, one of the earliest spring migrants. Horned larks breeding in Alaska and Canada winter south as far as the Gulf of Mexico; populations from the tundras of Northern Europe migrate annually to wintering areas around the North Sea.

FEEDING ECOLOGY AND DIET

Feeds mainly on insects during breeding season, but takes more seeds during winter months.

REPRODUCTIVE BIOLOGY

Monogamous. Nest cup-shaped, frequently surrounded with pebbles. Breeds March through July; one to two broods; three to five eggs, incubation takes 13 days; chicks leave nest after eight to 10 days. Both parents care for young.

CONSERVATION STATUS

Not threatened.

SIGNIFICANCE TO HUMANS

None known. ◆

Resources

Books

Burton, John F. *Birds and Climate Change*. London: Christopher Helm, 1995.

Cramp, Stanley, ed. *Handbook of the Birds of Europe, The Middle East and North Africa: The Birds of the Western Palearctic*. Vol. 5. Oxford and New York: Oxford University Press, 1988.

Glutz von Blotzheim, Urs N. (Hrsg.). *Handbuch der Vögel Mitteleuropas*. Bd. 10 Passeriformes (1. Teil). Wiesbaden: Aula-Verlag, 1985.

Keith, S., E. K. Urban, and C. H. Fry, eds. *The Birds of Africa*. Vol. 4. London: Academic Press, 1992.

Pätzold, Rudolf. *Die Lerchen der Welt*. Magdeburg: Westarp-Wissenschaften, 1994.

Sibley, Charles G., and Burt L. Monroe. *Distribution and Taxonomy of Birds of the World*. New Haven: Yale University Press, 1990.

Resources

Periodicals

Alström, Per. "Taxonomy of the *Mirafra assamica* Complex." *Forktail* 13 (1998): 97–107.

Baicich, Paul J., Steven C. Heinl, and Mike Toochin. "First Documented Breeding of the Eurasian Skylark in Alaska." *Western Birds* 27 (1996): 86–88.

Browne, Stephen, Juliet Vickery, and Dan Chamberlain. "Densities and Population Estimates of Breeding Skylarks *Alauda arvensis* in Britain in 1997." *Bird Study* 47 (2000): 52–65.

Ericson, Per G., Ulf S. Johansson, and Thomas J. Parsons. "Major Divisions in Oscines Revealed by Insertions in the Nuclear Gene *c-myc*: A Novel Gene in Avian Phylogenetics." *The Auk* 117 (2000): 1069–1078.

Lloyd, Penn. "Rainfall as a Breeding Stimulus and Clutch Size Determinant in South African Arid-Zone Birds." *Ibis* 141 (1999): 637–643.

Ratcliffe, Norman, Luis R. Monteiro, and Cornelis J. Hazevoet. "Status of Raso Lark *Alauda razae* with Notes on Threats and Foraging Behaviour." *Bird Conservation International* 9 (1999): 43–46.

Riley, Steve. "Crested Lark Using 'Anvil'." *British Birds* 82 (1989): 30–31.

Ryan, Peter G., Ian Hood, Paulette Bloomer, Joris Komen, and Timothy M. Crowe. "Barlow's Lark: A New Species in the Karoo Lark *Certhilauda albescens* Complex of Southwest Africa." *Ibis* 140 (1998): 605–619.

Ryan, Peter G., and Paulette Bloomer. "The Long-Billed Lark Complex: A Species Mosaic in Southwestern Africa." *Auk* 116 (1999): 194–208.

Spaepen, J. F. "A Study of the Migration of the Skylark *Alauda arvensis*, Based on European Ringing Data." *Gerfault* 85 (1995): 63–89.

Williams, Joseph B. "Lizard Burrows Provide Thermal Refugia for Larks in the Arabian Desert." *Condor* 101 (1999): 714–717.

Wilson, J. D., J. Evans, S. J. Browne, and J. R. King. "Territory Distribution and Breeding Success of Skylarks *Alauda arvensis* on Organic and Intensive Farmland in Southern England." *Journal of Applied Ecology* 34 (1997): 1–20.

Other

The European Commission. "Discover the Species of the Birds Directive." 1 January 2002. 13 March 2002 <http://europa.eu.int/comm/environment/nature/directive/birdshome_en.htm>

The IUCN Species Survival Commission. "2000 Red List of Threatened Species." 1 January 2002. 13 March 2002 <http://www.redlist.org/>

Albrecht Manegold
Walter Sudhaus, PhD

Swallows
(Hirundinidae)

Class Aves
Order Passeriformes
Suborder Passeri (Oscines)
Family Hirundinidae

Thumbnail description
Small to medium-sized, slender, streamlined birds with pointed wings, a small bill, and perching feet; excellent fliers that catch their insect prey on the wing

Size
Body length ranges from about 3.9 to 9.4 in (10–24 cm); weight 0.4–2.1 oz (10–60 g)

Number of genera, species
15 genera; about 88 species

Habitat
Usually found in open habitats near water, along forested rivers, in wooded savanna, and near cliffs and caves close to open areas with an abundant supply of insects

Conservation status
Critically Endangered: 1 species; Vulnerable: 4 species

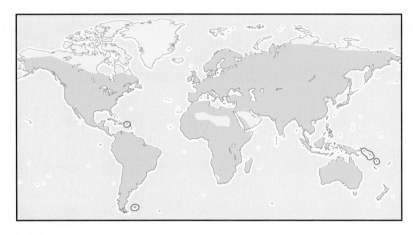

Distribution
Almost worldwide, with the exception of the high Arctic, Antarctica, and some remote oceanic islands.

Evolution and systematics

The evolutionary progenitor that gave rise to the modern swallow family probably separated from other primitive insectivorous birds in the Eocene (i.e., in the Lower Tertiary about 50 million years ago). In addition to many relatively recent adaptations related to their mode of aerial hunting, the swallows show some ancient evolutionary characters, such as the presence of bronchial rings, and so they are placed fairly low in the phylogenetic tree of the passerine birds (Passeriformes). Ornithologists consider the Hirundinidae to be closely related to larks (Alaudidae), pipits and wagtails (Motacilladae), and cuckoo-shrikes (Campephagidae).

The family Hirundinidae is separated into two major groups: the river-martins (subfamily Pseudochelidoninae) and the true swallows and martins (subfamily Hirundininae). The river-martins consist of only two species, while the true swallows include 14 genera and about 86 species. The river-martins have relatively robust legs and feet and a stouter bill than the other hirundines. They are thought to have diverged from the main hirundine line early in the evolution of the group.

During the past century, the classification of swallows and martins into genera has been modified several times. Initially, all the swallows (and also the swifts, now in the separate family, Apodidae) were placed into a single genus, *Hirundo*. With additional study, however, the swifts were recognized as being unrelated to the swallows and were assigned to their own family. More detailed research then resulted in the description of additional genera of swallows. Presently, 15 genera are

recognized, with separation based on aspects of their morphology, biochemistry, and ecological and behavioral characters. Some genera form subgroups of closely related species.

Modern ornithologists do not consider the swallows (Hirundinidae) and swifts (Apodidae) to be closely related—their morphological and behavioral similarities are due to convergent evolution. Such convergence is manifest when, in otherwise not closely related organisms, similar body forms and behavioral traits develop as a result of selective pressures associated with comparable ecological niches. Their similar adaptations account for the fact that swallows and swifts are often confused and were once regarded as close relatives. Since the end of the nineteenth century, however, they have been recognized as belonging to different orders of birds—the swifts (Apodiformes) are not even perching birds.

Physical characteristics

Species in the swallow family are delicately built birds that can fly swiftly with great maneuverability and endurance. Their body length ranges from about 3.9–9.4 in (10–24 cm) and the weight from about 0.4–2.1 oz (10–60 g). The wings are long, pointed, and have nine primary feathers. The tail has 12 feathers and may be deeply forked, somewhat indented, or square-ended. The term "martin" often refers to species with a slightly indented or squared tail. The body plumage is short and pressed close to the skin. The most usual coloration is earth-brown or dark-and-white, often with

Male and female tree swallow (*Tachycineta bicolor*). (Photo by Windland Rice. Bruce Coleman Inc. Reproduced by permission.)

an attractive green or purple iridescence; rust-brown or rust-red markings are also frequent. The sexes do not differ in external appearance in most species. The flight muscles are strong, and the legs are short and permit only a clumsy waddling gait. The feet are typical of the perching birds. Swallows are primarily insect hunters, catching their prey almost exclusively in flight. To assist in this mode of feeding their small beak has a broad gape, with a wide base that runs almost as far back as the eye.

Distribution

Species of swallows occur almost worldwide, with the exception of the high Arctic tundra, Antarctica, and some isolated oceanic islands. Almost 30 species of swallows breed in Africa, and others migrate to winter there after breeding in Europe or Asia. Australia is home to four endemic species, while North America has nine species in six genera, and Central and South America have about 20 species. On a local scale, however, the distribution of swallows is usually rather patchy, depending on the presence of key habitat elements, particularly nesting sites.

Habitat

Major habitats for swallows and martins include forested ecosystems close to lakes, rivers, streams, and wetlands, as well as wooded savanna and prairies. Their habitats are normally open areas with an abundant supply of flying insects as prey. The proximity of appropriate nest-site habitat is also critical. Some swallow species utilize tree cavities for nesting, while others require mud or sand banks in which nesting tunnels can be dug, and yet others demand access to soft mud as a material used to construct clinging nests on cliffs or other vertical structures. Swallows and martins have fewer habitat restrictions during the non-breeding season, and they may migrate or wander extensively at that time.

Specific habitat requirements differ greatly among genera in the swallow family. Species in the genus *Hirundo*, for example, have specialized habitat needs ranging from desert to tropical rainforest, and from highlands to seacoasts. Some swallow species have greatly expanded their range of habitat utilization by taking advantage of human-built structures for nesting. For instance, barn swallows (*Hirundo rustica*) commonly nest on or inside of farm buildings and cottages, whereas they were previously restricted to using caves and cliffs in coastal areas. This adaptive change to utilizing anthropogenic habitats has allowed barn swallows to expand their range and overall abundance. Similarly, cliff swallows (*H. pyrrhonota*) often build their nests in protected locations on bridges and buildings. Most species of hirundines, however, continue to use only natural nesting sites.

Behavior

Swallows are excellent fliers, and are capable of executing impressive aerial maneuvers when hunting or courting. Males typically choose and defend a nest site and then attract a female and guard their territory using song and inspiring flight patterns. The size of the defended territory may be quite small in colonial-nesting species, being restricted to the nest-site itself and a small surrounding area. It may, however, be considerably larger for species that do not nest in groups. Mated pairs of non-migratory species often stay in the vicinity of their breeding area all year, although they defend the nest site most vigorously during the breeding season. Migratory species often return to the same breeding area each year, and may even utilize the exact same nest site if they have been successful there previously. First-year breeders typically return to a breeding area close to where they were born and raised.

Behavioral displays have not been well studied for many species of swallows and martins, but they appear to be important in attracting a mate and defending the nest-site and breeding territory. Aggressive displays can include ruffling of the head and neck feathers, quivering of the wings, bill snap-

Barn swallow (*Hirundo rustica*) young at nest in Massachusetts. (Photo by Dwight Kuhn. Bruce Coleman Inc. Reproduced by permission.)

American cliff swallow (*Hirundo pyrrhonota*) nests at Chaco Canyon National Cultural Park in New Mexico. (Photo by Brenda Tharp. Photo Researchers, Inc. Reproduced by permission.)

ping, and lunging at an opponent. Swallows may also gape aggressively with their bill open and neck extended when other birds approach their nest site. Displays associated with pairing and copulation often include the male flying skillfully, quivering of the wings, and spreading of the tail. Females may also quiver their wings prior to mating. Plumage appears to play a key role in such displays—during singing or threat displays males often show prominent colored patches on their forehead and throat. The tail of some swallows, when spread, shows additional markings.

Swallows exhibit an array of calls, which are used when excited or agitated, to maintain contact with others of the same species, during courtship, or as an alarm indicating the presence of a predator. In addition, begging calls are used by the young when soliciting food from their parents. The typical song of swallows is a simple, sometimes musical twittering.

Feeding ecology and diet

Swallows feed almost exclusively on insects that they catch in flight. The tree swallow (*Tachycineta bicolor*) is the only species that eats a substantial amount of plant food, including berries and other small fruits. Specific information is lacking on the diet of many swallow species, and the size of insects consumed varies greatly. The largest species are able to catch and eat dragonflies and butterflies, but the majority of swallows consume medium-sized insects such as flies, beetles, and ants. They tend to avoid stinging insects such as bees and

wasps. Swallows also will sometimes eat spiders. Their diet typically changes during the year depending on which insect species are available.

Swallows usually feed in areas where insects are plentiful and often catch their prey when flying in open places. Consequently, swallows are often seen flying above areas of open habitat, such as over water or near a forest edge or above the canopy. During the breeding season, they typically feed rather close to the nesting site. During inclement weather, however, they may need to fly more than a kilometer to find suitable food. Swallows may feed in large groups or flocks, or they may hunt individually.

Reproductive biology

Hirundines form monogamous breeding pairs, but males are often promiscuous and will attempt to copulate with females other than their mate. This typically happens at locations where the birds congregate in flocks. In some species, the males closely guard their mate, although this is not always the case, as in the colonial cliff swallow in which guarding the nest site is more important than defending the mate.

Nest construction is highly variable among species of swallows and martins. Their nests are usually built of mud used to cement grasses and twigs, and are typically located in a tree cavity, in a tunnel dug in an earthen or sandy bank, on a rocky cliff, or on a wall or in an accessible building. Hole nesters may use pre-existing cavities such as a hollow

Tree swallows (*Tachycineta bicolor*) gathered in South Carolina. (Photo by J. Trott/VIREO. Reproduced by permission.)

tree, an abandoned woodpecker hole, or a crevice in a cliff or rocky slope. Other species dig a 3.3–6.6 ft (1–2 m) long burrow in the side of a riverbank or sandy slope. Nests made of mud are typically attached to a protected portion of a cliff, cave, or tree, or they may be attached to an artificial structure such as a bridge or a building wall beneath an overhang. The mud nest may be an open bowl, such as the barn swallow constructs, or an enclosed mud nest, such as the cliff swallow creates. Mud nests may be renovated and used several years in a row.

Hirundines that nest in cavities typically have white eggs, while those with open nests have eggs that are spotted and/or lightly colored. Clutch size is commonly four or five eggs, but can range from about two to eight. Clutches are typically smaller in the tropics at two or three eggs, and larger in temperate areas, as many as seven or eight eggs. The average clutch size tends to decrease for second and later clutches in re-nests during the same breeding season. Eggs are laid a day apart and the incubation period ranges from 11 to 20 days, with 14–16 days being typical. Incubation is by the female only in some species, and by both parents in others.

Eggs within a clutch hatch asynchronously over several days. Because the nestlings initially have little down, they need to be brooded by one of the parents until they become better feathered. The parents alternate the brooding chore. The parents feed the young as frequently as possible, often several times an hour. After the nestlings become sufficiently

feathered to regulate their own body temperature, both parents are able to search for food at the same time. Nestling survivorship is strongly tied to weather conditions, which affects the availability of insect prey. Many nestlings may die of starvation during prolonged bad weather. Ectoparasites and predation can also be important causes of nestling mortality. The nestling period is approximately three weeks, ranging from 17 to 30 days. After they fledge, juveniles will still be fed by their parents for several weeks, until they become self-sufficient.

Conservation status

Five species of hirundines are at risk. The white-eyed river martin (*Eurychelidon sirintarae*) of Thailand is listed as Critically Endangered, although it has not been seen for more than 20 years and may already be Extinct. Four other species are listed as Vulnerable. They are the blue swallow (*Hirundo atrocaerulea*) of tropical and southern Africa, the white-tailed swallow (*H. megaensis*) of Ethiopia, the Bahama swallow (*Tachycineta cyaneoviridis*) of the Bahamas, Cuba, and the extreme southeastern United States, and the golden swallow (*T. euchrysea*) of the Dominican Republic, Haiti, and Jamaica. Four additional species are listed as Data Deficient, meaning they are likely at risk but additional research is required to conclusively demonstrate this. These additional species are the Red Sea cliff-swallow (*Hirundo perdita*) of Sudan, Brazza's swallow (*Phedina brazzae*) of Angola, Congo, and the Democratic Republic of Congo, the Sinaloa martin (*Progne sinaloae*) of Guatemala and Mexico, and the African river-martin (*Pseudochelidon eurystomina*) of Congo, the Democratic Republic of Congo, and Gabon.

However, current research and monitoring of the distribution and abundance of rare swallow species is incomplete. Local climate changes can cause local population declines, and competition for nest sites from abundant non-native cavity nesters, such as house sparrows (*Passer domesticus*) and starlings (*Sturnus vulgaris*), has caused local declines in swallow and martin populations. Pollution and insecticides are also suspected to have caused declines in swallow populations due to reductions of insect populations, contamination of mud used to build nests, and direct poisoning of birds. Habitat destruction also is a concern for all threatened swallows.

Significance to humans

Typically, there is little direct interaction between humans and swallows. Sometimes, their nests on buildings are viewed as a nuisance and are removed by people. Only the African river-martin is hunted as a minor source of food. Humans may, however, have large indirect influences on swallow populations through the destruction of their habitat and the creation of toxic pollution. On the other hand, some species of swallows benefit from the presence of bridges, sheltered places beneath building overhangs, and other built structures that are used as nesting habitat. Some species are also aided by the provision of nesting boxes for their use. There are also indirect economic benefits of ecotourism and bird-watching focused on seeing swallows and other native birds in natural habitats. Species of swallows that live in residential and urbanized areas are often greatly appreciated by local people, and may even be considered harbingers of good fortune.

1. American cliff swallow (*Hirundo pyrrhonota*); 2. House swallow (*Hirundo tahitica*); 3. Square-tailed saw-wing (*Psalidoprocne nitens*); 4. Crag martin (*Ptyonoprogne rupestris*): 5. Mascarene martin (*Phedina borbonica*); 6. Sand martin (*Riparia riparia*); 7. Barn swallow (*Hirundo rustica*); 8. Purple martin (*Progne subis*); 9. African river martin (*Pseudochelidon eurystomina*); 10. House martin (*Delichon urbica*). (Illustration by Brian Cressman)

Species accounts

African river-martin
Pseudochelidon eurystomina

SUBFAMILY
Pseudochelidoninae

TAXONOMY
Pseudochelidon eurystomina Hartlaub, 1861, Gabon.

OTHER COMMON NAMES
English: Congo swallow; French: Hirondelle de rivière; German: Trugschwalbe; Spanish: Avión de Río Africano.

PHYSICAL CHARACTERISTICS
5.5 in (14 cm). The plumage is glossy black overall and the beak, feet, and eyes are colored red.

DISTRIBUTION
Breeds in coastal Gabon and inland along the Congo River and other large tropical rivers of central western Africa. After breeding it migrates far up the Congo River and other major watercourses in the region. It occurs in Congo, the Democratic Republic of Congo, and Gabon.

HABITAT
Breeds in sandy places along forested tropical rivers and in coastal tropical savanna. After breeding it migrates inland along large forested rivers, where it often roosts on sandy bars and embankments.

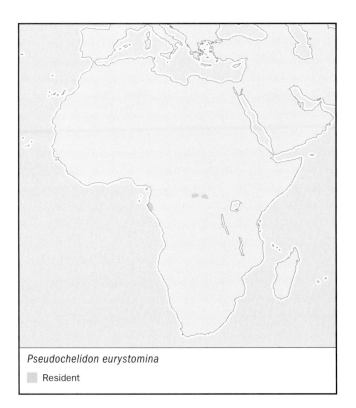

Pseudochelidon eurystomina
 Resident

BEHAVIOR
The species is highly gregarious when breeding and also during migration. It defends a local territory around the nesting site, and attracts a mate by aerial displays and singing.

FEEDING ECOLOGY AND DIET
Feeds on flying insects caught on the wing.

REPRODUCTIVE BIOLOGY
Monogamous. Populations breeding on large rivers breed during the drier season, when sandy riverbanks are exposed. Populations breeding in coastal savanna breed from about September to November. Colonial nesting sites are in sandy embankments, dunes, and islands of large tropical rivers and coastal savanna. It digs a slanting tunnel of 6.6–10 ft (2–3 m) into soft ground for its nesting site. The clutch size is two to four eggs. Both parents share in the incubation and care of the young.

CONSERVATION STATUS
A rare, little-known, but locally abundant species that is listed as Data Deficient.

SIGNIFICANCE TO HUMANS
African river-martins have been recently hunted in significant numbers by local people as a source of food. Otherwise, they are of no direct importance to humans, except for indirect economic benefits of ecotourism involving rare birds. ◆

Barn swallow
Hirundo rustica

SUBFAMILY
Hirundinae

TAXONOMY
Hirundo rustica Linnaeus, 1758. Six subspecies are recognized.

OTHER COMMON NAMES
English: Chimney swallow, European swallow, house swallow, swallow; French: Hirondelle de cheminée; German: Rauchschwalbe; Spanish: Golondrina Bermeja.

PHYSICAL CHARACTERISTICS
7.5 in (19 cm); 0.6 oz (17 g). The back is glossy blue-black, the throat and belly rusty brown, and there is a brighter rusty-red patch on the forehead. The tail is deeply forked. The six geographic subspecies vary somewhat in coloration.

DISTRIBUTION
The most widespread species of swallow. It breeds in northern regions of Eurasia, North America, and northern North Africa. It migrates to winter in more southern regions of its range, including northern and central South America, central and southern Africa, and South and Southeast Asia.

HABITAT
Forages in open areas, often close to water. Nests primarily on built structures, as well as natural cliffs and caves.

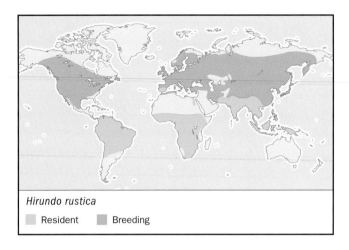

Hirundo rustica

Resident   Breeding

BEHAVIOR
Begins to migrates south at the end of August, but particularly in the first half of September. Before they depart and while resting during migration, they may aggregate into large flocks, often with other species of swallows. Late migrants may persist until late October or early November. Has a loud, twittering song and is perhaps the most talented singer in the family. Its contact call, used to attract and connect with others, is a high-pitched, loud, repeated weet.

FEEDING ECOLOGY AND DIET
Feeds on insects caught in flight.

REPRODUCTIVE BIOLOGY
Monogamous. Builds a cup-shaped nest of mud and some plant fibers, often attached to a building beneath a shading over-hang, and sometimes inside if there is easy access. Its natural nest sites are cliffs, cave walls, and clefts in the ground. It usually takes about eight days to build a new nest, but in unfavorable weather this may be as long as four weeks. The nest is lined mostly with feathers. A nest may be used for a decade or longer, being refurbished each year. The clutch is typically three to six eggs. However, clutches laid later in the season have fewer eggs. Generally only the female incubates, although the male may also participate, especially in the American sub-species. The young hatch asynchronously after 11–18 days of incubation and are fledged after 15–23 days. Generally breeds once or twice per season, rarely three times.

CONSERVATION STATUS
Not threatened. A widespread and abundant species.

SIGNIFICANCE TO HUMANS
The barn swallow is a popular bird for many people. In parts of Europe it is considered an omen of good luck and a harbinger of spring. It has lived for millennia in close association with humans, and has likely benefited from this relationship and become more abundant than in former times. This bird is welcomed by farmers because it eats many insect pests that affect livestock and crops. ◆

American cliff swallow
Hirundo pyrrhonota

SUBFAMILY
Hirundinae

TAXONOMY
Hirundo pyrrhonota Vieillot, 1817.

Hirundo pyrrhonota

Resident   Breeding

OTHER COMMON NAMES
English: Cliff swallow; French: Hirondelle à front blanc; German: Fahlstirnschwalbe; Spanish: Golondrina de las Rocas.

PHYSICAL CHARACTERISTICS
5.1 in (13 cm); 0.8 oz (22.7 g). It has a bluish brown back, tail, and wings, a reddish rump, blue on the crown of the head, a rusty-brown chin and sides of head, and a white forehead.

DISTRIBUTION
Breeds locally but widely in North America and migrates to winter in Central America and northern South America.

HABITAT
Occurs in open areas near suitable breeding sites, often near water.

BEHAVIOR
A migratory species that spends the non-breeding season in southern parts of its range. During migration it gathers in large flocks. The song is a simple, melodious note.

FEEDING ECOLOGY AND DIET
Feeds on flying insects caught on the wing.

REPRODUCTIVE BIOLOGY
Monogamous. Builds a bulb-shaped nest of clay beneath a sheltering projection on a cliff. They may also nest beneath overhanging eaves of a building, within the structure of a bridge, or on protected places on a dam. A social species that nests in colonies of various size. The clutch size is usually three to four eggs. The eggs are incubated by the female, but both parents feed the young.

CONSERVATION STATUS
Not threatened. A widespread and locally abundant species.

SIGNIFICANCE TO HUMANS
A familiar and popular bird to many people. ◆

House swallow
Hirundo tahitica

SUBFAMILY
Hirundinae

TAXONOMY
Hirundo tahitica Gmelin, 1789.

OTHER COMMON NAMES
English: Hill swallow, Pacific sea swallow, Pacific swallow, welcome swallow; French: Hirondelle de Tahiti; German: Südseeschwalbe; Spanish: Golondrina Pacífica.

PHYSICAL CHARACTERISTICS
5.1 in (13 cm). The back, wings, and tail are colored glossy purple-black, with a reddish face and chin and a brown-streaked belly. The tail is deeply forked.

DISTRIBUTION
Southern India, Southeast Asia, New Guinea, and many Pacific islands.

HABITAT
Occurs in open tropical habitats, usually in the vicinity of coastal water.

BEHAVIOR
A non-migratory species that uses song and aerial display to defend a breeding site and attract a mate. The song is a loud twittering.

FEEDING ECOLOGY AND DIET
Feeds on insects that are caught in flight.

REPRODUCTIVE BIOLOGY
Monogamous. Builds a cup-shaped nest of mud and some plant fibers that is attached to a cliff or building. The clutch size ranges from one to three eggs. The eggs are incubated by the female, but both parents feed the young.

CONSERVATION STATUS
Not threatened. A widespread and locally abundant species.

SIGNIFICANCE TO HUMANS
Not of much importance to humans, other than the indirect economic benefits of ecotourism focused on birding. ◆

House martin
Delichon urbica

SUBFAMILY
Hirundinae

TAXONOMY
Hirundo urbica Linnaeus, 1758. Six geographic subspecies are recognized.

OTHER COMMON NAMES
English: Common house-martin, northern house-martin; French: Hirondelle de fenêtre; German: Mehlschwalbe; Spanish: Avión Común.

PHYSICAL CHARACTERISTICS
5.5 in (14 cm). It is colored glossy purple-brownish on the back and white below, with a white rump and a forked tail.

DISTRIBUTION
Breeds widely in Eurasia, from Britain through to Japan and the Koreas, and including part of northern North Africa. It

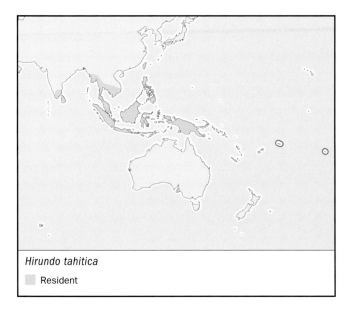

Hirundo tahitica

  Resident

Delichon urbica

  Resident   Breeding

migrates to spend the non-breeding season in central and southern Africa, South Asia, and southeastern China.

HABITAT
Occurs in open areas, usually close to water, and also in towns and residential areas.

BEHAVIOR
It migrates to spend the non-breeding season in southern parts of its range. Attracts a mate and defends a nest site by song and aerial displays.

FEEDING ECOLOGY AND DIET
Feeds on flying insects, which are caught on the wing.

REPRODUCTIVE BIOLOGY
Monogamous. The nest is a half-cup of mud with an oval entrance at the top and is built against a vertical wall under an overhanging roof. It is usually placed on a building in an open area, but the species also breeds on natural cliff faces. Both sexes incubate the four to five eggs and there are often two broods per season.

CONSERVATION STATUS
Not threatened. A widespread and abundant species.

SIGNIFICANCE TO HUMANS
The house martin is a popular bird for many people. It lives in close association with humans and has likely benefited from this relationship and become more abundant than in former times. ◆

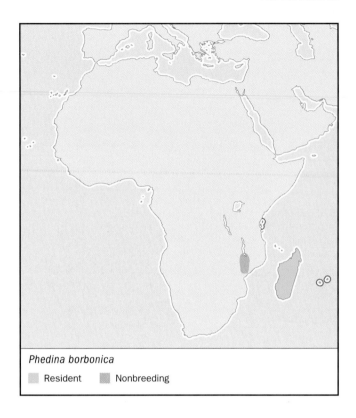

Phedina borbonica
▨ Resident ▨ Nonbreeding

Mascarene martin
Phedina borbonica

SUBFAMILY
Hirundinae

TAXONOMY
Hirundo borbonica Gmelin, 1860.

OTHER COMMON NAMES
English: Mascarene swallow; French: Hirondelle des Mascareignes; German: Maskarenenschwalbe; Spanish: Golondrina de Mascarene.

PHYSICAL CHARACTERISTICS
Colored brownish overall, with a whitish rump and white throat. Has long pointed wings and a slightly forked tail.

DISTRIBUTION
An endemic species that only breeds on the islands of Madagascar, Mauritius, and Reunion, all in the western Indian Ocean off eastern Africa. Migrants from Madagascar may occur in southern and eastern Africa during the non-breeding season.

HABITAT
Occurs in open areas, usually near water.

BEHAVIOR
It is generally a resident species, but birds from Madagascar may wander during the non-breeding season and can occur in continental Africa. Attracts a mate and defends a nest site by song and aerial displays.

FEEDING ECOLOGY AND DIET
Feeds on insects, which are caught in flight.

REPRODUCTIVE BIOLOGY
Monogamous. Builds a simple cup-shaped nest of twigs on a rocky cliff and sometimes on buildings. The clutch size is two to three eggs, and both parents feed the young.

CONSERVATION STATUS
Not threatened. An endemic species but locally abundant within its restricted range. Its populations are vulnerable to devastatation by monsoons.

SIGNIFICANCE TO HUMANS
Of no direct importance to humans, except for indirect economic benefits of ecotourism involving rare birds. ◆

Sand martin
Riparia riparia

SUBFAMILY
Hirundinae

TAXONOMY
Hirundo riparia Linnaeus, 1758.

OTHER COMMON NAMES
English: Bank swallow; French: Hirondelle de rivage; German: Uferschwalbe; Spanish: Avión Zapador.

PHYSICAL CHARACTERISTICS
5 in (12 cm); 0.5 oz (14 g). The back and wings are colored brownish, with a white chin and belly and a brown band across the chest.

DISTRIBUTION
Breeds widely throughout most of Eurasia and North America. It migrates to spend the non-breeding season in southern parts

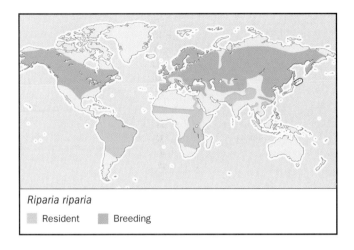

Riparia riparia
☐ Resident ☐ Breeding

Progne subis
☐ Resident ☐ Breeding

of its range, including central and northern South America and south and Southeast Asia.

HABITAT
Often occurs in the vicinity of high sandy banks and cliffs near water, but may also breed away from water in old sand quarries.

BEHAVIOR
A migratory species that winters in southern parts of its range. Attracts a mate and defends a nest site by song and aerial maneuvers. The most common call is a hoarse, whetstone-like sound.

FEEDING ECOLOGY AND DIET
Feeds on flying insects that are caught on the wing.

REPRODUCTIVE BIOLOGY
Monogamous. Builds a nest at the end of an approximately 3-ft (1-m) long passage dug into an earthen or sandy bank. The clutch consists of three to six white eggs. There are from one to three broods per season, depending on the local food availability.

CONSERVATION STATUS
Not threatened. A widespread and abundant species.

SIGNIFICANCE TO HUMANS
A familiar and popular bird to many people. ◆

Purple martin
Progne subis

SUBFAMILY
Hirundinae

TAXONOMY
Hirundo subis Linnaeus, 1758.

OTHER COMMON NAMES
French: Hirondelle noire; German: Purpurschwalbe; Spanish: Golondrina de Iglesias.

PHYSICAL CHARACTERISTICS
7 in (18 cm); 1.7 oz (48 g). The male is colored overall a glossy, iridescent purple-black, with darker wings, while the female is brownish with a lighter belly.

DISTRIBUTION
Breeds widely in North America, from southern Canada to northern Mexico. It migrates to spend the non-breeding season in central South America.

HABITAT
Inhabits open areas near suitable nesting sites, often near water.

BEHAVIOR
A long distance migrant, it winters in southern parts of its range (Venezuela to southeastern Brazil). During migration it often gathers in large flocks. Attracts a mate and defends a nest site by song and aerial maneuvers. The song is a low-pitched, bubbling twitter.

FEEDING ECOLOGY AND DIET
Feeds on insects that are caught in flight.

REPRODUCTIVE BIOLOGY
Monogamous. Breeds in colonies in special, apartment-style nest-boxes with individual compartments, or in a dead, hollow tree. It may be non-colonial at natural nesting sites. Typically

lays a clutch of four to six eggs. The eggs are incubated by the female, but both parents feed the young.

CONSERVATION STATUS

Not threatened. A locally abundant species, although declining over parts of its range. The practice of removing dead trees with cavities has reduced nesting sites, and introduced species also compete for nesting cavities.

SIGNIFICANCE TO HUMANS

A familiar and popular bird to many people. An occupied apartment nest-box is highly prized, because of the lively nature of the martins and the fact that they eat such large numbers of irritating insects, such as mosquitoes. ◆

Crag martin

Ptyonoprogne rupestris

SUBFAMILY

Hirundinae

TAXONOMY

Hirundo rupestris Scopoli, 1769.

OTHER COMMON NAMES

English: Eurasian crag martin; French: Hirondelle de rochers; German: Felsenschwalbe; Spanish: Avión Roquero.

PHYSICAL CHARACTERISTICS

5 in (14 cm). It has a brownish gray back, tail, and wings, a lighter colored throat and belly, and white spots on the tail.

DISTRIBUTION

Occurs in mountainous regions of southern and central Europe, western and central Asia, and northern North Africa. It migrates to winter in southern parts of its range in the Middle East and central and southern Africa.

HABITAT

A migratory species that inhabits open meadow habitat near cliffs and ravines, often near water, in mountainous areas at mid-elevation.

BEHAVIOR

Usually migrates in flocks. It defends a breeding site and attracts a mate by song and aerial displays.

FEEDING ECOLOGY AND DIET

Feeds on insects that are caught in flight.

REPRODUCTIVE BIOLOGY

Monogamous. The nest is built of mud and is typically located beneath a protecting overhang on a rocky cliff. Typically lays a clutch of three to four eggs, which are incubated by the female. Both parents feed the young.

CONSERVATION STATUS

Not threatened. A widespread and locally abundant species.

SIGNIFICANCE TO HUMANS

Not of much importance to humans, but sought after by birders and other ecotourists. ◆

Square-tailed saw-wing

Psalidoprocne nitens

SUBFAMILY

Hirundinae

TAXONOMY

Atticora nitens Cassin, 1857.

OTHER COMMON NAMES

English: Square-tailed saw-winged swallow; French: Hirondelle à queue courte; German: Glanzschwalbe; Spanish: Alas de Sierra de Cola Cuadrada.

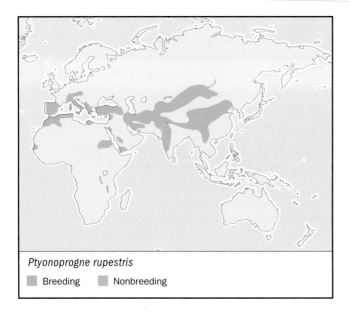

Ptyonoprogne rupestris
■ Breeding ■ Nonbreeding

Psalidoprocne nitens
■ Resident

PHYSICAL CHARACTERISTICS

4 in (11 cm). The body is overall colored glossy dark-brown and black, with sheens of green and purplish and a squared back of the tail. Males have a distinctive, hook-like thickening of the outer vane of the first primary, which may play a role in mating or a means of clinging to a vertical wall.

DISTRIBUTION

Central and western tropical Africa.

HABITAT

Occurs in open areas within tropical forested habitats, generally near water.

BEHAVIOR

A non-migratory species that uses aerial display and a relatively faint song to defend a breeding site and attract a mate.

FEEDING ECOLOGY AND DIET

Feeds on flying insects, which are caught on the wing.

REPRODUCTIVE BIOLOGY

Monogamous. Nests in a tunnel dug into a sandy bank. The clutch size is two eggs.

CONSERVATION STATUS

Not threatened. A locally abundant species.

SIGNIFICANCE TO HUMANS

Not of much importance to humans, other than the indirect economic benefits of ecotourism focused on birding. ◆

Resources

Books

Ali, Salim, and S. Dillon Ripley. *Handbook of the Birds of India and Pakistan.* 2nd ed. New York: Oxford University Press, 1981.

BirdLife International. *Threatened Birds of the World.* Barcelona, Spain and Cambridge, UK: Lynx Edicions and BirdLife International, 2000.

Brown, C., and M. Brown. *Coloniality in the Cliff Swallow.* Chicago: University of Chicago Press, 1996.

Cramp, Stanley, K.E.L. Simmons, and C.M. Perrins, eds. *Birds of the Western Palearctic.* Vol. 5, *Tyrant Flycatchers to Thrushes.* Oxford, UK: Oxford University Press, 1988.

Howell, S.N.G., and S. Webb. *A Guide to the Birds of Mexico and Northern South America.* Oxford, UK: Oxford University Press, 1995.

Keith, Stuart, et al. *The Birds of Africa.* Vol. 4. New York: Academic Press, 1992.

Meyer de Schauensee, R.A., and W.H. Phelps, Jr. *A Guide to the Birds of Venezuela.* Princeton, NJ: Princeton University Press, 1978.

Ridgely, R.S., and G. Tudor. *The Birds of South America.* Austin: University of Texas Press, 1989.

Sibley, D. *The Sibley Guide to Bird Life and Behavior.* New York: Alfred A. Knopf, 2001.

Turner, A., and C. Rose. *Swallows & Martins: An Identification Guide and Handbook.* Boston: Houghton Mifflin Company, 1989.

Periodicals

Bent, A.C. "Life Histories of North American Flycatchers, Larks, Swallows, and Their Allies." *U.S. National Museum Bulletin* 179 (1942).

Organizations

BirdLife International. Wellbrook Court, Girton Road, Cambridge, Cambridgeshire CB3 0NA United Kingdom. Phone: +44 1 223 277 318. Fax: +44-1-223-277-200. E-mail: birdlife@birdlife.org.uk Web site: <http://www.birdlife.net>

IUCN–The World Conservation Union. Rue Mauverney 28, Gland, 1196 Switzerland. Phone: +41-22-999-0001. Fax: +41-22-999-0025. E-mail: mail@hq.iucn.org Web site: <http://www.iucn.org>

Gregory J. Davis, PhD
Bill Freedman, PhD

Pipits and wagtails
(Motacillidae)

Class Aves

Order Passeriformes

Suborder Passeri (Oscines)

Family Motacillidae

Thumbnail description
Small passerines with slender body, short neck, medium to long tail and legs, and thin, pointed bill; mainly terrestrial and insectivorous

Size
4.7–8.3 in (12–21 cm); 0.025–0.14 lb (11–64 g).

Number of genera, species
5 genera; 63 species

Habitat
Predominantly open country, often near water: rocky shores, wet meadows, grassland, arid regions and tundra; also woodland

Conservation status
Endangered: 2 species; Vulnerable: 3 species; Near Threatened: 4 species; Data Deficient: 2 species

Distribution
Cosmopolitan; all continents, from Arctic tundra to Antarctic (South Georgia)

Evolution and systematics

The family Motacillidae is well defined and homogeneous but its relationships to other oscine passerine groups (singing birds such as larks, finches, and crows), as indicated by traditionally accepted morphological characters, are obscure. The family was once placed next to the larks (Alaudidae) but later widely accepted classifications placed it between the Hirundinidae (swallows and martins) and the Campephagidae (cuckoo-shrikes). However, egg-white protein evidence suggests ties with Old World warblers and flycatchers (Muscicapidae). The greatly reduced outermost primary feathers suggests affinities with nine-primaried oscines and this is supported by DNA hybridization evidence; which led researchers to treat the group as a subfamily (Motacillidae) within the family Passeridae, alongside the subfamilies Passerinae (sparrows), Prunellinae (accentors), Ploceinae (weavers), and Estrildinae (waxbills).

It has been proposed that two African endemic species, Sharpe's longclaw (*Macronyx sharpei*) and the yellow-breasted pipit (*Anthus chloris*) should be associated in the genus *Hemimacronyx* (or even that both should be placed in *Anthus*) on the basis of shared characteristics suggesting that they form a link between the pipits and the longclaws. However, this is not justified in terms of the many typical longclaw characteristics shown by Sharpe's longclaw. In structural characters and behavior, the yellow-breasted pipit resembles typical pipits more than it resembles Sharpe's longclaw.

The first fossil material for the family dates back to the Upper Oligocene, about 30 million years ago. During the Miocene epoch (26–7 million years ago), when drying conditions reduced forests and encouraged the spread of grasslands, this and other bird families radiated extensively into these more open habitats.

Physical characteristics

The members of this family are small birds. Pipits and wagtails are structurally very similar, having a slim, elongated body, a small rounded head and short neck, a slender pointed bill with rictal bristles, a medium to long tail (longest in the wagtails and shortest in pipits associated with trees), rather long legs and toes and, especially in pipits, a long hind claw. The wings have 10 primaries, the outermost being vestigial, and in most species the tertials (top feathers) are long, often reaching to the tips of the primaries in the folded wing. The wing formula is useful in identifying some pipits. The longclaws are larger and more robust, with relatively short tails and very long, curved hind claws that can reach 1.25 in (mean 0.8 in; 32 mm, mean 21 mm) in the yellow-throated longclaw (*Macronyx croceus*) and facilitate walking and perching on grass clumps. Wagtails have a horizontal stance on the ground and a more upright stance when perched, pipits usually have a less horizontal stance, and longclaws are even more upright, often having a lark-like appearance.

Wagtails are strikingly colored or patterned, at least in adult plumage, with black, white, gray, yellow, or green. Young birds and nonbreeding plumages are generally less conspicuous. The males and females of most species differ to

A South Georgia pipit (*Anthus antarcticus*) perches near the water. (Photo by G. Lasley/VIREO. Reproduced by permission.)

some extent in plumage. Most species have pale wing-bars or white wing-panels, the tertials have prominent pale outer edges and the tail is edged with white. Variation in the yellow wagtail (*Motacilla flava*) is complex, with up to 15 morphological types, which may be morphs, intergrading subspecies or sympatric species (those that occupy the same area but maintain their own identity).

Pipits are cryptically colored and patterned, usually having brown upperparts and whitish underparts, with dark streaking, although some species are almost unstreaked. The tail is edged with white, buff, or pale brown, the color and pattern being important in identification. In most species the sexes are identical in plumage and there is little or no seasonal variation. Many species are very similar in appearance and are very hard to identify, although vocalizations are usually diagnostic. The olive-backed pipit (*Anthus hodgsoni*) is unique with its greenish upperparts and white underparts beautifully decorated with lines of large black spots on the breast and flanks. Some pipits have brighter colors in the breeding plumage, with red or pink on the throat and breast in rosy pipits (*A. roseatus*) and red-throated pipits (*A. cervinus*), and yellow underparts on the yellow-breasted pipit. The golden pipit is peculiar because the lower part of the tibia is not feathered and the sexes are very distinct, the male being very brightly colored.

Longclaws have cryptically patterned upperparts and brightly colored underparts. Adults have a blackish or strongly dark-streaked "necklace" bordering a yellow, orange, or red chin and throat. This color extends to the rear underparts in

some species. The tail has white corners. Young birds have a less distinct necklace and duller underparts. The yellow-throated longclaw is the ecological equivalent of the American meadowlark (*Sturnella neglecta*), which it resembles very closely in plumage—a striking example of convergence.

Distribution

The Motacillidae family is cosmopolitan; some wagtails and pipits breed as far north as the Arctic tundra, while the South Georgia pipit (*Anthus antarcticus*) occurs on the sub-Antarctic island of South Georgia. Almost all holarctic wagtails and pipits are strongly migratory, many moving south to winter in Africa and Asia.

Wagtails occur throughout most of the Old World but are of limited occurrence in Australia, where the yellow wagtail regularly reaches the north and the yellow-hooded wagtail (*Motacilla citreola*), the gray wagtail (*M. cinerea*), and the white wagtail (*M. alba*) are vagrants. The yellow wagtail breeds from western Europe and North Africa east to Siberia, and the race *tschutschensis* is the only wagtail breeding in the western hemisphere—in the arctic tundra of western Alaska. Most wagtails are widely distributed, but the Mekong wagtail (*M. samveasnae*) has a very restricted distribution in Southeast Asia, while the Japanese wagtail (*M. grandis*) occurs only in Japan. Three species are endemic to Africa: the African pied wagtail (*M. aguimp*), the cape wagtail (*M. capensis*) and the mountain wagtail (*M. clara*). One species, *M. flaviventris*, is endemic to Madagascar.

The pipits are very widely distributed, breeding from the arctic tundra through Eurasia, Africa, the Americas and Australasia. The northernmost breeding Eurasian species, such as the red-throated pipit, the olive-backed pipit and the pechora pipit (*Anthus gustavi*) winter far to the south, in Africa and/or southern and southeastern Asia. The tawny pipit (*A. campestris*) and the tree pipit (*A. trivialis*), which breed in the western Palearctic and east into central Asia, winter in Africa and Asia south to the Indian subcontinent. The meadow pipit (*A. pratensis*) breeds mainly in the western Palearctic and winters in Europe, North Africa and southwest Asia. Africa boasts 13 endemic pipit species, some migratory but most resident, and several with restricted ranges.

The Australasian pipit (*Anthus novaeseelandiae*) is the only species occurring in Australia and New Zealand, while the alpine pipit (*A. gutturalis*) is endemic to New Guinea. North America has three pipits, all migratory: the endemic Sprague's pipit (*A. spraguei*); the American pipit (*A. rubescens*), which breeds in eastern Siberia, Alaska, northern Canada and south locally to New Mexico, and winters south to Central America; and the red-throated pipit, which breeds in western Alaska. South America has seven endemic pipits, including the widespread yellowish pipit (*A. lutescens*), the migratory correndera pipit (*A. correndera*), which occurs south from central Peru, and the very poorly known chaco pipit (*A. chacoensis*), which is recorded only from southern Paraguay and northern Argentina.

The longclaws are endemic to sub-Saharan grassland regions of Africa. The yellow-throated longclaw is very widespread, occurring through west, central and eastern Africa, its range hardly overlapping with that of the very similar Fuelleborn's longclaw (*Macronyx fuellebornii*) occurring in south-central Africa. The other species are less widespread.

Habitat

Most species in the family inhabit open country, but a few are associated with woodland, forest or riparian vegetation. The forest wagtail occurs in forests and woodlands, the mountain wagtail is found along the banks of fast-flowing streams in forests and the gray wagtail occupies well-vegetated waterside habitats but also sometimes occurs on streams with no vegetation cover. Eurasian woodland and forest pipits include the tree pipit, olive-backed pipit, and Pechora pipit. African species include the bush pipit (*Anthus caffer*), woodland pipit (*A. nyassae*), and Sokoke pipit (*A. sokokensis*).

Wagtails are found in a wide variety of open and semi-open habitats. They often occur in wet habitats, ranging from streams, rivers and open bodies of water to the edges of vegetated wetlands. Several species are associated with farmland, parks, gardens and human habitations, and may breed in buildings inside villages and towns. The yellow wagtail breeds in arctic tundra habitats as well as vegetated wetlands and meadows at lower latitudes, and occupies a wide range of open, short-vegetated, often wet habitats on its wintering grounds. On migration and in their wintering areas, migratory wagtail and pipit species often associate with each other.

Pipits occur in many habitat types, mostly open and especially grassland, from sea level to high altitudes, and the rosy pipit reaches 17,400 ft (5,300 m) in the Himalayas. Rocky shores attract the rock pipit (*Anthus petrosus*), while species such as the meadow pipit prefer wet meadows and grasslands.

A red-throated pipit (*Anthus cervinus*) nest with chick and eggs. (Photo by G. Olioso/VIREO. Reproduced by permission.)

Drier open grasslands are inhabited by many species, including almost all of those occurring in South America, while the tawny pipit inhabits dry, often sparsely vegetated regions, sometimes semi-desert and often sandy. In Africa, the yellow-tufted pipit (*A. crenatus*) prefers rocky hills with grass clumps and the striped pipit (*A. lineiventris*) rocky sites with trees. On its nonbreeding grounds in the Philippines, Borneo, and Wallacea, the Arctic tundra-breeding pechora pipit frequents moist grassy areas, forest trails and coastal forests.

The longclaws are predominantly grassland birds, often occurring on moist ground at wetland edges, although the pangani longclaw (*Macronyx aurantiigula*) also occurs widely in grassland with acacia bushes in semi-arid country, and in West Africa the yellow-throated longclaw is also found on the seashore. Longclaw species with overlapping ranges exhibit a wider ecological tolerance in areas where other longclaw species of the same habitats are not present.

Behavior

Members of this family are territorial when breeding, and males are often very aggressive, threatening and chasing intruding individuals of their own and other species. Commensal wagtails often show aggressive behavior to their reflection in car mirrors and hubcaps, threatening and attacking the image vigorously, often for prolonged periods.

Many species form flocks when on migration and in the nonbreeding quarters. Wagtails often roost communally, sometimes with other species, usually in reedbeds but also in bushes or scrub and sometimes at sewage works, factories and greenhouses. Some wagtail and pipit species defend feeding territories outside the breeding season. For example, rock pipits defend stretches of coastline and wagtails defend stretches of shoreline adjacent to open flowing or still water.

Wagtails are lively and attractive birds, and commensal species are usually fearless of humans. Wagtails are so named because they frequently wag their tails up and down. Pipits also perform this movement, though less strongly, but the olive-backed pipit pumps its tail as vigorously as a wagtail. Wagtails and pipits either walk with a rather deliberate gait or run at great speed over the ground. To escape detection, pipits and longclaws are adept at crouching in short vegetation and moving quietly through short ground vegetation, adopting an upright stance to look around. Forest and woodland species such as the Sokoke, striped and bush pipits are often unobtrusive and relatively difficult to locate.

The flight of wagtails and pipits is strong and is often undulating, especially in wagtails. Longclaws normally have a jerky flight, with alternating periods of flaps and glides. Songflights, launched either from the ground or a perch, are characteristic of the pipits and longclaws, while wagtails usually sing from the ground or from an elevated perch, more rarely in a short fluttering song-flight. Pipits are well known for their spectacular song flights, in which the song is delivered from high in the air, often as the bird parachutes back to the ground or a perch. The songs of pipits vary from short and simple (e.g., in the upland pipit [*Anthus gustavi*]) to extended,

Gray wagtail (*Motacilla cinerea*) with chicks at its nest. (Photo by Hans Reinhard. Bruce Coleman Inc. Reproduced by permission.)

complex and varied (e.g., in Sprague's pipit). Wagtails have simple, often quite melodious songs, while the longclaws have distinctive voices, giving whistled, melodious calls and simple songs. Flight calls are common in all species in the family, and are of great use in identifying pipits.

Most pipits and wagtails are migratory. The species that breed in temperate regions are usually medium- to long-distance migrants, whereas those breeding further south are usually short-distance migrants or residents. Many Palearctic wagtails and pipits migrate to tropical Africa and Asia for the winter. The correndera pipit is unique among South American endemic pipits in having some migratory movements.

Feeding ecology and diet

The principal food of all species is adult and larval insects of a very wide variety and a great size range, from tiny midges and chironomid larvae to locusts and dragonflies. Insects taken include Diptera (especially Scatophagidae, Tipulidae, and Chironomidae), beetles (including cockroaches), grasshoppers and locusts, crickets, Hemiptera (including aphids), mantids, ants, termites, and wasps. Virtually the full range of insect prey is taken by the members of each genus, and some of the most popular food items are beetles (including weevils) and grasshoppers, plus dragonfly larvae, adult and larval Diptera, and Lepidoptera larvae, while termites are usually taken whenever available. Wagtail species take a variety of larval aquatic insects, such as dragonfly and mayfly nymphs, and caddisfly and stonefly larvae.

Other invertebrates eaten include spiders, crustaceans such as Isopoda, Amphipoda and crabs, annelid worms, Myriapoda, and small terrestrial, freshwater and marine mollusks. Vertebrate prey includes small fish, small frogs, tadpoles and small chameleons. Plant material is sometimes eaten, especially in the winter, and includes grass seeds, weed seeds, berries, grass blades, pine tree seeds, and tree buds; even young vegetables are reportedly eaten by the Australasian pipit. The olive-

backed pipit feeds chiefly on insects in the summer and seeds in the winter. Some species occasionally take more unusual foods, including carrion, and the cape wagtail will forage around human habitation, eating raw meat, fat, cheese, maize meal, bread, and cake.

The birds forage on open ground, in grass and herbaceous vegetation, among domestic stock, at water margins, in shallow water, on floating vegetation, and in trees and bushes; some species even follow the plough. The red-throated pipit also forages in seaweed on beaches, while the rock pipit wades in seawater, following retreating waves on beaches. Foraging methods vary with species, and include the following main techniques: (1) picking from the ground or the water surface while walking; (2) run-picking by making darting runs to catch prey; (3) immersion, by plunging the head into water; (4) fly-catching and aerial pursuit, by making short to long flights in pursuit of prey; (5) hovering to catch airborne prey or prey on the water surface; (6) probing in ground vegetation, crevices or leaf-litter. The long tail of wagtails helps the birds' balance when run-picking and flycatching, and assists in the control of aerial maneuvers when pursuing insects in flight.

In the nonbreeding season, some wagtail and pipit species feed in flocks, exploiting large patches of food. Some wagtails maintain individual winter feeding territories to defend dependable but localized food supplies, especially adjacent to water. Territory boundaries are vigorously defended with displays involving head-bobbing and short jumps into the air. Territoriality may vary with food abundance, and individuals may switch between defending patchy resources and feeding communally at widespread patches. Winter pairs may occupy territories, or an adult may share the territory with an immature bird.

Reproductive biology

Courtship displays are given by some species. The mountain wagtail, which pairs permanently and defends a permanent territory, indulges in erratic aerial chases and an aerial "spiral dance." Other wagtails also pair monogamously and permanently, and the male cape wagtail displays by presenting the female with nesting material throughout the year. Carrying and presenting nesting material is also recorded in Berthelot's pipit (*Anthus berthelotii*) and other pipit species, while courtship feeding is practiced by wagtails and pipits. Breeding pipits perform aerial courtship chases, which sometimes precede copulation. Breeding pairs of Sharpe's longclaw (*Macronyx sharpei*) perform fluttering or circular flights together over the territory.

Pipits nest on the ground, often in grass. The red-throated pipit sometimes builds at the end of a short tunnel in a mossy hummock. Wagtails may also nest on the ground, in grass or reeds, on flood debris or below bushes, but they commonly nest in crevices or holes in rocks, cliffs, stream banks, and walls, under bridges, and in tree roots and hollow branches; the white wagtail sometimes uses old nests of other species. Longclaw nests are hidden in, or at the base of, a grass tussock or among herbaceous plants.

Nests are cup-shaped, sometimes placed in a depression or a shallow scrape, are usually neatly built of grass, stems, rootlets, twigs, or moss and are often lined with hair, wool, feathers, or plant fibers. The female builds, usually with the male in attendance and, in some species, with the help of the male.

Egg colors vary from white, cream, buff, or gray to (in pipits) olive, reddish, or dark brown, spotted or blotched (in wagtails sometimes also streaked) with brown, gray, mauve, purple, or black. Longclaws and the tree pipit sometimes lay pale blue, pink, or green eggs. The clutch size of wagtails is three to eight (usually four to six) in higher latitudes and one to seven (usually two to four) at lower latitudes; pipits lay two to nine (usually four to six) eggs at higher latitudes and only two to four (usually three) in the tropics. Longclaws lay two to five eggs, most commonly two or three. Incubation is usually by the female only, but by both sexes in some species; it takes 11–16 days. Both parents usually care for the young, which fledge after 10–17 days (exceptionally 19–20 days in the cape wagtail). Young often leave the nest before they are fully fledged and able to fly.

In temperate latitudes, wagtails and pipits breed from April to August (mostly April through June), but Berthelot's pipit has an extended season, from January to August. In the tropics, pipits and wagtails breed mainly at the end of the dry season and during the rains. Longclaws and the golden pipit breed during or just after the rains, the development of grass cover for nest concealment probably being an important factor in the timing of breeding. Some pipits and wagtails breed two or three times per year, including the African pied wagtail, which may sometimes breed continually throughout the year. High-latitude species are usually single-brooded because the breeding season is short.

Conservation status

The worldwide and increasing loss and degradation of grassland and wetland habitats has had its effect on the Motacillidae. Of the family's 63 species, two are Endangered, three are Vulnerable, four are Near Threatened and two are Data Deficient. In addition, the recently described Mekong wagtail, although existing in healthy numbers in Cambodia as of 2001, should probably be regarded as Near Threatened because of its very small known range and its susceptibility to habitat loss.

Three of the eight longclaw species are of global conservation concern. The Endangered Sharpe's longclaw, which is endemic to grassland in the Kenya highlands, has a very small and fragmented range and is threatened by rapid habitat loss and degradation through cultivation, tree-planting and heavy grazing. The Abyssinian longclaw (*Macronyx flavicollis*), a highland grassland species confined to Ethiopia, has apparently declined in numbers since the 1970s. It is likely to suffer further declines as a result of constantly increasing levels of cultivation and grazing, and is therefore classed as Near Threatened. Grimwood's longclaw (*M. grimwoodi*), which occurs in moist grasslands in southwest Democratic Republic of the Congo, central and eastern Angola and extreme northwestern Zambia, was formerly regarded as locally common but there is no recent information on its population or on potential threats, and it is regarded as Data Deficient.

A Richard's pipit (*Anthus richardi*) holds an insect in its bill. (Photo by M. Hale/VIREO. Reproduced by permission.)

The other eight species of conservation concern are all pipits, six of which are sedentary, with very restricted ranges. In Africa, the Sokoke pipit is restricted to coastal forest and thickets in Kenya and Tanzania, where its small population is seriously threatened by habitat loss and degradation; it is regarded as Endangered. The Near Threatened Malindi pipit (*Anthus melindae*) is endemic to the coastal strip from southern Kenya to southern Somalia, where it is locally common in seasonally flooded short grassland that is under pressure from intensive grazing and demands for arable land. The Vulnerable yellow-breasted pipit is restricted to highland grasslands in South Africa, where its small, declining population is threatened by habitat loss. The long-tailed pipit (*A. longicaudatus*), recently described from Kimberley, South Africa, is presumed to be a migrant and its range and status are unclear. It is regarded as Data Deficient.

The Near Threatened Nilgiri pipit (*A. nilghiriensis*) is endemic to southern India, where it has a small range on grassy upland slopes of the Western Ghats of Kerala and Tamil Nadu. Its habitats are being converted to tea, wattle and eucalyptus. In the Americas, the Vulnerable ochre-breasted pipit (*A. nattereri*) is confined to dry grasslands in southeast Brazil, southern Paraguay and northern Argentina. It has declined dramatically in Brazil and is threatened by extensive and continuing habitat destruction. The migratory Sprague's pipit of Canada and the United States is also Vulnerable. The Near Threatened South Georgia pipit has a population of 3,000–4,000 pairs confined to about 20 small rat-free offshore islands and a few mainland areas on South Georgia. An ongoing threat is the invasion of its habitats by rats.

Significance to humans

Wagtails, with their striking plumage, confiding nature and frequent association with humans and domestic stock, have become the subject of legends throughout their wide geographical range. In Japanese Shinto mythology, wagtails are sacred to the central deities of the creation myth, the brother and sister gods Izanagi and Izanami, who were taught love by them. In Ainu myth the wagtail was sent by the cuckoo to Earth, which was then a sterile quagmire. The wagtail beat the earth down, flattening rough places with its wings and tail until the ground hardened and became habitable for people. The "water wagtail" is the Ainu Cupid and its feathers and bones are love charms.

In Greek mythology, wagtails were seen as a gift from Aphrodite, the goddess of love, and the wagtail was a symbol of love. In India the wagtail is a bird of divination and bears a holy caste mark. The situation in which it appears is an omen: if it is near a lotus flower, elephants, cows, snakes or horses it is favorable; if near bones, ashes or refuse it presages evil and the gods should be placated.

Wagtails featured strongly in tribal life among the Xhosa people of South Africa in the early twentieth century. The wagtail (primarily the cape wagtail, possibly also the African pied wagtail) was widely known as "the bird of the cattle" and "the bird of good fortune." It was held in high regard and was protected because its presence was thought to assure the increase of stock, while its call was likened to a herd boy's whistle. The departure of cape wagtails from a region was seen as a sign that war was about to take place.

Despite their often striking display flights, the more cryptic and less approachable pipits hardly seem to have figured in myth and legend. However, young Zulu men in South Africa formerly manufactured love charms from pipits and it is interesting to note that the Xhosa people were aware of the close relationship between pipits and wagtails.

In the remote hinterland of Borneo, the Kelabit people determine their crucial rice-planting cycle by the arrival of a series of migratory bird species from far northern breeding grounds. These birds, which include the yellow wagtail, indicate the sequence of clearing, planting, bedding, weeding, protecting and harvesting the rice crop, and give their names to the months.

1. Forest wagtail (*Dendronanthus indicus*); 2. Golden pipit (*Tmetothylacus tenellus*); 3. Sprague's pipit (*Anthus spragueii*); 4. Yellow-breasted pipit (*Anthus chloris*); 5. Rosy-breasted longclaw (*Macronyx ameliae*); 6. Mekong wagtail (*Motacilla samveasnae*); 7. Gray wagtail (*Motacilla cinerea*); 8. Berthelot's pipit (*Anthus berthelotii*); 9. Red-throated pipit (*Anthus cervinus*); 10. Bush pipit (*Anthus caffer*). (Illustration by Bruce Worden)

Species accounts

Forest wagtail
Dendronanthus indicus

TAXONOMY
Motacilla indica Gmelin, 1798, India (Malabar).

OTHER COMMON NAMES
English: Tree wagtail; French: Bergeronnette de forêt; German: Baumstelze; Spanish: Lavandera de los Bosques.

PHYSICAL CHARACTERISTICS
6.3–6.9 in (16–17.5 cm); 0.5–0.6 oz (14–17 g). Grayish head and upperparts with white eye stripe; white bands on wing. Whitish throat and underparts with black bib-like collar with black stripes on breast.

DISTRIBUTION
Northeast China and southeast Russia south to Korea and southwest Japan; winters in south China, Bangladesh, India, Sri Lanka, and Indochina through Malay Peninsula to Sumatra, Java, Borneo, and the Philippines.

HABITAT
Deciduous and evergreen broadleaf forests, riverine forest, open woodland, and pinewoods; also winters in bamboo, plantations, and parks.

BEHAVIOR
Readily flies to trees when disturbed; sometimes roosts in reeds or mangroves. Sways tail and rear body laterally. Migrates south in August through October; return migration March through May.

FEEDING ECOLOGY AND DIET
Insects (including ants) and spiders. Forages on the ground, usually close to cover.

REPRODUCTIVE BIOLOGY
Monogamous; breeds May through June. Nest is a small, neat cup of small twigs, leaves, grass, rootlets, moss, cobwebs and lichen, on a horizontal tree branch, often near water; female builds. Lays four eggs; incubation 15–16 days, by female.

CONSERVATION STATUS
Not threatened; formerly regarded as locally distributed.

SIGNIFICANCE TO HUMANS
None known. ◆

Gray wagtail
Motacilla cinerea

TAXONOMY
Motacilla cinerea Tunstall, 1771, Yorkshire, England. Six subspecies.

OTHER COMMON NAMES
French: Bergeronnette des ruisseaux; German: Gebirgsstelze; Spanish: Lavandera Cascadeña.

PHYSICAL CHARACTERISTICS
7.1–7.5 in (18–19 cm); 0.5–0.8 oz (14–22 g). Gray upperparts; yellow underparts. Tail longer, more black-and-white than the yellow wagtail. In summer males develop a bold face pattern with white stripes and a black bib.

DISTRIBUTION
M. c. patriciae: Azores; *M. c. schmitzi*: Madeira; *M. c. canariensis*: Canary Islands; *M. c. cinerea*: Northwest Africa and Europe east

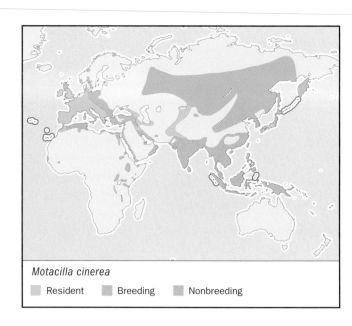

Dendronanthus indicus
■ Breeding   ■ Nonbreeding

Motacilla cinerea
□ Resident   ■ Breeding   ■ Nonbreeding

to Iran, winters in western Europe, Middle East and Africa south to Malawi; *M. c. melanope*: Ural mountains and Afghanistan east to Amur R.; *M. c. robusta*: east Asia from Kamchatka and Okhotsk Sea to northeast China and Japan; central and eastern Asian birds winter in Pakistan east to southeast China, and southeast Asia to New Guinea.

HABITAT
Fast-running, rocky upland streams and rivers; also canals, and lakeshores with stones, trees and dense herbage; in winter also in lowlands at waterbodies, estuaries and coasts near water.

BEHAVIOR
Territorial when breeding; some defend winter feeding territories; gregarious only at winter roosts.

FEEDING ECOLOGY AND DIET
Takes mainly aquatic insects; also tadpoles and small fish; forages on ground or in water; also flycatches.

REPRODUCTIVE BIOLOGY
Monogamous; breeds March through May. Cup nest placed on cliff ledge, in crevice, bank or tree roots; both sexes build. Three to seven eggs; incubation 11–14 days, by both sexes; fledging 11–17 days.

CONSERVATION STATUS
Not threatened. Uncommon to common and widespread. Some populations decreasing, but no significant threats noted.

SIGNIFICANCE TO HUMANS
None known. ◆

Motacilla samveasnae
▨ Resident

Mekong wagtail
Motacilla samveasnae

TAXONOMY
Motacilla samveasnae Duckworth *et al.*, 2001, Cambodia.

OTHER COMMON NAMES
French: Bergeronnette de Mekong; German: Mekongstelze; Spanish: Lavandera Mekong.

PHYSICAL CHARACTERISTICS
6.7 6.9 in (17–17.5 cm). Black head, breast, and upperparts with white eye stripe, underside, and chin to throat.

DISTRIBUTION
Mekong River and tributaries in Cambodia and adjacent east Thailand and south Laos, possibly also Vietnam.

HABITAT
Breeding habitat is swift-flowing rivers at braided channels with rocks and bushes.

BEHAVIOR
Territorial when breeding. Possibly resident; almost no observations outside breeding season.

FEEDING ECOLOGY AND DIET
Insectivorous; often forages within emergent bushes, walking along branches.

REPRODUCTIVE BIOLOGY
Monogamous; breeds February through May, late in river's low-flow season. No other information available.

CONSERVATION STATUS
Should be considered Near Threatened. Although its population density is locally high, it has a very restricted range and is permanently at risk of extinction by habitat loss. The greatest threat is dam-building; flow-modifying events, even in other countries and far upstream of its known range, could drastically affect its habitat.

SIGNIFICANCE TO HUMANS
None known. ◆

Golden pipit
Tmetothylacus tenellus

TAXONOMY
Macronyx tenellus Cabanis, 1878, Taita, southeast Kenya.

OTHER COMMON NAMES
French: Pipit doré; German: Goldpieper; Spanish: Bisbita Dorada.

PHYSICAL CHARACTERISTICS
5.1–6.3 in (13–16 cm); 0.6–0.8 oz (18–22 g). Mottled brown from forehead to rump; golden-yellow underparts, cheek, and brow with black bib across throat.

DISTRIBUTION
Western Somalia and Ethiopia, west to southeast Sudan and south to southern Tanzania; vagrant to the north in Oman and south to northern South Africa.

HABITAT
Arid or semi-arid grassland with bushes and small trees, usually up to about 3,300 ft (1,000 m) above sea level.

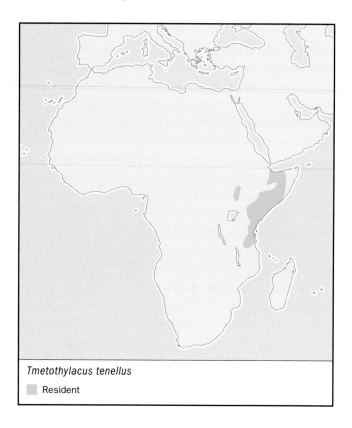

Tmetothylacus tenellus

◼ Resident

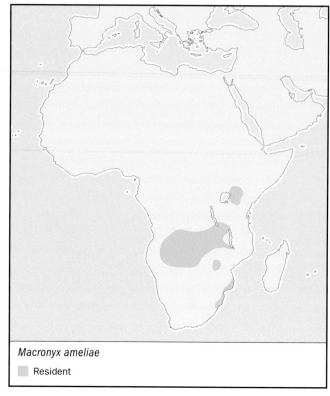

Macronyx ameliae

◼ Resident

BEHAVIOR
Territorial when breeding, otherwise often in family groups or parties. Rather shy; frequently perches on trees or bushes, wagging tail. Displays by fluttering up and then gliding to ground, showing yellow in spread wings and tail. Migratory; marked influxes during rains.

FEEDING ECOLOGY AND DIET
Insectivorous, taking prey from ground or grass stems.

REPRODUCTIVE BIOLOGY
Monogamous. Breeds November through May, during rains. Nest is a cup of grass and stems, lined rootlets, hidden low in a bush; lays two to four eggs.

CONSERVATION STATUS
Not threatened; sometimes locally abundant after rains.

SIGNIFICANCE TO HUMANS
None known. ◆

Rosy-breasted longclaw
Macronyx ameliae

TAXONOMY
Macronyx ameliae de Tarragon, 1845, Durban, South Africa.

OTHER COMMON NAMES
English: Pink-throated longclaw; French: Sentinelle à gorge rose; German: Rubinkehlpieper; Spanish: Bisbita de Pecho Rosado.

PHYSICAL CHARACTERISTICS
7.5–8 in (19–20 cm); 1.1–1.4 oz (30–40 g). Mottled upperparts with orange-red upper throat, blackish band across lower throat, and rosy breast.

DISTRIBUTION
Southwestern Kenya, north and southwest Tanzania, west to Angola and south to Botswana, Zimbabwe and coastal east South Africa.

HABITAT
Short or tussocky grassland, usually permanently or seasonally moist and near marshes or open water.

BEHAVIOR
Territorial when breeding; normally in pairs or in family groups. Usually shy. Male sings from top of bush or in song flight. Some movements in relation to seasonal rainfall.

FEEDING ECOLOGY AND DIET
Takes insects, sometimes small frogs. Forages in grass or on bare ground; sometimes catches prey in flight.

REPRODUCTIVE BIOLOGY
Monogamous; breeds mainly during or after rains. Nest is a cup of grass, lined rootlets or grass, placed in grass tuft; built by female. Lays two to four eggs; incubation is 13–14 days by female. Young leave nest after about 16 days.

CONSERVATION STATUS
Near Threatened in South Africa, where range has contracted significantly through loss of coastal habitat; threatened by habitat loss in coastal south Mozambique.

SIGNIFICANCE TO HUMANS
None known. ◆

Yellow-breasted pipit

Anthus chloris

TAXONOMY
Anthus chloris Lichtenstein, 1842, Vaal/Modder Rivers, South Africa.

OTHER COMMON NAMES
French: Pipit à gorge jaune; German: Gelbbrustpieper; Spanish: Bisbita de Pecho Amarillo.

PHYSICAL CHARACTERISTICS
6.3–7.1 in (16–18 cm); 0.9 oz (25 g). Mottled brown upperparts with yellowish eye stripe and yellow chin to belly.

DISTRIBUTION
Eastern South Africa and Lesotho.

HABITAT
Submontane, flat to undulating lush grasslands, usually tussocky; normally breeds at 4,600–7,900 ft (1,400–2,400 m); outside breeding season also in lower-elevation pastures and fallow lands.

BEHAVIOR
Territorial when breeding; usually in pairs but in small flocks when not breeding. Skulking and furtive. Sings from ground or in display flight. Some move to lower altitudes after breeding.

FEEDING ECOLOGY AND DIET
Forages on the ground for insects.

REPRODUCTIVE BIOLOGY
Monogamous; breeds November though January, during rains. Nest is a cup of stalks, grass and roots, lined rootlets and hair; built under tussock. Lays two to three eggs.

CONSERVATION STATUS
Vulnerable because habitat loss and range contraction suggest its small population (2,500–6,500 birds in 2000) is declining. Threatened by burning, grazing, agricultural intensification and commercial afforestation.

SIGNIFICANCE TO HUMANS
None known. ◆

Red-throated pipit

Anthus cervinus

TAXONOMY
Motacilla cervinus Pallas, 1811, Kolmya, Siberia.

OTHER COMMON NAMES
French: Pipit à gorge rousse; German: Rotkehlpieper; Spanish: Bisbita Gorgirrojo.

PHYSICAL CHARACTERISTICS
5.7–5.9 in (14.5–15 cm); 0.6–1 oz (16.5–29 g). Upperparts are dark brown with black and whitish streaks on the back. Underparts are buffy with dark streaks across the breast and onto the flanks. Wings are blackish with whitish bars and the tail is black with white outer tail feathers. In the breeding season, breeding males develop a reddish head, throat, and breast; the head and breast of females and nonbreeding males is duller.

DISTRIBUTION
Breeds in northern Scandinavia east through arctic Russia to Bering Strait and western Alaska; winters in east Africa south to Tanzania, Turkey, Middle East, and southeast China, southeast Asia, and northern Indonesia.

HABITAT
Shrubby or mossy tundra, and willow/birch swamps; also damp grassy flats. In winter, on short-grazed grassland, especially with mud or shallow water; also mudflats, ploughland and moorland.

BEHAVIOR
Territorial when breeding. In winter, often forms large, loose flocks. Often perches on rocks, bushes and fences. Male displays with horizontal or parachuting song flight.

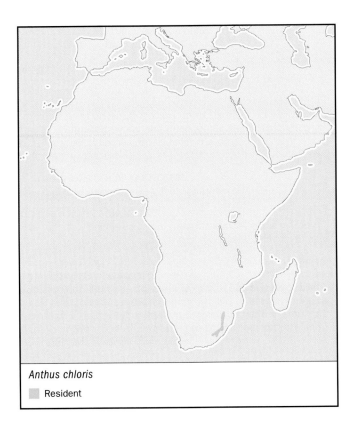

Anthus chloris

　▨　Resident

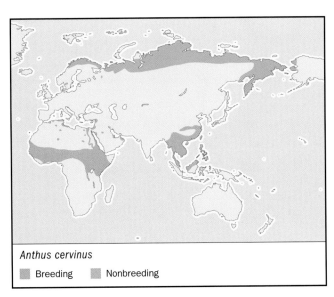

Anthus cervinus

　▨　Breeding　　▨　Nonbreeding

FEEDING ECOLOGY AND DIET
Forages on the ground, probing amongst vegetation for insects.

REPRODUCTIVE BIOLOGY
Monogamous; breeds May though August. Nest is a hollow (made by male) in moss or ground, with cup of grass, lined hair and feathers; female builds. Lays two to seven eggs; incubation is 11–14 days, by female; fledging after 11–15 days.

CONSERVATION STATUS
Not threatened. Locally abundant, but facing extinction in Finland; no major range changes recorded.

SIGNIFICANCE TO HUMANS
None known. ◆

Berthelot's pipit
Anthus berthelotii

TAXONOMY
Anthus berthelotii Bolle, 1862, Canary Islands. Two subspecies.

OTHER COMMON NAMES
French: Pipit de Berthelot; German: Kanarenpieper; Spanish: Bisbita caminero.

PHYSICAL CHARACTERISTICS
5.5 in (14 cm); 0.6 oz (16–17 g). Brownish-gray upperparts with white brow stripe. Underparts buff-gray with dark streaks.

DISTRIBUTION
A. b. berthelotii: Canary Islands and Ilhas Selvagens; *Anthus b. madeirensis*: Madeira.

HABITAT
Mostly island habitats; prefers dry, open areas (including rocky plains and slopes) with bushes, grass, and herbaceous vegetation; also open grasslands, cultivation, vine-clad slopes, dunes, areas of volcanic rock, and open pine forests.

BEHAVIOR
Territorial, possibly throughout the year, but forms small groups in winter. Often very tame. Runs rapidly; jumps nimbly over boulders. Sedentary; not recorded outside island breeding range.

FEEDING ECOLOGY AND DIET
Eats insects and seeds. Forages on the ground, climbing over small plants.

REPRODUCTIVE BIOLOGY
Monogamous; breeds January through August. Nest is a cup of stems, lined hair, wool and feathers; on ground under low plant, bushes or stone; female builds. Lays two to five eggs. Possibly double-brooded.

CONSERVATION STATUS
Not threatened. Locally common throughout most islands where it occurs, but has declined on Gran Canaria.

SIGNIFICANCE TO HUMANS
None known. ◆

Bush pipit
Anthus caffer

TAXONOMY
Anthus caffer Sundevall, 1850, Natal, South Africa. Five subspecies.

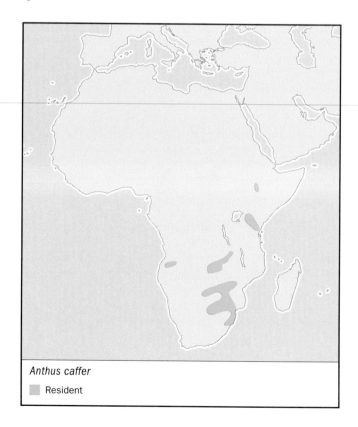

Anthus berthelotii
■ Resident

Anthus caffer
■ Resident

OTHER COMMON NAMES
English: Bushveld pipit; French: Pipit cafre; German: Busch-pieper; Spanish: Bisbita Negra.

PHYSICAL CHARACTERISTICS
4.9–5.5 in (12.5–14 cm); 0.6 oz (16 g). Buff underparts streaked with brown at the throat and breast. Dark brownish head and upperparts with lighter eyestripe and chin.

DISTRIBUTION
A. c. caffer: Southeast Botswana, southwest Zimbabwe, northern South Africa and west Swaziland; *A. c. traylori*: South Mozambique and adjacent northeast South Africa; *A. c. mzimbaensis*: northeast Botswana, central Zimbabwe, northern Zambia, extreme southeast Democratic Republic of the Congo (DRC) and western Malawi; *A. c. blaneyi*: Kenya and Tanzania; *A. c. australoabyssinicus*: Ethiopia.

HABITAT
Open woodland with patchy ground cover, woodland edges, and grassland with scattered trees.

BEHAVIOR
Occurs in pairs or small flocks, often with mixed-species bird parties. Flies to trees when disturbed. Some *A. c. mzimbaensis* undertake post-breeding movements from Botswana and Zimbabwe north to Zambia, DRC, and Malawi.

FEEDING ECOLOGY AND DIET
Forages for insects on the ground in grass and leaf-litter.

REPRODUCTIVE BIOLOGY
Monogamous. Breeds October through April, during rains. Nest is a thick-walled cup of grass, lined rootlets, on ground under tuft of grass. Lays two to three eggs.

CONSERVATION STATUS
Not threatened.

SIGNIFICANCE TO HUMANS
None known. ◆

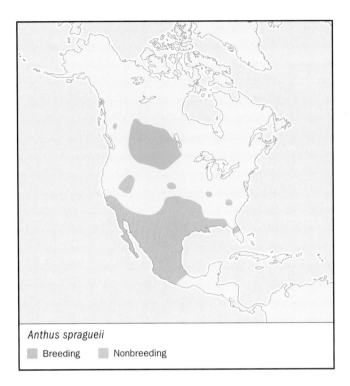

Anthus spragueii

■ Breeding ■ Nonbreeding

Sprague's pipit
Anthus spragueii

TAXONOMY
Alauda spragueii Audubon, 1844, North Dakota.

OTHER COMMON NAMES
French: Pipit de Sprague; German: Präriepieper; Spanish: Bisbita Llanera.

PHYSICAL CHARACTERISTICS
6.3–7 in (16–18 cm); 0.8–1 oz (22–29 g). Pale buff face; olive-tan upperparts streaked with buff and black; buff to whitish underparts with dark streaking; white outer tail feathers. Dark eyes; thin pale bill; creamy pink to yellowish legs and feet. Plumage camouflages the bird in prairie grasses.

DISTRIBUTION
Breeds in south central Canada (Alberta, Saskatchewan, Manitoba, British Columbia) and adjacent north central United States (Montana, North and South Dakota); winters in southern United States and northern Mexico south to Guerrero and Veracruz.

HABITAT
Tall grass prairies and short-grass plains; on migration uses stubble and fallow fields.

BEHAVIOR
Often flies high when flushed. Displays with a high, arcing song flight. Migrates south September through November, returns to breed in April through May.

FEEDING ECOLOGY AND DIET
Forages on the ground for insects; also takes some seeds.

REPRODUCTIVE BIOLOGY
Monogamous. Nest is a cup of grass and weed stems, built on the ground and often overarched with grass. Lays four to seven eggs; fledging period 10–11 days or more. May be double-brooded.

CONSERVATION STATUS
Vulnerable because of rapid population declines in the United States and Canada due to loss of prairie breeding habitat to crops, pasture and hayfields, and the introduction of alien plant species; intensive grazing is a threat throughout its range.

SIGNIFICANCE TO HUMANS
None known. ◆

Resources

Books

Ali, S., and S.D. Ripley. *Handbook of the Birds of India and Pakistan.* (Compact Edition). Delhi: Oxford University Press, 1983.

Blakers, M., S.J.J.F. Davies, and P.N. Reilly. *The Atlas of Australian Birds.* Melbourne: Melbourne University Press, 1984.

Cramp, S., ed. *The Birds of the Western Palearctic.* Vol. 5, *Tyrant Flycatchers to Thrushes.* Oxford: Oxford University Press, 1988.

Keith, S., E.K. Urban, and C.H. Fry, eds. *The Birds of Africa.* Vol. 4. London: Academic Press, 1992.

Ridgely, R.S., and G. Tudor. *The Birds of South America.* Vol. 1, *The Oscine Passerines.* Oxford: Oxford University Press, 1989.

Sibley, C.G., and B.L. Monroe. *Distribution and Taxonomy of Birds of the World.* New Haven: Yale University Press, 1990.

Sibley, C.G., and J.E. Ahlquist. *Phylogeny and Classification of Birds: A Study of Molecular Evolution.* New Haven: Yale University Press, 1990.

Stattersfield, A.J., and D.R. Capper, eds. *Threatened Birds of the World: The Official Source for Birds on the IUCN Red List.* Cambridge: BirdLife International, 2000.

Periodicals

Clancey, P.A. "A Review of the Indigenous Pipits (Genus *Anthus* Bechstein: Motacillidae) of the Afrotropics." *Durban Mus. Novit.* 15 (1990): 42–72.

Cooper, M.R. "A Review of the Genus *Macronyx* and Its Relationship to the Yellow-bellied Pipit." *Honeyguide* 31 (1985): 81–92.

Duckworth, J.W., P. Alstrom, P. Davidson, T.D. Evans, C.M. Poole, Tan Setha, and R.J. Timmins. "A New Species of Wagtail from the Lower Mekong Basin." *Bulletin of the British Ornithology Club* 121 (2001): 152–182.

Hall, B.P. "The Taxonomy and Identification of Pipits." *Bulletin of the British Museum of Natural History* 7 (1961): 245–289.

Voelcker, G., and S.V. Edwards. "Can Weighting Improve Bushy Trees? Models of Cytochrome b Evolution and the Molecular Dystematics of Pipits and Wagtails (Aves: Motacillidae)." *Systematic Biology* 47 (1998): 589–603.

Barry Taylor, PhD

Cuckoo-shrikes

(*Campephagidae*)

Class Aves
Order Passeriformes
Suborder Passeri (Oscines)
Family Campephagidae

Thumbnail description
Small to medium-sized birds with broad-based bills, moderately long tails, and erectile rump feathers; some species are brightly colored

Size
5.5–14.5 in (14–37 cm); 0.2–6.3 oz (6–180 g)

Number of genera, species
9 genera; 74 species

Habitat
Forest, woodland, savanna, scrub, and mangroves

Conservation status
Endangered: 1 species; Vulnerable: 3 species; Near Threatened: 9 species; Data Deficient: 1 species

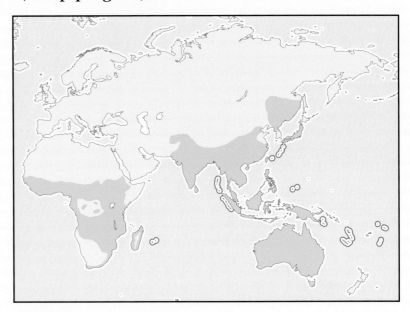

Distribution
Sub-Saharan Africa, southern and Southeast Asia to Australasia and the western Pacific islands

Evolution and systematics

Traditional classifications place the family Campephagidae between the wagtails and pipits (Motacillidae) and the bulbuls (Pycnonotidae). Nine genera are recognized: cuckoo-shrikes (*Pteropodocys*, *Coracina*, *Campochaera*, *Chlamydochaera*, and *Campephaga*), trillers (*Lalage*), minivets (*Pericrocotus*), flycatcher-shrikes (*Hemipus*), and woodshrikes (*Tephrodornis*).

DNA-DNA hybridization studies have suggested that the cuckoo-shrikes' closest relatives are the Old World orioles (Oriolidae). In 1990 Sibley and Monroe included both groups in the tribe Oriolinac within the expanded family Corvidae. This revision has not gained general acceptance, and in 1994 Christidis and Boles retained Campephagidae and Oriolidae as separate families pending further study. Sibley and Monroe also placed the genus *Tephrodornis* in the Corvidae, subfamily Malaconotidae, and tribe Vangini (helmet-shrikes). This radical rearrangement requires further investigation.

While the genus *Chlamydochaera* is retained within the Campephagidae, it has the distinctive syrinx morphology of true thrushes (formerly Turdidae) and muscicapine flycatchers (now united in the subfamily Muscicapinae), while DNA-DNA hybridization studies also suggest a turdine relationship. Storrs Olson in 1987 described skeletal features that confirm the proper placement of *Chlamydochaera* in the Muscicapidae.

Physical characteristics

Cuckoo-shrikes are small- to medium-sized birds. The bill is broad at the base, notched and slightly hooked, and rictal

bristles are well developed. The wings are rather long and pointed, and the tail is moderately long, rounded or graduated. In most genera the back and rump feathers have very stiff shafts and soft tips, and are partially erectile. These spine-like feathers are easily shed and may act as a means of defense.

The minivets are a distinctive group. Most species have brilliantly colored plumage; males are striking red and black, and females are yellow or orange and black or gray. Most other cuckoo-shrikes are less brightly colored, and the female is often a paler, washed-out version of the male. Male flycatcher-shrikes and trillers are typically black and white, while females have the black replaced by gray or brown; female trillers are often barred below. Most *Coracina* species (often called graybirds) are gray or gray and white.

In four *Campephaga* species, males are glossy black with brightly colored gape wattles, while females are olive-yellow and white, most with strong blackish barring. The other two species are sometimes separated into a different genus (*Lobotos*) in view of their predominantly green, yellow, and orange plumage, prominent facial wattles, and lack of strong sexual dimorphism.

Distribution

The family is confined to the Old World, from Africa through south and Southeast Asia to Australasia and the western Pacific islands. The genus *Campephaga* is endemic to Africa, where four endemic *Coracina* species also occur, while

A ground cuckoo-shrike (*Pteropodocys maxima*) at its nest. (Photo by R. Brown/VIREO. Reproduced by permission.)

the western Indian Ocean islands of Madagascar, Mauritius, and Reunion each have one endemic *Coracina* species. The remaining 34 *Coracina* species are distributed through the Oriental and Australasian regions. Twenty-four species occur only in Australasia, where several species have very restricted island distributions.

The ground cuckoo-shrike (*Pteropodocys maxima*) is endemic to Australia. Of the other monospecific genera, the golden cuckoo-shrike (*Campochaera sloetii*) is endemic to New Guinea, and the black-breasted triller (*Chlamydochaera jeffreyi*) to Borneo. The genus *Lalage* is predominantly Australasian in distribution. Only three of its nine species occur in the Oriental region, while two species inhabit Pacific islands east of Western Samoa.

The remaining three genera (*Hemipus*, *Tephrodornis*, and *Pericrocotus*) are predominantly birds of southern and Southeast Asia. The 11 minivet species are widely distributed in southern Asia. The rosy minivet (*Pericrocotus roseus*) and the long-tailed minivet (*Pericrocotus ethologus*) occur as far west as eastern Afghanistan, and the ashy minivet (*Pericrocotus divaricatus*) ranges east to eastern Siberia and Japan.

Habitat

All species except the ground cuckoo-shrike are predominantly or exclusively arboreal, and many are found mainly in the canopy of tall trees. Habitats include the interior and edge of forest (swampy, humid, or dry), woodland, savanna, and scrub. Some species, such as the golden cuckoo-shrike, and the wattled cuckoo-shrikes (*Campephaga lobata* and *oriolina*) of Africa, are restricted to the forest interior, but many are more typical of forest edge, secondary growth, riparian or gallery forest, gardens, or coastal vegetation (including mangroves).

Most *Campephaga* species inhabit forest and woodland areas, but the black cuckoo-shrike (*Campephaga flava*) also frequents acacia savanna, semi-arid bushland, scrub, and exotic plantations. *Coracina* species inhabit many forest types, as well as savanna, woodland, scrub, farmlands, gardens, and plantations, as do the trillers and the flycatcher-shrikes. Minivets

are predominantly birds of the treetops in forest, woodland, urban areas, and sometimes mangroves.

Behavior

Most species are seen singly, in pairs, or in family groups. Mixed-species bird parties of forest or woodland often contain one or more species from this family. In the nonbreeding season, many species, especially the minivets, associate in monospecific parties or flocks, and in Australia the yellow-eyed cuckoo-shrike (*Coracina lineata*) roosts communally.

Most cuckoo-shrikes tend to be unobtrusive and quite silent, although many have loud calls. These whistles or rather raucous squawks are not uttered frequently. However, flocks of minivets, with their bright colors and musical contact calls, are much more obvious. The cicadabird (*Coracina tenuirostris*) has a cicada-like call. Several species have male-female duets.

Some cuckoo-shrikes have a habit of perching motionless on a branch, often in an upright stance and sometimes for long periods. The larger species have the habit of shuffling or refolding their wings on alighting.

Many species, especially those of forested regions, are sedentary. However, other species, especially in Africa and Australia, show local or limited seasonal movements. Three minivet species of central/eastern Asia are long-distance migrants, wintering south to Southeast Asia. The ashy minivet reaches the Philippines.

A white-breasted cuckoo-shrike (*Coracina pectoralis*) will dispose of its chick's fecal sac away from the nest. (Photo by C. Laubscher. Bruce Coleman Inc. Reproduced by permission.)

A black-faced cuckoo-shrike (*Coracina novaehollandiae*) shades its young from the midday heat, and pants to keep its own body temperature down, in Currawinya National Park, Queensland, Australia. (Photo by Wayne Lawler. Photo Researchers, Inc. Reproduced by permission.)

Feeding ecology and diet

Food is predominantly insects and other arthropods, and many species take caterpillars, including hairy ones. Many species also eat fruit, while some take seeds and other vegetable matter. The black-breasted triller is apparently entirely frugivorous, while the varied triller (*Lalage leucomela*) and the yellow-eyed cuckoo-shrike are particularly attracted to figs. Minivets eat predominantly insects, buds, and berries.

Most species take insect prey by gleaning in the foliage of trees, bushes, and creepers, while some also search trunks and branches. Many species also make aerial flycatching sallies, while some occasionally take prey from the ground. The ground cuckoo-shrike feeds mainly on the ground on large insects such as grasshoppers, which it runs to catch.

Reproductive biology

Monogamy is prevalent, and some species are thought to be permanently territorial. The breeding and breeding seasons of many species are poorly known or undescribed. Most species nest solitarily but the white-winged triller (*Lalage sueurii*) of-

ten nests in loose colonies and the ground cuckoo-shrike also nests communally. Males of some of the larger cuckoo-shrike species have a courtship display in which the bird lifts each wing alternately, calling strongly while doing so.

The nest is a small, shallow cup of fine twigs, roots, bark, grasses, lichens, or moss, often bound together and sometimes lined with spider webs. It is usually placed on a fork or horizontal branch high in a tree and is difficult to see from below. In many species both sexes build the nest, but in the genera *Campephaga* and *Pericrocotus* nest-building is done chiefly by the female, assisted by the male.

The clutch is one to five eggs, usually two or three. In many species only the female incubates; in others both sexes share incubation duties. The incubation period varies from 14 to 25 days, and in many species is three weeks or more. The nestling period is of similar length (13–24 days). Both sexes care for the young. Most species for which information is available breed during or just after the rains.

Conservation status

Only four species are considered threatened. The Reunion cuckoo-shrike (*Coracina newtoni*) is Endangered. In 2000 it had a population of 120 pairs and a very small range of 6.2 mi² (16 km²) in increasingly degraded forest habitat. Of the three Vulnerable species, the Mauritius cuckoo-shrike (*Coracina typica*) has a very small range. However, it has responded well to rehabilitation of native ecosystems, and its small population is increasing. The white-winged cuckoo-shrike (*Coracina ostenta*) of the Philippines is declining through destruction of its forest habitat.

Nine species are Near Threatened and six of these occur only in Indonesia or the Philippines, where forest destruction has been severe.

Detailed information is lacking on the current status of most species, but many have suffered from habitat loss. A few species have adapted to modified or degraded habitats. In New Guinea, the white-bellied cuckoo-shrike (*Coracina papuensis*) has invaded some urban areas, where it is locally abundant.

Significance to humans

Although some cuckoo-shrikes are brightly colored and many are vocal, almost all are relatively unobtrusive and often overlooked by humans. They have little or no significance in either a cultural or agricultural context.

1. Varied triller (*Lalage leucomela*); 2. Blue cuckoo-shrike (*Coracina azurea*); 3. Mauritius cuckoo-shrike (*Coracina typica*); 4. Fiery minivet (*Pericrocotus igneus*); 5. Black-winged flycatcher-shrike (*Hemipus hirundinaceus*); 6. Golden cuckoo-shrike (*Campochaera sloetii*); 7. Western wattled cuckoo-shrike (*Campephaga lobata*); 8. Red-shouldered cuckoo-shrike (*Campephaga phoenicea*). (Illustration by Emily Damstra)

Species accounts

Western wattled cuckoo-shrike

Campephaga lobata

TAXONOMY

Ceblepyris lobatus Temminck, 1824, Gold Coast (Ghana).

OTHER COMMON NAMES

English: Ghana cuckoo-shrike; French: Echenilleur à barbillons; German: Westlicher lappenraupenfresser; Spanish: Oruguero Occidental.

PHYSICAL CHARACTERISTICS

8.3 in (21 cm); 1.0–1.3 oz (29–37 g). Male has glossy black head and black bill with orange wattles on gape. Breast and flanks orange-yellow, wings greenish yellow. Female slightly duller with greenish forehead and crown.

DISTRIBUTION

East Sierra Leone, Liberia, west Côte d'Ivoire, south Ghana, one record in southeast Nigeria.

HABITAT

Upper Guinea forest zone, in the canopy of primary or tall lowland rainforest, often near rivers, and in open swamp-forest. Also recorded in logged forest, *Terminalia* plantations, disturbed forest, and managed forest.

BEHAVIOR

Usually seen singly or in pairs, often with a mixed bird party; inconspicuous.

FEEDING ECOLOGY AND DIET

The diet includes caterpillars, grasshoppers, mantids, and seeds.

REPRODUCTIVE BIOLOGY

Descriptions of the nest and eggs are not available. Birds from Liberia are known to be in breeding condition in February and during the months of August to November.

CONSERVATION STATUS

Although it has apparently adapted to secondary habitat, this bird is seriously threatened by massive forest destruction throughout its range. It is therefore considered Vulnerable. In 2000, it was locally rare to uncommon over much of its range.

SIGNIFICANCE TO HUMANS

None known. ◆

Mauritius cuckoo-shrike

Coracina typica

TAXONOMY

Oxynotus typicus Hartlaub, 1865, Mauritius.

OTHER COMMON NAMES

French: Echenilleur de Maurice; German: Mauritiusraupenfänger; Spanish: Oruguero de Mauricio.

PHYSICAL CHARACTERISTICS

8.7 in (22 cm); one male 1.5 oz (43 g). Gray upperparts and grayish white below with blackish wings; females orange-brown above and rich orange rufous below.

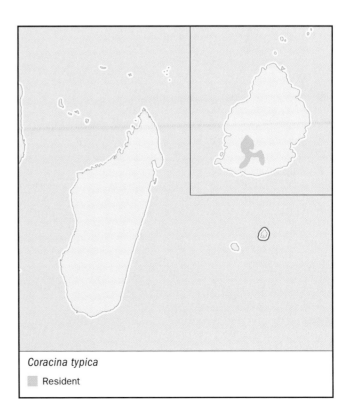

Campephaga lobata

Resident

Coracina typica

Resident

DISTRIBUTION
South Mauritius.

HABITAT
Canopy of moist tropical evergreen forest above 1,500 ft (460 m); also adjacent degraded or altered forest.

BEHAVIOR
Lives solitarily or in pairs; territorial throughout the year. Unobtrusive and secretive, but may be located by its melodic trill and harsh call-note.

FEEDING ECOLOGY AND DIET
It eats mainly large arthropods, including caterpillars, mantids, stick insects, and beetles, and also geckos.

REPRODUCTIVE BIOLOGY
Highest density 25 territories/km^2. Breeds from September to March, during the rains. Monogamous. The nest is a shallow cup of fine twigs, lichens, and spider webs, and is attached to a horizontal tree branch. Both sexes build the nest. The female lays two eggs; incubation is by both sexes for 24–25 days.

CONSERVATION STATUS
Habitat loss and degradation have caused a long-term decline in the population cinced human colonization and remains a long-term threat. The species is considered Vulnerable because of its very small range and population (300–350 pairs in 2000). Since 1975, its range and density have increased, thanks to conservation action to rehabilitate native ecosystems.

SIGNIFICANCE TO HUMANS
None known. ◆

Blue cuckoo-shrike

Coracina azurea

TAXONOMY
Graucalus azureus Cassin, 1852, Sierra Leone.

OTHER COMMON NAMES
French: Echenilleur bleu; German: Azurraupenfänger; Spanish: Oruguero Azul Africano.

PHYSICAL CHARACTERISTICS
8.5 in (21.5 cm); 1.5–1.8 oz (43–51 g). Bright blue plumage with black mask, flight feathers, and tail. Eyes red. Female has duller, green-blue forehead and wings.

DISTRIBUTION
Sierra Leone, Liberia, southwest Côte d'Ivoire, south Ghana and south Nigeria to southwest Cameroon, Gabon, southwest Central African Republic, north and central DRC and northwest Angola.

HABITAT
Canopy of lowland primary and secondary forest; also open woodland and sometimes clearings.

BEHAVIOR
Territorial. Unobtrusive and easily overlooked despite its bright blue color, but its powerful "peeeoo" calls advertise its presence. It joins mixed-species bird parties.

FEEDING ECOLOGY AND DIET
Eats caterpillars, grasshoppers, termites, beetles, snails, and occasionally small fruit. Forages like a flycatcher, snatching prey in the air and from foliage; also gleans.

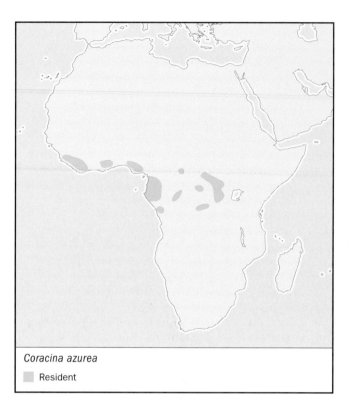

Coracina azurea
  ▨ Resident

REPRODUCTIVE BIOLOGY
Monogamous. Breeds November–March, when rainfall is relatively low. Nest is a loose bowl of lichens and spider webs on a horizontal tree branch.

CONSERVATION STATUS
Not threatened. Its known range is fragmented, and it has suffered considerably from habitat destruction. In the 1980s, it was still locally frequent to common in some parts of its range.

SIGNIFICANCE TO HUMANS
None known. ◆

Red-shouldered cuckoo-shrike

Campephaga phoenicea

TAXONOMY
Ampelis phoenicea Latham, 1790, Gambia.

OTHER COMMON NAMES
French: Echenilleur à épaulettes rouges; German: Mohrenraupenfresser; Spanish: Oruguero de Hombros Rojos.

PHYSICAL CHARACTERISTICS
8 in (20 cm); 0.8–1.3 oz (23–35.5 g). Male black with red or orange shoulder-patch; female heavily barred below.

DISTRIBUTION
South Mauritania, Senegal and Gambia east to Ethiopia and south to north DRC, Uganda, and west Kenya.

HABITAT
Bushes and small trees in forest patches, secondary growth, wooded grassland, and thickets in savanna.

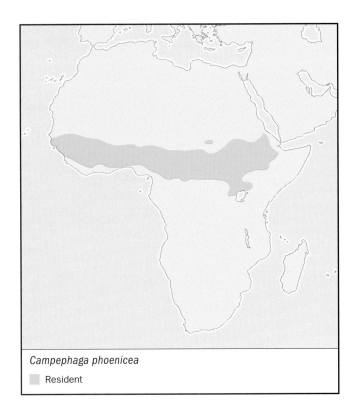

Campephaga phoenicea
■ Resident

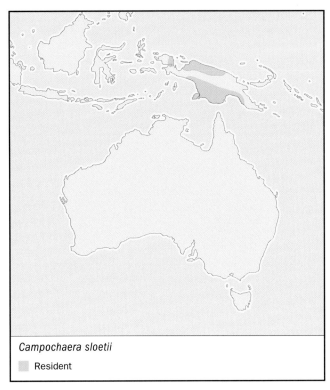

Campochaera sloetii
■ Resident

BEHAVIOR
Usually occurs singly or in pairs; unobtrusive and mainly silent. Flight is undulating and low between trees. Sedentary or nomadic. In Nigeria and Sudan, it moves north to breed during rains.

FEEDING ECOLOGY AND DIET
Eats caterpillars and other insects, especially Orthoptera and Hemiptera. Forages mainly by gleaning from leaves and branches; also takes prey on ground and in flycatching sallies.

REPRODUCTIVE BIOLOGY
Breeds May–September, during rains. Monogamous. Nest is a small, shallow cup of moss, lichens, and spider webs, and is well concealed in fork of leafless tree. Lays two eggs.

CONSERVATION STATUS
Widespread and uncommon to locally common; not threatened. There is no information on effects of habitat loss.

SIGNIFICANCE TO HUMANS
None known. ◆

Golden cuckoo-shrike
Campochaera sloetii

TAXONOMY
Campephaga sloetii Schlegel, 1866, New Guinea. Two subspecies.

OTHER COMMON NAMES
English: Golden triller; French: Echenilleur doré; German: Goldraupenfresser; Spanish: Oruguero Anaranjado.

PHYSICAL CHARACTERISTICS
7.9 in (20 cm); 1.3–1.6 oz (36–46 g). Grayish white forehead and browline with olive-gray crown. Neck to uppertail-coverts

are orange-yellow, as is the abdomen. Black wings with two white stripes; throat and upper breast black.

DISTRIBUTION
C. s. sloetii: Arfak Mountains (foothills) and north New Guinea lowlands east to Wewak area. *C. s. flaviceps*: South New Guinea lowlands from Mimika River east to Port Moresby area.

HABITAT
Tall tree canopy of forest interior and edge.

BEHAVIOR
Occurs in pairs or small parties. Active and noisy, with musical, high-pitched whistling calls. Duetting displays are recorded.

FEEDING ECOLOGY AND DIET
Eats fruit; reports of feeding on insects are unconfirmed.

REPRODUCTIVE BIOLOGY
Unknown; it is thought to breed during the rainy season (November–March).

CONSERVATION STATUS
Not threatened. The known distribution is rather patchy but the species was formerly regarded as not uncommon locally.

SIGNIFICANCE TO HUMANS
None known. ◆

Varied triller
Lalage leucomela

TAXONOMY
Campephaga leucomela Vigors and Horsefield, 1827, Queensland. Fifteen subspecies.

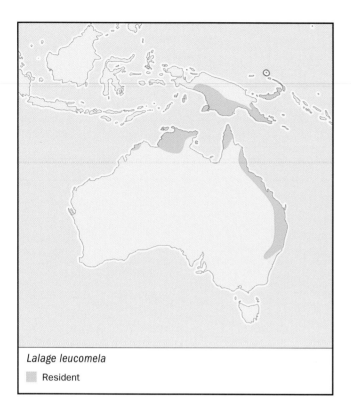

Lalage leucomela
▨ Resident

FEEDING ECOLOGY AND DIET
Feeds on insects (especially caterpillars), fruit, berries, and seeds; particularly attracted to figs.

REPRODUCTIVE BIOLOGY
Assumed monogamous. Breeds during rains. Nest is a very small shallow cup of dry grass, twigs, rootlets, and spider webs on a horizontal branch or fork. Lays one egg; incubation/fledging unknown.

CONSERVATION STATUS
Not threatened. Considered locally fairly common in New Guinea and North Australia.

SIGNIFICANCE TO HUMANS
None known. ◆

Fiery minivet
Pericrocotus igneus

TAXONOMY
Pericrocotus igneus Blyth, 1846, Malacca. Sometimes considered a race of small minivet (*P. cinnamomeus*) but is morphologically distinct and marginally sympatric.

OTHER COMMON NAMES
English: Small minivet; French: Minivet flamboyant; German: Feuerrotmennigvogel; Spanish: Minivete Chico.

PHYSICAL CHARACTERISTICS
6–6.5 in (15–16.5 cm); 0.5–0.6 oz (14–16 g). Male has black upperparts and throat with red breast, belly, rump, and outer tail feathers. Female has gray upperparts with orange rump, yellow underparts, and black tail.

OTHER COMMON NAMES
English: White-browed/pied triller; French: Echenilleur varié; German: Weißbrauenlalage; Spanish: Gorjeador de Cejas Blancas.

PHYSICAL CHARACTERISTICS
6.7–7.1 in (17–18 cm); 0.8–1.3 oz (24–37.5 g). Males have black upperparts, dark gray rumps, and white markings through the wings. Their underparts are white with fine dark barring and a cinnamon vent. Females have browner upperparts and barred, gray-buff underparts.

DISTRIBUTION
L. l. keyensis: Kei Islands; *L. l. rufiventer*: Melville Island and coastal Northern Territory; *L. l. leucomela*: North Bismarck Archipelago, East Queensland to North New South Wales; *L. l. yorki*: North Queensland; *L. l. polygrammica*: Aru Islands and east New Guinea; *L. l. obscurior*: D'Entrecasteaux Archipelago; *L. l. trobriandi*: Trobriand Islands; *L. l. pallescens*: Louisaide Archipelago; *L. l. falsa*: New Britain, Umboi Island, Duke of York Islands; *L. l. karu*: New Ireland; *L. l. albidior*: New Hanover; *L. l. ottomeyeri*: Lihir Island; *L. l. tabarensis*: Tabar Island; *L. l. conjuncta*: St. Matthias Island; *L. l. sumunae*: Djaul Island.

HABITAT
Canopy of tropical and subtropical rainforest, dense eucalyptus forest, forest edges, and secondary growth; sometimes in mangroves and dense savanna; also gardens.

BEHAVIOR
Normally seen singly or in pairs, sometimes in threes. Inconspicuous but vocal, with a loud, rolling, repeated "brreeer" call; several birds may call together. Sometimes sits quietly on exposed perch. Sedentary over most of its range; some movements in Australia.

Pericrocotus igneus
▨ Resident

DISTRIBUTION
Malay Peninsula (south Burma, south Thailand, Malaysia), Sumatra, Borneo, and southwest Philippines (Palawan).

HABITAT
Canopy of forest and forest edges; also pine plantations and casuarinas. Occurs in lowlands, but lives in montane forest up to 8,900 ft (2,700 m) in Sumatra.

BEHAVIOR
Has a rising "swee-eet" call. Joins mixed-species bird parties.

FEEDING ECOLOGY AND DIET
Not specifically described; probably insects, especially caterpillars and moths. Forages in the canopy.

REPRODUCTIVE BIOLOGY
Probably monogamous. Breeds in Palawan in December (dry season) and in Malaysia in May (start of rains). Nest is not described; probably as for small minivet: a cup of fine twigs, stems, leaves, lichens, and spider webs on a tree branch. Lays two eggs.

CONSERVATION STATUS
Not threatened. In the 1980s–90s regarded as locally quite common in Palawan and Sumatra but uncommon in Thailand.

SIGNIFICANCE TO HUMANS
None known. ◆

Hemipus hirundinaceus

▨ Resident

Black-winged flycatcher-shrike
Hemipus hirundinaceus

TAXONOMY
Muscicapa hirundinacea Temminck, 1822, Java.

OTHER COMMON NAMES
French: Echenilleur véloce; German: Schwarzflügel-raupenschmätzer; Spanish: Minivete de Alas Negras.

PHYSICAL CHARACTERISTICS
6 in (15 cm); 0.4 oz (10 g). Black plumage with dusky white underparts and white rump.

DISTRIBUTION
Extreme south Thailand, peninsular Malaysia; Riau and Lingga Archipelagos to Sumatra, Borneo, Java, and Bali.

HABITAT
Canopy of forest, including swamp forest and tall secondary growth; also mangroves, mature plantations, and wooded gardens. Lowlands to 3,600 ft (1,100 m), but to 4,900 ft (1,500 m) in Java and Bali.

BEHAVIOR
Moves around in pairs or small parties; unobtrusive except for its short trilling or twittering calls

FEEDING ECOLOGY AND DIET
Eats small insects.

REPRODUCTIVE BIOLOGY
In Sumatra, Borneo, and peninsular Malaysia, it breeds January–April and August–November. Monogamous. Nest is a truncated cone with a cup-shaped depression, consisting of lichens, bryophytes, and spider webs and placed on a horizontal branch. Lays two eggs; young are fed by both parents.

CONSERVATION STATUS
Not threatened. Probably locally frequent to common, but was regarded as local and uncommon in Thailand in the 1980s–90s.

SIGNIFICANCE TO HUMANS
None known. ◆

Resources

Books
Ali, S., and S.D. Ripley. *Handbook of the Birds of India and Pakistan.* Compact ed. Delhi: Oxford University Press, 1983.

Christidis, L., and W.E. Boles. *The Taxonomy and Species of Birds of Australia and Its Territories.* Royal Australasian Ornithologists Union Monograph 2. Hawthorn East: RAOU, 1994.

Keith, S., E.K. Urban, and C.H. Fry, eds. *The Birds of Africa.* Vol. 4. London: Academic Press, 1992.

Sibley, C.G., and B.L. Monroe. *Distribution and Taxonomy of Birds of the World.* New Haven, CT: Yale University Press, 1990.

Stattersfield, A.J., and D.R. Capper, eds. *Threatened Birds of the World: The Official Source For Birds on the IUCN Red List.* Cambridge: BirdLife International, 2000.

Resources

Periodicals

Olson, S.L. "More on the Affinities of the Black-Collared Thrush of Borneo (*Chlamydochaera jefferyi*)." *Journal of Ornithology* 128 (1987): 246–248.

Ripley, S.D. "Notes on the Genus *Coracina*." *Auk* 58 (1941): 381–395.

Voous, K.H., and J.G. van Marle. "The Distributional History of *Coracina* in the Indo-Australian Archipelago." *Bijdragen Tot De Dierkunde* 28 (1949): 513–529.

Barry Taylor, PhD

Bulbuls
(Pycnonotidae)

Class Aves
Order Passeriformes
Suborder Passeri (Oscines)
Family Pycnonotidae

Thumbnail description
Medium sized, short-necked songbirds with short round, wings; fluffy plumage, many with crested heads

Size
3.6–11.5 in (9.3–29 cm); 0.5–2 oz (15–59 g)

Number of genera, species
15 genera; 123 species

Habitat
Forest, forest edge, open woodland, scrub, cultivated lands

Conservation status
Endangered: 2 species; Vulnerable: 5 species

Distribution
Africa, Indian subcontinent, Asia, Indonesia, northwestern Australia, Malagasy region, introduced populations in United States

Evolution and systematics

In the introduction to his 1945 revision of the family Pycnonotidae, Jean Delacour states that "The bulbuls constitute one of the most clearly defined groups of perching birds (Passeres). This means that the different genera and subgenera which belongs to the group are obviously related to one another, and rather far removed from any others." Recognition of the bulbuls as a "natural" group has not stood the test of time, however, and the placement of the Pycnonotidae within Passeriformes remains unresolved, as do the relationships of these songbirds to each other.

Depending on the characters examined, the Pycnonotidae have been allied with a number of families. Early authors placed them next to cuckoo-shrikes (Campephagidae) based on shared anatomical and plumage characters, such as abundant rump feathers. Egg-white protein analysis in the early 1970s suggested a relationship to drongos (Dicruridae) or starlings and orioles (Sturnidae, Oriolidae). Current DNA evidence suggests that bulbuls belong in the superfamily Sylvioidea, between kinglets (genus *Regulus*, sometimes treated as a family, Regulidae) and the African warblers (genus *Cisticola*, sometimes treated as a family, Cisticolidae). More controversial than the family placement however, are the relationships of various bulbuls to each other. Several species will probably eventually be shuffled to different genera and even other families as research continues. DNA analyses in 2001 suggest that the African and Asiatic genus *Criniger*, united chiefly by the conspicuously colored throat, is not a natural group. This study recommends that the Asian *Criniger* take the genus name *Alophoixus*, and the African species remain as *Criniger*. Other research examining the relationships of 13 birds endemic to Madagascar, including some of the African greenbuls (*Phyllastrephus* spp.) found that the Mala-

gasy genera, normally distributed across three families, were actually all most closely related to each other. Some ancient relative of these birds probably arrived at the island at least nine million years ago, then evolved and diversified there rather than different birds of different lineages colonizing the island several times, as previously thought. This study recommends that the Malagasy bulbuls be removed from the genus.

Also controversial is the placement of two of the endemic African genera *Neolestes* and *Nicator*. The striking plumage of the black-collared bulbul (*Neolestes torquatus*) allies it with the shrikes (Malaconotidae, Laniidae, or Prionopidae), DNA data ally it with other bulbuls. Similarly, *Nicator* has also been allied with the shrikes, but feather protein and DNA evidence suggest the birds are bulbuls. Because the three *Nicator* species and *Neolestes* lack a thin sheet of nostril-covering bone that is present in the rest of the bulbuls, they are sometimes placed elsewhere.

Whatever their relationships, it is clear that birds considered bulbuls evolved in the last 10 million years and much of the family may have radiated more recently, probably during the Plio-Pleistocene, about 2.8 million years ago. During the Tertiary, northern Africa experienced general drying, and the eastern lowland forests became isolated as mountains swelled and rifted in the region. *Andropadus* probably colonized these "new" mountain habitats, which became centers of speciation for the group.

Physical characteristics

Most of the bulbuls are dressed in somber browns, olive tones, or grays, and are often heard before they are seen.

However, several species have distinctive face markings with a bright splash of red, yellow, orange, or white plumage. The plumage of *Pycnonotus* tends to be more variable than other genera, and these birds usually have are a red, yellow, orange, or white "vent" which contrasts with the rest of their underparts. Many species, especially among the *Pycnonotus*, *Criniger*, and some *Hypsipetes*, have long crown feathers that form a prominent head crest; the crest is absent in *Andropadus*. Almost all have at least a few bristles on the nape of their short necks and a small area without feathers, such that a bulbul with its neck stretched out shows a small bare patch between the nape and upper back. The tail can be fairly long, usually with a round tip, although it is slightly forked in some species.

Small to medium-sized birds, bulbuls range in size from 3.6 to 11.5 in (9.3–29 cm); and between 0.5 and 2 oz (15–59 g). Except for the finchbills (*Spizixos* spp.), the pycnonotids have medium-sized, slender, notched bills. Among the *Criniger*, the bill is usually strongly hooked at the tip. Bulbuls tend to have short, weak legs and short, rounded wings. Males and females usually look alike, though the female is often slightly smaller; this size difference is greatest in *Phyllastrephus*. Likewise, the juveniles look like the adults, but their plumage is duller and often more brown.

Distribution

Bulbuls occur mainly in forest and wooded areas across Africa and Madagascar, north to the Middle East and Japan, and east to the Philippines and Indonesia. Except for recently introduced species, there are no bulbuls in the New World. Of the roughly 120 species, there are 27 species in China. About 52 species (in 11 genera) are African, and bulbuls are widely distributed across that continent, being absent only from desert regions. Nine genera are found exclusively in Africa, and two of the larger genera (*Pycnonotus* and *Criniger*) are found in both Africa and Asia. The most ubiquitous of the *Pycnonotus* species is the common bulbul (*P. barbatus*). This bold and noisy bird is well adapted to man-made habitats and is one of the most widespread and abundant birds in Africa.

Many bulbuls are endemics and have quite restricted distributions, especially those on oceanic islands. The yellow-eared bulbul (*Pycnonotus pencillatus*) is found exclusively in the montane areas of Sri Lanka; the Nicobar bulbul (*Hypsipetes nicobariensis*) occurs only on the southern Nicobar Islands south of Burma. Others are widespread and common, the yellow-vented bulbul (*Pycnonotus goaivier*), occurring from southern Vietnam to the Malay Peninsula and the Philippines, is probably the most commonly seen bulbul of the region. An opportunistic bird, it is often seen around gardens, feeding on food scraps and using potted plants to nest in.

In 1960 a population of red-whiskered bulbuls (*Pycnonotus jocosus*) became established in Florida when a few birds escaped while being transported from one aviary to another. This population had increased to 500 birds by 1973, at which time it was still expanding in a southerly direction. Red-whiskered bulbuls also became established in Los Angeles County, California, in 1968. Both the red-whiskered and red-

A red-vented bulbul (*Pycnonotus cafer*) at its nest with a chick and eggs in India. (Photo by G.D. Dodge & D.R. Thompson. Bruce Coleman Inc. Reproduced by permission.)

vented bulbuls (*P. cafer*), were introduced to the Hawaiian island of O'ahu in the late 1960s. Populations of both birds have dramatically increased. Red-vented bulbuls are now found across the island, while the red-whiskered bulbul is found throughout southeastern areas. These birds are considered serious pests and threats to native bird populations.

Habitat

Forest, open woodland, gardens, and cultivated areas all constitute bulbul habitat. Essentially arboreal birds, the majority of pycnonotids live in or next to forested areas, but many are well adapted to human-made habitats. Many bulbuls show a preference for a particular level of the forest canopy. So as long as there is enough fruit and insects, a relatively small area of forest may support a large number of birds.

The majority of *Andropadus* species keep to the forest interior, as do many others, such as Finsch's bulbul (*Criniger finschii*) of Indonesia, and the African gray-headed greenbul (*Phyllastrephus poliocephalus*). Other species prefer more open areas and are found frequently in edge habitat. The gray-headed bristlebill (*Bleda canicapilla*) and the white-throated greenbul (*Phyllastrephus albigularis*), both found exclusively in western north-central Africa, are typically found at forest clearings and edges where there is tangled vegetation in which they can forage for insects. *Pycnonotus* species are well adapted to drier habitats, and are frequently found in areas that have been cultivated by humans. A few are confined to the forest, but they generally inhabit open country with scattered trees and shrubs. The yellow-throated leaf-love (*Chlorocichla flavicollis*) actually avoids large forest blocks, occurring mainly in brushy, more open areas. This bird also adapts extremely well to man-made habitats; and can be found in plantations, abandoned cultivated sites, orchards, parks, and gardens.

Several bulbuls show a preference for water and are found alongside rivers and forest streams. The gray-olive bulbul (*Phyllastrephus cerviniventris*) is one such bird. Infrequently entering the forest, this smallish bulbul frequently inhabits streamside thickets. Primarily an insect eater in Zambia, it is especially fond of feeding on logs that have fallen across streams or ravines. The swamp greenbul (*Thescelocichla leucopleura*) and the leaf-love (*Pyrrhurus scandens*) are also partial to water, both prefer swampy areas with luxuriant vegetation and palm trees, especially *Raphia* and the oil palm *Elaeis*.

Although most bulbuls prefer areas with lots of green vegetation, a small number are found in drier scrub habitats, especially *Pycnonotus*. The African red-eyed bulbul (*Pycnonotus nigricans*) occupies drier areas, including savanna, semiarid scrub, and bushy hillsides. The northern brownbul (*Phyllastrephus strepitans*) also prefers scrub, and is often the only bulbul present in the driest parts of its range.

Behavior

Many bulbuls are quite social; some readily join mixed-species groups, others flock with members of their own species. Perhaps the most gregarious bulbul is the spotted greenbul (*Ixonotus guttatus*). A distinctively colored bird, gray-olive with white spots on its wings and rump, the spotted greenbul is very social, traveling in monospecific groups of five to 50 birds. The groups work their way quickly through trees, never staying long, even if there is plenty of fruit. While briefly at rest they sit very close together on a branch, preening and flicking the wings and tail. The striated bulbul (*Pycnonotus striatus*), so named because of the yellow-streaking on its underparts and white-streaked upper body, lives in active noisy flocks of six to 15 birds, as does the yellow-browed bulbul (*Hypsipetes indicus*). The yellow-streaked greenbul (*Phyllastrephus flavostriatus*), Cabanis's greenbul (*P. cabanisi*), and Sjöstedt's honeyguide greenbul (*Baeopogon clamans*) often join mixed-species flocks, traveling right behind or mixed in with other bird species.

Bulbuls can be quite aggressive towards members of their own species, and other species as well. Some, such as the puff-throated bulbul (*Criniger pallidus*) and the mountain bulbul (*Hypsipetes mcclellandii*) will aggressively mob birds of prey. If the face-off is against a bulbul of the same species, the threat display may be different than against other birds. Among *Pycnonotus*, there are roughly three to seven threat displays. These include tail-flicking, tail-spreading, crest-raising, undertail-covert spreading, wing-flicking-and-spreading, and crouch display (the latter may also be an appeasement display). The red-vented bulbul has been observed attacking birds by poking with its bill. The red-tailed greenbul (*Criniger calurus*) and other *Criniger* bulbuls will puff out their fluffy beard-like throat feathers, both as a preening gesture and as an aggressive display.

A cape bulbul (*Pycnonotus capensis*) at its nest in South Africa. (Photo by W. Tarboton/VIREO. Reproduced by permission.)

Most bulbuls have distinctive voices, and often the best way of distinguishing similar species is by their song. All over the map, their calls may be jolly phrases of three to six notes (*Pycnonotus flavescens*), cacophonous explosions of loud discordant babbles (*Pycnonotus luteolus*), or mewing cat-like calls (*Hypsipetes mcclellandii* and *H. leucocaphalus*). Typically, their voices have a gravelly quality, many are chattery and noisy, some with whistles. Very few are actually musical. Most sing in the morning or evening, and many of the more social species will chatter as they forage. The somber greenbul (*Pycnonotus importunus*) constantly advertises its presence with its distinctive song, a series of clear strong notes "ti-ti-wer cheeo-cheeo cheeo-wer chi-wee chi-wer chi-wee." A most persistent vocalist, this bird sings all day, even at midday, and all year. The name *importunus* is from "importunate," named because it sings persistently to the point of annoyance.

Sadly, the species with the most celebrated song, the straw-headed bulbul (*Pycnonotus zeylanicus*) is now threatened as a result of being highly prized and traded for its voice. Described as "a prolonged series of magnificently warbled notes, richer and more powerful by far than the songs of such celebrated performers as the Nightingale and the blackbird," the song of the straw-headed bulbul is by no means typical of the Pycnonotidae.

The majority of the Pycnonotidae are nonmigratory, either sedentary or only locally nomadic. Banding and recapture records from Asia indicate that some bulbuls remain in the same few hundred yard area for several years. A handful of the cooler-climate, temperate-zone species, such as the black bulbul (*Hypsipetes madagascariensis*), are partly migratory. Flocks of several hundred of these birds move to southern China in winter.

Feeding ecology and diet

The bulbul diet spans the range of fruits and berries to insects and other arthropods, as well as small vertebrates such as frogs, snakes, and lizards. A few eat nectar and pollen. The jaw apparatus of Pycnonotids is rather generalized compared to other Passeriform birds, and while some Pycnonotids eat mainly fruit *or* insects, most can and do have a mixed diet. This flexibility may be critical during the dry season: since most bulbuls are non-migratory, they must take advantage of the food sources available within in their range, which can mean shifting to feeding on more plant matter when insects are not as abundant.

Among *Phyllastrephus* and *Criniger*, diets tend more toward insects such as caterpillars, dragonflies, wood lice, and ants. The diets of *Chlorocichla* and *Hypsipetes* include more fruit, and many of these bulbuls are important for dispersing the seeds of forest plants and as pollinators. Fruit-eating species typically forage in trees, shrubs, and bushes, gathering fruits and berries while perched on twigs and stems. They will often consume smaller fruits whole, and will repeatedly peck fruits with hard, thick walls until they have torn a hole in the outer coat to get at the pulp. Figs (*Ficus* spp.), are present in the diet of most fruit-eating bulbuls, as well as *Schefflera*, *Musanga*, and *Lantana* berries. These birds can do serious damage to orchards and other cultivated fruit crops. Indiscriminate in their preference for native or exotic berries, fruit-eating bulbuls often disperse noxious, weedy plant seeds.

Most insect-eating bulbuls forage on and among vegetation, but will also sally for insects in the air and hunt along the ground on fallen logs and branches. Many favor caterpillars and dragonflies, and several species have been found attending swarms of army ants. The yellow-bellied greenbul (*Chlorocichla flaviventris*) frequently forages on antelopes, landing on the animal and grooming its head, ears, and even eyes, presumably searching for small insects in the antelope's coat.

Although most bulbuls are omnivorous, a few specialize in certain foods. Sjöstedt's honeyguide greenbul is closely associated with the small, black wasps *Polybioides melaina* that build large paper nests in riverbank trees. The bird will tear apart the nests, despite the vicious retaliation of the wasps, and the young birds are fed exclusively on the wax, larvae, and pupae of the insect.

Reproductive biology

Most bulbuls are found in pairs, or in small groups that tend to be family parties and often include juveniles. Mostly monogamous and territorial, except for the yellow-whiskered greenbul (*Pycnonotus latirostris*), a lekking species, some bulbuls will form groups that defend a large home range together. Both the leaf-love and the swamp greenbul will gather and chorus to defend communal territory. In the case of the swamp greenbul, the loud vocalizations are accompanied by displays of spread wings and tail.

The timing of bulbul reproduction varies greatly, depending on the climate and region. In some areas breeding appears to be tied to rainfall, and many species have two broods per year, usually before and after the monsoon season. Breeding is common year-round in other species, and some African species may breed throughout the rainy season, or after the rains. Most bulbuls are monogamous and territorial, often the pair-bond is maintained year after year. One species, the yellow-whiskered greenbul, has a quite flexible social system, and in high density areas uses leks, but is monogamous and territorial in lower-density areas. There is evidence of cooperative breeding in a handful of species including, the spotted greenbul, the yellow-throated leaf-love, and *Ixonotus*. Groups of four to six birds will feed the young, usually both in the nest and after leaving it.

Pre-copulatory displays have been observed in some bulbuls, in which the birds chase each other while softly calling. Little is known about the selection of the nest. It is usually built by both parents, although in some cases just the female. Nests tend to be an untidy cup nestled in the fork of a tree. The construction materials vary, but usually include a variety of twigs, rootlets, plant stems, grasses, cobwebs, and hairs. Some species "decorate" the outside of the nest with fern fronds or bark strips. There are usually two eggs per clutch, although there may be as many as five, and there is great variation in the egg color and markings, even within a single species. Some eggs are elongated ovals, others truncated ovate, they may be glossy white, pinkish, lilac, gray, brown, and mauve with scratchy markings, blotches, or spots, sometimes concentrated at one end so the egg appears to be "capped." In some species both parents incubate the eggs, in others the female only, usually for 10–14 days. Young are born naked and are usually cared for by both parents. They are often just given insects at first, even among the species that primarily eat fruit. They generally fledge at 14 days, but as early as seven days in some species.

Conservation status

The primary threat to the Endangered (two species) and Vulnerable (five species) bulbuls is habitat loss, though hybridization and trapping for the caged bird trade are also problematic. The effect of destruction of habitat is especially pronounced because so many of these birds have quite restricted ranges. The Endangered streak-breasted bulbul (*Hypsipetes siquijorensis*) is endemic to four islands of the Philippines, and while it lives in open areas, forest in some condition appears to be essential to its survival. Forest destruction has severely affected population numbers, and although on the island of Siquijor the remaining four forest patches are now reserves, suitable protection of the rest of its habitat may be critical to its survival. Habitat loss is also a problem for the Nicobar bulbul (*Hypsipetes nicobarensis*), found exclusively on the Nicobar islands of India. This bird also suffers from competition with the recently introduced Andaman red-whiskered bulbul (*Pycnonotus jocosus*), which has flourished throughout the islands. Expansion of the Great Nicobar Biosphere Reserve would provide adequate protection for this bird, and controlling the populations of introduced species might also help. The Vulnerable Styan's bulbul (*Pycnonotus taivanus*), endemic to the lowlands of Taiwan, is also threatened by habitat loss. Also problematic is hybridization with the Chinese bulbul (*Pycnonotus sinensis*), whose range has expanded to overlap with *Pycnonotus taivanus*. The only genetically distinct Taiwan bulbuls left live in the coastal mountains. A protected species since 1995, captive-breeding programs and protected areas from which Chinese bulbuls are excluded have been proposed. Habitat loss is also a concern for the hook-billed bulbul (*Setornis criniger*), confined to the islands of Borneo, Sumatra, and Bangka in the greater Sundas, and the yellow-throated bulbul (*Pycnonotus xantholaemus*), found exclusively on the southern peninsula of India. Although both birds have some generic protection, preservation of suitable habitat is recommended for both.

The straw-headed bulbul (*Pycnonotus zeylanicus*) is now threatened as a result of being highly prized for its voice and hence traded as a caged bird. Listed on Appendix II of CITES, some measures have been taken to protect this bird, but it is still widely traded, and captive-breeding programs are subject to theft. Habitat protection might also help as long as areas are guarded.

In addition to habitat protection, the ecology and behavior of these birds must be studied further to better understand survival rates and ecological needs. Efforts to protect these birds by local governments and international groups will do little without the involvement of local people.

Significance to humans

It is not surprising that these vocal birds, many well adapted to human-made habitats, often figure in the folk-tales and songs of the regions in which they live. In each region where bulbuls are common, local lore is often associated with the more conspicuous birds. In Ghana, the noisy swamp greenbul is known as the "talky-talky bird," and children purportedly refuse to eat the flesh of these birds, for fear if they do they will never stop talking. The red-whiskered bulbul is mentioned in many local songs of the Lanna people in Thailand, and is considered a symbol of the old Lanna kingdom. Sadly, like the highly prized straw-headed bulbul, the red-whiskered bulbul is seen less and less in Thailand, as more and more are captured for the southern Thailand bird market. Known as "Nok Krong Hua Juk" (the caged bird with a crest), the red-whiskered bulbul was frequently entered in fighting contests in the early 1960s. Owners would put two of these fiercely territorial birds in a cage until one almost killed the other. In the early 1970s these contests were replaced with the less bloody singing contests, in which two caged birds are placed next to each other and both sing effusively, as though defending their territory. Although the Thai government began requiring permits for owning these birds in 1992, they are still widely traded. As recently as June 2001, 500 captured bulbuls were discovered in transport to the south of Thailand. Seen as a symbol of wealth and prestige, owning these birds has been described as a way of broadcasting the owner's status to the neighborhood, and unless attitudes surrounding their ownership change, they will continue to be captured for the caged bird trade.

1. Crested finchbill (*Spizixos canifrons*); 2. Black-collared bulbul (*Neolestes torquatus*); 3. Yellow-throated nicator (*Nicator vireo*); 4. Eastern bearded greenbul (*Criniger chloronotus*); 5. Ashy bulbul (*Hypsipetes flavala*); 6. Icterine greenbul (*Phyllastrephus icterinus*); 7. Red-tailed greenbul (*Criniger calurus*); 8. White-throated bulbul (*Criniger flaveolus*); 9. Black bulbul (*Hypsipetes madagascariensis*). (Illustration by Brian Cressman)

1. Red-whiskered bulbul (*Pycnonotus jocosus*); 2. Red-vented bulbul (*Pycnonotus cafer*); 3. Common bulbul (*Pycnonotus barbatus*); 4. Yellow-vented bulbul (*Pycnonotus goiaver*); 5. Leaf-love (*Phyllastrephus scandens*); 6. Straw-headed bulbul (*Pycnonotus zeylanicus*); 7. Joyful greenbul (*Chlorocichla laetissima*); 8. Yellow-whiskered greenbul (*Pycnonotus latirostris*); 9. Shelley's greenbul (*Pycnonotus masukuensis*). (Illustration by Brian Cressman)

Species accounts

Common bulbul
Pycnonotus barbatus

TAXONOMY
Turdus barbatus Desfontaine, 1789, "Côtes de Barbarie" (= near Algiers).

OTHER COMMON NAMES
English: Yellow-vented bulbul, dark-capped bulbul, black-eyed bulbul, white-eared bulbul, garden bulbul; French: Bulbul commun, Bulbul des jardins; German: Graubülbül; Spanish: Bulbul Naranjero.

PHYSICAL CHARACTERISTICS
3.6–4.2 in (93–107 mm), 0.8–2.1 oz (23–60 g). Thrush-sized with dark, slightly crested head, dark eye-ring and black bill. Grayish brown upperparts and breast, white belly and white or yellow undertail. Sexes alike, female slightly smaller. Juvenile duller than adult with rusty tones.

DISTRIBUTION
Widespread and common, almost everywhere in Africa south of 20°N, except in dry southwest and the Cape.

HABITAT
Any wooded or bushy habitat, especially near water.

BEHAVIOR
Usually in pairs, congregates in fruiting trees with other birds. Not territorial outside breeding season. Song abrupt "quick, chop, toquick"; one of earliest birds to sing in the morning, starting before dawn and continuing for up to two hours. Communal singing noted following feeding. Fond of bathing and scolding, but not overly aggressive.

FEEDING ECOLOGY AND DIET
Eats wide variety of wild and cultivated fruits, also flowers, termites, and other insects, sometimes small lizards.

REPRODUCTIVE BIOLOGY
Monogamous, said to mate for life. Pairs bond with preening ceremony and duet singing. Lays two to five eggs in shallow, thin cuplike nest in bush or shrub, lays twice per season. Incubation 12–14 days, usually by female only. Naked young cared for by both parents, fledging 10–17 days.

CONSERVATION STATUS
Not threatened; widespread and common.

SIGNIFICANCE TO HUMANS
None known. ◆

Red-vented bulbul
Pycnonotus cafer

TAXONOMY
Pycnonotus cafer Linnaeus, 1766. Nine subspecies recognized, extensive hybridization between races. Forms superspecies with six other bulbuls: common *P. barbatus*, black-fronted *P. nigricans*, Cape *P. capensis*, white-speckled *P. xanthopygos*, white-eared *P. leucotis*, and Himalayan and sooty-headed *P. aurigaster*.

OTHER COMMON NAMES
French: Bulbul à ventre rouge; German: Russbülbül; Spanish: Bulbul Ventrirrojo.

PHYSICAL CHARACTERISTICS
7.8–9 in (20–23 cm), 1.0–2.0 oz (31–59 g). Glossy black chin and throat, slightly tufted head. Back and breast feathers

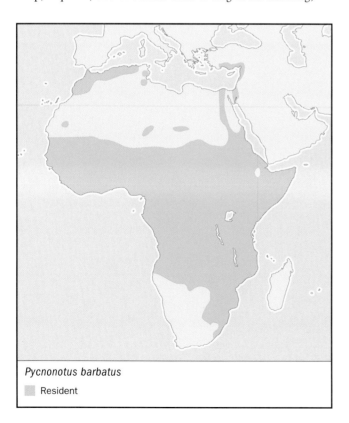

Pycnonotus barbatus
■ Resident

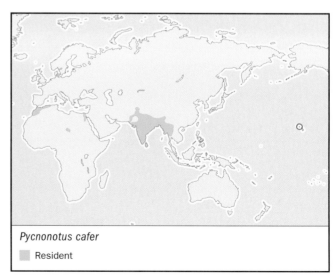

Pycnonotus cafer
■ Resident

brownish black edged with gray, appearing scalelike. Rump white, tail brownish black with white tip, undertail ("vent") crimson. Sexes alike. Juvenile resembles adult, but paler; undertail pinkish.

DISTRIBUTION
Native range covers the Indian subcontinent to southwestern China and Polynesia. Established populations in O'ahu, Hawaii.

HABITAT
Deciduous forest, gardens, and light scrub.

BEHAVIOR
Keeps in pairs or loose flocks. Bold and tame, can be quarrelsome. Aggressive behavior includes "crest-raising," whereby crest feathers arch over bill and body feathers fluff out. Call a "be-care-ful" or "be-quick-quick," alarm a sharp "peep." Nonmigratory.

FEEDING ECOLOGY AND DIET
Mainly eats fruits and berries, including figs (*Ficus* spp.), nightshade (*Solanum* spp.), and *Lantana*. Also eats insects, often caught on the wing.

REPRODUCTIVE BIOLOGY
Appears to be monogamous. Male courtship display involves showing erect crimson undertail. Two broods per season, male and female build nest, and both may incubate eggs 10–14 days. Fledge 12 days.

CONSERVATION STATUS
Not threatened. Widespread and common throughout range, well adapted to human environments. In Pakistan, building of canals and plantations has increased range.

SIGNIFICANCE TO HUMANS
Fruit-eating a threat to nurseries and agricultural orchards; management taken in Hawaii to prevent spread. Also disperses noxious weed seeds. ◆

Red-whiskered bulbul
Pycnonotus jocosus

TAXONOMY
Pycnonotus jocosus Linnaeus, 1758. Nine subspecies recognized.

OTHER COMMON NAMES
English: Crested bulbul, red-eared bulbul; French: Bulbul orphée; German: Rotohrbülbül; Spanish: Bulbul de Bigotes Rojos.

PHYSICAL CHARACTERISTICS
6.6–9 in (17–23 cm), 0.8–1.0 oz (24–31 g). Sooty black crown and erect pointed crest. White chick patch encircled with black line. Brown upperparts, white underparts. Glossy crimson feathers behind eye ("whiskers"). Sexes alike. Juvenile lacks whiskers and crest; head is brown.

DISTRIBUTION
Native range Saudi Arabia to Indian subcontinent to southern China. Introduced in Australia, Singapore. Established populations in southeastern Florida, O'ahu, Hawaii, sighted in southern California.

HABITAT
Open forest, scrub jungle, cultivated gardens, and orchards.

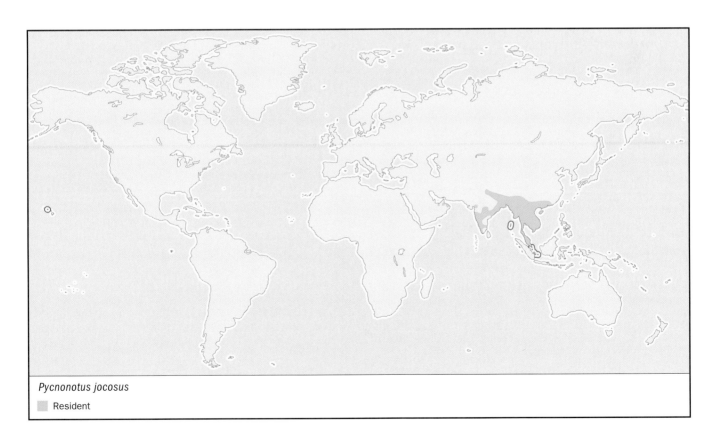

Pycnonotus jocosus

■ Resident

BEHAVIOR
Active and noisy, especially in morning and evening. May flock outside breeding season. Somewhat territorial, adult pairs frequently observed with third adult foraging nearby. Call a "pettigrew-kick-pettigrew." Nonmigratory.

FEEDING ECOLOGY AND DIET
Fruit, also insects, spiders, nectar, and flower buds. Will forage on ground, along buildings, and on tree trunks.

REPRODUCTIVE BIOLOGY
Probably monogamous. During courtship male lowers head in a bow, while emitting small croaks. Nests well hidden in low vegetation and often left unattended until full clutch (two to five eggs) is laid. Both parents feed young, insects at first, then fruits and berries.

CONSERVATION STATUS
Not threatened. Common and widespread throughout range. Florida populations grew from 40 to 50 birds in 1964, to 500 in 1973. Handful of birds escaped in southern California in late 1960s; 15 counted there in Christmas bird count of 1995.

SIGNIFICANCE TO HUMANS
Fruit-eating a threat to nurseries and agricultural orchards, 75% of some orchid plantations destroyed in Hawaii because of bud and flower damage. Blamed for drastic reduction in populations of native Hawaiian white-eyes (*Zosterops* spp.) on Mauritius I. Management taken in Hawaii to prevent spread. When southern California populations increased so dramatically that they became a threat to citrus crops, the California Department of Agriculture initiated an eradication program that has been partially successful. Also problematic out of range as it disperses noxious weed seeds. ◆

Yellow-vented bulbul
Pycnonotus goiaver

TAXONOMY
Pycnonotus goiaver Scopoli, 1786. Forms superspecies with *P. barbatus, P. nigricans, P. capensis, P. leucotis, P. leucogenys* and possibly *P. cafer, P. aurigaster.*

OTHER COMMON NAMES
French: Bulbul d'Arabie; German: Gelbsteißbülbül; Spanish: Bulbul Capirotado;.

PHYSICAL CHARACTERISTICS
7.4 in (19 cm), 1.2–1.6 oz (35–46 g). Black head with white eye-ring. Brownish underparts, white in center of belly. Conspicuous yellow undertail (vent). Sexes alike. Juvenile resembles adult, but head brown.

DISTRIBUTION
Turkey, Lebanon, Israel, western Syria, Jordan, Sinai and Arabian peninsulas.

HABITAT
Open forest, cultivated gardens, orchards, plantations—anywhere with trees, bushes or scrub; will nest quite close to humans.

BEHAVIOR
Thought to spend time in "duo" with a sibling of same or different sex. Gregarious where food is abundant, hundreds may

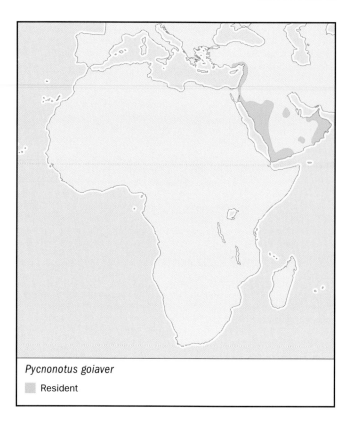
Pycnonotus goiaver
■ Resident

flock together outside of breeding season. Active and noisy, squabbling common. Calls various chirping, bubbling, whistling, scolding, occasional mimic of other birds. Nonmigratory.

FEEDING ECOLOGY AND DIET
Mainly feeds on fleshy fruits, also insects; flying ants, bees, wasps, mole-crickets, worms, and snails.

REPRODUCTIVE BIOLOGY
Monogamous, pairs remain together all year, often for several years. Two to three broods, nests in bushes or low palms. Two to four subelliptical light violet to pink eggs, with violet or red-brown and gray speckles. Incubation 14 days, fledge 13–15 days.

CONSERVATION STATUS
Not threatened. Fairly common throughout range, largest numbers in Israel where range is expanding (few hundred thousand pairs).

SIGNIFICANCE TO HUMANS
None known. ◆

Straw-headed bulbul
Pycnonotus zeylanicus

TAXONOMY
Pycnonotus zeylanicus Gmelin, 1789.

OTHER COMMON NAMES
English: Straw-crowned bulbul; French: Bulbul á tête jaune; German: Gelbscheitelbülbül; Spanish: Bulbul Bigotudo.

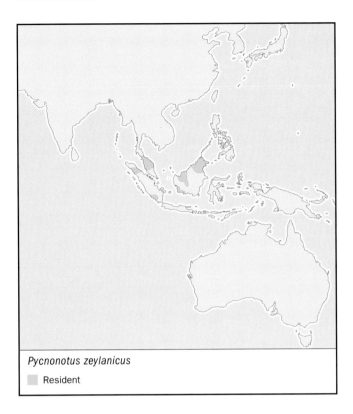

Pycnonotus zeylanicus

Resident

Shelley's greenbul
Pycnonotus masukuensis

TAXONOMY
Pycnonotus masukuensis Shelley, 1897, Masuku Range, east-central Africa. Species level taxonomy unclear. Two groups of subspecies recognized: gray-headed western group and green-headed nominate group.

OTHER COMMON NAMES
English: Shelley's bulbul; French: Bulbul des Monts Masukus; German: Shelleybülbül; Spanish: Bulbul de Shelley.

PHYSICAL CHARACTERISTICS
6.3 in (16 cm); 0.9–1.1 oz (27–31 g). Bright green upperparts, underparts dull olive. Head and neck variable; gray, olive-gray, or olive. Pale eye-ring. Sexes alike. Juvenile resembles adult, but olive parts deeper colored, upperparts duller.

DISTRIBUTION
Endemic resident of middle elevation highlands in east-central Africa. Eastern Zaire to Uganda, Rwanda, western Kenya, eastern Tanzania into Malawi.

HABITAT
Montane rainforest, riverside scrub.

BEHAVIOR
Typically found singly or in pairs, may forage in larger groups. Unusually silent for a bulbul, has soft song of "chip, wa-da-tee, chee-tu, ti-wew." Nonmigratory.

FEEDING ECOLOGY AND DIET
Eats berries and insects. Often uses "treecreeper" foraging method, whereby it starts at the base of a tree looking for insects, works its way up the tree in short hops for 18–30 ft (6–10 m), then flies to base of next tree and starts again.

PHYSICAL CHARACTERISTICS
11.5 in (29 cm). Large bulbul with orange-yellow crown and cheek, white throat lined with black on top, and large bill. Whitish belly and orange rump, olive-green back and wings. Sexes alike. Juvenile duller with brownish head.

DISTRIBUTION
Sundaic range; Myanmar and Thailand through Malaysia, Singapore, Sumatra, Java, Borneo.

HABITAT
Secondary forest and edge, near water, disturbed areas.

BEHAVIOR
Found in pairs or family groups up to six birds. Have large territories designated by singing duets (male and female alternating). Very vocal, especially at dawn and dusk. Song a strong, clear, melodious warble.

FEEDING ECOLOGY AND DIET
Mixed diet of insects, snails, fruits, and berries.

REPRODUCTIVE BIOLOGY
Little known, may breed year round. Nests of plant material often on boughs overhanging water. Both adults incubate eggs and care for young. Fledge 16 days.

CONSERVATION STATUS
Vulnerable. Listed on Appendix II of CITES. Highly prized caged birds; poaching and trapping has led to dramatic population declines and local extinction, habitat loss also a problem. Habitat preservation offers some protection, but capture and trade still permitted in some areas.

SIGNIFICANCE TO HUMANS
Prized songbird worldwide. Known as *maki boyah* or "alligator bird" from its supposed habit of annoying "alligators," and extending as far up the river crocodiles do. ◆

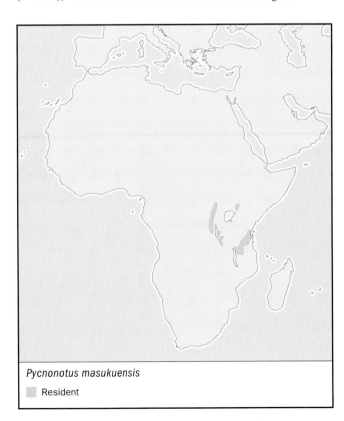

Pycnonotus masukuensis

Resident

REPRODUCTIVE BIOLOGY
Not well known. Nest a strong, shallow cup; lays one or two pinkish, spotted eggs.

CONSERVATION STATUS
Not threatened. Fairly common to locally abundant.

SIGNIFICANCE TO HUMANS
None known. ◆

Yellow-whiskered greenbul
Pycnonotus latirostris

TAXONOMY
Pycnonotus latirostris Strickland, 1844, Fernando Po (now Bioko, Equatorial Guinea). Four races recognized on basis of plumage variation.

OTHER COMMON NAMES
English: Yellow-whiskered bulbul; French: Bulbul à moustaches jaunes; German: Gelbbartbülbül; Spanish: Bulbul de Bigotes Amarillos.

PHYSICAL CHARACTERISTICS
6.3–7.5 in (16–19 cm); 0.6–1.1 oz (19–32 g). Upperparts and head sooty olive, rump with rufous tinge, back and wings brownish, tail dark reddish brown. Bright yellow moustache stripes on sides of throat. Sexes alike. Juvenile resembles adult, but more dingy brown, no moustache.

DISTRIBUTION
Endemic; central and West Africa.

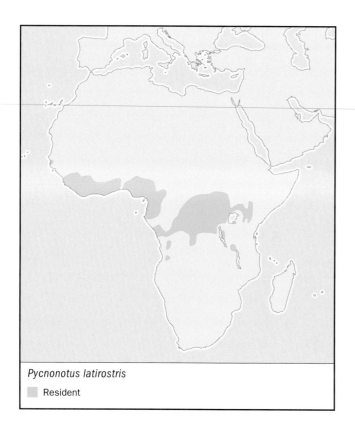

Pycnonotus latirostris
◼ Resident

HABITAT
Primary and secondary forest, interior and edge.

BEHAVIOR
Solitary and generally unsociable, usually travels singly. Bathes frequently. Sings continuously, dry jumbled notes; "chop, chip, chirrop, chup, prip, prip" repeated, volume increases throughout song, ends with loud "kick kick." Nonmigratory.

FEEDING ECOLOGY AND DIET
Omnivorous, eats fruits, berries, many invertebrates including mollusks, woodlice, spiders, frogs, and geckos. Shy when feeding, will fly from undergrowth, gorge on fruits, then return to cover. Often hovers while plucking fruit or gleaning insects from vegetation. Will dig in loose soil for insects.

REPRODUCTIVE BIOLOGY
Quite variable. In high-density areas uses leks; polygamous and non-territorial with some evidence of cooperative breeding. Before breeding season, males gather and sing at lek sites. Females gather at leks and choose male, female then takes care of young. At lower densities, monogamous and territorial. One to four eggs, incubation by female only 12–14 days. Outside of leks, both care for young.

CONSERVATION STATUS
Not threatened. Common and abundant, often most abundant bird in species in range.

SIGNIFICANCE TO HUMANS
None known. ◆

Joyful greenbul
Chlorocichla laetissima

TAXONOMY
Andropadus laetissimus Sharpe, 1899, Kenya. Two races recognized based on plumage.

OTHER COMMON NAMES
English: Joyful bulbul; French: Bulbul joyeux; German: Dotterbülbül; Spanish: Bulbul Feliz.

PHYSICAL CHARACTERISTICS
7.8 in (20 cm); 1.5–1.9 oz (43–55 g). One of the most brightly colored bulbuls, yellow-green upperparts, bright yellow chin and throat, golden-green underparts. Olive tail edged with yellow. Sexes alike. Juvenile washed-out brown, greenish underparts.

DISTRIBUTION
Endemic; Sudan, eastern Zaire, western Uganda, northern Zambia.

HABITAT
Open parts of primary and secondary forest, forest edges at altitudes of 3,150–7,000 ft (1,050–2,300 m).

BEHAVIOR
Sociable, forages in small flocks of four to eight birds, with noisy, bubbling, chatter. Song described as pleasant and energetic, call a sharp "chik" or "chak."

FEEDING ECOLOGY AND DIET
Eats berries and seeds, often forages in groups. Prefers low and intermediate levels, less often in canopy and undergrowth.

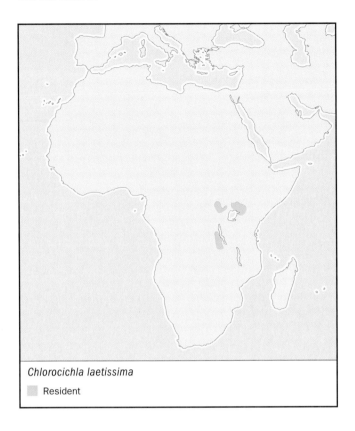

Chlorocichla laetissima

■ Resident

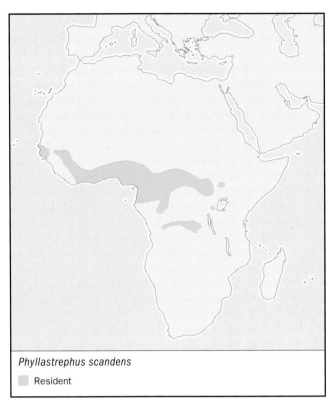

Phyllastrephus scandens

■ Resident

REPRODUCTIVE BIOLOGY
Unknown in wild. In captivity builds typical cuplike nest of rootlets, grasses, and Spanish moss. Young cared for by both parents, fed live insects and fruit.

CONSERVATION STATUS
Not threatened. Common locally but patchy distribution.

SIGNIFICANCE TO HUMANS
None known. ◆

Leaf-love
Phyllastrephus scandens

TAXONOMY
Phyllastrephus scandens Swainson, 1837, West Africa. Closest allies *Thescelocichla* and *Chlorocichla flavicollis*, based on bill shape. Two races recognized based on plumage and wing length.

OTHER COMMON NAMES
English: Leaflove; French: Bulbul à queue rousse; German: Uferbülbül; Spanish: Amante de Hojas.

PHYSICAL CHARACTERISTICS
5.9 in (15 cm); 1.1–1.9 oz (33–53 g). Gray head, back gray-olive, bright, rusty tail, feathers of tail and rump fluffy. Some black bristles on nape of neck and near bill. Belly creamy whitish yellow. Sexes alike. Juvenile mostly olive-gray with rusty wash, chin and underparts white, undertail pale rust.

DISTRIBUTION
Endemic to east central Africa; Sudan, western Gambia, Senegal, Guinea, Sierra Leone, Liberia, Mali, Ivory Coast, Ghana,

Togo, Nigeria, Cameroon, Gabon, southern Congo, Central African Republic, and Zaire.

HABITAT
Forest and thickets near water.

BEHAVIOR
Moves in pairs or small flocks; will defend communal territory with chorus. Drops from high perch into streams to bathe, flies back to perch to shake and preen, then drops again. Groups produce loud, raucous chorus.

FEEDING ECOLOGY AND DIET
Forages in canopy, on ground, and in vegetation for insects and their larvae, also eats small snails, seeds, and berries.

REPRODUCTIVE BIOLOGY
Territorial during breeding season. Cup-shaped nest suspended in twigs by cobwebs, appears too small for the bird. Incubation by female only.

CONSERVATION STATUS
Not threatened. Locally common, though fragmented distribution.

SIGNIFICANCE TO HUMANS
None known. ◆

Icterine greenbul
Phyllastrephus icterinus

TAXONOMY
Trichophorus icterinus Bonaparte, 1850, Guinea.

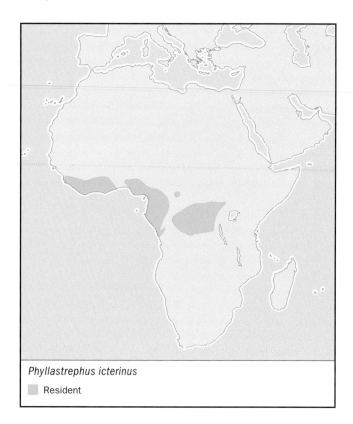

Phyllastrephus icterinus
■ Resident

ally two eggs, incubation 14 days, by female only. When surprised on nest, female will fall to ground and run to distract predator. Both parents feed young.

CONSERVATION STATUS
Not threatened.

SIGNIFICANCE TO HUMANS
None known. ◆

Eastern bearded greenbul
Criniger chloronotus

TAXONOMY
Trichophorus chloronotus Cassin, 1860, Gabon. Forms superspecies with *C. barbatus*.

OTHER COMMON NAMES
French: Bulbul crinon oriental; German:Haarbülbül; Spanish: Bulbul de Lomo Verde.

PHYSICAL CHARACTERISTICS
7.5–8.3 in (19–21 cm); 1.3–1.5 oz (38–45 g). Large bulbul; breast and head gray, few black bristles on hindneck and mantle. Mantle, rump, and wings yellow-green olive. Tail bright rust-maroon. Chin and throat white, appears puffy and beardlike, creamy belly. Sexes alike. Juvenile resembles adult.

DISTRIBUTION
Endemic central Africa; Cameroon, Gabon, southern Congo, western Zaire, Angola, Central African Republic, western Uganda.

OTHER COMMON NAMES
English: Lesser icterine bulbul; French: Bulbul ictérin; German: Zeisigbülbül; Spanish: Bulbul Icterino.

PHYSICAL CHARACTERISTICS
5.9 in (15 cm); 0.5–0.8 oz (15–25 g). Top of head and upperparts olive green, uppertail rusty, rump feathers long and fluffy. Chin and throat sulfur yellow, breast and belly yellow washed with green. Reddish tail. Sexes alike. Juvenile resembles adult but upperparts greener and washed brownish breast and throat.

DISTRIBUTION
Endemic east central Africa; Sierra Leone, Liberia, Ivory Coast, Ghana, Nigeria, Cameroon, Equatorial Guinea, Gabon, Congo, Central African Republic, and Zaire.

HABITAT
Forest, including patchy and swampy areas, plantations.

BEHAVIOR
Moves in family parties of three to five, up to 12. Group stays together by using nasal call. Group will defend territory and fight with other groups if confrontation occurs. Call a repeated "gur-guk," or nasal "gur-gur-gaaa." Will mob potential predators such as owls.

FEEDING ECOLOGY AND DIET
Mainly eats insects; often forages in mixed species flocks. Follows small mammals such as squirrels and antelopes, catching insects flushed out by mammals.

REPRODUCTIVE BIOLOGY
Territorial and monogamous, pairs staying together for several years. Nest a small cup of dry leaves held together by the fungus *Marasmius*, slung like a hammock in fork of branch. Usu-

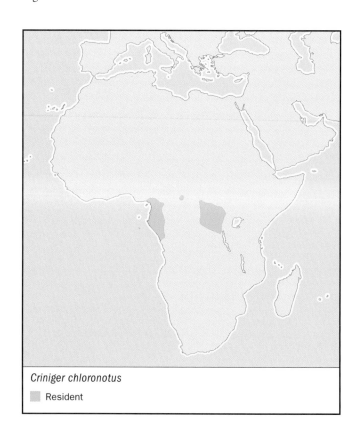

Criniger chloronotus
■ Resident

HABITAT
Lowland rainforest, lower montane forest to 5,900 ft (1,800 m).

BEHAVIOR
Often occurs in groups; three to five birds, seem to unite by song of leader. Group territory defended by chorus and displays in which "beard" is puffed out. Mournful song of two quavering notes.

FEEDING ECOLOGY AND DIET
Mainly eats insects and their eggs, some fruit.

REPRODUCTIVE BIOLOGY
Territorial, possibly a cooperative breeder. Solid cuplike nest decorated with tropical epiphytic fern *Microgramma owariensis*, which stays green through fledging. Usually two eggs, incubation 14 days by female only.

CONSERVATION STATUS
Not threatened. Fairly common in range.

SIGNIFICANCE TO HUMANS
None known. ◆

Red-tailed greenbul
Criniger calurus

TAXONOMY
Trichophorus calurus Cassin, 1857, Gabon. Two races recognized based on plumage variation.

OTHER COMMON NAMES
English: Red-tailed bulbul; French: Bulbul à barbe blanche; German: Swainsonbülbül; Spanish: Bulbul de Cola Roja.

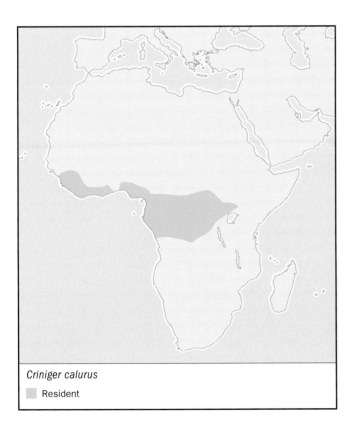

Criniger calurus

  Resident

PHYSICAL CHARACTERISTICS
7.1 in (18 cm); 0.7–1.2 oz (22–35 g). Head and hindneck olive brown, long black bristles on hindneck. Bright yellow underparts contrast with olive flanks. White conspicuous "beard," often puffed out. Sexes alike. Juvenile resembles adult but is dull cinnamon on wings.

DISTRIBUTION
Endemic central Africa; Senegal, Guinea, Mali, Sierra Leone, Liberia, Ivory coast, Ghana, Nigeria, Cameroon, Zaire, Gabon, Congo, Angola, Central African Republic, Sudan, and Uganda.

HABITAT
Forest, forest-grassland mosaic.

BEHAVIOR
Territorial groups of three through 12. Common in mixed bird flocks, often the leader. Call a weak "chit, chiro-chiro" or whistle "peeyu." Frequently flicks wings and fans tail.

FEEDING ECOLOGY AND DIET
Mainly insects and insect larvae, also fruits and seeds.

REPRODUCTIVE BIOLOGY
Territorial breeder, usually two eggs. Incubation by female only; young fed by both parents, fledge 14 days.

CONSERVATION STATUS
Not threatened.

SIGNIFICANCE TO HUMANS
None known. ◆

White-throated bulbul
Criniger flaveolus

TAXONOMY
Criniger flaveolus Gould, 1836. Two races recognized.

OTHER COMMON NAMES
English: Ashy-fronted bearded bulbul; French: Bulbul flavéole; German: Weisskehlbülbül; Spanish: Bulbul Barbudo de Frente Ahumado.

PHYSICAL CHARACTERISTICS
8.6 in (22 cm). Brownish crest, back olive-brown, lemon-yellow breast and belly, white fluffy throat. Wings and tail have rusty tinge. Sexes alike. Juvenile resembles adult but crest not as prominent and browner belly.

DISTRIBUTION
Himalayas to northeastern Myanmar.

HABITAT
Bushes and undergrowth in dense forest.

BEHAVIOR
Noisy, heard more than seen. Often in groups of up to 15 birds which are aggressive in mobbing birds of prey. Prefers lower story of forest. Song a nasal "cheer" or loud clear "teek, da-te-ek, da-te-ek." Frequently puffs out throat and fans tail. Resident, may move altitudinally.

FEEDING ECOLOGY AND DIET
Mainly berries, also insects.

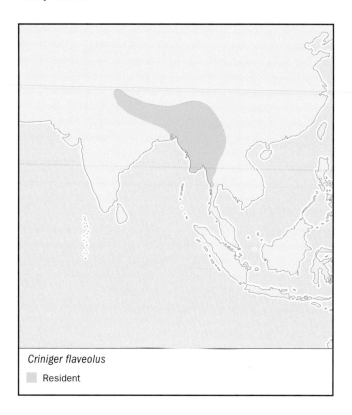

Criniger flaveolus
■ Resident

Hypsipetes madagascariensis
■ Resident ■ Breeding

REPRODUCTIVE BIOLOGY
Breeds April through July. Builds low nest in undergrowth, usually within 3 ft (1 m) of ground.

CONSERVATION STATUS
Not threatened. Locally common.

SIGNIFICANCE TO HUMANS
None known. ◆

Black bulbul

Hypsipetes madagascariensis

TAXONOMY
Hypsipetes madagascariensis Gmelin, 1789. Four races recognized based on plumage variation.

OTHER COMMON NAMES
English: Madagascar bulbul, Madagascar black bulbul, Comoro bulbul, Comoro black bulbul; French: Bulbul malgache, Bulbul des Comores; German: Madagaskarfluchtvogel, Rotschnabel-Fluchtvogel; Spanish: Bulbul Negro.

PHYSICAL CHARACTERISTICS
7.8 in (20 cm). Black, with slight crest and forked tail. Bright red legs and feet. Some races have white head, western races have grayer plumage. Sexes alike. Juvenile has less prominent crest, whitish throat and grayish brown plumage.

DISTRIBUTION
Pakistan, India, Bangladesh, Sri Lanka, India, southern China, Taiwan, Hainan, Myanmar, and Indochina.

HABITAT
Tall forest, shade trees in plantations.

BEHAVIOR
Gregarious, very noisy, will gather in flocks of several hundred birds. Has a variety of loud screeching, mewing notes, also repeated "pa-chit-chit" or "pip-per-tree." Flight is strong and swift. One of the few migrating bulbuls, resident in some areas, also moves altitudinally.

FEEDING ECOLOGY AND DIET
Mainly eats insects, also berries. Forages mainly in treetops, moving restlessly from tree to tree.

REPRODUCTIVE BIOLOGY
Breeds March through September, nest built high in forest trees.

CONSERVATION STATUS
Not threatened. Locally common.

SIGNIFICANCE TO HUMANS
None known. ◆

Ashy bulbul

Hypsipetes flavala

TAXONOMY
Hypsipetes flavala Blyth, 1845. Some authors consider races of chestnut bulbul (*Hemixos castanonotus*) as races of the ashy bulbul.

OTHER COMMON NAMES
English: Ashy bulbul, brown-eared bulbul, chestnut bulbul; French: Bulbul à ailes vertes; German: Braunohrbülbül; Spanish: Bulbul Ahumado.

PHYSICAL CHARACTERISTICS
7.8 in (20 cm). Distinctive bird, black crest and mask, brown cheeks and white throat. Back, chest, and tail gray, white belly,

Hypsipetes flavala
☐ Resident

Spizixos canifrons
☐ Resident

olive-yellow patch on wings. Sexes alike. Juvenile similar, but browner upperparts.

DISTRIBUTION
Himalayas, southwestern China, Southeast Asia, and Greater Sundas.

HABITAT
Broadleaved forest and edge plantations.

BEHAVIOR
Found in noisy parties, pairs during breeding season. Puffs out throat feathers like *Alophoixos* bulbuls. Voice a loud ringing call of four to five notes, also harsh "trrk." Resident, may move altitudinally.

FEEDING ECOLOGY AND DIET
Forages among trees for berries, nectar, and insects; the latter it catches on the wing.

REPRODUCTIVE BIOLOGY
Breeds in May and June. Deep, cuplike nest.

CONSERVATION STATUS
Not threatened. Locally common.

SIGNIFICANCE TO HUMANS
None known. ◆

Crested finchbill
Spizixos canifrons

TAXONOMY
Spizixos canifrons Blyth, 1845. Two races recognized.

OTHER COMMON NAMES
English: Finch-billed bulbul; French: Bulbul á gros bec; German: Fimkenbülbül; Spanish: Pico de Pinzón Copetón.

PHYSICAL CHARACTERISTICS
8.6 in (22 cm). Large, olive-green bulbul with prominent blackish crest and stout, pale-yellow, finchlike bill. Gray forehead, blackish throat, broad blackish green tail. Sexes alike. Juvenile resembles adult, but browner head and throat.

DISTRIBUTION
Myanmar, Bangladesh, Assam, and south-central China.

HABITAT
Forest and semicultivated areas up to 9,800 ft (3,000 m).

BEHAVIOR
Flocks outside the breeding season, often perched on telephone wires. Chattering, bubbling voice; "purr-purr-prruit-prruit-prruit." Resident, nonmigratory.

FEEDING ECOLOGY AND DIET
Forages at all levels for insects, fruit, and seeds.

REPRODUCTIVE BIOLOGY
Breeds April through July. Distinctive cuplike nest of vine tendrils, usually low in a shrub.

CONSERVATION STATUS
Not threatened.

SIGNIFICANCE TO HUMANS
None known. ◆

Black-collared bulbul

Neolestes torquatus

TAXONOMY

Neolestes torquatus Cabanis, 1875, Chinchoxo, Loango Coast. Taxonomy controversial, some morphological characters ally *Neolestes* with bulbuls, some ally it with shrikes (Malaconotidae, Laniidae, or Prionopidae). DNA and behavioral data suggest a closer relationship to bulbuls.

OTHER COMMON NAMES

French: Bulbul à collier noir; German: Rüttelbülbül; Spanish: Bulbul de Collar Negro.

PHYSICAL CHARACTERISTICS

6.3 in (16 cm), 0.7–1 oz (19–27 g). Distinctive, forehead to hindneck gray, black mask continues down neck to form a broad black band across the white breast. Back and tail olive-greenish brown, wings with yellow stripe. Sexes alike. Juvenile resembles adult but duller, crown and neck greenish.

DISTRIBUTION

South central Africa, Gabon, Congo, Angola, Zambia, and Zaire.

HABITAT

Wooded savanna, grassland.

BEHAVIOR

Territorial. Found singly, in pairs, or in small groups of three to four birds, will perch on fence posts. Fluid song of "tji-li-li" and "dee-de-de-de-de." Resident.

FEEDING ECOLOGY AND DIET

Mainly eats insects and fruit, often foraging in tall grasses.

REPRODUCTIVE BIOLOGY

Solitary nester with some distance between territories (out of range of song). Two eggs laid in untidy nest; young fed by both parents, mainly fruit. Feigns injury to distract predators from nest.

CONSERVATION STATUS

Not threatened. Locally common, though distribution spotty.

SIGNIFICANCE TO HUMANS

None known. ◆

Yellow-throated nicator

Nicator vireo

TAXONOMY

Nicator vireo Cabanis, 1876, Chinchoxo, Portuguese Congo. Taxonomy unresolved. The genus has been placed in the bush-shrikes (Malaconotidae), other authors place it in the Pycnonotidae.

OTHER COMMON NAMES

French: Bulbul à gorge jaune; German: Gelbkehlnicator; Spanish: Bulbul de Garganta Amarilla.

PHYSICAL CHARACTERISTICS

5.5 in (14 cm); 0.88 oz (25 g). Upperparts dark green, head tinged gray. Short yellow stripe above the eye, bordered on top with a black line, cheeks grayish. Yellow throat, undertail greenish yellow, yellow tip on tail. Conspicuous yellow wing spots. Sexes alike. Juvenile resembles adult, but greener forehead and less yellow on throat.

DISTRIBUTION

Central Africa; Cameroon, Gabon, Congo, Angola, and Zaire.

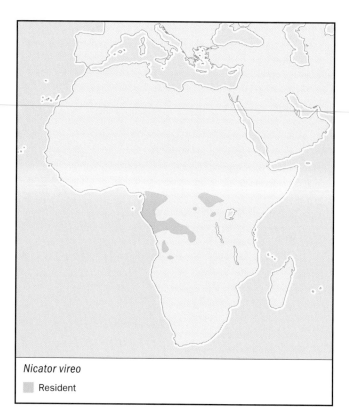

Neolestes torquatus
◻ Resident

Nicator vireo
◻ Resident

HABITAT
Lowland forest, especially vegetation near clearings.

BEHAVIOR
Territorial, lives in pairs, sedentary. Very shy, sings from dense cover. Song of up-and-down whistles with soft in-between notes "po-tyoo-ho-ho-ho-whee, tyoo-ho-ho-ho." Nonmigratory.

FEEDING ECOLOGY AND DIET
Mainly insects, especially large dragonflies, also caterpillars, beetles, and mantids. Forages in dense clumps of vegetation near clearings.

REPRODUCTIVE BIOLOGY
Apparently monogamous; holds same breeding territory for several years. Nest often exposed, two grayish eggs spotted with yellow, red, and gray; spots form "cap" on end. Incubation by female, while male stays nearby and sings. Young fed by both parents.

CONSERVATION STATUS
Not threatened. Locally common.

SIGNIFICANCE TO HUMANS
None known. ◆

Resources

Books

BirdLife International. *Threatened Birds of Asia: BirdLife International Red Data Book.* Cambridge: BirdLife International, 2001.

Grimmett, Richard, et al. *Birds of the Indian Subcontinent.* London: Christopher Helm Ltd., 1998.

Hagemeijer, Ward J.M., and Michael J. Blair, eds. *The EBCC Atlas of European Breeding Birds.* London: T.& A.D. Poyser, 1997.

Islam, K., and R.N. Williams. "Red-Vented Bulbul (*Pycnonotus cafer*), Red-Whiskered Bulbul (*Pycnonotus jocosus*)." In *The Birds of North America.* 520 (2000).

Jeyarajasingam, A., and A. Pearson. *A Field Guide to the Birds of West Malaysia and Singapore.* New York: Oxford University Press, 1999.

Keith, S., E. Urban, and C.H. Fry, eds. *The Birds of Africa.* Vol. IV. San Diego: Academic Press, 2000.

MacKinnon, J., and K. Phillips. *A Field Guide to the Birds of China.* New York: Oxford University Press, 2000.

Periodicals

Brosset, A. "The Social Life of the African Forest Yellow-Whiskered Greenbull *Andropadus latirostris.*" *Z. Tierpsychol.* 60 (1981): 239–255.

Cibois, A., B. Slikas, T.S. Schulenberg, and E. Pasquet. "An Endemic Radiation of Malagasy Songbirds is Revealed by Mitochondrial Sequence Data." *Evolution* 55, no. 6 (2001).

Dowsett, R.J., S.L. Olson, M.S. Roy, and F. Dowsett-Lemaire. "Systematic Status of the Black-Collared Bulbul *Neolestes torquatus.*" *Ibis* 141 (1999): 22–28.

Narang, M.L., R.S. Rana, and P. Mukesh. "Avian Species Involved in Pollination and Seed Dispersal of Some Forestry Species in Himachal Pradesh." *Journal of the Bombay Natural History Society* 97, no. 2 (2000): 215–222.

Pasquet, E., L. Han, O. Khobkhet, and A. Cibois. "Towards a Molecular Systematics of the Genus *Criniger*, and a Preliminary Phylogeny of the Bulbuls (Aves, Passeriformes, Pycnonotidae)." *Zoosystema* 23, no. 4 (2001): 857–863.

Roy, M.S. "Recent Diversification in African Greenbuls (Pycnonotidae: *Andropadus*) Supports a Montane Speciation Model." *Proceedings of the Royal Society of London B,* 264 (1997): 1337–1344.

Organizations

The Bird Conservation Society of Thailand. 69/12 Rarm Intra 24, Jarakhebua Lat Phrao, Bangkok 10230 Thailand. Phone: 943-5965 or 519-3385. E-mail: bcst@box1.a-net.net.th Web site: <http://www.geocities.com/TheTropics/Harbor/7503/ruang_nok/princess_bird.html>

Other

"Zoonomen Nomenclatural Data." January 19, 2002. 18 March 18, 2002. <http://www.zoonomen.net/avtax/frame.html>

Rachel Ehrenberg, MS

Fairy bluebirds and leafbirds
(Irenidae)

Class Aves

Order Passeriformes

Suborder Passeri (Oscines)

Family Irenidae

Thumbnail description
Medium-sized, mostly brightly-colored, thrush-shaped birds, with melodious voices

Size
5.4–10 in (13–25 cm); 0.5–2.6 oz (13.5–75 g)

Number of genera, species
3 genera, 14 species

Habitat
Primarily forests, as well as orchards, gardens and mangrove swamps

Conservation status
Vulnerable: 1 species; Near Threatened: 3 species

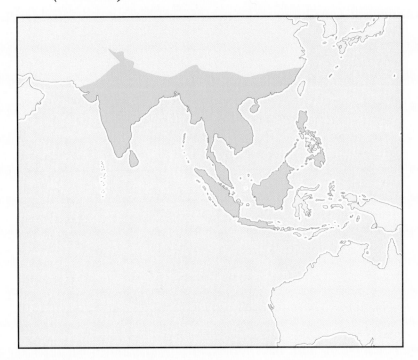

Distribution
South and Southeast Asia

Evolution and systematics

In familiar classification systems of the 1950s through the 1970s, the Irenidae were the only bird family entirely restricted to tropical Asia. Traditionally, this family has comprised three genera: the ioras (*Aegithina*), the leafbirds (*Chloropsis*), and the fairy bluebirds (*Irena*). As each is markedly distinct from the other, their inclusion in one family has long been questioned. For years, the leafbirds were commonly considered bulbuls (Pycnonotidae). The status of the fairy bluebirds was particularly controversial. At one time, these birds were popularly known as "blue drongos" (Dicruridae). More recently, various ornithologists have reassigned them to the Old World orioles (Oriolidae), based on skeletal studies. They have also been designated the sole genus in the family Irenidae.

However, the extensive DNA hybridization conducted in the 1980s by Sibley, Ahlquist, and Monroe indicates *Irena* and *Chloropsis* comprise a natural grouping, as the great systematist Jean Delacour concluded decades before. On the other hand, the same DNA research, published in 1990, suggests *Aegithina* is not part of this family, but instead forms a subfamily in the great assemblage Sibley, Ahlquist and Monroe classify under the family Corvidae. No fossils can be attributed to the Irenidae, but DNA hybridization suggests the divergence of these birds as distinct lines in the early Oligocene (roughly 30 million years ago), as well as an origin in Australia (where no modern forms have ever occurred).

Physical characteristics

According to the ornithologist Oliver Austin, two features shared by *Aegithina*, *Chloropsis*, and *Irena* are also typical of bulbuls (Pycnonotidae): the upper tail coverts are long and fluffy, and a patch of hair-like veinless feathers is present on the nape. All are excellent vocalists.

Varying in length from 6 to 7 in (13–18 cm) and averaging half an ounce (13 g) in weight, the four iora species are quietly patterned in shades of yellow and dull green, with some black in the males. With the exception of the great iora (*Aegithina lafresnayei*), the wings are dark, with boldly contrasting white bars, in both males and females. They have pale gray eyes and thin, uncurved black beaks. Breeding males are generally dark and more brightly colored than females. Unlike leafbirds or fairy bluebirds, nonbreeding male ioras have a comparatively dull eclipse plumage.

Brilliant green plumage in both sexes is a striking feature of all leafbirds. Excluding the Philippine leafbird (*Chloropsis flavipennis*), the species are distinguished by contrasting head patterns, often involving black throats, and bright blue "moustaches" (malar stripes). Most females do not have black throats, but maintain either the "moustaches" or a blue throat. Another distinction of the males of most species are shiny malachite-green "epaulets" at the bend of the wing. The build manages to be at once graceful and compact. Tails are of moderate length. The feet are notably small in comparison to body

An Asian fairy bluebird (*Irena puella*) perches on a branch. (Photo by S. Lipschutz/VIREO. Reproduced by permission.)

size, and the tarsi are short and fairly strong. The beak is slender and slightly down-curved, and the tongue appears to be specialized for nectar feeding. The size range is 7–8 in (17–20 cm), and weights are 1–1.5 oz (25–40 g).

The plumage of the male *Irena puella* can be described as a meeting of large fields of blue and velvet-black, accented by brilliant red eyes. The female's feathers are largely greenish cobalt, with black primaries and bright blue elongate rump feathers, as well as an equally red eye. Both sexes of *I. cyanogaster* are patterned in a comparatively indistinct pattern of deep blue and black. The feet of both are remarkably small in proportion to the generally robust, thrush-like body. At roughly 10 in (25 cm), and 2.5 oz (75 g), the fairy bluebirds are by far the largest members of the Irenidae.

Distribution

Irenids occur throughout the Oriental zoogeographical realm. Each of the three genera is represented by at least one species of very wide zoogeographical range, including the Indian subcontinent, southern China, Indochina, the Malay Peninsula, and the Greater Sundas. Each also includes species of far more limited distribution. Examples are Marshall's iora (*Aegithina nigrolutea*), found only in tropical Pakistan and northwestern India, and the yellow-throated leafbird (*Chloropsis palawanensis*), confined to the western Philippine island of Palawan, and some of its tiny satellites. The Philippines represent the eastern boundry of all three genera's range, and none occur in the Lesser Sundas.

Habitat

None of these three genera include habitual ground or undergrowth dwellers, though they may bathe in streams, and leafbirds may rarely forage on the ground or in low bushes. All species of *Chloropsis* and *Irena* are birds of the forest canopy, though some leafbirds are also regular garden visitors. Ioras have a far broader habitat tolerance, occurring everywhere from beaches and mangrove swamps to secondary forests, and are a common feature of gardens and orchards.

Behavior

When not nesting, ioras and fairy bluebirds roam the forest in flocks, generally in association with other species. Fairy bluebirds are often found in the company of feeding fruit pigeons, hornbills, bulbuls, and others. Ioras are generally associated with species of similar sized insect-eaters. Leafbirds are likewise nomadic outside the breeding season, but are generally in smaller numbers, or pairs, and are noted for frequent aggression towards other birds. All three genera are known for tuneful songs. Leafbirds are excellent natural mimics of other species. All of these birds form pairs that remain monogamous during the breeding season, and then become territorial.

Feeding ecology and diet

Though ioras have been observed to eat some fruit, they are avid arthropod hunters. Salim Ali describes them foraging through trees, hunting "in the foliage in pairs for caterpillars, moths and spiders, hopping from twig to twig and often hanging upside down in other acrobatic postures among the leaves." Studying bird communities of the dipterocarp forests of north-central Myanmar in the 1990s, David King and John Rappole found common ioras (*Aegithina tiphia*) were regular components in what they designated "woodshrike flocks," otherwise typically consisting of wood-shrikes, fantails, tits, minivets, nuthatches, and woodpeckers. With the wood-shrikes and minivets, ioras are species that most often initiate a relocation of a flock.

The members of the genus *Chloropsis* are well-known nectar-feeders. Leafbirds are highly effective pollinators. Groves of appropriate trees around villages have proved very attractive to leafbirds, making them familiar to people. They also hunt insects and spiders and eat small fruits, especially those of the oriental mistletoes (*Loranthus* sp.).

The abundance of wild fig species that are so important a feature of the southeast Asian forest ecology, are a major component of the diet of fairy bluebirds, flocks of which join the great assemblages of hornbills, fruit pigeons, orangutans, and other animals feasting on one tree after another. Fairy bluebirds will also eat berries from fairly low bushes. They also consume nectar and insects.

Reproductive biology

In India, the common iora may nest from May to September varying slightly with locale, while Marshall's iora, with its dryer range in Pakistan and northwest India, only breeds from June to August. Ioras have a well-known courtship display, which makes a vibrant spectacle of the normally inconspicuous male. During the display the male begins by chasing the female. Then it perches, wings lowered, and fluffs up its lower back feathers, lifts its tail and issues its call, long-drawn hissing phrases. Periodically it jumps up above its perch with its white back feathers fluffed and glides slowly back to its perch in a spiral while calling, sounding like a cricket or tree frog. Iora nests are deep, well constructed cups covered on the outside with spider webs, situated three or four meters up in a tree fork. The clutch is usually two to four gray, brown, or purplish streaked pinkish white eggs.

In India, the blue-winged leafbird (*Chloropsis cochinchinensis*) nests chiefly from April to August, while the golden-fronted leafbird (*C. aurifrons*) commences a month later, but also breeds until August. The courtship of the latter appears confrontational, with the male and female chasing and screeching at each other. The nest is a loosely formed cup composed of runners, moss, and rootlets weakly held together with spider webs. *Chloropsis* eggs are typically pale, speckled or flecked with black and various other colors.

Wild fairy bluebirds (*Irena puella*) have a short breeding period, typically from February to April (captive birds may breed from April through October). The nest is a loose platform of twigs, overlaid by moss, roots, and detritus, hidden in leafy trees at least five meters up. The usual clutch is of two eggs, olive-gray, irregularly splotched with brown.

Conservation status

The green iora (*Aegithina viridissima*) and the lesser green leafbird (*Chloropsis cyanopogon*) have nearly identical ranges, in

Thailand to the Malay Peninsula, Sumatra, and Borneo. They are listed as Near Threatened, as is the blue-masked leafbird (*C. venusta*), a Sumatran endemic. The Philippine leafbird (*C. flavipennis*) is Vulnerable. In addition to being restricted to threatened habitats, it has always been considered uncommon, in contrast to the other three IUCN listed species, which are considered locally common in parts of their range. Forest destruction has certainly caused losses of optimal habitat for a number of species, but as they are widespread, they are not considered globally threatened.

There is also concern about the cage bird trade. Several species of *Chloropsis* and fairy bluebirds were found to be popular among the markets of Singapore and Hong Kong in the 1990s. As of 2001, none of these species have been listed in any appendix of the Convention on International Trade in Endangered Species (CITES).

Significance to humans

The highly insectivorous ioras are recognized as a form of natural pest control. The value of leafbirds as pollinators of flowering trees can certainly not be overlooked, though this may be balanced against their spreading the seeds of the parasitic oriental mistletoe from one tree to another.

Appreciation for the beauty of members of the genus *Chloropsis* is centuries old. Leafbirds have appeared in Chinese art as early as the fifteenth century. The London Zoo obtained an orange-bellied leafbird as early as 1879, and by its centennial, in 1927, had exhibited two other species. In the twentieth century, various leafbirds (traditionally called fruitsuckers) arrived in large commercial shipments to Europe and America. Prior to the Second World War, India was the major source, succeeded by Thailand in the 1950s and 1960s, then by Indonesia and the People's Republic of China in the 1990s. China imposed an export ban on cage birds in 2001, making, for the time being, Indonesia the only commercial source. Due to their pugnacious behavior, they had a justified reputation as being unsuited for small mixed aviaries, and were traditionally kept as pets in small cages.

With the increasing role of ecotourism in the international economy, the potential of fairy bluebirds, leafbirds, and ioras to attract visitors to wildlife areas makes them a valuable resource to preservationists.

1. Blue-winged leafbird, Sumatran subspecies (*Chloropsis cochinchinensis icterocephalus*); 2. Orange-bellied leafbird (*Chloropsis hardwickii hardwickii*); 3. Blue-masked leafbird (*Chloropsis venusta*); 4. Common iora (*Aegithina tiphia tiphia*); 5. Green iora (*Aegithina viridissima*); 6. Philippine or yellow-quilled leafbird (*Chloropsis flavipennis*); 7. Golden-fronted leafbird (*Chloropsis aurifrons*); 8. Lesser green leafbird (*Chloropsis cyanopogon*); 9. Asian fairy-bluebird (*Irena puella sikkimensis*); 10. Philippine fairy bluebird (*Irena cyanogaster*). (Illustration by Amanda Humphrey)

Species accounts

Common iora
Aegithina tiphia

TAXONOMY
Motacilla tiphia Linnaeus, 1758, Bengal.

OTHER COMMON NAMES
English: Iora, black-winged iora; French: Petit iora; German: Schwarzflügelaegithina; Spanish: Iora Común.

PHYSICAL CHARACTERISTICS
5.5–6 in (13–17 cm); 0.5 oz (13.5 g). Males have dark green to black upperparts, bright yellow underparts, black wings with white bars, dark tails, and black crowns. Females have olive-green upperparts, duller yellow underparts, foreheads, and eyebrows, and olive-green crowns.

DISTRIBUTION
Almost all of Indian subcontinent, Sri Lanka, southern Yunnan and southwestern Guanxi, all of Myanmar, Indochina and the Malay Peninsula, Sumatra, Java, Borneo, and Palawan.

HABITAT
Open woodlands, secondary forest, gardens, orchards, mangroves, and beach forests.

BEHAVIOR
Outside of breeding season, travels in small flocks or pairs, continuously hunting for small arthropods. Contact is maintained through frequent vocalizations. Distinctive melodious songs and whistles.

FEEDING ECOLOGY AND DIET
Though some fruit is consumed, diet predominantly small arthropods (spiders, moths, caterpillars, etc.) gleaned from leaves.

REPRODUCTIVE BIOLOGY
Monogamous. Distinctive courtship behavior: Male repeatedly leaps one or two meters above perch, then glides back down with feathers erected, assuming a spherical appearance. The nest is deep and cup-shaped. Clutch size two to four. Eggs are pinkish white with brownish or purplish blotches.

CONSERVATION STATUS
Not threatened. Has likely expanded its range due to creation of orchards and gardens.

SIGNIFICANCE TO HUMANS
Controls caterpillars and other harmful insects in fruit orchards. ◆

Green iora
Aegithina viridissima

TAXONOMY
Jora viridissima Bonaparte, 1851, Sumatra.

OTHER COMMON NAMES
French: Iora émeraude; German: Smaragdaegithina; Spanish: Iora Verde.

PHYSICAL CHARACTERISTICS
5.5 in (13 cm). Dark olive-green plumage with black bill, wings, and tail. Yellow eye ring. Wing bars are white in males and yellow in females.

DISTRIBUTION
Malay Peninsula, including southern Thailand, Sumatra, Borneo, and nearby small islands.

HABITAT
Lowland primary and tall secondary forest.

BEHAVIOR
Confined to forest canopy, behavior, including vocalizations, otherwise similar to common iora's.

Aegithina tiphia
▨ Resident

Aegithina viridissima
▨ Resident

FEEDING ECOLOGY AND DIET
Similar to that of common iora, but apparently restricted to forest canopy.

REPRODUCTIVE BIOLOGY
Similar to common iora.

CONSERVATION STATUS
Near Threatened, due to drastic continuing reduction of forest habitat throughout range. Occurs in several important national parks.

SIGNIFICANCE TO HUMANS
None traditionally, but a target species for ecotourists. ◆

Philippine leafbird
Chloropsis flavipennis

TAXONOMY
Phyllornis flavipennis Tweeddale, 1878, Cebu.

OTHER COMMON NAMES
English: Yellow-quilled leafbird; French: Verdin à ailes jaunes; German: Philippinenblattvogel; Spanish: Verdín Amarillento.

PHYSICAL CHARACTERISTICS
7.5 in (19 cm). Unique among genus in both sexes lacking clearly defined head pattern. Green plumage with lighter, yellowish throat and yellow eye ring.

DISTRIBUTION
Philippine Islands of Mindanao, Cebu, and Leyte. Possibly extinct on Cebu.

HABITAT
Canopy of forest and forest edge, from sea level to 4,900 ft (1,500 m).

BEHAVIOR
Unlike some other leafbirds, very inconspicuous. Individuals or pairs, restricted to forest canopies.

FEEDING ECOLOGY AND DIET
Little recorded, presumed similar to other species of *Chloropsis*.

REPRODUCTIVE BIOLOGY
Nest and eggs unrecorded as of 2000. Breeding season appears to be from June through August.

CONSERVATION STATUS
Vulnerable, due to extensive habitat loss over limited range. Probably extinct on Cebu. Occurs in some Mindanao parks.

SIGNIFICANCE TO HUMANS
None known. ◆

Lesser green leafbird
Chloropsis cyanopogon

TAXONOMY
Phyllornis cyanopogon Temminck, 1829, Sumatra.

OTHER COMMON NAMES
French: Verdin barbe-bleue; German: Blaubart-Blattvogel; Spanish: Verdín Chico.

PHYSICAL CHARACTERISTICS
6.75 in (17cm). Green plumage with lighter underparts. Black bill and throat with small blue cheek patch. Yellowish tinge around face and throat.

DISTRIBUTION
Thailand, Tenasserim (Peninsular Malaysia), the remainder of the Malay Peninsula, Sumatra, Borneo, and nearby small islands.

HABITAT
Dense canopies of lowland primary and tall secondary forest.

BEHAVIOR
Forages from one tree to another individually, in pairs, or in flocks with other species.

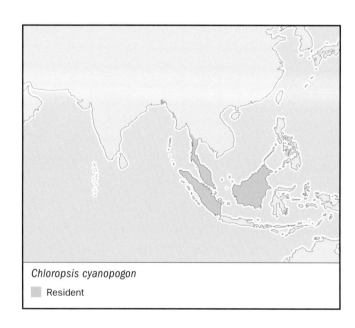

Chloropsis flavipennis
■ Resident

Chloropsis cyanopogon
■ Resident

FEEDING ECOLOGY AND DIET
Continuously moves from the top of one tree to another, drinking nectar from blossoms and systematically searching branches for arthropods, often as parts of mixed species "hunting parties."

REPRODUCTIVE BIOLOGY
Monogamous. Nesting behavior presumed similar to that of other leafbirds.

CONSERVATION STATUS
Near Threatened, due to dependence on tall trees in increasingly deforested areas, as well as vigorous exploitation for the bird trade.

SIGNIFICANCE TO HUMANS
Most specimens in the international bird trade become individually housed cage birds in Asia, highly regarded for melodious singing. ◆

Blue-winged leafbird
Chloropsis cochinchinensis

TAXONOMY
Turdus cochinchinensis Gmelin, 1788, Cochinchina (Vietnam).

OTHER COMMON NAMES
English: Jerdon's leafbird, yellow-headed leafbird, gold-mantled chloropsis; French: Verdin à tête jaune; German: Blauflügel-Blattvogel; Spanish: Verdín de Alas Azules.

PHYSICAL CHARACTERISTICS
6.5–7 in (16–17.5 cm). Most subspecific variation in family: head color varies from green to yellow and primaries may be blue or green. Distinctive blue outer tail and patch on cheek. Black throat patch in males.

DISTRIBUTION
Peninsular India, Sri Lanka, Bangladesh, Assam, southern Yunnan, Myanmar, all of Indochina and the Malay Peninsula, Sumatra, Borneo, Java, and smaller islands.

HABITAT
Favors groves and trees around villages and fields in India. Indonesian and Malaysian populations occur in woodland, and primary and tall secondary forest, up to 4,900 ft (1500 m).

BEHAVIOR
Indian specimens are aggressive and territorial. Indonesian varieties are more social among other species, appearing singly or in pairs, sometimes in groups.

FEEDING ECOLOGY AND DIET
Enthusiastic nectar feeders, especially from red flowers, serving as major pollinators. Also eat insects and small fruits, especially mistletoe.

REPRODUCTIVE BIOLOGY
Monogamous. Breeding season more or less from April to August. Nest is a loose, shallow cup composed of fine plant material, plaster on the exterior with cobwebs. Two or three pinkish or creamy-white eggs with variously colored specks, blotches and hair-streaks.

CONSERVATION STATUS
Not threatened, but some southern subspecies are in areas of intense habitat loss. Species fairly popular in the cage bird trade.

SIGNIFICANCE TO HUMANS
Important pollinator of flowering trees, but may also spread mistletoe. Traditional caged songbird in India and other Asian countries. Significant international trade, especially in Sumatran yellow-headed subspecies. ◆

Golden-fronted leafbird
Chloropsis aurifrons

TAXONOMY
Phyllornis aurifrons Temminck, 1829, India.

OTHER COMMON NAMES
English: Gold-fronted chloropsis, green bulbul, gold-fronted fruitsucker; French: Verdin à front d'or; German: Goldstirnblattvogel; Spanish: Verdín de Frente Dorado.

Chloropsis cochinchinensis
◻ Resident

Chloropsis aurifrons
◻ Resident

PHYSICAL CHARACTERISTICS
7.5 in (19 cm). Green plumage with golden-orange forehead, black border around blue throat, and bright blue wing patch.

DISTRIBUTION
Himalayan foothills, larges areas of the Indian subcontinent, Sri Lanka, southwest China, Myanmar, and Indochina, with an isolated population in Sumatra.

HABITAT
Prefers more forested conditions than blue-winged leafbird, but also more likely to be found in middle canopy than other leafbirds.

BEHAVIOR
Usually in pairs or small parties. A highly accomplished mimic.

FEEDING ECOLOGY AND DIET
Vigorously hunts insects and spiders, and equally busy nectar feeder. Also eats some fruits.

REPRODUCTIVE BIOLOGY
Monogamous. Noisy, frantic courtship involving chasing, screeching, and hanging upside down. Nest is loose shallow cup of tendrils, roots, etc, lined with soft plant material, near tip of branch, but concealed by foliage, in tall tree. Two eggs.

CONSERVATION STATUS
Not threatened, though heavily exploited for cage bird trade, and subject to habitat loss in many parts of range.

SIGNIFICANCE TO HUMANS
Important pollinator. Traditionally popular cage bird, both for appearance and as songbird and mimic. ◆

Orange-bellied leafbird
Chloropsis hardwickii

TAXONOMY
Chloropsis hardwickii Jardine & Selby, 1830, Nepal.

OTHER COMMON NAMES
French: Verdin de Hardwick; German: Blaubartblattvogel; Spanish: Verdín de Pico Anaranjado.

PHYSICAL CHARACTERISTICS
7.5 in (19 cm). Olive-green upperparts with distinctive yellow-orange belly and undertail-coverts. Black bill and throat patch with blue cheek. Bright blue shoulder patch and dark flight feathers.

DISTRIBUTION
Eastern Himalayas, Myanmar, southern China, Hainan Island, Indochina, Malay Peninsula.

HABITAT
Highland forests, occurs at higher elevations than golden-fronted leafbird.

BEHAVIOR
Usually in pairs or small groups. Frequently mimics other species.

FEEDING ECOLOGY AND DIET
Active, acrobatic, arthropod hunter and nectar feeder.

REPRODUCTIVE BIOLOGY
Monogamous. Nest very similar to gold-fronted leafbird. Two eggs, spotted with various colors, slightly larger than gold-fronted.

CONSERVATION STATUS
Not threatened, but experiencing some habitat loss, though less than lowland species.

SIGNIFICANCE TO HUMANS
Avid pollinator, but also spreads mistletoe. Popular cage bird for centuries. ◆

Blue-masked leafbird
Chloropsis venusta

TAXONOMY
Phyllornis venusta Bonaparte, 1850, Sumatra.

Chloropsis hardwickii
◼ Resident

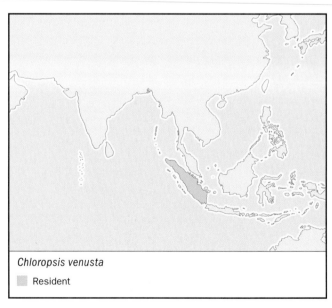

Chloropsis venusta
◼ Resident

OTHER COMMON NAMES
French: Verdin à front bleu; German: Blaustirn-Blattvogel;
Spanish: Verdín Enmascarado.

PHYSICAL CHARACTERISTICS
5.5 in (14 cm). Green plumage with blue forehead, cheek, and
upper throat.

DISTRIBUTION
Sumatra.

HABITAT
Hill forests.

BEHAVIOR
Presumed similar to other leafbirds.

FEEDING ECOLOGY AND DIET
Presumed typical of genus.

REPRODUCTIVE BIOLOGY
Not known.

CONSERVATION STATUS
Near Threatened, due primarily to deforestation. Locally common in some parts of habitat.

SIGNIFICANCE TO HUMANS
Potential "flagship" species to promote ecotourism in Sumatran elevated forest. ◆

Fairy bluebird
Irena puella

TAXONOMY
Coracias puella Latham, 1790, Travancore.

OTHER COMMON NAMES
English: Blue-mantled fairy bluebird, common fairy bluebird,
Asian fairy-blue-bird; French: Irène vierge; German: Elfen-
blauvogel; Spanish: Ave Flor de Espalda Negra.

PHYSICAL CHARACTERISTICS
Solidly built. 10 in (25 cm), 2.5 oz (75 g). Males have black underparts, wings, and tail; upperparts and undertail-coverts are a
distinctive ultramarine blue. Females are a uniform dark
turquoise-blue with black flight feathers. Both sexes have red
eyes.

DISTRIBUTION
Coastal southern India, eastern Himalayas, Myanmar, Yunnan,
Indochina, Malay Peninsula, Java, Sumatra, Borneo, and west-
ern Philippine island of Palawan.

HABITAT
Primary and tall secondary forests.

BEHAVIOR
May occur in flocks of up to thirty birds, largely staying in upper canopy, though descending to bath in streams. Loud melodious whistles are typical.

FEEDING ECOLOGY AND DIET
While known to eat insects and nectar, primarily fruit eaters,
specializing in figs.

REPRODUCTIVE BIOLOGY
Monogamous. Nest is a loose platform of twigs, lined with fine
plant materials, in fork of leafy tree. Clutch is two olive-gray,
brown splotched eggs.

CONSERVATION STATUS
Not threatened, due to enormous range, but many populations
are at risk due to forest destruction.

SIGNIFICANCE TO HUMANS
None known. ◆

Philippine fairy bluebird
Irena cyanogaster

TAXONOMY
Irena cyanogastra Vigors, 1831, Manilla.

Irena puella
■ Resident

Irena cyanogaster
■ Resident

OTHER COMMON NAMES
English: Black-mantled fairy bluebird; French: Irène à ventre bleu; German: Kobalt-Irene; Spanish: Ave Azul de Manto Negro.

PHYSICAL CHARACTERISTICS
9.75 in (24 cm). Black plumage with dark blue belly, tail, and wing edges. Blue from bill to nape; eyes red.

DISTRIBUTION
Most of the major eastern Philippine islands, from Luzon, south to Mindanao.

HABITAT
Forest canopy.

BEHAVIOR
Similar to fairy bluebird.

FEEDING ECOLOGY AND DIET
As in fairy bluebird.

REPRODUCTIVE BIOLOGY
Not known.

CONSERVATION STATUS
Not threatened. Still considered common on all islands as of 2000.

SIGNIFICANCE TO HUMANS
A target species for ecotourists. ◆

Resources

Books

Ali, S. *Field Guide to the Birds of the Eastern Himalayas.* New York: Oxford University Press, 1977.

Ali, S. *The Book of Indian Birds.* New York: Oxford University Press, 1996.

Austin, O.L. *Birds of the World.* New York: Golden Press, 1961.

Collar, N.J., A.V. Andreev, S. Chan, M.J. Crosby, S. Subramanya, and J.A. Tobias. *Threatened Birds of Asia.* Barcelona and Cambridge: Lynx Edicions and BirdLife International, 2001.

Kennedy, R.S., P.C. Gonzales, E.C. Dickinson, H.C. Miranda, and T.H. Fisher. *A Guide to the Birds of the Philippines.* New York: Oxford University Press, 2000.

MacKinnon, J., and K. Phillipps. *A Field Guide to the Birds of Borneo, Sumatra, Java and Bali.* New York: Oxford University Press, 1993.

Sibley, C.G., and J.E. Ahlquist. *Phylogeny and Classification of Birds: A Study in Molecular Evolution.* New Haven: Yale University Press, 1990.

Organizations

International Species Inventory System. Web site: <http://www.isis.org>

Josef Harold Lindholm, III, BA

Shrikes
(Laniidae)

Class Aves
Order Passeriformes
Suborder Passeri (Oscines)
Family Laniidae

Thumbnail description
Small to relatively large birds with a more or less sharply hooked bill and strong legs; feet and claws adapted for catching prey

Size
5.7–19.6 in (14.5–50 cm); 0.63–3.52 oz (18–100 g)

Number of genera, species
12 genera; 74 species

Habitat
Forest, woodlands, savannas, cultivated areas

Conservation status
Critically Endangered: 2 species; Endangered: 5 species; Vulnerable: 2 species; Near Threatened: 5 species

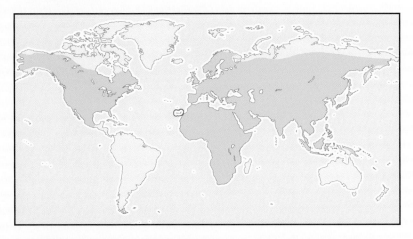

Distribution
Africa, Europe, Russia, Asia, Indonesia, New Guinea, North America

Evolution and systematics

The origin of shrikes is not known for certain. A fossil of a shrike-like bird (*Lanius miocaenus*) found in Lower Miocene deposits in France, indicates that the genus *Lanius* existed in Europe 25–30 million years ago. African origins have been suggested by various authors who consider that, out of the 74 species classically retained in the shrike family, only 10 have no link with Africa. These birds all belong to the genus *Lanius*, which still has nine endemic species on the African continent and may well have invaded Eurasia, where it also radiated.

The theory of African shrike origins, however, has recently been questioned, as molecular studies suggest that shrike ancestors originated in the Australasian region. From there, as Australia drifted northwards some 20–30 million years ago in the Tertiary period, shrikes might have emigrated to Asia.

Shrikes probably arrived much later in North America, where the loggerhead shrike (*L. ludovicianus*) may represent a first wave of shrike immigrants to the New World. *L. ludovicianus* may have been pushed south by a glacial epoch and was later followed by another wave of birds in the northern parts of the continent. It may be these birds that are now regarded as a race of the northern shrike (*L. excubitor*), which has a large holarctic distribution. The more common and widespread west-palearctic red-backed shrike (*L. collurio*) may be a much more recent invader, or re-invader, of northern latitudes after the retreat of the last glaciers about 12,000 years ago. Its typical migration pattern may reflect the route of its progressive northward spread. Its resident ancestor in Africa could be the very similar Emin's shrike (*L. gubernator*), known today from a belt spreading from central-west Africa to central-east Africa.

The most widely accepted standard sequence for recent birds dates back to the 1960s. It recognizes four subfamilies in the Laniidae, with 74 species in all. The subfamilies and the genera concerned are the following:

- Prionopinae (helmet-shrikes): *Eurocephalus* (two species); *Prionops* (seven species)

- Malaconotinae (bush-shrikes): *Lanioturdus* and *Nilaus* (monotypic); *Dryoscopus* (six species); *Tchagra* (six species); *Laniarius* (10 species); *Telophorus* (10 species); *Malaconotus* (five species)

- Laniinae ("true" shrikes): *Corvinella* (two species); *Lanius* (23 species)

- Pityriasinae (Bornean bristlehead): *Pityriasis* (one species)

Recent studies of genetic material and behavior, as well as observations of morphological and plumage characteristics, however, suggest that the above classifications may not reflect the true relationships among shrikes. Lefranc (1997) and Harris (2000) have suggested alternate taxonomies.

Physical characteristics

Bush-shrikes are a mixed group with at least seven genera. *Malaconotus* bush-shrikes are large birds (9.1–9.8 in [23–25 cm]) with a strong bill and beautiful, bright colors. Their upperparts are mainly olive-green and their underparts may be green or, more often, yellow with red or orange. A marked polymorphism occurs in the fiery-breasted bush-shrike (*M. cruentus*). Some species of the genus *Telophorus* are very similar,

Loggerhead shrike (*Lanius ludovicianus*) with a vole at its larder. (Illustration by John Megahan)

but are smaller and have a weaker bills. The many-colored bush-shrike (*T. multicolor*) shows yellow, orange, red, buff, and even black morphs; the very rare Mount Kupé bush-shrike (*T. kupeensis*) measures about 8.7 in (22 cm), is mainly gray with a distinct black face-mask, a white throat, and a yellow vent. The birds of the genus *Laniarius* are robust, medium to large (6.7–9.8 in [17–25 cm]), with black or mainly black upper-parts; their underparts may be black, white, red, orange, or yellow. *Dryoscopus* species are somewhat similar, often with a pied coloration; contrary to all the species mentioned above, they show a more or less strong sexual dimorphism. Their English name, puffback, is due to the fact that they can erect the thick, fluffy feathers of their back and rump into a puff. Tchagras are medium to large in size (6.3–9.1 in [16–23 cm]) with gray or gray-brown upper-parts, chestnut wings, black-streaked head, and a rather long, graduated tail; the sexes are similar, except in the marsh tchagra (*Tchagra minuta*), which is sometimes placed in its own genus. The brubru (*Nilaus afer*) is small (about 5.5 in [14 cm]) with a pied plumage; its white eyebrow is characteristic, and most races show rufous flanks. The strange white-tailed shrike (*Lanioturdus torquatus*) (5.9 in [15 cm]) is also pied with long legs and a very short, white tail.

Helmet shrikes (*Prionops*) are small to relatively large (7.5–9.8 in [19–25 cm]) with a fairly stout bill. Their plumage is generally contrasted, and white, black, and gray are often the dominant colors. The bill is most often white, but can be red in some species. The group is characterized by the presence of stiff, bristle-like feathers on the forehead and, in most species, by a colored wattle around the eye. The two very similar *Eurocephalus* species are plump and relatively large

(7.5–9.8 in [19–25 cm]); their plumage is pied and buffy, and their snow-white crown has given them their English name of white-crowned shrikes.

In the true shrikes, the two *Corvinella* species are large and long-tailed. The yellow-billed shrike (*C. corvina*) (12 in [30 cm] including tail, or 7 in [18 cm]) has a brown, vermiculated plumage and resembles a giant, juvenile *Lanius* shrike. The magpie-shrike (*C. melanoleuca*) (18 in [45 cm] including tail, or about 12 in [30 cm]) is almost completely black with white scapulars and wing patches. *Lanius* shrikes are small to relatively large birds (5.9–11.8 in [15–30 cm]). The smallest is the central African Emin's shrike; the largest is the high-elevation race *giganteus* of the Chinese gray shrike (*L. sphenocercus*). All *Lanius* bear the "highwayman's mask," which can extend well over the bill. The most common colors in the plumages are black, white, and various shades of gray and brown; however, deep chestnut, yellow, or even orange, can occur. One of the most brightly colored *Lanius* is the race *tricolor* of the Asian long-tailed shrike (*L. schach*). Males and females are often quite similar; sexual dimorphism is only very obvious in a few species and particularly in the red-backed shrike. Young birds are typically brown and heavily vermiculated.

Distribution

Almost all the bush-shrikes are found in Africa, south of the Sahara. The only exception is the black-crowned tchagra (*Tchagra senegala*), which is widespread on the African continent and has two isolated races, one in north Africa and the other on the southwest Arabian peninsula, in Yemen and

Oman. Among the most widespread species are the gray-headed bush-shrike (*Malaconotus blanchoti*) and its smaller replica, the orange-breasted bush-shrike (*Telophorus sulfureopectus*); the two species have almost the same vast breeding range in sub-Saharan Africa. Among the very rare species are the Mount Kupé bush shrike, known only from two or three tiny areas in Cameroon, and the Bulo Berti boubou, known from only one individual in central Somalia.

Helmet-shrikes are confined to sub-Saharan Africa. The most widespread species found in west, east, and southern Africa is the white-crested helmet-shrike (*Prionops plumata*).

True shrikes are also well represented in sub-Saharan Africa. The genus *Corvinella* occurs only there, but the most widespread species is the common fiscal (*Lanius collaris*); with about 10 races, it can be seen almost everywhere in adequate habitats. The rarest species is the rather similar Sao Tomé fiscal (*L. newtoni*), confined to an island in the gulf of Guinea. Two *Lanius* species breed both in north Africa and Eurasia: the southern gray shrike (*L. meridionalis*) and the woodchat (*L. senator*). Populations of the latter also migrate from Eurasia to Africa, as populations of three other *Lanius* do. Other shrikes of the same genus are common in Asia; a race of the long-tailed shrike breeds in New Guinea. Only two species have reached North America: the holarctic northern shrike, which has some breeding populations beyond the Arctic circle and breeds in Alaska and northern Canada, and the very similar endemic loggerhead shrike (*L. ludovicianus*).

Habitat

The large bush-shrikes of the genus *Malaconotus* and the generally smaller species of the genus *Telophorus* are birds of the forest, occurring in lowland and montane woodland up to 9,800 ft (3,000 m); they are arboreal, often skulking in canopies and sometimes in undergrowth. The puffbacks (*Dryoscopus*) most often inhabit canopies of forest or other habitats rich in high trees, including suburban gardens. Gonoleks and boubous (*Laniarius*) occur in all kinds of thickets and dense shrub; they generally live rather low in the vegetation and spend much of the time foraging on the ground. The tchagras are more terrestrial and inhabit a variety of semi-open habitats, generally arid and dominated by dense shrub; the marsh tchagra, however, lives in humid areas dominated by reeds or papyrus. The white-tailed shrike is also a terrestrial bird, typically found in the semi-arid scrub savanna extending from Namibia to southwestern Angola.

Helmet shrikes are found in a wide range of open wooded savannas or woodlands. Highly gregarious groups fly from one tree to another in the search of insects. The most common species, the white-crested helmet shrike, may occur in suburban gardens outside the breeding season.

True shrikes are typical birds of semi-open habitats; they need perches of some kind with a good view on the ground, where most of their prey is taken. Most species have benefited from deforestation and have adapted well to low intensity types of farming. A few species are forest birds, however. The rarest, the São Tomé fiscal, is restricted to primary low-

A white helmet-shrike (*Prionops plumata*) at its nest. Family parties assist in tending the nest and young in this species—five birds helped tend this nest. (Photo by C. Laubscher. Bruce Coleman Inc. Reproduced by permission.)

land and mid-altitude forest; it has never been recorded in secondary forest or in cultivated areas.

Behavior

Bush-shrikes are generally seen singly or in pairs, but small parties, probably family groups, have been recorded. They appear to be monogamous and territorial, but very little is known about the social organization of many species, as their tendency to keep to thick cover makes them difficult to observe. One of the easiest to spot is the beautiful, long-tailed, rosy-patched shrike (*Tchagra cruentus*), which lives in dry bush areas in eastern Africa. Its noisy family groups are constantly active, flying low or hopping on the ground. Members of a pair may be conspicuous in breeding time, when they project their melodious whistles from the top of a bush. Song is an important means of communication in tropical forests or dense shrub and often betrays the presence of bush-shrikes. *Malaconotus* species project many sounds, including distinctive, far-carrying whistles and bell-like phrases. *Laniarius* are well-known for their extraordinary duet-songs; what appears to be the call of a single bird is often a kind of whistle by the male immediately followed by generally harsher notes by the female. The brubru bears its name well; its call is onomatopoeic and recalls the trilling song of the male. Some tchagras produce clear lilting, melodious, almost human-sounding

A northern shrike (*Lanius excubitor*) decapitates its prey—a white-footed mouse—before eating it. (Photo by Ron Austing. Photo Researchers, Inc. Reproduced by permission.)

phrases that may accompany typical display flights. These displays start with a steep climb, during which the wings are extended and the tail is fanned out, and they end with a gliding descent. The puffbacks may be seen for a few seconds during their short butterfly flights, made with their puff expanded. Most bush-shrikes are apparently sedentary, although local movements are obvious in a few species. Altitudinal movements have been recorded in the olive bush-shrike (*Telophorus olivaceus*) in Malawi and Zimbabwe.

Helmet-shrikes are easier to observe; they are highly sociable with up to 30 birds in a single group; they forage and roost communally and breed cooperatively, with each group under the dominance of the only breeding female. Most species have local or altitudinal movements.

Two African true shrikes are gregarious: the long-tailed fiscal (*L. cabanisi*) and the gray-backed fiscal (*L. excubitoroides*). The latter, and possibly also the former, breed cooperatively, with only one breeding pair that benefits from the presence of a varying number of helpers. All the other species are found singly or in solitary pairs. However, loose colonies are known, particularly in the case of the lesser gray shrike (*L. minor*), and concentrations may also occur during migration. All *Lanius* are normally monogamous; that, however, does not exclude extra pair paternity, which has been proven in at least three species. They are also highly territorial, with territories covering on average 3.7 acres (1.5 ha) in the small red-backed shrike and up to 250 acres (100 ha) or more in the much larger northern shrike. In many *Lanius* species, songs are relatively rare and not far-carrying, although territorial calls are important. Almost all or all *Lanius* shrikes impale their prey, but the regularity of this behavior varies between and even within species. Pair formation takes place inside the territory, display-flights are known in some species, and curious group meetings have been recorded in the northern shrike and in the loggerhead shrike. Courtship feeding is regular. Depending on latitude and feeding habits, *Lanius* species can be resident; or partial, altitudinal, or long-distance migrants. African species are probably sedentary, while some populations of the

Eurasian red-backed and lesser gray shrikes may travel 6,200 mi (10,000 km) or more.

Feeding ecology and diet

All types of shrikes feed on a large variety of arthropods and are largely insectivorous. Small vertebrates and birds' eggs are also on the menu for the larger *Malaconotus* species and the true shrikes. At least some of the true shrikes occasionally indulge in small fruits and berries.

Bush-shrikes most often feed by gleaning the vegetation inside branches, trunks, and foliage, and have been compared to oversized warblers. Prey is taken at various forest levels according to the species. Some, like the rare Uluguru bush-shrike (*Malaconotus alius*), appear to be confined to the canopy; while others, like the gray-headed bush-shrike, forage at all levels. Where two related species come into contact, there might be an ecological segregation, and each shrike keeps to its preferred level. Some species, like the tchagras, some boubous, gonoleks, and the white-tailed shrike, pick much of their captures off the ground, where they hop about in the vegetation.

Helmet-shrikes search for food in noisy, sometimes mixed, groups. They may forage from ground to canopy, but seem to prefer middle or lower levels; at least some *Prionops*, when perched on a branch and looking for prey, tilt their heads on one side as if they were listening to their future victims.

True shrikes are "sit and wait" predators that spend a lot of time on a variety of perches. Prey is mainly caught on the ground, but in fair weather a lot of insects are hawked in the air. Impaling or wedging of prey is regular in *Lanius*, but has not been recorded in all species. It is regular in larger species like the northern shrike, which must anchor its small vertebrate victims in order to dismember and eat them. Impaled prey can serve as a larder available in bad weather, when insects are not very active. The habit of impaling for a few days may enable species like the loggerhead shrike to consume toxic prey once it has degraded. Impaling may also serve as mate attraction. The more prey that is impaled in a territory, the better the male may appear to a female, but this point needs further research. Impaling or wedging is not entirely confined to *Lanius*; it is regularly recorded in the gray-headed bush-shrike and may occur in other *Malaconotus*; it has occasionally been recorded in *Laniarius*, but so far only in captivity.

Reproductive biology

All bush-shrikes appear to be monogamous and territorial. The most remarkable aspects of courtship behavior are dueting, mainly found in *Laniarius*; the "puffback" displays of *Dryoscopus*; and the flight songs of *Tchagra*. Courtship feeding is known in the gray-headed bush-shrike; it may occur in the brubru. Nests, built in trees or bushes, are tidy cups made up of fine rootlets, twigs, and grass. Some species like to incorporate spider web, but the medal of originality goes to the marsh tchagra, which often decorates its nest with snake skin. A few nests, like those built by the rosy-patched shrike, are flimsy, almost transparent, and recall those of doves. Bush-shrikes lay few eggs, most often only two or three. The breeding season appears to be favored by the start of a rainy season.

Helmet-shrikes are cooperative breeders, with a dominant breeding pair assisted by helpers. The nest is a compact cup, generally built on a horizontal branch, often made of bark and decorated with spider web. *Prionops* lay two to five eggs, often four. The breeding biology of five of seven *Prionops* helmet-shrike species remains virtually unknown.

Cooperative breeding also occurs in true shrikes, in the two species of *Corvinella*, and in at least one species of African *Lanius*, the gray-backed fiscal. *Lanius* nests are of the classical cup-shaped type, generally well structured, but not always neatly made. Eggs number between three and eight; clutch-size varies with latitude, both within the genus and within populations of the same species. The African *Lanius* lay generally few eggs, for instance, apparently never more than three in Sousa's shrike (*L. souzae*). However, in Alaska the modal clutch-size of the northern shrike is eight eggs. Young *Lanius* stay in the nest for two to three weeks, depending on the species, weather conditions, and availability of food.

Conservation status

Nine species of shrikes have the unhappy privilege of being on the list of the globally threatened species, including six bush-shrikes, two helmet-shrikes, and one true shrike. They all live in sub-Saharan Africa, where they are still poorly known and have a very small range. Another common characteristic is that they almost all, including the *Lanius* shrike, live in forest habitats. The possible exception is the recently discovered Bulo Berti boubou; the only known individual was trapped in acacia shrub. All these birds have habitats that face deforestation for the benefit of various types of agriculture.

Population estimates must be taken with extreme caution. It is assumed that the Bulo Berti boubou population totals fewer than 50 pairs in the only area where it is known in Somalia. The Gabela bush-shrike (*Laniarius amboimensis*) and the orange-breasted bush-shrike (*L. brauni*) both live in tiny areas in western Angola and have populations that are thought to be between 250 and 1,000 pairs. The Mt. Kupé bush-shrike and the green-breasted bush-shrike (*Malaconotus gladiator*) are mainly confined to western Cameroon; the latter species also has a tiny population in eastern Nigeria. They are respectively thought to have populations numbering between 50 and 250 pairs and between 2,500 and 10,000 pairs. The population of the Uluguru bush-shrike (*Malaconotus alius*), endemic to the mountains of the same name in eastern Tanzania, benefited from a detailed survey in 2000. The results were rather encouraging, as about 1,200 pairs were located, thus doubling previous estimates. The two helmet-shrikes on the Red List are the Endangered Gabela helmet-shrike (*Prionops gabela*) with 1,000–2,500 pairs in western Angola, and the Vulnerable yellow-crested helmet-shrike (*Prionops alberti*) with 2,500–10,000 pairs in the east of the Democratic Republic of Congo. The São Tomé fiscal is endemic to the island of the same name, which lies in the gulf of Guinea, 160 mi (255 km) off the coast of Gabon. It is the only true shrike on the IUCN

A loggerhead shrike (*Lanius ludovicianus*) has its lizard prey impaled on a barbed wire fence in California. Shrikes may stash their prey by impaling it on fences or thorned branches. (Photo by Maslowski. Photo Researchers, Inc. Reproduced by permission.)

Red List. It is judged to be Critically Endangered with perhaps fewer than 50 pairs; other estimates, however, give a few hundred pairs.

Another five species are classified as Near Threatened. Three are in Africa: two bush-shrikes and one helmet-shrike; two are in Asia: one *Lanius* shrike and the Bornean bristlehead.

Almost all the species mentioned above are confronted with a major problem: deforestation. That is, however, not the general cause explaining the present worrying decline of *Lanius* shrikes in North America and Europe; on the contrary, these semi-open habitat species have certainly benefited from the clearance of forests. They have adapted extremely well to "untidy" open landscapes associated with low-intensity farming. A strong association exists between them and the lifestyle based around cultivation and domestic stock. The golden age for many such shrikes has, however, come to an end with the industrialization of agriculture. Large-scale production techniques have involved a high level of mechanization and an increase in field size. This has led to the disappearance of large areas of non-productive habitats, such as hedgerows, bushes, isolated trees, ponds, marshes, banks, ditches and even of grassy paths, which are favored by shrikes. Widespread use of pesticides has reduced their food resources.

Significance to humans

True shrikes are miniature birds of prey and were generally regarded as harmful up to about the middle of the twentieth century, both in Europe and in North America. This negative reputation was reflected in the writings of hunters and gamekeepers, but also in those of some ornithologists. Shrikes are still hunted by humans in many parts of the world, particularly when they migrate. Many are killed in the Middle East and Greece. In Turkey, numerous red-backed shrikes are caught, blindfolded, and used as decoys to attract and net sparrowhawks (*Accipiter nisus*); the raptors are then trained and used to catch common quail (*Coturnix coturnix*). In Taiwan, the brown shrike (*Lanius cristatus*) is caught with bamboo foot traps, killed, sold, and barbecued for tourists visiting a national park.

In the western world, shrikes have gained a better reputation; but the birds are confronted with drastic habitat changes. The decline of shrikes, their beauty, and the fact that they are generally conspicuous and easy to study, has prompted the creation of an International Shrike Working Group. It met for the first time in Florida in 1991 and organizes regular meetings that are followed by the publication of proceedings. So far it has only taken a strong interest in true shrikes, but the other shrike-like birds should not be forgotten. Very little is known about them, but it is obvious that a few species are already endangered. Much of their future will depend on the development of ornithology and conservation measures in Africa.

1. Loggerhead shrike (*Lanius ludovicianus*); 2. Female red-backed shrike (*Lanius collurio*); 3. Male red-backed shrike (*Lanius collurio*); 4. Long-tailed shrike (*Lanius schach*); 5. Yellow-crowned gonolek (*Laniarius barbarus*); 6. Northern puffback (*Dryoscopus gambensis*); 7. Gray-headed bush-shrike (*Malaconotus blanchoti*); 8. White helmet-shrike (*Prionops plumatus*); 9. Black-crowned tchagra (*Tchagra senegala*). (Illustration by John Megahan)

Species accounts

White helmet-shrike
Prionops plumatus

SUBFAMILY
Prionopinae

TAXONOMY
Prionops plumata Shaw, 1809, Senegal. Up to nine races described. Variation affects size, amount of white in wings, and characteristics of frontal feathers and crest.

OTHER COMMON NAMES
English: White-crested helmet-shrike, curly-crested helmet-shrike; French: Bagadais casqué German: Brillenwürger; Spanish: Alcaudón de Copete Yelcobé.

PHYSICAL CHARACTERISTICS
7.4–9.8 in (19–25 cm); 0.88–1.3 oz (25–37 g). A relatively large species; sexes similar. Mainly black on upperparts, but with white wing-stripe; wholly white on underparts; head whitish with stiff frontal feathers and long, straight crest; eyes yellow surrounded by yellow wattle. In flight, obvious white patch in wings and white outer feathers in tail. Juveniles are similar, but duller, with no crest and no wattle. Race *cristatus* (Ethiopia) has no white in the wings and its long crest curls forward. Race *talacoma* (parts of eastern and southern Africa) shows a grayish head and has no long crest.

DISTRIBUTION
Most common and widespread helmet-shrike, exclusively in sub-Saharan Africa. Absent from some parts of south, central-western and eastern Africa.

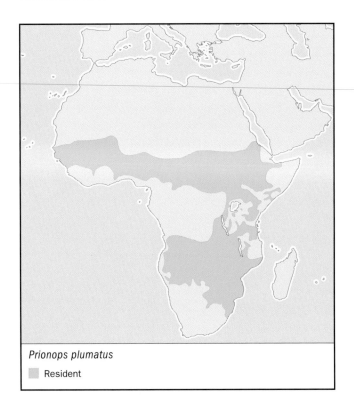

Prionops plumatus

▬ Resident

HABITAT
Woodlands and wooded savannas, sometimes in suburban gardens outside breeding season. Occurs up to 5,900 ft (1,800 m) in Kenya.

BEHAVIOR
Highly gregarious in all seasons. A given group generally numbers up to seven birds, but up to 22 have been recorded together outside the breeding season. The group defends a home range covering about 50 acres (20 ha) (10–75 acres [5–30 ha]). When searching food, birds fly from tree to tree and may be seen anywhere between ground and canopy. Local movements are known, but not yet fully understood; they appear to be favored by drought years.

FEEDING ECOLOGY AND DIET
All kinds of arthropods, mainly insects and particularly butterflies and moths, as well as their caterpillars. Small reptiles are sometimes taken.

REPRODUCTIVE BIOLOGY
Monogamous cooperative breeder. Only one dominant pair breeds and is assisted by what are thought to be closely related birds. Over the vast breeding area, egg-laying may occur almost in any month. Lays two to five eggs, most often four, in a cup-shaped nest made of bark that is cemented and decorated with spider web. It is placed a few yards (meters) above the ground in a tree. Incubation done by all the birds, but probably mostly by the dominant pair, for about 18 days. Young stay in the nest about 20 days; they are still fed by the group when about two months old.

CONSERVATION STATUS
Not threatened.

SIGNIFICANCE TO HUMANS
None known. ◆

Northern puffback
Dryoscopus gambensis

SUBFAMILY
Malaconotinae

TAXONOMY
Lanius gambensis Lichtenstein, 1823, Senegambia. Taxonomy still unclear; up to five races described; variation relatively well marked, but only in color of underparts and upperparts of females. Appears to be very close both to Pringle's puffback (*D. pringli*), an uncommon resident in eastern Africa, and to the black-backed puffback (*D. cubla*), which is widespread south of the equator.

OTHER COMMON NAMES
English: Gambian puffback-shrike; French: Cubla de Gambie; German: Gambia-Schneeballwürger; Spanish: Obispillo Común.

PHYSICAL CHARACTERISTICS
7–7.5 in (18–19 cm); 0.95–1.37 oz (27–39 g). The red-eyed male has mainly glossy black upperparts, but scapulars, rump,

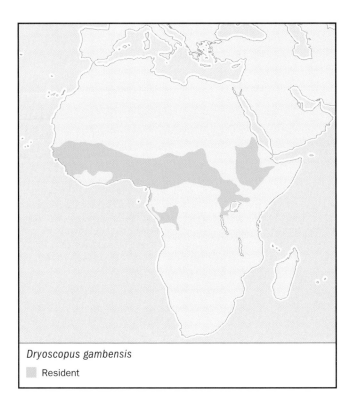

Dryoscopus gambensis
▪ Resident

wing-coverts, and edges of wing-feathers are pale gray; underparts are creamy white. Female is duller with orange eyes, gray-brown upperparts, and mainly creamy reddish underparts; the intensity of these colors varies with races. Young birds are similar to the adult female, but have reddish tips on upperparts and brown eyes. In the eastern race *erythrea*, the female strongly resembles the male; her upperparts are, however, dark brown instead of black, and her underparts are generally creamier.

DISTRIBUTION
Most common puffback north of the equator; found in a broad belt stretching from Senegambia to Eritrea and Ethiopia; almost absent from the Horn. Nominate from Senegal to Cameroon; most distinct race, *erythrea* from eastern Sudan to northwestern Somalia.

HABITAT
All types of savanna woodland and areas with large trees, including gardens; avoids closed forests. In Eritrea occurs up to about 4,900 ft (1,500 m).

BEHAVIOR
Generally solitary or occurring in pairs, keeping to the tree canopy. Only relatively easily spotted and seen for a few seconds in breeding season when very demonstrative. Males fly from tree to tree with back and rump feathers fluffed out. Thought to be sedentary, but local movements are possible.

FEEDING ECOLOGY AND DIET
Arthropods, mainly insects and particularly caterpillars gleaned from foliage.

REPRODUCTIVE BIOLOGY
Territorial and probably monogamous. Little known. Few nests found; generally high up in trees; plastered with spider web. Appears to breed in any month of the year.

CONSERVATION STATUS
Not threatened.

SIGNIFICANCE TO HUMANS
None known. ◆

Black-crowned tchagra
Tchagra senegala

SUBFAMILY
Malaconotinae

TAXONOMY
Lanius senegala Linnaeus, 1776, Senegal. About 12 races described, but a few poorly differentiated. Variation concerns size, coloration of back and underparts, and coloration and shape of superciliary stripe.

OTHER COMMON NAMES
French: Tchagra à tête noire; German: Senegaltschagra; Spanish: Chagra.

PHYSICAL CHARACTERISTICS
7.8–9 in (20–23 cm); on average 1.9 oz (54 g). Relatively large with a long, graduated tail and a heavy black bill; sexes are similar. Characteristic black crown and black stripe through eye; white, prominent supercilium. In nominate, upperparts are gray brown; tail is dark fringed and tipped white; underparts are pale gray. Juveniles are similar to adult, but duller with a mottled crown, buff eyebrow, and paler bill. The Arabian race *percivali* is the darkest one, relatively similar to the North African race *cucullata*, also darker than nominate; on the contrary *remigialis* from Chad to Sudan is very pale with almost pure white underparts.

Tchagra senegala
▪ Resident

DISTRIBUTION
Rather a coastal species in northwestern Africa from Morocco to northern Libya (race *cucullata*). Nominate occurs from Senegambia to Sierra Leone. Other races widespread in sub-Saharan Africa, except in the Horn, the Congo basin, and the southwestern area of the continent.

HABITAT
Wide range of semi-open habitats dotted with bushes, thickets, and isolated trees; sometimes occurs in forested habitats, in Ethiopia, up to 9,800 ft (3,000 m); also in plantations, parks, and gardens.

BEHAVIOR
The male defends a rather small territory covering about 10 acres (4 ha). Generally secretive; likes to keep to inside of bushes, but song of warbled whistles betrays its presence. Also conspicuous during display flights, when it can climb up to 50 ft (15 m) before gliding downward in full song toward another perch. Spends much of its time hopping on the ground where most prey is caught, particularly near bases of bushes and trees. Only local movements are known and only in a few areas.

FEEDING ECOLOGY AND DIET
Arthropods, mainly insects, particularly grasshoppers and beetles. Small reptiles and amphibians, as well as berries, are also taken.

REPRODUCTIVE BIOLOGY
Monogamous, occurs singly or in pairs. The nest is a shallow cup, built in a bush or small tree generally between 1.6 and 6.5 ft (0.5 and 2 m) above the ground. A normal clutch contains two or three eggs (rarely four), which are apparently incubated by both sexes, but probably mainly by female, for about 14 days. Over the vast range, laying may begin in all months with local peaks, possibly favored by early rains. Normally only one clutch. Young leave the nest when they are about 16 days old.

CONSERVATION STATUS
Not threatened. A few isolated races like *percivali* in the southwestern Arabian peninsula, might, however, deserve to attract the attention of conservationists.

SIGNIFICANCE TO HUMANS
None known. ◆

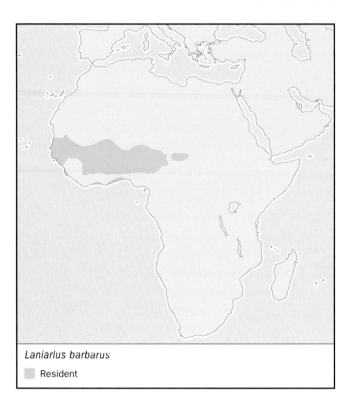

Laniarius barbarus

◻ Resident

Yellow-crowned gonolek
Laniarius barbarus

SUBFAMILY
Malaconotinae

TAXONOMY
Lanius barbarus Linnaeus, 1766, Senegal. Two races; slight differences mainly affecting coloration of crown and nape.

OTHER COMMON NAMES
French: Gonolek de Barbarie; German: Goldscheitelwürger; Spanish: Gonolek Común.

PHYSICAL CHARACTERISTICS
9.0–9.8 in (23–25 cm); on average 1.7 oz (49 g). A robust, relatively large, and brightly colored species. The similar sexes have black upperparts contrasting with a golden crown, brown

eyes, and vermilion red underparts. Juveniles are similar, but have duller upperparts and buffy underparts with heavy dark barring. Race *helenae* has a deeper reddish crown.

DISTRIBUTION
Western Africa from southern Mauritania to northern Cameroon. *Helenae* is confined to coastal areas in Cameroon.

HABITAT
Dense, woody undergrowth in savannas; the Sierra Leone race inhabits mangroves.

BEHAVIOR
Occurs singly or in pairs. Spends much time in low vegetation and on the ground, but may also look for prey on small branches up to about 16 ft (5 m). Its flight is short and appears heavy. More often heard than seen; produces remarkable duets that sound as if they were produced by a single bird. Sedentary as far as known.

FEEDING ECOLOGY AND DIET
Arthropods, mainly insects and particularly caterpillars and grasshoppers. Said to predate on birds' nests.

REPRODUCTIVE BIOLOGY
Monogamous and territorial. Few precise data available. The nest is hidden relatively low, between 4.9 and 14.8 ft (1.5–4.5 m) in a dense bush or a small tree. It receives two, or more rarely three, eggs. Eggs may be laid at any time of the year.

CONSERVATION STATUS
Not threatened. The precise status of the Sierra Leone race might require further investigations.

SIGNIFICANCE TO HUMANS
None known. ◆

Gray-headed bush-shrike
Malaconotus blanchoti

SUBFAMILY
Malaconotinae

TAXONOMY
Malaconotus blanchoti Stephens, 1826, Senegal. Up to seven races described with regular intergradation at common boundaries. Main differences concern color of underparts.

OTHER COMMON NAMES
French: Gladiateur de Blanchot; German: Graukopfwürger; Spanish: Gladiador de Cabeza Gris.

PHYSICAL CHARACTERISTICS
9–10.2 in (23–26 cm) on average 2.7 oz (77 g). One of the largest is the *Malaconotus* shrike, which has a heavy bill. Sexes are similar. Head and nape is grayish; eyes are pale yellow and lores are white. Upperparts and wings are mainly olive green; wings show yellow spots to coverts. Underparts are yellow with a varying amount of orange on breast. Juveniles are similar, but duller and with brown eyes. Adult of nominate form, in West Africa, has no or little orange on its underparts, which are very dark in the eastern African race *approximans*. Race *hypopyrrhus* from Tanzania southwards has less intensively colored underparts.

DISTRIBUTION
Widespread in sub-Saharan Africa, but absent from the Horn, the Congo basin, and the southwestern area of the continent.

HABITAT
Various types of woodlands, including riverine woods; can also occur in large parks or suburban gardens.

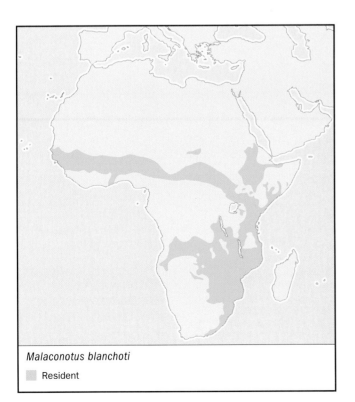

Malaconotus blanchoti

▨ Resident

BEHAVIOR
Seen solitary, in pairs, or in family groups. Territory covers about 124 acres (50 ha) and is advertised by various sounds, particularly by mournful, far-carrying whistles. Generally elusive, it hides in dense thickets, but may be relatively conspicuous during courtship activities, which include a display flight. Looks for food at all levels of vegetation, also occasionally on ground. Remarkably, caches food like *Lanius* shrikes; small vertebrates can be wedged into forks. Local movements have been suspected; they might be related to the onset of rains.

FEEDING ECOLOGY AND DIET
Arthropods: scorpions, worms, centipedes, but mainly insects. Small birds and reptiles are also regularly taken.

REPRODUCTIVE BIOLOGY
Monogamous, though little is known. The nest is an untidy cup, generally placed at about 13 ft (4 m) above the ground in a small, deciduous tree; it receives two to four, and most often three eggs. The laying season varies with the geographical area. Female alone appears to incubate for about 16 days. Nestling season is about 20 days.

CONSERVATION STATUS
Not threatened.

SIGNIFICANCE TO HUMANS
None known. ◆

Red-backed shrike
Lanius collurio

SUBFAMILY
Laniinae

TAXONOMY
Lanius collurio Linnaeus, 1758, Sweden. Up to five races described, but probably best considered monotypic as the supposed and slight differences between races also occur within given local populations all over the breeding range.

OTHER COMMON NAMES
French: Pie-grièche écorcheur; German: Neuntöter; Spanish: Alcaudon Dorsirrojo.

PHYSICAL CHARACTERISTICS
6.2–7 in (16–18 cm); on average 1.05–1.12 oz (30–32 g). One of the smallest *Lanius* shrikes; strong sexual dimorphism. The brightly colored male is unmistakable with his gray head, reddish brown upperparts, gray rump, black tail fringed white, and pinkish underside. Female is much duller, but her ground color is variable; brown and gray are dominant; her under-parts are generally heavily vermiculated. Juveniles are very similar to the female but with strong barring, (black crescents), also on upperparts.

DISTRIBUTION
Most widespread and common shrike in the western Palearctic, almost reaching the Arctic circle in the north. In some hot Mediterranean areas, like Spain and southern France, it is regarded as a bird of low mountains. This long-distance migrant winters in eastern and southern Africa, from southwestern Kenya southwards.

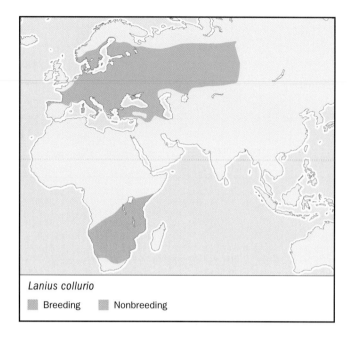

Lanius collurio

☐ Breeding ☐ Nonbreeding

HABITAT

Semi-open habitats dotted with low perches and thorny bushes like hawthorn and blackthorn. Benefits from low-intensity farming, particularly traditional pastures; also found in young plantations, forest clearings, etc. Breeds up to about 6,600 ft (2,000 m) in the French Alps and even up to about 10,500 ft (3,200 m) on meadows in Caucasus. In Africa, it favors arid savannas and particularly dry *Acacia* thornveld.

BEHAVIOR

In breeding season, it defends a rather small territory covering 3.7–7.4 acres (1.5–3 ha). At the end of April or beginning of May, when they have returned from Africa, perched males advertise their territories by typical, far-carrying, and somewhat nasal calls. It is a "sit and wait" predator that takes most of its prey on the ground; however, many insects are also caught in the air in fair weather. Larders are kept, but that habit varies among individuals and is less common in warmer climates, where insects are plentiful and easier to locate. Leaves its breeding sites from the beginning of July onward; October observations are rare in Europe. It is a long-distance and a loop migrant; in autumn, west European populations converge towards Greece and its islands before crossing the sea to Egypt, and then to Sudan and Ethiopia; in spring, these two last countries are almost completely avoided as the species passes more to the east, toward the Arabian peninsula.

FEEDING ECOLOGY AND DIET

Arthropods, mainly insects, particularly beetles and grasshoppers. Small vertebrates constitute about 5% of the captures, but may even be more important in certain years when moles are abundant.

REPRODUCTIVE BIOLOGY

Monogamous. Densities can be locally high with up to seven pairs, almost loose colonies, per 25 acres (10 ha) in good habitats. Isolated pairs are not rare, however. Nests are built soon after arrival from wintering quarters at an average height of about 4.2 ft (1.3 m) above the ground, often in a thorny bush. In most areas, laying begins at the end of May and peaks in the first two weeks of June. Second broods are very rare, but re-

placement clutches are frequent. Female usually lays four to six eggs (range one through eight); incubation is 13–15 days exclusively by female, who is regularly fed by male. Nestling period is about 15 days. Young are completely independent when they are about six weeks old.

CONSERVATION STATUS

Not threatened, but declining in many countries, particularly at lower altitudes where they have been eliminated by the intensification of agriculture. Climatic fluctuations may also play a role; has disappeared as a regular breeding bird from Britain, which lies on the north-western limits of its range.

SIGNIFICANCE TO HUMANS

In Europe, it is now protected and regarded as a significant bio-indicator of the health of the wider environment. Thousands still are killed on migration, particularly in Greece and in the Middle East. ◆

Long-tailed shrike

Lanius schach

SUBFAMILY
Laniinae

TAXONOMY
Lanius schach Linnaeus, 1758, China. Generally nine races recognized; intermediate forms exist. Variation concerns body and bill size, tail-length, and color of head and upperparts. Closely allied with the gray-backed (or Tibetan) shrike (*Lanius tephronotus*), but the latter is obviously a distinct species with two races.

OTHER COMMON NAMES
English: Black-headed shrike, Schach shrike, rufous-backed shrike; French: Pie-grièche schach; German: Schachwürger; Spanish: Alcaudon Cabecinegro.

Lanius schach

☐ Resident

PHYSICAL CHARACTERISTICS

9–10.6 in (23–27 cm); on average 1.76–1.86 oz (50–53 g) for nominate and about 1.3 oz (37 g) for *erythronotus*. Nominate, which inhabits China, is the largest race. Sexes similar or nearly so. Head and mantle are dark gray, back and rump are rufous; tail is long, black, and graduated; wings are dark with conspicuous white primary patches. Underparts are whitish, strongly tinged with rufous on the sides of breast and flanks. *Erythronotus*, widespread in central Asia and in the Indian subcontinent, is similar, but distinctly smaller, somewhat duller, and with a narrower black band on the forehead. *Caniceps* from southern India and Sri Lanka is paler, with less rufous on its upperparts. Race *tricolor* is a superb Himalayan bird; it bears a black cap, shows a small grayish area on upper mantle, and has mainly deep rufous upperparts. It is rather similar to the three insular races. Race *longicaudatus* from Thailand has a very long tail. Remarkably, in certain areas, nominate has a melanistic form called *fuscatus*; mixed pairs have been recorded.

DISTRIBUTION

Has a vast breeding area, from central Asia, Turkmenistan, and possibly Iran, to the Chinese Pacific coast, Southeast Asia, and New Guinea.

HABITAT

A scrub jungle bird, but also associated with lightly wooded country, cultivated areas, and gardens. Generally a bird of lowlands, but in the Himalayas, *tricolor* populations have been found up to 9,800 ft (3,000 m) and occasionally up to 14,000 ft (4,300 m). Nominate breeds up to 9,800 ft (3,000 m) in China.

BEHAVIOR

Solitary in habits, and highly territorial. Densities can, however, be very high locally with up to 1.6 pairs/acre (4 pairs/ha) in suburban areas in Afghanistan. It is very vocal in the breeding season during pair formation. It takes most of its prey on the ground, but also hawks insects in the air and occasionally pirates other birds. Impales some of its victims. Most populations are resident; however, local movements, including altitudinal ones, are known. The western part of the breeding range, covering central Asia, is almost completely vacated by *erythronotus* between August and November; the birds return in late February.

FEEDING ECOLOGY AND DIET

Arthropods, mainly insects; also small vertebrates and occasionally fruits.

REPRODUCTIVE BIOLOGY

Monogamous. The cup-shaped nest is rather bulky and placed in a thorny bush or in a tree; it is hidden 9.8–39.4 ft (3–12 m) above the ground. There appears to be a geographical variation in clutch size: four to six eggs in China, four in Sri Lanka, three in the Malay Peninsula, and two in New Guinea. Breeding season varies with geographical areas; the western race lays eggs between the end of March and July. Locally double-brooded; replacement clutches are frequent. Incubation by female lasts 13–16 days, and the young fledge after 14–19 days.

CONSERVATION STATUS

Not threatened.

SIGNIFICANCE TO HUMANS

In Nepal, its bill is used to "feed" newborn babies; this ceremony is supposed to bring luck to the young children. ◆

Loggerhead shrike
Lanius ludovicianus

SUBFAMILY
Laniinae

TAXONOMY
Lanius ludovicianus Linnaeus, 1766, Louisiana. Up to 12 races described; some of them poorly differentiated. Variation affects size and plumage coloration, particularly of the back and underparts.

OTHER COMMON NAMES
English: Migrant shrike; French: Pie-grièche migratrice; German: Louisianawürger; Spanish: Alcaudon Yanqui.

PHYSICAL CHARACTERISTICS
About 8.2 in (21 cm); on average 1.7 oz (48 g). The loggerhead is relatively large, large-headed, gray, black, and white shrike; sexes are similar or nearly so. Facial mask extends just over the eyes. Upperparts are gray with white scapulars. Wings and tail are mainly black, but with a white primary patch and white outer tail feathers. Underparts are white, sometimes with faint indications of barring. Juveniles are similar, but paler, brownish gray, and barred overall. Nominate, common in southeastern United States, is dark gray above, including on rump, and almost pure white below, whereas *excubitorides* from the Great Plains region of the west is a pale race with a white rump. The endangered race *mearnsi*, confined to the San Clemente Island of California, is a darker gray than any other subspecies on it upperparts. The loggerhead shrike resembles the northern shrike, but the latter is about 25% larger, paler gray above, and with a facial mask not extending over the eyes, and particularly narrowed in lores.

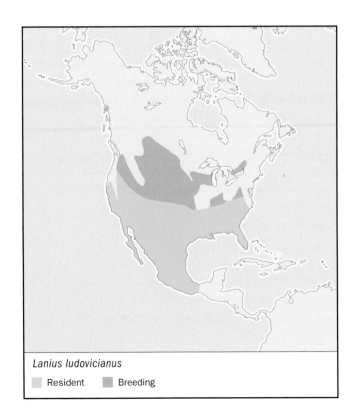

Lanius ludovicianus
☐ Resident ☐ Breeding

DISTRIBUTION
Only endemic shrike in North America, from southern Canada to Mexico. Northern part of breeding range is vacated in winter; contacts then possible with the relatively similar northern shrike, which also migrates further south from its breeding grounds in Alaska and northern Canada.

HABITAT
Various types of semi-open habitats with short vegetation; pastures are favored in many areas in Missouri, Illinois, and New York. In the western part of its range, it also occurs in semi-arid sagebrush areas, desert scrub, and pinyon-juniper woodlands; may occur in residential areas that are well dotted with perches. May be present in mountainous areas up to about 6,600 ft (2,000 m).

BEHAVIOR
Usually solitary or in pairs, but curious group meetings have been reported. Size of territory varies with habitat quality and averages about 25 acres (10 ha), but appears to be three times as large in the rare San Clemente race. Territory is defended with loud song and harsh territorial calls that may be produced in flight. It hunts from an elevated perch (6.6–33 ft [2–10 m]) and catches most of its prey on the ground. Impaling is regular; prey weighing more than about 1.1 oz (30 g) may be carried in the feet. Populations nesting north of about 40°N migrate and may move into areas with resident birds; most leave between September and November and return in March or April. In some southern areas, pairs maintain territories throughout the year; but in others, they separate and often defend adjacent territories.

FEEDING ECOLOGY AND DIET
All kinds of arthropods, mainly insects, most commonly beetles and grasshoppers. Vertebrate prey is regularly taken, particularly in winter; it includes small birds, lizards, mice, and occasionally bats and fish.

REPRODUCTIVE BIOLOGY
Territorial and normally monogamous. Even in well-populated areas, nests are normally several hundred yards (meters) apart, but small colonies have been reported occasionally. Northern populations are single-brooded, whereas second broods are common and third broods occasional in southern areas, such as Florida. Replacement clutches are frequent everywhere. Female usually lays four or five eggs (ranges one through seven); clutch size seems to vary with latitude. Female alone incubates for about 16 days; she is fed in the nest by the male. Nestling period is 16–21 days according to weather conditions. The young are independent four to five weeks after fledging, but they may stay up to three months with the parents.

CONSERVATION STATUS
Not threatened yet, but populations have experienced a marked decline in many regions, particularly in the northern part of the range. It is classified as endangered in Quebec. The highly endangered San Clemente race benefits from strong conservation measures; these measures include eradication of predators, removal of herbivores that tend to destroy the habitat, and breeding in captivity in the San Diego Zoo for reinforcement operations.

SIGNIFICANCE TO HUMANS
Of high significance to North American conservationists, as it is declining dramatically in most parts of range. ◆

Resources

Books

BirdLife International. *Threatened birds of the world*. Barcelona and Cambridge: Lynx Edicions and BirdLife International, 2000.

Fry, C.H., S. Keith, and E.K. Urban, eds. *The Birds of Africa*. Vol. 6. London: Academic Press, 2000.

Harris, T. *Shrikes and Bush-shrikes*. Princeton and Oxford: Princeton University Press, 2000.

Lefranc, N., and T. Worfolk. *Shrikes. A Guide to the Shrikes of the World*. New Haven and London: Yale University Press, 1997.

Yosef, R., F. Lohrer, D. Van Nieuwenhuyse, and P. Busse, eds. *Proceedings of the Third International Shrike Symposium, 15-18 September 2000, Gdansk, Poland*. Choczewo: The Ring; Polish Zoological Society, 2001.

Periodicals

Burgess, N., T.S. Romdal, and M. Rahner. "Forest Loss in the Ulugurus, Tanzania and the Status of the Uluguru Bush-Shrike *Malaconotus alius*." *Bull. ABC* 8 (2001): 89–90.

Kristin, A., H. Hoi, F. Valera, and C. Hoi. "Breeding Biology and Breeding Success of the Lesser Grey Shrike *Lanius minor* in a Stable and Dense Population." *Ibis* 142 (2000): 305–311.

Schollaert, V., and G. Willem. "A New Site for Newton's Fiscal *Lanius newtoni*." *Bull. ABC* 8 (2001): 21–22.

Van Nieuwenhuyse, D. "Global Shrike Conservation; Problems, Methods and Opportunities." *Aves* 36 (1999): 193–204.

Organizations

International Shrike Working Group. "Het Speihuis," Speistraat, 17, Sint-Lievens-Esse (Herzele), B-9550 Belgium. Phone: +32 54 503 789. E-mail: dries_van_nieuwenhuyse@hotmail.com

Norbert Lefranc, PhD

▲
Vanga shrikes
(Vangidae)

Class Aves

Order Passeriformes

Suborder Passeri (Oscines)

Family Vangidae

Thumbnail description
Small to medium-sized shrike-like birds with a wide variety of body forms and lifestyles, essentially restricted to forest

Size
5–13 in (12–32 cm); 1.2–11 oz (35–300 g)

Number of genera, species
11 genera; 16 species

Habitat
Primary eastern rainforest, western deciduous forest, and southern spiny bush; some species locally in plantations and wooded areas not far from primary forest

Conservation status
Endangered: 1 species; Vulnerable: 3 species; Near Threatened: 1 species; Data Deficient: 1 species

Distribution
Madagascar; one species locally on the Comoro Islands

Evolution and systematics

The startling range of body forms found in the vangas has provoked endless taxonomic confusion. Currently they are considered a monophyletic group within the large shrike assemblage, but members of the group have variously been considered nuthatches, shrikes, orioles, and bulbuls. In common with many other Malagasy birds, diversity is comparatively rich at the generic level but there are few species in each genus. All are at present included in the same sub-family.

Physical characteristics

The vangas display a remarkable variety of body formats over a relatively small size range. They start with the small vireo-like *Calicalicus* species, which glean small insects from foliage, through the nuthatch-like *Hypositta*, via the wood-swallow-like Chabert's vanga (*Leptopterus chabert*), and the electrified blue vanga (*Cyanolanius madagascarinus*), the thick, powerful-billed hook-billed vanga (*Vanga curvirostris*), *Xenopirostris*, and white-headed *Leptopterus viridis* vangas, to the larger and more extravagantly adorned sicklebilled vanga (*Falculea palliata*) and helmet vanga (*Euryceros prevostii*). The sicklebilled vanga is a large vigorous bird with a bill that is evidently suited to winkling invertebrates from their hiding places. The helmet vanga sports a huge flattened pale-blue

instrument that has no apparent foraging advantage, given that the species prefers small insects to opening coconuts.

In general, apart from their bills, vangas are similar in form in that they have rather stocky solid bodies, short wings, medium-length tails, short and rather strong legs, and large eyes. About the only exception to this is the Chabert's vanga, which has rather long wings suited to its largely aerial life. It is in the bill and head structure that the vangas really distinguish themselves. The 16 species of vanga can be divided up into three main groups—relatively slender, narrow bills for gleaning and general insectivory (red-tailed [*Calicalicus madagascariensis*], red-shouldered [*C. rufocarpalis*], nuthatch [*Hypositta corallirostris*], Tylas [*Tylas eduardi*], Chabert's, and blue vangas), laterally flattened bills for breaking open insect refuges or killing large invertebrates or small vertebrates (white, Pollen's [*Xenopirostris polleni*], Van Dam's [*X. damii*], Lafresnaye's [*X. xenopirostris*], and hook-billed), and the oddities, sicklebilled and helmet vangas.

Just over half the species are sexually dichromatic (red-tailed, red-shouldered, rufous [*Schetba rufa*], the *Xenopirostris* species, white-headed, blue and Bernier's [*Oriolia bernieri*], and the nuthatch vanga). The Tylas seems to be most unusual in having one form (the nominate) monochromatic, and the other (*T. e. albigularis*) possibly dichromatic, although the

Pollen's vanga (*Xenopirostris polleni*) is an endangered bird that lives in Madagascar. (Photo by A. Forbes-Watson/VIREO. Reproduced by permission.)

latter form has been seen so few times the question is not yet resolved. Most species are some combination of black and white, with some showing larger areas of red. The smaller vangas are brightly colored, the red-shouldered and red-tailed with black bibs, gray caps, and reddish wingpatches, the blue vanga with an ultramarine back and bill base. Chabert's vanga is unique in the family in having brightly colored blue skin around the eye. Rufous and helmet vangas are black, red, and white with pale blue bills (though of dramatically different form), while the other species are essentially gray, black, and white. Juveniles of many species and females of some (Pollen's, Van Dam's) show variable amounts of orange underneath, while in the Tylas vanga both sexes are orange-bellied. The nuthatch vanga is the only species with an orange as opposed to grey, blue, or black bill.

Distribution

With one strange exception, vangas are endemic to the mainland and adjacent small islands of Madagascar. They are found all over the island, being most scarce on the central plateau region where there is scarcely any natural habitat left. The curious exception to this rule is that of two populations of blue vangas in the Comoro islands, 125 mi (200 km) from the Malagasy mainland. The presence of this species in the Comoros is all the more remarkable given the fact that it is one of the more forest-limited vangas, never being seen far from the forest canopy in the west and east of Madagascar. It is very difficult to imagine such an apparently sedentary and habitat-limited species suddenly setting out over the open ocean, particularly when there are so many other more mobile species (in particular Chabert's vanga) that might have been expected to make the journey.

Habitat

Vangas are essentially birds of primary forest, like most other endemic Malagasy species. This means that they are ab-

sent from most of the high plateau region, and in coastal areas, essentially absent from secondary forest, plantations, or built-up areas. The two exceptions to the rule are hook-billed and Chabert's vangas. The latter may be found in degraded and secondary plantations a fair way from primary forest, where it appears to be able to hold territories and breed. Chabert's vanga is very mobile and may visit plantations (especially eucalyptus or mango) in search of pollinating insects tens of miles from primary forest. But neither has ever been recorded, for instance, from the centers of the larger cities.

The more adaptable species (hook-billed, Chabert's, white-headed) are found in all types of primary forest (eastern rainforest, western deciduous forest, and southern spiny subdesert), although populations in different forest types are subspecifically distinct in all three species. Blue, rufous, and Tylas vangas are found in both east and west; they are all restricted entirely to primary forest, and rufous and Tylas vangas have different subspecies in the west. The sicklebill occurs only in the west and south, although it manages to get by in the northern parts of the east, where forests are transitional. All the other species are limited either to southern spiny subdesert (Lafresnaye's, red-shouldered), western deciduous (Van Dam's), or eastern rainforest (Pollen's, Bernier's vangas, helmet vanga, and nuthatch vanga). The latter three species seem to be limited to the lower elevations in the rainforest, and helmet and Bernier's vangas to the northern half of the rainforest belt. Pollen's vanga seems to be much more common in the southern half of the rainforest than the north, and may be more common at higher altitudes. *Hypositta perdita*, described recently from juvenile specimens, is only known from rainforest at low altitude in the far southeast.

Behavior

Most vangas are found in pairs (during the breeding season), small groups, and often in large multi-species feeding flocks, which generally consist of vangas but often include other small passerines such as jerys, drongos, and cuckoo-shrikes. Two species (the sicklebilled and Chabert's vangas) are quite gregarious and can be seen in flocks of 6–32. Vanga shrikes are almost strictly forest dwellers, with the exception of a few species that will reside in both forest and savanna. Because of this limited habitat, long-range flight of vangas is often awkward and slow, even in the most powerful of short-range flyers. Some species (notably the sicklebilled, Lafresnaye's, Van Dam's, helmet, and white-headed vangas) are known for a distinctive loud whirring noise from their wings in flight. As a whole, vangas are a highly vocal group with diverse calls within the family; this is beneficial for tracking the birds while in feeding flocks with other vanga species.

Feeding ecology and diet

All vangas are essentially active seekers after exposed or hidden invertebrate or vertebrate prey. In line with bill form, the species can be separated into three basic groups—gleaners, predators/substrate strippers, and other species. The gleaners include the smaller *Calicalicus* and *Hypositta* species, which take small insect prey from substrates, the latter genus directly from trunks and large branches while moving in the manner of a nuthatch. The *Calicalicus* vangas glean from leaves and twigs. Blue vangas have quite conical-shaped bills that they use to extract insects from the tips of narrow canopy branches. Even longer and more conical is the bill of Chabert's vanga, which is used as a gleaning instrument in the canopy as well as for catching flying insects, particularly around flowers. Tylas vangas take larger invertebrates such as caterpillars from leaves. The heavy-billed vangas (Pollen's, Van Dam's, Lafresnaye's, rufous, white-headed, Bernier's, and hook-billed) exploit the mass of their bills to great advantage. The *Xenopirostris* vangas (Van Dam's, Pollen's, and Lafresnaye's) are capable of breaking off bits of dead branch, levering off mats of moss adhering to branches, opening insect cocoons, and crunching up heavily armored *Gasteracantha* spiders. The longer and more raptor-like bill of the hook-billed vanga is used for tearing apart small vertebrates, often through the use of a horizontal tree-fork in which to wedge the prey. Chameleons, bats, geckos, frogs, and even other birds are dealt with in this way. Rufous vangas eat rather similar kinds of food, but hunt for it in a different way. They spot prey by its movement, and so the actual species eaten are very different from those consumed by the hook-billed vanga. The Rufous vanga takes cockroaches, scorpions, and ants from the forest floor, and katydids and beetles from the undersides of understory leaves. The white-headed vanga, having a rather less substantial bill than the latter species, is more of a generalist, and while quite capable of killing and eating small chameleons, will also occasionally take fruit. Blue and Chabert's vangas have also been recorded eating fruits. The Bernier's vanga uses its triangular-shaped bill to dig into rotting masses of vegetation, especially those in the bases of *Pandanus* trees, from which it then energetically flings debris in search for invertebrates.

The two odd-billed species (sicklebilled vanga and helmet vanga) provide an object lesson in the interpretation of structural adaptation. For sicklebills, the bills are used as one might predict, to probe deeply into holes and recover well-hidden invertebrate larvae. The long powerful neck of this species is a great advantage in twisting the bill around curves in holes. By contrast, the helmet vanga uses its remarkable bill for little more than crunching up medium-sized invertebrates. It often displays great dexterity in sally-gleaning some invertebrate from a leaf or branch in a manner very similar to a rufous vanga, and likewise takes much prey from the ground.

Reproductive biology

Aside from studies on one or two species, not much is known of vanga reproduction. Like most Malagasy passerines, the breeding season is from October to December, in the early part of the warm rainy season. Most species lay two or three eggs, whitish or pinkish with dull reddish or brown speckles, in a cup-shaped nest, constructed by both male and female. Hook-billed, rufous, helmet, sicklebilled, and the *Xenopirostris* vangas place their nests in a tree fork between 3 and 16 ft (1–5 m) from the ground. The smaller species, especially Tylas, blue, Chabert's, and nuthatch vangas, place their nests higher in the canopy, and so are more difficult to study.

A male Van Dam's vanga (*Xenopirostris damii*) unzips an insect cocoon with its bill, while holding the cocoon in place with its foot. (Illustration by Marguette Dongvillo)

Nuthatch vangas attach their nest to a main trunk, often in a slight crevice. Blue vangas choose a site near the end of a branch near the crown, as do Chabert's vangas.

By far the best studied species is the rufous vanga, which has a cooperative breeding system where young males contribute to the upbringing of the young.

Several species of vanga produce beautiful calls. The rufous vanga produces a wide array of echoing and bell-like noises, many as a duet between male and female of a pair. Helmet vangas, apparently the closest relation to the rufous vanga, make similar calls, while the *Xenopirostris* vangas are easily detectable via their piercing descending whistle. Social species like the sicklebilled vanga make a lot of contact calls, and the Malagasy name of this species, *Voronzaza* (baby bird), reflects accurately a particular call that sounds like a baby crying. The red-shouldered and red-tailed vangas sing loud simple songs from the canopy, while the nuthatch vanga barely makes any discernible sound at all. The hook-billed vanga is unusual in clapping its bill together as part of a threat display.

Conservation status

Given that most vangas are tied closely to primary forest, and that Malagasy primary forests are under considerable threat, the future of many members of the family is uncertain. The most threatened species are Van Dam's vanga, which is found only in two widely separated populations in highly threatened western deciduous forest, and the lowland rainforest specialists helmet and Bernier's vangas. The latter two species are also only found in the northern half of the rainforest, in which zone lowland forest is fast disappearing due to slash-and-burn cultivation. Bernier's vanga in particular seems to have rather a patchy distribution, being absent from some areas of apparently suitable habitat. The newly described red-shouldered vanga is limited to dense *Euphorbia* scrub in the south-west of the country, and while this forest occurs on poor soils that are difficult to cultivate, pressure for land from a rising human population has led to recent significant clearance, putting this species under threat in its limited range. There are also some species that have regionally distinct populations that are much more threatened. The Comoros populations of blue vangas are very rare, in particular *Cyanolanius madagascarinus bensoni,* known from very few sightings and a single specimen from Grande Comore (Ngazidja) and Moheli (Mweli). The western form of the Tylas vanga, *Tylas eduardi albigularis,* seems very rare and sparsely distributed, being apparently most common in or near mangroves.

Significance to humans

Most vangas are primary forest species and, being too small to hunt, are not particularly well-known to local people. Some species, notably the sicklebilled vanga, are large and noisy enough to make themselves conspicuous wherever they occur, and some desultory hunting of this and other species occurs when they cross paths with small boys equipped with catapults. However the main relevance these species have to humans is probably in their amazing diversity of form and function, one of the main attractions for wildlife tourists and birdwatchers as well as a fertile source of research for scientists.

Species accounts

Rufous vanga
Schetba rufa

TAXONOMY
Schetba rufa Linnaeus, 1766. The rufous vanga is the sole occupant of its genus. It has two subspecies, the nominate and long-billed form in eastern Madagascar and *S. r. occidentalis*, somewhat shorter-billed, in the central parts of the west.

OTHER COMMON NAMES
French: Artamie rousse; German: Rotvanga; Spanish: Vanga Rufa.

PHYSICAL CHARACTERISTICS
The rufous vanga is a medium-sized vanga, with short wings and a medium-length tail, and a strong head and neck. It has a thick and slightly hook-tipped bill, and large eyes. The head, throat, and breast of the male is glossy black, while the female has a black cap and a ghost of the male pattern in pale gray on the throat and breast. The under-parts of both sexes

Schetba rufa

are white, and the backs and tails bright rusty red. The bill is pale blue and the legs blackish.

DISTRIBUTION
The eastern form of the species is found in lowland forest from Marojejy National Park in northern Madagascar to Andohahela National Park in the south. The western subspecies occurs in primary western deciduous forest from just north of the Mangoky river to the Sambirano rainforest belt in the northwest.

HABITAT
In the eastern rainforests, the rufous vanga is rather patchily distributed in lowland forests. It seems to be exclusively limited to forests with large trees and a fairly open understory; in these conditions it is fairly abundant. It has not been recorded at higher altitudes than about 3,300 ft (1,000 m), and is most common in forests about sea-level.

In the west, rufous vangas are found only in areas of primary deciduous forest. It is rarely found in degraded or open areas.

BEHAVIOR
The rufous vanga is characteristically perched on a low liana or branch. Often the birds travel in family groups, and individuals may sit on particular perches for several minutes. In the early

Schetba rufa

☐ Resident

morning and late afternoon birds move into the canopy to sing. Much food is taken from the ground.

FEEDING ECOLOGY AND DIET
Rufous vangas sit still for long periods looking for movement of potential prey. Finding such prey is difficult if the substrate against which prey might be detected is itself moving, so foraging outside the calm forest interior is likely to be unproductive. In addition, the stable territories of rufous vangas need to contain food resources all year round, and, during the dry season, only shaded and cool areas maintain enough humidity to permit enough insect life to survive on the forest floor.

REPRODUCTIVE BIOLOGY
Rufous vangas occupy apparently stable territories in the forest interior. The nest is built in a tree fork or exceptionally in a rock crevice, of spiders webs, lichens, small flakes of bark, etc. Both sexes contribute to construction, resulting in a neat hemispheric or inverted conical bowl. Usually two or three eggs are laid, off-white with darker reddish markings. During the nestling phase, young males from the previous year, distinguishable by the black spotting on the breast, help feed the young.

CONSERVATION STATUS
Despite being limited to the interior of lowland rainforest and dry deciduous forest, the rufous vanga is not currently considered threatened, as it has a fairly large range (including many protected areas) and is common where it occurs. However this situation could change, particularly in the west where forest destruction, even in reserves, is occurring rapidly.

SIGNIFICANCE TO HUMANS
None known. ◆

Hook-billed vanga
Vanga curvirostris

TAXONOMY
Vanga curvirostris Linnaeus, 1766. The hook-billed vanga is in a genus of its own. There are two subspecies, the nominate, which is widely distributed in rainforest and west of Madagascar, and *V. c. cetera*, limited to the southern spiny forests.

OTHER COMMON NAMES
French: Vanga écorcheur; German: Hakenvanga; Spanish: Vanga de Pico Curvo.

PHYSICAL CHARACTERISTICS
The hook-billed vanga has a slim body, short wings, long tail, a long neck, a relatively heavy head, and a thick black bill. Male, female, and juvenile are similar. The underparts are pure white, the head has a black nape-band. The mantle is also black, with wide pale fringes to the greater and median coverts. The base of the tail is pale gray.

Vanga curvirostris

Vanga curvirostris
■ Resident

DISTRIBUTION
The hook-billed vanga occurs all over the east in plantations not too far from primary forest, gardens, lowland, and mid-altitude forest. It is relatively scarce in rainforest. It is absent from the high plateau except in large areas of primary forest. In the west and south it is more common, especially in the fringes of primary deciduous and spiny forest.

HABITAT
In the east, the hook-billed vanga is largely a bird of forest and forest edge, though occasionally found some way from the forest. In the west, it is found most commonly in dense forest, particularly around regenerating gaps.

BEHAVIOR
The hook-billed vanga is often difficult to find, as it feeds high in dense vegetation. The song is a short high single whistle, very difficult to locate. Imitiation of the song will often bring the bird flying in overhead to investigate, as they are very terri-torial. They do not really follow mixed-species flocks, but they are sometimes seen on the periphery of groups. They are usually seen in pairs, although the couple may be widely separated.

FEEDING ECOLOGY AND DIET
The hook-billed vanga forages mostly in dense vegetation where it will tear open leaf-clumps or loose bark in search of invertebrates. When looking for chameleons, the hook-billed vangas move rapidly through the understory, hopping from vertical stem to vertical stem, looking intently for the shapes of chameleons. Hook-billed vangas take prey up to the size of a medium-sized chameleon or a bird or bat.

REPRODUCTIVE BIOLOGY
The pair constructs a nest in the fork of a tree, often quite low down, and the female lays in it two or three whitish or reddish eggs. Pairs seem to be very territorial, and singing birds from adjacent territories may end up fighting or loudly bill-clapping at each other.

CONSERVATION STATUS
Not being limited to primary forests and having a wide distribution, the hook-billed vanga is not considered threatened.

SIGNIFICANCE TO HUMANS
None known. ◆

Resources

Books
Dee, T. J. *The Endemic Birds of Madagascar.* Cambridge, United Kingdom: ICBP, 1986.

Langrand, O. *Guide des oiseaux de Madagascar.* Paris: Delachaux et Niestlé, 1995.

Milon, P., J. J. Petter, and G. Randrianasolo. *Faune de Madagascar 35: Oiseaux.* Tananarive and Paris: ORSTOM and CNRS, 1973.

Morris, P., and F. Hawkins. *A Photographic Fieldguide to the Birds of Madagascar.* Robertsbridge, East Sussex: Pica Press, 1998.

Periodicals
Appert, O. "La répartition géographique des vangidés dans la région du Mangoky et la question de leur présence aux différentes époques de l'année." *Oiseaux et R.F.O.* 38 (1968): 6–19.

Resources

Appert, O. "Zur biologie der Vangawurger (Vangidae) sudwest-Madagaskars." *Orn. Beob.* 67: 101–133.

Eguchi, K., S. Yamagishi, and V. Randrianasolo. "The Composition and Foraging Behaviour of Mixed Species Flocks of Forest-living Birds in Madagascar." *Ibis* 134 (1993): 91–96.

Goodman, S. M., A. F. A. Hawkins, and C. A. Domergue. "A New Species of Vanga (Vangidae) from South-western Madagascar." *Bulletin of the British Ornithologist's Club* 117 (1997): 5–10.

Goodman, S. M., M. Pidgeon, A. F. A. Hawkins, and T. S. Schulenberg. "The Birds of South-eastern Madagascar." *Fieldiana* Zoology 87 (1997): 1–132.

Griveaud, P. "Le peuplement ornithologique de Madagascar-origines-biogéographie." *Cahiers ORSTOM série Biologie* 4 (1967): 53–69.

Hawkins, A. F. A. "Recent Observations of the Western Tylas vanga, (*Tylas (eduardi) albigularis*, Vangidae)." *Bulletin of the African Bird Club* 2 (1995): 13–16.

Hawkins, A. F. A. "Altitudinal and Latitudinal Distribution of East Malagasy Forest Bird Communities." *Journal of Biogeography* 26 (1999): 447–458.

Hawkins, A. F. A. "The Importance of Foraging Strategy to the Habitat Selection of the Rufous vanga (*Schetba rufa*, Vangidae) in the western forests of Madagascar." *Ibis* (submitted).

Hawkins, A. F. A., and L. Wilmé. "Effects of logging on forest birds." In *Economy and Ecology of a Tropical Dry Forest in Madagascar*, edited by J. U. Ganzhorn and J. P. Sorg, 203–213. *Primate Report* (Special Issue) 46 (1996): 1.

Hawkins A. F. A., M. Rabenandrasana, C. V. Marie, O. M. Rabeony, R. Mulder, R. E. Emahalala, and R. Ramariason. "Field Observations of the Red-shouldered Vanga *Calicalicus rufocarpalis*: A Newly Described Malagasy Endemic." *Bulletin of the African Bird Club* 5 (1997): 30–32.

La Marca, G., and R. Thorstrom. "Breeding Biology, Diet and Vocalisation of the Helmet Vanga, (*Euryceros prevostii*), on the Masoala Peninsula, Madagascar." *Ostrich* 71 (2000): 400–403.

Rand, A. L. "The Distribution and Habits of Madagascar Birds." *Bull. Amer. Mus. Nat. Hist.* 72 (1936): 143–499.

Yamagishi, S., and K. Eguchi. "Comparative Foraging Ecology of Madagascar Vangids (Vangidae)." *Ibis* 138 (1996): 283–290.

Yamagishi, S., E. Urano, and K. Eguchi. "Group Composition and Contributions to Breeding by Rufous Vangas (*Schetba rufa*) in Madagascar." *Ibis* 137 (1995): 157–161.

Organizations

African Bird Club, c/o BirdLife International. Wellbrook Court, Girton Road, Cambridge, Cambridgeshire CB3 0NA United Kingdom. Phone: +44 1 223 279 800. Fax: +44-1-223-277-200. E-mail: birdlife@birdlife.org.uk Web site: <www.africanbirdclub.org>

Working Group on Birds in Madagascar and the Indian Ocean Islands. World Wide Fund for Nature. Antananarivo 101, BP 738 Madagascar. Phone: +261 3207 80806. E-mail: wfrep@dts.mg

Frank Hawkins, PhD

Waxwings and silky flycatchers
(Bombycillidae)

Class Aves
Order Passeriformes
Suborder Passeri (Oscines)
Family Bombycillidae

Thumbnail description
Sleek, medium-sized, berry-eating songbirds, with plump bodies, short bills and crested heads

Size
5.9–9.4 in (15–24 cm); 1–2.1 oz (30–60 g)

Number of genera, species
5 genera; 8 species

Habitat
Forest, open woodland, semi-arid scrubland, and desert

Conservation status
Most species widespread and common, none are targeted for conservation efforts

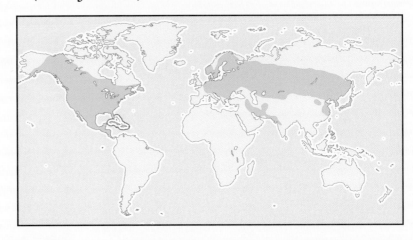

Distribution
North and Central Americas, north temperate Eurasia, Middle East, and Indian subcontinent

Evolution and systematics

Despite the ubiquity of some members of the Bombycillidae, the taxonomic status of the waxwings, their relationship to the gray and silky flycatchers, and indeed the relationships among these flycatchers themselves, remains controversial. Most modern taxonomic treatments place the waxwings proper (Bombycillinae) within the order Passeriformes together with the silky flycatchers (Ptilogonatinae) and the palmchat (Dulidae). This classification is based primarily on coloration, nesting, skeletal features, musculature, and results of DNA-DNA hybridization studies. Here the consistency ends. Waxwings have also been allied with several other Passeriformes, including thrushes (Turdidae), Old World flycatchers (Muscicapidae), pipits (Motacillidae), and starlings (Sturnidae). More controversial is the placement of the gray hypocolius (*Hypocolius ampelinus*). This Middle Eastern bird is frequently allied with the other bombycillids, either as a subfamily or as a monotypic family closely related to the waxwings and silky flycatchers. However, *Hypocolius* has also been allied with the bulbuls (Pycnonotidae) and thrushes. Further research is needed to resolve these relationships.

This treatment recognizes three subfamilies. They are the waxwings: Bombycillinae, comprised of one genus and three species: the cedar waxwing (*Bombycilla cedrorum*), bohemian waxwing (*B. garrulus*), and Japanese waxwing (*B. japonica*). The silky flycatchers, subfamily Ptilogonatinae, with three genera and four species: the phainopepla (*Phainopepla nitens*), black and yellow silky flycatcher (*Phainoptila melanoxantha*), long-tailed silky flycatcher (*Ptilogonys caudatus*), and gray silky flycatcher (*Ptilogonys cinereus*). The latter is not to be confused with the only member of the third subfamily—the gray hypocolius (*Hypocolius ampelinus*) of the Hypocolinae.

Physical characteristics

Bombycillids are medium-sized songbirds with short bills, sleek plumage, head crests, and tails of varying length. The waxwings have fawn-gray bodies, and black chin and eye masks. Their wings have contrasting plumage with white, crimson, or yellow patches and their common name refers to the red, wax-like tips present on the secondary flight feathers. (These are absent however in the Japanese waxwing.) All have a bright band of orange or yellow at the base of the tail. The cedar waxwing's tail band is usually yellow, but in the past 30 years many have been sighted with orange tail bands, apparently due to pigments in the fruits of the invasive European honeysuckle (*Lonicera* spp.), which has successfully established itself in the United States and is eaten by the birds.

Silky flycatchers have longer tails than waxwings and their head crests (present in three of the four species) often appear as erect bristles. The plumage is black, gray, or brown, some species have brightly colored patches of yellow or white. The gray hypocolius is gray with a black tail band and face mask.

Distribution

The ranges of the three subfamilies are quite different. Waxwings, whose fossils have been reported from the Pleistocene in California, are distributed across the temperate Old and New Worlds; the cedar waxwing winters as far south as Guatemala. Silky flycatchers occur from the southern United States into Central America. The gray hypocolius is found in the Middle East and on the Indian subcontinent.

A cedar waxwing (*Bombycilla cedrorum*) adult feeds its young. (Photo by Hal H. Harrison. Photo Researchers, Inc. Reproduced by permission.)

Habitat

The habitat preferences of the bombycillids vary, although all occur where they can find fruits and berries. Waxwings prefer open woodlands and hedgerows, and increasingly are found in suburban areas. Silky flycatchers and *Hypocolius* inhabit desert and arid scrub-land.

Behavior

All bombycillids are gregarious, social birds, although to varying degrees. Silky flycatchers nest in loose colonies and are somewhat territorial. Waxwings are non-territorial, and will form flocks of thousands, migrating erratically to areas of high fruit density. Bent (1950) wrote of the Bohemian waxwing "We never know when or where we may see these roving bands of gypsies. They come and they go, we know not whence or whither, in the never-ending search for a bounteous food supply on which to gorge themselves." Bombycillids are migratory, the phainopeplas moving altitudinally to moister habitats after breeding. They are quite vocal birds, their calls a mixture of chatters, warbles, and whistles.

Feeding ecology and diet

Bombycillids feed predominantly on small fruits, which are the mainstay of the diet for the north temperate species for seven months of the year. They also eat insects, plucking them off of vegetation and tree bark or swooping down from high perches and taking them in flight. The cedar waxwing can store ingested fruits in a portion of the esophagus, presumably to maximize the amount of food ingested while foraging. Unlike fruit-eating thrushes, they have the enzymes to digest sucrose. In recent years, waxwings have come to rely increasingly on crops and ornamental fruits planted in suburban areas. While it will eat a variety of fruits, the phainopepla specializes on mistletoe (*Phoradendron* spp.) berries and is closely associated with these plants that grow on the trunks and branches of mesquite.

Reproductive biology

Waxwings are socially monogamous and the same is believed of silky flycatchers, though their habits are less well known.

The breeding season of waxwings is one of the latest of North American birds, with eggs laid from early June through August. Adults tend to pair with similar-aged birds, and pairs of older birds nest earlier and raise more young than younger birds. Copulation in waxwings is preceded by courtship-hopping, which often involves passing a small item such as an insect or flower petal between male and female, and touching the bills together which results in a clicking noise. All bombycillids make a small, cup-shaped nest, usually at a strong fork in a tree. The female does most nest building in the waxwings, while the reverse is true for the phainopepla. Clutch sizes range from four to six in waxwings, from two to four in silky flycatchers. In both groups the young hatch naked and blind and are fed crushed berries and insects by both parents. The breeding biology of the gray hypocolius is not well known.

Conservation status

No conservation measures have been taken for the family, and none appear needed. According to Breeding Bird Survey data, cedar waxwing populations have increased across North America, and phainopepla populations remain stable. The status of populations of Middle Eastern and Central American species is not well known, but as of 2001, none are listed as Endangered or Threatened.

Phainopepla (*Phainopepla nitens*) male at nest with mistletoe berries. (Photo by Anthony Mercieca. Photo Researchers, Inc. Reproduced by permission.)

Significance to humans

The irregular mass invasion of Europe and temperate North America by waxwings has caused them to be considered pests at times. As they have become more prominent in suburban areas, more cedar waxwings are killed by flying into windows.

1. Female gray hypocolius (*Hypocolius ampelinus*); 2. Male gray silky flycatcher (*Ptilogonys cinereus*); 3. Female phainopepla (*Phainopepla nitens*); 4. Male Bohemian waxwing (*Bombycilla garrulus*); 5. Male cedar waxwing (*Bombycilla cedrorum*). (Illustration by Jacqueline Mahannah)

Species accounts

Cedar waxwing
Bombycilla cedrorum

SUBFAMILY
Bombycillinae

TAXONOMY
Bombycilla cedrorum Vieillot, 1808. Two subspecies (*B. c. cedrorum* and *B. c. larifuga*) are recognized by some researchers based on geographic variation in plumage.

OTHER COMMON NAMES
French: Jaseur des cèdres; German: Zederseidenschwanz; Spanish: Ampelis Americano.

PHYSICAL CHARACTERISTICS
6.1 in (15.5 cm), 1.1 oz (32 g). Smaller of two North American waxwings. Sleek, crested birds with small bill, overall plumage grayish brown with pale yellow belly. Adults have black face mask with white edge and black chin patch. Named for red, wax-like tips on the secondary flight feathers of many adults. Pointed wings, tail square with distinctive yellow band at tip. Female chin patch may be smaller and lighter colored. Red wax-tips absent and plumage more gray than brown in juveniles.

DISTRIBUTION
North to southeast Alaska, throughout Canadian provinces, east to Newfoundland, throughout United States and Central America to Panama, east to Bermuda, occasionally winters in West Indies and the Bahama Islands.

HABITAT
Uses various open woodland forests and old fields; avoids forest interior; also found in riparian areas of grasslands, farms, orchards, conifer plantations, and suburban gardens.

BEHAVIOR
Very social species, flocking throughout year. Rarely ventures to ground, frequent preening at high exposed sites. Non-territorial, but may show aggressive behavior near nest. Short flights are direct with steady wing beats. Two basic calls; rapidly repeated buzzy or trilled high-pitch notes and high-pitched hissy whistles.

FEEDING ECOLOGY AND DIET
Diet consists mainly of fleshy fruits, but also includes insects caught in air or gleaned from vegetation. Forages in branches of fruiting trees, typically plucks fruit while grasping a branch. Fleshy, berry-like cones of cedar (*Juniperus* spp.) historically dominated winter diet. In spring, this bird will hang from maple (*Acer* spp.) branches to feed on suspended drops of sap. Instances have been recorded of cedar waxwings becoming drunk from alcohol in overripe fruits. This often results in them falling to the ground, hitting windows, being hit by vehicles, and dying from injuries.

REPRODUCTIVE BIOLOGY
Appears to be monogamous within a breeding season. Among latest-nesting birds in North America, apparently cued to mid-summer ripening of fruit. Lays two to six sparsely spotted pale blue-gray eggs, in woven cup-like nest. Female incubates, 12–15 days. One to two broods per season. Young hatch naked and blind; both parents feed nestlings. Fledge 14–17 days. Occasional brood parasitism by brown-headed cowbirds (*Molothrus ater*).

CONSERVATION STATUS
Not threatened. No conservation measures have been taken for cedar waxwings and none appear needed. Sharp population increases occurred in late 1970s, in apparent rebound from elimination of DDT in agriculture and increase in edge habitats conducive to fruit-bearing shrubs.

SIGNIFICANCE TO HUMANS
Commonly killed by hitting windows, perhaps because many ornamental fruit-bearing shrubs are planted near homes. ◆

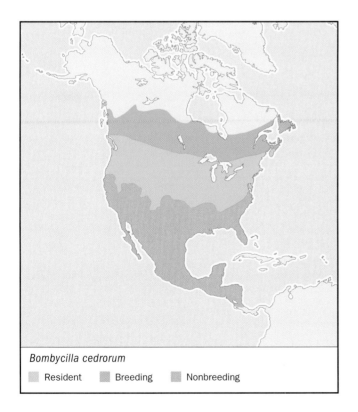

Bombycilla cedrorum

Resident   Breeding   Nonbreeding

Bohemian waxwing
Bombycilla garrulus

SUBFAMILY
Bombycillinae

TAXONOMY
Bombycilla garrulus Linnaeus, 1758.

OTHER COMMON NAMES
English: Waxwing; French: Jaseur boréal; German: Seidenschwanz; Spanish: Ampelis Europeo.

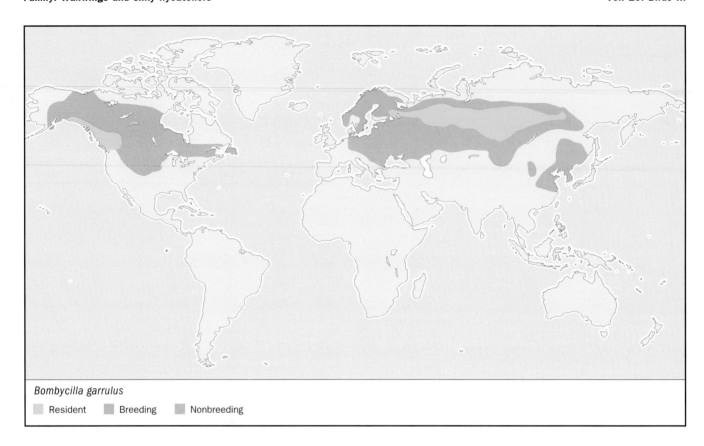

Bombycilla garrulus

Resident   Breeding   Nonbreeding

PHYSICAL CHARACTERISTICS
6.7–8.3 in (17–21 cm), wingspan: 14.5 in (37 cm), 2 oz (56 g). Larger and slightly shorter-tailed than the cedar waxwing, appears small-headed and round-bodied. Sleek, crested bird, overall plumage grayish with rusty tinge on forehead and cheeks. Black eye-mask and chin patch, white edge between eye-patch and chin, little to no white on forehead. Undertail rust colored. Yellow band on terminal end of tail, yellow and white bands on wing. Characteristic red wax droplets on some secondaries of many adults. Juvenile resembles adult, but lacks wax droplets and rust coloring on head.

DISTRIBUTION
Holarctic, northern Eurasia, to northeast and north-central China, northwest and north-central North America.

HABITAT
Old stands of coniferous trees, with open canopy and rich field layer.

BEHAVIOR
Very social species, flocking throughout year. Characteristic silvery buzzing *sirr* of flock call. Nomadic and irruptive; occasionally large flocks invade Europe and the United States, presumably due to fruit shortages in northern regions. Non-territorial, but may show aggressive behavior near nest.

FEEDING ECOLOGY AND DIET
Diet mainly sugary, fleshy fruits, especially rose-hips (*Rosa* spp.) and mountain-ash berries (*Sorbus* spp.); also insects in spring and summer. Birds may gorge themselves until they can hardly fly.

REPRODUCTIVE BIOLOGY
Monogamous. Late breeder, from early June through August. Lays four to six pale bluish gray eggs spotted with black in woven cup-like nest. Female incubates, 12–15 days. Young hatch naked and blind; both parents feed nestlings; fledging 14–17 days.

CONSERVATION STATUS
Not threatened. Widespread but can be intermittent, numbers of breeding pairs may vary considerably from year to year.

SIGNIFICANCE TO HUMANS
Has been considered a pest in some areas during irregular mass-invasions in Europe. ◆

Phainopepla
Phainopepla nitens

SUBFAMILY
Ptilogonatinae

TAXONOMY
Phainopepla nitens Swainson, 1838. This species is often considered a distinct family.

OTHER COMMON NAMES
French: Phénopèple luisant; German: Trauerseidenschnäpper; Spanish: Papamoscas Sedoso.

PHYSICAL CHARACTERISTICS
6.3–9.4 in (16–24 cm) Sleek, long-tailed, crested birds with rounded wings. Adult male has a short black bill and ragged crest, glossy black with white patches on wings (white not visible when perching). Female is ashy-gray with whitish edges on all wing feathers. Juvenile resembles female, gradually acquires black feathers throughout first year.

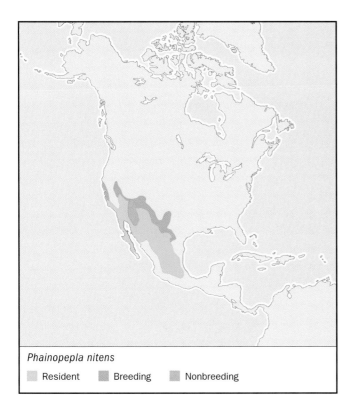

Phainopepla nitens

☐ Resident ■ Breeding ■ Nonbreeding

DISTRIBUTION
Americas only. North to central California, south to Panama.

HABITAT
Arid scrub-land and open woodland within reach of water courses. Believed to breed in lowland deserts and then move to higher, moister habitats.

BEHAVIOR
Somewhat social, nests in loose colonies and forms small flocks in non-breeding season. Territory size varies with food abundance. Call a soft-rising whistle *hoi*; song a mellow, gurgled warble, series of short phrases with long pauses between *krrtii-ilwa*. Also mimics other species.

FEEDING ECOLOGY AND DIET
Diet mainly fruit and berries, some insects often caught on the wing. Phainopepla is closely associated with mistletoe, a parasitic plant that grows on many desert trees especially mesquite (*Prosopis* spp.). Phainopeplas have a specialized digestive system for consuming mistletoe berries. In the gizzard they remove the seed and pulp from the seed coat of the berries; they then digest the pulp and defecate the seeds and seed coat separately, usually on the branch of the tree where the bird was perched. The seeds sprout in the tree, continuing their parasitic lifestyle, having been dispersed by the bird.

REPRODUCTIVE BIOLOGY
Monogamous, with large breeding territories. Breeding pairs form loose colonies where food sources are concentrated. In a courtship display a male flies as high as 300 ft (90 m) over his territory, chasing the female while making circles and erratic zig-zag patterns. Two to three grayish blue, mottled eggs laid in cup-shaped nest of plant matter, spider webs, and hair; nest built in central fork of a tree. Both parents incubate 14–16 days; young hatch naked and helpless. Fledge at 18–25 days. Phainopeplas can product two to three broods per year.

CONSERVATION STATUS
Not threatened. Breeding Bird Survey data suggest population levels have been stable since the 1960s, however, reliance on mesquite and mistletoe makes the phainopepla vulnerable to habitat loss.

SIGNIFICANCE TO HUMANS
None known. ◆

Gray silky flycatcher
Ptilogonys cinereus

SUBFAMILY
Ptilogonatinae

TAXONOMY
Ptilogonys cinereus Swainson, 1827.

OTHER COMMON NAMES
English: Gray silky; French: Ptilogon cendré German: Grauseidenschnäpper; Spanish: Capulinero Gris.

PHYSICAL CHARACTERISTICS
7.3–8.3 in (18.5–21 cm). Sleek, long-tailed birds with crests and small bills. Male's head, crest, and upperparts blue-gray, with white eye crescents and black wings. Belly white, black tail has bright yellow undertail with a white band. Female head and crest grayish, upperparts, throat, and underparts dusky wine-colored with yellow undertail. Whitish belly. Juvenile resembles female, but belly dull yellow.

DISTRIBUTION
Middle Americas. Throughout Mexico, south into Guatemala at 3,300–11,500 ft (1,000–3,500 m), lower in winter. Documented twice in Texas, sight records for southern Arizona. Also seen in southern California, though may be an escape.

HABITAT
Humid to semi-arid pine-oak and evergreen forest, wanders to adjacent habitats in winter.

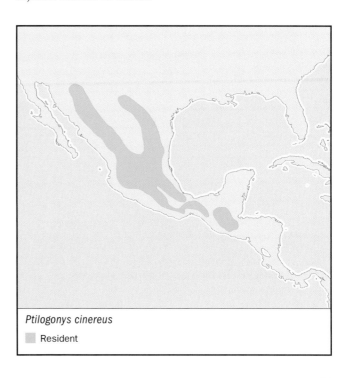

Ptilogonys cinereus

☐ Resident

BEHAVIOR
Commonly seen in pairs or small flocks; often perches conspicuously atop tall trees; flies high. Call is chattering and nasal: *chi-che-rup*, *che-chep* and *k-lik*, *k-li-lik*. Song a pleasant, soft, warbled series of clucks with intermittent quiet whistles.

FEEDING ECOLOGY AND DIET
Diet mainly insects, often caught in flight, and fruits and berries.

REPRODUCTIVE BIOLOGY
Monogamous. Lays two bluish white eggs, speckled and spotted with brown and gray, in cup-like nest of fine plant material and lichen; nests in bushes or trees. Female incubates for 12–14 days, young hatch naked and blind. Nestlings are fed by both parents and fledge in 18–20 days.

CONSERVATION STATUS
Not threatened.

SIGNIFICANCE TO HUMANS
None known. ◆

Gray hypocolius
Hypocolius ampelinus

SUBFAMILY
Hypocoliinae

TAXONOMY
Hypocolius ampelinus Bonaparte, 1850. Single species in subfamily, taxonomy contoversial. Allied with thrushes, Turdidae (Lowe, 1947); throughout latter half of 1900s usually considered a subfamily within the waxwings. Sibley and Monroe (1990) and Clements (1991) suggest it might be closely related to bulbuls (Pycnonotidae), although affinities uncertain.

OTHER COMMON NAMES
English: Gray flycatcher; French: Hypocolius gris; German: Seidenwürger; Spanish: Hipocolino Gris.

PHYSICAL CHARACTERISTICS
Crested, long-tailed birds with white on primaries. Male has black mask that encircles the head and black band on tail, in flight shows black primaries with prominent white tips. Female sandy-brown with creamy throat, lacks black mask, tail tip dark. Juvenile resembles female.

Hypocolius ampelinus
■ Resident   ■ Breeding   ■ Nonbreeding

DISTRIBUTION
Breeding range centered in Iran and Iraq, east to Pakistan, winters in Saudi Arabia. Rare and irregular visitor to India.

HABITAT
Semi-desert with scattered thorn scrub of berried bushes or around oases and date palm plantations.

BEHAVIOR
Gregarious, in winter seen in flocks up to 20 birds. Flight is strong and direct with rapid wing beats and occasional swooping glides. Raises ear coverts and nape-feathers when excited. Call a mellow, liquid *tre-tur-tur*, and *whee-oo*.

FEEDING ECOLOGY AND DIET
Feeds mainly on berries, occasionally insects. Chiefly forages in trees and bushes, hopping about, occasionally descends to the ground to pick up insects.

REPRODUCTIVE BIOLOGY
Lays three to six pale bluish gray eggs with black or brown speckles which may merge around the broader end of the egg to form a distinct ring.

CONSERVATION STATUS
Status uncertain, but not currently a target of conservation efforts.

SIGNIFICANCE TO HUMANS
None known. ◆

Resources

Books

Grimmett, Richard, et al. *Birds of the Indian Subcontinent.* London: Christopher Helm Ltd., 1998.

Helle, Pekka, and Timo Pakkala. "Waxwing." In *The EBCC Atlas of European Breeding Birds*, edited by Ward J. M. Hagemeijer and Michael J. Blair. London: T. & A. D. Poyser, 1997.

Sibley, Charles G., and Burt L. Monroe. *Distribution and Taxonomy of Birds of the World.* New Haven: Yale University Press, 1991.

Periodicals

Witmer, M. C., et al. "Cedar Waxwing." *The Birds of North America* 309 (1997).

Witmer, M. C. "Consequences of an Alien Shrub on the Plumage, Coloration and Ecology of Cedar Waxwings." *Auk* 113 (1996): 735-43.

Other

"Zoonomen Nomenclatural Data." Alan P. Peterson. 2000 (9 April 2002). <http://www.zoonomen.net/>

Rachel Ehrenberg, MS

▲
Palmchats
(Dulidae)

Class Aves

Order Passeriformes

Suborder Passeri (Oscines)

Family Dulidae

Thumbnail description
A highly vocal bird that flocks in small bands, generally around royal palm trees; brown above and have yellowish white underparts with heavy brown streaking

Size
7.5–8 in (19–20 cm)

Number of genera, species
1 genus; 1 species

Habitat
Distributed throughout the low to mid-elevations, especially in royal palm savannas

Conservation status
Not threatened

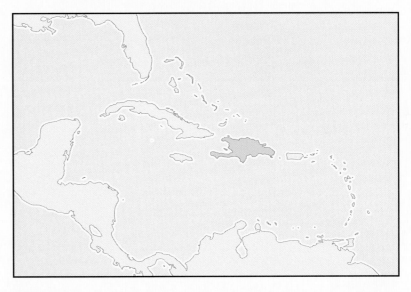

Distribution
Endemic to Hispaniola (Dominican Republic and Haiti), including Gonave and the Saona Islands

Evolution and systematics

The palmchat (*Dulus dominicus*) is the only member of its genus and family. Todies and palmchats are the only two endemic bird families of the Caribbean. Because of the palmchat's evolutionary and taxonomic importance, the Dominican Republic (which occupies two thirds of the island of Hispaniola) has declared it the national bird.

The palmchat was first described in 1766 by Linnaeus as a member of the genus *Tanagra*. In 1847 Haurtlaub classified it under its present genus, *Dulus*.

Surprisingly, palmchats are unrelated to any other Hispaniolan bird species, but are distantly related to North American waxwings (Bombycillidae), with whom they share similar skeletal structures.

Physical characteristics

The palmchat is 7.5–8 in (19–20 cm) long with a moderately long tail and strong bill. The bird is olive brown above with a yellowish green wash on the rump and edges of the primary wing feathers. The underside is yellowish white with sooty brown streaking. Its bill is yellow, its iris brownish red. The sexes have similar plumage; immature birds have darker throats and forenecks.

Distribution

The palmchat is endemic to the island of Hispaniola, including Gonave and the Saona Islands.

Habitat

Palmchat are common and widely distributed throughout the low to mid elevations, although occasionally they extend into altitudes of 4,900 ft (1,500 m) or more. They are absent from the tallest peaks and mountains, and from dense rainforests; they prefer royal palm savannas and valleys.

Behavior

Palmchats are gregarious and live in small bands, which have a communal nest as the center of activity. Groups consist of several pairs that tend to be sociable and affectionate; they continuously sidle up to and perch beside their companions.

When they are not searching for food, the birds rest on palm fronds or perch on the projected spikes above palm trees. They often use their nests as resting spots, and as common roosts at night. In the morning, groups of up to eight individuals can be seen perched on limbs to preen and dry themselves in the sun.

Palmchats often seem alert and vivacious because they stand with bodies erect and tails pointing straight down. They are very vocal and noisy, especially around their nests. They utter a wide array of call notes, but nothing that can be considered a song.

Feeding ecology and diet

Palmchats are fruit eaters. They prefer a variety of berries, including those from palm trees. Undoubtedly they help

Palmchat (*Dulus dominicus*). (Illustration by Jacqueline Mahannah)

disperse the seeds of royal palms, gumbo-limbo trees, and many other fruiting tree species. Their diet includes flower buds and blossoms, especially orchid tree blooms, but the birds are not considered harmful to plants or trees.

Palmchats feed by searching through trees in twos or threes. Insects provide an alternate source of dietary protein, and at times the birds hunt insects by pursuing them in flight.

Reproductive biology

The nest is a large structure of twigs built around the crown of a palm and supported by lower fronds. Twigs are interlaced loosely to form a disorderly looking nest that measures 3–6.5 ft (1–2 m) in diameter. Twigs chosen for their nests are large (10–18 in [25–45 cm]), so it is impressive to see these relatively small birds struggling along with large twigs in their bills.

The nest is built by several pairs of birds that have individual nest chambers with separate entrances. The inner chamber is lined with shredded bark, on which eggs are laid. The female lays two to four purplish gray eggs that are heavily spotted at the broad end. Breeding is mainly between March and June but has been recorded throughout the year.

Old nests tend to fall to the ground as supporting palm fronds mature and break. Immediately afterward, if the nest

is still active, twigs from the fallen nest are usually reused. In certain habitats in Haiti and the Dominican Republic, where palms are uncommon (particularly in dry scrub and pine forests), palmchats are reported to build nests on telephone poles and in pine trees, and sometimes in mature broadleaf trees.

Palmchats are parasitized by a fly of the *Philornis* genus, whose eggs are generally laid on nestling birds and develop in a sac under the skin on the head or under the wing. The parasites apparently do no harm and when they complete the larval cycle abandon their hosts without complications. Up to 90% of all birds examined were hosts to the fly larvae.

Conservation status

Not threatened.

Significance to humans

None known.

A palmchat (*Dulus dominicus*) holds a berry in its bill. (Photo by Doug Wechsler/VIREO. Reproduced by permission.)

Resources

Books

Dod, A.S. *Aves de La Republica Dominicana.* Santo Domingo: Museo Nacional de Historia Natural, 1978.

Raffaele, Herbert, James Wiley, Orlando Garrido, Allan Keith, and Janis Raffaele. *A Guide to the Birds of the West Indies.* Princeton: Princeton University Press, 1998.

Wetmore, A., and B.H. Swales. *The Birds of Haiti and The Dominican Republic.* U.S. National Museum, Bulletin 155. Washington, DC: Smithsonian Institution, 1931.

Other

I-bird.com. International Birding Information Resource Data, 2000. (29 Jan. 2002). <http://www.i-bird.com/FamilyOfOne/Palmchat/Palmchat.htm>

Eladio M. Fernandez

Hedge sparrows
(Prunellidae)

Class Aves

Order Passeriformes

Suborder Passeri (Oscines)

Family Prunellidae

Thumbnail description
Sparrow-like birds, but with a relatively slender and pointed beak; the legs and feet are sturdy, the tail short, and the coloration a drab gray or brown patterned with streaks of white, black, and dull-red

Size
Body length of 5–7 in (13–18 cm) and weight 0.5–1.4 oz (18–40 g)

Number of genera, species
1 genus; 12 species

Habitat
Most species live in shrubby places or meadows in alpine habitats. One species occurs in open woodland and cultivated lands at lower altitude

Conservation status
Near Threatened: 1 species

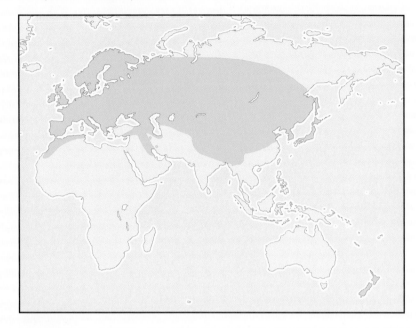

Distribution
Occur widely in Eurasia and in the northern edge of Africa

Evolution and systematics

Hedge sparrows or accentors are a small group of rather similar-looking birds, all of which are placed in the genus *Prunella*. All of these species may have evolved from a single, alpine-dwelling progenitor. Recent studies of the phylogeny of the Passeriformes suggest that the accentors are most closely related to the wagtails and pipits (family Motacillidae), sunbirds (Nectariniidae), and finches (Fringillidae).

Physical characteristics

Accentors are small, plainly colored birds. They are similar to sparrows in general appearance, but their beak is much more slender and pointed. Their legs and feet are sturdy, and the wing has 10 functional primary feathers, which may be rounded or pointed at their tips. The tail is relatively short and squared at the end. The plumage is rather drab, being colored in inconspicuous patterns of brownish and grayish, with spots and streaks of white, black, and dull red. Males and females are similar in appearance, but males have somewhat longer wings, are heavier, and are slightly brighter in coloration.

Distribution

Accentors are widely distributed over the Palearctic region, including much of Europe, Asia north of the Himalayan mountains, and Africa north of the Sahara desert.

Habitat

Primarily occur in the thick undergrowth of shrubby habitats and alpine meadows.

Behavior

Accentors are relatively quiet, unobtrusive birds. They are somewhat gregarious during the non-breeding season. They spend most of their time on or close to the ground, where they move about by running and hopping, often with small flicks of the wings and tail. Their flight is rapid and often undulating. They have high-pitched call notes, and a warbling song that is often given in flight.

Feeding ecology and diet

Accentors forage for invertebrates on the ground, especially beetles, flies, aphids, ants, spiders, and worms. During the winter, they mostly eat seeds and berries.

Reproductive biology

Accentors construct an open, cup-shaped nest of plant fibers, mosses, and feathers. The nest is built in a hollow in the ground, in a crevice among rocks, or in low scrub as high as about 6 ft (2 m) above the ground. The breeding season lasts from late March to August, but this varies among species and latitude. Accentors lay a clutch of three to six

A dunnock (*Prunella modularis*) feeding its chicks in the nest. (Photo by D. Hadden/VIREO. Reproduced by permission.)

eggs, which are uniformly colored light bluish green to blue. Typically, two broods are raised each year. The incubation period is 11–15 days, and the nestlings are raised for another 12–14 days until they fledge. Both partners share in the care of the young.

The breeding territories established vary among species of accentors. Solitary males, male-female pairs, and trios of a fe-male and two males have been observed to defend a territory. The presence of more than one male in a breeding group, a breeding relationship known as polyandry, is somewhat unusual among passerines. There may, however, be some advantage to having several males participate in feeding and defense of the young, particularly in a predator-rich environment. Typically, the territory occupied by a polyandrous breeding group is larger than that of a male-female pair.

Conservation status

The IUCN lists the Yemen accentor (*Prunella fagani*) as Near Threatened. Its range and abundance have declined precipitously because of degradation of its shrubland habitat by grazing animals and other damages caused by human activities. Populations of some other species have also declined significantly, but not to the point of being considered at risk. The population of dunnocks (*Prunella modularis*) breeding in Britain declined during the 1980s, but has since maintained itself.

Significance to humans

Accentors are not of direct importance to humans, other than the indirect economic benefits of ecotourism focused on seeing birds in natural habitats. In 2001, ticks living on dunnocks and other migratory birds were implicated in the spread of a bacterial pathogen, *Ehrlichia phagocytophila*, the causative agent of the rare disease ehrlichiosis in humans.

1. Siberian accentor (*Prunella montanella*); 2. Robin accentor (*Prunella rubeculoides*); 3. Alpine accentor (*Prunella collaris*); 4. Dunnock (*Prunella modularis*). (Illustration by Barbara Duperron)

Species accounts

Dunnock
Prunella modularis

TAXONOMY
Prunella modularis Linnaeus, 1758.

OTHER COMMON NAMES
English: Hedge-sparrow, hedge accentor; French: Accenteur mouchet; German: Heckenbrounelle; Spanish: Acentor Común.

PHYSICAL CHARACTERISTICS
6 in (15 cm); 0.7 oz (19 g). The beak is pointed and slender, and the feet and legs are sturdy. Upperparts are brownish gray and usually streaked with black and lighter colors; underparts are more uniformly gray with a few apricot markings.

DISTRIBUTION
The dunnock occurs in suitable habitat throughout Europe, ranging as far east as the far western regions of Russia. It is migratory in northern parts of its range, but may be resident in more southern parts in France and Spain. The dunnock was introduced to New Zealand between 1860 and 1880, where it persists today.

HABITAT
Dunnocks occur in woods with abundant undergrowth, and in hedges and shrubbery near forest edges, typically at a breeding density of less than about 2.6 pairs per m² (1 pair per km²). In well-vegetated farmland and gardens of midland Britain, however, their density can be up to 30 times greater.

BEHAVIOR
Dunnocks are shy and unassuming birds. In many areas, they are characteristic but little-known garden birds because of their secretive habits. Most populations are migratory. Dunnocks are normally seen as individuals or pairs during the breeding season, but during winter they may occur in large foraging flocks. A particularly good local food source may encourage a flock of a hundred or more birds to gather. Dunnocks sing from the tops of conifers of medium height. The song is a short, bright, soft, metallic twittering.

FEEDING ECOLOGY AND DIET
During the growing season, dunnocks predominantly feed on invertebrates, including insects, spiders, and worms. During the winter they eat seeds and small berries. They will readily consume food put out for songbirds at winter feeders.

REPRODUCTIVE BIOLOGY
Dunnocks build a cup-shaped nest in a shrub or low tree. They breed from about the beginning of April to the end of July, and typically raise two broods per year. The incubation period is 12–14 days, and the young are fully fledged at 11–13 days after hatching. Both parents care for their young. Dunnocks are often polyandrous breeders, that is, a female will mate with several males within her breeding territory, and all may cooperate to raise the young.

CONSERVATION STATUS
Not threatened. The dunnock is a widespread and relatively abundant bird. In some parts of their range, however, they have significantly decreased in abundance. In Britain, for example, the species has declined by 45–60% between 1975 and 2001. The most precipitous population decline was during 1975 through 1986, with maintenance of the lower population thereafter. The reasons for the decline are not known, and it did not occur throughout the British Isles (Wales had a population increase).

SIGNIFICANCE TO HUMANS
Dunnocks are of no direct significance to people, except for the economic benefits of bird-watching. Ticks living on dunnocks and other migratory birds have been recently (2001) implicated in the spread of a bacterial pathogen, *Ehrlichia phagocytophila*, the causative agent of the rare disease ehrlichiosis in humans. The dunnock is mentioned by the English novelist, Emily Brontë, in her famous novel, *Wuthering Heights*. ◆

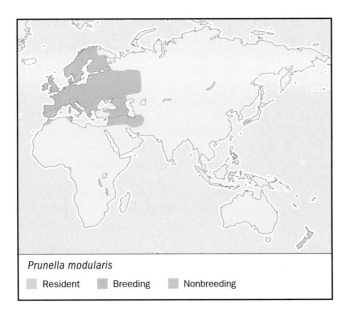

Prunella modularis

☐ Resident ☐ Breeding ☐ Nonbreeding

Alpine accentor
Prunella collaris

TAXONOMY
Prunella collaris Scopoli, 1769.

OTHER COMMON NAMES
French: Accenteur Alpin; German: Alpenbraunelle; Spanish: Acentor Alpino.

PHYSICAL CHARACTERISTICS
Length is about 7.5 in (18 cm). A drab songbird, with a slender, pointed beak and stout feet and legs. Upperparts are brownish gray streaked with black and lighter colors; underparts are more uniformly gray, with chestnut on the flanks.

DISTRIBUTION
A widespread species in the mountains of Europe and Asia. Also has a minor presence in North Africa.

HABITAT
Inhabits high alpine meadows and rocky slopes above the treeline. A short-distance migrant that breeds at high altitude and winters in lower valleys.

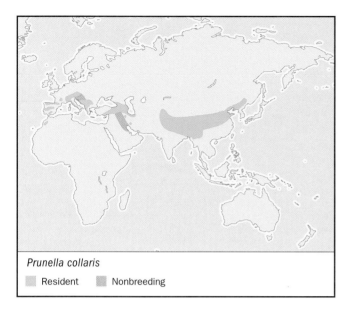

Prunella collaris

■ Resident ■ Nonbreeding

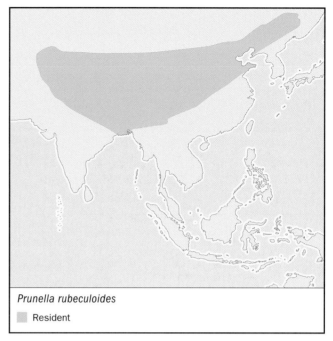

Prunella rubeculoides

■ Resident

BEHAVIOR
Possibly because of a lack of large predators and scarcity of food at high altitude, the alpine accentor is rather unafraid to approach humans for food. Its song is a high-pitched warbling, sometimes given in flight.

FEEDING ECOLOGY AND DIET
Forages on the ground for insects, spiders, and other invertebrates. Feeds on seeds and fruits during the winter.

REPRODUCTIVE BIOLOGY
Builds a cup-shaped nest of moss, grass stalks, and fine roots in holes and rocky clefts on the ground. Lays a clutch of four to five eggs. Because there are two clutches each year, the eggs may be found from late May to July. Females display a complex breeding behavior, involving mating with several males in succession. The polyandrous breeding system may benefit the female by encouraging the tending of the nest and protection of the young by one than one male.

CONSERVATION STATUS
Not threatened. A widespread and abundant species within its habitat.

SIGNIFICANCE TO HUMANS
None known. ◆

Robin accentor
Prunella rubeculoides

TAXONOMY
Prunella rubeculoides Moore, 1854.

OTHER COMMON NAMES
French: Accenteur rougegorge; German: Rotbrustbraunelle; Spanish: Acentor de Cola Roja.

PHYSICAL CHARACTERISTICS
Length is about 6.5 in (16–17 cm). The back is colored gray streaked with brown, the head grey, the belly whitish, and the breast rusty-orange.

DISTRIBUTION
A widespread species in the Himalayas, the Tibetan Plateau, and other highlands of eastern Asia.

HABITAT
Mostly occurs in high montane forests of rhododendron and in willows and sedge in meadowlands, but also ventures above the tree-line to scrub habitats and wet meadows. A local, short-distance migrant that winters in lower-altitude valleys, often around human habitation. Breeds from about 9,850 to 14,000 ft (3,000 to 4,260 m) and winters as low as 4,900 ft (1,500 m).

BEHAVIOR
Usually occurs singly, in pairs, or in small groups. A relatively tame bird. The song is a high-pitched, repeated phrase.

FEEDING ECOLOGY AND DIET
Forages on the ground for insects, spiders, and other invertebrates. Feeds on seeds and fruits during the winter.

REPRODUCTIVE BIOLOGY
Polyandrous. Builds a nest on the ground, often near a stream.

CONSERVATION STATUS
Not threatened. A widespread species within its habitat.

SIGNIFICANCE TO HUMANS
None known. ◆

Siberian accentor
Prunella montanella

TAXONOMY
Prunella montanella Pallas, 1776.

OTHER COMMON NAMES
English: Mountain accentor; French: Accenteur montanelle; German: Bergbraunelle; Spanish: Acentor de Pallas.

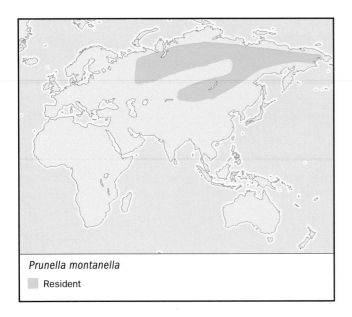

Prunella montanella
▨ Resident

PHYSICAL CHARACTERISTICS
Length is about 6 in (15 cm). Upperparts are colored dark brown, the belly whitish with rufous flanks, the throat lighter, and the head with black and rufous stripes.

DISTRIBUTION
Ranges through the higher-altitude taiga regions of Russia from the Ural Mountains to the Bering Sea. It migrates to spend the winter in southern parts of its range in eastern China and Korea, and occurs as a vagrant in eastern Europe.

HABITAT
Breeds in shrubby thickets along watercourses within the boreal forest region. Winters in montane forest.

BEHAVIOR
Usually occurs singly, in pairs, or in small groups. The song is a high-pitched trill.

FEEDING ECOLOGY AND DIET
Forages on the ground for insects, spiders, and other invertebrates. Feeds on seeds and fruits during the winter.

REPRODUCTIVE BIOLOGY
Polyandrous. Nests in a tree low to the ground. Lays 4–6 light-blue-colored eggs.

CONSERVATION STATUS
Not threatened. A widespread and abundant species within its habitat.

SIGNIFICANCE TO HUMANS
None known. ◆

Resources

Books

BirdLife International. *Threatened Birds of the World.* Barcelona and Cambridge: Lynx Edicions, 2000.

Davies, N.B. *Dunnock Behaviour and Social Evolution,* edited by R.M. May and P.H. Harvey. Oxford: Oxford University Press, 1992.

Forshaw, J. M., and D. Kirshner. "Accentors." In *Encyclopedia of Birds.* San Diego: Academic Press, 1998.

Sibley, C. G., and B. L. Monroe Jr. *Distribution and Taxonomy of Birds of the World.* New Haven: Yale University Press, 1990.

Periodicals

Bjoersdorff, A., S. Bergstrom, R.F. Massung, P.D. Haemig, and B. Olsen. "*Ehrlichia*-infected ticks on migrating birds." *CDC Journal of Emerging Infectious Diseases* 7, 5 (2001): 1.

Burke, T., N.B. Davies, M.W. Bruford, and B.J. Hatchwell. "Parental care and mating behaviour of polyandrous dunnocks *Prunella modularis* related to paternity and DNA fingerprinting." *Nature* 338: (1989): 249–251.

Organizations

BirdLife International. Wellbrook Court, Girton Road, Cambridge, Cambridgeshire CB3 0NA United Kingdom. Phone: +44 1 223 277 318. Fax: +44-1-223-277-200. E-mail: birdlife@birdlife.org.uk Web site: <http://www.birdlife.net>

British Trust for Ornithology. The Nunnery, Thetford, Norfolk IP24 2PU United Kingdom. Phone: +44 (0) 1842 750050. Fax: +44 (0) 1842 750030. E-mail: info@bto.org Web site: <www.bto.org>

IUCN–The World Conservation Union. Rue Mauverney 28, Gland, 1196 Switzerland. Phone: +41-22-999-0001. Fax: +41-22-999-0025. E-mail: mail@hq.iucn.org Web site: <http://www.iucn.org>

Bill Freedman, PhD
Brian Douglas Hoyle, PhD

Thrashers and mockingbirds
(Mimidae)

Class Aves
Order Passeriformes
Suborder Passeri
Family Mimidae

Thumbnail description
Medium-sized songbirds, usually with long graduated tails and rounded wings, and long, often decurved bills; frequently with loud, varied, and sometimes imitative, songs

Size
8.2–12.2 in (20.5–30.5 cm)

Number of genera, species
10 genera; 35 species

Habitat
Forest edges, scrubland, suburban areas and abandoned farmland, desert scrub, and waterless desert

Conservation status
Critically Endangered: 2 species; Endangered: 2 species; Near Threatened: 2 species

Distribution
North America from southern Canada south, through Central America and West Indies south to Patagonia; Galápagos Islands, Bermuda

Evolution and systematics

The mimids (a collective term which covers mockingbirds, thrashers, catbirds, and tremblers) are a group of New World oscine passerines with many common structural features. The relationships of this group to other passerine families has been the subject of some debate over the years. Originally they were considered to be closest to the thrushes, a family that breeds on every continent except Antarctica (indeed, Linnaeus, in 1758, put both the brown thrasher [*Toxostoma rufum*] and the northern mockingbird [*Mimus polyglottos*] in the same genus as the Eurasian blackbird [*Turdus merula*]); however, recent work based on DNA hybridization studies suggests that mimids are a sister-group to the starlings (Sturnidae), an exclusively Old World family. The geographic origin of the mimids is clearly New World, since they occur nowhere else, but exactly where in the Americas is less clear. The greatest diversity of breeding species of mimids is found in the southwestern United States and Mexico. South America, the continent richest in bird species, has only seven mimids, all rather similar members of the one genus *Mimus*, arguing for a relatively recent colonization with little time for widely divergent evolution into the dif-

ferent habitats found in South America. By contrast, the West Indies has eight species. Some have clearly been there for a substantial time, since they are members of three very diverse genera endemic to the region. Based purely on considerations of species abundance and diversity, an origin in Central America is argued for, with an early colonization of the West Indies and a later invasion of South America. This remains speculative in the absence of hard evidence, since the fossil record is unhelpful.

Physical characteristics

The 35 species of mimid (Peter's checklist lists 13 genera and 31 species, but a more recent classification is used here) share many common characteristics. All are, for passerines, quite large, typically of a size similar to a Eurasian blackbird (*Turdus merula*) or American robin (*Turdus migratorius*). Legs and feet tend to be stout and sturdy, the tail long and graduated, the wings generally rather short and rounded. The bill is always relatively large, sometimes long and decurved. Coloration is usually fairly muted; most species are various shades of gray or brown (often brightly rufous), or in some cases

A northern mockingbird (*Mimus polyglottos*) bathes in south Texas. (Photo by Joe McDonald. Bruce Coleman Inc. Reproduced by permission.)

blue. Bright greens, yellows, and reds are absent, although in several species the irides are a startling red or orange. Sexes are similar. The two largest groups within the Mimidae are the thrashers, genus *Toxostoma*, with 11 or 12 (according to accepted taxonomy) species, and the mockingbirds, genus *Mimus*, with 9 or 10 species. Thrashers tend to be heavy, often largely terrestrial species, sometimes heavily spotted below, with strong and often markedly decurved bills. In plumage, mockingbirds tend to be various shades of gray, often with conspicuous white flashes on wings and tail; the bill is more thrush-like, not strongly decurved. A closely allied genus, *Nesomimus*, is endemic to the Galápagos Islands; the four species are very similar, a generally brown-gray plumage with darker markings. Catbirds comprise two species; one, highly migratory, occurs all over eastern North America, the second, sedentary, is confined to Guatemala, Belize, and the Yucatán Peninsula of Mexico.

The greatest diversity of physical form is found in the West Indies where there are several unique island forms. The bizarre tremblers, found only in the Lesser Antilles, are the most aberrant mimids, with a fine, decurved bill (longer in the female) and an extraordinary habit of shaking and trembling the wings. Also endemic to the Lesser Antilles (two islands only) is the striking and endangered white-breasted thrasher (*Ramphocinclus brachyurus*,) with blackish upperparts, white underparts, and a red eye.

Distribution

Mimids are exclusively a New World family, apart from vagrant records of three migratory North American species in Europe. Mimids occupy most of North and South America, being absent only from the boreal forest and tundra regions of North America (Newfoundland, Labrador, and the northern sections of the Canadian provinces across to Alaska), the Falkland Islands and extreme southern South America, and as breeding species, some parts of Central America. Mimids occur throughout the West Indies, Bermuda, and the Galápagos Islands. The greatest species diversity occurs in the southwestern United States and adjacent Mexico, but the largest generic diversity is found in the Caribbean, where there are a number of endemic, highly restricted genera. The mimids found throughout much of mainland South America are all closely related and very similar members of the genus *Mimus*. The Galápagos is home to a genus, *Nesomimus*, closely related to *Mimus*.

Habitat

The habitats required by mimids are varied, but essentially involve bushland. Apart from some West Indian species, they are absent from dense forest (though found in forest edge) and from savanna or prairie that lack bush. Within these requirements, habitats are quite diverse, from well-watered bushland in eastern North America (being especially fond of abandoned farmland) to cactus desert in the American Southwest, including some extremely arid zones such as Death Valley in California. Some West Indian species, notably the genus *Cinclocerthia*, inhabit wet epiphytic rainforest, but the family is generally absent from such habitat-types in South America.

Behavior

Most mimids are not excessively shy and some species are noted for their demonstrative and rambunctious behavior. Many species, especially of the genus *Mimus*, have exploited very modified habitats and live in comfortable proximity to man in suburban locations. Some others, such as the Crissal thrasher (*Toxostoma crissale*), an inhabitant of riparian brush-

A brown thrasher (*Toxostoma rufum*) with eggs at its nest in Arizona. (Photo by F. Truslow/VIREO. Reproduced by permission.)

Hood mockingbird (*Nesomimus macdonaldi*) attacks a waved albatross egg on Hood Island, one of the Galápagos Islands. (Photo by George Holton. Photo Researchers, Inc. Reproduced by permission.)

land in the American Southwest, are very secretive, rushing into dense cover at the slightest disturbance. Most members of the family spend much of their time on the ground or in the lower levels of vegetation, even though they may sing from exposed perches. Many species are notably aggressive, not hesitating to harass cats, dogs, or corvids by physical attack.

The songs of most mimids are loud, liquid, and generally very attractive. In some cases, especially the mockingbirds, there are strong imitative elements of other species. Some species sing at night.

Feeding ecology and diet

Mimids typically have long, pointed bills; in many cases these have a pronounced downward curvature and in some species are extremely decurved. Species with especially long curved bills, such as the California thrasher (*Toxostoma redivivum*), spend much time probing in loose soil. Food is varied; for most species it is predominantly invertebrate, but small frogs, lizards, and crawfish may be taken. Substantial amounts of vegetable matter including berries and seed-pods are taken, especially, in northern species, in fall and winter. The mockingbirds of the Galápagos Islands have developed some extraordinary and unique feeding traits in response to their harsh and variable environment. On islands where there are seabird colonies, such as Hood (Española), the resident species will eat seabird eggs, cracking open eggs of smaller species (such as the swallow-tailed gull, *Creagrus furcatus*). The more robust eggs of boobies and albatrosses cannot be broken into, but those with cracks are immediately attacked. An even more bizarre food source is blood. Marine mammals such as sea lions, wounded in territorial battles, soon attract groups of mockingbirds, which actively drink blood from open wounds and peck up congealed blood from the ground. One visiting ornithologist recorded that he twice had to fend off mockingbirds that attempted to pick off blood spots from scratches on his legs. Other food sources include ectoparasites (often engorged with blood) and loose skin from marine iguanas (*Amblyrhynchus subcristatus*), disgorged fish, and other carrion and marine mammal feces. Species found on islands

without seabird colonies such as Chatham (San Cristóbal) are more conventional in diet.

Reproductive biology

Most mimid species build open-cup nests, usually situated at heights of 3–6 ft (1–2 m) in bushes. The exceptions are the pearly-eyed thrasher (*Margarops fuscatus*) of the West Indies, which builds a bulky stick nest in cavities of trees and the two tremblers, found in the Lesser Antilles, whose nests are domed structures with side entrances. Desert species, such as Leconte's thrasher (*Toxostoma lecontei*), take great care to position their nests in such a way as to take greatest advantage of shade. Egg color is variable within the family; most thrashers (*Toxostoma*) and mockingbirds (*Mimus*) lay speckled or spotted eggs, with either a whitish or greenish background, but catbirds (*Dumetella* and *Melanoptila*), the Central American blue mockingbird (*Melanotis caerulescens*), and most of the endemic West Indian genera lay unmarked bluish or greenish eggs. Incubation is generally by both sexes in the thrashers, but by the female alone in mockingbirds and catbirds. In most species breeding is conventional (i.e. one male and one female per nest), but in the mockingbirds of the Galápagos Islands a complex system of social breeding has evolved, with a "territorial group" of birds, which in the case of the Hood mockingbird (*Nesomimus macdonaldi*) may number up to 40, defending a group territory. Within the group is a set hierarchy, the dominant male being the oldest bird. Up to four breeding females occur in a group,

A common or gray catbird (*Dumetella carolinensis*) at its nest. (Photo by F. Truslow/VIREO. Reproduced by permission.)

A group of Galápagos mockingbirds (*Nesomimus trifasciatus*). (Photo by Sam Fried/VIREO. Reproduced by permission.)

and non-breeding birds may assist in the rearing of broods of related pairs.

Conservation status

Most species of mimids pose no conservation concerns, and some populations seem to have benefited from human conversions of forest to bushland. However, five species, four of which are island endemics, are the source of serious anxiety. The Socorro mockingbird (*Mimodes graysoni*), formerly abundant on Socorro Island in the Revillagigedo group west of Mexico is Critically Endangered due to predation by introduced cats and habitat destruction by introduced goats. Rather puzzlingly, the Cozumel thrasher (*Toxostoma guttatum*) of Cozumel Island off the coast of the Yucatán Peninsula is also Critically Endangered, having suffered catastrophic population declines following hurricanes in 1988 and 1995; curiously none of the other endemic Cozumel species has suffered and all seem as abundant as previously. In the West Indies the two races of white-breasted thrasher, found on St Lucia and Martinique, are both Endangered, possibly critically, with shrinking populations of fewer than 100 pairs each. In these cases the causes seem to be habitat destruction and predation by introduced mongooses. The various species of mockingbirds

in the Galápagos Islands seem to be doing quite well, with the exception of the Charles mockingbird (*Nesomimus trifasciatus*), formerly found on Charles (Floreana) Island, now restricted to two tiny islets off Charles. In this case the culprit appears to be the black rat (*Rattus rattus*), though other species in the archipelago appear to be able to coexist with this introduced species.

Because of its limited range, the black catbird (*Melanoptila glabrirostris*) of the Yucatán Peninsula is classified as Near Threatened; it is, however, quite common in several locations. Recent information is lacking on the San Andres mockingbird (*Mimus magnirostris*), confined to Isla San Andrés (St. Andrew Island) in the western Caribbean; the island has been the subject of much recent development as a resort.

Significance to humans

None of the mimids has great economic significance apart from sporadic reports of damage to berry crops by some species. They do, however, include a number of well-known and very popular species, so much so that the brown thrasher has been adopted as the state bird of Georgia, and the northern mockingbird by no fewer than five southern U.S. states from Tennessee to Texas.

1. Brown trembler (*Cinclocerthia ruficauda*); 2. Blue-and-white mockingbird (*Melanotis hypoleucus*); 3. Hood mockingbird (*Nesomimus macdonaldi*); 4. Northern mockingbird (*Mimus polyglottos*); 5. Brown thrasher (*Toxostoma rufum*); 6. Gray catbird (*Dumetella carolinensis*). (Illustration by Emily Damstra)

Species accounts

Gray catbird
Dumetella carolinensis

TAXONOMY
Muscicapa carolinensis Linnaeus, 1766, the Carolinas.

OTHER COMMON NAMES
English: (Bermuda) blackbird; French: Moqueur chat; German: Katzendrossel; Spanish: Sinsonte Maullador.

PHYSICAL CHARACTERISTICS
8.1–8.5 in (20.5–21.5 cm); 0.94–2.0 oz (26.6–56.5 g). Plumage generally dark leaden-gray, crown dull black, crissum deep chestnut-brown. Eye deep brown, bill and legs black.

DISTRIBUTION
Breeds across eastern North America from Nova Scotia to east Texas, west to British Columbia, north New Mexico; Bermuda. Winters from southern New England through southeast states including Florida, coastal Texas, east Mexico through Central America to Panama, Bahamas, Cuba, Jamaica, rarely in Lesser Antilles. Vagrant Colombia, western Europe.

HABITAT
Scrubland, forest edge, well-vegetated suburban areas, abandoned orchards.

BEHAVIOR
Usually in lower levels of vegetation. Rather retiring; frequently sings from dense cover. Song is a pleasant series of whistled and gurgled notes, interspersed with harsher notes. Female sometimes sings. Calls include a cat-like mewing and more grating notes.

FEEDING ECOLOGY AND DIET
Food is varied, including many invertebrates (ants, caterpillars, spiders, grasshoppers) and some vegetable matter, including wild grapes, especially in fall; takes eggs of other small passerine species up to the size of American robin.

REPRODUCTIVE BIOLOGY
Nest is a cup-shaped construction of fine grasses etc, lined with horsehair and fine materials, built mostly by female, from ground level to 48 ft (15 m), usually about 4–5 ft (1.5 m) up. Eggs three to four, sometimes one to six, bright blue-green. Incubation by female alone, 12–14 days. Young fed by both parents, fledging period 8–12 days. Double-brooded over most of range, triple-brooded in southern areas.

CONSERVATION STATUS
Not threatened. Common over much of its range; has probably benefited from clearance of forest and subsequent abandonment of marginal farmland.

SIGNIFICANCE TO HUMANS
No economic significance; a popular and widespread species. ◆

Blue-and-white mockingbird
Melanotis hypoleucus

TAXONOMY
Melanotis hypoleucus Hartlaub, 1852, Guatemala.

OTHER COMMON NAMES
English: White-breasted blue mockingbird; French: Moqueur bleu et blanc; German: Lazurspottdrossel; Spanish: Sinsonte Matorralejo.

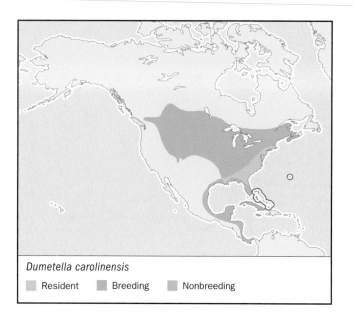

Dumetella carolinensis
■ Resident ■ Breeding ■ Nonbreeding

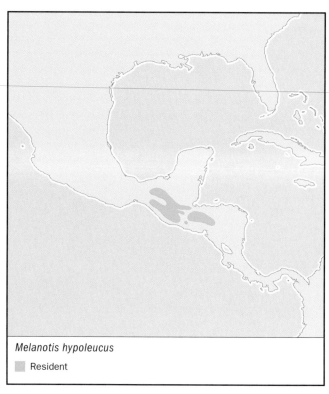

Melanotis hypoleucus
■ Resident

PHYSICAL CHARACTERISTICS
10–11 in (25.5–28 cm); 2.1–2.3 oz (60–63 g). Upperparts deep blue, underparts white apart from dark blue-gray flanks and vent. Facial mask black. Eyes deep red, bill and legs black. Juvenile dull blackish gray above and below with some white on underparts.

DISTRIBUTION
Southern Mexico (Chiapas), highlands of Guatemala, southern Honduras, northern El Salvador.

HABITAT
Open pine and oak woodland and scrub with dense under-storey, fairly humid at higher altitudes; at lower altitudes drier scrubland. 3,300–9,800 ft (1,000–3,000 m).

BEHAVIOR
Shy and retiring, usually found in pairs or family parties feeding on or near the ground. Song has a variety of repeated short musical notes, or a medley of whistles and harsher notes.

FEEDING ECOLOGY AND DIET
Feeds mostly on the ground, sweeping leaf-litter aside with sideways movements of the bill. Food omnivorous; invertebrates, also berries.

REPRODUCTIVE BIOLOGY
Little known. Nest is a shallow untidy cup of coarse sticks, lined with fibers, placed 3–15 ft (1–5 m) above ground in dense thickets or saplings. Eggs light blue, unmarked, average two at higher altitudes, three at lower altitudes. Incubation by female, period not recorded; young fed by both parents, fledging period 14–15 days.

CONSERVATION STATUS
Not threatened. In suitable habitat quite common; susceptible to habitat destruction and now very restricted in some parts of range (e.g., El Salvador).

SIGNIFICANCE TO HUMANS
None known. ◆

Northern mockingbird
Mimus polyglottos

TAXONOMY
Turdus polyglottos Linnaeus, 1758, Carolina. Three subspecies.

OTHER COMMON NAMES
English: Mockingbird; mocking thrush; French: Moqueur polyglotte; German: Spottdrossel; Spanish: Sinsonte Común.

PHYSICAL CHARACTERISTICS
9–10 in (23–25.5 cm); 1.3–2 oz (36.2–55.7 g). Plumage generally gray, wings and tail darker gray with pale edgings to coverts, white flash on primaries and white outer rectrices. Eye yellow, bill black with paler base, legs dusky. Juvenile with obscure spots on chest.

DISTRIBUTION
M. p. polyglottos: eastern North America from east Nebraska to Nova Scotia, south to east Texas and Florida; *M. p. leucopterus*: western North America from northwest Nebraska and west

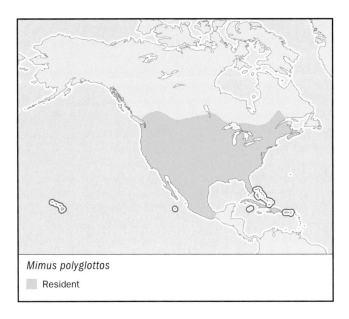

Mimus polyglottos
▨ Resident

Texas to the Pacific coast, south throughout Mexico to the Isthmus of Tehuantepec; *M. p. orpheus*: Bahama Islands, Greater Antilles east to Virgin Is. Vagrant to Great Britain; introduced to Hawaiian Islands.

HABITAT
Open bushland, well-vegetated suburban areas, abandoned farmland, orchards.

BEHAVIOR
Self-assured and conspicuous. Spends most of its time on the ground or in low vegetation. During breeding season very aggressive, not hesitating to attack predators such as cats and crows. Song is a loud, very varied series of strong musical notes, frequently including imitations of other species, sometimes repeated over and over; often sings at night, especially in urban settings.

FEEDING ECOLOGY AND DIET
Food includes both animal (mostly invertebrate, but including small lizards) and vegetable matter (especially berries).

REPRODUCTIVE BIOLOGY
Nest is an open cup, the base of dead twigs, built mainly by the male, the lining of grasses added by the female, usually less than 10 ft (3 m) from the ground in shrubs. Eggs two to six, base-color bluish or greenish white or somewhat darker, with brownish markings. Incubation by female alone, 12–13 days. Fed by both parents, fledging period 12–15 days. Two or three broods per year.

CONSERVATION STATUS
Not threatened. Over most of its range common or abundant. Has adapted well to human situations and probably expanded as forest was cleared; has benefited from plantings of such species as multiflora roses.

SIGNIFICANCE TO HUMANS
Little economic significance. A popular species, widely celebrated in literature and the state bird of Texas, Arkansas, Mississippi, Florida, and Tennessee. ◆

Hood mockingbird
Nesomimus macdonaldi

TAXONOMY
Nesomimus macdonaldi Ridgway, 1890, Hood Island (Isla Española), Galápagos Archipelago.

OTHER COMMON NAMES
French: Moqueur d'Espanola; German: Galapagos-Spottdrossel; Spanish: Sinsonte de los Galápagos.

PHYSICAL CHARACTERISTICS
11 in (28 cm). Weight not available. Generally variegated dull gray-brown above, off-white below with variable amount of darker spotting on chest. Eye yellow, bill and legs black.

DISTRIBUTION
Confined entirely to Hood Island (Española) in the Galápagos, an island measuring approximately 8 mi long and 4 mi wide (13 by 6.5 km).

HABITAT
Arid scrub; spends much time in seabird colonies.

BEHAVIOR
Totally fearless of humans. The mockingbirds of the Galápagos Islands live in groups, defending communal territories; in the Hood mockingbird the groups are larger than in the other species, at times up to 40 individuals. Groups rapidly adopt any temporary human encampment, investigating unfamiliar objects inquisitively and snapping up any offerings found therein.

FEEDING ECOLOGY AND DIET
Hood is the site of major marine mammal and seabird colonies and the Hood mockingbird, in contrast to species that live on other islands without such colonies, has developed some unique and bizarre feeding habits. The birds will eat blood from animals' fresh wounds or eat congealed blood from the sand. Engorged ticks and skin are picked from live iguanas. The Hood mockingbird has a more powerful bill than other members of the genus, and uses this to predate the eggs of swallow-tailed gulls (*Creagrus furcatus*), but larger eggs of boobies and albatrosses are only attacked if already damaged. Other food items include invertebrates, vegetable matter, carrion and feces. Groups whose communal territories do not include a seabird or mammal colony have less varied diets.

REPRODUCTIVE BIOLOGY
Groups of up to 40 individuals occupy and mutually defend a group territory, intruders being intimidated by consolidated displays of aggression. In less optimal territories (e.g., those without access to colonies), group size is smaller. Little published information on nest and eggs; presumably similar to other species of the genus; the Galapagos mockingbird (*N. parvulus*) builds a cup-shaped nest in a shrub and lays three to four eggs which are blue-green with brown markings.

CONSERVATION STATUS
Not threatened. Within its small range, abundant and not in any obvious hazard. However, introduction of any alien predatory species such as cats would very rapidly change this situation.

SIGNIFICANCE TO HUMANS
Of very great significance with regard to evolutionary and behavioral studies. The Galápagos fauna also has a major economic impact on the finances of Ecuador by virtue of the significant influx of foreign capital from tourism and from visiting zoologists. ◆

Brown thrasher
Toxostoma rufum

TAXONOMY
Turdus rufus Linnaeus, 1758, Carolina. Two subspecies.

OTHER COMMON NAMES
French: Moqueur roux; German: Rotrücken-Spottdrossel; Spanish: Cuitlacoche Rojizo.

PHYSICAL CHARACTERISTICS
10–11 in (25.5–28 cm); 2–3.2 oz (57.6–89 g) Upperparts deep rufous-brown, two whitish wing-bars on covert edges, underparts buffy-white with conspicuous blackish tear-drop shaped markings, eye yellow, bill brownish with flesh-colored base, legs brownish.

DISTRIBUTION
T. r .rufum breeds in eastern North America from New Brunswick and Maine south to Florida and west to central Ontario. *T. r. longicauda* breeds from western Ontario to southern Alberta, south to northeast New Mexico and central Oklahoma. The species winters from coastal Massachusetts to Texas, north to southern Illinois and Indiana. Vagrant to England, Bermuda, northern Alaska, Cuba, Bahamas, and northern Mexico.

HABITAT
Brushy areas, field-edges with hedges, well-vegetated suburbs, abandoned farmland.

BEHAVIOR
Not particularly shy. Spends most of its time on or near ground, but frequently sings from high exposed perches. Song

Nesomimus macdonaldi

■ Resident

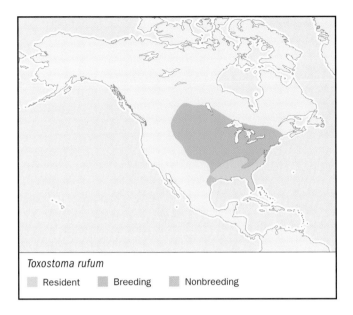

Toxostoma rufum

▢ Resident   ▢ Breeding   ▢ Nonbreeding

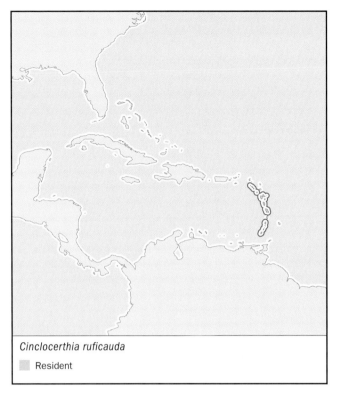

Cinclocerthia ruficauda

▢ Resident

is an attractive series of loud musical phrases, each phrase usually repeated, the whole song often continuing for several minutes at a time.

FEEDING ECOLOGY AND DIET
Food is very varied; many invertebrates including caterpillars, spiders, grasshoppers, and crayfish, also small frogs, snakes, and lizards; vegetable matter, especially berries, but also acorns, corn etc. Feeds mostly on the ground, probing into soft soil, sweeping aside leaf litter with sideways movements of the bill.

REPRODUCTIVE BIOLOGY
Nest is a bulky open cup, built of thorny twigs and lined with grass stems etc., located as high as 20 ft (6 m) in trees or bushes, but also frequently on the ground; built by both sexes. Eggs three to five, rarely only two, pale bluish white with fine reddish dots. Incubation by both sexes, 11–14 days; young fed by both sexes. Usually double-brooded.

CONSERVATION STATUS
Not threatened. Probably benefited from the spread of agriculture and especially from the abandonment and re-vegetation of marginal cropland. However, recent significant declines in overall population in Ontario and by implication elsewhere.

SIGNIFICANCE TO HUMANS
A popular and well-known species, the state bird of Georgia. ◆

Brown trembler
Cinclocerthia ruficauda

TAXONOMY
Stenorhynchus ruficauda Gould, 1835. Three subspecies.

OTHER COMMON NAMES
English: Brown bird, gray trembler, (Caribbean) trembling thrush; French: Trembleur brun; German: Zitterdrossel; Spanish: Temblador Castaño.

PHYSICAL CHARACTERISTICS
9–10 in (23–25.5 m); 1.8 oz (50 g) (average). Upperparts dark olivaceous-brown, with blackish brown crown and rufescent

rump. Underparts dull gray-brown, becoming orange-brown on belly. Eyes light yellow. Bill long, slender and decurved, black; legs brown. Females have longer bill than males; juveniles faintly spotted on chest.

DISTRIBUTION
C. r. ruficauda Dominica, West Indies; *C. r. tremula* Saba south to Guadeloupe, W.I.; *C. r. tenebrosa*, St. Vincent, W.I.

HABITAT
Wet forest with epiphytes, more rarely in grapefruit, cacao, and banana plantations.

BEHAVIOR
Occurs at all levels from the ground to tree-crowns. Occurs in pairs or small groups. Has peculiar habit of trembling, consisting of drooping both wings then quickly bringing them upwards while jerking the tail nervously. On ground it hops with cocked tail. Song is varied; loud notes uttered at intervals, some rich and warbled, others harsh or squeaky.

FEEDING ECOLOGY AND DIET
Food very varied; much invertebrate matter including insects, spiders, scorpions, and snails, also small tree-frogs and lizards. Fruit and berries including seedpods. Feeds at all levels; probes into ground and epiphytes with long bill; clings to tree trunks while probing into cavities in the manner of a woodcreeper.

REPRODUCTIVE BIOLOGY
Nest is a cup of rootlets, lined with finer material, in the base of a palm frond, in a tree cavity, or in a hollow stump; may use nest-boxes. Some nests reported to be domed with a side entrance. Eggs two to three, unmarked greenish blue. Incubation and fledging periods not recorded.

CONSERVATION STATUS
Not threatened. On several islands, populations are much reduced due to habitat destruction, but still abundant on Dominica and common on Guadeloupe and Saba. Extinct on St. Eustatius.

SIGNIFICANCE TO HUMANS
None known. ◆

Resources

Books

Bent, A.C. *Life Histories of North American Nuthatches, Wrens, Thrashers and their Allies.* Bulletin 195, U.S. Nat. Mus. 1948.

Brewer, David, and B.K. MacKay. *Wrens, Dippers and Thrashers.* London: Christopher Helm; and New Haven: Yale University Press, 2001.

Collar, N.J., L.P. Gonzaga, N. Krabbe, A. Madroño Nieto, L.G. Naranjo, and T.A. Parker. *Threatened Birds of the Americas: the ICBP/IUCN Red data Book.* Washington, DC: Smithsonian Inst. Press, 1992.

Periodicals

Bowman, R.I., and A. Carticr. "Egg-pecking behaviour in Galápagos mockingbirds." *Living Bird* 10 (1971): 243–270.

Castellanos, A., and R. Rodriguez-Estrella. "Current status of the Socorro Mockingbird." *Wilson Bull.* 105 (1993): 167–171.

Curry, R.L., and D.J. Anderson. "Interisland variation in blood-drinking by Galápagos Mockingbirds, *Nesomimus*

parvulus, in a climatically variable environment." *Auk* 104 (1987): 517–521.

Grant, P.R. and N. Grant. "Breeding and feeding of Galápagos Mockingbirds *Nesomimus parvulus.*" *Auk* 90 (1979): 723–736.

Markowsky, J.K., W. Glanz, and M. Hunter. "Why do Brown Tremblers tremble?" *Journal of Field Ornithology* 65 (1994): 247–249.

Skutch, A.F. "Life history of the White-breasted Blue Mockingbird" *Condor* 52 (1950): 220–227.

Zusi, R.L. "Ecology and adaptations of the Trembler on the island of Dominica." *The Living Bird*, eighth annual of the Cornell Laboratory of Ornithology, (1969): 137–164. Cornell University, Ithaca, NY.

Other

Hardy, J.W., J.C. Barlow, and B.B. Coffey Jr. *Voices of all the mockingbirds, thrashers and their allies.* Gainesville, FL: ARA Records, 1987.

David Brewer, PhD

Dippers
(Cinclidae)

Class Aves
Order Passeriformes
Suborder Passeri (Oscines)
Family Cinclidae

Thumbnail description
Medium-sized passerines with plump bodies, rounded wings, short tails, and sturdy legs, adapted to an exclusively aquatic existence in rushing mountain streams

Size
5.5–9 in (14–23 cm); 1.3–3.1 oz (38–88 g)

Number of genera, species
1 genus; 5 species

Habitat
Mountain streams

Conservation status
Vulnerable: 1 species

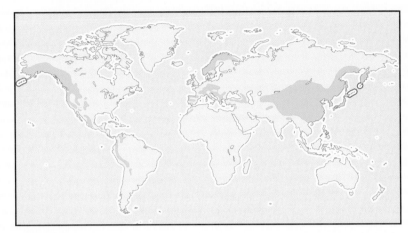

Distribution
Eurasia, from Ireland, Scandinavia to east Siberia, Japan, west China, north India, central Asia, Caucasus, Turkey, southern Europe, northern Africa, western North America, Central America, western South America to northern Argentina

Evolution and systematics

Dippers are a most unusual family; there are only five species, but all five are very uniform in shape, habits, habitat, and nesting behavior, yet are spread across five continents and a huge geographic area. Linnaeus in 1758 named the Eurasian dipper, the only species that he could have observed personally, as a starling, but for most of their history dippers have been regarded as an off-shoot of the wren family. Given their rounded body-shape, short wings and tail, and the domed nests with side-entrance, all also typical of wrens, this would seem a very reasonable classification. Recent DNA studies, however, indicate that the closest relatives of dippers are, in fact, the thrushes, and resemblance to wrens is coincidental.

There are widely different views as to the geographic origins of dippers. Wrens are undoubtedly of New World origin, and when dippers were regarded as wrens a similar geographic origin was unarguable. However, with the overturning of this relationship the facts become very cloudy, with different authorities arguing for origins in eastern Asia, followed by colonization of the New World via the Bering land-bridge, and almost the precise opposite: an origin in western North America followed by colonization of Eurasia and South America as land-bridges became available. The fossil record is of little help since it is essentially non-existent; the mountain-stream habitat of dippers is hardly conducive to deposition of remains for fossilization.

Physical characteristics

All the species of dipper are extremely similar in physical form. All are stocky birds, with short, rounded wings with a concave lower surface (the ninth to seventh, or eighth to sixth,

primary feathers being the longest), a short tail with broad, round-tipped feathers, sturdy legs and powerful feet, and a short, compressed bill without rictal bristles. The preen gland is unusually large for a passerine bird and the body-feathers are water-repellant. Some species, such as the American and brown dippers, have an almost uniformly colored plumage; others, like the Eurasian and white-capped dippers, are strikingly marked with white. Bright colors are absent; plumage is usually gray or brown, sometimes strongly rufous, but blues, greens, and yellows are not found. All species have white surfaces to the eyelids, conspicuous when the eyes are blinking, and prominent nictitating membranes.

Distribution

Dippers occur on five continents: the Americas, Eurasia, and marginally, in Africa. Because of their habitat requirements they are largely confined to mountainous regions with rushing streams. Hence they are absent from large lowland areas of all the continents on which they breed. In North America they occur throughout the western states and provinces, from the Arctic Circle in Alaska, south through Central America to Panama, but are, curiously, absent from the eastern two-thirds of the continent, including areas of apparently suitable habitat in eastern Canada, New England, and the Appalachians. In South America they occur exclusively in the Andes and adjacent mountains, from Venezuela and Colombia south to northern Argentina. In Europe, they occur from Scandinavia and the British Isles across to Greece, Turkey, Syria, and the Caucasus; they are also found in suitable habitat in North Africa. In Asia, they occur in isolated areas of the Ural mountains, central Asia from Afghanistan to eastern Siberia and Japan, south to Burma and north-western

Mexican (or American) dipper (*Cinclus mexicanus*) with food for its chicks. (Photo by Jeff Foott. Bruce Coleman Inc. Reproduced by permission.)

Thailand. Dippers have not succeeded in colonizing suitable habitat in Africa south of the Sahara, southern India, or Australasia; in these areas their ecological niche is exploited by birds of several families, none of which, however, has evolved the unique structural and behavioral characteristics of dippers.

Habitat

The habitat requirements of all the dippers are extremely uniform throughout the genus; they need fast-flowing, well-oxygenated streams, not usually very deep, with generally clear water with little sediment or turbidity. So long as there remain large enough areas of unfrozen water they can withstand very severe conditions, being found in winter in central Asia in temperatures dropping to −40°F (−40°C) or lower. Where too much of their habitat freezes they may move to lower altitudes or migrate to warmer areas. Dippers also occur, especially outside the breeding season, on the shores of mountain lakes, in beaver ponds, and occasionally on rocky sea-shores.

A second habitat requirement is rock-faces with crevices for siting of nests; artificial sites such as bridge trestles frequently suffice for this purpose.

Behavior

Dipper behavior is very uniform throughout the genus. Dippers are rarely found more than a few feet from a stream and are usually seen perched on a boulder surrounded by rushing water, or flying rapidly (with a peculiar resemblance to an enormous bumble-bee) with whirring wing-beats a few inches above the surface. A characteristic activity of a perched dipper is a rapid series of bowing movements, which in white-chested species emphasizes that feature; bowing, or, "dipping," becomes more frequent and intense when birds are agitated or during territorial disputes. Dippers also blink rapidly, showing the white upper surface of the eyelid.

Uniquely among the passerines, dippers spend much of their time in water, frequently totally submerged for several seconds at a time. Birds may wade in shallow water; in deeper water they may dive head-first from a boulder or even from flight. Birds may also escape pursuing hawks by diving straight into water from mid-flight. Under water, birds overcome their natural buoyancy with rapid beating movements of the wings; although pebbles are occasionally grasped with the feet, wing movements are the main way that birds keep themselves on the bottom. Submerged birds have a silvery appearance due to entrapped air on their waterproof plumage.

Songs of dippers tend to be loud bubbling warbles; in the absence of loud water noise, they are audible at considerable distances; both sexes sing. Calls are sharp "zitting" sounds, audible above the sound of rushing water.

Feeding ecology and diet

Food consists almost entirely of aquatic invertebrates such as may fly and caddis fly larvae; the latter build themselves a substantial protective case of tiny pebbles, which dippers break by hammering them on a rock. Some small fish such as minnows are consumed, as are fish eggs and some vegetable matter. Some species have been observed to glean torpid insects from snow-banks.

Reproductive biology

The nesting habits of all the dipper species are very similar. Nests are domed structures with a wide side-entrance,

A white-throated dipper (*Cinclus cinclus*) rests on a rock. (Photo by S. Young/VIREO. Reproduced by permission.)

Mexican (or American) dipper (*Cinclus mexicanus*) at its nest. (Photo by Jeff Foott. Bruce Coleman Inc. Reproduced by permission.)

made of mosses, grass-stems, and leaves, with an inner cup of finer material and hair. Both sexes build the nest but the female usually completes it and builds the cup. Nests are usually situated close to water, most often actually above it, and sometimes behind waterfalls, in crevices in rock-faces, or in artificial situations such as under bridges. Eggs are white, up to seven in temperate populations but frequently only two in the tropics. Incubation is mainly or entirely done by the female, usually for about 16 days; young are fed by both parents. Fledging period (where known) is usually about 22 days. Temperate populations frequently produce double broods.

Conservation status

Dippers require clear flowing water and consequently are rather vulnerable to many human-made changes. They can tolerate only a modest amount of pollution, whether domestic or farm, industrial or mining, and frequently abandon degraded habitats; even the planting of conifers close to streams can be detrimental. On the other hand, man-made structures such as bridges may enhance dipper populations by providing nest-sites. Generally, dipper populations have diminished in the face of human activity over most of their ranges, though in some cases (for example, in the United Kingdom) some populations have recovered as the decline of the mining industry has reduced stream pollution. One race of the Eurasian dipper (*C. c. olympicus*), formerly of Cyprus, is extinct, and several other populations, for example those in North Africa, are very small. One species, the rufous-fronted dipper of northern Argentina and southern Bolivia, is classified as Vulnerable with a total world population in the probable range of 1,000 pairs.

Significance to humans

Dippers have no economic significance *per se* apart from infrequent reports of nuisance predation in fish-hatcheries. Their main significance is as an indicator species for an undegraded habitat.

1. White-capped dipper (*Cinclus leucocephalus leuconotus*); 2. Eurasian dipper (*Cinclus cinclus aquaticus*); 3. American dipper (*Cinclus mexicanus*); 4. Rufous-throated dipper (*Cinclus schulzi*); 5. Brown dipper (*Cinclus pallasii*). (Illustration by Dan Erickson)

Species accounts

Eurasian dipper
Cinclus cinclus

TAXONOMY
Sturnus cinclus Linnaeus, 1758, Sweden.

OTHER COMMON NAMES
English: English dipper, white-throated dipper, white-breasted dipper, water ousel; French: Cincle plongeur; German: Wasseramsel; Spanish: Mirlo Acuático.

PHYSICAL CHARACTERISTICS
6.7–7.9 in (17–20 cm); weight, male 1.9–2.7 oz (53–76 g), female 1.6–2.5 oz (46–72 g). Dark brown above. Throat and chest pure white, belly dark brown or rufous according to race, lower belly dark brown. Some Asiatic races have largely white underparts.

DISTRIBUTION
Numerous disjunct populations in hilly areas of Europe from Ireland, northern Norway, Iberia, central and southern Europe to Italy and Greece, Urals and Caucasus, northern Africa, Turkey, central Asia to Tibet and western China. Largely sedentary, but some Scandinavian birds winter in the Baltic States, Poland, and adjacent Russia, and the Urals population appears to be partially migratory.

HABITAT
Rushing mountain streams, occasionally rocky lakeshores.

BEHAVIOR
Behaves as others of the genus; territorial, lives in pairs along streams. Hunts in and under water. Song is a sweet, rippling warble; both sexes sing. Call is a loud penetrating "zit-zit."

FEEDING ECOLOGY AND DIET
Bulk of prey aquatic invertebrates, but also fish eggs and fry, almost all taken underwater from streambeds.

REPRODUCTIVE BIOLOGY
Nest typical of genus, domed with a side entrance, built by both sexes, of mosses, grass, etc., situated in rock crevices or in artificial sites such as bridges, always close to and frequently directly over water. Nests early; eggs unmarked white, one to seven, usually four or five, incubated by female alone, 12–18 days, usually about 16. Young fed by both sexes, fledging 20–24 days, usually about 22. Frequently double—occasionally triple—brooded. Generally monogamous, but polygamy does occur.

CONSERVATION STATUS
Not threatened. Generally widespread in suitable habitat, but most populations have shown declines, some serious. Will recolonize previous habitat if pollution ceases or water quality improves. Race *olympicus* in Cyprus Extinct.

SIGNIFICANCE TO HUMANS
No economic significance; has great appeal as a symbol of beautiful and unspoiled mountain countryside. ◆

Brown dipper
Cinclus pallasii

TAXONOMY
Cinclus pallasii Temminck, 1820, "Crimea," actually Okhota River, eastern Siberia.

OTHER COMMON NAMES
English: Asian dipper, Pallas's dipper; French: Cincle de Pallas; German: Flusswasseramsel; Spanish: Mirlo Aquático Castaño.

PHYSICAL CHARACTERISTICS
8.3–9.1 in (21–23 cm); weight (sexes not distinguished) 2.3–3.1 oz (66–88 g). Plumage uniformly dark brown.

DISTRIBUTION
Central Asia from Afghanistan, Kazakhstan to Tibet, Nepal, North Burma, North Vietnam; disjunctly northern India,

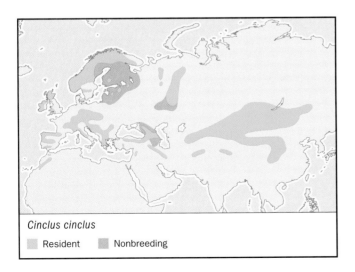

Cinclus cinclus
■ Resident ■ Nonbreeding

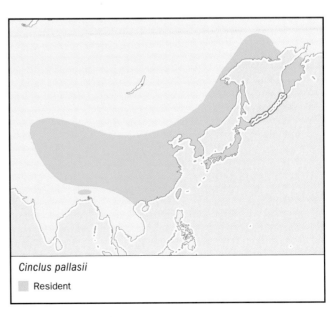

Cinclus pallasii
■ Resident

Siberia to Kamchatka, Japan from Kyushu northwards. Over most of its range largely sedentary, in some cases moves to lower altitudes in winter; some apparent migration from northern edge of range in China.

HABITAT
Rushing mountain streams and shores of mountain lakes.

BEHAVIOR
Much as other members of the genus; spends all of its time in close proximity to fast-flowing water, diving and swimming down to the bed. Bobs and curtsies in a manner similar to the Eurasian species. Song is a loud bubbling warble; call a sharp "zit-zit."

FEEDING ECOLOGY AND DIET
Aquatic invertebrates, caddis fly and may fly larvae, etc., taken mostly underwater.

REPRODUCTIVE BIOLOGY
Nest is spherical with a side entrance, the outer layer of moss, the inner of rootlets and leaf-webs, usually located in crevices in rock faces above flowing water; more rarely in artificial sites such as bridges. Both sexes build. Eggs three to six, usually five, incubation period 19–20 days, fledging period 23–24 days. Probably mostly single-brooded.

CONSERVATION STATUS
Not threatened. Widespread and common; susceptible to habitat degradation, but much of its range has very sparse human habitation.

SIGNIFICANCE TO HUMANS
None known. ◆

American dipper
Cinclus mexicanus

TAXONOMY
Cinclus mexicanus Swainson, 1827, Temescaltepic, Mexico.

OTHER COMMON NAMES
English: Mexican dipper, water ouzel, waterthrush; French: Cincle d'Amérique, cincle américain; German: Grauwasseramsel; Spanish: Cinclo Norteamericano.

PHYSICAL CHARACTERISTICS
6–7 in (15–17.5 cm); weight, male 2.0–2.3 oz (57–66 g); female, 1.5–2.3 oz (43–65 g). Plumage overall dark gray, paler on chin.

DISTRIBUTION
Western North America from Alaska (north to the Arctic Circle) and east Aleutians, south through western Canada to Arizona and Colorado; numerous disjunct populations from Mexico through Central America to Panama.

HABITAT
Rushing mountain streams; in winter also the fringes of lakes and beaver ponds, sometimes sea-shores. Sea-level in north to 11,000 ft (3,500 m); in Costa Rica 2,600–8,200 ft (800–2,500 m).

BEHAVIOR
Very similar to Eurasian dipper; dives into and swims in fast-flowing water, usually to be seen perched on a mid-stream boulder or flying low above the water. Song is a medley of single notes, audible for long distances; call a sharp "dzik."

FEEDING ECOLOGY AND DIET
Most food is taken from boulders or stream-beds. Birds usually only remain submerged for 10 seconds or less. Prey almost en-

Cinclus mexicanus
▨ Resident

tirely aquatic invertebrates. Will occasionally fly-catch; has been seen to pick frozen insects off stream-side snowbanks and beach-hoppers out of cast-up seaweed.

REPRODUCTIVE BIOLOGY
Nest is a spherical or elliptical ball with a side entrance, the outer layer grass or moss, inside a woven cup of grass, leaves and bark, usually located near or above flowing water in rock crevices; also, increasingly, in artificial sites such as bridges or nest boxes. Both sexes build. Eggs white, in North America usually four or five, in Costa Rica two to four. Incubation by female alone, 14–17 days, young fed by both sexes, 24–26 days. Sometimes double-brooded. Usually monogamous; males may, rarely, be polygamous.

CONSERVATION STATUS
Not threatened. Frequently common in pristine habitat but susceptible to water pollution from activities such as mining. Some populations apparently augmented by provision of nest sites such as bridges or suitable nest boxes.

SIGNIFICANCE TO HUMANS
Occasionally accused of causing significant damage to fish hatcheries; otherwise no other direct economic significance. ◆

White-capped dipper
Cinclus leucocephalus

TAXONOMY
Cinclus leucocephalus Tschudi, 1844, Junín, Peru.

OTHER COMMON NAMES
French: Cincle à tête blanche; German: Weisskopf-Wasseramsel; Spanish: Pájaro de Agua.

Cinclus leucocephalus

■ Resident

PHYSICAL CHARACTERISTICS
6 in (15 cm); weight (sexes not distinguished) 1.3–2.1 oz (38–59 g). Plumage varies according to race; back black, chest white, eyestripe black, crown gray-white with blackish streaks (Peru and Bolivia), or back black with white center, chest and belly white, lower belly blackish, eyestripe blackish, crown gray with darker streaks, eyestripe blackish (Venezuela to Ecuador).

DISTRIBUTION
Andes of western South America, from northwest Venezuela, Colombia through Ecuador and Peru to northern Bolivia.

HABITAT
Fast-flowing mountain streams, 3,300–12,800 ft (1,000–3,900 m).

BEHAVIOR
Typical of genus; territorial, living in pairs along mountain streams. Less inclined to dive with rushing water than Eurasian or American species; also forages in stream-side vegetation. Song is a loud musical trill; call a sharp "zeet-zeet."

FEEDING ECOLOGY AND DIET
Bulk of prey items are aquatic invertebrates. Prey is picked off wet boulders as well as from stream bottoms; will also take items such as earthworms from sides of streams.

REPRODUCTIVE BIOLOGY
Relatively little information available. Nest is a roughly spherical construction with a circular entrance hole at the side, built of mosses with an inner cup of dry leaves, strips of bark, etc., situated in crevices in rock-faces above flowing water. Eggs two, color not recorded but presumably white. Incubation and fledging data unknown.

CONSERVATION STATUS
Not threatened. Generally common and widely distributed in suitable habitat; however, it is susceptible to habitat degradation and has disappeared from some watersheds, e.g., near Quito, Ecuador, as a result of pollution.

SIGNIFICANCE TO HUMANS
None known. ◆

Rufous-throated dipper
Cinclus schulzi

TAXONOMY
Cinclus schulzi Cabanis, 1883, Cerro Bayo, Tucumán, Argentina.

OTHER COMMON NAMES
English: Rufous-throated dipper; French: Cincle à gorge rousse; German: Rostkehl-Wassermasel; Spanish: Mirlo de Agua Gorjirufo.

PHYSICAL CHARACTERISTICS
5.6–6 in (14–15 cm); weight, no data. Plumage uniformly gray-brown above and below, with contrasting orange-brown throat and upper chest. Inner webs of primaries partially white, giving a white flash on open wing.

DISTRIBUTION
Very restricted range in northern Argentina (Salta, Jujuy, Catamarca, and Tucumán) and extreme southern Bolivia (Tarija).

Cinclus schulzi

■ Resident

HABITAT

Swift-flowing mountain streams, apparently mostly in areas where the dominant forest-type is alder (4,600–8,200 ft; 1,400–2,500 m).

BEHAVIOR

Sometimes grouped with *Cinclus leucocephalus*. Generally similar to other dipper species; however, apparently it does not dive into rushing water or swim as frequently as the others. The bowing and dipping behavior also seems to be absent, replaced by a wing-flicking which displays the white flash on the primary feathers. Song is similar to that of white-capped dipper, but more thrush-like; call a sharp "zeet-zeet."

FEEDING ECOLOGY AND DIET

Little data; aquatic invertebrates including beetles.

REPRODUCTIVE BIOLOGY

Only a few nests have been described. Nest is typical globular dipper structure with side entrance, the outer part moss, with an inner bowl of grasses, etc., located in rock crevices or under bridges. Eggs two, unmarked glossy white. Incubation and fledging periods unknown; young fed by both parents.

CONSERVATION STATUS

Classified as Vulnerable. Total range is in an area about 60 by 375 mi (100 by 600 km), but within this area not all suitable habitat is occupied. Susceptible to stream quality degradation; only a small part of its total range is protected.

SIGNIFICANCE TO HUMANS

No economic significance; accessible populations give some financial benefit for local towns from ecotourism. ◆

Resources

Books

Brewer, David, and Barry Kent MacKay. *Wrens, Dippers and Thrashers.* London: Christopher Helm; and New Haven: Yale University Press, 2001.

Cramp, S. *Handbook of the Birds of Europe and North Africa; Birds of the Western Palaearctic.* Vol. 5. Oxford: Oxford University Press, 1988.

Kingery, H. E. "American Dipper (*Cinclus mexicanus*)." In *The Birds of North America*, edited by A. Poole and F. Gill. No. 229. Philadelphia: The Academy of Natural Sciences; and Washington, DC: The American Ornithologists' Union, 1996.

Tyler, S. J., and S. J. Ormerod. *The Dippers.* London: T and A. D. Poyser, 1994.

Periodicals

Salvador, S., S. Narosky, and R. Fraga. "First description of the nest and eggs of the Rufous-throated Dipper *Cinclus schulzi* in northwestern Argentina." *Gerfaut* 76 (1986): 63–66.

Skutch, A. F. *Studies of Tropical American Birds.* Publications of the Nuttall Ornithological Society. 10 (1972).

Tyler, S. J. "The Yungas of Argentina; in search of Rufous-throated Dippers (*Cinclus schulzi*)." *Cotinga* 2 (1994): 38–41.

Tyler, S. J., and L. Tyler. "The Rufous-throated Dipper (*Cinclus schulzi*) on rivers in north-west Argentina and southern Bolivia." *Bird Conservation International* 6 (1996): 103–116.

David Brewer, PhD

Thrushes and chats

(*Turdidae*)

Class Aves
Order Passeriformes
Suborder Passeri (Oscines)
Family Turdidae

Thumbnail description
Small to medium-sized songbirds with stocky build and short, strong but slender bills, stout legs, and relatively short wings; great variety in color, many larger thrushes more or less spotted beneath, smaller chats varying from plain and brown to boldly patterned and colorful

Size
Smaller chats are 5 in (12.5 cm) long and weigh 0.5 oz (15 g); larger thrushes are 13 in (33 cm) long, weighing 4.9 oz (140 g)

Number of genera, species
54 genera; 300 species

Habitat
From desert to rainforest, and from lowlands to high mountain slopes; most live in woodland or scrub, some entirely terrestrial

Conservation status
Critically Endangered: 7 species; Endangered: 5 species; Vulnerable: 18 species; Near Threatened: 22 species; Data Deficient: 3 species; Extinct: 3 species

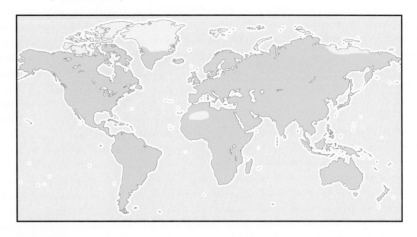

Distribution
Widespread in Old and New Worlds, including Pacific, Atlantic, and Indian Ocean islands, from far northern America and Eurasia to southern tips of America, Africa, and Australia

Evolution and systematics

Thrushes and chats are found almost throughout the world's land masses, with their greatest diversity in the tropics of Africa and Asia. Many are much-loved species in their native lands. The relationship of some groups such as the solitaires (*Myadestes*) and forktails (*Enicurinae*) within the family is contentious. Discussion of groups is somewhat hampered by the complication of English names, especially the use of the word robin to describe species that have some red in the plumage (thus resembling the Eurasian robin, *Erithacus rebecula*) or groups of species that have a robin-like shape (hence bush robins, also sometimes termed bush chats and robin chats).

Chats, generally smaller, stockier birds than most thrushes, include some well-marked groups such as 10 species of bush robins, three typical robins (*Erithacus*), a dozen nightingales and similar species, 13 robin chats, 11 redstarts, 11 stonechats, three bluebirds, and 19 wheatears. All of these are well-defined and recognizable groups sharing distinctive features. Some 164 species of chats are found in the Old World, and 87 in Eurasia, while only the solitaires, the three bluebirds, a bluethroat, and a wheatear are native to the New World.

Thrushes comprise 15 genera with a dozen rock thrushes, 26 mountain and ground thrushes (*Zoothera*), 10 small so-called nightingale-thrushes (*Catharus*), and approximately 64 typical

thrushes (*Turdus*). Of all of these, 55 species are Eurasian, 21–25 are found in the Afrotropics, and 46 in the Americas.

This distribution of species implies that the Palearctic, basically Europe and much of Asia, is the center from which thrushes evolved and spread, with a secondary center in Africa, perhaps particularly so for chats. New World thrushes probably evolved after invasions from the Old World. Most recently, the northern wheatear (*Oenanthe oenanthe*) and the bluethroat (*Luscinia svecica*) have expanded their range into North America. Similarly, thrushes have expanded from Asia into Oriental regions and Australia and, in the far distant past, reached a number of isolated island groups, including the Seychelles in the Indian Ocean and Hawaii in the Pacific, where unique species have evolved. Some isolated forms have since become among the most threatened bird species in the world. Long-distance flights to isolated islands are not difficult for this strong-flying group: American and eastern Asian species turn up each fall in northwest Europe, for example, far from their normal range.

Physical characteristics

Thrushes are compact, alert, rather upright birds in general, and very varied in appearance. Nevertheless, they do not look particularly dramatic or exceptional in terms of size,

Blackbird (*Turdus merula*) display. (Illustration by Barbara Duperron)

overall shape, bill shape, leg length, or tail length or shape. They do not exhibit any great variety of crests, ruffs, or other feathered ornament, but many are richly colored and beautifully patterned. Indeed, even the most plain have some exquisite subtlety of color visible in a close view.

The bill is strong and slightly decurved, sometimes with a very slightly developed hook. The stout tarsi (legs) have fused horny plates, which are often stronger and larger than in related birds, in keeping with the thrushes' mode of life, which is carried out more on the ground than in trees. Typically, thrushes run and hop or combine the two in a kind of hurried shuffle across the ground, but they are also adept at perching in trees, on overhead wires, on rocks, and on all kinds of artificial structures.

The wings, which have 10 primaries, the outermost often very short, are somewhat rounded except in species that are true migrants, which tend to have longer, more pointed wingtips as might be expected from the journeys they undertake. Some species even show subspecific differences in this regard, with those that migrate farthest having longer wings than more resident species.

The tail has 12 feathers. Usually it is only moderately long, sometimes rather short, and square-tipped, conforming to the thrushes' common conformation.

In many species, adults of both sexes are alike, but in others there is a marked difference. Within true thrushes, blackbirds (*Turdus merula*) include jet-black males and plain brown females, while male and female song thrushes (*T. philomelos*) are identical, as are both sexes of the related American robin (*T. migratorius*), although most females tend to be a little plainer than males. There is little seasonal variation, although paler feather fringes giving duller colors in winter may wear off to reveal stronger patterning in spring, and bill colors may intensify. More usual is a marked difference between adults and young, with a much more spotted effect common on juveniles. In some smaller chats such as the redstart (*Phoenicurus phoenicurus*) and northern wheatear, juveniles are very like freshly molted fall adults, if not inseparable.

Most wheatears (*Oenanthe*) have strikingly white rumps and upper tail coverts, usually combined with white beneath the tail and a bold black bar across the tail tip. In some, the center of the tail is black, too, creating a black T shape against an eye-catching blaze of white. This is usually more or less hidden on a perched or standing bird, but revealed in a sudden flash of white as the bird takes flight.

Many typical thrushes and smaller North American thrushes are brown above and pale below with dark spots. Separating them may rely on rather subtle differences: the wood thrush (*Catharus mustelina*) is brightest and reddest around the head, grayer towards the rear, while Swainson's (*C. ustulatus*) and gray-cheeked (*C. minimus*) thrushes are duller overall, and the hermit thrush (*C. guttatus*) is dull on the head but redder on the tail. Rock thrushes (*Monticola*) are richly colored, with variable blue or blue-gray on the head and back, and deep orange to rusty-red on the underside and tail. The tropical whistling-thrushes (*Myophonus*) have some spectacular deep blue plumages, highlighted by small spots or crescents of electric blue or white, while the *Zoothera* thrushes include several glorious black, rufous, and white species, a number that are largely rich orange or orange and gray (often with white, spotted wing bars, and black and white face patterns), and some more typically spotted species. The rich orange combined with striking black, white, and gray is repeated in a number of the African robin chats (*Cossypha*). Nightingales (*Luscinia*) are rather dull and plain, but the related bluethroats include males with stunning deep sky-blue gorgets (neck ruffs), some marked with a central spot of red or white.

Some larger African chats, the small nightingales and bluethroats, and many true thrushes are especially prized for their vocalizations. In many parts of the world, the dawn chorus of territory-holding birds in woodland or suburban gardens is largely made up of thrushes and chats singing loud and long, each species with its characteristic sound and pattern. In Europe, the blackbird and song thrush are common and expert songsters, while in much of Africa the songs (and in some cases, duets) of such birds as the white-browed robin chat (*Cossypha heuglini*) and olive thrush (*Turdus olivaceus*) are essential to the aesthetic enjoyment of each dawn and dusk.

Distribution

Thrushes and chats can be found on most land masses except for Antarctica, although only two artificially introduced species occupy New Zealand. A number are extremely restricted in their range. Three species of rock thrush (*Monticola*) are found only in Madagascar; whistling-thrushes (*Myophonus*) include one species restricted to Sri Lanka, another to Malaysia, and another to Taiwan. The rarely seen geomalia (*Geomalia heinrichi*) is a rufous and brown thrush found only in Sulawesi rainforests, but is not the only thrush restricted to Sulawesi. The Amami thrush (*Zoothera major*) is found on just one small island in the Ryukyu Islands south of Japan, while other *Zoothera* thrushes are also confined to one or two islands and several *Myadestes* thrushes are found only within the Hawaiian group. Even within *Turdus*, there are species with remarkably restricted ranges, including various African islands, while Tristan da Cunha, a remote South Atlantic island group, the Tres Marias Islands off Mexico, and several West Indies islands also have their own unique thrushes. The Seychelles magpie-robin (*Copsychus sechellarum*) was, until the 1990s, restricted to just one Seychelles island with only 20 individuals, before translocation to other islands in the group helped a recovery to more than 100 individuals by 2001.

These, however, are the exceptions to the general rule, with most species having quite extensive geographical ranges. Some are long-distance migrants such as the northern wheatear. Birds of this species breeding in Greenland set off on a massive trans-oceanic flight direct to Africa each fall, and return in spring via western Europe. The whinchat (*Saxicola rubetra*) spends each winter in Africa but moves north in spring to breed in Europe, often alongside the similar stonechat (*S. torquata*), which is an all-year resident in Europe and also in much of Africa. American robins breed throughout North America as far as the northern coasts of Alaska and mainland Canada and south to Mexico, but vast numbers from the northern two-thirds of this huge range move south in winter to join their more southerly relatives. Of the spotted thrushes in North America, the hermit thrush (*Catharus guttatus*) is the most widespread and the only one that commonly winters in much of the United States. Several thrushes breed in an extensive range across Europe and northern Asia, including the redwing (*Turdus iliacus*), hundreds of thousands of which pour out of cold Asia and northern Europe in autumn to find milder conditions in western and southern Europe each winter. The dusky thrush (*T. naumanni*), however, moves south instead of west, to winter in Japan and south-east Asia, while the dark-throated thrush (*T. ruficollis*) moves south from central Asia to spend the winter months in a belt just north of the Indian subcontinent. The blackbird has one of the most extensive ranges of all, from northwest Europe eastwards and south across Eurasia to the Oriental region, and into Australia and New Zealand. The stonechat has one of the widest ranges of Old World chats, breeding from Britain and Ireland south and east through Eu-

A gray-cheeked thrush (*Catharus minimus*) sings while perched on a branch. (Photo by C. Witt/VIREO. Reproduced by permission.)

rope, through East Africa south as far as the Cape, in Madagascar and Arabia and, in a separate area of distribution, through a vast area of central and eastern Asia and Japan.

Habitat

Most thrushes are forest birds. Rock thrushes (*Monticola*) are least tied to woodland, although some are very much woodland birds in their wintering areas, even though they breed in open places. In some species, part of the population may be relatively tame and confiding and live close to human settlements, while others remain shy.

Each species has a preferred habitat that may be extremely narrow or very broad. The American robin has been called one of the most adaptable of North American birds as it has such a wide habitat choice: it lives in just about any kind of forest, in parks, gardens, and farmland, showing a similar ability to accept a wide variety of habitat in winter and in summer. The eastern bluebird (*Sialia sialis*) is found in open woodland, orchards, gardens, and farmland, whereas the western bluebird (*S. mexicana*) prefers high-altitude slopes and recently burned areas. Many similar pairs or groups of species are separated by such differences in habitat choice.

An example of a rather narrow habitat requirement is the dense thicket needed by the nightingale (*Luscinia megarhynchos*) in Europe. Conservationists repeatedly cut trees down to the stump to encourage the growth of new, slender poles from ground level; a dense canopy is of no use for the nightingale,

A mountain bluebird (*Sialia currucoides*) fluffs its feathers to keep warm in cold weather. (Photo by R. Curtis/VIREO. Reproduced by permission.)

which requires thick growth right down to the ground, where it feeds and nests. Changes in forest management over the years threatened nightingales until wardens of nature reserves started to replicate out-dated forest practices.

The bluethroat (*L. svecica*) has much closer affinity to the edges of wetlands, where it skulks in reeds and dense willow thickets. The redstart (*Phoenicurus phoenicurus*) is principally a bird of oak woodland with a thick, dense canopy above an open forest floor, but it also enjoys the light of woodland glades and forest edge. An all-year resident in Europe is the black redstart (*P. ochruros*), a bird of cliffs and crags that drops down in winter to rocky shores and the floors of quarries. It has increased and expanded its range by occupying many European towns and villages, finding holes under tiles and in old stonewalls that make an ideal alternative to a hole in a cliff for nesting. It showed a particularly interesting expansion of range into southern England after World War II, when it moved into the vast areas of rubble and crumbling brickwork left by bombing raids on London and elsewhere. With the clearance of such sites from English cities, the black redstart has found it difficult to survive as a British breeding bird, although some still spend the winter in semi-derelict places near the coast where concrete and building rubble replicate a rocky habitat. The blackstart (*Cercomela melanura*) from the Middle East, another species with a narrow habitat, is a desert rock specialist, living in wadis and along the foot of crags with hot, bare rock and nothing more than a few scrubby acacia bushes. In parts of Africa, the mocking chat (*Thamnolaea cinnamomeiventris*) lives on exposed rocky outcrops on otherwise wooded cliffs. It finds that tourist lodges around places such as Lake Nakuru in Kenya are useful alternative sites, with stone-walled rooms with tiled roofs mimicking the rock outcrops quite well.

Behavior

Many thrushes feed on the ground most of the time and keep under cover in forest or scrub. They scratch with their feet and turn over dead leaves and other litter with their bills to get at invertebrate food. Chats include species that do much the same, and others that live in more open places, especially the wheatears, which hop over open grass, stony or gravelly ground, scree slopes, and rocky places in desert or semi-arid areas. They are territorial birds and sing to attract mates and to warn other males that they are present and claiming their territory. Most sing from perches, but some smaller chats also have a song flight. In some species such as the robin, both sexes may sing and hold winter territories that they defend vigorously. This makes springtime courtship even more imperative to break down territorial tendencies between the sexes. Wood thrushes (*Hylocichla mustelina*) wintering in the United States return to the same spot year after year and defend territories with great determination. Experienced birds that can fight to defend a territory do better than young ones that have not yet found themselves a vacant space: these floating individuals, moving from place to place and losing out in territorial fights, are less likely to survive.

The mistle thrush is well known for defending individual bushes with plentiful berries in fall and winter, fending off competition with great expenditure of energy in order to maintain exclusive access to this food supply all winter. Yet flocks of mistle thrushes feed happily together in late summer and early fall when there is such an abundance of fresh berries that defending them is both impossible and unnecessary: All can feed without competition. Other species remain in flocks throughout the winter, and in western Europe mixed groups of fieldfares and redwings, often with blackbirds and song thrushes at least temporarily joining them, live a nomadic life, wandering and pausing wherever there is food. They are often driven far to the west and south by the onset of severe winter weather.

Keeping plumage in good condition is essential to any bird and thrushes and chats preen frequently. This includes scratching of the head and neck by drooping one wing and reaching above it with the foot, called the indirect method, as opposed to reaching up directly in a method with the foot beneath the wing. They bathe regularly in shallow water but, unlike many smaller species such as sparrows and larger ones such as the game birds, they do not dust-bathe. These birds are, however, noted for their frequent sun bathing and they adopt strange positions in order to soak up the sun as much as possible. The sun may help in keeping feathers in good condition, or reducing infestation by parasites. Some species, including the American robin and blackbird, have also been seen anting, which is placing ants into their feathers or sinking down to let ants run over them; it is likely that this also helps reduce parasites.

At night, thrushes seek a safe, warm, dry, sheltered spot to sleep. Many roost alone, but some do so communally, at least outside the breeding season. Blackbirds call loudly at dusk, sorting out their social hierarchies as they seek the best roost sites. They may form roosts hundreds strong; fieldfares (*Turdus pilaris*) have been known to roost in flocks of 20,000, and one mixed-thrush roost in winter in France held 200,000 birds. These roosts are usually in dense thickets, where temperature and humidity are much the same inside as outside. However, exposure and wind speed are greatly reduced. Rock thrushes, in contrast, are solitary, roosting in rock crevices or in high tree branches, occasionally in the roofs of old, secluded buildings, while ring ouzels (*T. torquata*) roost alone among rocks and boulders.

Feeding ecology and diet

Thrushes are mostly generalists: They eat a wide variety of food, concentrating on what is most easily available at the time. This may include earthworms and larvae, beetles or other insects, berries, and fruit. Rock thrushes, however, entirely feed on animals.

The song thrush has become a specialist in finding snails, and smashes them open against a stone (an "anvil") to get at the soft, nutritious fleshy body. Song thrushes in Scotland look for snails on the rocky seashore, which may help the thrushes to survive in dry, hot summers when the ground is too hard for earthworms to be found and easily extracted. The use of anvils is restricted to a handful of thrush species worldwide, including the African thrush (*Turdus pelios*) and Malabar whistling-thrush (*Myophonus horsfieldii*).

American robins feed on the ground or in bushes; in winter they forage in large flocks. They advance across the ground in a series of short, bouncy hops, pausing to look and listen. They tilt their heads from side to side, looking for tell tale signs of a worm moving under the surface, then swing forwards to get at the prey with their bills. This is a typical thrush feeding habit. Blackbirds and the various *Zoothera* thrushes of North America and Asia are great diggers in leaf litter, scratching and tossing away leaves in a noisy performance. The *Zoothera* species has a curious up-and-down bobbing action while feeding. Blackbirds sometimes take tadpoles, and sometimes newts and tiny fish, from garden ponds. Bluebirds, however, take insects from leaves, usually while perched but sometimes in a hover, and also take flying insects with an aerial foray. They typically use a technique of dropping from a perch to the ground to pick up prey.

Some thrushes eat vast amounts of fruit and berries and excrete the seeds whole. These bird droppings are often responsible for the spread of shrubs and even parasitic plants such as mistletoe, whose seeds are deposited in tree bark in bird droppings. Townsend's solitaire helps disperse several species of juniper, mountain ash, and serviceberry.

In Europe, robins follow wild boars and other animals that forage in forests, and hop around the heaps of earth that are thrown up by moles, hoping to catch a worm or two. In exactly the same way, tame British robins have learned to follow gardeners and often literally hop at the feet of people as they turn over the soil of a flowerbed or prepare a vegetable patch. Robins remain territorial in winter and feed solitarily, even at bird tables and feeders, but these territorial barriers may be broken down by prolonged spells of severe weather, which force the robins to forget their differences for a time

Eastern bluebird (*Sialia sialis*) with young. (Photo by Laura Riley. Bruce Coleman Inc. Reproduced by permission.)

and get on with the business of surviving. In spring, robins, as do some other chats, may be seen feeding one another. This is an important part of the courtship behavior, probably with two main functions: It helps break down inhibitions and allows the pair to bond together and accept close contact, and it also helps the female to get sufficient nutritious food to develop a large and healthy clutch.

The flycatcher-thrushes and ant-thrushes (*Neocossyphus*) of South America are four species that perch for long spells on a branch, watching for prey, or hop on the ground ready to pick up prey flushed by columns of ants. The whistling-thrushes (*Myophonus*) forage at the edge of flowing streams. They eat more mollusks and crustaceans than most thrushes. Some island thrushes occasionally eat young birds and even the occasional small rodent.

Reproductive biology

Thrushes are monogamous, but both males and females will take the chance to mate with others if an opportunity arises. Families are, however, reared by both members of the pair. Indeed, it is generally the case that male and female share nest building, incubation, and feeding duties.

Larger thrushes make large nests, in the form of a deep bowl of thin twigs, thick stems, grasses, and straw, variously lined and shaped into a neater central cup. European blackbirds and American robins strengthen the nest with a layer of mud and dung, but line it with a soft layer of fine grass. Song thrushes make do with the mud and the dung, laying their eggs directly onto this strong, hard lining. Mistle thrushes make a larger, less tidy nest, often earlier in the year before leaves are on the trees so the nest is rather conspicuous. These

An American robin (*Turdus migratorius*) feeds its young at its nest. (Photo by T. Fink/VIREO. Reproduced by permission.)

While other species may lay a single large clutch to coincide with a seasonal flush of food, and still others live much longer, lay just one or two eggs, and put all their effort into rearing a single chick, thrushes typically lay several small clutches through the course of a season. The end result is the same, but it is a different strategy: more hit and miss. From each clutch, only one or two chicks might survive; eventually, almost all will die in the winter. Perhaps a dozen eggs will be laid, 10 may hatch, and nine of the resultant chicks will die, through disease, starvation, accident, or predation. So long as one adult survives the winter, the single surviving chick is enough to keep the population stable.

Chicks hatch naked and blind and develop a coat of down before growing their first feathers. Chicks of larger thrushes tend to leave the nest several days before they can fly. They are probably too exposed to predators in the nest, which becomes easy to locate as the large adults constantly fly in and out carrying bills full of food, and the chicks themselves beg for food with loud, far-carrying calls. So, while the chicks might seem extremely vulnerable on the ground and unable to fly, they must have a better chance by splitting up and hiding away in the bushes than by staying in the nest.

Smaller chats nest in holes or make domed nests, some with a more or less developed entrance tunnel, carefully hidden in thick vegetation at or close to ground level. They have larger clutches and rear just one or two broods, putting more effort into each than the big thrushes. They risk being around longer as they feed the chicks in the nest until they are ready to fly, but then the nests are harder to find. Birds such as the stonechat, however, impart a remarkably agitated, worried impression as they perch nearby with a bill full of caterpillars or flies, undecided whether to give the game away or dive in to feed the hungry chicks.

Conservation status

Many thrushes are numerous and secure. Others are the subject of intensive research, such as the song thrush in Britain, where numbers have fallen dramatically in recent decades in some areas. The problem appears to involve the survival of fledged chicks and the ability to produce a second brood. Some habitats, especially intensive farmland, have too few nest sites and too little food for a pair of thrushes to rear enough young to replace themselves when they die. Nightingales in Britain are declining, primarily through habitat loss and neglect, but numbers remain high in much of Europe.

Other species are genuinely super-rarities. The Seychelles magpie-robin was pushed to the brink of extinction as its habitats and food supplies were reduced, and introduced predators took their toll. Coconut plantations replaced natural forest, cats and rats caught adult and young birds, nest sites and big insect food disappeared. Only an intensive campaign has achieved a recovery.

The Usambara robin-chat (*Alethe montana*) is restricted to a tiny area of one Tanzanian forest that is subject to clearance for agriculture and tea plantations. Several island species are at risk from habitat loss compounded by the effects of in-

are, however, large and aggressive birds and usually able to drive away most potential predators, swooping at cats and squirrels and even striking the heads of people if they get too close. Mostly, thrushes are much less demonstrative.

American robins and blackbirds, among others, nest in some unusual places; thrushes have been found nesting between the lights of traffic signals, for example. Blackbirds are likely to build inside a garden shed or garage, finding a shelf or ledge somewhere to accommodate the nest, but risking failure if they become shut out (or, indeed, shut in). Occasionally, blackbirds and European robins have succeeded in rearing young from a nest built inside a vehicle or boat that is in regular use.

Northern wheatears nest in holes in the ground or among piles of rocks or in dry-stone walls. A small heap of tiny pebbles or sand may make a kind of porch at the entrance to a nest hole. Redstarts are particularly likely to use wooden nest boxes, either with a small round entrance hole or an open-fronted design. Eastern bluebirds have been helped by the provision of large numbers of nest boxes that make up for the loss of natural cavities.

troduced cats and rats while others such as the Comoro thrush (*Turdus bewsheri*) that, despite their very restricted range, remain common. The kamao (*Myadestes myadestinus*), common in Hawaii a century ago, and the olomao (*M. lanaiensis*), also of Hawaii, are both probably Extinct, while the omao (*M. obscurus*) occupies less than a third of its past Hawaiian range and is greatly reduced in numbers. The puaiohi (*M. palmeri*) is Critically Endangered. These are a sad reflection of the inability of conservationists and governments to save the several endangered species on these magical islands.

Significance to humans

In much of southern Europe, especially in Spain, Italy, Greece, and many Mediterranean islands, thrushes are still caught and killed in vast numbers. They are offered in restaurants and sold in supermarkets, bottled or as thrush paté. The overall effect of this long-term exploitation is difficult to determine, but, as it has been happening for so long, it is reasonable to assume that it is not threatening the species concerned. The ethical position, however, is a matter for the individual, and many people find the situation highly distasteful.

Most western Europeans prefer to enjoy their thrushes in a different way. Nightingales, with their wonderful song, have a special place in the poetry of some countries. Certainly, however, the presence of a singing song thrush or blackbird in the garden in spring is a great pleasure to tens of thousands of people. In North America, a nesting pair of bluebirds has a similar effect, and thrushes and chats include some of the best-loved garden birds in the world.

1. Wheatear (*Oenanthe oenanthe*); 2. White-browed robin chat (*Cossypha heuglini*); 3. Townsend's solitaire (*Myadestes townsendi*); 4. Eastern bluebird (*Sialia sialis*); 5. White-crowned forktail (*Enicurus leschenaulti*); 6. Anteater chat (*Myrmecocichla aethiops*); 7. Magpie-robin (*Copsychus saularis*); 8. Nightingale (*Luscinia megarhynchos*); 9. Stonechat (*Saxicola torquata*); 10. Rufous bush chat (robin) (*Cercotrichas galactotes*); 11. Blackstart (*Cercomela melanura*). (Illustration by Barbara Duperron)

1. Blackbird (*Turdus merula*); 2. Capped wheatear (*Oenanthe pileata*); 3. Hermit thrush (*Catharus guttatus*); 4. Red-legged thrush (*Turdus plumbeus*); 5. American robin (*Turdus migratorius*); 6. Black redstart (*Phoenicurus ochruros*); 7. Spotted palm thrush (*Cichladusa guttata*); 8. Olive thrush (*Turdus olivaceus*); 9. Rock thrush (*Monticola saxatilis*); 10. Siberian ruby throat (*Erithacus calliope*); 11. White's thrush (*Zoothera dauma*). (Illustration by Barbara Duperron)

Species accounts

Nightingale
Luscinia megarhynchos

TAXONOMY
Luscinia megarhynchos C. L. Brehm, 1831.

OTHER COMMON NAMES
English: Common nightingale; French: Rossignol philomèle;
German: Nachtigall; Spanish: Ruiseñor Común.

PHYSICAL CHARACTERISTICS
6.5 in (16.5 cm); male 0.6–0.8 oz (17–23 g); female 0.6–0.85 oz
(17–24 g). Brown upperparts, gray-buff underparts (throat
paler), rusty-red rump and tail.

DISTRIBUTION
Breeds southeast England eastwards through central and south-
ern Europe, into central Asia; locally North Africa. Winters in
Africa south of Sahara.

HABITAT
Low, dense thickets, woodland, bushes beside heaths. In winter, bushy, dry savanna.

BEHAVIOR
Skulking, feeds on or near ground, sings from hidden perch,
sometimes more open on bush or tree; territorial, solitary.

FEEDING ECOLOGY AND DIET
Eats beetles, ants, other invertebrates, some berries in summer;
insectivorous in winter.

REPRODUCTIVE BIOLOGY
Monogamous; nest on or near ground. Lays four to five eggs
April–June, incubation 13 days, fledging 11 days; one or two
broods.

CONSERVATION STATUS
Not threatened, though declining in north and west of range,
secure in south and east.

SIGNIFICANCE TO HUMANS
Exceptional song greatly revered but actually less well known
than may be suspected; frequent allusions in literature, poetry
and music. ◆

Magpie-robin
Copsychus saularis

TAXONOMY
Copsychus saularis Linnaeus, 1758, Bengal.

OTHER COMMON NAMES
English: Asian magpie-robin, Oriental magpie-robin; French:
Shama dayal; German: Dajal; Spanish: Robín la Gazza.

PHYSICAL CHARACTERISTICS
9 in (23 cm); male 1.1–1.5 oz (31–42 g); female 1.1–1.4 oz
(32–40 g). In males, upperparts, head, and breast are black; underparts are white; the tail is black with white outer feathers;
the wings have white bars. In females, upperparts, head and
breast are dull dark gray. Juveniles resemble adults but have
mottled brown breasts.

DISTRIBUTION
Pakistan, India, Bangladesh, Thailand, and Indochina, Andaman
Islands, Sri Lanka, Malaysia, Indonesia, and the Philippines.

HABITAT
Gardens, woodland edge, and forest clearings, open
broadleaved forest.

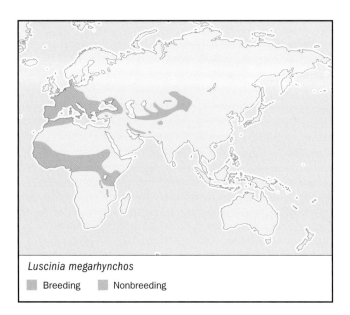

Luscinia megarhynchos
▮ Breeding ▮ Nonbreeding

Copsychus saularis
▮ Resident

BEHAVIOR
Pairs or family groups, lively and easily visible, feeds on ground, perches on branches, high wires, poles.

FEEDING ECOLOGY AND DIET
Large insects taken from ground, spiders, centipedes, earthworms, small lizards, seeds, and nectar.

REPRODUCTIVE BIOLOGY
Breeds February–August, untidy grassy nest in hole in tree, bank, or wall; four to five eggs, incubation 12–13 days.

CONSERVATION STATUS
Not threatened.

SIGNIFICANCE TO HUMANS
None known. ◆

White-browed robin chat
Cossypha heuglini

TAXONOMY
Cossypha heuglini Hartlaub, 1866, Sudan.

OTHER COMMON NAMES
English: Heuglin's robin; French: Cossyphe de Heuglin; German: Weissbrauenrötel.

PHYSICAL CHARACTERISTICS
7.9 in (20 cm); male 1.1–1.6 oz (30–44 g); female 1.0–1.3 oz (29–36 g). Brownish upperparts with black head and white brow from bill to nape. Orange throat, neck collar, and underparts.

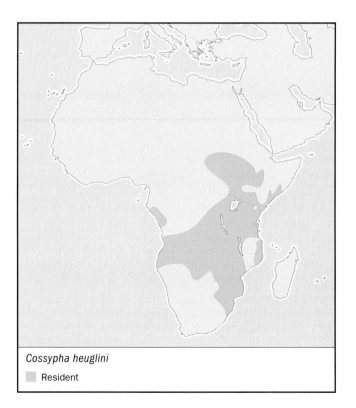

Cossypha heuglini
■ Resident

DISTRIBUTION
Africa south of Sahara, chiefly southerb Sudan through East Africa, Central Africa west to Angola and south to northeastern South Africa and Mozambique.

HABITAT
Riverside forest and evergreen thickets, with dense understory, garden shrubberies.

BEHAVIOR
Solitary or in pairs, feeds mostly on ground but sings from high perch.

FEEDING ECOLOGY AND DIET
Insects, especially beetles and ants, also caterpillars, moths, grasshoppers, and other small invertebrates.

REPRODUCTIVE BIOLOGY
Monogamous and territorial; strong pair bond throughout year. Nest in cavity in tree or stump. Two to three eggs, incubation by female for 12–17 days.

CONSERVATION STATUS
Not threatened. Common and secure in core range, rare in some peripheral regions.

SIGNIFICANCE TO HUMANS
None known. ◆

Rufous bush chat
Cercotrichas galactotes

TAXONOMY
Cercotrichas galactotes Temminck, 1820, Algeciras.
other common names
English: Rufous-tailed scrub-robin; French: Agrobate roux; German: Heckensänger; Spanish: Alzacola Español.

PHYSICAL CHARACTERISTICS
5.9 in (15 cm); 0.7–0.9 oz (21–25 g). Rufous-brown upperparts with reddish tail and buff underparts. Wings are streaked brown and buff. Rufous head with white brow and cheek stripes.

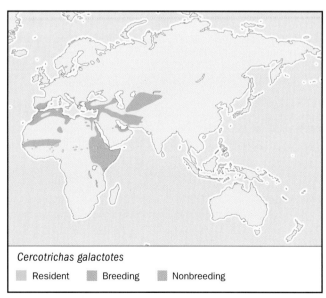

Cercotrichas galactotes
■ Resident ■ Breeding ■ Nonbreeding

DISTRIBUTION
Southwest Europe including Spain, Portugal, southeast Europe, southwest Asia, and Arabia, North Africa from Morocco to Egypt, West Africa south of Sahara, locally Sudan, Ethiopia, Somalia.

HABITAT
Cactus thickets, dry scrub in gullies, warm, bushy slopes, orchards, gardens.

BEHAVIOR
Solitary or in pairs; mostly on ground, secretive, but also demonstrative to intruders and potential predators, flirting boldly-marked tail in display, threat, or distraction activities.

FEEDING ECOLOGY AND DIET
Caterpillars, beetles, grasshoppers, earwigs, and other invertebrates, a few seeds, and fruits.

REPRODUCTIVE BIOLOGY
Monogamous and territorial; males sing to proclaim territory; nest loose and untidy, in palm thicket or thorn bush; three to five eggs incubated by female for 13 days, young fledge in 12–13 days.

CONSERVATION STATUS
Not threatened. Scarce in much of European range, but numbers stable.

SIGNIFICANCE TO HUMANS
None known. ◆

Eastern bluebird
Sialia sialis

TAXONOMY
Sialia sialis Linnaeus, 1758.

OTHER COMMON NAMES
French: Merlebleu de l'est; German: Rotkehl-Hüttensánger; Spanish: Azulejo.

PHYSICAL CHARACTERISTICS
7.1 in (18 cm); 1.1 oz (31 g). Males have bright blue upperparts; reddish brown chin, throat, breast, sides, and flanks; and a white belly and undertail coverts. Females have gray upperparts; blue wings, rump, and tail; and paler chestnut where the male is reddish brown. Juveniles have gray-brown upperparts with white spotting on the back, a brownish chest with white scalloping, bluish tail and wings, and white belly and undertail coverts.

DISTRIBUTION
Eastern North America, north to Hudson Bay, west to Arizona, south to Bermuda, Florida, and Mexico.

HABITAT
Open woodland, farmland with scattered trees, orchards, gardens with trees and shrubberies.

BEHAVIOR
In pairs or family groups, perching upright on exposed branch or post or tree top. Gregarious in winter, often forming large flocks and roosting communally.

FEEDING ECOLOGY AND DIET
Feeds largely on small insects from ground, foliage, or caught in the air; also eats fruits and berries.

Sialia sialis

■ Resident ■ Breeding ■ Nonbreeding

REPRODUCTIVE BIOLOGY
Nests in tree hole or hollow branch, increasingly in artificial nest boxes; nest constructed of grass, weeds, pine needles, and twigs by the female; three to seven eggs; incubation 12–14 days; chicks fledge after 15–19 days; two or three broods.

CONSERVATION STATUS
Not threatened. Decreased by up to 90% in twentieth century after competition for nest holes from introduced house sparrows and starlings; increased locally after nest box provision became popular.

SIGNIFICANCE TO HUMANS
None known. ◆

White-crowned forktail
Enicurus leschenaulti

TAXONOMY
Enicurus leschenaulti Hartert, 1909, Assam.

OTHER COMMON NAMES
English: Leschenault's forktail; French: Énicure de Leschenault; German: Weissscheitel-Scherenschwanz; Spanish: Enicurino de Corona Blanca.

PHYSICAL CHARACTERISTICS
11 in (28 cm); 1.8–1.9 oz (50–55 g). Black mantle and breast with white from bill to nape. Wings are dark with white bar; tail is banded black and white. Rump and abdomen white, legs pinkish, and bill black.

DISTRIBUTION
Northeast India, Bangladesh, Myanmar, Thailand, Indochina.

Enicurus leschenaulti
◻ Resident

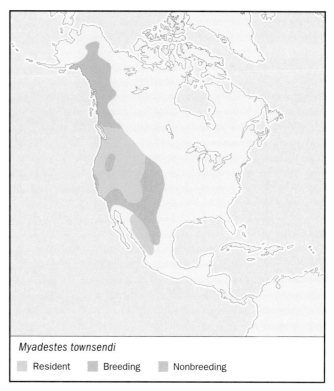

Myadestes townsendi
◻ Resident   ◼ Breeding   ◼ Nonbreeding

HABITAT
Fast-flowing streams and rivers, sometimes tiny rivulets in dense evergreen forest.

BEHAVIOR
Perches on rocks and stumps in or beside streams and waterfalls, sometimes on forest tracks, constantly bobbing long tail up and down and flicking it open; sometimes pursues insects in hovering flight, also submerges in water and walks on river bottom.

FEEDING ECOLOGY AND DIET
Aquatic insects and their larvae.

REPRODUCTIVE BIOLOGY
Monogamous; breeds April–June; nest of moss, leaves and rootlets, in overgrown bank, rocks or boulders; three or four eggs.

CONSERVATION STATUS
Not threatened. Fairly common and secure.

SIGNIFICANCE TO HUMANS
None known. ◆

Townsend's solitaire
Myadestes townsendi

TAXONOMY
Ptiliogonys townsendi Audubon, 1838, Oregon.

OTHER COMMON NAMES
French: Solitaire de Townsend; German: Bergklarino; Spanish: Clarín Norteño.

PHYSICAL CHARACTERISTICS
7.9–8.7 in (20–22 cm); 1.1–1.2 oz (30–35 g). Adults are gray overall; black tails with white outer feathers that show during

flight; buff wing patches near the base of blackish flight feathers; white eye rings. Juveniles are brownish gray overall marked with buff and white scalloping on upperparts and underparts.

DISTRIBUTION
North America from Alaska south to Mexico, east to southeast Wyoming, central Arizona, western South Dakota, and Montana.

HABITAT
Open stands of conifers, edges of extensive conifer forest, often near streams and with little or no undergrowth; also high mountain slopes, cliffs, and ravines up to and above the tree line.

BEHAVIOR
Often conspicuous; solitary or in loose parties, typically perched upright on exposed perch on tree or post, flying up to catch prey or dropping to the ground to forage.

FEEDING ECOLOGY AND DIET
Insects, including moths, beetles, caterpillars, ants, bees, and wasps; also eats a wide range of berries.

REPRODUCTIVE BIOLOGY
Breeds May–July, nest of bark, grass, and roots built by female in bush or shrub or on the ground, often near a stream; three to four eggs incubated for 11–14 days, chicks fledge after 10–12 days; two broods.

CONSERVATION STATUS
Not threatened, though uncommon in much of range.

SIGNIFICANCE TO HUMANS
None known. ◆

Stonechat
Saxicola torquata

TAXONOMY

Saxicola torquata Linnaeus, 1766.

OTHER COMMON NAMES

English: Common stonechat; French: Traquet pâtre; German: Schwartzkehlchen; Spanish: Tarabilla Común.

PHYSICAL CHARACTERISTICS

4.9 in (12.5 cm); 0.46–0.60 oz (13–17 g). Males have black heads, orange breasts, and large white patches on the sides of the neck. Females and juveniles have a similar plumage pattern, but have brown (rather than black) heads and less ponounced orange and white areas.

DISTRIBUTION

Britain and Ireland, Europe from Denmark south to Iberia and east to Black Sea, Middle East, locally Arabia; Asia east to Japan, south to China; scattered through Africa south to the Cape.

HABITAT

Heath and rough grassland with thorny scrub, young plantations, forest clearings with bushy undergrowth, open coastal strip above rocky shore and cliffs.

BEHAVIOR

In pairs or family groups, perching on open bush tops or tall stems, overhead wires, giving frequent harsh, scolding calls.

FEEDING ECOLOGY AND DIET

Insects and other small invertebrates.

REPRODUCTIVE BIOLOGY

Monogamous and territorial; nest on or close to ground in dense vegetation, well hidden, sheltered from sun, loosely woven from grass stems, with entrance tunnel; four to six eggs incubated for 13–14 days by female; fledging period 13 days.

CONSERVATION STATUS

Not threatened.

SIGNIFICANCE TO HUMANS

None known. ◆

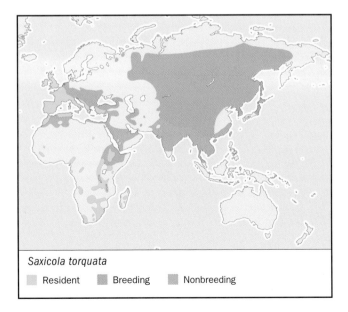

Saxicola torquata

▢ Resident   ▣ Breeding   ▣ Nonbreeding

Wheatear
Oenanthe oenanthe

TAXONOMY

Oenanthe oenanthe Linnaeus, 1758.

OTHER COMMON NAMES

English: Northern wheatear; French: Traquet motteaux; German: Steinschmätzer; Spanish: Collalba Gris.

PHYSICAL CHARACTERISTICS

5.5–5.9 in (14–15 cm); male 0.6–1.0 oz (18–28 g); female 0.7–1.0 oz (19–28 g). Males have dark gray upperparts, white underparts, and black wings. Their tails have a white base and a black tip. Their throats, breasts, and flanks are covered in a buff wash. Females have brown upperparts and darker brown wings. A rusty wash covers their throats and breasts.

DISTRIBUTION

Alaska, northern coast of Canada, Baffin Island, coasts of Greenland, Iceland, and wide band from northwest Europe south to Iberia and eastwards across Asia; locally Morocco, Tunisia.

HABITAT

Open grassy ground near rocks, stone walls or crags, sandy heaths, coastal grassland with outcrops of rock, boulders, or scree.

BEHAVIOR

Singly or in pairs, terrestrial, running or hopping over open spaces or perching on small eminences; male sings from perch or in song flight.

FEEDING ECOLOGY AND DIET

Insects and other small invertebrates; a few berries.

REPRODUCTIVE BIOLOGY

Monogamous and territorial; nest in cavity in rocks or wall or in hole in ground, loosely made of grass and stems; four to seven eggs incubated only by female for 13 days; young fledge after 15 days.

CONSERVATION STATUS

Not threatened.

SIGNIFICANCE TO HUMANS

None known. ◆

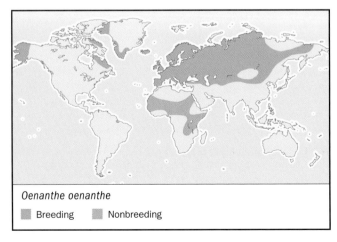

Oenanthe oenanthe

▣ Breeding   ▣ Nonbreeding

Capped wheatear
Oenanthe pileata

TAXONOMY
Oenanthe pileata Gmelin, 1789, Cape of Good Hope.

OTHER COMMON NAMES
French: Traquet du Cap; German: Erdschmätzer; Spanish: Collalba Encapotada.

PHYSICAL CHARACTERISTICS
7.1 in (18 cm); 1.1 oz (32.5 g). Brown upperparts with black tail and rufous-tinged flight feathers. Black crown, bill, cheek, and chest collar. White brow stripe, throat, and breast with rosy-buff belly.

DISTRIBUTION
Africa from Kenya and Angola southwards to the Cape.

HABITAT
Dry grassy plains, especially overgrazed or burnt areas with a few bushes or termite mounds.

BEHAVIOR
Solitary but often common, scattered over open ground, perching on small mounds or hopping over short grass; wags tail and flicks wings, bobs on landing after low, fast flight.

FEEDING ECOLOGY AND DIET
Insects, especially ants, also flies, beetles, locusts, termites, and caterpillars.

REPRODUCTIVE BIOLOGY
Monogamous, territorial, nesting in hole in ground or termite mound; nest of straw, grass, and leaves; three to four or six eggs, incubation not established.

CONSERVATION STATUS
Not threatened.

SIGNIFICANCE TO HUMANS
None known. ◆

Blackstart
Cercomela melanura

TAGFXONOMY
Cercomela melanura Temminck, 1824, Arabia.

OTHER COMMON NAMES
French: Traquet à queue noir; German: Schwarzschwanz; Spanish: Colinegro Real.

PHYSICAL CHARACTERISTICS
5.5 in (14 cm); 0.49–0.53 oz (14–15 g). Gray-brown upperparts with black tail. White eye ring. Buff cheek, underparts, and wing edges.

DISTRIBUTION
Locally in Africa along southern edge of Sahara, Red Sea coast, Israel, Jordan, Arabia.

HABITAT
Rocky, hot hills with or without acacia scrub, thorny bushes in desert ravines and wadis, scree slopes, dry river beds.

BEHAVIOR
In pairs; tame and fearless; flits between rocks and bushes, often perching deep inside bush, or drops to ground and hops about; flicks and flirts black tail, half-opens wings.

Oenanthe pileata
▨ Resident

Cercomela melanura
▨ Resident

FEEDING ECOLOGY AND DIET
Insects and their larvae; a few berries.

REPRODUCTIVE BIOLOGY
Monogamous, territorial, solitary breeder, nesting in cavity in rocks, scree, or wall, under eaves of buildings; three to four eggs.

CONSERVATION STATUS
Not threatened.

SIGNIFICANCE TO HUMANS
None known. ◆

Anteater chat
Myrmecocichla aethiops

TAXONOMY
Myrmecocichla aethiops Cabanis, 1850, Senegal.

OTHER COMMON NAMES
English: Northern anteater chat; French: Traquet brun; German: Ameisenschmätzer; Spanish: Hormiguero.

PHYSICAL CHARACTERISTICS
7.1 in (18 cm); male 1.8–2.3 oz (51–66 g); female 1.7–2.0 oz (47–58 g). Dark, sooty-brown plumage with black bill and legs. White wing patches are conspicous during flight.

DISTRIBUTION
Narrow band across Africa south of Sahara from Senegal to Sudan, locally southern Kenya and extreme northern Tanzania.

HABITAT
Open grassy ground with termite mounds and scattered bushes.

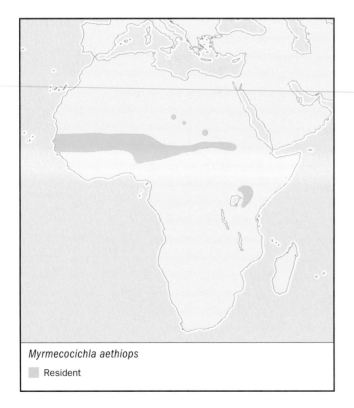

Myrmecocichla aethiops
▨ Resident

BEHAVIOR
Usually in pairs or small groups, often 5–15 scattered over a small area, perched on bushes, mounds of earth or termite mounds.

FEEDING ECOLOGY AND DIET
Insects, especially moths, and termites, beetles, spiders, and some fruits.

REPRODUCTIVE BIOLOGY
Mostly monogamous but cooperative groups assist at some nests; pairs remain together for several years. Nests in tunnel up to 5 ft (1.5 m) long, dug by both sexes in the side of an earth bank, termite mound, or within an animal burrow; two to five eggs, incubated only by female for 14–16 days; young fledge after 21–23 days.

CONSERVATION STATUS
Not threatened.

SIGNIFICANCE TO HUMANS
None known. ◆

Black redstart
Phoenicurus ochruros

TAXONOMY
Phoenicurus ochruros Gmelin, 1774, Iran.

OTHER COMMON NAMES
French: Rougequeue nior; German: Hausrotschwanz; Spanish: Colirrojo Tizón.

PHYSICAL CHARACTERISTICS
5.5 in (14 cm); 0.5–0.7 oz (13–20 g). Black head and upperparts with dark gray forehead and reddish tail. Throat is black; underparts are rufous. In females, plumage is brown.

DISTRIBUTION
Widespread in Europe north to Baltic Sea, Middle East, Morocco, Algeria, Tunisia, coast of Libya, Egypt, and Red Sea coasts, Iran, and Afghanistan.

Phoenicurus ochruros
▨ Resident   ▨ Breeding   ▨ Nonbreeding

HABITAT
Cliffs and rocky outcrops from sea level to high altitude in Alps, Pyrenees, Caucasus, and Atlas Mountains, also industrial sites, city buildings, towns, and villages with old buildings, tiled roofs, chimneys; sometimes woodland glades with rocky slopes.

BEHAVIOR
Singly or in pairs, perching on high rock or rooftop, frequently flirting reddish tail; mostly terrestrial.

FEEDING ECOLOGY AND DIET
Beetles, ants, grasshoppers, and other insects, many spiders, also earthworms, millipedes, and some fruit.

REPRODUCTIVE BIOLOGY
Monogamous in most cases, some males with two females; territorial; nest a loose cup of grass, on ledge in roof or outbuilding, in cavity in rocks or hole in wall or pipe; four to six eggs incubated only by female for 12–13 days; young leave nest after 12–17 days, sometimes before they can fly.

CONSERVATION STATUS
Not threatened.

SIGNIFICANCE TO HUMANS
None known. ◆

Siberian rubythroat
Erithacus calliope

TAXONOMY
Erithacus calliope Pallas, 1776.

OTHER COMMON NAMES
French: Calliope Sibérienne; German: Rubinkehlchen; Spanish: Ruiseñor Caliope.

PHYSICAL CHARACTERISTICS
5.5 in (14 cm); male 0.74–1.0 oz (21–29 g); female 0.56–0.78 oz (16–22 g). Brownish upperparts with white eyebrow and malar stripe. Throat is red (male) or white (female). Males have light brown beast fading to buff, while females have whitish buff underparts with a brown breast band.

DISTRIBUTION
Northern and central Siberia from Urals to Kamchatka.

HABITAT
Lowland forest, also up to tree line or above in subalpine scrub; usually in thickets, boggy clearings, riverine glades, or meadows.

BEHAVIOR
Solitary or in small, loose parties, usually on or close to the ground in thick undergrowth, but runs rapidly over open spaces.

FEEDING ECOLOGY AND DIET
Mostly beetles and other insects, but also various aquatic invertebrates from riversides and shorelines.

REPRODUCTIVE BIOLOGY
Apparently monogamous and territorial; nests from June–July, in thick bush or tussock near ground; four to six eggs, incubated by female, period undetermined.

CONSERVATION STATUS
Not threatened.

SIGNIFICANCE TO HUMANS
None known. ◆

Spotted palm-thrush
Cichladusa guttata

TAXONOMY
Cichladusa guttata Heuglin, 1862, White Nile.

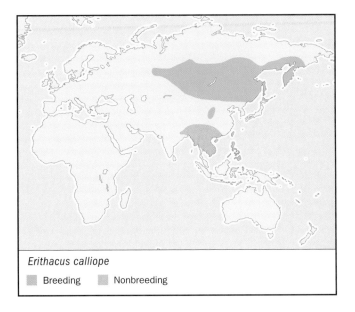

Erithacus calliope

▨ Breeding ▨ Nonbreeding

Cichladusa guttata

▨ Resident

OTHER COMMON NAMES
English: Collared palm-thrush, morning warbler; French: Cichladuse à collier; German: Morgenrötel; Spanish.

PHYSICAL CHARACTERISTICS
6.7 in (17 cm); male 0.6–1.1 oz (17–30 g); female 0.6–1.1 oz (16–30 g). Rufous upperparts and tail. Buff brow and underparts with black spotting on breast, fading to mottled rufous on flanks and belly.

DISTRIBUTION
East Africa, from southern Sudan and Ethiopia to northern Tanzania.

HABITAT
Thick scrub and dense riverside thickets, bushes in dry savanna, mostly at low altitude.

BEHAVIOR
In pairs or family groups, shy, keeping well hidden, but sometimes more visible in gardens and hotel grounds; mostly terrestrial, flicks wings and slowly waves tail as it hops about.

FEEDING ECOLOGY AND DIET
Insects, snails, and other small invertebrates.

REPRODUCTIVE BIOLOGY
Monogamous, territorial; nest of mud, bound with a few strands of grass and lined with roots, on tree branch; two to three eggs incubated mostly by female for 12 days; both adults feed young.

CONSERVATION STATUS
Not threatened.

SIGNIFICANCE TO HUMANS
None known. ◆

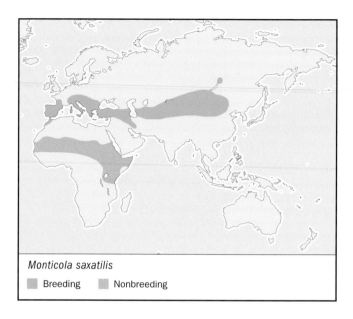

Monticola saxatilis
▮ Breeding   ▮ Nonbreeding

FEEDING ECOLOGY AND DIET
Large insects, also centipedes, millipedes, spiders, small lizards, snails, and some fruits.

REPRODUCTIVE BIOLOGY
Monogamous, territorial; breeds from April onwards; nest in crevice in rock face or wall, under boulder; four to five eggs incubated only by female for 14–15 days; young fly after 14–16 days.

CONSERVATION STATUS
Not threatened.

SIGNIFICANCE TO HUMANS
None known. ◆

Rock thrush

Monticola saxatilis

TAXONOMY
Monticola saxatilis Linnaeus, 1776.

OTHER COMMON NAMES
English: Rufous-tailed rock thrush, European rock-thrush; French: Merle de roche; German: Steinrötel; Spanish: Roquero Rojo.

PHYSICAL CHARACTERISTICS
7.3 in (18.5 cm); male 1.4–2.3 oz (40–65 g); female 1.5–2.3 oz (42–65 g). Gray head, upperparts, and throat with dull orange breast to undertail. Grayish bill and legs.

DISTRIBUTION
Locally from Iberia and Morocco, eastwards through southern Europe, Turkey, and Iran to central Asia; winters in Africa.

HABITAT
Sunny, dry, stony, or rocky slopes, upland meadows, and pastures with scattered bushes, barren stony hillsides; winters in wooded savanna.

BEHAVIOR
Pairs or family groups, mostly terrestrial or perching on low trees or bushes; hops over ground, stands upright like a wheatear, wags tail; mostly shy and solitary.

White's thrush

Zoothera dauma

TAXONOMY
Turdus dauma Latham, 1790.

OTHER COMMON NAMES
English: Scaly thrush; French: Grive dorée; German: Erddrossel; Spanish: Zorzal Dorado.

PHYSICAL CHARACTERISTICS
10.6 in (27 cm); male 3.5–6.9 oz (100–195 g); female 3.5–6.3 oz (100–180 g). Distinctive black scales above and below. Upperparts are olive-brown; underparts are white with buffy breast.

DISTRIBUTION
Central and eastern Siberia, southwest India, northern India east to China, Japan, Taiwan, Ryukyu Islands, Solomon Islands.

HABITAT
Coniferous forest, often in river valleys or close to water, wooded hillsides with mossy rocks, dense undergrowth with deep leaf litter.

BEHAVIOR
Mostly terrestrial, in deep cover, but flies up into trees when disturbed; shy and reclusive.

Zoothera dauma

■ Breeding ■ Nonbreeding

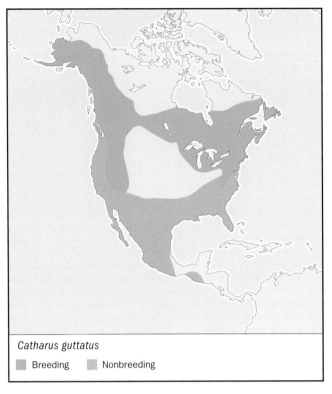

Catharus guttatus

■ Breeding ■ Nonbreeding

FEEDING ECOLOGY AND DIET
Insects, worms, and berries.

REPRODUCTIVE BIOLOGY
Monogamous, territorial, and solitary, nesting in June, in tree fork; four to five eggs incubated by female; little information available.

CONSERVATION STATUS
Not threatened. Presumed stable, but range and numbers poorly known.

SIGNIFICANCE TO HUMANS
None known. ◆

Hermit thrush
Catharus guttatus

TAXONOMY
Muscicapa guttata Pallas, 1811, Alaska.

OTHER COMMON NAMES
French: Grive solitaire; German: Einsiedlerdrossel: Spanish: Zorzal de hermit.

PHYSICAL CHARACTERISTICS
6.3–7.1 in (16–18 cm); male 1.0–1.3 oz (27–37 g); female 1.0–1.1 oz (27–32 g). Rich brown to grayish brown upperparts; reddish tail; whitish underparts with buff-washed breast and gray- or brownish-washed flanks; dark spots on breast and sides of throat. There are size and color variations across the wide breeding range of this species.

DISTRIBUTION
North America, breeding from Alaska to Newfoundland across Canada and south to California, New Mexico; Long Island; winters in southern United States and Central America.

HABITAT
Coniferous and mixed woodlands and thickets, forest bogs and clearings, also very dry areas but prefers neighborhood of water.

BEHAVIOR
More secretive than shy, usually solitary, terrestrial or flitting through low vegetation, hopping about on open grass or in deep cover and flying into higher canopy if disturbed; flicks wings and tail and quickly raises and slowly lowers tail on landing.

FEEDING ECOLOGY AND DIET
Worms, insects, and fruits.

REPRODUCTIVE BIOLOGY
Breeds May–August, nest of twigs, bark, grass, and roots in tree; three to four eggs incubated only by female for 11–13 days, chicks fly after 10–15 days; two broods.

CONSERVATION STATUS
Not threatened.

SIGNIFICANCE TO HUMANS
None known. ◆

Olive thrush
Turdus olivaceus

TAXONOMY
Turdus olivaceus Linnaeus, 1766, Cape of Good Hope.

OTHER COMMON NAMES
English: African thrush, West African thrush; French: Grive olivâtre; German: Kapdrossel; Spanish: Zorzal olivo.

PHYSICAL CHARACTERISTICS
8.3–9.4 in (21–24 cm); 1.9–2.9 oz (54–81 g). Dull olive-brown upperparts and tail, with orange underparts and white vent. Throat is speckled white. Bill and legs are yellow-orange.

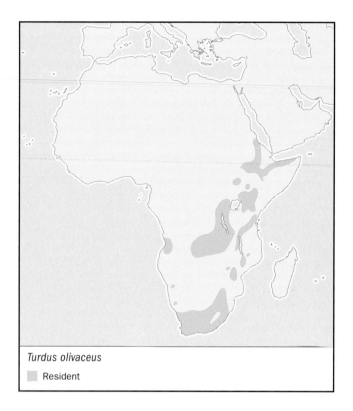

Turdus olivaceus

Resident

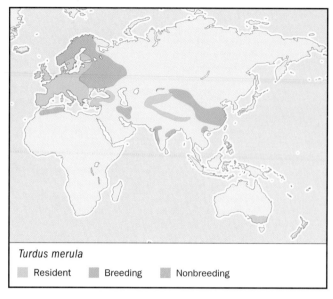

Turdus merula

Resident   Breeding   Nonbreeding

DISTRIBUTION
Africa, from Eritrea and Ethiopia discontinuously south to the
Cape, west to Angola.

HABITAT
Upland and lowland forest, gardens, and hotel grounds.

BEHAVIOR
Usually solitary, in trees or on ground beneath, foraging with
steady, hopping or walking action, often close to buildings in
parks and ornamental grounds.

FEEDING ECOLOGY AND DIET
Spiders, ants, termites, grasshoppers, millipedes, and other
small invertebrates, various household scraps, and many fruits
and berries.

REPRODUCTIVE BIOLOGY
Nests almost throughout the year in some parts of its range;
nest is large, untidy cup of leaves, grass, bark and roots, lined
with mud, in tree fork, built by female; two to three eggs incubated mostly by female for 14–15 days; young fledge after
16 days.

CONSERVATION STATUS
Not threatened.

SIGNIFICANCE TO HUMANS
None known. ◆

Blackbird
Turdus merula

TAXONOMY
Turdus merula Linnaeus, 1758.

OTHER COMMON NAMES
English: Eurasian blackbird, common blackbird; French: Merle
noir; German: Amsel; Spanish: Mirlo Comúun.

PHYSICAL CHARACTERISTICS
9.4–11.4 in (24–29 cm); male 2.1–5.3 oz (60–149 g); female
3.0–3.7 oz (85–106 g). Males have black plumage and a yellow
bill; females have brown plumage and a dark bill.

DISTRIBUTION
Europe from Iceland eastwards.

HABITAT
Mainly damp forest and woodlands, from tundra to golf
courses, gardens, parks, and town shrubberies, farmland with
hedges, and scattered woods.

BEHAVIOR
Bold and tame, feeding on ground where walks, hops, or runs;
large roosts after breeding season. Flocks in winter.

FEEDING ECOLOGY AND DIET
Fruits, berries, grass seeds, many invertebrates including beetles, caterpillars, grasshoppers, snails, spiders, and earthworms.

REPRODUCTIVE BIOLOGY
Breeds April–August, nest large and untidy, of grass, twigs,
stems, and string, lined with mud and fine grass. Three to
four eggs, incubation 11–14 days, fledging 15–16 days. Two
broods.

CONSERVATION STATUS
Not threatened.

SIGNIFICANCE TO HUMANS
None known. ◆

American robin
Turdus migratorius

TAXONOMY
Turdus migratorius Linnaeus, 1766, America.

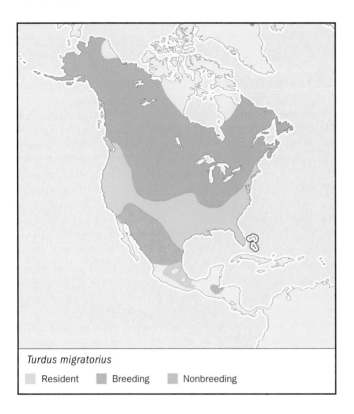

Turdus migratorius

☐ Resident ☐ Breeding ☐ Nonbreeding

OTHER COMMON NAMES
French: Merle d'Amérique; German: Wanderdrossel; Spanish: Robín Americano.

PHYSICAL CHARACTERISTICS
9.8–11.0 in (25–28 cm); male 2.1–3.2 oz (59–91 g); female 2.5–3.3 oz (72–94 g). Dark brownish gray upperparts; black head (brownish gray in female) with broken white eye ring and yellow bill; brick-red breast (chestnut-orange in female); white lower belly and undertail coverts; dark tail with white outer corners. Juveniles are similar to adults but have white markings on the back and shoulders, and heavy spotting on the underparts.

DISTRIBUTION
Throughout Canada, Alaska, United States, Mexico; winters in south of breeding range, Bahamas, Guatemala.

HABITAT
Mainly damp forest and woodlands, from tundra to golf courses, gardens, parks, and town shrubberies, farmland with hedges, and scattered woods.

BEHAVIOR
Bold and tame, feeding on ground where walks, hops, or runs; large roosts after breeding season. Flocks in winter.

FEEDING ECOLOGY AND DIET
Fruits, berries, grass seeds, many invertebrates, including beetles, caterpillars, grasshoppers, snails, spiders, and earthworms.

REPRODUCTIVE BIOLOGY
Breeds April–August, nest, large and untidy, of grass, twigs, stems, and string, lined with mud and fine grass. Three to four eggs, incubation 11–14 days, fledging 15–16 days. Two broods.

CONSERVATION STATUS
Not threatened.

SIGNIFICANCE TO HUMANS
None known. ◆

Red-legged thrush
Turdus plumbeus

TAXONOMY
Turdus plumbeus Linnaeus, 1758, Bahamas.

OTHER COMMON NAMES
French: Merle vantard; German: Rotfussdrossel; Spanish: Zorzal cubano.

PHYSICAL CHARACTERISTICS
10.2–10.6 in (26–27 cm); 1.8– 2.9 oz (50–82 g). Gray upperparts, reddish legs and bill, red eye ring, white chin, and large white tail tips. Belly color varies between races from white to rufous; throat can be dark to spotted white.

DISTRIBUTION
Bahamas, Cuba, Hispaniola, Puerto Rico, Dominica.

HABITAT
Woodlands, mangrove, and scrub, coffee plantations, cactus, and thickets in large gardens.

BEHAVIOR
Shy, solitary or in pairs, forages low in vegetation or on ground, noisy at roost.

FEEDING ECOLOGY AND DIET
Insects in summer, fruit in winter, including royal palm fruit and various berries.

REPRODUCTIVE BIOLOGY
Breeds March–November; three to five eggs in bulky nest of stems, grass, paper, in tree fork or crown of tall palm.

CONSERVATION STATUS
Not threatened.

SIGNIFICANCE TO HUMANS
None known. ◆

Turdus plumbeus

☐ Resident

Resources

Books

Campbell, B., and E. Lack. *A Dictionary of Birds.*: Harrell Books, 1985.

Clement, P., and R. Hathway. *Thrushes.* London: Christopher Helm, 2000.

Cramp, S. *The Birds of the Western Palearctic.* Oxford: Oxford University Press, 1988.

Keith, S., K. Urban, and C. H. Fry. *The Birds of Africa.* Vol. 4. London: Academic Press, 1992.

Simms, E. *British Thrushes.* London: Collins, 1978.

Robert Arthur Hume, BA

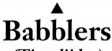

Babblers
(Timaliidae)

Class Aves

Order Passeriformes

Suborder Passeri (Oscines)

Family Timaliidae

Thumbnail description
Small to medium-sized birds of highly variable shapes and colors

Size
3–16 in (7.6–40 cm); 0.35–5.25 oz (10–150 g)

Number of genera, species
Roughly 50 genera, around 280 species

Habitat
Generally forests, but some semi-desert, scrub, grassland, and wetlands

Conservation status
Endangered: 5 species; Vulnerable: 22 species; Near Threatened: 39 species; Data Deficient: 1 species

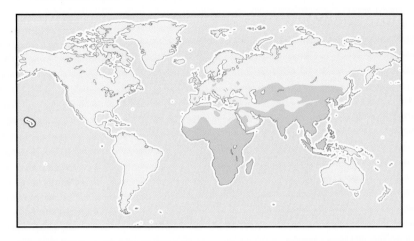

Distribution
Eurasia (excluding Japan), Wallacia, Africa, Madagascar, the Pacific Coast of North America (Introduced in Hawaii)

Evolution and systematics

Nearly a century ago, Ernst Hartert, Curator of the incredible Rothschild collection of preserved birds, observed: "What can't be classified is regarded as a babbling thrush." In the opinions of many of today's ornithologists, Hartert then proceeded to confuse matters further. The convoluted history of babbler classification is well treated in Sibley and Ahlquist's *Phylogeny and Classification of Birds*, and the reader is best referred to that book. Suffice it to say that exactly what a babbler is, and how it should be classified, has been, and remains, a controversy among ornithologists.

Jean Delacour's masterful arrangement of babblers, proposed in the 1940s and 1950s, has worked remarkably well, even as 1980s experiments in DNA-hybridization completely reorganized bird classification, and the rationale behind it. Delacour considered the babblers a vast subfamily, Timaliinae, one among many in the huge family Muscicapidae, created by Hartert in 1910, and now determined to be a hodgepodge of unrelated birds. Delacour divided his Timaliinae into a series of tribes, including one for both the wrentit (*Chamaea*) and bearded reedling (*Panurus*), both of which were traditionally placed in different families. He also gave tribe status within the babblers to the parrotbills (*Paradoxornis* and *Conostoma*) and to the genus *Picathartes*. *Picathartes* had long been especially controversial, being assigned to the starling and crow families by various authorities.

The DNA research of Charles Sibley and his associates has led them to place the babblers in the family Sylviidae, together with many of the birds traditionally called Old World warblers. Within this family, they have divided the babblers into two subfamilies. Two genera, the laughing thrushes (*Garru-*

lax) and liocichlas (*Liocichla*), are given their own subfamily, Garrulacinae. The other babblers, including the parrotbills, which have traditionally been placed in a subfamily, Paradoxornithinae (or Panurinae), are included by Sibley in the subfamily Sylviinae, divided into three tribes. All but one of these are in the tribe Timaliini, the exception being the wrentit, comprising its own tribe, Chamaeini. The third tribe, Sylviini, consists of the warblers of the genus *Sylvia*, including such familiar European birds as the blackcap, whitethroats, and the garden warbler. One controversial inclusion in the tribe Timaliini, as Sibley defines it, is the genus *Rhabdornis*, the Philippine creepers, traditionally placed near the nuthatches and creepers. On the other hand, the genera *Pomatostomus* and *Garritornis*, considered by Delacour and others to be close relatives of the scimitar babblers, now constitute their own family, Pomatostomidae, which Sibley and colleagues placed near the crow family, Corvidae. The rail babblers (*Eupetes* and *Ptilorrhoa*), which Delacour initially assigned to a tribe within the Timaliinae, then later considered a separate subfamily, are classified by Sibley as a subfamily within the Corvidae. It is not surprising that Sibley removed *Picathartes*, long a source of disagreement, from the Timaliinae, nor is his refusal to place their family in close proximity to any other. What is truly startling is his inclusion, in the family Picathartidae, of the rock jumpers (*Chaetops*), traditionally classified as thrushes (Turdidae).

Paleontology has, thus far, not served to clarify the origins and relationships of babblers, the only fossils being a Middle Pleistocene example of the modern species, the Arabian babbler (*Turdoides squamiceps*). DNA research seems to support Asian origins in the Oligocene, about 40 million years ago.

Jungle babblers (*Turdoides striatus*) perched on a branch. (Photo by R. Saldino/VIREO. Reproduced by permission.)

Physical characteristics

Almost any book published over the last 50 years is likely to present a different definition of exactly what constitutes "the babblers." Over 40 years ago, Oliver Austin, in his now classic *Birds of the World*, summed up the situation by describing this group as "poorly defined and loosely delimited." All of these birds have 10 primary feathers, the outermost shorter than the rest. The tail is composed of 12 retrices. The plumage is soft and loose. Juvenal plumage is never speckled. The legs and feet are proportionately large and powerful, while flight potential is comparatively small (leading, in part, to the great number of species with small ranges). In appearance babblers range from the utterly nondescript to the unforgettable.

Despite the great disagreements among most books, one usually finds the babblers arranged thusly. To begin with are several genera of "small brown birds," otherwise referred to as "somber," "dull" or "nondescript." Often their most recognizable feature, at least when they're alive, are their voices. The number of genera is in dispute, but a few more than 30 species compose these "jungle-babblers" (*Pellorneum, Trichastoma, Illadopsis, Kakamega, Malacopteron*, etc.), all of them tropical, most of them forest-dwellers, occurring through large areas of Africa and Asia. There follow around a dozen scimitar babblers (*Pomatorhinus, Xiphirhynchus*, and *Jaboulleia*), named for their long curved bills, some quite spectacular. They appear to be closely related to the around 20 very different-looking wren-babblers (*Rimator, Ptilocichla, Napothera, Pnoepyga, Spelaeornis*, etc.). These are short-tailed, mostly plump-bodied birds, some tiny, none with brilliant colors, but some with vivid patterns. Some strongly resemble New World antbirds. Scimitar babblers and wren-babblers are Asian, with one species extending down the Lesser Sundas to Timor. What comes next depends on the book one is reading, and here things may differ wildly. Suffice to say that there follows the majority of the babblers, in many genera great and small. Towards the end, one usually finds the

parrotbills (*Paradoxornis* and *Conostoma*) of Palearctic and Oriental Asia, and the primarily Palearctic bearded reedling (*Panurus biarmicus*), mostly compact-bodied and long-tailed, many with distinctive bills, one, (*Paradoxornis paradoxus*), uniquely with three toes instead of four. Finally, one may or may not find the unique and magnificent *Picathartes*. What they are remains a question.

Distribution

Asia is the center of evolution for babblers. With the exception of the unique picathartes, which DNA research suggests may not be babblers at all, but an ancient African lineage, most of the 33 African babblers appear to be derivative of large Asian genera. Only recently have several of them been removed from Asian genera, and placed in ones exclusive to Africa, such as *Illadopsis* and *Pseudoalcippe*, while 15 are clearly members of *Turdoides*, widespread in the Indian subcontinent. Typical of the island, the six species in Madagascar, all endemic, have no connection to the African mainland.

China, where the Palearctic and Oriental zoogeographical realms meet, is a great center of distribution, and at least 143 (including four Taiwanese endemics) occur there, many nowhere else. Chinese babblers include 35 of the 49 species of laughing thrushes (*Garrulax*), 14 of the 16 fulvettas (*Alcippe*), and every one of the 19 parrotbills (Paradoxornithinae). Through Southeast Asia, there continues a great diversity of babblers, especially in Indochina, where several were only recently discovered. Fifty-five species have been recorded for Java, Sumatra, and Borneo, many of them endemic, or otherwise found only on the Malay Peninsula. While the "babblers" inhabiting Australia and New Guinea are no longer considered Timaliids, a few species occur in Wallacea, including the enigmatic malia (*Malia grata*), found only on Sulawesi, and remarkably, the fragile-looking pygmy wren-babbler (*Pnoepyga pusilla*), found all the way from Nepal to Timor. It is remarkable that Japan and Okinawa have no native babblers. On the other hand, 19 species occur in the Philippines, including 10 of the 24 species of *Stachyris*, and all but one are endemic. The Philippines and Timor are the western-most of the Pacific islands where babblers occur naturally. (Four Asian species have been successfully introduced in Hawaii.)

The Indian subcontinent, including Sri Lanka, is another babbler stronghold, and 131 species are recorded there. The majority are birds of the Himalayas, shared with China, but a number are endemic to Peninsular India or Sri Lanka. In the Middle East, babblers are represented by three species of *Turdoides*. Europe, including the United Kingdom, share the bearded reedling (*Panurus biarmicus*) with temperate Asia. Finally, on the Pacific Coast of North America is the wrentit (*Chamaea fasciata*).

Habitat

Forests are the stronghold of babblers, and most are dependent on one type or another. However, natural selection being what it is, there are babblers adapted to scrublands, near-desert, grassy savannas, and marshes, and some species have become partial to orchards, fields, and gardens.

Behavior

For so diverse a group of birds, generalizations are difficult. As did Bertram Smythies in the first edition of this encyclopedia, one cannot do better than to quote Jean Delacour, from his monumental 1946 monograph of the babblers: "They move restlessly among twigs and on the ground; they hop about and dig among fallen leaves. Usually they live in the undergrowth, sometimes on the ground among dense plant growth, fallen branches, climbers and evergreen trees, where they can be observed searching for berries and insects. While doing so, they move busily, flutter the wings a great deal, wag their tails and utter noisy calls. As rule, they are loud and varied vocally, hence the name babbler, for they are virtually never quiet. Some sing very well and their full melodies ring out far. Outside the breeding season, they move about in small troops. Often they join with other birds into the mixed flocks characteristic of tropical forest, all seeking food together."

Delacour's description fits most of the forest-dwelling babblers fairly well. It is in other sorts of habitat that particularly interesting variations have evolved. The Arabian babbler (*Turdoides squamiceps*), of the semi-desert, with its highly developed "tribal" social system, and the semi-aquatic bearded reedling (*Panurus biarmicus*) of the marshes, are two striking examples.

Feeding ecology and diet

Insects are the core of babbler diets. Some species appear to feed exclusively on them, while many also eat fruit, other invertebrates, and small frogs and reptiles. As with any huge family, there have evolved some peculiar specialists. While the bearded reedling feeds vigorously on insects for most of the year, and rears its young entirely on them, it subsists on seeds during the winter, its digestive tract making remarkable adjustments to this change in diet. As one might expect, the desert-adapted Arabian babbler will eat practically anything. On the other hand, the fire-tailed myzornis (*Myzornis pyyrhoura*) has come to resemble a hummingbird or sunbird, consuming nectar with its insects, and becoming a pollinator in the process.

Reproductive biology

In general, babblers form pairs in the breeding season, establish territories, raise one or two broods, then reassemble as flocks. Again, this seems to be the case for forest species, with most exceptions being birds of other habitat. The bearded reedling has evolved to produce up to four broods in quick succession each breeding season, with the first potentially able to breed themselves by season's end. The Arabian babbler's reproductive system, however, is remarkably different. Its highly regimented social units, with birds waiting as long as six years for their first opportunity to breed, is only the most extreme situation in its large and widespread genus *Turdoides*. Researchers have confirmed that at least 14, and possibly 26, of the 29 species in this genus practice some sort of cooperative breeding, with groups defending territories where only a few members will breed at any given time. The *Turdoides* are primarily open-country birds. In India, the tendency to always remain in their social unit has earned some species the name "Seven Sisters," and India's revered ornithologist Salim Ali refers to their groups as "sisterhoods." Another departure from the norm are the colonial picathartes, who use their unique, mud pottery-like nests, built on rock faces, as residences, and raise chicks in them year after year. Such nests are unique among babblers, who, in general, construct cup or bowl-shaped, or spherical, nests of plant materials, usually not far from the ground. Eggs run a spectrum from white and patternless, to various beautiful colors and intricate patterns of spots and streaks.

Conservation status

As of 2002, the IUCN and BirdLife International designated five babblers as Endangered, 22 as Vulnerable, and 39 as Near Threatened. In addition, it appears that subspecies of two otherwise globally non-threatened species may be extinct: The southern Turkish bearded reedling (*Panurus biarmicus kosswigi*) appears to have been a victim of wetlands destruction, while the mysterious "Astley's leiothrix" (*Leiothrix lutea astleyi*), known only from the bird trade more than 80 years ago, may have been exterminated through trapping.

All five Endangered babblers are primarily threatened by habitat loss. All are forest birds. Two are Philippine endemics: the flame-templed babbler (*Stachyris speciosa*) occurs only on Negros and Panay, while the Negros striped-babbler (*S. negrorum*) is entirely restricted to that severely deforested island. The remaining three depend on high-elevation forests: the Nilgiri laughing thrush (*Garrulax cachinnnans*), one of many imperiled inhabitants of India's Nilgiri Hills, the white-throated mountain babbler (*Kupeornis gilberti*), known to science only since 1949, restricted to several places in Nigeria and Cameroon, and the gray-crowned crocias (*Crocias langbianis*), rediscovered in 1994 after 56 years of no records, from a few locations in Vietnam.

Three Vulnerable species, the Omei Shan liocichla (*Liocichla omeiensis*), found only around Mt. Emei, in southwestern China, and the white-necked and gray-necked picathartes (*Picathartes gymnocephalus* and *P. oreas*), of the Guinea forests of West Africa, were vigorously exploited by the cage-bird trade, resulting in their listing by the Convention on International Trade in Endangered Species (CITES). Like the 19 other Vulnerable babblers, they are threatened by habitat destruction as well.

Fourteen island endemics, seven restricted to the Philippines, are included among the 39 Near Threatened species, and of the remainder, nine occur only in the Malay Peninsula, Sumatra, and Borneo. Again, loss of habitat, in often restricted ranges, is the cause for their designation. Finally, there is one categorized as Data Deficient, the miniature tit-babbler (*Micromacronus leytensis*), at 3 in (7.6 cm), the smallest babbler. Restricted to the Philippine islands of Leyte, Samar, and Mindanao, it is a forest-dependent species in a land of increasing deforestation, and has remained rare, and little known since its discovery in 1961.

Significance to humans

Babblers have long been admired for their appearance, songs, and behavior. One aspect of this admiration has been the compulsion to keep them in captivity. The enormous volume of commercial trade in living babblers has caused increasing concern. At the same time, over 30 species have been hatched in captivity.

Some Chinese babblers, especially laughing thrushes (*Garrulax* sp.) and parrotbills (*Paradoxornis* sp.) cause some damage to crops, but this appears to be minor, and offset by insect control by these same birds. On the other hand, babblers are playing an increasingly important role in the developing economy of ecotourism. Such species as the fire-tailed myzornis are specifically featured in advertisements enticing trekkers to Nepal and Bhutan, already attracted simply by the potential of seeing great numbers of species in the foraging "bird waves" that sweep across the Himalayas. As ecotourism continues to grow, increasing numbers of people from around the world will enjoy the magnificent mixed choruses of otherwise unobtrusive brown babblers in Borneo or Myanmar, search for Madagascar's peculiar endemics, observe the colonial nesting of fantastic-looking picathartes in Ghana or Gabon, and search for laughing thrushes only recently unknown to science in the highlands of Vietnam. At the same time, vigorous efforts are being made by environmentalists in these countries to instill pride in a precious natural heritage.

1. Crossley's babbler (*Mystacornis crossleyi*); 2. Yellow-naped yuhina (*Yuhina flavicollis*); 3. Red-billed leiothrix (*Leiothrix lutea*); 4. Golden-breasted fulvetta (*Alcippe chrysotis*); 5. Black-crowned barwing (*Actinodura sodangorum*); 6. Bearded reedling (*Panurus biarmicus*); 7. Fire-tailed myzornis (*Myzornis pyrrhoura*); 8. Vinous-throated parrotbill (*Paradoxornis webbianus*); 9. Yellow-headed rockfowl (*Picathartes gymnocephalus*); 10. Red-headed rockfowl (*Picathartes oreas*). (Illustration by Bruce Worden)

1. Hwamei (*Garrulax canorus*); 2. Chestnut-backed scimitar-babbler (*Pomatorhinus montanus*); 3. Wrentit (*Chamaea fasciata*); 4. Pygmy wren-babbler (*Pnoepyga pusilla*); 5. Arabian babbler (*Turdoides squamiceps*); 6. Flame-templed babbler (*Stachyris speciosa*); 7. White-crested laughing thrush (*Garrulax leucolophus*); 8. Yellow-throated laughing thrush (*Garrulax galbanus*); 9. Rufous-winged akalat (*Trichastoma rufescens*); 10. Omei Shan liocichla (*Liocichla omeinsis*). (Illustration by Bruce Worden)

Species accounts

Rufous-winged akalat
Trichastoma rufescens

SUBFAMILY
Timaliinae

TAXONOMY
Turdirostris rufescens Reichenow, 1878, Liberia. Monotypic. Endemic African genus *Illadopsis* has been merged by various authors into Asian genera *Malacocincla* or *Trichastoma*.

OTHER COMMON NAMES
English: Rufous-winged illadopsis; French: Akalat à ailes rousses; German: Rostschwingen-Buchdrossling.

PHYSICAL CHARACTERISTICS
7.2 in (18 cm); 1.2 oz (35 g). Profoundly nondescript thrush-like bird, as are most of the other six African members of the genus. Sexes alike. Rufous-brown dorsally, whitish ventrally, with indistinct brown wash on chest and brownish gray sides. Eyes dark; legs and feet pale.

DISTRIBUTION
Endemic to Upper Guinea forest, from southern Senegal to Togo.

HABITAT
Undergrowth in primary and disturbed forest.

BEHAVIOR
Ground dweller. Found in pairs or small flocks, sometimes associates with other species of insectivorous passerines. Territorial.

FEEDING ECOLOGY AND DIET
Actively searches the ground for insects and tiny frogs, scratching at substrate with feet.

REPRODUCTIVE BIOLOGY
Undescribed.

CONSERVATION STATUS
Near Threatened due to former paucity of observations, and continuous destruction of forests throughout narrow range. Recent surveys have found species common in some localities, and occurring in areas where previously unrecorded. Population densities in logged forest are a fraction of those in primary habitat, but has been found in plantations and open thickets.

SIGNIFICANCE TO HUMANS
None known. ◆

Chestnut-backed scimitar-babbler
Pomatorhinus montanus

SUBFAMILY
Timialiinae

TAXONOMY
Pomatorhinus montanus Horsfield, 1821, Java. Four subspecies.

Trichastoma rufescens

◼ Resident

Pomatorhinus montanus

◼ Resident

OTHER COMMON NAMES
English: Yellow-billed scimitar babbler; French: Moineau friquet; German: Rotrückensäbler.

PHYSICAL CHARACTERISTICS
8 in (20 cm). Striking bird reminiscent of New World thrashers (Mimidae). Bright chestnut mantle, flanks, and vent. Brownish primaries and tail. White chest and throat. Remainder of head black, except for well-defined white eyebrow. Long curving bill bright yellow, with black at rear of upper mandible, extending along part of culmen. Legs gray. Melodious voice.

DISTRIBUTION
Malay Peninsula, Sumatra, Java, Bali, and Borneo.

HABITAT
Lower and middle stories of forests, at all altitudes.

BEHAVIOR
Often associated with laughing thrushes (*Garrulax* sp.). Single birds may be observed, but usually in small flocks. Vocal, but usually hard to see.

FEEDING ECOLOGY AND DIET
Continually hunting for snails, spiders, and insects, as well as small seeds and fruits.

REPRODUCTIVE BIOLOGY
Monogamous. Presumed to be similar to that of other members of genus, with cone or bowl-shaped nest of plant material constructed by both sexes, and containing four white eggs.

CONSERVATION STATUS
Not threatened but endemic in areas subject to increasing habitat loss due to logging.

SIGNIFICANCE TO HUMANS
For many years, fairly small numbers have been exported through the cage-bird trade. ◆

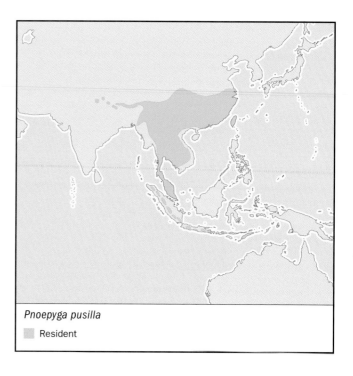

Pnoepyga pusilla

▨ Resident

Pygmy wren-babbler

Pnoepyga pusilla

SUBFAMILY
Timaliinae

TAXONOMY
Pnoepyga pusillus Hodgson, 1845, Nepal. Eight subspecies; some may constitute separate species.

OTHER COMMON NAMES
English: Lesser scaly-breasted wren-babbler, brown wren-babbler; French: Turdinule maillée; German: Moostimalie.

PHYSICAL CHARACTERISTICS
3.5 in (9 cm). Tiny, almost tail-less brown bird with speckled underparts, and dark eyes. Light and dark-breasted phases are present. Short, thin pointed bill.

DISTRIBUTION
Himalayas, east from Nepal, southern China, Taiwan, Indochina, Malay Peninsula, Sumatra, Java, Flores, and Timor.

HABITAT
Highland forest floor thickets and moist ravines.

BEHAVIOR
Highly furtive and mouse-like, staying concealed in ferns and other plants, but surprisingly loud, piercing whistle, followed by softer note, is heard often, apparently a contact call. Distinctive voice.

FEEDING ECOLOGY AND DIET
Presumably small invertebrates.

REPRODUCTIVE BIOLOGY
Undescribed.

CONSERVATION STATUS
Not threatened, but would be vulnerable to any logging of montane forests.

SIGNIFICANCE TO HUMANS
Target species for ecotourists. ◆

Flame-templed babbler

Stachyris speciosa

TAXONOMY
Dasycrotapha speciosa Tweeddale, 1878, Negros. Listed by some authorities in the monotypic genus *Dasycrotapha*.

OTHER COMMON NAMES
English: Rough-templed tree-babbler; French: Lobélia spéciosa; German: Goldstirn-Buschtimalie.

PHYSICAL CHARACTERISTICS
5.25 in (13.2 cm). Unique, rather bizarre appearance: black head with golden yellow forehead, lores, eye ring, and chin; bright reddish orange tufts on side of head. Streaked olive mantle with brown wings and tail. Breast yellow with black spots.

Stachyris speciosa

■ Resident

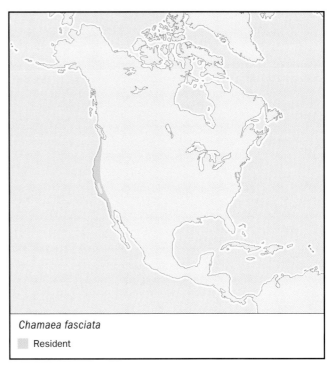

Chamaea fasciata

■ Resident

DISTRIBUTION
Restricted to central Philippine islands of Negros and Panay.

HABITAT
Lowland forest understory.

BEHAVIOR
Forages in small flocks, individually, or with other species. Very responsive to contact calls.

FEEDING ECOLOGY AND DIET
Presumed to be insects and seeds.

REPRODUCTIVE BIOLOGY
Nest and eggs undiscovered as of 2000.

CONSERVATION STATUS
Endangered due to severe reduction and fragmentation of range resulting from deforestation.

SIGNIFICANCE TO HUMANS
Negros and Panay "flagship" potential. ◆

Wrentit
Chamaea fasciata

SUBFAMILY
Timaliinae

TAXONOMY
Parus fasciatus Gambel, 1845, Monterey, California. Six subspecies. While Delacour long ago considered this species a babbler, many authorities have persisted in assigning it its own family or subfamily. However, the DNA hybridization research

of Sibley and Ahlquist suggests the wrentit is the only New World babbler, whose ancestors crossed the Bering land bridge in the mid-Miocene (15–20 million years ago). Sibley and Ahlquist do place it in its own tribe, Chamaeini.

OTHER COMMON NAMES
French: Cama brune; German: Chaparraltimalie.

PHYSICAL CHARACTERISTICS
6.3 in (16 cm); 0.5 oz (14 g). Gray (southern) or brown (northern) compact bird with long tail (usually held up), pale eyes, and small, sharp bill. Song is a very distinctive accelerating series of high notes.

DISTRIBUTION
Pacific Coast of North America, from northern Oregon to northern Baja California.

HABITAT
Chaparral and coniferous scrub.

BEHAVIOR
Generally live in pairs, communicating continuously with unique calls. Frequents heavy low vegetation.

FEEDING ECOLOGY AND DIET
Forages for insects and berries in undergrowth.

REPRODUCTIVE BIOLOGY
Monogamous, paired for life. Sexes share in all aspects of nesting and chick rearing. Nest is located deep in bushes, a tightly woven cup of twigs, bark, and feathers, lined with spider webs. Three to five greenish blue, unpatterned eggs compose clutch.

CONSERVATION STATUS
Not threatened, though significant portions of range have been developed.

SIGNIFICANCE TO HUMANS
Popular with birdwatchers. ◆

Arabian babbler
Turdoides squamiceps

SUBFAMILY

TAXONOMY
Malurus squamiceps Cretzschmar, 1827, Transjordan. Three subspecies.

OTHER COMMON NAMES
French: Cratérope écaillé; German: Graudrossling; Spanish: Tordalino Arábigo.

PHYSICAL CHARACTERISTICS
11 in (28 cm). Slender grayish brown, long-tailed bird, which may remind Americans of thrashers (Mimidae). Plumage identical in both sexes, but male's eyes pale yellow, female's brown.

DISTRIBUTION
Israel (Negev, Dead Sea Depression, Arava Valley, Sinai), Jordan, coastal Arabian Peninsula (except for Red Sea Coast).

HABITAT
Dependent on vegetation and water sources, but otherwise a dry country bird, favoring acacias, tamarisk, saltbush, and date-palm groves and gardens.

BEHAVIOR
Complex social system, reminiscent of wolves. Lives in groups of six to a dozen, but sometimes more than 20, composed of an alpha breeding pair, usually the oldest birds, which maintain a strict dominance hierarchy over rest of unit. Groups compete with others for fixed territories, and may displace them, sometimes violently. Most unit members related to breeders, but some recruitment of unrelated birds from nearby colonies. When not actively foraging, generally secreted in bushes.

FEEDING ECOLOGY AND DIET
Opportunistic feeders, energetically foraging in groups for anything edible. Main items include very wide variety of arthropods, small reptiles, berries, flowers, nectar, seeds, and garbage.

REPRODUCTIVE BIOLOGY
In some units, only the alpha male mates with any of the females. In others, subordinate males may breed with subordinate females, but the alpha female is always the one most fiercely defended by the alpha male. Courtship behavior appears to be initiated by females soliciting subordinate males in the group, causing the alpha male to aggressively prevent copulations. No copulations take place if other members of the group are present. (As a rule, they actively interfere with impending matings, only dominant males being able to prevent other group members from following potential breeding pair). Successful copulation takes place in seclusion of bushes. Nest is large, untidy, cup-shaped structure, composed of coarse plant materials, with no clearly defined lining. Generally, only one is constructed in a group's territory. Up to three females may lay eggs in group's nest. Eggs are glossy turquoise. Clutch size per female varies from three to five. At night, only the alpha female incubates, but any female in the group may do so in daytime. Eggs take 13–14 days to hatch, and hatching may occur continuously through breeding season, so that nest may be occupied by large number of chicks of disparate ages. All members of group feed chicks, often competing to do so. Unusually long reproductive longevity for a passerine bird. Sexual maturity not attained until two years, some may remain non-breeders in group as long as six years. Individuals may still be group's alpha breeders at 13 years of age.

CONSERVATION STATUS
Globally non-threatened. Range has expanded, especially in Israel, due to irrigation and abundance of food (including garbage) around residences.

SIGNIFICANCE TO HUMANS
This species may pose some harm to certain types of agriculture, eating such products as sorghum. On the other hand, they consume vast numbers of potentially harmful insects, such as termites and weevils, and are especially avid tick eaters. Subject of continuous research for decades. ◆

Turdoides squamiceps
Resident

White-crested laughing thrush
Garrulax leucolophus

SUBFAMILY
Timaliinae

TAXONOMY
Corvus leucolophus Hardwicke, 1815, Uttar Pradesh. Five subspecies. DNA hybridization study suggests *Garrulax* and *Liocichla* compose a separate subfamily (Garrulacinae) from the other babblers, which belong in another subfamily, together with some warblers (Sylviinae).

OTHER COMMON NAMES
English: White-crested jay thrush; French: Garrulaxe à huppe blanche; German: Weisshaubenhäherling.

PHYSICAL CHARACTERISTICS
12 in (30 cm). Instantly recognizable dark-bodied, white-headed bird with bushy crest and black mask and beak. Sexes monomorphic. Subspecific variation fairly pronounced: Himalayan nominate subspecies has brown chest and underparts. These areas are white in southeastern Yunnan and Indo-Chinese

Garrulax leucolophus
■ Resident

G. l. diardi, which also has a much grayer nape. *G. l. patkaicus* of Assam, northern Myanmar, and western Yunnan, has brown underparts, but a pure white nape. Sumatran *G. l. bicolor* has underparts, back, and tail blackish brown, a very different shade from that of other subspecies.

DISTRIBUTION
The length of the Himalayas, with some gaps, Assam, Myanmar, southern China, Indochina, and an isolated subspecies in Sumatra.

HABITAT
Evergreen forests, especially secondary, where heavy undergrowth and bamboo stands abound.

BEHAVIOR
Frequents lower storeys of forest, roaming in flocks of varying sizes (up to 40), often with other bird species. Distinctive ringing, antiphonal vocalizations are frequent.

FEEDING ECOLOGY AND DIET
Groups continuously forage through plants and leaf litter for insects, lizards, fruit, nectar, and seeds.

REPRODUCTIVE BIOLOGY
Species observed in flocks during the breeding season. From captive observations, however, it may prove that only one pair actually breeds. Clutch size is usually three to five. Eggs are white. The wide, shallow nest is usually only about 6 ft (1.8 m) above ground. Bamboo leaves are preferred material for nest. In China, two broods are usually raised each year, any time from March through August. In zoos, breedings have been recorded all year.

CONSERVATION STATUS
Not threatened. Preference for secondary forest decreases vulnerability to habitat loss. Traditionally exploited for cage-bird trade; as of 2002, only Indonesia exporting.

SIGNIFICANCE TO HUMANS
Very popular in zoos; more than 400 hatched in United States alone since 1968. ◆

Yellow-throated laughing thrush
Garrulax galbanus

SUBFAMILY
Timaliinae

TAXONOMY
Garrulax galbanus Godwin-Austin, 1874, Manipur. Three subspecies. Chinese populations may constitute separate species.

OTHER COMMON NAMES
English: Yellow-bellied laughing thrush; French: Garrulaxe à gorge jaune; German: Gelbbauchhäherling.

PHYSICAL CHARACTERISTICS
9.1 in (23 cm); 1.75 oz (50 g). Elegant thrush-shaped bird with brown mantle, black mask, and yellow throat and underparts. Sexes monomorphic. Nominate subspecies, from the western part of the range, has an olive green nape and crown, while the two isolated eastern subspecies have brilliant dark blue napes and crowns instead. Jiangxi population has a clear, brilliant yellow chest, while birds from Yunan have yellowish gray chests.

DISTRIBUTION
Western population in eastern India (the Manipur Valley, Nagaland, Assam, and Mizoram), the Chin Hills of Mynamar, and possibly Bangladesh. Isolated Chinese populations: one in Jiangxi Province, another in Yunnan, and perhaps another yet to be ascertained, from which captive specimens have arrived.

HABITAT
Dense scrub.

BEHAVIOR
Lives in flocks, stays near ground.

FEEDING ECOLOGY AND DIET
Hunts through leaf-litter and other substrate for insects. Also consumes fruit.

Garrulax galbanus
■ Resident

REPRODUCTIVE BIOLOGY
Captive birds produce clutches of two or three eggs, in smallish cup-shaped nests of twigs and plant fibers. Incubation, lasting 13 days, is performed by both parents. As with other *Garrulax*, while a flock may be present during nesting, only one pair actually breeds.

CONSERVATION STATUS
Though previously listed as Near Threatened, IUCN designated it as globally non-threatened as of 2002. However, isolated Chinese populations, possibly constituting a separate species, appear to be very small and occupy limited ranges. Not listed by CITES. Commercial live bird exports banned by People's Republic of China in 2001.

SIGNIFICANCE TO HUMANS
International captive breeding projects hatched over 100 since 1990. ◆

Hwamei
Garrulax canorus

SUBFAMILY
Timaliinae

TAXONOMY
Turdus canorus Linnaeus, 1758, Amoy. Three subspecies.

OTHER COMMON NAMES
English: Melodious jay thrush, spectacled thrush, Chinese thrush; French: Garrulaxe hoamy; German: Augenbrauen-häherling.

PHYSICAL CHARACTERISTICS
9 in (22 cm). Thrush-shaped. Uniformly brown with fine darker streaking. Distinctive white eye ring, with long rearwards extension lending an "Egyptian" look. Bill pale. Taiwanese subspecies more heavily streaked with much reduced eye ring.

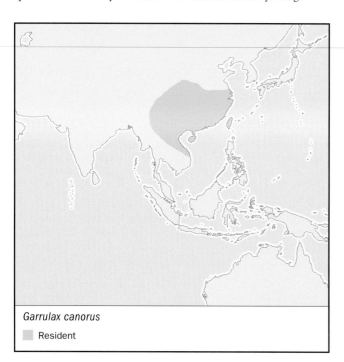

Garrulax canorus
 Resident

DISTRIBUTION
Central and southeast China, Taiwan, Hainan, and northern Indochina. Feral populations established on the Hawaiian Islands of Oahu, Maui, Hawaii, Molokai, and Kaua'i.

HABITAT
Scrub, secondary forest, and nearby farmland.

BEHAVIOR
A notably shy bird, yet famous for male's robust song. Stays near the ground, a weak flyer. Usually in pairs or small groups.

FEEDING ECOLOGY AND DIET
Forages in leaf litter for insects, fruit, seeds, and insects.

REPRODUCTIVE BIOLOGY
Monogamous. Large bowl or oval-shaped nest situated in thickets, made of leaves, pine needles, grasses, fine twigs, vines, and other plant material. May or may not be lined. Three to five bluish eggs, sometimes spotted. Female alone incubates for 15 days. Chicks fledge after 13 days of being fed by both parents. Two breeds raised between February and September.

CONSERVATION STATUS
Not threatened, though Taiwanese *G. canorus taewanus* may be threatened due to introduction of mainland nominate subspecies. As disturbed and agricultural habitat is favored, still quite common on Chinese mainland, despite massive exploitation for the cage-bird trade. Added to Appendix II of CITES in 2001.

SIGNIFICANCE TO HUMANS
Most revered songbird in China. Apparently inflicts limited damage to peanut and pea crops in China, but negative presence offset by insect control. ◆

Omei Shan liocichla
Liocichla omeiensis

SUBFAMILY
Timaliinae

TAXONOMY
Liocichla omeiensis Riley, 1926, Mt. Emei, Sichuan. Only recently recognized as full species, distinct from Taiwanese endemic Steere's babbler (*L. steerii*).

OTHER COMMON NAMES
English: Mount Omei babbler; French: Garrulaxe de l'Omei; German: Omei-Haeherling.

PHYSICAL CHARACTERISTICS
7 in (17 cm). Gray with bright reddish orange pattern on wings, squared-off, red-tipped tail, red-and-black vent, yellowish ear coverts, dark eyes and bill.

DISTRIBUTION
The vicinity of Mt. Emei, southern Sichuan and northeastern Yunnan.

HABITAT
Mountain forests with bamboo stands.

BEHAVIOR
Noted for being not as shy as other babblers, but rather curious.

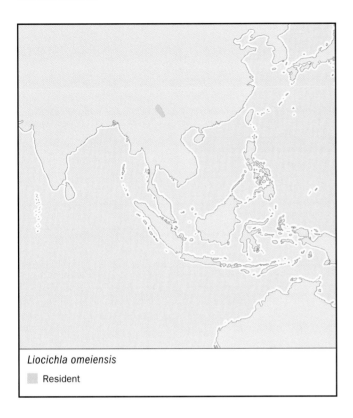

Liocichla omeiensis

Resident

Leiothrix lutea

Resident

throat, olive back, orange and yellow wings, and uniquely forked tail.

FEEDING ECOLOGY AND DIET
Small flocks forage for insects and fruits both in trees and on the ground.

REPRODUCTIVE BIOLOGY
Male displays to female by exposing bright colors on wings while vocalizing. Nest and eggs undescribed.

CONSERVATION STATUS
Vulnerable. Still occurs in some numbers, but threatened both by bamboo shoot harvesting and other forms of habitat destruction, and by the cage-bird trade. Accorded CITES Appendix II status in 1997. Export from China prohibited in 2001.

SIGNIFICANCE TO HUMANS
Target species for ecotourists. ◆

Red-billed leiothrix
Leiothrix lutea

SUBFAMILY
Timaliinae

TAXONOMY
Sylvia lutea Scopoli, 1786, Anhui. Five subspecies generally recognized.

OTHER COMMON NAMES
English: Pekin robin, red-billed hill tit; French: Léiothrix jaune; German: Sonnenvogel.

PHYSICAL CHARACTERISTICS
6 in (15.5 cm); 0.8 oz (22 g). Unmistakable combination of red bill, dark eyes with white ring, orange chest, yellow

DISTRIBUTION
The length of the Himalayas, Assam, western and northern Myanmar, southern China, northern Vietnam. Introduced in Hawaiian islands of Oahu, Hawaii, Maui, Molokai, and Kaua'i, though now possibly extinct on last island.

HABITAT
Undergrowth of secondary forests and bamboo stands, gardens.

BEHAVIOR
Found in groups, which, when perched, typically maintain close bodily contact with frequent mutual preening. In winter, moves to lower elevations.

FEEDING ECOLOGY AND DIET
Groups mostly forage near ground for insects, fruits, seeds, and flowers.

REPRODUCTIVE BIOLOGY
Apparent group courtship, followed by pairs separating off and setting up territories. Distinctive male territorial song. Both sexes build cup-shaped nest from a variety of plant materials and cocoon silk. Usual clutch three to four greenish white, variously speckled eggs.

CONSERVATION STATUS
Not threatened. Massive exploitation for cage bird trade led to listing as CITES Appendix II in 1997. China banned commercial export of songbirds in 2001. Mysterious, brightly colored *L. lutea astleyi* may represent a population trapped to extinction.

SIGNIFICANCE TO HUMANS
Enormously popular cage bird. Implicated in spreading malaria to native Hawaiian birds. ◆

Black-crowned barwing
Actinodura sodangorum

SUBFAMILY
Timaliinae

TAXONOMY
Actinodura sodangorum Eames, Le, Nguyen, and Eve, 1999, Ngoc Linh, Vietnam. Monotypic.

OTHER COMMON NAMES
French: Actinodure de Vietnam; German: Schwarzkronensibia.

PHYSICAL CHARACTERISTICS
9.6 in (24 cm). Like the other six members of the genus, a compact-bodied, brownish bird, with a short bushy crest, a long tail, finely barred with black, and edged with white, and primary feathers barred like tail. Unique among genus in possessing a black crown, bold streaking on the throat, and very dark wings. Sexes monomorphic.

DISTRIBUTION
Discovered in 1996 on Mount Ngoc Linh, Kon Tum province, in the western highlands of Vietnam. Since found in six further Kon Tum sites, and on the Laotian Dakchung plateau.

HABITAT
Montane evergreen and pine forest, as well as grasslands and cultivated land.

BEHAVIOR
Only observed as individuals or pairs, rarely with other species.

FEEDING ECOLOGY AND DIET
Gleans insects from foliage, primarily high in trees, sometimes in scrub.

REPRODUCTIVE BIOLOGY
Monogamous. Pairs often duet antiphonally. Nest undescribed.

CONSERVATION STATUS
Vulnerable. Habitat increasingly encroached upon by agriculture, though able to tolerate at least some cultivation.

SIGNIFICANCE TO HUMANS
Target for ecotourists. ◆

Golden-breasted fulvetta
Alcippe chrysotis

SUBFAMILY
Timaliinae

TAXONOMY
Proparus chrysotis Blyth, 1845, Nepal. Five subspecies.

OTHER COMMON NAMES
English: Golden-breasted tit babbler; French: Alcippe à poitrine dorée; German: Gold-Alcippe.

PHYSICAL CHARACTERISTICS
4.3 in (11 cm). The most brilliantly colored member of a large genus of mostly brown birds. Chest orange or yellow; head black, with gray throat and whitish cheeks and crown. Wings black with brilliant orange and yellow highlights. Tail black, edged with orange or yellow. Mantle grayish and rump yellow. Sexes similar. Shape typical of genus: rounded, like a titmouse or kinglet, with a short, sharp bill, and tail of moderate length.

DISTRIBUTION
Himalayas, east from Nepal, Assam, northeastern Myanmar, southwestern China, and northwestern Vietnam.

HABITAT
Highland evergreen forest and scrub.

BEHAVIOR
Member of perpetually moving "bird waves," mixed-species flocks sweeping from one feeding site to another, continuously vocalizing.

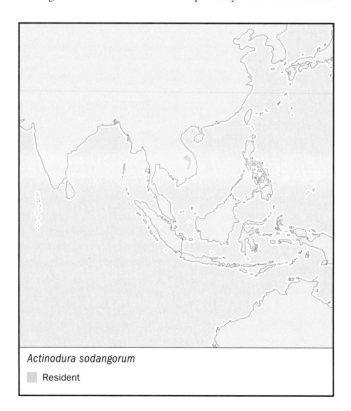

Actinodura sodangorum
☐ Resident

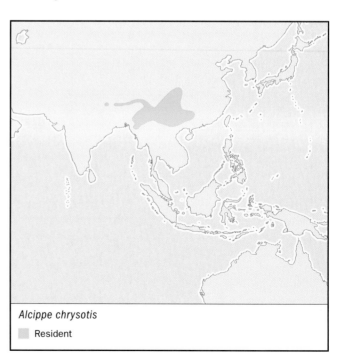

Alcippe chrysotis
☐ Resident

FEEDING ECOLOGY AND DIET
Typical mixed-species flock foliage gleaner, moving from plant to plant on a perpetual search for insects and spiders on bark, twigs, and under leaves. May also eat seeds and fruit.

REPRODUCTIVE BIOLOGY
Not known.

CONSERVATION STATUS
Not threatened, though parts of range subject to deforestation.

SIGNIFICANCE TO HUMANS
A popular target for ecotourists, many of whom come to the Himalayas to momentarily spot as many participants in "bird waves" as possible. ◆

Whiskered yuhina
Yuhina flavicollis

SUBFAMILY
Timaliinae

TAXONOMY
Yuhina flavicollis Hodgson, 1836, central Nepal. Seven subspecies.

OTHER COMMON NAMES
English: Yellow-naped yuhina, yellow-collared ixulus; French: Yuhina à cou roux; German: Gelbnackenyuhina.

PHYSICAL CHARACTERISTICS
5 in (13 cm). Like other nine members of the genus, a compact, quietly colored bird with small, pointed bill and short, but well-defined, crest. Distinguished by combination of black malar stripe ("moustache") and white streaking on brown flanks. Wings, mantle, and tail dark brown. Chest and throat white. Back of head gray, bordered by orange-brown, then white collars.

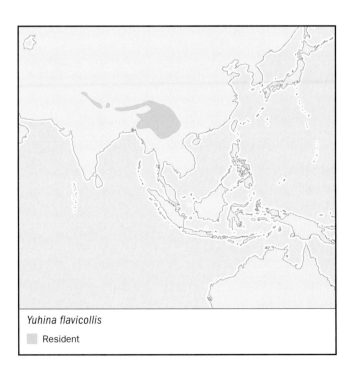

Yuhina flavicollis
▨ Resident

DISTRIBUTION
The length of the Himalayas, western China, Myanmar, and northern Indochina.

HABITAT
Montane evergreen and deciduous forests.

BEHAVIOR
Usually in large, noisy mixed-species flocks, with birds of similar size. Altitudinal migrant, moving downhill in winter.

FEEDING ECOLOGY AND DIET
Continuously foraging in interspecies groups for insects, berries, and nectar.

REPRODUCTIVE BIOLOGY
Monogamous, pairs become territorial during breeding season. Bowl-shaped nest of fine plant materials tied on to branches of shrubs and climbing plants. Two to three speckled eggs, incubated by both male and female.

CONSERVATION STATUS
Not threatened.

SIGNIFICANCE TO HUMANS
While considered "uncommon" in China, it is one of the core species in "bird wave" multi-species flocks which attract birders from around the world to the Himalayas. ◆

Fire-tailed myzornis
Myzornis pyrrhoura

SUBFAMILY
Timaliinae

TAXONOMY
Myzornis pyrrhoura Blyth, 1843, Nepal. Monotypic and only member of genus. Position within family remains unclear.

OTHER COMMON NAMES
French: Myzorne queue-de-feu; German: Feuerschwänzchen.

PHYSICAL CHARACTERISTICS
5 in (12.5 cm). Startling bright green bird with brilliant green, red, and black tail, reddish undertail coverts, black and orange pattern on wings, red wash on chest, scaly crown, red eyes, and slender, black, slightly down-turned bill.

DISTRIBUTION
Himalayas, east from Nepal, and south, along the Salween Divide, into western Sichuan, Yunan, and northeastern Myanmar.

HABITAT
Rhododendron, bamboo, juniper, and other montane forests, as well as tree farms.

BEHAVIOR
May be seen singly, in groups of three to four, and in parties of sunbirds, warblers, and other small babblers. Altitudinal migrant; from 9,800–12,000 ft (3,000–3,660 m) in summer, to 6,600 ft (2,000 m) in winter.

FEEDING ECOLOGY AND DIET
Remarkably similar to unrelated sunbirds (Nectariniidae), with whom it often associates. Continuously probes flowers, especially rhododendrons, becoming covered with pollen, while extracting nectar and insects. Observed by Salim Ali to hover at

Myzornis pyrrhoura
■ Resident

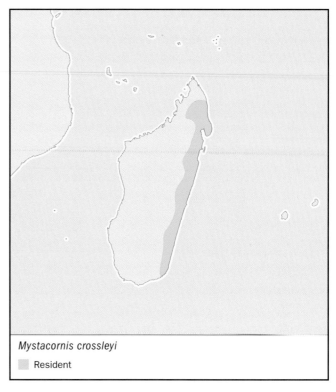

Mystacornis crossleyi
■ Resident

flowers like a sunbird, also to run up moss-covered trunks, creeper-like, in search of insects and spiders. Also eats berries.

REPRODUCTIVE BIOLOGY
Eggs undescribed. Observed feeding chicks at nest in May.

CONSERVATION STATUS
Not threatened.

SIGNIFICANCE TO HUMANS
The "Living Emerald" of the Himalayas, a major draw for ecotourists trekking in Nepal and Bhutan. ◆

Crossley's babbler
Mystacornis crossleyi

SUBFAMILY
Timaliinae

TAXONOMY
Bernieria crossleyi Grandidier, 1870, Madagascar. Monotypic. Position within family unclear.

OTHER COMMON NAMES
French: Mystacornis; German: Mystacornis.

PHYSICAL CHARACTERISTICS
6.3 in (16 cm). Compact-bodied, with relatively long bill. Sexes strongly dimorphic. Male distinguished by black throat and gray crown, while female's white throat and brown crown gives it a markedly different appearance. Both sexes have a broad white malar stripe ("moustache"), far more noticeable in male, as well as black band through eye, and small white spot above eye. Mantle, wings, and tail brown, belly gray, off-white undertail coverts. Eyes dark.

DISTRIBUTION
Eastern Madagascar.

HABITAT
Understory of primary evergreen humid forests.

BEHAVIOR
Found either individually or in family units. Adult males do not tolerate the presence of others, and respond aggressively to others' vocalizations. (Unlike many other babblers, fledged juveniles do not resemble adults, but are instead uniform rufous.)

FEEDING ECOLOGY AND DIET
Walks around on forest floor, in fairly open places, looking for insects.

REPRODUCTIVE BIOLOGY
Monogamous. Cup-shaped nest built from twigs, close to the ground.

CONSERVATION STATUS
Not threatened, but habitat dependent. Range includes several reserves.

SIGNIFICANCE TO HUMANS
A target species for ecotourists. ◆

Bearded reedling
Panurus biarmicus

SUBFAMILY
Paradoxornithinae

TAXONOMY
Parus biarmicus Linnaeus, 1758, Holstein. Three subspecies. Some classification systems have placed this species and other

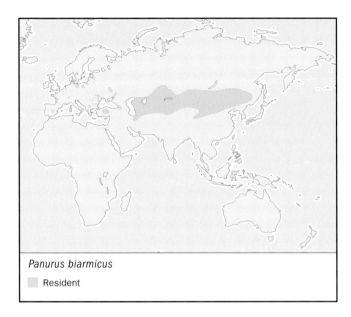

Panurus biarmicus
▢ Resident

parrotbills in separate family (Panuridae or Paradoxornithidae), but Delacour considered them subfamily within babblers. DNA hybridization research suggests they may belong in the same subfamily (Timaliinae) as most babblers.

OTHER COMMON NAMES
English: Bearded tit, marsh pheasant (East Anglia); French: Mesange a moustaches; German: Bartmeise; Spanish: Bigotudo.

PHYSICAL CHARACTERISTICS
5 in (12.7 cm). Highly sexually dimorphic. Male unmistakable: unique elongated black feathers in front of eye droop down to chest, forming a "moustache" that contrasts strikingly with orange eyes and bill; gray head, white throat, warm brown mantle, rump, and tail, black-white-and-brown-patterned wings, and black vent. Female: long-tailed, compact-bodied bird like male, with similarly patterned wings, and orange eyes and bill, but otherwise generally buffy-brown. Juvenile male very distinct: gray eyes set in black mask, black back, black on tail, and overall golden color. Legs black and unusually long.

DISTRIBUTION
Scattered range in British Isles and western and central Europe, then broadly across central and northeastern Asia to the Pacific (Bokai Sea). Accidental to Japan.

HABITAT
Wetlands, especially *Phragmites* reed beds.

BEHAVIOR
European populations essentially nonmigratory, but inclined to wander in winter, and severe weather may cause wholesale relocations. East Asian populations appear to regularly migrate from interior to coast in winter. Highly gregarious and social outside of breeding season. Complex series of communication calls and displays. Usual flock size 10–20. Noted for skillful negotiation through reed beds, typically grasping a different stalk in each foot. Also capable of what Otto Koenig termed "flutter-swimming," should bird fall in water.

FEEDING ECOLOGY AND DIET
Remarkable seasonal changes in digestive tract. Winter diet largely *Phragmites*, *Typha* and rush (*Juncus*) seeds. Stomach then muscular and hard-walled, and bird ingests numerous small stones. During remainder of year, when diet is mostly insects, spiders, and snails, stomach is smaller, flaccid, and contains no, or fewer, stones. Forages both on the ground and in vegetation.

REPRODUCTIVE BIOLOGY
Adapted to produce large numbers of offspring quickly, in response to relatively unstable environment. Regularly hatches three broods a year, sometimes four, and young from earliest clutch may themselves breed before end of season, in fall. Pairs mate for life, but nest in loose colonies without territorial boundaries. Nest is a deep cup of reed leaves and other plants, lined with reed flower-heads and feathers, located deep amidst plant stalks, above water, or on land, constructed by both members of pair, who continue to add material throughout 10–14 day incubation period. Four to eight pale, streaked, and speckled eggs incubated by both parents, who also raise chick together.

CONSERVATION STATUS
Not threatened, but distinctive, pink-bellied southern Turkish *P. biarmicus kosswigi* appears to be extinct. In early nineteenth century common along the Thames, from the estuary to Oxford, but British Isles population reduced to two to four pairs in East Anglia by 1948 due to marsh and fen draining, hard winters, and private and commercial egg collectors. Since then British population has increased to more than 500 pairs, at least partially recolonized from continental Europe during hard winter dispersals. German range has expanded since the 1950s. All European populations are closely monitored.

SIGNIFICANCE TO HUMANS
Since the 1990s large numbers were available cheaply from People's Republic of China for the captive-bird trade. Export prohibited by China in 2001, but may continue from Russian Federation. ◆

Vinous-throated parrotbill
Paradoxornis webbianus

SUBFAMILY
Paradoxornithinae

TAXONOMY
Suthora webbiana Gould, 1852, Shanghai. Seven subspecies.

OTHER COMMON NAMES
English: Rufous-headed crowtit; French: Paradoxornis de Webb; German: Papageischnabel.

PHYSICAL CHARACTERISTICS
5 in (12 cm). Tiny, long-tailed, brownish bird with rufous accents, finch-like yellow bill, and dark eyes. Sexes similar.

DISTRIBUTION
Greater portion of eastern China, north to southeastern Siberia, south to northeastern Myanmar and northwestern Vietnam, as well as Korean Peninsula and Taiwan.

HABITAT
Summer: undergrowth of mixed conifer and broad-leafed secondary forest. Winter: grassy hillsides, reed beds, thickets, farmlands.

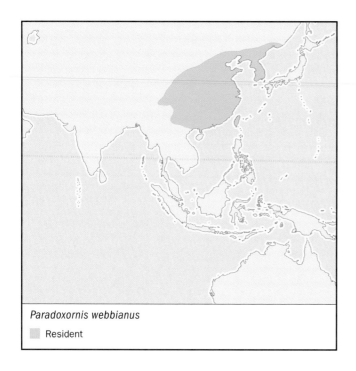

Paradoxornis webbianus

Resident

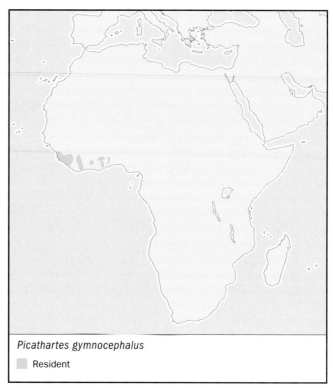

Picathartes gymnocephalus

Resident

BEHAVIOR

Some populations migrate south, most only change habitat in winter. Flocks usually contain about 10 birds, may include up to 80. Continuous contact calls. Noted for great agility among dense vegetation.

FEEDING ECOLOGY AND DIET

Continuously forages through vegetation in flocks for seeds, insects, spiders, and occasionally cultivated grain.

REPRODUCTIVE BIOLOGY

Monogamous. Male constructs outer framework of deep bowl-shaped nest of grass and various fibers, in small trees, bamboo, dense grass, or thickets. Female assists in lining with moss, cobwebs, hair, etc. Thirteen-day incubation of 3–5 blue-green or white eggs shared by parents.

CONSERVATION STATUS

Not threatened. Habitat may have expanded through recent human activity. Recent massive commercial exportation from China ended by ban in 2001, though shipments may arrive from Siberia.

SIGNIFICANCE TO HUMANS

May pose minor threat to millet, sorghum, wheat, and rice. Continuous acrobatic activity has made it a popular traditional cagebird in China and Japan. Before the 1949 Chinese Revolution, males popular for wagered bird fights. ◆

White-necked picathartes

Picathartes gymnocephalus

SUBFAMILY

Picathartinae

TAXONOMY

Corvus gymnocephalus Temminck, 1825, Coast of Guinea. Mono-

typic. Classification long controversial. Prior to Delacour's placement as a subfamily within babblers, often considered in crow family (Corvidae). DNA melting curve experiments show no close relation to babblers, but instead suggest *Picathartes* belongs in the same family with Southern African rock jumpers (*Chaetops* sp.), with possible relationship to corvids.

OTHER COMMON NAMES

English: White-necked bald crow, yellow-headed rockfowl, Guinea picathartes; French: Picatharte de Guinée; German: Weisshals Stelzenkraehe; Spanish: Picatartes Cuelliblanco.

PHYSICAL CHARACTERISTICS

16 in (40 cm), 7 oz (200 g). Elegantly proportioned bird of unmistakable appearance. Head unfeathered, with unique black and orange-yellow skin pattern. Eyes and powerful bill dark. Mantle, wings, and tail black, or nearly so. Neck and underparts creamy white.

DISTRIBUTION

Guinea, south to Ghana.

HABITAT

Primary and mature secondary forest with boulders or rock formations.

BEHAVIOR

Keeps to understory, mostly stays in vicinity of nest sites throughout the year. May be found singly, in pairs, or in small groups. Moves about as much by leaping and hopping as by flying. Noted to be inquisitive of humans entering habitat.

FEEDING ECOLOGY AND DIET

Forages on or near ground for wide variety of insects, as well as other arthropods, snails, earthworms, frogs, and lizards. Follows army ant columns in company of other insectivorous birds, snatching fleeing insects.

REPRODUCTIVE BIOLOGY
Monogamous. Usually nests in colonies of several pairs. Thick-walled, cup-shaped nests are composed largely of mud, with incorporated plant material, constructed on rock faces (cave mouths, boulders, cliffs), or sometimes stream banks or large fallen trees. Separate roosting and breeding nests thought to be routine. Nests are reused over periods of at least several years. Only one or two blotched, variously colored eggs laid, with rather long incubation period of 23–28 days. Usually two clutches a year.

CONSERVATION STATUS
Vulnerable. One of Africa's most well-known threatened birds. As of 2002, main threats are from forest clearance. Prior to 1973 CITES Appendix I listing, Liberian animal dealers appeared to have destroyed entire colonies. Recent fieldwork has located previously unknown populations, such as one in Mont Peko National Park, Ivory Coast, discovered in 1998. A number of protected areas include breeding colonies.

SIGNIFICANCE TO HUMANS
Chicks and adults sometimes taken for food. Once popular zoo bird, as of 2002, none outside Africa since 1998. Future "flagship" and ecotourism potential. ◆

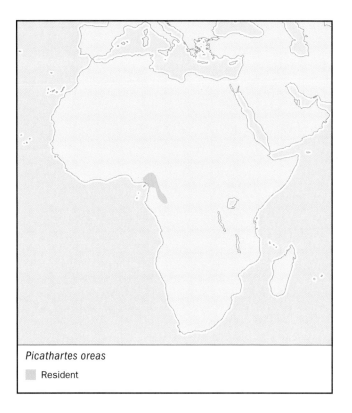

Picathartes oreas

█ Resident

Gray-necked picathartes
Picathartes oreas

SUBFAMILY
Picathartinae

TAXONOMY
Picathartes oreas Reichenow, 1899, Cameroon. Monotypic.

OTHER COMMON NAMES
English: Red-headed rockfowl, blue-headed picathartes, gray-necked bald crow, Cameroon picathartes; French: Picatharte du Cameroun; German: Blaustirn Stelzenkrahe; Spanish: Picatartes Cuelligris.

PHYSICAL CHARACTERISTICS
14 in (35 cm), 7.7 oz (220 g). Appears obviously related to preceding species, but markedly distinct. Uniquely beautiful blue, black, and red pattern of bare skin on head, blue extending to base of bill. Neck, mantle, back, and tail gray. A patch of black bristles on crown, and short ruff at base of bald head, can be erected when bird is excited. Primaries black; underparts pale yellow.

DISTRIBUTION
Primarily Cameroon, and contiguous Nigeria and Gabon. Recently discovered to occur on the island of Bioko, Gulf of Guinea.

HABITAT
Closed canopy, undisturbed rainforest, often hilly, always in vicinity of rock formations.

BEHAVIOR
Found on or near ground, singly, in pairs, or in groups of up to 10. Flocks have been observed bounding along, almost in unison. Shares previous species' curiosity towards humans. Groups gather to roost in communal nesting sites at night.

FEEDING ECOLOGY AND DIET
Actively hunts, poking through leaf litter, pouncing on prey, primarily arthropods (including crabs), snails, worms, frogs, and lizards. Often follows army ants, and may poke through bat guano.

REPRODUCTIVE BIOLOGY
Monogamous. Mud and fiber nests similar to that of white-necked picathartes, similarly plastered to rocks, built by both sexes. Construction may take months, sometime a year. Clutch of usually two variously colored speckled eggs incubated by both parents for 24 days.

CONSERVATION STATUS
Vulnerable due to dependence on primary forest. Still abundant in various locations, and new populations have been recently discovered. Limited commercial exploitation as zoo birds from 1968 to 1970. CITES Appendix I status, awarded in 1973, prevents further international trade.

SIGNIFICANCE TO HUMANS
Future "flagship" and ecotourism potential. ◆

Resources

Books

Ali, S. *The Book of Indian Birds.* New York: Oxford University Press, 1996.

Collar, N. J., A. V. Andreev, S. Chan, M. J. Crosby, S. Subramanya, and J. A. Tobias. *Threatened Birds of Asia.* Barcelona and Cambridge: Lynx Edicions and BirdLife International, 2001.

Cramp, S., C. M. Perrins, and D. J. Brooks, eds. *Handbook of the Birds of Europe, the Middle East, and North Africa,* Vol. 7. New York: Oxford University Press, 1993.

Fry, C. H., S. Keith, and E. K. Urban, eds. *The Birds of Africa,* Vol. 6. London: Academic Press, 2000.

Kennedy, R. S., P. C. Gonzales, E. C., Dickinson, H. C. Miranda, and T. H. Fisher. *A Guide to the Birds of the Philippines.* New York: Oxford University Press, 2000.

MacKinnon, J. R., K. Phillipps, and Fen-qi He. *A Field Guide to the Birds of China.* New York: Oxford University Press, 2000.

Morris, P., and F. Hawkins. *Birds of Madagascar.* New Haven and London: Yale University Press, 1998.

Sibley, C. G., and B. L. Monroe. *Distribution and Taxonomy of Birds of the World.* New Haven and London: Yale University Press, 1990.

Sibley, C. G., and J. E. Ahlquist. *Phylogeny and Classification of Birds: A Study in Molecular Evolution.* New Haven and London: Yale University Press, 1990.

Periodicals

Eames, J. C. "On the Trail of Vietnam's Endemic Babblers." *Oriental Bird Club Bulletin* 33 (June 2001): 20–26.

Lindholm, J. H. "The Laughing Thrushes." *A.F.A. Watchbird* 24 (January/February and March/April 1997): 42–47, 53–58.

Other

African Bird Club. <http://www.africanbirdclub.org>. (29 March 2002).

International Species Inventory System. <http://www.isis.org>. (29 March 2002).

Oriental Bird Club. <http://www.orientalbirdclub.org>. (29 March 2002).

Josef Harold Lindholm, III, BA

Wrens
(Troglodytidae)

Class Aves
Order Passeriformes
Suborder Passeri (Oscines)
Family Troglodytidae

Thumbnail description
Very small to medium-sized passerines, plumage usually predominantly brown, sometimes with striking black or white markings; frequently with superbly beautiful songs that are often the product of both sexes singing in concert

Size
3.6–8.8 in (9–22 cm); 0.26–2.25 oz (7.3–62.8 g)

Number of genera, species
14 genera; 60 species

Habitat
Forests, scrubland, suburban areas, marshes, grassland, rocky areas

Conservation status
Critical: 1 species; Endangered: 2 species; Vulnerable: 3 species; Near Threatened: 3 species

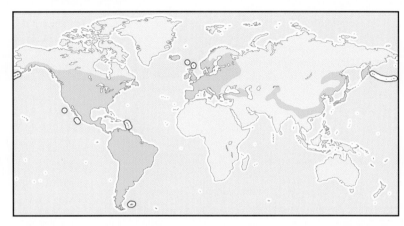

Distribution
North and South America, except for northern Canada, Alaska, and most of the West Indies; one species in Europe, North Africa, Middle East, central Asia to Japan, eastern Siberia, and Taiwan

Evolution and systematics

The classification of the wren family has long been an area of dispute among scientists. The commonly held view over much of the twentieth century was that wrens were very closely related to dippers and more distantly related to mockingbirds and thrushes. Given the obvious physical similarities between wrens and dippers—the stumpy shape, usually short tail, and rounded wings—as well as the fact that both groups build domed nests with a side entrance-hole, this seemed eminently reasonable. However, more recent work based on DNA studies has radically changed this picture. Wrens appear to be most closely allied to the New World gnatcatchers and gnatwrens, and rather more distantly to the creepers (treecreepers) and nuthatches; dippers, by contrast, are evolutionarily closer to thrushes than to wrens.

The ancestral seat of the wrens is in the New World, but precisely where is a source of debate. At the time when most present-day passerine families were evolving, there was no continuous land bridge between North and South America. One theory postulates an ancestral center somewhere in present southwestern North America, followed by the invasion of South America. This viewpoint is by no means universally held, and an origin in northern South America has some advocates. The fossil record of wrens is extremely scanty and so recent as to be of little help in elucidating the geographic origins of the family. The greatest abundance of modern wrens is in southern Central America and northwestern South America.

The precise number of species of wrens is a fluid and debatable quantity. Although only four totally new species have

been described to science since 1945 (the most recent in 1985), different authorities have wildly differing opinions as to the taxonomy of the family. Thus the house wren group, occupying the Americas from Canada to Tierra del Fuego, has been variously treated as one species or as many as ten. Peters checklist contains 60 species in 14 genera; more recently, Clements has 78 species in 16 genera, while the most recent studies suggests 83 species in 16 genera. Wren taxonomy is currently in a state of great flux, and both the total number of species and their allocation into genera will almost certainly change in the near future. The bizarre and aberrant *Donacobius*, a raucous and rambunctious inhabitant of South American marshes, is sometimes classified as a wren, a viewpoint that varies among taxonomies.

Physical characteristics

In size, wrens are among the smaller birds. In North America, the only lighter birds are the kinglets, hummingbirds, some smaller warblers, and the gnatcatchers. The largest wrens are about the same size and weight as a small thrush, and the majority weigh less than a house sparrow. Plumage is generally brown, buff, or gray-brown, usually with prominent barring on wings and tail. Some of the tropical species, especially of the genus *Thryothorus*, have bright reddish brown plumage, frequently with attractive and striking patterns of black and white on the face and underparts. A few terrestrial South American species have bare blue skin around the eye.

A marsh wren (*Cistothorus palustris*) sings from the top of a cattail. (Photo by Lee Rentz. Bruce Coleman Inc. Reproduced by permission.)

Wrens spend most of their time foraging in thick vegetation and consequently tend to have short, rounded wings, more suited to brief flights between adjacent thickets than to sustained flight. However, several short-winged species—such as the winter, house, and sedge wrens—are long-distance migrants, undertaking migrations of 600–1,200 mi (1,000–2,000 km) in some cases. Migration occurs after dark. Many wrens have short, stumpy tails (usually with 12 feathers) that are frequently carried cocked up over the back, although the unusual *Odontorchilus* (tooth-billed wrens) of South America are long-tailed gnatcatcher-like birds.

Wrens are largely insectivorous birds, and their bills reflect this diet. The bills are long and fine, never heavy or conical. Within the family there is substantial variation in feeding specialties with corresponding variation in the bill. For example, the "cactus" wrens of the genus *Campylorhynchus*, species that make their livings by probing into epiphytes, tend to have finer bills than arid-country species, which take coarser prey. The cliff-face and rock specialists of the genus *Hylorchilus* have exceptionally long and fine bills to aid in probing narrow crannies among rocks.

The plumage of the sexes is always very similar in wrens. Juvenal plumage may, in some cases, be substantially different from that of adults, but in other cases, very similar.

Distribution

The wrens are an American family. With the notable exception of the winter or northern wren, they are confined entirely to the Nearctic and Neotropical zoogeographic regions. Within this vast area, ranging from 62° north in Alaska to 55° south in Tierra del Fuego, the abundance of wren species varies greatly, with the maximum diversity in southern Central and South America. Canada, with an area of 3.85 million sq mi (10 million km²), has eight species; the United States (3.8 million sq mi, 9.8 million km²) has ten; Mexico (760,000 square miles, 1.98 million km²) has 35, with 11 species that are endemic; Panama (30,000 sq mi, 77,700 km²) has 21; and Colombia (440,000 sq mi, 1.14 million km²) has 30. Species abundance remains high in the Andean chain, but drops off sharply in the lowlands of the Amazonian basin.

The high diversity in the mountainous regions of Central and South America is at least partially the result of the varied terrain; frequently several species are found in close proximity in the different habitats created by different altitudes and the varying precipitation levels caused by mountains and rain shadows. Conversely, the Amazonian basin is of almost uniform altitude. Furthermore, in lowland areas the families of the antbirds and ovenbirds reach their maximum abundance, doubtless competing for food resources with wrens. In South America south of Bolivia, the number of species diminishes rapidly, with essentially only two species south of the tropic of Capricorn.

Curiously, wrens are almost absent from the Caribbean subregion; the southern house wren extends to some of the Windward Islands and the peculiar and unique Zapata wren (*Ferminia cerverai*) occupies a few square miles of swampland in Cuba, but much apparently excellent habitat on the large islands is inexplicably wrenless.

One species, known in North America as the winter wren (*Troglodytes trogodytes*) but in Britain simply as the wren, crossed the Bering Strait into the Old World. Lacking competition, it has expanded across three further continents, from Kamchatka and Taiwan in the east to Morocco and Iceland in the west, occupying a diverse range of habitats, from remote sea-girt islands to Himalayan scrubland and suburban gardens.

The different genera of wrens have different centers of abundance. By far the most widespread genus is *Troglodytes*. It is both the northernmost and the southernmost genus in the Americas, and the only one in the Old World. It includes a modest number of species, some of which have very restricted distributions in the mountains of Costa Rica, Panama, and Colombia. The genus *Thryothorus*, at 21 species, is by far the largest, and has an almost exclusively tropical distribution. The genera *Salpinctes*, *Catherpes* and *Hylorchilus*, comprising rock, canyon, and alllied wrens, occur from western Canada to Costa Rica and specialize in living in rocky habitats and cliff-faces. From Arizona to Ecuador and Brazil are found the largest wrens, the rambunctious and boisterous members of the genus

Campylorhynchus, which includes the familiar cactus wren (*Campylorhynchus brunneicapittus*) of the arid American southwest as well as several highly restricted species found in limited areas of Mexico. In the thick tropical forests from Mexico to Brazil and Bolivia is found the genus *Henicorhina*, which includes two widely distributed species, one highland and one lowland, as well as one that was only recently discovered in a few mountains in Peru and Ecuador. The genus *Microcerculus* is a specialist of dense tropical forest, usually at low elevations, and occurs in these habitats from Mexico to Brazil and Ecuador. In Andean South America the genus *Cinnycerthia* is found exclusively in wet mountain forest, often at considerable elevations, while the genus *Cyphorhinus* is found at lower elevations from Honduras to Bolivia and the Guianas. The distribution of the four species of the marsh-specializing genus *Cistothorus* is peculiar. Two are found in restricted areas of Venezuela and Colombia. One occurs across the whole of North America, from California to Florida and New Brunswick. The fourth, the sedge wren (*Cistothorus platensis*), is found discontinuously from the Canadian prairies through Central and South America to Cape Horn and the Falkland Islands. The other wren genera are almost all confined to the New World tropics, with the exception of Bewick's wren (*Thryomanes bewickii*), which occurs across western and central North America.

Habitat

Wrens have evolved to take advantage of virtually all types of habitat in their geographic range. However, certain genera tend to specialize in particular kinds of habitat. The genus *Cistothorus*, the marsh wrens, is predominantly found in marsh-edge vegetation such as reeds and cattails, though one species also occurs in wet high-altitude Andean grassland. The most arid areas are occupied by the large wrens of the genus *Campylorhynchus*, though again, some members occur in lowland tropical forest. The largest genus is *Thryothorus*, the majority of whose species live in dense forest or forest-edge, often at considerable elevations. Some wrens have become very specialized; the canyon wren of western North America is essentially confined to canyons, more rarely sea-cliffs. The two members of the genus *Hylorchilus* occur exclusively in tall forest on limestone karst outcrops. The majority of species in the tropics are found in forest. The highly terrestrial members of the genus *Microcerculus* are found in wet lowland forest, and the genus *Cinnycerthia* occurs in wet montane forest. Wrens do not usually adapt well to gross habitat modification, but a few species have developed a reasonable coexistence with humans. Northern and southern house wrens are abundant in abandoned farmland, clearings, and well-treed suburbs, and undoubtedly have expanded their range to take advantage of forest clearing. The Bewick's wren expanded greatly into eastern North America as agriculture moved westward in the nineteenth century. The winter or Eurasian wren occupies a range of habitats. In North America, it tends to be restricted to cold, wet northern forest. However, after it crossed the Bering Strait into three new continents, it was without competition from any other wrens, and it expanded into a great variety of habitats, from low brush in remote oceanic islands to high-altitude bushland in Central Asia, cedar forest in North Africa, and suburban gardens in England.

A house wren (*Troglodytes aedon*) in flight in Pennsylvania. (Photo by Joe McDonald. Bruce Coleman Inc. Reproduced by permission.)

Behavior

Wrens tend to be a retiring and secretive family. This is by no means universal—the large cactus wrens of the genus *Campylorhynchus* are noisy, uninhibited, and conspicuous—but most wrens tend to spend most of their time in the lower levels of dense vegetation, going about their daily lives with an immense busyness, but liable to disappear from view at the slightest disturbance. This is especially true of the nightingale wren (*Microcerculus marginatus*), which is legendarily difficult to observe. Wrens are almost always first detected by song. Wren vocalizations are loud beyond all proportion to the size of the bird; some tropical wrens are stated to have a vocal production ten times louder, weight for weight, than the crowing of a cockerel. Indeed, the name for the house wren in the Ojibwa language of western Ontario means "he who makes a lot of noise for his size." Many species of tropical wrens, especially in the genera *Campylorhynchus* and *Thryothorus*, have developed elaborate, mutual songs by both sexes. In some cases these are duets, with both birds singing simultaneously. More often, though, they are antiphonal, with each sex singing a different part. The contributions often are so tightly interwoven that the casual observer would not guess that more than one bird was involved.

Another peculiar fact of wren behavior, widely spread among several genera, is the destruction of eggs of other birds, sometimes of their own species, but frequently of others. In the marshland communities of Canada and the United States, the marsh wren (*Cistothorus palustris*) may destroy enough eggs of the much larger red-winged and yellow-headed blackbirds as to have a significant effect on their breeding success. Marsh wrens can, in fact, be caught in traps baited with small eggs.

A Carolina wren (*Thryothorus ludovicianus*) with chicks at its nest. (Photo by R. & S. Day/VIREO. Reproduced by permission.)

The decline of the Bewick's wren in eastern North America has been correlated to the increase in the population of the house wren (aided, ironically, by the provision of nest-boxes). Two closely related *Campylorhynchus* wrens, the giant wren of southern Mexico and the bicolored wren of northern South America have the local name *Chupahuevo*, literally, egg-sucker, acquired apparently by their depredations in hen houses. The function of egg predation is not clear; it may reduce competition on the nesting-ground. Sometimes destroyed eggs are eaten, but frequently they are simply punctured and left.

Feeding ecology and diet

Surprisingly little is known about the feeding ecology of many—in fact the majority—of wren species. The 10 North American species are well-studied, but for many tropical species the only information available comes from collectors' notes on the labels of museum specimens, which are often more cryptic than enlightening. Generally speaking, almost all species eat largely or wholly an arthropod diet. The large wrens of the genus *Campylorhynchus* are something of an exception, since they do take substantial quantities of vegetable matter, such as cactus seeds. In some instances small frogs or, in the case of the Zapata wren, substantial lizards may be taken. Egg-destruction occurs in a variety of species, but eggs so attacked are not always eaten. Feeding techniques of wrens vary among different genera.

Food items are generally taken from a perched position, not caught in mid-air. *Microcerculus* feed almost exclusively among forest-floor litter; the wing-banded wren specializes in foraging in rotted logs. Most other wren species feed in the lower levels of tangled vegetation, but some species range higher up; the aberrant genus *Odontorchilus* is unusual for feeding mostly in the forest canopy. Some species of tropical wren will briefly join mixed flocks following ant-swarms, but no species is truly a habitual ant-follower. In Arizona, cactus wrens have learned to exploit a novel food source, the squashed and conveniently dehydrated insects on the radiators of parked cars.

Reproductive biology

There are three unusual features in the reproductive biology of some wren species: multiple nest building, polygamy, and cooperative nesting. The building of superfluous nests occurs in many genera of wrens, but is most pronounced in the marsh-living genus *Cistothorus*. Male marsh wrens may build up to 20 nests in a season. Typically, nest construction is by the male, with the female apparently selecting the nest to be used for breeding and adding the lining. The sheer amount of energy that goes into building apparently surplus and useless nests suggests that it has a strong evolutionary advantage, but there is debate as to the actual nature of that advantage. Doubtless it is useful to have back-up nests that can be rapidly refurbished in the event of damage to the breeding nest, but after one or two back-ups that advantage must surely be played out. It has also been suggested that extra nests act as decoys to predators, or that the number of nests that a male makes is an indication of his reproductive vigor; indeed, if their surplus nests are experimentally removed, male marsh wrens have difficulty attracting mates. Surplus nests are also useful for roosting and may be constructed exclusively for that purpose. In the white-breasted wood wren, breeding nests are substantial and well concealed. Nests specifically built for roosting tend to be flimsy, situated higher up, and have less camouflage, since the ability to make a rapid exit on disturbance is apparently of more importance.

Most wren nests are domed structures with a side entrance, although nests built by some cavity-nesting species such as the northern house wren have no roof. In many species—for example in the genera *Thryothorus* and *Uropsila*—the nest is an elaborate, often a two-chambered, beautifully built structure; in others, for example, the song wren (*Cyphorhinus aradus*), it is tattered and untidy. Sometimes, as in the winter wren, nests are artfully concealed. By contrast, the cactus wren usually builds in savagely spiny cholla cacti with little effort at concealment. Many tropical wrens specifically nest in acacia trees that harbor aggressive colonies of symbiotic ants that attack any intruder, and the birds gain vicarious protection from predators such as monkeys. In the absence of such ants, some wren species deliberately site their nests next to pendant hornets' nests for the same purpose.

Clutch size in wrens varies from two in many tropical species to up to ten in temperate zones. Eggs are white, immaculate to heavily speckled, or rarely, blue. Incubation is by females alone, but young are fed by both sexes, except in polygamous species, where male help may be rare.

Polygamy is especially pronounced in the genus *Cistothorus*, reaching its peak with the marsh wren. In some populations of that species, more than half of the breeding males are bigamists, and a significant number are trigamists. The sex ratio among marsh wrens also seems to be skewed, with females outnumbering males almost two to one.

Cooperative breeding occurs in a number of species, but reaches its greatest development in the stripe-backed wren (*Campylorhynchus nuchalis*) of northern South America. There, groups of up to 14 birds will defend a territory, but only one dominant pair actually breeds. Nesting success and the rearing of two broods a year are both strongly correlated to group size; without at least two helpers a pair will not attempt a second brood. Birds in a group are usually blood relations, but after a year or so all females and some males disperse to other groups, thus minimizing incest.

Large gaps remain in human knowledge of wren breeding biology. For example, of the wrens found only south of the United States-Mexican border, the nest and eggs have never been described in about a third of species, and for a good many others available information is sparse.

Conservation status

Most of the 60 species of wren are in no immediate danger of extinction, though with forest destruction the ranges of many have been fragmented and the total population doubtless substantially reduced. In some instances human activity has helped wren populations, notably in the case of the northern and southern house wrens, which do not occur in undisturbed thick woodland but readily take to second growth and bushland, the amount of which greatly increased as European settlement spread across the Americas. Bewick's wren expanded eastward in the nineteenth century for the same reason, but may now be retreating due to competition and egg destruction by the northern house wren.

There are several species of wren for which there is cause for concern. The unique Zapata wren of southwestern Cuba inhabits a very small area at a rather low density. Discovered in 1926, it was feared extinct in the 1970s but appears to have a current population of about 75–100 pairs. The main threats are the burning of its marshland habitat and predation by introduced mongooses. Its current status is Endangered. Niceforo's wren (*Thryothorus nicefori*) was discovered in a tiny area of Colombia in 1945 and not seen again until 1989. The population is obviously quite small, and the habitat currently unprotected and at risk; Niceforo's wren was classified as critically Endangered in 2001. Apolinar's wren (*Cistothorus apolinari*), which is Endangered, also occurs in a very restricted range in Colombia. It has very specific habitat requirements, namely lakeside reed-beds, and much of the habitat lies in well-populated areas; several previously known areas are now lost. The two species of the genus *Hylorchilus* each occupy very specific and restricted habitat in southern Mexico, being tied to open forest on karst limestone outcrops. Sumichrast's wren (*H. sumichrasti*) has an overall range of about 2,000 sq mi (5,200 km²), but only occurs in small isolated pockets in that area; Nava's wren (*H. navai*) occurs in isolated pockets in an even smaller overall range. Some of the range of Nava's wren is protected, but none of that of Sumichrast's wren; they were classified in 2001 as Vulnerable and Near Threatened, respectively.

Although no full species of wren have been lost, several island races have become extinct in recent times. Thus two races of Bewick's wren, formerly found on islands off the Californias, are gone, as is the Martinique race of the southern house wren. Two other West Indian races of the same species, on St. Lucia and Guadeloupe, are in very parlous states; and finally, an isolated race of the rock wren on San Benedicto in the Revellagigedo islands off western Mexico became extinct in a very spectacular manner in 1952, when its island home erupted catastrophically.

There are several other species or subspecies of wren found only in very small ranges: the Clarion (*Troglodytes tanneri*), Socorro (*Thryomanes sissonii*), and Cozumel wrens on the Mexican islands of the same names; the giant wren, occurring only in Chiapas state of Mexico; and the Yucatan wren (*Campylorhynchus yucatanicus*) in the Yucatán peninsula. Although the first two are classified as Near Threatened due to their small island range, most recent observers have found them to be fairly common. The other three species seem to be common to abundant in their limited ranges. One further South American species, the bar-winged wood wren (*Henicorhina leucoptera*), has a very limited distribution in Ecuador and Peru, where it is probably protected by the remoteness of its mountain habitat.

Significance to humans

Wrens have no economic impact on agriculture and probably little on forestry. Nevertheless, it is hardly surprising that birds so abundant and so vehement of song should figure as prominently in legend and folklore as wrens do. In Celtic mythology, the wren was the king of the oak tree and the symbol of the old year, while the robin was the symbol of the new year. This gave rise to a tradition of hunting the wren at the winter solstice to make way for the robin. On the Isle of Man and in southern Ireland, groups of boys caught wrens on St. Stephens Day (December 26), which were then paraded around with accompanying verse and solicitations for modest funds. "Wrenning" was not, in fact, confined to Celtic regions of the British Isles, but up until the nineteenth century was widespread over much of England as well. The verses chanted often made reference to the wren as the king of the birds, a belief widespread in the mythologies of very diverse peoples in Europe. In fact, in several European languages the name of the wren implies royalty—Winterkonig (Winter King) in Dutch and Zaunkönig (King of the Fencerow) in German, for example.

Although the origins are obscure, there is a common theme in cultures as various as Ojibwa and Scottish Gaelic of the wren attaining kingship by outwitting the eagle. In one well-known story, all the birds agreed to decide their king by holding a competition to see who could fly highest. When the eagle, who had outflown all the other birds, could climb no more, the wren, which had concealed itself among the eagle's back feathers, popped out and flew just a little higher. The origin of this story may be an Aesop fable that has now been lost. In Native American folklore the wren appears frequently, often as a busybody; in Cherokee belief the wren is supposed to observe women in labor, rejoicing in the birth of a girl and lamenting the appearance of a boy.

1. Winter wren (*Troglodytes troglodytes*); 2. Gray-breasted wood wren (*Henicorhina leucophrys*); 3. Canyon wren (*Salpinctes mexicanus*); 4. Marsh wren (*Cistothorus palustris*); 5. Black-capped donacobius (*Donacobius atricapillus*); 6. Northern house wren (*Troglodytes aedon*); 7. Bay wren (*Thryothorus nigricapillus*); 8. Sumichrast's wren (*Hylorchilus sumichrasti*); 9. Zapata wren (*Ferminia cerverai*); 10. Cactus wren (*Campylorhynchus brunneicapillus*). (Illustration by Barbara Duperron)

Species accounts

Cactus wren
Campylorhynchus brunneicapillus

TAXONOMY
Picocolaptes brunneicapillus Lafresnaye 1835, Guaymas, Mexico. Eight subspecies recognized.

OTHER COMMON NAMES
French: Troglodyte des cactus; German: Kaktuszaunkönig; Spanish: Matraca Desértica.

PHYSICAL CHARACTERISTICS
7.2–7.6 in (18–19 cm); 1.2–1.6 oz, mean 1.4 oz (33.4–46.9 g, mean 38.9 g). The largest species of wren in the United States. The bird is chocolate-brown above, with a plain cap. The back is heavily streaked with black and white, the wings prominently barred with buff and blackish, the tail feathers with alternating blackish brown and gray-brown bars, the outer tail feathers conspicuously barred black and white. Underparts are buff-white and heavily spotted with black, especially on chest. Lower flanks are buff. It has a conspicuous white supercilium. Eyes are reddish brown, bill dull black with paler base, legs pinkish brown. Sexes are similar. The juvenile has less well-defined streaks and spots; eye color is muddy gray-brown.

DISTRIBUTION
Resident from southeast California, southwest Nevada, southern Arizona and New Mexico, southwest Texas through central Mexico as far south as Michoacán and Hidalgo; Baja California.

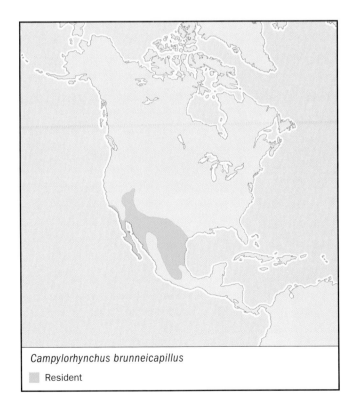

Campylorhynchus brunneicapillus
▨ Resident

HABITAT
Semi-desert from sea level to 4,500 ft (1,400 m), rarely to 6,500 ft (2,000 m), in various vegetation-types, provided that there are spiny cacti such as cholla for nesting. Will adapt to badly degraded habitat so long as some spiny cactus nesting sites remain.

BEHAVIOR
A rambunctious and noisy bird, usually found in pairs or family parties. Song is a loud, harsh series of "jar-jar-jar" notes, frequently delivered from the top of a cactus or other perch. Roosts in nests that are often built for that purpose; old birds may roost alone, fledged broods are usually together.

FEEDING ECOLOGY AND DIET
Majority of food is invertebrate (ants, wasps, spiders, caterpillars, etc.); also eats small frogs and lizards. Vegetable matter includes cactus seeds and fruit; may visit bird feeders. Can exist without drinking, but will drink if water is available. Tends to feed on the ground, overturning litter and stones for prey.

REPRODUCTIVE BIOLOGY
Monogamous. Nest is a conspicuous ovoid ball with a side entrance hole, made of dry grasses and fibers and lined with feathers. Nests are almost invariably located in spiny cacti; little effort is made at concealment. Eggs usually number three to five, sometimes two to seven, are buff or pinkish in color and finely speckled with reddish brown. Populations in Baja California tend to lay smaller clutches. Incubation is by the female alone, about 16 days in length. Young are fed by both sexes for 19–23 days. In Arizona, nesting may begin as early as January, more usually February. Multibrooded; may attempt up to six broods a year, but only three successful broods are reared. Unlike tropical members of its genus, additional birds (other than the breeding pair) rarely help at the nest.

CONSERVATION STATUS
Not threatened. In suitable habitat one of the most abundant species. Can withstand significant habitat modification provided some spiny cactus remain for nesting sites.

SIGNIFICANCE TO HUMANS
A familiar and popular local species, it is the state bird of Arizona. ◆

Canyon wren
Salpinctes mexicanus

TAXONOMY
Thryothorus mexicanus Swainson, 1829, Real del Monte, Hidalgo, Mexico.

OTHER COMMON NAMES
French: Troglodyte des canons; German: Schluchtenzaunkönig; Spanish: Saltaparad Barranquero.

PHYSICAL CHARACTERISTICS
5–5.6 in (12.5–14 cm). Male 0.35–0.52 oz (9.9–14.8 g), female 0.35–0.43 oz (9.9–12.2 g). A slender, long-tailed wren with a

Salpinctes mexicanus
■ Resident

fine decurved bill, quite unlike any other species in its range or habitat. Crown and nape are gray-brown, back is reddish, rump is chestnut. Throat and upper chest are white, contrasting with chestnut-brown belly and rich red-brown lower belly. Crown and back have numerous white speckles; wing and tail feathers have blackish transverse bars. Eyes are dark brown, bill is grayish black and paler at base, legs are dull gray-black. Sexes are similar. In the juvenile, pale speckles on upperparts are obscure, and underparts are less brightly colored.

DISTRIBUTION
Mountainous regions of western North America from southern British Columbia, east to Montana, Wyoming, South Dakota, and western Texas; south through Mexico to Oaxaca and disjunctly in Chiapas. Largely sedentary; northern populations descend in winter. Some vagrant records outside breeding range.

HABITAT
Confined to areas with rock faces, canyons, bluffs, and, rarely, sea-coasts; also occurs in ancient ruins, especially in Mexico. Sea-level to 9,800 ft (3,000 m), lower in northern parts of range.

BEHAVIOR
Forages on rock faces, over which it crawls much in the manner of a wallcreeper; the tail is not used as a prop as in the true creepers. Frequently enters narrow crevices and cracks in rock face; will sometimes hawk for aerial prey, more rarely forages on ground. Song is a superb descending trill, ending in a series of six or seven beautiful clear notes.

FEEDING ECOLOGY AND DIET
Specially adapted to gain access to narrow cracks in rock; the long bill, flattened head, special articulation of the skull and spine, and widely spaced legs are all adapted for this purpose. Food is entirely invertebrate, including beetles, spiders, ter-

mites, etc. Has been seen to steal paralyzed spiders from nests of mud-dauber wasps.

REPRODUCTIVE BIOLOGY
Monogamous. Nest, built by both sexes, is an open cup of hair, feathers, and wool with a base of twigs situated in crevices in rock faces; sometimes in artificial cavities in ruins or buildings. Eggs number three to seven, usually six; they are glossy white with fine reddish spots. Clutch size is smaller in Mexican populations. Incubation is by the female alone, for 12–18 days. Young are fed by both parents for 12–17 days. Does not build special roosting nests.

CONSERVATION STATUS
Not threatened; the remoteness and ruggedness of its habitat gives it substantial protection.

SIGNIFICANCE TO HUMANS
None known. ◆

Slender-billed wren
Hylorchilus sumichrasti

TAXONOMY
Catherpes sumichrasti Lawrence, 1871, Mato Bejuco, Veracruz, Mexico.

OTHER COMMON NAMES
English: Sumichrast's wren; French: Troglodyte de Sumichrast; German: Schmalschnabel-Zaunkönig; Spanish; Saltaparad de Sumichrast.

PHYSICAL CHARACTERISTICS
6–6.6 in (15–16.5 cm), 1.1 oz (29–30g). A bulky and dark-plumaged wren with a notably long and slender bill. Upperparts including tail are a deep chocolate-brown. Wings are dull blackish brown. Throat is whitish brown, chest orange-brown,

Hylorchilus sumichrasti
■ Resident

belly rich chocolate-brown. Lower belly and flanks have small white spots. Eyes are brown; bill is very long, slender, and decurved, blackish with dull orange-yellow base; legs are dark gray. Sexes are identical. The juvenile has a dull buff throat with diffuse scaling and whitish flecks on the belly.

DISTRIBUTION
Restricted area in the Mexican states of Veracruz, Oaxaca, and Puebla; total area is about 2,300 mi² (6,000 km²), but actual range within this area is much less due to exacting habitat requirements.

HABITAT
Confined entirely to karst limestone country with large mature forest canopy.

BEHAVIOR
Typically hops over boulders, tail cocked, bobbing like a dipper. Is not excessively wary, but rather curious. Generally occurs low to the ground, rarely at any height in vegetation, and most frequently on rocks and boulders. Normally found alone or in pairs. Both sexes sing in a distinguishable manner. Males have two song types: one an arresting series of loud clear notes, finishing with a longer series of slow, descending notes; the other is a song of shorter notes alternating in pitch. Female song is a single note repeated in a series.

FEEDING ECOLOGY AND DIET
Most often forages in mosses and lichens on boulders and rock faces, frequently disappearing into crevices. Food is almost entirely invertebrate.

REPRODUCTIVE BIOLOGY
Little known; only three nests described, two in crevices in rocks and one in roof of limestone cave. Eggs are white, number three, and are laid in May. Incubation and fledging periods are unknown. Not known if the species is multiple-brooded, or whether polygamy or nest-helping occur.

CONSERVATION STATUS
Near threatened. Highly restricted both in geographic range and in habitat. Does, however, seem to be able to tolerate some disturbance to habitat, such as the planting of coffee bushes, provided the disturbance is moderate. No part of its range is under formal protection.

SIGNIFICANCE TO HUMANS
None known. ◆

Marsh wren
Cistothorus palustris

TAXONOMY
Certhia palustris Wilson, 1810, eastern Pennsylvania. 15 subspecies recognized.

OTHER COMMON NAMES
English: Long-billed marsh wren; French: Troglodyte de marais; German: Sumpfzaunkönig; Spanish; Saltapared Pantañero.

PHYSICAL CHARACTERISTICS
4.5–5 in (11.5–12.5 cm). Male 0.43–0.5 oz, mean 0.45 oz (12.1–14 g, mean 12.7 g). Populations in western North America weigh slightly less. Crown is blackish brown; back is

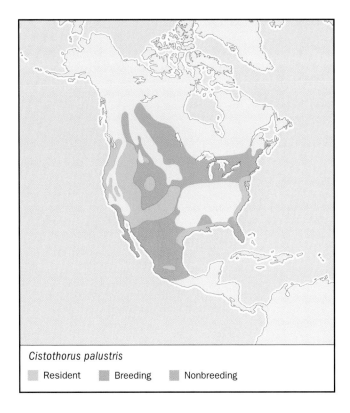

Cistothorus palustris

Resident · Breeding · Nonbreeding

medium-brown with broad collar of blackish; streaked white on upper back. Wings and tail are reddish brown with darker bars. Supercilium is whitish, contrasting with cap and brown eyestripe and ear-coverts. Underparts are gray, becoming buff on lower belly and flanks. Eyes are brown; bill is dark brown above and yellowish brown below; legs are pale brown. Sexes are identical. The juvenile rather duller than an adult with reduced white streaking on back; supercilium is less obvious.

DISTRIBUTION
Breeds across North America from New Brunswick and Virginia to British Columbia, north to northern Alberta, south to central Midwestern states and California; also coastal marshes along the eastern United States seaboard and Gulf of Mexico. Populations in eastern and western North America may be two distinct species.

HABITAT
Freshwater marshes with cattails, bulrushes, or phragmites; coastal marshes with *Spartina*. Winter migrants occur frequently in coastal brackish marshes.

BEHAVIOR
Active and noisy, moving rapidly among reeds, often clinging spread-eagled to two separate stems. The bird is highly territorial and predatory on eggs, both of its own species and of others, such as red-winged blackbirds. Song is a characteristic metallic rattle, sometimes uttered at night. There are many distinctive patterns of songs, with major differences between eastern and western populations.

FEEDING ECOLOGY AND DIET
Food is mainly invertebrate, including beetles, bugs, dragonfly larvae, spiders, etc. Predated eggs may be partially eaten. Feeds from the water level to the tops of reeds, but usually stays fairly low.

REPRODUCTIVE BIOLOGY

Exhibits two features found in many other wren species to a high degree, multiple nest-building and polygamy. The nest is a domed structure with a side entrance usually about 3 ft (1 m) above water level. It is made of strips of leaves and grass and lined with down. Eggs number three to 10, usually four to six, fewer in southern populations. They are brown with darker speckles, rarely white. Incubation is by the female, for 13–16 days; young fledge in 13–15 days. Rate of polygamy varies among different populations; may be more than 50% in central Canada, with a significant number of trigamists.

CONSERVATION STATUS

Not threatened. Can be common in suitable habitat. Can reach high densities of up to 75 males per acre (190 per ha) in prime habitat. Much habitat loss is due to swamp drainage; conversely, recent water impounds may create habitat.

SIGNIFICANCE TO HUMANS

Has been and continues to be an intense object of research because of its complex song patterns and interesting breeding behavior. ◆

Zapata wren

Ferminia cerverai

TAXONOMY

Ferminia cerverai Barbour, 1926, Santo Tomás, Ciénaga de Zapata, Cuba.

OTHER COMMON NAMES

English: Cuban marsh wren, Cervera's wren; French: Troglodyte de Zapata; German: Kubazaunkönig; Spanish: Fermina.

PHYSICAL CHARACTERISTICS

6.2–6.5 in (15.5–16 cm). Weight: no data. Quite unlike any other wren, with very short wings, long tail, and sturdy legs. Crown is blackish brown; back is brownish with numerous fine bars on back and tertial feathers. Tail is long and fluffy with numerous fine dark bars; underparts are whitish buff, darker

below, the flanks with dark barring. Eyes are clear brown; bill is dark brown with paler base; legs are brownish. Sexes are identical. Juveniles are similar to adults, but with fine speckles on throat and more diffuse barring on flanks.

DISTRIBUTION

Confined entirely to the Ciénaga de Zapata, southwest Cuba. Originally entirely in the vicinity of Santo Tomás; recently small additional populations discovered 9–12 mi (15–20 km) from Santo Tomás.

HABITAT

Restricted to savanna swampland, with saw grass, rushes, and scattered bushes. Seems to prefer drier areas where it can forage on the ground.

BEHAVIOR

A very poor flyer. Is quarrelsome and prone to drive off much larger species. Is rather secretive, though it frequently sings from low bushes. Song is a series of gurgling whistles mixed with harsher churring notes. Some dispute as to whether females sing; the female song is stated to be shorter and higher-pitched than that of the male.

FEEDING ECOLOGY AND DIET

Food is varied for a wren; invertebrates such as crickets, caterpillars, flies and spiders, snail eggs, and slugs are eaten. Also takes *Anolis* lizards, some up to 4 in (10 cm) in length, and some vegetable matter. Forages low in vegetation, but also takes much from the ground.

REPRODUCTIVE BIOLOGY

A nest was not discovered until 1986, and by 2001 only four had been described. The nest is a ball of saw grass leaves situated in saw grass, with a side entrance hole. It occurs 20–28 in (50–70 cm) above the ground. Eggs number two and are white. Incubation is by the female alone; incubation and fledging periods are unknown. Breeding season extends from January to July; the species may be double-brooded. It is not known whether polygamy occurs.

CONSERVATION STATUS

Endangered. The bird is extremely restricted in range. Original range was only 5 sq mi (13 km²), but recently other populations were discovered up to 12 mi (20 km) away. Even within this range, the birds are rather sparse. In the late 1970s no birds could be located and it was feared extinct; present best estimates (April 2000) are of 80–100 pairs. Threats to survival include grass fires and predation by introduced mongooses.

SIGNIFICANCE TO HUMANS

None known. ◆

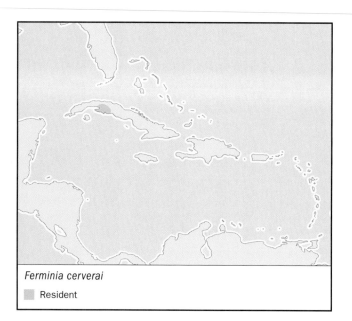

Ferminia cerverai

▨ Resident

Bay wren

Thryothorus nigricapillus

TAXONOMY

Thryothorus nigricapillus Sclater, 1860, Nanegal, Pichincha, Ecuador.

OTHER COMMON NAMES

French: Troglodyte á tête noir; German: Kastanienzaunkönig; Spanish: Cucarachero Cabecinegro, (Panama) El Guerrero.

PHYSICAL CHARACTERISTICS

5.8 in (14.5 cm). 0.78 oz (21.9 g). Strikingly marked, with black head, white facial markings, chestnut back, and heavily

Thryothorus nigricapillus

Resident

it occurs from February to November, in Colombia from January to August.

CONSERVATION STATUS
Not threatened; substantial areas of habitat have been lost, however.

SIGNIFICANCE TO HUMANS
None known. ◆

Gray-breasted wood wren
Henicorhina leucophrys

TAXONOMY
Troglodytes leucophrys Tschudi 1844, Montaña del Vitoc, Junín, Peru. 18 subspecies recognized.

OTHER COMMON NAMES
English: Highland wood wren; French: Troglodyte à poitrine gris; German: Einsiedlerzaunkönig; Spanish: Saltapared-selvatico Pechigris.

PHYSICAL CHARACTERISTICS
4–4.5 in (10–11.5 cm). The bird is a large-headed plump wren with a short tail and sturdy legs, chestnut-brown above, and gray below with brownish buff lower flanks. Sides of face are strikingly marked with black and white streaks; eyestripe is black; supercilium is white; crown is blackish; throat is grayish white. Flight feathers are chestnut-brown with narrow blackish bars. Eyes are reddish brown; bill is black with dark gray base; legs are brownish black. Male and female are identical in

barred wings and tail. Central American populations have a white throat and unbarred chestnut belly; the South American races are heavily barred with black from the upper chest to the lower flanks. Eyes are brown; bill is blackish with pale orange-yellow base; legs are blackish. Sexes are similar. Juveniles have less clear-cut facial markings, and colors are generally somewhat less rich.

DISTRIBUTION
Caribbean slope of Central America from central Nicaragua and Costa Rica to Panama; both drainages from central Panama to Colombia, western Colombia along Pacific slope to southern Ecuador.

HABITAT
Edges of humid forest, generally absent from forest interior; *Heliconia* thickets. Occurs from sea level to 3,600 ft (1,100 m).

BEHAVIOR
Found in pairs, low in thick vegetation; rather secretive. It is territorial throughout year; both sexes sing and defend territory. Song is a series of varied, loud, ringing whistles. Males may sing on their own, but antiphonal song with female is frequent. The female always initiates a song duet.

FEEDING ECOLOGY AND DIET
Food is entirely invertebrate, including spiders, beetles, roaches, earwigs, etc.

REPRODUCTIVE BIOLOGY
Nest, built by both sexes, is an elbow-shaped construction made of grass and vine tendrils, placed in vines or crotches of shrubs at heights of 3–16 ft (1–5 m). Eggs number three and are white with cinnamon speckles. Incubation and fledging periods are not recorded. Breeding season protracted. In Panama

Henicorhina leucophrys

Resident

plumage; juvenile is similar to adults but with the facial pattern less well-defined. There is considerable variation in plumage between different subspecies.

DISTRIBUTION
Disjunct distribution in highlands of Central and South America, separated by unsuitable lowland areas, from central to southern Mexico through Guatemala, Honduras, Nicaragua, Costa Rica, and Panama; Andes and adjacent mountain ranges from northern and central Venezuela through Colombia south to central Bolivia.

HABITAT
Humid mountain forests, usually above 4,900 ft (1,500 m) but rarely as low as 1,300 ft (400 m), up to 12,500 ft (3,800 m). Occupies several different forest types, including mixed oak and pine woodland, bamboo thickets, and second growth, up to the lower edge of páramo (high altitude heathland).

BEHAVIOR
Territorial; lives in pairs or family groups. Is very secretive and usually keeps to lower levels of vegetation. Song is a magnificent series of loud ringing whistled phrases, uttered by both sexes in duet. Roosts in pairs or family groups, in nests built for that purpose. Form of roosting nest is identical to that of breeding nest.

FEEDING ECOLOGY AND DIET
Feeds in lower levels of dense vegetation, actively moving around and probing into leaf-litter and tangles. Diet apparently entirely invertebrate; no evidence that any vegetable matter is taken.

REPRODUCTIVE BIOLOGY
Nest, built of roots, plant fibers, and moss, is a globular structure with an antechamber leading to a rounded chamber, placed in low vegetation including bamboo thickets. Both sexes build the nest. Eggs number two and are immaculate white. Incubation is by the female only, for 19–20 days. Young are fed by both sexes for 17–18 days. Breeding season depends on geographic location. In Costa Rica breeding occurs late March to early June, in Colombia December to June. Probably multiple-brooded.

CONSERVATION STATUS
Not threatened. Much habitat has been lost to agriculture and logging, but large areas remain; some are protected by reserve status.

SIGNIFICANCE TO HUMANS
None known. ◆

House wren
Troglodytes aedon

TAXONOMY
Silvia domestica Wilson, 1808; suppressed in favor of *Troglodytes aedon*, Vieillot, 1809. Three subspecies recognized (Clements).

OTHER COMMON NAMES
French: Troglodyte familier; German: Hauszaunkönig; Spanish: Saltapared Cucarachero.

PHYSICAL CHARACTERISTICS
4.6–5 in (11.5–12.5 cm). Weight 0.28–0.40 oz (8–11 g). A rather plain gray-brown wren without prominent markings.

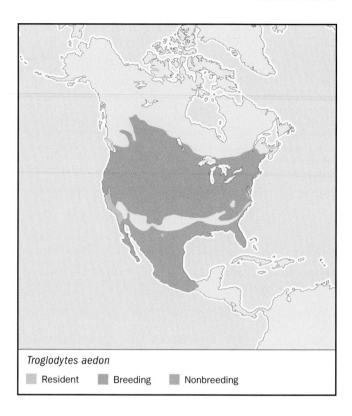

Troglodytes aedon

■ Resident   ■ Breeding   ■ Nonbreeding

Crown and nape are medium-brown; back and rump are reddish brown; wings and tail have narrow dark barrings. Face is gray-brown with ill-defined pale supercilium; throat and chest are buffy-white; belly and flanks are buffy-red. Eyes are brown; bill is dusky-brown above and pale horn-gray below; legs are dark brown. Sexes are similar. Juvenile has dusky mottles on breast.

DISTRIBUTION
Breeds across North America from New Brunswick south to California, west to central Alberta and southern British Columbia and south to California. Winters almost entirely south of breeding range, from South Carolina westwards to southern Arizona, south through Mexico to Oaxaca.

HABITAT
Partially open country with brushy areas, abandoned farmland, forest edges, and well-vegetated suburban areas; in western North America in open deciduous and coniferous forest.

BEHAVIOR
Very territorial in breeding season. Not particularly secretive, especially when singing. Frequently destroys the eggs of cavity-nesting birds, such as the prothonotary warbler and Bewick's wren. Song is a long series of rapid notes, not particularly tuneful, mostly on one pitch then descending at end. Both sexes sing.

FEEDING ECOLOGY AND DIET
Most often forages low in tangled vegetation. Food is mainly invertebrate, including spiders, caterpillars, bugs, etc. Also eats a small amount of vegetable matter.

REPRODUCTIVE BIOLOGY
Is a cavity nester, using old woodpecker holes, tree cavities, disused hornets' nests, old nests of northern oriole or cliff swallow, but increasingly using nest boxes and other artificial sites. Nest

has a coarse twig base with a cup lined with hair, wool, and feathers. Eggs number four to eight, rarely three to 10. They are whitish with small spots of reddish brown. Incubation is by the female alone for 12–14 days; young are fed by both sexes for 16–18 days. Single-brooded in Canada, double-brooded or rarely triple-brooded in southern parts of the range. Generally monogamous, but polygyny does occur regularly.

CONSERVATION STATUS
Not threatened; populations augmented by provision of nest boxes and abandonment of marginal farmland.

SIGNIFICANCE TO HUMANS
Well-known and popular. Due to the readiness with which it takes to nest boxes, is one of the best-studied of North American passerines. ◆

Winter wren
Troglodytes troglodytes

TAXONOMY
Motacilla troglodytes Linnaeus, 1758, Sweden. 44 subspecies recognized.

OTHER COMMON NAMES
English: Wren, northern wren, holarctic wren; French: Troglodyte mignon; German: Zaunkönig; Spanish: Saltapared Invernal, Cochín.

PHYSICAL CHARACTERISTICS
3.6–4 in (9–10 cm). Weight 0.26–0.39 oz (7.5–11.0 g). A very small, short-tailed, and heavily barred wren. Upperparts are warm dark brown, with conspicuous narrow dark bars on wing and tail feathers. Chin and throat are grayish brown, becoming more reddish lower; flanks are deep reddish brown with darker bars. Eyes are brown; bill is brown with paler base; legs brown. Sexes are similar. Juvenile has obscure mottles on chest, and flank bars are less distinct.

DISTRIBUTION
The most widespread of the wrens, found in four continents. North America, breeding range from Alaska, the Aleutian Islands south in mountains to California, eastward across Canada to Newfoundland and south in mountains to Georgia; winter to northern Mexico. Old World from Iceland and Scandinavia

south to Spain, Morocco, Algeria, and Libya; eastward to Russia, Caucasus, Turkey, and Iran; central Asia from Afghanistan to eastern Siberia, Japan, China, and Taiwan, including many offshore islands in Europe and east Asia.

HABITAT
Extremely varied habitat. In North America, occurs most often in cool forested areas, especially coniferous. In Europe and Asia, occurs in various types of bush and woodland, including deciduous woods with extensive undergrowth; also in suburban areas and treeless offshore islands with low scrubby vegetation. Sea level to 13000 ft (4000 m) in central Asia.

BEHAVIOR
Territorial in breeding season, but will sometimes roost communally in winter, with several dozen birds crammed in layers into a suitable cavity. Spends most of time down in vegetation, hopping mouse-like through dense tangles; flights are always low and short, between adjacent cover. Song is loud and vehement, a prolonged series of trills and clear notes, much sweeter and purer than song of the house wren.

FEEDING ECOLOGY AND DIET
Not known.

REPRODUCTIVE BIOLOGY
Nest is a domed structure with side entrance, usually in dense vegetation or cavities; island races may nest in crannies in rock-faces or ruined buildings. Eggs number five to eight, rarely three to nine. They are white with red-brown speckles; some Aleutian races may lay white eggs. Incubation is by the female for 16 days; young are fed by both parents for 14–19 days. Usually double-brooded. Polygyny is frequent but varies among races. Males build many surplus nests; the female selects the one to be used for breeding and adds the lining.

CONSERVATION STATUS
Not threatened. Generally common or abundant over much of its range; populations in Britain fluctuate wildly, being decimated by a severe winter but bouncing back very rapidly. Some island races have very small populations that may fluctuate quickly.

SIGNIFICANCE TO HUMANS
A familiar and popular bird, it is the subject of extensive folklore in many countries, well enough known to be among the few birds with a given name in English, Jenny Wren. ◆

Troglodytes troglodytes

☐ Resident   ☐ Breeding   ☐ Nonbreeding

Black-capped donacobius
Donacobius atricapillus

TAXONOMY
Turdus atricapilla Linnaeus, 1766, location in error, actually eastern Brazil.

OTHER COMMON NAMES
English: Black-capped mockingthrush; French: Troglodyte à miroirs; German: Rohrspottdrossel; Spanish: Paraulata de Agua, Donacobio.

PHYSICAL CHARACTERISTICS
8.5–9 in (21–22 cm); 1.1–1.5 oz (31–42 g). Unique and unmistakable. Crown, nape, and shoulders are glossy black; back is browner; rump is olive-brown. Tail feathers are black with conspicuous white tips. Wings are blackish with conspicuous

Donacobius atricapillus

☐ Resident

HABITAT
Brushy vegetation over slow-moving rivers and ponds; occurs at sea level to rarely 2,000 ft (750 m), usually lower.

BEHAVIOR
Demonstrative and noisy. Pairs indulge in loud ritualized displays, advertising the white wing-flashes and tail edgings by spreading the wings and tail. Song is a series of loud whistles. Both sexes sing in antiphonal style. Female song has a lower, grating quality not found in the male's contribution. Other members of a nesting group may also join in. The bird is territorial, with the territories often being linear along a marsh edge.

FEEDING ECOLOGY AND DIET
Not known.

REPRODUCTIVE BIOLOGY
A cooperative breeder; a nesting pair may have up to two additional helpers, which are usually young from the previous year or two. Pairs with no assistants rear only one young bird, while helpers can increase the number to two young. Nest is an open cup, usually built over or near water. Eggs usually number two in Venezuela, in Brazil frequently three. They are purplish white, covered with reddish or purplish spots and blotches. Incubation is by the female alone for 16–18 days; young are fed by both parents and by helpers, fledging at 17–18 days. Single-brooded. Adults keep young in nest cool by soaking their body feathers in water and wetting the nestlings.

CONSERVATION STATUS
Not threatened. Common or abundant over much of its range. Although marsh-drainage may destroy habitat, the species will colonize suitable vegetation arising around artificial water impoundments.

SIGNIFICANCE TO HUMANS
For two hundred years has provided much diversion and entertainment for taxonomic ornithologists, a process that will doubtless continue. ◆

white flash at the base of primaries. Underparts warm yellow buff with black bars on flanks. Eyes are bright yellow; legs are dusky green. Has a distendable yellow cheek pouch.

DISTRIBUTION
Panama (Darién) through lowland South America east of the Andes to coastal Brazil and northern Argentina

---

Resources

Books

Anderson, A.H., and A. Anderson. *The Cactus Wren.* Tucson: University of Arizona Press, 1973.

Armstrong, E.A. *The Wren.* London: Collins, 1955.

Brewer, D., and B.K. MacKay. *Wrens, Dippers and Thrashers.* New Haven and London: Yale University Press and A&C Black, 2001.

Hejl, S., and J.A. Holmes. "Winter Wren (*Troglodytes troglodytes*)." In *The Birds of North America*, ed. F. Gill and A. Poole. Philadelphia and Washington, DC: The Academy of Natural Sciences and the American Ornithologists' Union, 2000.

Johnson, L.S. "House Wren (*Troglodytes aedon*)." In *The Birds of North America*, ed. F. Gill and A. Poole. Philadelphia and Washington, DC: The Academy of Natural Sciences and the American Ornithologists' Union, 1998.

Raffaele, H., J. Wiley, O. Garrido, A. Keith, and J. Raffaele. *A Guide to the Birds of the West Indies.* Princeton: Princeton University Press, 1998.

Ridgely, R.S., and G.E. Tudor. *The Birds of South America.* Vol. 1. Austin: University of Texas Press, 1989.

Periodicals

Atkinson, P.W., M.J. Whittingham, H. Gómez de Silva Garza, A.M. Kent, and R.T. Maier. "Notes on the Ecology, Conservation and Taxonomic Status of *Hylorchilus* Wrens." *Bird Conservation International* 3 (1993): 75–85.

Joyce, F.J. "Nesting Success of Rufous-Naped Wrens (*Campylorhynchus rufinucha*) is Greater Near Wasps' Nests." *Behavioral Ecology and Sociobiology* 32 (1993): 71–77.

Kiltie, R.A., and J.W. Fitzgerald. "Reproduction and Social Organisation of the Black-Capped Donacobius (*Donacobius atricapillus*) in Southern Peru." *Auk* 101 (1993): 804–811.

Verner, J. "Evolution of Polygamy in the Long-billed Marsh Wren." *Evolution* 18 (1964): 252–261.

Woods, R.W. "Cobb's Wren *Troglodytes (aedon) cobbi* of the Falkland Islands." *Bulletin of the British Ornithologists' Club* 113 (1993): 195–207.

David Brewer, PhD

For further reading

Ali, S. and Ripley, S. D. *Handbook of the Birds of India and Pakistan.* 2nd edition. 10 Vols. New York: Oxford University Press, 1978-1999.

American Ornithologists' Union. *Check-list of North American Birds: the Species of Birds of North America from the Arctic through Panama, including the West Indies and Hawaiian Islands.* 7th ed. Washington, DC: The American Ornithologists' Union, 1998.

Bennett, Peter M. and I. P. F. Owens. *Evolutionary Ecology of Birds: Life Histories, Mating Systems, and Extinction.* Oxford Series in Ecology and Evolution. Oxford: Oxford University Press, 2002.

Berthold, P. *Bird Migration: A General Survey.* Translated by H.-G. Bauer and V. Westhead. 2nd edition. Oxford Ornithology Series, no. 12. Oxford: Oxford University Press, 2001.

Boles, W. E. *The Robins and Flycatchers of Australia.* Sydney: Angus & Robertson, 1989.

Borrow, Nik, and Ron Demey. *Birds of Western Africa: An Identification Guide.* London: Christopher Helm, 2001.

Brewer, David, and B. K. Mackay. *Wrens, Dippers, and Thrashers.* New Haven: Yale University Press. 2001.

Brown, L. H., E. K. Urban and K. Newman, eds. *The Birds of Africa.* Vol. 1, *Ostriches to Falcons.* London: Academic Press, 1982.

Bruggers, R. L., and C. C. H. Elliott. *Quelea quelea: Africa's Bird Pest.* Oxford: Oxford University Press, 1990.

Burger, J., and Olla, B. I., eds. *Shorebirds: Breeding Behavior and Populations.* New York: Plenum Press, 1984.

Burton, J.A., Ed. *Owls of the World: Their Evolution, Structure and Ecology.* 2nd edition. London: Peter Lowe, 1992.

Byers, Clive, Jon Curson, and Urban Olsson. *Sparrows and Buntings: A Guide to the Sparrows and Buntings of North America and the World.* Boston: Houghton Mifflin Company, 1995.

Castro, I., and A. A. Phillips. *A Guide to the Birds of the Galápagos Islands.* Princeton: Princeton University Press, 1997.

Chantler, P., and G. Driessens. *Swifts: A Guide to the Swifts and Treeswifts of the World.* Sussex: Pica Press, 1995.

Cheke, R. A., and C. Mann. *Sunbirds: A Guide to the Sunbirds, Flowerpeckers, Spiderhunters and Sugarbirds of the World.* Christopher Helm, 2001.

Cleere, N., and D. Nurney. *Nightjars: A Guide to the Nightjars and Related Nightbirds.* Sussex: Pica Press, 1998.

Clement, P. *Finches and Sparrows: An Identification Guide.* Princeton: Princeton University Press, 1993.

Clement, P., et al. *Thrushes.* London: Christopher Helm, 2000.

Clements, J. F., and N. Shany. *A Field Guide to the Birds of Peru.* Temecula, California: Ibis Pub. 2001.

Coates, B. J. *The Birds of Papua New Guinea: Including the Bismarck Archipelago and Bougainville.* 2 vols. Alderley, Queensland: Dove Publications, 1985, 1990.

Coates, B. J., and K. D. Bishop. *A Guide to the Birds of Wallacea: Sulawesi, The Moluccas and Lesser Sunda Islands, Indonesia.* Alderley, Queensland: Dove Publications, 1997.

Cooke, Fred, Robert F. Rockwell, and David B. Lank. *The Snow Geese of La Pérouse Bay: Natural Selection in the Wild.* Oxford Ornithology Series, no. 4. Oxford: Oxford University Press, 1995.

Cooper, W. T., and J. M. Forshaw. *The Birds of Paradise and Bowerbirds.* Sydney: Collins, 1977.

Cramp, S., ed. *Handbook of the Birds of Europe the Middle East and North Africa. The Birds of the Western Palearctic.* Vol. 1, *Ostrich to Ducks.* Oxford: Oxford University Press, 1977.

Cramp, S., ed. *Handbook of the Birds of Europe the Middle East and North Africa. The Birds of the Western Palearctic.* Vol. 2, *Hawks to Bustards.* Oxford: Oxford University Press, .

Cramp, S., ed. *Handbook of the Birds of Europe the Middle East and North Africa. The Birds of the Western Palearctic.* Vol. 3, *Waders to Gulls.* Oxford: Oxford University Press, 1983.

Cramp, S., ed. *Handbook of the Birds of Europe the Middle East and North Africa. The Birds of the Western Palearctic.* Vol. 4, *Terns to Woodpeckers.* Oxford: Oxford University Press, 1985.

Cramp, S., ed. *Handbook of the Birds of Europe the Middle East and North Africa. The Birds of the Western Palearctic.* Vol. 6, *Warblers.* Oxford: Oxford University Press, 1992.

Cramp, S., ed. *Handbook of the Birds of Europe the Middle East and North Africa. The Birds of the Western Palearctic.* Vol. 7, *Flycatchers to Shrikes.* Oxford: Oxford University Press, 1993.

Cramp, S., ed. *Handbook of the Birds of Europe the Middle East and North Africa. The Birds of the Western Palearctic.* Vol. 8, *Crows to Finches.* Oxford: Oxford University Press, 1994.

Cramp, S., ed. *Handbook of the Birds of Europe the Middle East and North Africa. The Birds of the Western Palearctic.* Vol. 9, *Buntings and New World Warblers.* Oxford: Oxford University Press, 1994.

Davies, N. B. *Dunnock Behaviour and Social Evolution.* Oxford: Oxford University Press, 1992.

Davies, N. B. *Cuckoos, Cowbirds and Other Cheats.* London: T. & A. D. Poyser, 2000.

Davis, L. S., and J. Darby, eds. *Penguin Biology.* New York: Academic Press, 1990.

Deeming, D. C., ed. *Avian Incubation: Behaviour, Environment, and Evolution.* Oxford Ornithology Series, no. 13. Oxford: Oxford University Press, 2002.

del Hoyo, J., A. Elliott and J. Sargatal, eds. *Handbook of the Birds of the World.* Vol. 1, *Ostrich to Ducks.* Barcelona: Lynx Edicions, 1992.

del Hoyo, J., A. Elliott and J. Sargatal, eds. *Handbook of the Birds of the World.* Vol. 2, *New World Vultures to Guineafowl.* Barcelona: Lynx Edicions, 1994.

del Hoyo, J., A. Elliott and J. Sargatal, eds. *Handbook of the Birds of the World.* Vol. 3, *Hoatzin to Auks.* Barcelona: Lynx Edicions, 1996.

del Hoyo, J., A. Elliott and J. Sargatal, eds. *Handbook of the Birds of the World.* Vol. 4, *Sandgrouse to Cuckoos.* Barcelona: Lynx Edicions, 1997.

del Hoyo, J., A. Elliott and J. Sargatal, eds. *Handbook of the Birds of the World.* Vol. 5, *Barn-owls to Hummingbirds.* Barcelona: Lynx Edicions, 1999.

del Hoyo, J., A. Elliott and J. Sargatal, eds. *Handbook of the Birds of the World.* Vol. 6, *Mousebirds to Hornbills.* Barcelona: Lynx Edicions, 2001.

del Hoyo, J., A. Elliott and J. Sargatal, eds. *Handbook of the Birds of the World.* Vol. 7, *Jacamars to Woodpeckers.* Barcelona: Lynx Edicions, 2002.

Delacour, J., and D. Amadon. *Currasows and Related Birds.* New York: American Museum of Natural History, 1973.

Diamond, J., and A. B. Bond. *Kea, Bird of Paradox: The Evolution and Behavior of a New Zealand Parrot.* Berkeley: University of California Press, 1999.

Erritzoe, J. *Pittas of the World: a Monograph on the Pitta Family.* Cambridge: Lutterworth, 1998.

Feare, C., and A. Craig. *Starlings and Mynahs.* Princeton: Princeton University Press, 1999.

Feduccia, A. *The Origin and Evolution of Birds.* 2nd edition. New Haven: Yale University Press, 2001.

Ferguson-Lees, J., and D. A. Christie. *Raptors of the World.* Boston: Houghton Mifflin, 2001.

Fjeldsa, J., and N. Krabbe. *Birds of the High Andes.* Svendborg, Denmark: Apollo Books, 1990.

Forshaw, J. M. *Parrots of the World.* 3rd rev. edition. Melbourne: Lansdowne Editions, 1989.

Forshaw, J., ed. *Encyclopedia of Birds.* 2nd edition. McMahons Point, N.S.W.: Weldon Owen, 1998.

Frith, C. B., and B. M. Beehler. *The Birds of Paradise: Paradisaeidae.* Bird Families of the World, no. 6. Oxford University Press, 1998.

Fry, C. H., and K. Fry. *Kingfishers, Bee-eaters and Rollers: A Handbook.* Princeton: Princeton University Press, 1992.

Fry, C. H., S. Keith, and E. K. Urban, eds. *The Birds of Africa.* Vol. 3, *Parrots to Woodpeckers.* London: Academic Press, 1988.

Fry, C H., and S. Keith, eds. *The Birds of Africa.* Vol. 6. *Picathartes to Oxpeckers.* London: Academic Press, 2000.

Fuller, Errol. *Extinct Birds.* Rev. ed. Ithaca, N.Y.: Comstock Pub., 2001.

Gehlbach, Frederick R. *The Eastern Screech Owl: Life History, Ecology, and Behavior in the Suburbs and Countryside.* College Station: Texas A & M University Press, 1994.

Gibbs, D., E. Barnes, and J. Cox. *Pigeons and Doves: A Guide to the Pigeons and Doves of the World.* Robertsbridge: Pica Press, 2001.

Gill, F. B. *Ornithology.* 2nd edition. New York: W. H. Freeman, 1995.

Grant, P. R. *Ecology and Evolution of Darwin's Finches.* Princeton: Princeton University Press, 1986.

Hagemeijer, Ward J. M., and Michael J. Blair, eds. *The EBCC Atlas of European Breeding Birds: Their Distribution and Abundance.* London: T. & A. D. Poyser, 1997.

Hancock, J. A., J. A. Kushlan, and M. P. Kahl. *Storks, Ibises, and Spoonbills of the World.* New York: Academic Press, 1992.

Hancock, J. A., and J. A. Kushlan. *The Herons Handbook.* New York: Harper and Row, 1984.

Harris, Tony, and Kim Franklin. *Shrikes and Bush-shrikes: Including Wood-shrikes, Helmet-shrikes, Flycatcher-shrikes, Philentomas, Batises and Wattle-eyes.* London: Christopher Helm, 2000.

Harrap, S., and D. Quinn. *Chickadees, Tits, Nuthatches, and Treecreepers.* Princeton: Princeton University Press, 1995.

Heinrich, B. *Ravens in Winter.* New York: Summit Books, 1989.

Higgins, P. J., and S. J. J. F. Davies, eds. *The Handbook of Australian, New Zealand and Antarctic Birds.* Vol. 3, *Snipe to Pigeons.* Melbourne: Oxford University Press, 1996.

Higgins, P. J., ed. *The Handbook of Australian, New Zealand and Antarctic Birds.* Vol. 4, *Parrots to Dollarbirds.* Melbourne: Oxford University Press, 1999.

Hilty, S. L., and W. L. Brown. *A Guide to the Birds of Colombia.* Princeton, N. J.: Princeton University Press, 1986.

Holyoak, D.T. *Nightjars and Their Allies: The Caprimulgiformes. Bird Families of the World,* no. 7. Oxford: Oxford University Press, 2001.

Howard, R., and A. Moore. *A Complete Checklist of the Birds of the World.* 2nd edition. London: Macmillan, 1991.

Howell, S. N. G., and S. Webb. *A Guide to the Birds of Mexico and Northern Central America.* Oxford: Oxford University Press, 1995.

Howell, S. N. G. *Hummingbirds of North America: The Photographic Guide.* San Diego: Academic Press, 2002.

Isler, M. L., and P. R. Isler. *The Tanagers: Natural History, Distribution, and Identification.* Washington, D.C.: Smithsonian Institution Press, 1987.

Jaramillo, A., and P. Burke. *New World Blackbirds: The Icterids.* Princeton: Princeton University Press, 1999.

Jehl, Joseph R., Jr. *Biology of the Eared Grebe and Wilson's Phalarope in the Nonbreeding Season: A study of Adaptations to Saline Lakes. Studies in Avian Biology,* no. 12. San Diego: Cooper Ornithological Society, 1988.

Johnsgard, Paul A. *Bustards, Hemipodes, and Sandgrouses, Birds of Dry Places.* Oxford: Oxford University Press, 1991.

Johnsgard, Paul A. *Cormorants, Darters, and Pelicans of the World.* Washington, D.C.: Smithsonian Institution Press, 1993.

Johnsgard, Paul A. *Cranes of the World.* Bloomington: Indiana University Press, 1983.

Johnsgard, Paul A. *Diving Birds of North America.* Lincoln: University of Nebraska Press, 1987.

Johnsgard, Paul A. *The Hummingbirds of North America.* 2nd ed. Washington, D.C.: Smithsonian Institution Press, 1997.

Johnsgard, P. A. *The Plovers, Sandpipers, and Snipes of the World.* Lincoln: University of Nebraska Press, 1981.

Johnsgard, Paul A. *Trogons and Quetzals of the World.* Washington, D.C.: Smithsonian Institution Press, 2000.

Johnsgard, Paul A., and Montserrat Carbonell. *Ruddy Ducks and other Stifftails: Their Behavior and Biology.* Norman: University of Oklahoma Press, 1996.

Jones, D. N., R. W. R. J. Dekker, and C. S. Roselaar. *The Megapodes. Bird Families of the World,* no. 3. Oxford: Oxford University Press, 1995.

Juniper, Tony, and Mike Parr. *Parrots: A Guide to the Parrots of the World.* Sussex: Pica Press, 1998.

Kear, J., and N. Düplaix-Hall, eds. *Flamingos.* Berkhamsted: T. & A. D. Poyser, 1975.

Keith, S., E. K. Urban, and C. H. Fry, eds. *The Birds of Africa.* Vol. 4, *Broadbills to Chats.* London: Academic Press, 1992.

Kemp, A. *The Hornbills. Bird Families of the World,* no. 1. Oxford: Oxford University Press, 1995.

Kennedy, Robert S., et al. *A Guide to the Birds of the Philippines.* Oxford: Oxford University Press, 2000.

Lambert, F., and M. Woodcock. *Pittas, Broadbills and Asities.* Sussex: Pica Press, 1996.

Lefranc, Norbert, and Tim Worfolk. *Shrikes: A Guide to the Shrikes of the World.* Robertsbridge: Pica Press, 1997.

Lenz, Norbert. *Evolutionary Ecology of the Regent Bowerbird Sericulus chrysocephalus.* Special Issue of *Ecology of Birds,* Vol. 22, Supplement. Ludwigsburg, 1999.

MacKinnon, John R., and Karen Phillipps. *A Field Guide to the Birds of China.* Oxford Ornithology Series. Oxford; New York: Oxford University Press, 2000.

Madge, S. *Crows and Jays: A Guide to the Crows, Jays, and Magpies of the World.* New York: Houghton Mifflin, 1994.

Madge, Steve, and Phil McGowan. *Pheasants, Partridges and Grouse: A Guide to the Pheasants, Partridges, Quails, Grouse, Guineafowl, Buttonquails and Sandgrouse of the World.* London: Christopher Helm, 2002.

Marchant, S., and P. Higgins, eds. *The Handbook of Australian, New Zealand and Antarctic Birds.* Vol. 1, parts A , B, *Ratites to Ducks.* Melbourne: Oxford University Press, 1990.

Marchant, S., and P. Higgins, eds. *The Handbook of Australian, New Zealand and Antarctic Birds.* Vol. 2, *Raptors to Lapwings.* Melbourne: Oxford University Press, 1993

Higgins, P. J., J. M. Peter, and W. K. Steele. *The Handbook of Australian, New Zealand and Antarctic Birds.* Vol. 5, *Tyrant-Flycatcher to Chats.* Melbourne: Oxford University Press, 2001.

Marzluff, J. M., R. Bowman, and R. Donnelly, eds. *Avian Ecology and Conservation in an Urbanizing World.* Boston: Kluwer Academic Publishers, 2001.

Matthysen, E. The *Nuthatches.* London: T. A. & D. Poyser, 1998.

Mayfield, H. *The Kirtland's Warbler.* Bloomfield Hills, Michigan: Cranbrook Institute of Science, 1960.

Mayr, E. *The Birds of Northern Melanesia: Speciation, Ecology, and Biogeography.* Oxford: Oxford University Press, 2001.

McCabe, Robert A. *The Little Green Bird: Ecology of the Willow Flycatcher.* Madison, Wis.: Rusty Rock Press, 1991.

Mindell, D. P., ed. *Avian Molecular Evolution and Systematics.* New York: Academic Press, 1997.

Morse, D. H. *American Warblers, An Ecological and Behavioral Perspective.* Cambridge: Harvard University Press, 1989.

Mundy, P., D., et al. *The Vultures of Africa.* London: Academic Press, 1992.

Nelson, J. B. *The Sulidae: Gannets and Boobies.* Oxford: Oxford University Press, 1978.

Nelson, Bryan. *The Atlantic Gannet.* 2nd ed. Great Yarmouth: Fenix, 2002.

Olsen, Klaus Malling, and Hans Larsson. *Skuas and Jaegers: A Guide to the Skuas and Jaegers of the World.* New Haven: Yale University Press, 1997.

Olsen, Klaus Malling, and Hans Larsson. *Gulls of Europe, Asia and North America.* London: Christopher Helm, 2001.

Ortega, Catherine P. *Cowbirds and Other Brood Parasites.* Tucson: University of Arizona Press, 1998.

Padian, K., ed. *The Origin of Birds and the Evolution of Flight.* Memoirs of the California Academy of Sciences, no. 8. San Francisco: California Academy of Sciences, 1986.

Paul, Gregory S. *Dinosaurs of the Air: the Evolution and Loss of Flight in Dinosaurs and Birds.* Baltimore: Johns Hopkins University Press, 2001.

Poole, Alan F., P. Stettenheim, and Frank B. Gill, eds. *The Birds of North America.* 15 vols. to date. Philadelphia: The Academy of Natural Sciences; Washington, D.C.: American Ornithologists' Union, 1992-.

Raffaele, H., et al. *Birds of the West Indies.* London: Christopher Helm, A. & C. Black, 1998.

Ratcliffe, D. *The Peregrine Falcon.* 2nd ed. London: T. & A. D. Poyser, 1993.

Restall, R. *Munias and Mannikins.* Sussex: Pica Press, 1996.

Ridgely, R. S., and G. A. Gwynne, Jr. *A Guide to the Birds of Panama: with Costa Rica, Nicaragua, and Honduras.* 2nd ed. Princeton: Princeton University Press. 1989.

Ridgely, R. S., and G. Tudor. *The Birds of South America.* Vol. 1, *The Oscine Passerines.* Austin: University of Texas Press, 1989.

Ridgely, R. S., and G. Tudor. *The Birds of South America.* Vol. 2, *The Suboscine Passerines.* Austin: University of Texas Press, 1994.

Ridgely, Robert S., and Paul J. Greenfield. *The Birds of Ecuador.* Vol. 1, *Status, Distribution, and Taxonomy.* Ithaca, NY: Comstock Pub., 2001.

Ridgely, Robert S., and Paul J. Greenfield. *The Birds of Ecuador.* Vol. 2, *Field Guide.* Ithaca, NY: Comstock Pub., 2001.

Rising, J. D. *Guide to the Identification and Natural History of the Sparrows of the United States and Canada.* New York: Academic Press, 1996.

Rowley, I., and E. Russell. *Fairy-Wrens and Grasswrens. Bird Families of the World,* no. 4. Oxford: Oxford University Press, 1997.

Scott, J. M., S. Conant, and C. van Riper, III, eds. *Evolution, Ecology, Conservation, and Management of Hawaiian Birds: A Vanishing Avifauna. Studies in Avian Biology,* no. 22. Camarillo, CA : Cooper Ornithological Society, 2001.

Searcy, W. A., and K. Yasukawa. *Polygyny and Sexual Selection in Red-winged Blackbirds. Monographs in Behavior and Ecology.* Princeton: Princeton University Press, 1995.

Shirihai, H., G. Gargallo, and A. J. Helbig. *Sylvia Warblers.* Edited by G. M. Kirwan and L. Svensson. Princeton: Princeton University Press, 2001.

Short, L. L. *Woodpeckers of the World.* Greenville, Del.: Delaware Natural History Museum, 1984.

Short, L. L., and J. F.M. Horne. *Toucans, Barbets and Honeyguides: Ramphastidae, Capitonidae, and Indicatoridae. Bird Families of the World,* no. 8. Oxford: New York: Oxford University Press, 2001.

Sibley, C. G., and J. E. Ahlquist. *Phylogeny and Classification of Birds: A Study in Molecular Evolution.* New Haven: Yale University Press, 1990.

Sibley, C. G., and B. L. Monroe, Jr. *Distribution and Taxonomy of Birds of the World.* New Haven: Yale University Press, 1990.

Sibley, C. G., and B. L. Monroe, Jr. *A Supplement to Distribution and Taxonomy of Birds of the World.* New Haven: Yale University Press, 1993.

Sibley, David Allen. *The Sibley Guide to Birds.* New York: Knopf, 2000.

Sick, H. *Birds in Brazil, a Natural History.* Princeton University Press, 1993.

Simmons, Robert Edward. *Harriers of the World: Their Behaviour and Ecology. Oxford Ornithology Series,* no. 11. Oxford: Oxford University Press, 2000.

Sinclair, I., and O. Langrand. *Birds of the Indian Ocean Islands: Madagascar, Mauritius, Reunion, Rodrigues, Seychelles and the Comoros.* New Holland. 1998.

Skutch, A. F. *Antbirds and Ovenbirds: Their Lives and Homes.* Austin: University of Texas Press, 1996.

Skutch, A. F. *Orioles, Blackbirds, and Their Kin: A Natural History.* Tucson: University of Arizona Press, 1996.

Smith, S. M. *The Black-capped Chickadee: Behavioral Ecology and Natural History.* Ithaca: Cornell University Press, 1991.

Snow, D. W. *The Cotingas, Bellbirds, Umbrella-birds and Their Allies in Tropical America.* London: Brit. Museum (Nat. Hist.), 1982.

Snyder, Noel F. R., and Helen Snyder. *The California Condor: A Saga of Natural History and Conservation.* San Diego, Calif.: Academic Press, 2000.

Stacey, P. B., and W. D. Koenig, eds. *Cooperative Breeding in Birds.* Cambridge: Cambridge University Press, 1989.

Stattersfield, Alison J., et al. eds. *Threatened Birds of the World: The Official Source for Birds on the IUCN Red List.* Cambridge: BirdLife International, 2000.

Stiles, F. G., and A. F. Skutch. *A Guide to the Birds of Costa Rica.* Ithaca: Comstock, 1989.

Stokes, Donald W., and Lillian Q. Stokes. *A Guide to Bird Behavior.* 3 vols. Boston: Little, Brown, 1979-1989.

Stolz, D.F., J. W. Fitzpatrick, T. A. Parker III, and D. K. Moskovits. *Neotropical Birds, Ecology and Conservation.* Chicago: University of Chicago Press, 1996.

Summers-Smith, J. D. *The Sparrows: A Study of the Genus Passer.* Calton: T. & A. D. Poyser, 1988.

Taylor, B., and B. van Perlo. *Rails: A Guide to the Rails, Crakes, Gallinules and Coots of the World.* Sussex: Pica Press, 1998.

Terres, John K. *The Audubon Society Encyclopedia of North American Birds.* New York: Knopf, 1980.

Tickell, W. L. N. *Albatrosses.* Sussex: Pica Press, 2000.

Todd, F. S. *Natural History of the Waterfowl.* Ibis Publishing Co., San Diego Natural History Museum, 1996.

Turner, A., and C. Rose. *A Handbook to the Swallows and Martins of the World.* London: Christopher Helm, 1989.

Tyler, Stephanie J., and Stephen J. *The Dippers.* London: T. & A. D. Poyser, 1994.

Urban, E. K., C. H. Fry, and S. Keith, eds. *The Birds of Africa.* Vol. 2, *Gamebirds to Pigeons.* London: Academic Press, 1986.

Urban, E. K., C. H. Fry, and S. Keith, eds. *The Birds of Africa.* Vol. 5, *Thrushes to Puffback Flycatchers.* London: Academic Press, 1997.

van Rhijn, J. G. *The Ruff: Individuality in a Gregarious Wading Bird.* London: T. & A. D. Poyser, 1991.

Voous, Karel H. *Owls of the Northern Hemisphere.* Cambridge, Mass.: MIT Press, 1988.

Warham, J. *Behaviour and Population Ecology of the Petrels.* New York: Academic Press, 1996.

Williams, T. D. *The Penguins. Bird Families of the World,* no. 2. Oxford: Oxford University Press, 1995.

Winkler, H., D. A. Christie, and D. Nurney. *Woodpeckers: A Guide to the Woodpeckers, Piculets and Wrynecks of the World.* Sussex: Pica Press, 1995.

Woolfenden, G. E., and J. W. Fitzpatrick. *The Florida Scrub Jay: Demography of a Cooperative-breeding Bird. Monographs in Population Biology,* no. 20. Princeton: Princeton University Press, 1984.

Zimmerman, D. A., D. A. Turner, and D. J. Pearson. *Birds of Kenya and Northern Tanzania.* London: Christopher Helm, 1996.

Compiled by Janet Hinshaw, Bird Division Collection Manager, University of Michigan Museum of Zoology

Organizations

African Bird Club
  Wellbrook Court, Girton Road
  Cambridge, Cambridgeshire CB3 0NA
  United Kingdom
  Phone: +44 1 223 277 318
  Fax: +44-1-223-277-200
  <http://www.africanbirdclub.org>

African Gamebird Research, Education and Development
(AGRED)
  P.O. Box 1191
  Hilton, KwaZulu-Natal 3245
  South Africa
  Phone: +27-33-343-3784

African-Eurasian Migratory Waterbird Agreement (AEWA)
  UN Premises in Bonn, Martin Luther-King St.
  Bonn D-53175 Germany
  <http://www.wcmc.org.uk/AEWA>

American Ornithologists' Union
  Suite 402, 1313 Dolley Madison Blvd
  McLean, VA 22101
  USA
  <http://www.aou.org>

American Zoo and Aquarium Association
  8403 Colesville Road
  Suite 710
  Silver Spring, Maryland 20910
  <http://www.aza.org>

Association for BioDiversity Information
  1101 Wilson Blvd., 15th Floor
  Arlington, VA 22209
  USA
  <http://www.infonatura.org/>

Association for Parrot Conservation
  Centro de Calidad Ambiental iTESM Sucursal de Correos
  J., C.P. 64849
  Monterrey, N.L.
  Mexico

Australasian Raptor Association
  415 Riversdale Road
  Hawthorn East, Victoria 3123
  Australia
  Phone: +61 3 9882 2622
  Fax: +61 3 9882 2677
  <http://www.tasweb.com.au/ara/index.htm>

Australian National Wildlife Collection
  GPO Box 284
  Canberra, ACT 2601
  Australia
  Phone: +61 2 6242 1600
  Fax: +61-2-6242-1688

The Bird Conservation Society of Thailand
  69/12 Rarm Intra 24
  Jarakhebua Lat Phrao, Bangkok 10230
  Thailand
  Phone: 943-5965
  <http://www.geocities.com/TheTropics/Harbor/7503/
  ruang_nok/princess_bird.html>

BirdLife International
  Wellbrook Court, Girton Road
  Cambridge, Cambridgeshire CB3 0NA
  United Kingdom
  Phone: +44 1 223 277 318
  Fax: +44-1-223-277-200
  <http://www.birdlife.net>

BirdLife International Indonesia Programme
  P. O. Box 310/Boo
  Bogor
  Indonesia
  Phone: +62 251 357222
  Fax: +62 251 357961
  <http://www.birdlife-indonesia.org>

BirdLife International, Panamerican Office
  Casilla 17-17-717
  Quito
  Ecuador
  Phone: +593 2 244 3261
  Fax: +593 2 244 3261
  <http://www.latinsynergy.org/birdlife.html>

BirdLife South Africa
  P. O. Box 515
  Randburg 2125
  South Africa
  Phone: +27-11-7895188
  <http://www.birdlife.org.za>

Birds Australia
  415 Riversdale Road
  Hawthorn East, Victoria 3123
  Australia
  Phone: +61 3 9882 2622
  Fax: +61 3 9882 2677
  <http://www.birdsaustralia.com.au>

Birds Australia Parrot Association, Birds Australia
   415 Riversdale Road
   Hawthorn East, Victoria 3123
   Australia
   Phone: +61 3 9882 2622
   Fax: +61 3 9882 2677
   <http://www.birdsaustralia.com.au>

The Bishop Museum
   1525 Bernice Street
   Honolulu, HI 96817-0916
   Phone: (808) 847-3511
   <http://www.bishopmuseum.org>

British Trust for Ornithology
   The Nunnery
   Thetford, Norfolk IP24 2PU
   United Kingdom
   Phone: +44 (0) 1842 750050
   Fax: +44 (0) 1842 750030
   <http://www.bto.org>

Center for Biological Diversity
   P.O. Box 710
   Tucson, AZ 85702-0701
   USA
   Phone: (520) 623-5252
   Fax: (520) 623-9797
   <http://www.biologicaldiversity.org/swcbd/index.html>

Coraciiformes Taxon Advisory Group
   <http://www.coraciiformestag.com>

The Cracid Specialist Group
   PO Box 132038
   Houston, TX 77219-2038
   USA
   Phone: (713) 639-4776
   <http://www.angelfire.com/ca6/cracid>

Department of Ecology and Environmental Biology, Cornell University
   E145 Corson Hall
   Ithaca, NY 14853-2701
   USA
   Phone: (607) 254-4201
   <http://www.es.cornell.edu/winkler/botw/fringillidae.html>

Department of Ecology and Evolutionary Biology, Tulane University
   310 Dinwiddie Hall
   New Orleans, LA 70118-5698
   USA
   Phone: (504) 865-5191
   <http://www.tulane.edu/
   eeob/Courses/Heins/Evolution/lecture17.html>

Department of Zoology, University of Toronto
   25 Harbord Street
   Toronto, Ontario M5S 3G5
   Canada
   Phone: (416) 978-3482
   Fax: (416) 978-8532
   <http://www.zoo.utoronto.ca>

Ducks Unlimited, Inc
   One Waterfowl Way
   Memphis, TN 38120
   USA
   Phone: (800) 453-8257
   Fax: (901) 758-3850
   <http://www.ducks.org>

Emu Farmers Federation of Australia
   P.O Box 57
   Wagin, Western Australia 6315
   Australia
   Phone: +61 8 9861 1136

Game Conservancy Trust
   Fordingbridge, Hampshire SP6 1EF
   United Kingdom
   Phone: +44 1425 652381
   Fax: +44 1425 651026
   <http://www.gct.org.uk>

Gamebird Research Programme, Percy FitzPatrick Institute, University of Cape Town
   Private Bag
   Rondebosch, Western Cape 7701
   South Africa
   Phone: +27 21 6503290
   Fax: +27 21 6503295
   <http://www.uct.ac.za/depts/fitzpatrick>

Haribon Foundation for the Conservation of Natural Resources
   9A Malingap Cot, Malumanay Streets, Teachers' Village, 1101 Diliman
   Quezon City
   Philippines
   Phone: +63 2 9253332
   <http://www.haribon.org.ph>

The Hawk and Owl Trust
   11 St Marys Close
   Newton Abbot, Abbotskerswell, Devon TQ12 5QF
   United Kingdom
   Phone: +44 (0)1626 334864
   Fax: +44 (0)1626 334864
   <http://www.hawkandowltrust.org>

Herons Specialist Group
   Station Biologique de la Tour du Valat
   Le Sambuc, Arles 13200
   France
   Phone: +33-4-90-97-20-13
   Fax: 33-4-90-97-29-19
   <http://www.tour-du-valat.com>

Hornbill Research Foundation
   c/o Department of Microbiology, Faculty of Science, Mahidol University, Rama 6 Rd
   Bangkok 10400
   Thailand
   Phone: +66 22 460 063, ext. 4006

International Crane Foundation
   P.O. Box 447
   Baraboo, WI 53913-0447
   USA
   Phone: (608) 356-9462
   Fax: (608) 356-9465
   <http://www.savingcranes.org>

International Shrike Working Group
   "Het Speihuis," Speistraat, 17
   Sint-Lievens-Esse (Herzele), B-9550

Belgium
Phone: +32 54 503 789

International Species Inventory System
<http://www.isis.org>

International Touraco Society
Brackenhurst, Grange Wood
Netherseal, Nr Swadlincote, Derbyshire DE12 8BE
United Kingdom
Phone: +44 (0)1283 760541

International Waterbird Census
<http://www.wetlands.org>

IUCN Species Survival Commission
219c Huntingdon Road
Cambridge, Cambridgeshire CB3 0DL
United Kingdom
<http://www.iucn.org/themes/ssc>

IUCN–The World Conservation Union
Rue Mauverney 28
Gland 1196
Switzerland
Phone: +41-22-999-0001
Fax: +41-22-999-0025
<http://www.iucn.org>

IUCN–World Conservation Union, USA Multilateral Office
1630 Connecticut Avenue
Washington, DC 20009
USA
Phone: (202) 387-4826
<http://www.iucn.org/places/usa/inter.html>

IUCN/SSC Grebes Specialist Group
Copenhagen DK 2100
Denmark
Phone: +45 3 532 1323
Fax: +45-35321010
<http://www.iucn.org>

Japanese Association for Wild Geese Protection
Minamimachi 16
Wakayangi 989-5502
Japan
Phone: +81 228 32 2004
Fax: +81 228 32 2004
<http://www.japwgp.org>

Ligue pour la Protection des Oiseaux
La Corderie Royale, B.P. 263
17305 Rochefort cedex
France
Phone: +33 546 821 234
Fax: +33 546 839 586
<http://www.lpo-birdlife.asso.fr>

Loro Parque Fundación
Loro Parque S.A. 38400 Puerto de la Cruz
Tenerife, Canary Islands
Spain

National Audubon Society
700 Broadway
New York, NY 10003
USA

Phone: (212) 979-3000
Fax: (212) 978-3188
<http://www.Audubon.org>

National Audubon Society Population & Habitat Program
1901 Pennsylvania Ave. NW, Suite 1100
Washington, DC 20006
USA
Phone: (202) 861-2242
<http://www.audubonpopulation.org>

Neotropical Bird Club
c/o The Lodge
Sandy, Bedfordshire SG19 2DL
United Kingdom

Oriental Bird Club, American Office
4 Vestal Street
Nantucket, MA 02554
USA
Phone: (508) 228-1782
<http://www.orientalbirdclub.org>

Ornithological Society of New Zealand
P.O. Box 12397
Wellington, North Island
New Zealand
<http://osnz.org.nz>

Pacific Island Ecosystems Research Center
3190 Maile Way, St. John Hall, Room 408
Honolulu, HI 96822
USA
Phone: (808) 956-5691
Fax: (808) 956-5687
<http://biology.usgs.gov/pierc/piercwebsite.htm>

ProAves Peru
P.O. Box 07
Piura
Peru

Raptor Conservation Group, Endangered Wildlife Trust
Private Bag X11
Parkview, Gauteng 2122
South Africa
Phone: +27-11-486-1102
Fax: +27-11-486-1506
<http://www.ewt.org.za>

Raptor Research Foundation
P.O. Box 1897, 810 E. 10th Street
Lawrence, KS 66044-88973
USA
<http://biology.biosestate.edu/raptor>

Research Centre for African Parrot Conservation Zoology and
Entomology Department
Private Bag X01
Scottsville 3201
Natal Republic of South Africa

Roberts VII Project, Percy FitzPatrick Institute of African
Ornithology, University of Cape Town
Rondebosch 7701
South Africa
Fax: (021) 650 3295
<http://www.uct.ac.za/depts/fitzpatrick/docs/r549.html>

Royal Society for the Protection of Birds
Admail 975 Freepost ANG 6335, The Lodge
Sandy, Bedfordshire SG19 2TN
United Kingdom
<http://www.rspb.org.uk>

Ruffed Grouse Society
451 McCormick Rd
Coraopolis, PA 15108
Phone: (888) 564-6747
Fax: (412) 262-9207
<http://www.ruffedgrousesociety.org>

Smithsonian Migratory Bird Center, Smithsonian National
Zoological Park
3001 Connecticut Avenue, NW
Washington, DC 20008
USA
Phone: (202) 673-4800
<http://www.natzoo.si.edu>

The Songbird Foundation
2367 Eastlake Ave. East
Seattle, WA 98102
USA
Phone: (206) 374-3674
Fax: (206) 374-3674
<http://www.songbird.org>

University of Michigan
3019 Museum of Zoology, 1109 Geddes Ave
Ann Arbor, MI 48109-1079
USA
Phone: (734) 647-2208
Fax: (734) 763-4080
<http://www.ummz.lsa.umich.edu/birds/index.html>

Wader Specialist Group, Mr. David Stroud
Monkstone House, City Road
Peterborough PE1 1JY
United Kingdom
Phone: +44 1733 866/810
Fax: +44 1733 555/448

Wader Study Group, The National Centre for Ornithology
The Nunnery
Thetford, Norfolk JP24 2PU
United Kingdom

Waterbird Society
National Museum of Natural History, Smithsonian
Institution
Washington, DC 20560
USA
<http://www.nmnh.si.edu/BIRDNET/cws>

Western Hemisphere Shorebird Reserve Network (WHSRN)
Manomet Center for Conservation Science, P O Box 1770
Manomet, MA 02345
USA
Phone: (508) 224-6521
Fax: (508) 224-9220
<http://www.manomet.org/WHSRN/index.html>

Wetlands International
Droevendaalsesteeg 3A

Wageningen 6700 CA
The Netherlands
Phone: +31 317 478884
Fax: +31 317 478885
<http://www.wetlands.agro.nl>

Wetlands International (the Americas)
7 Hinton Avenue North, Suite 200
Ottawa, Ontario K1Y 4P1
Canada
Phone: (613) 722-2090
<http://www.wetlands.org>

Wetlands International/Survival Service Commission Flamingo
Specialist Group
c/o Station Biologique de la Tour du Valat
Le Sambuc, Arles 13200
France

Wildfowl and Wetlands Trust
Slimbridge, Glos GL2 7BT
United Kingdom
Phone: +44 01453 891900

Woodcock and Snipe Specialist Group
Director, European Wildlife Research Institute
Bonndorf, Glashuette D-79848
Germany
Phone: 949 7653 1891
Fax: 949 7653 9269

Woodhoopoe Research Project, FitzPatrick Institute of African
Ornithology, University of Cape Town
P.O. Rondebosch
Cape Town, Western Cape 7700
South Africa
Phone: +27 (0)21 650-3290
Fax: +27-21-650-3295
<http://www.fitztitute.uct.ac.za>

Working Group on Birds in Madagascar and the Indian Ocean
Islands. World Wide Fund for Nature
Antananarivo 101 BP 738
Madagascar
Phone: +261 3207 80806

Working Group on International Wader and Waterfowl
Research (WIWO)
Stichting WIWO, c/o P O Box 925
Zeist 3700 AX
The Netherlands
<http://www.wiwo-international.org>

World Center for Birds of Prey, The Peregrine Fund
566 West Flying Hawk Lane
Boise, ID 83709
USA
Phone: (208) 362-3716
Fax: (208) 362-2376
<http://www.peregrinefund.org>

World Parrot Trust
Glanmor House
Hayle, Cornwall TR27 4HB
United Kingdom
<http://www.worldparrottrust.org>

World Pheasant Association
  P. O. Box 5, Lower Basildon St
  Reading RG8 9PF
  United Kingdom
  Phone: +44 1 189 845 140
  Fax: +44 118 984 3369
  <http://www.pheasant.org.uk>

World Working Group on Birds of Prey and Owl
  P.O. Box 52
  Towcester NN12 7ZW
  United Kingdom
  Phone: +44 1 604 862 331
  Fax: +44 1 604 862 331
  <http://www.Raptors-International.de>

WPA/BirdLife/SSC Megapode Specialist Group
  c/o Department of Ornithology, National Museum of
  Natural History, P.O. Box 9517
  Leiden 2300 RA
  The Netherlands

WPA/BirdLife/SSC Partridge, Quail, and Francolin Specialist
Group
  c/o World Pheasant Association, PO Box 5
  Lower Basildon, Reading RG8 9PF
  United Kingdom
  Phone: +44 1 189 845 140
  Fax: +118 9843369
  <http:/www.pheasant.org.uk>
  PQF: <http://www.gct.org.uk/pqf>

Contributors to the first edition

The following individuals contributed chapters to the original edition of Grzimek's Animal Life Encyclopedia, *which was edited by Dr. Bernhard Grzimek, Professor, Justus Liebig University of Giessen, Germany; Director, Frankfurt Zoological Garden, Germany; and Trustee, Tanzanian National Parks, Tanzania.*

Dr. Michael Abs
Curator, Ruhr University
Bochum, Germany

Dr. Salim Ali
Bombay Natural History Society
Bombay, India

Dr. Rudolph Altevogt
Professor, Zoological Institute,
University of Münster
Münster, Germany

Dr. Renate Angermann
Curator, Institute of Zoology,
Humboldt University
Berlin, Germany

Edward A. Armstrong
Cambridge University
Cambridge, England

Dr. Peter Ax
Professor, Second Zoological Institute
and Museum, University of Göttingen
Göttingen, Germany

Dr. Franz Bachmaier
Zoological Collection of the State of
Bavaria
Munich, Germany

Dr. Pedru Banarescu
Academy of the Roumanian Socialist
Republic, Trajan Savulescu Institute of
Biology
Bucharest, Romania

Dr. A. G. Bannikow
Professor, Institute of Veterinary
Medicine
Moscow, Russia

Dr. Hilde Baumgärtner
Zoological Collection of the State of
Bavaria
Munich, Germany

C. W. Benson
Department of Zoology, Cambridge
University
Cambridge, England

Dr. Andrew Berger
Chairman, Department of Zoology,
University of Hawaii
Honolulu, Hawaii, U.S.A.

Dr. J. Berlioz
National Museum of Natural History
Paris, France

Dr. Rudolf Berndt
Director, Institute for Population
Ecology, Hiligoland Ornithological
Station
Braunschweig, Germany

Dieter Blume
Instructor of Biology, Freiherr-vom-
Stein School
Gladenbach, Germany

Dr. Maximilian Boecker
Zoological Research Institute and A.
Koenig Museum
Bonn, Germany

Dr. Carl-Heinz Brandes
Curator and Director, The Aquarium,
Overseas Museum
Bremen, Germany

Dr. Donald G. Broadley
Curator, Umtali Museum
Mutare, Zimbabwe

Dr. Heinz Brüll
Director; Game, Forest, and Fields
Research Station
Hartenholm, Germany

Dr. Herbert Bruns
Director, Institute of Zoology and the
Protection of Life
Schlangenbad, Germany

Hans Bub
Heligoland Ornithological Station
Wilhelmshaven, Germany

A. H. Chrisholm
Sydney, Australia

Herbert Thomas Condon
Curator of Birds, South Australian
Museum
Adelaide, Australia

Dr. Eberhard Curio
Director, Laboratory of Ethology,
Ruhr University
Bochum, Germany

Dr. Serge Daan
Laboratory of Animal Physiology,
University of Amsterdam
Amsterdam, The Netherlands

Dr. Heinrich Dathe
Professor and Director, Animal Park
and Zoological Research Station,
German Academy of Sciences
Berlin, Germany

Dr. Wolfgang Dierl
Zoological Collection of the State of
Bavaria
Munich, Germany

Dr. Fritz Dieterlen
Zoological Research Institute, A.
Koenig Museum
Bonn, Germany

Dr. Rolf Dircksen
Professor, Pedagogical Institute
Bielefeld, Germany

Josef Donner
Instructor of Biology
Katzelsdorf, Austria

Dr. Jean Dorst
Professor, National Museum of
Natural History
Paris, France

Dr. Gerti Dücker
Professor and Chief Curator,
Zoological Institute, University of
Münster
Münster, Germany

Dr. Michael Dzwillo
Zoological Institute and Museum,
University of Hamburg
Hamburg, Germany

Dr. Irenäus Eibl-Eibesfeldt
Professor and Director, Institute of
Human Ethology, Max Planck
Institute for Behavioral Physiology
Percha/Starnberg, Germany

Dr. Martin Eisentraut
Professor and Director, Zoological
Research Institute and A. Koenig
Museum
Bonn, Germany

Dr. Eberhard Ernst
Swiss Tropical Institute
Basel, Switzerland

R. D. Etchecopar
Director, National Museum of
Natural History
Paris, France

Dr. R. A. Falla
Director, Dominion Museum
Wellington, New Zealand

Dr. Hubert Fechter
Curator, Lower Animals, Zoological
Collection of the State of Bavaria
Munich, Germany

Dr. Walter Fiedler
Docent, University of Vienna, and
Director, Schönbrunn Zoo
Vienna, Austria

Wolfgang Fischer
Inspector of Animals, Animal Park
Berlin, Germany

Dr. C. A. Fleming
Geological Survey Department of
Scientific and Industrial Research
Lower Hutt, New Zealand

Dr. Hans Frädrich
Zoological Garden
Berlin, Germany

Dr. Hans-Albrecht Freye
Professor and Director, Biological
Institute of the Medical School
Halle a.d.S., Germany

Günther E. Freytag
Former Director, Reptile and
Amphibian Collection, Museum of
Cultural History in Magdeburg
Berlin, Germany

Dr. Herbert Friedmann
Director, Los Angeles County
Museum of Natural History
Los Angeles, California, U.S.A.

Dr. H. Friedrich
Professor, Overseas Museum
Bremen, Germany

Dr. Jan Frijlink
Zoological Laboratory, University of
Amsterdam
Amsterdam, The Netherlands

Dr. H.C. Karl Von Frisch
Professor Emeritus and former
Director, Zoological Institute,
University of Munich
Munich, Germany

Dr. H. J. Frith
C.S.I.R.O. Research Institute
Canberra, Australia

Dr. Ion E. Fuhn
Academy of the Roumanian Socialist
Republic, Trajan Savulescu Institute of
Biology
Bucharest, Romania

Dr. Carl Gans
Professor, Department of Biology,
State University of New York at
Buffalo
Buffalo, New York, U.S.A.

Dr. Rudolf Geigy
Professor and Director, Swiss Tropical
Institute
Basel, Switzerland

Dr. Jacques Gery
St. Genies, France

Dr. Wolfgang Gewalt
Director, Animal Park
Duisburg, Germany

Dr. H.C. Dr. H.C. Viktor Goerttler
Professor Emeritus, University of Jena
Jena, Germany

Dr. Friedrich Goethe
Director, Institute of Ornithology,
Heligoland Ornithological Station
Wilhelmshaven, Germany

Dr. Ulrich F. Gruber
Herpetological Section, Zoological
Research Institute and A. Koenig
Museum
Bonn, Germany

Dr. H. R. Haefelfinger
Museum of Natural History
Basel, Switzerland

Dr. Theodor Haltenorth
Director, Mammalology, Zoological
Collection of the State of Bavaria
Munich, Germany

Barbara Harrisson
Sarawak Museum, Kuching, Borneo
Ithaca, New York, U.S.A.

Dr. Francois Haverschmidt
President, High Court (retired)
Paramaribo, Suriname

Dr. Heinz Heck
Director, Catskill Game Farm
Catskill, New York, U.S.A.

Dr. Lutz Heck
Professor (retired), and Director,
Zoological Garden, Berlin
Wiesbaden, Germany

Dr. H.C. Heini Hediger
Director, Zoological Garder
Zurich, Switzerland

Dr. Dietrich Heinemann
Director, Zoological Garden, Münster
Dörnigheim, Germany

Dr. Helmut Hemmer
Institute for Physiological Zoology,
University of Mainz
Mainz, Germany

Dr. W. G. Heptner
Professor, Zoological Museum,
University of Moscow
Moscow, Russia

Dr. Konrad Herter
Professor Emeritus and Director
(retired), Zoological Institute, Free
University of Berlin
Berlin, Germany

Dr. Hans Rudolf Heusser
Zoological Museum, University of
Zurich
Zurich, Switzerland

Dr. Emil Otto Höhn
Associate Professor of Physiology,
University of Alberta
Edmonton, Canada

Dr. W. Hohorst
Professor and Director,
Parasitological Institute, Farbwerke
Hoechst A.G.
Frankfurt-Höchst, Germany

Dr. Folkhart Hückinghaus
Director, Senckenbergische Anatomy,
University of Frankfurt a.M.
Frankfurt a.M., Germany

Francois Hüe
National Museum of Natural
History
Paris, France

Dr. K. Immelmann
Professor, Zoological Institute,
Technical University of Braunschweig
Braunschweig, Germany

Dr. Junichiro Itani
Kyoto University
Kyoto, Japan

Dr. Richard F. Johnston
Professor of Zoology, University of
Kansas
Lawrence, Kansas, U.S.A.

Otto Jost
Oberstudienrat, Freiherr-vom-Stein
Gymnasium
Fulda, Germany

Dr. Paul Kähsbauer
Curator, Fishes, Museum of Natural
History
Vienna, Austria

Dr. Ludwig Karbe
Zoological State Institute and
Museum
Hamburg, Germany

Dr. N. N. Kartaschew
Docent, Department of Biology,
Lomonossow State University
Moscow, Russia

Dr. Werner Kästle
Oberstudienrat, Gisela Gymnasium
Munich, Germany

Dr. Reinhard Kaufmann
Field Station of the Tropical Institute,
Justus Liebig University, Giessen,
Germany
Santa Marta, Colombia

Dr. Masao Kawai
Primate Research Institute, Kyoto
University
Kyoto, Japan

Dr. Ernst F. Kilian
Professor, Giessen University and
Catedratico Universidad Austral,
Valdivia-Chile
Giessen, Germany

Dr. Ragnar Kinzelbach
Institute for General Zoology,
University of Mainz
Mainz, Germany

Dr. Heinrich Kirchner
Landwirtschaftsrat (retired)
Bad Oldesloe, Germany

Dr. Rosl Kirchshofer
Zoological Garden, University of
Frankfort a.M.
Frankfurt a.M., Germany

Dr. Wolfgang Klausewitz
Curator, Senckenberg Nature
Museum and Research Institute
Frankfurt a.M., Germany

Dr. Konrad Klemmer
Curator, Senckenberg Nature
Museum and Research Institute
Frankfurt a.M., Germany

Dr. Erich Klinghammer
Laboratory of Ethology, Purdue
University
Lafayette, Indiana, U.S.A.

Dr. Heinz-Georg Klös
Professor and Director, Zoological
Garden
Berlin, Germany

Ursula Klös
Zoological Garden
Berlin, Germany

Dr. Otto Koehler
Professor Emeritus, Zoological
Institute, University of Freiburg
Freiburg i. BR., Germany

Dr. Kurt Kolar
Institute of Ethology, Austrian
Academy of Sciences
Vienna, Austria

Dr. Claus König
State Ornithological Station of Baden-
Württemberg
Ludwigsburg, Germany

Dr. Adriaan Kortlandt
Zoological Laboratory, University of
Amsterdam
Amsterdam, The Netherlands

Dr. Helmut Kraft
Professor and Scientific Councillor,
Medical Animal Clinic, University of
Munich
Munich, Germany

Dr. Helmut Kramer
Zoological Research Institute and A.
Koenig Museum
Bonn, Germany

Dr. Franz Krapp
Zoological Institute, University of
Freiburg
Freiburg, Switzerland

Dr. Otto Kraus
Professor, University of Hamburg,
and Director, Zoological Institute and
Museum
Hamburg, Germany

Dr. Hans Krieg
Professor and First Director (retired),
Scientific Collections of the State of
Bavaria
Munich, Germany

Dr. Heinrich Kühl
Federal Research Institute for
Fisheries, Cuxhaven Laboratory
Cuxhaven, Germany

Dr. Oskar Kuhn
Professor, formerly University
Halle/Saale
Munich, Germany

Dr. Hans Kumerloeve
First Director (retired), State
Scientific Museum, Vienna
Munich, Germany

Dr. Nagamichi Kuroda
Yamashina Ornithological Institute,
Shibuya-Ku
Tokyo, Japan

Dr. Fred Kurt
Zoological Museum of Zurich
University, Smithsonian Elephant
Survey
Colombo, Ceylon

Dr. Werner Ladiges
Professor and Chief Curator,
Zoological Institute and Museum,
University of Hamburg
Hamburg, Germany

Leslie Laidlaw
Department of Animal Sciences,
Purdue University
Lafayette, Indiana, U.S.A.

Dr. Ernst M. Lang
Director, Zoological Garden
Basel, Switzerland

Dr. Alfredo Langguth
Department of Zoology, Faculty of
Humanities and Sciences, University
of the Republic
Montevideo, Uruguay

Leo Lehtonen
Science Writer
Helsinki, Finland

Bernd Leisler
Second Zoological Institute, University
of Vienna
Vienna, Austria

Dr. Kurt Lillelund
Professor and Director, Institute for
Hydrobiology and Fishery Sciences,
University of Hamburg
Hamburg, Germany

R. Liversidge
Alexander MacGregor Memorial
Museum
Kimberley, South Africa

Dr. Konrad Lorenz
Professor and Director, Max Planck
Institute for Behavioral Physiology
Seewiesen/Obb., Germany

Dr. Martin Lühmann
Federal Research Institute for the
Breeding of Small Animals
Celle, Germany

Dr. Johannes Lüttschwager
Oberstudienrat (retired)
Heidelberg, Germany

Dr. Wolfgang Makatsch
Bautzen, Germany

Dr. Hubert Markl
Professor and Director, Zoological
Institute, Technical University of
Darmstadt
Darmstadt, Germany

Basil J. Marlow, B.SC. (Hons)
Curator, Australian Museum
Sydney, Australia

Dr. Theodor Mebs
Instructor of Biology
Weissenhaus/Ostsee, Germany

Dr. Gerlof Fokko Mees
Curator of Birds, Rijks Museum of
Natural History
Leiden, The Netherlands

Hermann Meinken
Director, Fish Identification Institute,
V.D.A.
Bremen, Germany

Dr. Wilhelm Meise
Chief Curator, Zoological Institute
and Museum, University of Hamburg
Hamburg, Germany

Dr. Joachim Messtorff
Field Station of the Federal Fisheries
Research Institute
Bremerhaven, Germany

Dr. Marian Mlynarski
Professor, Polish Academy of
Sciences, Institute for Systematic and
Experimental Zoology
Cracow, Poland

Dr. Walburga Moeller
Nature Museum
Hamburg, Germany

Dr. H.C. Erna Mohr
Curator (retired), Zoological State
Institute and Museum
Hamburg, Germany

Dr. Karl-Heinz Moll
Waren/Müritz, Germany

Dr. Detlev Müller-Using
Professor, Institute for Game
Management, University of Göttingen
Hannoversch-Münden, Germany

Werner Münster
Instructor of Biology
Ebersbach, Germany

Dr. Joachim Münzing
Altona Museum
Hamburg, Germany
Dr. Wilbert Neugebauer
Wilhelma Zoo
Stuttgart-Bad Cannstatt, Germany

Dr. Ian Newton
Senior Scientific Officer, The Nature
Conservancy
Edinburgh, Scotland

Dr. Jürgen Nicolai
Max Planck Institute for Behavioral
Physiology
Seewiesen/Obb., Germany

Dr. Günther Niethammer
Professor, Zoological Research
Institute and A. Koenig Museum
Bonn, Germany

Dr. Bernhard Nievergelt
Zoological Museum, University of Zurich
Zurich, Switzerland

Dr. C. C. Olrog
Institut Miguel Lillo San Miguel de Tucuman
Tucuman, Argentina

Alwin Pedersen
Mammal Research and aRctic Explorer
Holte, Denmark

Dr. Dieter Stefan Peters
Nature Museum and Senckenberg Research Institute
Frankfurt a.M., Germany

Dr. Nicolaus Peters
Scientific Councillor and Docent, Institute of Hydrobiology and Fisheries, University of Hamburg
Hamburg, Germany

Dr. Hans-Günter Petzold
Assistant Director, Zoological Garden
Berlin, Germany

Dr. Rudolf Piechocki
Docent, Zoological Institute, University of Halle
Halle a.d.S., Germany

Dr. Ivo Poglayen-Neuwall
Director, Zoological Garden
Louisville, Kentucky, U.S.A.

Dr. Egon Popp
Zoological Collection of the State of Bavaria
Munich, Germany

Dr. H.C. Adolf Portmann
Professor Emeritus, Zoological Institute, University of Basel
Basel, Switzerland

Hans Psenner
Professor and Director, Alpine Zoo
Innsbruck, Austria

Dr. Heinz-Siburd Raethel
Oberveterinärrat
Berlin, Germany

Dr. Urs H. Rahm
Professor, Museum of Natural History
Basel, Switzerland

Dr. Werner Rathmayer
Biology Institute, University of Konstanz
Konstanz, Germany

Walter Reinhard
Biologist
Baden-Baden, Germany

Dr. H. H. Reinsch
Federal Fisheries Research Institute
Bremerhaven, Germany

Dr. Bernhard Rensch
Professor Emeritus, Zoological Institute, University of Münster
Münster, Germany

Dr. Vernon Reynolds
Docent, Department of Sociology, University of Bristol
Bristol, England

Dr. Rupert Riedl
Professor, Department of Zoology, University of North Carolina
Chapel Hill, North Carolina, U.S.A.

Dr. Peter Rietschel
Professor (retired), Zoological Institute, University of Frankfurt a.M.
Frankfurt a.M., Germany

Dr. Siegfried Rietschel
Docent, University of Frankfurt; Curator, Nature Museum and Research Institute Senckenberg
Frankfurt a.M., Germany

Herbert Ringleben
Institute of Ornithology, Heligoland Ornithological Station
Wilhelmshaven, Germany

Dr. K. Rohde
Institute for General Zoology, Ruhr University
Bochum, Germany

Dr. Peter Röben
Academic Councillor, Zoological Institute, Heidelberg University
Heidelberg, Germany

Dr. Anton E. M. De Roo
Royal Museum of Central Africa
Tervuren, South Africa

Dr. Hubert Saint Girons
Research Director, Center for National Scientific Research
Brunoy (Essonne), France

Dr. Luitfried Von Salvini-Plawen
First Zoological Institute, University of Vienna
Vienna, Austria

Dr. Kurt Sanft
Oberstudienrat, Diesterweg-Gymnasium
Berlin, Germany

Dr. E. G. Franz Sauer
Professor, Zoological Research Institute and A. Koenig Museum, University of Bonn
Bonn, Germany

Dr. Eleonore M. Sauer
Zoological Research Institute and A. Koenig Museum, University of Bonn
Bonn, Germany

Dr. Ernst Schäfer
Curator, State Museum of Lower Saxony
Hannover, Germany

Dr. Friedrich Schaller
Professor and Chairman, First Zoological Institute, University of Vienna
Vienna, Austria

Dr. George B. Schaller
Serengeti Research Institute, Michael Grzimek Laboratory
Seronera, Tanzania

Dr. Georg Scheer
Chief Curator and Director, Zoological Institute, State Museum of Hesse
Darmstadt, Germany

Dr. Christoph Scherpner
Zoological Garden
Frankfurt a.M., Germany

Dr. Herbert Schifter
Bird Collection, Museum of Natural History
Vienna, Austria

Dr. Marco Schnitter
Zoological Museum, Zurich University
Zurich, Switzerland

Dr. Kurt Schubert
Federal Fisheries Research Institute
Hamburg, Germany

Eugen Schuhmacher
Director, Animals Films, I.U.C.N.
Munich, Germany

Dr. Thomas Schultze-Westrum
Zoological Institute, University of
Munich
Munich, Germany

Dr. Ernst Schüt
Professor and Director (retired), State
Museum of Natural History
Stuttgart, Germany

Dr. Lester L. Short Jr.
Associate Curator, American Museum
of Natural History
New York, New York, U.S.A.

Dr. Helmut Sick
National Museum
Rio de Janeiro, Brazil

Dr. Alexander F. Skutch
Professor of Ornithology, University
of Costa Rica
San Isidro del General, Costa Rica

Dr. Everhard J. Slijper
Professor, Zoological Laboratory,
University of Amsterdam
Amsterdam, The Netherlands

Bertram E. Smythies
Curator (retired), Division of Forestry
Management, Sarawak-Malaysia
Estepona, Spain

Dr. Kenneth E. Stager
Chief Curator, Los Angeles County
Museum of Natural History
Los Angeles, California, U.S.A.

Dr. H.C. Georg H.W. Stein
Professor, Curator of Mammals,
Institute of Zoology and Zoological
Museum, Humboldt University
Berlin, Germany

Dr. Joachim Steinbacher
Curator, Nature Museum and
Senckenberg Research Institute
Frankfurt a.M., Germany

Dr. Bernard Stonehouse
Canterbury University
Christchurch, New Zealand

Dr. Richard Zur Strassen
Curator, Nature Museum and
Senckenberg Research Institute
Frankfurt a.M., Germany

Dr. Adelheid Studer-Thiersch
Zoological Garden
Basel, Switzerland

Dr. Ernst Sutter
Museum of Natural History
Basel, Switzerland

Dr. Fritz Terofal
Director, Fish Collection, Zoological
Collection of the State of Bavaria
Munich, Germany

Dr. G. F. Van Tets
Wildlife Research
Canberra, Australia

Ellen Thaler-Kottek
Institute of Zoology, University of
Innsbruck
Innsbruck, Austria

Dr. Erich Thenius
Professor and Director, Institute of
Paleontolgy, University of Vienna
Vienna, Austria

Dr. Niko Tinbergen
Professor of Animal Behavior,
Department of Zoology, Oxford
University
Oxford, England

Alexander Tsurikov
Lecturer, University of Munich
Munich, Germany

Dr. Wolfgang Villwock
Zoological Institute and Museum,
University of Hamburg
Hamburg, Germany

Zdenek Vogel
Director, Suchdol Herpetological
Station
Prague, Czechoslovakia

Dieter Vogt
Schorndorf, Germany

Dr. Jiri Volf
Zoological Garden
Prague, Czechoslovakia
Otto Wadewitz
Leipzig, Germany

Dr. Helmut O. Wagner
Director (retired), Overseas Museum,
Bremen
Mexico City, Mexico

Dr. Fritz Walther
Professor, Texas A & M University
College Station, Texas, U.S.A.

John Warham
Zoology Department, Canterbury
University
Christchurch, New Zealand

Dr. Sherwood L. Washburn
University of California at Berkeley
Berkeley, California, U.S.A.

Eberhard Wawra
First Zoological Institute, University
of Vienna
Vienna, Austria

Dr. Ingrid Weigel
Zoological Collection of the State of
Bavaria
Munich, Germany

Dr. B. Weischer
Institute of Nematode Research,
Federal Biological Institute
Münster/Westfalen, Germany

Herbert Wendt
Author, Natural History
Baden-Baden, Germany

Dr. Heinz Wermuth
Chief Curator, State Nature Museum,
Stuttgart
Ludwigsburg, Germany

Dr. Wolfgang Von Westernhagen
Preetz/Holstein, Germany

Dr. Alexander Wetmore
United States National Museum,
Smithsonian Institution
Washington, D.C., U.S.A.

Dr. Dietrich E. Wilcke
Röttgen, Germany

Dr. Helmut Wilkens
Professor and Director, Institute of
Anatomy, School of Veterinary
Medicine
Hannover, Germany

Dr. Michael L. Wolfe
Utah, U.S.A.

Hans Edmund Wolters
Zoological Research Institute and A.
Koenig Museum
Bonn, Germany

Dr. Arnfrid Wünschmann
Research Associate, Zoological Garden
Berlin, Germany

Dr. Walter Wüst
Instructor, Wilhelms Gymnasium
Munich, Germany

Dr. Heinz Wundt
Zoological Collection of the State of
Bavaria
Munich, Germany

Dr. Claus-Dieter Zander
Zoological Institute and Museum,
University of Hamburg
Hamburg, Germany

Dr. Fritz Zumpt
Director, Entomology and
Parasitology, South African Institute
for Medical Research
Johannesburg, South Africa

Dr. Richard L. Zusi
Curator of Birds, United States
National Museum, Smithsonian
Institution
Washington, D.C., U.S.A.

Glossary

The following glossary is not intended to be exhaustive, but rather includes primarily terms that (1) have some specific importance to our understanding of birds, (2) have been used in these volumes, (3) might have varying definitions relative to birds as opposed to common usage, or (4) are often misunderstood.

Accipiter—This is the genus name for a group of bird-eating hawks (Accipitridae; e.g., sharp-shinned hawk, Cooper's hawk). These birds show similar behavior and appearance and extreme sexual dimorphism. Females are much larger than males and the female of the sharp-shinned hawk often seems as large as the male of the Cooper's hawk, leading to some confusion on the part of birders. In the face of uncertainty, these birds are often just referred to as "Accipiters" and the name is now firmly ensconced in "birding" terminology.

Adaptive radiation—Diversification of a species or single ancestral type into several forms that are each adaptively specialized to a specific niche.

Aftershaft—A second rachis (= shaft) arising near the base of a contour feather, creating a feather that "branches." Aftershafts can be found in many birds (e.g., pheasants) but in most the aftershaft is much smaller than the main shaft of the feather. In ratites (ostrich-like birds), the aftershaft is about the same size as the main shaft. Sometimes the term "aftershaft" is restricted to the rachis that extends from the main rachis and the whole secondary structure is referred to as the "afterfeather."

Agonistic—Behavioral patterns that are aggressive in context. Most aggressive behavior in birds is expressed as song (in songbirds) or other vocal or mechanical sound (e.g. see Drumming). The next level of intensity is display, and only in extreme circumstances do birds resort to physical aggression.

Air sac—Thin-walled, extensions of the lungs, lying in the abdomen and thorax, and extending even into some bones of birds. Air sacs allow an increased respiratory capacity of birds and the removal of oxygen both as air passes in through the lungs and also as it passes back through the lungs as the bird exhales. The flow of air through the air sacs also helps dissipate the heat produced through muscle activity and increases a bird's volume while only minimally increasing weight—thus effectively making birds lighter relative to their size and more efficient in flying. Air sacs are best developed in

the strongest flying birds and least developed in some groups that are flightless.

Alcid—Referring to a member of the family Alcidae; including puffins, auks, auklets, murres, razorbills, and guillemots.

Allopatric—Occurring in separate, nonoverlapping geographic areas.

Allopreening—Mutual preening; preening of the feathers of one bird by another; often a part of courtship or pair bond maintenance.

Alpha breeder—The reproductively dominant member of a social unit.

Alternate plumage—The breeding plumage of passerines, ducks, and many other groups; typically acquired through a partial molt prior to the beginning of courtship.

Altricial—An adjective referring to a bird that hatches with little, if any, down, is unable to feed itself, and initially has poor sensory and thermoregulatory abilities.

Alula—Small feathers at the leading edge of the wing and attached to the thumb; also called bastard wing; functions in controlling air flow over the surface of the wing, thus allowing a bird to land at a relatively slow speed.

Anatid—A collective term referring to members of the family Anatidae; ducks, geese, and swans.

Anisodactyl— An adjective that describes a bird's foot in which three toes point forward and one points backwards, a characteristic of songbirds.

Anserine—Goose-like.

Anting—A behavior of birds that involves rubbing live ants on the feathers, presumably to kill skin parasites.

Antiphonal duet—Vocalizations by two birds delivered alternately in response to one another; also known as responsive singing.

AOU—American Ornithologists' Union; the premier professional ornithological organization in North America; the organizational arbiter of scientific and standardized common names of North American birds as given in the periodically revised Check-list of North American Birds.

Arena—See Lek.

Aspect ratio—Length of a wing divided by width of the wing; High aspect wings are long and narrow. These are characteristic of dynamic soaring seabirds such as albatrosses. These birds have tremendous abilities to soar over the open ocean, but poor ability to maneuver in a small area. In contrast, low aspect ratio wings are short and broad, characteristic of many forest birds, and provide great ability to quickly maneuver in a small space.

Asynchronous—Not simultaneous; in ornithology often used with respect to the hatching of eggs in a clutch in which hatching occurs over two or more days, typically a result of initiation of incubation prior to laying of the last egg.

Auricular—An adjective referring to the region of the ear in birds, often to a particular plumage pattern over the ear.

Austral—May refer to "southern regions," typically meaning Southern Hemisphere. May also refer to the geographical region included within the Transition, Upper Austral, and Lower Austral Life Zones as defined by C. Hart Merriam in 1892–1898. These zones are often characterized by specific plant and animal communities and were originally defined by temperature gradients especially in the mountains of southwestern North America.

Autochthonous—An adjective that indicates that a species originated in the region where it now resides.

Barb—One of the hair-like extensions from the rachis of a feather. Barbs with barbules and other microstructures can adhere to one another, forming the strong, yet flexible vane needed for flight and protection and streamlining of body surfaces.

Barbules—A structural component of the barbs of many feathers; minute often interlocking filaments in a row at each side of a barb. As a result of their microstructure, barbules adhere to one another much like "Velcro®" thus assuring that feathers provide a stiff, yet flexible vane.

Basic plumage—The plumage an adult bird acquires as a result of its complete (or near complete) annual molt.

Bergmann's rule—Within a species or among closely related species of mammals and birds, those individuals in colder environments often are larger in body size. Bergmann's rule is a generalization that reflects the ability of warm-blooded animals to more easily retain body heat (in cold climates) if they have a high body surface to body volume ratio, and to more easily dissipate excess body heat (in hot environments) if they have a low body surface to body volume ratio.

Bioacoustics—The study of biological sounds such as the sounds produced by birds.

Biogeographic region—One of several major divisions of the earth defined by a distinctive assemblage of animals and plants. Sometimes referred to as "zoogeographic regions or realms" (for animals) or "phytogeographic regions or realms" (for plants). Such terminology dates from the late nineteenth century and varies considerably. Major biogeographic regions each have a somewhat distinctive flora and fauna. Those generally recognized include Nearctic, Neotropical, Palearctic, Ethiopian, Oriental, and Australian.

Biomagnification—Sometimes referred to as "bioaccumulation." Some toxic elements and chemical compounds are not readily excreted by animals and instead are stored in fatty tissues, removing them from active metabolic pathways. Birds that are low in a food chain (e.g., sparrows that eat seeds) accumulate these chemicals in their fatty tissues. When a bird that is higher in the food chain (e.g., a predator like a falcon) eats its prey (e.g., sparrows), it accumulates these chemicals from the fatty tissue of each prey individual, thus magnifying the level of the chemical in its own tissues. When the predator then comes under stress and all of these chemicals are released from its fat into its system, the effect can be lethal. Chemicals capable of such biomagnification include heavy metals such as lead and mercury, and such manmade compounds as organochlorine pesticides and polychlorinated biphenyls (PCBs).

Booming ground—See Lek.

Booted—An adjective describing a bird tarsus (leg) that has a smooth, generally undivided, rather than scaly (= scutellate) appearance. The extent of the smooth or scaly appearance of a bird tarsus varies among taxonomic groups and there are many different, more specific, patterns of tarsal appearance that are recognized.

Boreal—Often used as an adjective meaning "northern"; also may refer to the northern climatic zone immediately south of the Arctic; may also include the Arctic, Hudsonian, and Canadian Life Zones described by C. Hart Merriam.

Bristle—In ornithology, a feather with a thick, tapered rachis and no vane except for a remnant sometimes found near the bristle base.

Brood—As a noun: the young produced by a pair of birds during one reproductive effort. As a verb: to provide warmth and shelter to chicks by gathering them under the protection of breast and/or wings.

Brood parasitism—Reproductive strategy where one species of bird (the parasite) lays its eggs in the nests of another species (the host). An acceptable host will incubate the eggs and rear the chicks of the brood parasite, often to the detriment or loss of the host's own offspring.

Brood patch—A bare area of skin on the belly of a bird, the brood patch is enlarged beyond the normal apterium (bare area) as a result of loss of feathers. It becomes highly vascularized (many blood vessels just under the surface). The brood patch is very warm to the touch and the bird uses it to cover and warm its chicks. In terms of structure, the brood patch is the same as the incubation patch and the two terms are often used synonymously. Technically the brood patch and incubation patch differ in function: the incubation patch is used in incubating eggs, the brood patch is used to brood the young after the eggs hatch.

Brood reduction—Reduction in the number of young in the nest. Viewed from an evolutionary perspective, mechanisms that allow for brood reduction may assure that at least some offspring survive during stressful times and that during times of abundant resources all young may survive. Asynchronous hatching results in young of different ages and sizes in a nest and is a mechanism that facilitates brood reduction: the smallest chick often dies if there is a shortage of food. The barn owl (*Tyto alba*; Tytonidae) depends on food resources that vary greatly in availability from year to year and it often experiences brood reduction.

Buteo—This is the genus name for a group of hawks that have broad wings and soar. These hawks are often seen at a distance and are easily recognized as "Buteos" although they may not be identifiable as species. Hence the genus name has come into common English usage.

Caecum (pl. caeca)—Blindly-ending branch extending from the junction of the small and large intestine. Most birds have two caeca, but the number and their development in birds is highly variable. Caeca seem to be most highly developed and functional in facilitating microbial digestion of food in those birds that eat primarily plant materials.

Caruncle—An exposed, often brightly colored, fleshy protuberance or wrinkled facial skin of some birds.

Casque—An enlargement at the front of the head (e.g., on cassowaries, Casuariidae) or sometimes of the bill (e.g., on hornbills, Bucerotidae) of a bird. A casque may be bony, cartilaginous, or composed of feathers (e.g., Pri-

onopidae). A casque is often sexual ornamentation, but may protect the head of a cassowary crashing through underbrush, may be used for vocal amplification, or may serve a physiological function.

Cavity nester—A species that nests in some sort of a cavity. Primary cavity nesters (e.g., woodpeckers, Picidae; kingfishers, Alcedinidae; some swallows, Hirundinidae) are capable of excavating their own cavities; secondary cavity nesters (e.g., starlings, Sturnidae; House Sparrows, Passeridae; bluebirds, Turdidae) are not capable of excavating their own cavities.

Cere—The soft, sometimes enlarged, and often differently colored basal covering of the upper bill (maxilla) of many hawks (Falconiformes), parrots (Psitaciformes), and owls (Strigiformes). The nostrils are often within or at the edge of the cere. In parrots the cere is sometimes feathered.

Cladistic—Evolutionary relationships suggested as "tree" branches to indicate lines of common ancestry.

Cleidoic eggs—Cleidoic eggs are simply ones that are contained, hence protected, inside of a somewhat impervious shell—such as the eggs of birds. The presence of a shell around an egg freed the amphibian ancestors of reptiles from the need to return to the water to lay eggs and provided greater protection from dying.

Cline—A gradient in a measurable character, such as size and color, showing geographic differentiation. Various patterns of geographic variation are reflected as clines or clinal variation, and have been described as "ecogeographic rules."

Clutch—The set of eggs laid by a female bird during one reproductive effort. In most species, a female will lay one egg per day until the clutch is complete; in some species, particularly larger ones (e.g., New World vultures, Cathartidae), the interval between eggs may be more than one day.

Colony—A group of birds nesting in close proximity, interacting, and usually aiding in early warning of the presence of predators and in group defense.

Commensal—A relationship between species in which one benefits and the other is neither benefited nor harmed.

Congeneric—Descriptive of two or more species that belong to the same genus.

Conspecific—Descriptive of two or more individuals or populations that belong to the same species.

Conspecific colony—A colony of birds that includes only members of one species.

Contact call—Simple vocalization used to maintain communication or physical proximity among members of a social unit.

Contour feather—One of those feathers covering the body, head, neck, and limbs of a bird and giving rise to the shape (contours) of the bird.

Convergent evolution—When two evolutionarily unrelated groups of organisms develop similar characteristics due to adaptation to similar aspects of their environment or niche. The sharply pointed and curved talons of hawks and owls are convergent adaptations for their predatory lifestyle.

Cooperative breeding—A breeding system in which birds other than the genetic parents share in the care of eggs and young. There are many variants of cooperative breeding. The birds that assist with the care are usually referred to as "helpers" and these are often offspring of the same breeding pair, thus genetically related to the chicks they are tending. Cooperative breeding is most common among tropical birds and seems most common in situations where nest sites or breeding territories are very limited. Several studies have demonstrated that "helping" increases reproductive success. By helping a helper is often assuring survival of genes shared with the related offspring. The helper also may gain important experience and ultimately gain access to a breeding site.

Coracoid—A bone in birds and some other vertebrates extending from the scapula and clavicle to the sternum; the coracoid serves as a strut supporting the chest of the bird during powerful muscle movements associated with flapping flight.

Cosmopolitan—Adjective describing the distribution pattern of a bird found around the world in suitable habitats.

Countershading—A color pattern in which a bird or other animal is darker above and lighter below. The adaptive value of the pattern is its ability to help conceal the animal: a predator looking down from above sees the darker back against the dark ground; a predator looking up from below sees the lighter breast against the light sky; a predator looking from the side sees the dark back made lighter by the light from above and the light breast made darker by shading.

Covert—A feather that covers the gap at the base between flight feathers of the wing and tail; coverts help create smooth wing and tail contours that make flight more efficient.

Covey—A group of birds, often comprised of family members that remain together for periods of time; usually applied to game birds such as quail (Odontophorinae).

Crepuscular—Active at dawn and at dusk.

Crèche—An aggregation of young of many colonially-nesting birds (e.g., penguins, Spheniscidae; terns, Laridae). There is greater safety from predators in a crèche.

Crissum—The undertail coverts of a bird; often distinctively colored.

Critically Endangered—A technical category used by IUCN for a species that is at an extremely high risk of extinction in the wild in the immediate future.

Cryptic—Hidden or concealed; i.e., well-camouflaged patterning.

Dichromic—Occurring in two distinct color patterns (e.g., the bright red of male and dull red-brown of female northern cardinals, *Cardinalis cardinalis*)

Diurnal—Active during the day.

Dimorphic—Occurring in two distinct forms (e.g., in reference to the differences in tail length of male and female boat-tailed grackles, *Cassidix major*).

Disjunct—A distribution pattern characterized by populations that are geographically separated from one another.

Dispersal—Broadly defined: movement from an area; narrowly defined: movement from place of hatching to place of first breeding.

Dispersion—The pattern of spatial arrangement of individuals, populations, or other groups; no movement is implied.

Disruptive color—A color pattern such as the breast bands on a killdeer (*Charadrius vociferus*) that breaks up the outline of the bird, making it less visible to a potential predator, when viewed from a distance

DNA-DNA hybridization—A technique whereby the genetic similarity of different bird groups is determined based on the extent to which short stretches of their DNA, when mixed together in solution in the laboratory, are able to join with each other.

Dominance hierarchy—"Peck order"; the social status of individuals in a group; each animal can usually dominate those animals below it in a hierarchy.

Dummy nest—Sometimes called a "cock nest." An "extra" nest, often incomplete, sometimes used for roosting, built by aggressive males of polygynous birds. Dummy nests may aid in the attraction of additional mates, help define a male's territory, or confuse potential predators.

Dump nest—A nest in which more than one female lays eggs. Dump nesting is a phenomenon often linked to young, inexperienced females or habitats in which nest sites are scarce. The eggs in dump nest are usually not incubated. Dump nesting may occur within a species or between species.

Dynamic soaring—A type of soaring characteristic of oceanic birds such as albatrosses (Diomedeidae) in which the bird takes advantage of adjacent wind currents that are of different speeds in order to gain altitude and effortlessly stay aloft.

Echolocation—A method of navigation used by some swifts (Apodidae) and oilbirds (Steatornithidae) to move in darkness, such as through caves to nesting sites. The birds emit audible "clicks" and determine pathways by using the echo of the sound from structures in the area.

Eclipse plumage—A dull, female-like plumage of males of Northern Hemisphere ducks (Anatidae) and other birds such as house sparrows (*Passer domesticus*) typically attained in late summer prior to the annual fall molt. Ducks are flightless at this time and the eclipse plumage aids in their concealment at a time when they would be especially vulnerable to predators.

Ecotourism—Travel for the primary purpose of viewing nature. Ecotourism is now "big business" and is used as a non-consumptive but financially rewarding way to protect important areas for conservation.

Ectoparasites—Relative to birds, these are parasites such as feather lice and ticks that typically make their home on the skin or feathers.

Emarginate—Adjective referring to the tail of a bird that it notched or forked or otherwise has an irregular margin as a result of tail feathers (rectrices) being of different lengths. Sometimes refers to individual flight feather that is particularly narrowed at the tip.

Endangered—A term used by IUCN and also under the Endangered Species Act of 1973 in the United States in reference to a species that is threatened with imminent extinction or extirpation over all or a significant portion of its range.

Endemic—Native to only one specific area.

Eocene—Geological time period; subdivision of the Tertiary, from about 55.5 to 33.7 million years ago.

Erythrocytes—Red blood cells; in birds, unlike mammals, these retain a nucleus and are longer lived. Songbirds tend to have smaller, more numerous (per volume) erythrocytes that are richer in hemoglobin than are the erythrocytes of more primitive birds.

Ethology—The study of animal behavior.

Exotic—Not native.

Extant—Still in existence; not destroyed, lost, or extinct.

Extinct—Refers to a species that no longer survives anywhere.

Extirpated—Referring to a local extinction of a species that can still be found elsewhere.

Extra-pair copulation—In a monogamous species, refers to any mating that occurs between unpaired males and females.

Facial disc—Concave arrangement of feathers on the face of an owl. The facial discs on an owl serve as sound parabolas, focusing sound into the ears around which the facial discs are centered, thus enhancing their hearing.

Fecal sac—Nestling songbirds (Passeriformes) and closely related groups void their excrement in "packages"—enclosed in thin membranes—allowing parents to remove the material from the nest. Removal of fecal material likely reduces the potential for attraction of predators.

Feminization—A process, often resulting from exposure to environmental contaminants, in which males produce a higher levels of female hormones (or lower male hormone levels), and exhibit female behavioral or physiological traits.

Feral—Gone wild; i.e., human-aided establishment of non-native species.

Fledge—The act of a juvenile making its first flight; sometimes generally used to refer to a juvenile becoming independent.

Fledgling—A juvenile that has recently fledged. An emphasis should be placed on "recently." A fledgling generally lacks in motor skills and knowledge of its habitat and fledglings are very vulnerable, hence under considerable parental care. Within a matter of a few days, however, they gain skills and knowledge and less parental care is needed.

Flight feathers—The major feathers of the wing and tail that are crucial to flight. (See Primary, Secondary, Tertial, Alula, Remex, Rectrix)

Flyway—A major pathway used by a group of birds during migration. The flyway concept was developed primarily with regard to North American waterfowl (Anatidae) and has been used by government agencies in waterfowl management. Major flyways described include the Atlantic, Mississippi, Central, and Pacific flyways. While the flyway concept is often used in discussions of other groups of birds, even for waterfowl the concept is an oversimplification. The patterns of movements of migrant waterfowl and other birds vary greatly among species.

Frugivorous—Feeds on fruit.

Galliform—Chicken-like, a member of the Galliformes.

Gape—The opening of the mouth of a bird; the act of opening the mouth, as in begging.

Gizzard—The conspicuous, muscular portion of the stomach of a bird. Birds may swallow grit or retain bits of bone or hard parts of arthropods in the gizzard and these function in a manner analogous to teeth as the strong muscles of the gizzard contract, thus breaking food into smaller particles. The gizzard is best developed in birds that eat seeds and other plant parts; in some fruit-eating birds the gizzard is very poorly developed.

Glareolid—A member of the family Glareolidae.

Gloger's rule—Gloger's rule is an ecogeographic generalization that suggests that within a species or closely related group of birds there is more melanin (a dark pigment) in feathers in warm humid parts of the species' or groups' range, and less melanin in feathers in dry or cooler parts of the range.

Gorget—Colorful throat patch or bib (e.g., of many hummingbirds, Trochilidae).

Graduated—An adjective used to describe the tail of a bird in which the central rectrices are longest and those to the outside are increasingly shorter.

Granivorous—Feeding on seeds.

Gregarious—Occuring in large groups.

Gular—The throat region.

Hallux—The innermost digit of a hind or lower limb.

Hawk—Noun: a member of the family Accipitridae. Verb: catching insects by flying around with the mouth open (e.g. swallows, Hirundinidae; nightjars, Caprimulgidae).

Heterospecific colony—A colony of birds with two or more species.

Heterothermy—In birds, the ability to go into a state of torpor or even hibernation, lowering body temperature through reduced metabolic activity and thus conserving energy resources during periods of inclement weather or low food.

Hibernation—A deep state of reduced metabolic activity and lowered body temperature that may last for weeks; attained by few birds, resulting from reduced food supplies and cool or cold weather.

Holarctic—The Palearctic and Nearctic bigeographic regions combined.

Homeothermy—In birds the metabolic ability to maintain a constant body temperature. The lack of development of homeothermy in new-hatched chicks is the underlying need for brooding behavior.

Hover-dip—A method of foraging involving hovering low over the water, and then dipping forward to pick up prey from the surface (e.g., many herons, Ardeidae).

Hybrid—The offspring resulting from a cross between two different species (or sometimes between distinctive subspecies).

Imprinting—A process that begins with an innate response of a chick to its parent or some other animal (or object!) that displays the appropriate stimulus to elicit the chick's response. The process continues with the chick rapidly learning to recognize its parents. Imprinting typically occurs within a few hours (often 13–16 hours) after hatching. Imprinting then leads to learning behavioral characteristics that facilitate its survival, including such things as choice of foraging sites and foods, shelter, recognition of danger, and identification of a potential mate. The most elaborate (and best studied) imprinting is associated with precocial chicks such as waterfowl (Anatidae).

Incubation patch—See Brood patch.

Indigenous—See Endemic.

Innate—An inherited characteristic; e.g., see Imprinting.

Insectivorous—In ornithology technically refers to a bird that eats insects; generally refers to in birds that feed primarily on insects and other arthropods.

Introduced species—An animal or plant that has been introduced to an area where it normally does not occur.

Iridescent—Showing a rainbow-like play of color caused by differential refraction of light waves that change as the angle of view changes. The iridescence of bird feathers is a result of a thinly laminated structure in the barbules of those feathers. Iridescent feathers are made more brilliant by pigments that underlie this structure, but the pigments do not cause the iridescence.

Irruptive—A species of bird that is characterized by irregular long-distance movements, often in response to a fluctuating food supply (e.g., red crossbill, *Loxia curvirostra*, Fringillidae; snowy owl, *Nyctea scandiaca*, Strigidae).

IUCN—The World Conservation Union; formerly the International Union for the Conservation of Nature, hence IUCN. It is the largest consortium of governmental and nongovernmental organizations focused on conservation issues.

Juvenal—In ornithology (contrary to most dictionaries), restricted to use as an adjective referring to a characteristic (usually the plumage) of a juvenile bird.

Juvenile—A young bird, typically one that has left the nest.

Kleptoparasitism—Behavior in which one individual takes ("steals") food, nest materials, or a nest site from another.

Lachrymal—Part of the skull cranium, near the orbit; lachrymal and Harderian glands in this region lubricate and protect the surface of the eye.

Lamellae—Transverse tooth-like or comb-like ridges inside the cutting edge of the bill of birds such as ducks (Anatidae) and flamingos (Phoenicopteridae). Lamellae serve as a sieve during feeding: the bird takes material into its mouth, then uses its tongue to force water out through the lamellae, while retaining food particles.

Lek—A loose to tight association of several males vying for females through elaborate display; lek also refers to the specific site where these males gather to display. Lek species include such birds as prairie chickens (Phasianidae) and manakins (Pipridae).

Lobed feet—Feet that have toes with stiff scale-covered flaps that extend to provide a surface analogous to webbing on a duck as an aid in swimming.

Lore—The space between the eye and bill in a bird. The loral region often differs in color from adjacent areas of a bird's face. In some species the area is darker, thus helping to reduce glare, serving the same function as the dark pigment some football players apply beneath each eye. In predatory birds, a dark line may extend from the eye to the bill, perhaps decreasing glare, but also serving as a sight to better aim its bill. The color and pattern of plumage and skin in the loral region is species-specific and often of use in helping birders identify a bird.

Malar—Referring to the region of the face extending from near the bill to below the eye; markings in the region are often referred to as "moustache" stripes.

Mandible—Technically the lower half of a bird's bill. The plural, mandibles, is used to refer to both the upper and lower bill. The upper half of a bird's bill is technically the maxilla, but often called the "upper mandible."

Mantle—Noun: The plumage of the back of the bird, including wing coverts evident in the back region on top of the folded wing (especially used in describing hawks (Accipitridae) and gulls (Laridae). Verb: The behavior in which a raptor (typically on the ground) shields its acquired prey to protect it from other predators.

Mesoptile—On chicks, the second down feathers; these grow attached to the initial down, or protoptile.

Metabolic rate—The rate of chemical processes in living organisms, resulting in energy expenditure and growth. Hummingbirds (Trochilidae), for example, have a very high metabolic rate. Metabolic rate decreases when a bird is resting and increases during activity.

Miocene—The geological time period that lasted from about 23.8 to 5.6 million years ago.

Migration—A two-way movement in birds, often dramatically seasonal. Typically latitudinal, though in some species is altitudinal or longitudinal. May be short-distance or long-distance. (See Dispersal)

Mitochondrial DNA—Genetic material located in the mitochondria (a cellular organelle outside of the nucleus). During fertilization of an egg, only the DNA from the nucleus of a sperm combines with the DNA from the nucleus of an egg. The mitochondrial DNA of each offspring is inherited only from its mother. Changes in mitochondrial DNA occur quickly through mutation and studying differences in mitochondrial DNA helps scientists better understand relationships among groups.

Mobbing—A defensive behavior in which one or more birds of the same or different species fly toward a potential predator, such as a hawk, owl, snake, or a mammal, swooping toward it repeatedly in a threatening manner, usually without actually striking the predator. Most predators depend on the element of surprise in capturing their prey and avoid the expenditure of energy associated with a chase. Mobbing alerts all in the neighborhood that a potential predator is at hand and the predator often moves on. Rarely, a predator will capture a bird that is mobbing it.

Molecular phylogenetics—The use of molecular (usually genetic) techniques to study evolutionary relationships between or among different groups of organisms.

Molt—The systematic and periodic loss and replacement of feathers. Once grown, feathers are dead structures that continually wear. Birds typically undergo a complete or near-complete molt each year and during this molt feathers are usually lost and replaced with synchrony between right and left sides of the body, and gradually, so that the bird retains the ability to fly. Some species, such as northern hemisphere ducks, molt all of their flight feathers at once, thus become flightless for a short time. Partial molts, typically involving only contour feathers, may occur prior to the breeding season.

Monophyletic—A group (or clade) that shares a common ancestor.

Monotypic—A taxonomic category that includes only one form (e.g., a genus that includes only one species; a species that includes no subspecies).

Montane—Of or inhabiting the biogeographic zone of relatively moist, cool upland slopes below timberline dominated by large coniferous trees.

Morphology—The form and structure of animals and plants.

Mutualism—Ecological relationship between two species in which both gain benefit.

Nail—The horny tip on the leathery bill of ducks, geese, and swans (Anatidae).

Nectarivore—A nectar-eater (e.g., hummingbirds, Trochilidae; Hawaiian honeycreepers, Drepaniidae).

Near Threatened—A category defined by the IUCN suggesting possible risk of extinction in the medium term future.

Nearctic—The biogeographic region that includes temperate North America faunal region.

Neotropical—The biogeographic region that includes South and Central America, the West Indies, and tropical Mexico.

Nestling—A young bird that stays in the nest and needs care from parents.

New World—A general descriptive term encompassing the Nearctic and Neotropical biogeographic regions.

Niche—The role of an organism in its environment; multidimensional, with habitat and behavioral components.

Nictitating membrane—The third eyelid of birds; may be transparent or opaque; lies under the upper and lower eyelids. When not in use, the nictitating membrane is held at the corner of the eye closest to the bill; in use it moves horizontally or diagonally across the eye. In flight it keeps the bird's eyes from drying out; some aquatic birds have a lens-like window in the nictitating membrane, facilitating vision underwater.

Nidicolous—An adjective describing young that remain in the nest after hatching until grown or nearly grown.

Nidifugous—An adjective describing young birds that leave the nest soon after hatching.

Nocturnal—Active at night.

Nominate subspecies—The subspecies described to represent its species, the first described, bearing the specific name.

Nuclear DNA—Genetic material from the nucleus of a cell from any part of a bird's body other than its reproductive cells (eggs or sperm).

Nuptial displays—Behavioral displays associated with courtship.

Oligocene—The geologic time period occurring from about 33.7 to 23.8 million years ago.

Old World—A general term that usually describes a species or group as being from Eurasia or Africa.

Omnivorous—Feeding on a broad range of foods, both plant and animal matter.

Oscine—A songbird that is in the suborder Passeri, order Passeriformes; their several distinct pairs of muscles within the syrinx allow these birds to produce the diversity of sounds that give meaning to the term "songbird."

Osteological—Pertaining to the bony skeleton.

Palearctic—A biogeographic region that includes temperate Eurasia and Africa north of the Sahara.

Paleocene—Geological period, subdivision of the Tertiary, from 65 to 55.5 million years ago.

Pamprodactyl—The arrangement of toes on a bird's foot in which all four toes are pointed forward; characteristic of swifts (Apodidae).

Parallaxis—Comparing the difference in timing and intensity of sounds reaching each ear (in owls).

Passerine—A songbird; a member of the order Passeriformes.

Pecten—A comb-like structure in the eye of birds and reptiles, consisting of a network of blood vessels projecting inwards from the retina. The main function of the pecten seems to be to provide oxygen to the tissues of the eye.

Pectinate—Having a toothed edge like that of a comb. A pectinate claw on the middle toe is a characteristic of nightjars, herons, and barn owls. Also known as a "feather comb" since the pectinate claw is used in preening.

Pelagic—An adjective used to indicate a relationship to the open sea.

Phalloid organ—Penis-like structure on the belly of buffalo weavers; a solid rod, not connected to reproductive or excretory system.

Philopatry—Literally "love of homeland"; a bird that is philopatric is one that typically returns to nest in the same area in which it was hatched. Strongly philopatric species (e.g., hairy woodpecker, *Picoides borealis*) tend to accumulate genetic characteristics that adapt them to local conditions, hence come to show considerable geographic variation; those species that show little philopatry tend to show little geographic variation.

Phylogenetics—The study of racial evolution.

Phylogeny—A grouping of taxa based on evolutionary history.

Picid—A member of the family Picide (woodpeckers, wrynecks, piculets).

Piscivorous—Fish-eating.

Pleistocene—In general, the time of the great ice ages; geological period variously considered to include the last 1 to 1.8 million years.

Pliocene—The geological period preceding the Pleistocence; the last subdivision of what is known as the Tertiary; lasted from 5.5 to 1.8 million years ago.

Plumage—The complete set of feathers that a bird has.

Plunge-diving—A method of foraging whereby the bird plunges from at least several feet up, head-first into the water, seizes its prey, and quickly takes to the wing (e.g., terns, Laridae; gannets, Sulidae).

Polygamy—A breeding system in which either or both male and female may have two or more mates.

Polyandry—A breeding system in which one female bird mates with two or more males. Polyandry is relatively rare among birds.

Polygyny—A breeding system in which one male bird mates with two or more females.

Polyphyletic—A taxonomic group that is believed to have originated from more than one group of ancestors.

Powder down—Specialized feathers that grow continuously and break down into a fine powder. In some groups (e.g., herons, Ardeidae) powder downs occur in discrete patches (on the breast and flanks); in others (e.g., parrots, Psitacidae) they are scattered throughout the plumage. Usually used to waterproof the other feathers (especially in birds with few or no oil glands).

Precocial—An adjective used to describe chicks that hatch in an advanced state of development such that they generally can leave the nest quickly and obtain their own food, although they are often led to food, guarded, and brooded by a parent (e.g., plovers, Charadriidae; chicken-like birds, Galliformes).

Preen—A verb used to describe the behavior of a bird when it cleans and straightens its feathers, generally with the bill.

Primaries—Unusually strong feathers, usually numbering nine or ten, attached to the fused bones of the hand at the tip of a bird's wing.

Protoptile—The initial down on chicks.

Pterylosis—The arrangement of feathers on a bird.

Quaternary—The geological period, from 1.8 million years ago to the present, usually including two subdivisions: the Pleistocene, and the Holocene.

Quill—An old term that generally refers to a primary feather.

Rachis—The shaft of a feather.

Radiation—The diversification of an ancestral species into many distinct species as they adapt to different environments.

Ratite—Any of the ostrich-like birds; characteristically lack a keel on the sternum (breastbone).

Rectrix (pl. rectrices)—A tail feather of a bird; the rectrices are attached to the fused vertebrae that form a bird's bony tail.

Remex (pl. remiges)—A flight feather of the wing; remiges include the primaries, secondaries, tertials, and alula).

Reproductive longevity—The length of a bird's life over which it is capable of reproduction.

Resident—Nonmigratory.

Rhampotheca—The horny covering of a bird's bill.

Rictal bristle— A specialized tactile, stiff, hairlike feather with elongated, tapering shaft, sometimes with short barbs at the base. Rictal bristles prominently surround the mouth of birds such as many nightjars (Caprimulgidae), New World flycatchers (Tyrannidae), swallows (Hirundinidae), hawks (Accipitridae) and owls (Strigidae). They are occasionally, but less precisely referred to as "vibrissae," a term more appropriate to the "whiskers" on a mammal.

Rookery—Originally a place where rooks nest; now a term often used to refer to a breeding colony of gregarious birds.

Sally—A feeding technique that involves a short flight from a perch or from the ground to catch a prey item before returning to a perch.

Salt gland—Also nasal gland because of their association with the nostrils; a gland capable of concentrating and excreting salt, thus allowing birds to drink saltwater. These glands are best developed in marine birds.

Scapulars—Feathers at sides of shoulders.

Schemochrome—A structural color such as blue or iridescence; such colors result from the structure of the feather rather than from the presence of a pigment.

Scutellation—An arrangement or a covering of scales, as that on a bird's leg.

Secondaries—Major flight feathers of the wing that are attached to the ulna.

Sexual dichromatism—Male and female differ in color pattern (e.g., male hairy woodpecker [*Picoides villosus*, Picidae] has a red band on the back of the head, female has no red).

Sexual dimorphism—Male and female differ in morphology, such as size, feather size or shape, or bill size or shape.

Sibling species—Two or more species that are very closely related, presumably having differentiated from a common ancestor in the recent past; often difficult to distinguish, often interspecifically territorial.

Skimming—A method of foraging whereby the skimmers (Rhynchopidae) fly low over the water with the bottom bill slicing through the water and the tip of the bill above. When the bird hits a fish, the top bill snaps shut.

Slotting—Abrupt narrowing of the inner vane at the tip of some outer primaries on birds that soar; slotting breaks up wing-tip turbulence, thus facilitating soaring.

Sonagram—A graphic representation of sound.

Speciation—The evolution of new species.

Speculum—Colored patch on the wing, typically the secondaries, of many ducks (Anatidae).

Spur—A horny projection with a bony core found on the tarsometatarsus.

Sternum—Breastbone.

Structural color—See Schemochrome.

Suboscine— A songbird in the suborder Passeri, order Passeriformes, whose songs are thought to be innate, rather than learned.

Sympatric—Inhabiting the same range.

Syndactyl—Describes a condition of the foot of birds in which two toes are fused near the base for part of their length (e.g., kingfishers, Alcedinidae; hornbills, Bucerotidae).

Synsacrum—The expanded and elongated pelvis of birds that is fused with the lower vertebrae.

Syrinx (pl. syringes)—The "voice box" of a bird; a structure of cartilage and muscle located at the junction of the trachea and bronchi, lower on the trachea than the larynx of mammals. The number and complexity of muscles in the syrinx vary among groups of birds and have been of value in determining relationships among groups.

Systematist—A specialist in the classification of organisms; systematists strive to classify organisms on the basis of their evolutionary relationships.

Tarsus—In ornithology also sometimes called Tarsometatarsus or Metatarsus; the straight part of a bird's foot immediately above its toes. To the non-biologist, this seems to be the "leg" bone—leading to the notion that a bird's "knee" bends backwards. It does not. The joint at the top of the Tarsometatarsus is the "heel" joint, where the Tarsometatarsus meets the Tibiotarsus. The "knee" joint is between the Tibiotarsus and Femur.

Taxon (pl. taxa)—Any unit of scientific classification (e.g., species, genus, family, order).

Taxonomist— A specialist in the naming and classification of organisms. (See also Systematist. Taxonomy is the older science of naming things; identification of evolutionary relationships has not always been the goal of taxonomists. The modern science of Systematics generally incorporates taxonomy with the search for evolutionary relationships.)

Taxonomy—The science of identifying, naming, and classifying organisms into groups.

Teleoptiles—Juvenal feathers.

Territory—Any defended area. Typically birds defend a territory with sound such as song or drumming. Territorial defense is typically male against male, female against female, and within a species or between sibling species. Area defended varies greatly among taxa, seasons, and habitats. A territory may include the entire home range, only the area immediately around a nest, or only a feeding or roosting area.

Tertiary—The geological period including most of the Cenozoic; from about 65 to 1.8 million years ago.

Tertial—A flight feather of the wing that is loosely associated with the humerus; tertials fill the gap between the secondary feathers and the body.

Thermoregulation—The ability to regulate body temperature; can be either behavioral or physiological. Birds can regulate body temperature by sunning or moving to shade or water, but also generally regulate their body temperature through metabolic processes. Baby birds initially have poor thermoregulatory abilities and thus must be brooded.

Threatened—A category defined by IUCN and by the Endangered Species Act of 1973 in the United States to refer to a species that is at risk of becoming endangered.

Tomium (pl. tomia)—The cutting edges of a bird's bill.

Torpor—A period of reduced metabolic activity and lowered body temperature; often results from reduced availability of food or inclement weather; generally lasts for only a few hours (e.g., hummingbirds, Trochilidae; swifts, Apodidae).

Totipalmate—All toes joined by webs, a characteristic that identifies members of the order Pelecaniformes.

Tribe—A unit of classification below the subfamily and above the genus.

Tubercle—A knob- or wart-like projection.

Urohydrosis—A behavior characteristic of storks and New World vultures (Ciconiiformes) wherein these birds excrete on their legs and make use of the evaporation of the water from the excrement as an evaporative cooling mechanism.

Uropygial gland—A large gland resting atop the last fused vertebrae of birds at the base of a bird's tail; also known as oil gland or preen gland; secretes an oil used in preening.

Vane—The combined barbs that form a strong, yet flexible surface extending from the rachis of a feather.

Vaned feather—Any feather with vanes.

Viable population—A population that is capable of maintaining itself over a period of time. One of the major conservation issues of the twenty-first century is determining what is a minimum viable population size. Population geneticists have generally come up with estimates of about 500 breeding pairs.

Vibrissae—See Rictal bristle.

Vulnerable—A category defined by IUCN as a species that is not Critically Endangered or Endangered, but is still facing a threat of extinction.

Wallacea—The area of Indonesia transition between the Oriental and Australian biogeographical realms, named after Alfred Russell Wallace, who intensively studied this area.

Wattles—Sexual ornamentation that usually consists of flaps of skin on or near the base of the bill.

Zoogeographic region—See Biogeographic region.

Zygodactyl—Adjective referring to the arrangement of toes on a bird in which two toes project forward and two to the back.

Compiled by Jerome A. Jackson, PhD

Aves species list

Struthioniformes [Order]
Struthionidae [Family]
Struthio [Genus]
S. camelus [Species]

Rheidae [Family]
Rhea [Genus]
R. Americana [Species]
Pterocnemia [Genus]
P. pennata [Species]

Casuaridae [Family]
Casuarius [Genus]
C. bennetti [Species]
C. casuarius
C. unappendiculatus

Dromaiidae [Family]
Dromaius [Genus]
D. novaehollandiae [Species]
D. diemenianus

Apterygidae [Family]
Apteryx [Genus]
A. australis [Species]
A. owenii
A. haastii

Tinamiiformes [Order]
Tinamidae [Family]
Tinamus [Genus]
T. tao [Species]
T. solitarius
T. osgoodi
T. major
T. guttatus
Nothocercus [Genus]
N. bonapartei [Species]
N. julius
N. nigrocapillus
Crypturellus [Genus]
C. berlepschi [Species]
C. cinereus
C. soui
C. ptaritepui
C. obsoletus
C. undulatus
C. transfasciatus

C. strigulosus
C. duidae
C. erythropus
C. noctivagus
C. atrocapillus
C. cinnamomeus
C. boucardi
C. kerriae
C. variegatus
C. brevirostris
C. bartletti
C. parvirostris
C. casiquiare
C. tataupa
Rhynchotus [Genus]
R. rufescens [Species]
Nothoprocta [Genus]
N. taczanowski [Species]
N. kalinowskii
N. omata
N. perdicaria
N. cinerascens
N. pentlandii
N. curvirostris
Nothura [Genus]
N. boraquira [Species]
N. minor
N. darwinii
N. maculosa
Taoniscus [Genus]
T. nanus [Species]
Eudromia [Genus]
E. elegans [Species]
E. formosa
Tinamotis [Genus]
T. pentlandii [Species]
T. ingoufi

Procellariiformes [Order]
Diomedidae [Family]
Diomedea [Genus]
D. exulans [Species]
D. epomophora
D. irrorata
D. albatrus
D. nigripes
D. immutabilis

D. melanophrys
D. cauta
D. chrysostoma
D. chlororhynchos
D. bulleri
Phoebetria [Genus]
P. fusca [Species]
P. palpebrata
Macronectes [Genus]
M. giganteus [Species]
M. halli
Fulmarus [Genus]
F. glacialoides [Species]
F. glacialis
Thalassoica [Genus]
T. antarctica [Species]
Daption [Genus]
D. capense [Species]
Pagodroma [Genus]
P. nivea [Species]
Pterodroma [Genus]
P. macroptera [Species]
P. lessonii
P. incerta
P. solandri
P. magentae
P. rostrata
P. macgillivrayi
P. neglecta
P. arminjoniana
P. alba
P. ultima
P. brevirostris
P. mollis
P. inexpectata
P. cahow
P. hasitata
P. externa
P. baraui
P. phaeopygia
P. hypoleuca
P. nigripennis
P. axillaris
P. cookii
P. defilippiana
P. longirostris
P. leucoptera

Halobaena [Genus]
  H. caerulea
Pachyptila [Genus]
  P. vittata [Species]
  P. desolata
  P. belcheri
  P. turtur
  P. crassirostris
Bulweria [Genus]
  B. bulwerii [Species]
  B. fallax
Procellaria [Genus]
  P. aequinoctialis [Species]
  P. westlandica
  P. parkinsoni
  P. cinerea
Calonectris [Genus]
  C. diomedea [Species]
  C. leucomelas
Puffinus [Genus]
  P. pacificus [Species]
  P. bulleri
  P. carneipes
  P. creatopus
  P. gravis
  P. griseus
  P. tenuirostris
  P. nativitatis
  P. puffinus
  P. gavia
  P. huttoni
  P. lherminieri
  P. assimilis
Oceanites [Genus]
  O. oceanicus [Species]
  O. gracilis
Garrodia [Genus]
  G. nereis [Species]
Pelagodroma [Genus]
  P. marina [Species]
Fregetta [Genus]
  F. tropica [Species]
  F. grallaria
Nesofregetta [Genus]
  N. fuliginosa [Species]
Hydrobates [Genus]
  H. pelagicus [Species]
Halocyptena [Genus]
  H. microsoma [Species]
Oceanodroma [Genus]
  O. tethys [Species]
  O. castro
  O. monorhis
  O. leucorhoa
  O. macrodactyla
  O. markhami
  O. tristami
  O. melania
  O. matsudairae
  O. homochroa

  O. hornbyi
  O. furcata
Pelecanoides [Genus]
  P. garnotii [Species]
  P. magellani
  P. georgicus
  P. urinator

Sphenisciformes [Order]
Spheniscidae [Family]
  Aptenodytes [Genus]
    A. patagonicus [Species]
    A. forsteri
  Pygoscelis [Genus]
    P. papua [Species]
    P. adeliae
    P. antarctica
  Eudyptes [Genus]
    E. chrysocome [Species]
    E. pachyrhynchus
    E. robustus
    E. sclateri
    E. chryoslophus
  Megadyptes [Genus]
    M. antipodes [Species]
  Eudyptula [Genus]
    E. minor [Species]
  Spheniscus [Genus]
    S. demersus [Species]
    S. humboldti
    S. magellanicus
    S. mendiculus

Gaviiformes [Order]

Gaviidae [Family]
  Gavia [Genus]
    G. stellata [Species]
    G. arctica
    G. immer
    G. adamsii

Podicipediformes [Order]

Podicipedidae [Family]
  Rollandia [Genus]
    R. rolland [Species]
    R. microptera
  Tachybaptus [Genus]
    T. novaehollandiae [Species]
    T. ruficollis
    T. rufolavatus
    T. pelzelnii
    T. dominicus
  Podilymbus [Genus]
    P. podiceps [Species]
    P. gigas
  Poliocephalus [Genus]
    P. poliocephalus [Species]
    P. rufopectus
  Podiceps [Genus]
    P. major [Species]

    P. auritus
    P. grisegena
    P. cristatus
    P. nigricollis
    P. occipitalis
    P. taczanowskii
    P. gallardoi
  Aechmophorus [Genus]
    A. occidentalis [Species]

Pelecaniformes [Order]
Phaethontidae [Family]
  Phaethon [Genus]
    P. aethereus [Species]
    P. rubricauda
    P. lepturus

Fregatidae [Family]
  Fregata [Genus]
    F. magnificens [Species]
    F. minor
    F. ariel
    F. andrewsi

Phalacrocoracidae [Family]
  Phalacrocorax [Genus]
    P. carbo [Species]
    P. capillatus
    P. nigrogularis
    P. varius
    P. harrisi
    P. auritus
    P. olivaceous
    P. fuscicollis
    P. sulcirostris
    P. penicillatus
    P. capensis
    P. neglectus
    P. punctatus
    P. aristotelis
    P. perspicillatus
    P. urile
    P. pelagicus
    P. gaimardi
    P. magellanicus
    P. bouganvillii
    P. atriceps
    P. albiventer
    P. carunculatus
    P. campbelli
    P. fuscescens
    P. melanoleucos
    P. niger
    P. pygmaeus
    P. africanus
  Anhinga [Genus]
    A. anhinga [Species]
    A. melanogaster

Sulidae [Family]
  Sula [Genus]
    S. bassana [Species]

S. capensis
S. serrator
S. nebouxii
S. variegata
S. dactylatra
S. sula
S. leucogaster
S. abbotti

Pelecanidae [Family]
  Pelecanus [Genus]
    P. onocrotalus [Species]
    P. rufescens
    P. philippensis
    P. conspicillatus
    P. erythrorhynchos
    P. occidentalis

Ciconiiformes [Order]
  Ardeidae [Family]
    Syrigma [Genus]
      S. sibilatrix [Species]
    Pilherodius [Genus]
      P. pileatus [Species]
    Ardea [Genus]
      A. cinerea [Species]
      A. herodias
      A. cocoi
      A. pacifica
      A. melanocephala
      A. hombloti
      A. imperialis
      A. sumatrana
      A. goliath
      A. purpurea
      A. alba
    Egretta [Genus]
      E. rufescens [Species]
      E. picata
      E. vinaceigula
      E. ardesiaca
      E. tricolor
      E. intermedia
      E. ibis
      E. novaehollandiae
      E. caerulea
      E. thula
      E. garzetta
      E. gularis
      E. dimorpha
      E. eulophotes
      E. sacra
    Ardeola [Genus]
      A. ralloides [Species]
      A. grayii
      A. bacchus
      A. speciosa
      A. idae
      A. rufiventris
      A. striata

Agamia [Genus]
  A. agami [Species]
Nyctanassa [Genus]
  N. violacea [Species]
Nycticorax [Genus]
  N. nycticorax [Species]
  N. caledonicus
  N. leuconotus
  N. magnificus
  N. goisagi
  N. melanolophus
Cochlearius [Genus]
  C. cochlearius [Species]
Tigrisoma [Genus]
  T. mexicanum [Species]
  T. fasciatum
  T. lineatum
Zonerdius [Genus]
  Z. heliosylus [Species]
Tigriornis [Genus]
  T. leucolophus [Species]
Zebrilus [Genus]
  Z. undulatus [Species]
Ixobrychus [Genus]
  I. involucris [Species]
  I. exilis
  I. minutus
  I. sinensis
  I. eurhythmus
  I. cinnamomeus
  I. sturmii
  I. flavicollis
Botaurus [Genus]
  B. pinnatus [Species]
  B. lentiginosus
  B. stellaris
  B. poiciloptilus

Scopidae [Family]
  Scopus [Genus]
    S. umbretta [Species]

Ciconiidae [Family]
  Mycteria [Genus]
    M. americana [Species]
    M. cinerea
    M. ibis
    M. leucocephala
  Anastomus [Genus]
    A. oscitans [Species]
    A. lamelligerus
  Ciconia [Genus]
    C. nigra [Species]
    C. abdimii
    C. episcopus
    C. maguari
    C. ciconia
  Ephippiorhynchus [Genus]
    E. asiaticus [Species]
    E. senegalensis
  Jabiru [Genus]
    J. mycteria [Species]

Leptoptilos [Genus]
  L. javanicus [Species]
  L. dubius
  L. crumeniferus

Balaenicipitidae [Family]
  Balaeniceps [Genus]
    B. rex [Species]

Threskiornithidae [Family]
  Eudocimus [Genus]
    E. albus [Species]
    E. ruber
  Phimosus [Genus]
    P. infuscatus [Species]
  Plegadis [Genus]
    P. falcinellus [Species]
    P. chihi
    P. ridgwayi
  Cercibis [Genus]
    C. oxycerca [Species]
  Theristicus [Genus]
    T. caerulescens [Species]
    T. caudatus
    T. melanopsis
  Mesembrinibis [Genus]
    M. cayennensis [Species]
  Bostrychia [Genus]
    B. hagedash [Species]
    B. carunculata
    B. olivacea
    B. rara
  Lophotibis [Genus]
    L. cristata [Species]
  Threskiornis [Genus]
    T. aethiopicus [Species]
    T. spinicollis
  Geronticus [Genus]
    G. eremita [Species]
    G. calvus
  Pseudibis [Genus]
    P. papillosa [Species]
    P. gigantea
  Nipponia [Genus]
    N. nippon [Species]
  Platalea [Genus]
    P. leucorodia [Species]
    P. minor
    P. alba
    P. flavipes
    P. ajaja

Phoenicopteriformes [Order]
  Phoenicopteridae [Family]
    Phoenicopterus [Genus]
      P. ruber [Species]
      P. chilensis
    Phoeniconaias [Genus]
      P. minor [Species]
    Phoenicoparrus [Genus]
      P. andinus [Species]
      P. jamesii

Falconiformes [Order]
Cathartidae [Family]
Coragyps [Genus]
C. atratus [Species]
Cathartes [Genus]
C. burrovianus [Species]
C. melambrotus
Gymnogyps [Genus]
G. californianus [Species]
Vultur [Genus]
V. gryphus [Species]
Sarcoramphus [Genus]
S. papa [Species]

Accipitridae [Family]
Pandion [Genus]
P. haliaetus [Species]
Aviceda [Genus]
A. cuculoides [Species]
A. madagascariensis
A. jerdoni
A. subcristata
A. leuphotes
Leptodon [Genus]
L. cayanensis [Species]
Chondrohierax [Genus]
C. uncinatus [Species]
Henicopernis [Genus]
H. longicauda [Species]
H. infuscata
Pernis [Genus]
P. apivorus [Species]
P. ptilorhynchus
P. celebensis
Elanoides [Genus]
E. forficatus [Species]
Macheiramphus [Genus]
M. alcinus [Species]
Gampsonyx [Genus]
G. swainsonii [Species]
Elanus [Genus]
E. leucurus [Species]
E. caeruleus
E. notatus
E. scriptus
Chelictinia [Genus]
C. riocourii [Species]
Rostrhamus [Genus]
R. sociabilis [Species]
R. hamatus
Harpagus [Genus]
H. bidentatus [Species]
H. diodon
Ictinia [Genus]
I. plumbea [Species]
I. misisippiensis
Lophoictinia [Genus]
L. isura [Species]
Hamirostra [Genus]
H. melanosternon [Species]

Milvus [Genus]
M. milvus [Species]
M. migrans
Haliastur [Genus]
H. sphenurus [Species]
H. indus
Haliaeetus [Genus]
H. leucogaster [Species]
H. sanfordi
H. vocifer
H. vociferoides
H. leucoryphus
H. albicilla
H. leucocephalus
H. pelagicus
Ichthyophaga [Genus]
I. humilis [Species]
I. ichthyaetus
Gypohierax [Genus]
G. angolensis [Species]
Gypaetus [Genus]
G. barbatus [Species]
Neophron [Genus]
N. percnopterus [Species]
Necrosyrtes [Genus]
N. monachus [Species]
Gyps [Genus]
G. bengalensis [Species]
G. africanus
G. indicus
G. rueppellii
G. himalayensis
G. fulvus
Aegypius [Genus]
A. monachus [Species]
A. tracheliotus
A. occipitalis
A. calvus
Circaetus [Genus]
C. gallicus [Species]
C. cinereus
C. fasciolatus
C. cinerascens
Terathopius [Genus]
T. ecaudatus [Species]
Spilornis [Genus]
S. cheela [Species]
S. elgini
Dryotriorchis [Genus]
D. spectabilis [Species]
Eutriorchis [Genus]
E. astur [Species]
Polyboroides [Genus]
P. typus [Species]
P. radiatus
Circus [Genus]
C. assimilis [Species]
C. maurus
C. cyaneus
C. cinereus

C. macrourus
C. melanoleucos
C. pygargus
C. ranivorus
C. aeruginosus
C. spilonotus
C. approximans
C. maillardi
C. buffoni
Melierax [Genus]
M. gabar [Species]
M. metabates
M. canorus
Accipiter [Genus]
A. poliogaster [Species]
A. trivirgatus
A. griseiceps
A. tachiro
A. castanilius
A. badius
A. brevipes
A. butleri
A. soloensis
A. francesii
A. trinotatus
A. fasciatus
A. novaehollandiae
A. melanochlamys
A. albogularis
A. rufitorques
A. haplochrous
A. henicogrammus
A. luteoschistaceus
A. imitator
A. poliocephalus
A. princeps
A. superciliosus
A. collaris
A. erythropus
A. minullus
A. gularis
A. virgatus
A. nanus
A. cirrhocephalus
A. brachyurus
A. erythrauchen
A. rhodogaster
A. ovampensis
A. madagascariensis
A. nisus
A. rufiventris
A. striatus
A. bicolor
A. cooperii
A. gundlachi
A. melanoleucus
A. henstii
A. gentilis
A. meyerianus
A. buergersi

A. radiatus
A. doriae
Urotriorchis [Genus]
  U. macrourus [Species]
Butastur [Genus]
  B. rufipennis [Species]
  B. liventer
  B. teesa
  B. indicus
Kaupifalco [Genus]
  K. monogrammicus [Species]
Geranospiza [Genus]
  G. caerulescens [Species]
Leucopternis [Genus]
  L. schistacea [Species]
  L. plumbea
  L. princeps
  L. melanops
  L. kuhli
  L. lacernulata
  L. semiplumbea
  L. albicollis
  L. polionota
Asturina [Genus]
  A. nitida [Species]
Buteogallus [Genus]
  B. aequinoctialis [Species]
  B. subtilis
  B. anthracinus
  B. urubitinga
  B. meridionalis
Parabuteo [Genus]
  P. unicinctus [Species]
Busarellus [Genus]
  B. nigricollis [Species]
Geranoaetus [Genus]
  G. melanoleucus [Species]
Harpyhaliaetus [Genus]
  H. solitarius [Species]
  H. coronatus
Buteo [Genus]
  B. magnirostris [Species]
  B. leucorrhous
  B. ridgwayi
  B. lineatus
  B. platypterus
  B. brachyurus
  B. swainsoni
  B. galapagoensis
  B. albicaudatus
  B. polyosoma
  B. poecilochrous
  B. albonotatus
  B. solitarius
  B. ventralis
  B. jamaicensis
  B. buteo
  B. oreophilus
  B. brachypterus
  B. rufinus

B. hemilasius
B. regalis
B. lagopus
B. auguralis
B. rufofuscus
Morphnus [Genus]
  M. guianensis [Species]
Harpia [Genus]
  H. harpyja [Species]
Pithecophaga [Genus]
  P. jeffreyi [Species]
Ictinaetus [Genus]
  I. malayensis [Species]
Aquila [Genus]
  A. pomarina [Species]
  A. clanga
  A. rapax
  A. heliaca
  A. wahlbergi
  A. gurneyi
  A. chrysaetos
  A. audax
  A. verreauxii
Hieraaetus [Genus]
  H. fasciatus [Species]
  H. spilogaster
  H. pennatus
  H. morphnoides
  H. dubius
  H. kienerii
Spizastur [Genus]
  S. melanoleucus [Species]
Lophaetus [Genus]
  L. occipitalis [Species]
Spizaetus [Genus]
  S. africanus [Species]
  S. cirrhatus
  S. nipalensis
  S. bertelsi
  S. lanceolatus
  S. philippensis
  S. alboniger
  S. nanus
  S. tyrannus
  S. ornatus
Stephanoaetus [Genus]
  S. coronatus [Species]
Oroaetus [Genus]
  O. isidori [Species]
Polemaetus [Genus]
  P. bellicosus [Species]

Sagittariidae [Family]
  Sagittarius [Genus]
    S. serpentarius [Species]

Falconidae [Family]
  Daptrius [Genus]
    D. ater [Species]
    D. americanus

Phalcoboenus [Genus]
  P. megalopterus [Species]
  P. australis
Polyborus [Genus]
  P. plancus [Species]
Milvago [Genus]
  M. chimachima [Species]
  M. chimango
Herpetotheres [Genus]
  H. cachinnans [Species]
Micrastur [Genus]
  M. ruficollis [Species]
  M. gilvicollis
  M. mirandollei
  M. semitorquatus
  M. buckleyi
Spiziapteryx [Genus]
  S. circumcinctus [Species]
Polihierax [Genus]
  P. semitorquatus [Species]
  P. insignis
Microhierax [Genus]
  M. caerulescens [Species]
  M fringillarius
  M. latifrons
  M. erythrogerys
  M. melanoleucus
Falco [Genus]
  F. berigora [Species]
  F. naumanni
  F. sparverius
  F. tinnunculus
  F. newtoni
  F. punctatus
  F. araea
  F. moluccensis
  F. cenchroides
  F. rupicoloides
  F. alopex
  F. ardosiaceus
  F. dickinsoni
  F. zoniventris
  F. chicquera
  F. vespertinus
  F. amurensis
  F. eleonorae
  F. concolor
  F. femoralis
  F. columbarius
  F. rufigularis
  F. subbuteo
  F. cuvieri
  F. severus
  F. longipennis
  F. novaeseelandiae
  F. hypoleucos
  F. subniger
  F. mexicanus
  F. jugger
  F. biarmicus

F. cherrug
F. rusticolus
F. kreyenborgi
F. peregrinus
F. deiroleucus
F. fasciinucha

Anseriformes [Order]
Anatidae [Family]
  Anseranas [Genus]
    A. semipalmata [Species]
  Dendrocygna [Genus]
    D. guttata [Species]
    D. eytoni
    D. bicolor
    D. arcuata
    D. javanica
    D. viduata
    D. arborea
    D. autumnalis
  Thalassornis [Genus]
    T. leuconotus [Species]
  Cygnus [Genus]
    C. olor [Species]
    C. atratus
    C. melanocoryphus
    C. buccinator
    C. cygnus
    C. bewickii
    C. columbianus
  Coscoroba [Genus]
    C. coscoroba [Species]
  Anser [Genus]
    A. cygnoides [Species]
    A. fabalis
    A. albifrons
    A. erythropus
    A. anser
    A. indicus
    A. caerulescens
    A. rossii
    A. canagicus
  Branta [Genus]
    B. sandvicensis [Species]
    B. canadensis
    B. leucopsis
    B. bernicla
    B. ruficollis
  Cereopsis [Genus]
    C. novaehollandiae [Species]
  Stictonetta [Genus]
    S. naevosa [Species]
  Cyanochen [Genus]
    C. cyanopterus [Species]
  Chloephaga [Genus]
    C. melanoptera [Species]
    C. picta
    C. hybrida
    C. poliocephala
    C. rubidiceps

Neochen [Genus]
  N. jubata [Species]
Alopochen [Genus]
  A. aegyptiaca [Species]
Tadorna [Genus]
  T. ferruginea [Species]
  T. cana
  T. variegata
  T. cristata
  T. tadornoides
  T. tadorna
  T. radjah
Tachyeres [Genus]
  T. pteneres [Species]
  T. brachypterus
  T. patachonicus
Plectropterus [Genus]
  P. gambensis [Species]
Cairina [Genus]
  C. moschata [Species]
  C. scutulata
Pteronetta [Genus]
  P. hartlaubii
Sarkidiornis [Genus]
  S. melanotos [Species]
Nettapus [Genus]
  N. pulchellus [Species]
  N. coromandelianus
  N. auritus
Callonetta [Genus]
  C. leucophrys [Species]
Aix [Genus]
  A. sponsa [Species]
  A. galericulata
Chenonetta [Genus]
  C. jubata [Species]
Amazonetta [Genus]
  A. brasiliensis [Species]
Merganetta [Genus]
  M. armata [Species]
Hymenolaimus [Genus]
  H. malacorhynchos [Species]
Anas [Genus]
  A. waigiuensis [Species]
  A. penelope
  A. americana
  A. sibilatrix
  A. falcata
  A. strepera
  A. formosa
  A. crecca
  A. flavirostris
  A. capensis
  A. gibberifrons
  A. bernieri
  A. castanea
  A. aucklandica
  A. platyrhynchos
  A. rubripes
  A. undulata

A. melleri
A. poecilorhyncha
A. superciliosa
A. luzonica
A. sparsa
A. specularioides
A. specularis
A. acuta
A. georgica
A. bahamensis
A. erythrorhyncha
A. versicolor
A. hottentota
A. querquedula
A. discors
A. cyanoptera
A. platalea
A. smithii
A. rhynchotis
A. clypeata
Malacorhynchus [Genus]
  M. membranaceus [Species]
Marmaronetta [Genus]
  M. angustirostris [Species]
Rhodonessa [Genus]
  R. caryophyllacea [Species]
Netta [Genus]
  N. rufina [Species]
  N. peposaca
  N. erythrophthalma
Aythya [Genus]
  A. valisineria [Species]
  A. ferina
  A. americana
  A. collaris
  A. australis
  A. baeri
  A. nyroca
  A. innotata
  A. novaeseelandiae
  A. fuligula
  A. marila
  A. affinis
Somateria [Genus]
  S. mollissima [Species]
  S. spectabilis
  S. fischeri
Polysticta [Genus]
  P. stelleri [Species]
Camptorhynchus [Genus]
  C. labradorius [Species]
Histrionicus [Genus]
  H. histrionicus [Species]
Clangula [Genus]
  C. hyemalis [Species]
Melanitta [Genus]
  M. nigra [Species]
  M. perspicillata
  M. fusca

Bucephala [Genus]
  B. clangula [Species]
  B. islandica
  B. albeola
Mergus [Genus]
  M. albellus [Species]
  M. cucullatus
  M. octosetaceous
  M. serrator
  M. squamatus
  M. merganser
  M. australis
Heteronetta [Genus]
  H. atricapilla [Species]
Oxyura [Genus]
  O. dominica [Species]
  O. jamaicensis
  O. leucocephala
  O. maccoa
  O. vittata
  O. australis
Biziura [Genus]
  B. lobata [Species]

Anhimidae [Family]
  Anhima [Genus]
    A. cornuta [Species]
  Chauna [Genus]
    C. chavaria [Species]
    C. torquata

Galliformes [Order]
Megapodiidae [Family]
  Megapodius [Genus]
    M. nicobariens [Species]
    M. tenimberens
    M. reinwardt
    M. affinis
    M. eremita
    M. freycinet
    M. laperouse
    M. layardi
    M. pritchardii
  Eulipoa [Genus]
    E. wallacei [Species]
  Leipoa [Genus]
    L. ocellata [Species]
  Alectura [Genus]
    A. lathami [Species]
  Talegalla [Genus]
    T. cuvieri [Species]
    T. fuscirostris
    T. jobiensis
  Aepypodius [Genus]
    A. arfakianus [Species]
    A. bruijnii
  Macrocephalon [Genus]
    M. maleo [Species]

Cracidae [Family]
  Nothocrax [Genus]
    N. urumutum [Species]

Mitu [Genus]
  M. tomentosa [Species]
  M. salvini
  M. mitu
Pauxi [Genus]
  P. pauxi [Species]
Crax [Genus]
  C. nigra [Species]
  C. alberti
  C. fasciolata
  C. pinima
  C. globulosa
  C. blumenbachii
  C. rubra
Penelope [Genus]
  P. purpurascens [Species]
  P. ortoni
  P. albipennis
  P. marail
  P. montagnii
  P. obscura
  P. superciliaris
  P. jacu-caca
  P. ochrogaster
  P. pileata
  P. argyrotis
Ortalis [Genus]
  O. motmot [Species]
  O. spixi
  O. araucuan
  O. superciliaris
  O. guttata
  O. columbiana
  O. wagleri
  O. vetula
  O. ruficrissa
  O. ruficauda
  O. garrula
  O. canicollis
  O. erythroptera
Penelopina [Genus]
  P. nigra [Species]
Chamaepetes [Genus]
  C. goudotii [Species]
  C. unicolor
Pipile [Genus]
  P. pipile [Species]
  P. cumanensis
  P. jacutinga
Aburria [Genus]
  A. aburri [Species]
Oreophasis [Genus]
  O. derbianus [Species]

Tetraonidae [Family]
  Tetrao [Genus]
    T. urogallus [Species]
    T. parvirostris
  Lyrurus [Genus]
    L. tetrix [Species]
    L. mlokosiewiczi

Dendragapus [Genus]
  D. obscurus [Species]
Lagopus [Genus]
  L. scoticus [Species]
  L. lagopus
  L. mutus
  L. leucurus
Canachites [Genus]
  C. canadensis [Species]
  C. franklinii
Falcipennis [Genus]
  F. falcipennis [Species]
Tetrastes [Genus]
  T. bonasia [Species]
  T. sewerzowi
Bonasa [Genus]
  B. umbellus [Species]
Pedioecetes [Genus]
  P. phasianellus [Species]
Tympanuchus [Genus]
  T. cupido [Species]
  T. palladicinctus
Centrocercus [Genus]
  C. urophasianus [Species]

Phasianidae [Family]
  Dendrortyx [Genus]
    D. barbatus [Species]
    D. macroura
    D. leucophrys
    D. hypospodius
  Oreortyx [Genus]
    O. picta [Species]
  Callipepla [Genus]
    C. squamota [Species]
  Lophortyx [Genus]
    L. californica [Species]
    L. gambelli
    L. leucoprosopon
    L. douglasii
  Philortyx [Genus]
    P. fasciatus [Species]
  Colinus [Genus]
    C. virginianus [Species]
    C. nigrogularis
    C. leucopogon
    C. cristatus
  Odontophorus [Genus]
    O. gujanensis [Species]
    O. capueira
    O. erythrops
    O. hyperythrus
    O. melanonotus
    O. speciosus
    O. loricatus
    O. parambae
    O. strophium
    O. atrifrons
    O. leucolaemus
    O. columbianus
    O. soderstromii

O. balliviani
O. stellatus
O. guttatus
Dactylortyx [Genus]
   D. thoracicus [Species]
Cyrtonyx [Genus]
   C. montezumae [Species]
   C. sallei
   C. ocellatus
Rhynchortyx [Genus]
   R. cinctus [Species]
Lerwa [Genus]
   L. lerwa [Species]
Ammoperdix [Genus]
   A. griseogularis [Species]
   A. heyi
Tetraogallus [Genus]
   T. caucasicus [Species]
   T. caspius
   T. tibetanus
   T. altaicus
   T. himalayensis
Tetraophasis [Genus]
   T. obscurus [Species]
   T. szechenyii
Alectoris [Genus]
   A. graeca [Species]
   A. rufa
   A. barbara
   A. melanocephala
Anurophasis [Genus]
   A. monorthonyx [Species]
Francolinus [Genus]
   F. francolinus [Species]
   F. pictus
   F. pintadeanus
   F. pondicerianus
   F. gularis
   F. lathami
   F. nahani
   F. streptophorus
   F. coqui
   F. albogularis
   F. sephaena
   F. africanus
   F. shelleyi
   F. levaillantii
   F. finschi
   F. gariepensis
   F. adspersus
   F. capensis
   F. natalensis
   F. harwoodi
   F. bicalcaratus
   F. icterorhynchus
   F. clappertoni
   F. hartlaubi
   F. swierstrai
   F. hildebrandti
   F. squamatus

F. ahantensis
F. griseostriatus
F. camerunensis
F. nobilis
F. jacksoni
F. castaneicollis
F. atrifrons
F. erckelii
Pternistis [Genus]
   P. rufopictus [Species]
   P. afer
   P. swainsonii
   P. leucoscepus
Perdix [Genus]
   P. perdix [Species]
   P. barbata
   P. hodgsoniae
Rhizothera [Genus]
   R. longirostris [Species]
Margaroperdix [Genus]
   M. madagarensis [Species]
Melanoperdix [Genus]
   M. nigra [Species]
Coternix [Genus]
   C. coturnix [Species]
   C. coromandelica
   C. delegorguei
   C. pectoralis
   C. novaezelandiae
Synoicus [Genus]
   S. ypsilophorus [Species]
Excalfactoria [Genus]
   E. adansonii [Species]
   E. chinensis
Perdicula [Genus]
   P. asiatica [Species]
Cryptoplectron [Genus]
   C. erythrorhynchum [Species]
   C. manipurensis
Arborophila [Genus]
   A. torqueola [Species]
   A. rufogularis
   A. atrogularis
   A. crudigularis
   A. mandellii
   A. brunneopectus
   A. rufipectus
   A. gingica
   A. davidi
   A. cambodiana
   A. orientalis
   A. javanica
   A. rubrirostris
   A. hyperythra
   A. ardens
Tropicoperdix [Genus]
   T. charltonii [Species]
   T. chloropus
   T. merlini
Caloperdix [Genus]
   C. oculea [Species]

Haematortyx [Genus]
   H. sanguiniceps [Species]
Rollulus [Genus]
   R. roulroul [Species]
Ptilopachus [Genus]
   P. petrosus [Species]
Bambusicola [Genus]
   B. fytchii [Species]
   B. thoracica
Galloperdix [Genus]
   G. spadicea [Species]
   G. lunulata
   G. bicalcarata
Ophrysia [Genus]
   O. superciliosa [Species]
Ithaginis [Genus]
   I. cruentus [Species]
Tragopan [Genus]
   T. melanocephalus [Species]
   T. satyra
   T. blythii
   T. temminckii
   T. caboti
Lophophorus [Genus]
   L. impejanus [Species]
   L. sclateri
   L. lhuysii
Crossoptilon [Genus]
   C. mantchuricum [Species]
   C. auritum
   C. crossoptilon
Gennaeus [Genus]
   G. leucomelanos [Species]
   G. horsfieldii
   G. lineatus
   G. nycthemerus
Hierophasis [Genus]
   H. swinhoii [Species]
   H. imperialis
   H. edwardsi
Houppifer [Genus]
   H. erythrophthalmus [Species]
   H. inornatus
Lophura [Genus]
   L. rufa [Species]
   L. ignita
Diardigallus [Genus]
   D. diardi [Species]
Lobiophasis [Genus]
   L. bulweri [Species]
Gallus [Genus]
   G. gallus [Species]
   G. lafayetii
   G. sonneratii
   G. varius
Pucrasia [Genus]
   P. macrolopha [Species]
Catreus [Genus]
   C. wallichii [Species]
Phasianus [Genus]
   P. colchicus [Species]

Syrmaticus [Genus]
   S. reevesii [Species]
   S. soemmerringii
   S. humiae
   S. ellioti
   S. mikado
Chrysolophus [Genus]
   C. pictus [Species]
   C. amherstiae
Chalcurus [Genus]
   C. inopinatus [Species]
   C. chalcurus
Polyplecton [Genus]
   P. bicalcaratum [Species]
   P. germaini
   P. malacensis
   P. schleiermacheri
   P. emphanum
Rheinardia [Genus]
   R. ocellata [Species]
Argusianus [Genus]
   A. argus [Species]
Pavo [Genus]
   P. cristatus [Species]
   P. muticus

Numididae [Family]
Phasidus [Genus]
   P. niger [Species]
Agelastes [Genus]
   A. meleagrides [Species]
Numida [Genus]
   N. meleagris [Species]
Guttera [Genus]
   G. plumifera [Species]
   G. edouardi
   G. pucherani
Acryllium [Genus]
   A. vulturinum [Species]

Meleagridae [Family]
Meleagris [Genus]
   M. gallopavo [Species]
Agriocharis [Genus]
   A. ocellata [Species]

Opisthocomidae [Family]
Opisthocomus [Genus]
   O. hoazin [Species]

Gruiformes [Order]
Mesoenatidae [Family]
Mesoenas [Genus]
   M. variegata [Species]
   M. unicolor
Monias [Genus]
   M. benschi [Species]

Turnicidae [Family]
Turnix [Genus]
   T. sylvatica [Species]
   T. worcesteri

T. nana
T. hottentotta
T. tanki
T. suscitator
T. nigricollis
T. ocellata
T. melanogaster
T. varia
T. castanota
T. pyrrhothorax
T. velox
Ortyxelos [Genus]
   O. meiffrenii [Species]
Pedionomus [Genus]
   P. torquatus [Species]

Gruidae [Family]
Grus [Genus]
   G. grus [Species]
   G. nigricollis
   G. monacha
   G. canadensis
   G. japonensis
   G. americana
   G. vipio
   G. antigone
   G. rubicunda
   G. leucogeranus
Bugeranus [Genus]
   B. carunculatus [Species]
Anthropoides [Genus]
   A. virgo [Species]
   A. paradisea
Balearica [Genus]
   B. pavonina [Species]

Aramidae [Family]
Aramus [Genus]
   A. scolopaceus [Species]

Psophiidae [Family]
Psophia [Genus]
   P. crepitans [Species]
   P. leucoptera
   P. viridis

Rallidae [Family]
Rallus [Genus]
   R. longirostris [Species]
   R. elegans
   R. limicola
   R. semiplumbeus
   R. aquaticus
   R. caerulescens
   R. madagascariensis
   R. pectoralis
   R. muelleri
   R. striatus
   R. philippensis
   R. ecaudata
   R. torquatus
   R. owstoni
   R. wakensis

Nesolimnas [Genus]
   N. dieffenbachii [Species]
Cabalus [Genus]
   C. modestus [Species]
Atlantisia [Genus]
   A. rogersi [Species]
Tricholimnas [Genus]
   T. conditicius [Species]
   T. lafresnayanus
   T. sylvestris
Ortygonax [Genus]
   O. rytirhynchos [Species]
   O. nigricans
Pardirallus [Genus]
   P. maculatus [Species]
Dryolimnas [Genus]
   D. cuvieri [Species]
Rougetius [Genus]
   R. rougetii [Species]
Amaurolimnas [Genus]
   A. concolor [Species]
Rallina [Genus]
   R. fasciata [Species]
   R. eurizonoides
   R. canningi
   R. tricolor
Rallicula [Genus]
   R. rubra [Species]
   R. leucospila
Cyanolimnas [Genus]
   C. cerverai [Species]
Aramides [Genus]
   A. mangle [Species]
   A. cajanea
   A. wolfi
   A. gutturalis
   A. ypecaha
   A. axillaris
   A. calopterus
   A. saracura
Aramidopsis [Genus]
   A. plateni [Species]
Nesoclopeus [Genus]
   N. poeciloptera [Species]
   N. woodfordi
Gymnocrex [Genus]
   G. rosenbergii [Species]
   G. plumbeiventris
Gallirallus [Genus]
   G. australis [Species]
   G. troglodytes
Habropteryx [Genus]
   H. insignis [Species]
Habroptila [Genus]
   H. wallacii [Species]
Megacrex [Genus]
   M. inepta [Species]
Eulabeornis [Genus]
   E. castaneoventris [Species]
Himantornis [Genus]
   H. haematopus [Species]

AVES SPECIES LIST

Canirallus [Genus]
  C. oculeus [Species]
Mentocrex [Genus]
  M. kioloides [Species]
Crecopsis [Genus]
  C. egregria [Species]
Crex [Genus]
  C. crex [Species]
Anurolimnas [Genus]
  A. castaneiceps [Species]
Limnocorax [Genus]
  L. flavirostra [Species]
Porzana [Genus]
  P. parva [Species]
  P. pusilla
  P. porzana
  P. fluminea
  P. carolina
  P. spiloptera
  P. flaviventer
  P. albicollis
  P. fusca
  P. paykullii
  P. olivieri
  P. bicolor
  P. tabuensis
Porzanula [Genus]
  P. palmeri [Species]
Pennula [Genus]
  P. millsi [Species]
  P. sandwichensis
Nesophylax [Genus]
  N. ater [Species]
Aphanolimnas [Genus]
  A. monasa [Species]
Laterallus [Genus]
  L. jamaicensis [Species]
  L. spilonotus
  L. exilis
  L. albigularis
  L. melanophaius
  L. ruber
  L. levraudi
  L. viridis
  L. hauxwelli
  L. leucopyrrhus
Micropygia [Genus]
  M. schomburgkii [Species]
Coturnicops [Genus]
  C. exquisita [Species]
  C. noveboracensis
  C. notata
  C. ayresi
Neocrex [Genus]
  N. erythrops [Species]
Sarothura [Genus]
  S. rufa [Species]
  S. lugeus
  S. pulchra
  S. elegans

S. bohmi
S. antonii
S. lineata
S. insularis
S. watersi
Aenigmatolimnas [Genus]
  A. marginalis [Species]
Poliolimnas [Genus]
  P. cinereus [Species]
Porphyriops [Genus]
  P. melanops [Species]
Tribonyx [Genus]
  T. ventralis [Species]
  T. mortierii
Amaurornis [Genus]
  A. akool [Species]
  A. olivacea
  A. isabellina
  A. phoenicurus
Gallicrex [Genus]
  G. cinerea [Species]
Gallinula [Genus]
  G. tenebrosa [Species]
  G. chloropus
  G. angulata
Porphyriornis [Genus]
  P. nesiotis [Species]
  P. comeri
Pareudiastes [Genus]
  P. pacificus [Species]
Porphyrula [Genus]
  P. alleni [Species]
  P. martinica
  P. parva
Porphyrio [Genus]
  P. porphyrio [Species]
  P. madagascariensis
  P. poliocephalus
  P. albus
  P. pulverulentus
Notornis [Genus]
  N. mantelli [Species]
Fulica [Genus]
  F. atra [Species]
  F. cristata
  F. americana
  F. ardesiaca
  F. armillata
  F. caribaea
  F. leucoptera
  F. rufrifrons
  F. gigantea
  F. cornuta

Heliornithidae [Family]
  Podica [Genus]
    P. senegalensis [Species]
  Heliopais [Genus]
    H. personata [Species]
  Heliornis [Genus]
    H. fulica [Species]

Rhynochetidae [Family]
  Rhynochetos [Genus]
    R. jubatus [Species]
Eurypygidae [Family]
  Eurypyga [Genus]
    E. helias [Species]
Cariamidae [Family]
  Cariama [Genus]
    C. cristata [Species]
  Chunga [Genus]
    C. burmeisteri [Species]
Otidae [Family]
  Tetrax [Genus]
    T. tetrax [Species]
  Otis [Genus]
    O. tarda [Species]
  Neotis [Genus]
    N. cafra [Species]
    N. ludwigii
    N. burchellii
    N. Nuba
    N. heuglinii
  Choriotius [Genus]
    C. arabs [Species]
    C. kori
    C. nigriceps
    C. australis
  Chlamydotis [Genus]
    C. undulata [Species]
  Lophotis [Genus]
    L. savilei [Species]
    L. ruficrista
  Afrotis [Genus]
    A. atra [Species]
  Eupodotis [Genus]
    E. vigorsii [Species]
    E. ruppellii
    E. humilis
    E. senegalensis
    E. caerulescens
  Lissotis [Genus]
    L. melanogaster [Species]
    L. hartlaubii
  Houbaropsis [Genus]
    H. bengalensis [Species]
  Sypheotides [Genus]
    S. indica [Species]

Charadriiformes [Order]
Jacanidae [Family]
  Microparra [Genus]
    M. capensis [Species]
  Actophilornis [Genus]
    A. africana [Species]
    A. albinucha
  Irediparra [Genus]
    I. gallinacea [Species]
  Hydrophasianus [Genus]
    H. chirurgus [Species]

Metopidius [Genus]
    M. indicus [Species]
Jacana [Genus]
    J. spinosa [Species]

Rostratulidae [Family]
    Rostratula [Genus]
        R. benghalensis [Species]
    Nycticryphes [Genus]
        N. semicollaris [Species]

Haematopodidae [Family]
    Haematopus [Genus]
        H. ostralegus [Species]
        H. leucopodus
        H. fuliginosus
        H. ater
    Chettusia [Genus]
        C. leucura [Species]
        C. gregaria

Charadriidae [Family]
    Vanellus [Genus]
        V. vanellus [Species]
    Belonopterus [Genus]
        B. chilensis [Species]
    Hemiparra [Genus]
        H. crassirostris [Species]
    Tylibyx [Genus]
        T. melanocephalus [Species]
    Microsarcops [Genus]
        M. cinereus [Species]
    Lobivanellus [Genus]
        L. indicus [Species]
    Xiphidiopterus [Genus]
        X. albiceps [Species]
    Rogibyx [Genus]
        R. tricolor [Species]
    Lobibyx [Genus]
        L. novaehollandiae [Species]
        L. miles
    Afribyx [Genus]
        A. senegallus [Species]
    Stephanibyx [Genus]
        S. lugubris [Species]
        S. melanopterus
        S. coronatus
    Hoplopterus [Genus]
        H. spinosus [Species]
        H. armatus
        H. duvaucelii
    Hoploxypterus [Genus]
        H. cayanus [Species]
    Ptilocelys [Genus]
        P. resplendens [Species]
    Zonifer [Genus]
        Z. tricolor [Species]
    Anomalophrys [Genus]
        A. superciliosus [Species]
    Lobipluvia [Genus]
        L. malabarica [Species]

Sarciophorus [Genus]
    S. tectus [Species]
Squatarola [Genus]
    S. squatarola [Species]
Pluvialis [Genus]
    P. apricaria [Species]
    P. dominica
Pluviorhynchus [Genus]
    P. obscurus [Species]
Charadrius [Genus]
    C. rubricollis [Species]
    C. hiaticula
    C. melodus
    C. dubius
    C. alexandrinus
    C. venustus
    C. falklandicus
    C. alticola
    C. bicinctus
    C. peronii
    C. collaris
    C. pecuarius
    C. sanctaehelenae
    C. thoracicus
    C. placidus
    C. vociferus
    C. tricollaris
    C. mongolus
    C. wilsonia
    C. leschenaultii
Elseyornis [Genus]
    E. melanops [Species]
Eupoda [Genus]
    E. asiatica [Species]
    E. veredus
    E. montana
Oreopholus [Genus]
    O. ruficollis [Species]
Erythrogonys [Genus]
    E. cinctus [Species]
Eudromias [Genus]
    E. morinellus [Species]
Zonibyx [Genus]
    Z. modestus [Species]
Thinornis [Genus]
    T. novaeseelandiae [Species]
Anarhynchus [Genus]
    A. frontalis [Species]
Pluvianellus [Genus]
    P. socialis [Species]
Phegornis [Genus]
    P. mitchellii [Species]

Scopacidae [Family]
    Aechmorhynchus [Genus]
        A. cancellatus [Species]
        A. parvirostris
    Prosobonia [Genus]
        P. leucoptera [Species]
    Bartramia [Genus]
        B. longicauda [Species]

Numenius [Genus]
    N. minutus [Species]
    N. borealis
    N. phaeopus
    N. tahitiensis
    N. tenuirostris
    N. arquata
    N. madagascariensis
    N. americanus
Limosa [Genus]
    L. limosa [Species]
    L. haemastica
    L. lapponica
    L. fedoa
Tringa [Genus]
    T. erythropus [Species]
    T. totanus
    T. flavipes
    T. stagnatilis
    T. nebularia
    T. melanoleuca
    T. ocrophus
    T. solitaria
    T. glareola
Pseudototanus [Genus]
    P. guttifer [Species]
Xenus [Genus]
    X. cinereus [Species]
Actitis [Genus]
    A. hypoleucos [Species]
    A. macularia
Catoptrophorus [Genus]
    C. semipalmatus [Species]
Heteroscelus [Genus]
    H. brevipes [Species]
    H. incanus
Aphriza [Genus]
    A. virgata [Species]
Arenaria [Genus]
    A. interpres [Species]
    A. melanocephala
Limnodromus [Genus]
    L. griseus [Species]
    L. semipalmatus
Coenocorypha [Genus]
    C. aucklandica [Species]
Capella [Genus]
    C. solitaria [Species]
    C. hardwickii
    C. nemoricola
    C. stenura
    C. megala
    C. nigripennis
    C. macrodactyla
    C. media
    C. gallinago
    C. delicata
    C. paraguaiae
    C. nobilis
    C. undulata

Chubbia [Genus]
  C. imperialis [Species]
  C. jamesoni
  C. stricklandii
Scolopax [Genus]
  S. rusticola [Species]
  S. saturata
  S. celebensis
  S. rochussenii
Philohela [Genus]
  P. minor [Species]
Lymnocryptes [Genus]
  L. minima [Species]
Calidris [Genus]
  C. canutus [Species]
  C. tenuirostris
Crocethia [Genus]
  C. alba [Species]
Ereunetes [Genus]
  E. pusillus [Species]
  E. mauri
Eurynorhynchus [Genus]
  E. pygmeus [Species]
Erolia [Genus]
  E. ruficollis [Species]
  E. minuta
  E. temminckii
  E. subminuta
  E. minutilla
  E. fuscicollis
  E. bairdii
  E. melanotos
  E. acuminata
  E. maritima
  E. ptilocnemis
  E. alpina
  E. testacea
Limicola [Genus]
  L. falcinellus [Species]
Micropalama [Genus]
  M. himantopus [Species]
Tryngites [Genus]
  T. subruficollis [Species]
Philomachus [Genus]
  P. pugnax [Species]

Recurvostridae [Family]
  Ibidorhyncha [Genus]
    I. struthersii [Species]
  Himantopus [Genus]
    H. himantopus [Species]
  Cladorhynchus [Genus]
    C. leucocephala [Species]
  Recurvirostra [Genus]
    R. avosetta [Species]
    R. americana
    R. novaehollandiae
    R. andina

Phalaropodidae [Family]
  Phalaropus [Genus]
    P. fulicarius [Species]

Steganopus [Genus]
  S. tricolor [Species]
Lobipes [Genus]
  L. lobatus [Species]

Dromadidae [Family]
  Dromas [Genus]
    D. ardeola [Species]

Burhinidae [Family]
  Burhinus [Genus]
    B. oedicnemus [Species]
    B. senegalensis
    B. vermiculatus
    B. capensis
    B. bistriatus
    B. superciliaris
    B. magnirostris
  Esacus [Genus]
    E. recurvirostris [Species]
  Orthoramphus [Genus]
    O. magnirostris [Species]

Glareolidae [Family]
  Pluvianus [Genus]
    P. aegyptius [Species]
  Cursorius [Genus]
    C. cursor [Species]
    C. temminckii
    C. coromandelicus
  Rhinoptilus [Genus]
    R. africanus [Species]
    R. cinctus
    R. chalcopterus
    R. bitorquatus
  Peltohyas [Genus]
    P. australis [Species]
  Stiltia [Genus]
    S. isabella [Species]
  Glareola [Genus]
    G. pratincola [Species]
    G. maldivarum
    G. nordmanni
    G. ocularis
    G. nuchalis
    G. cinerea
    G. lactea
  Attagis [Genus]
    A. gayi [Species]
    A. malouinus
  Thinocorus [Genus]
    T. orbignyianus [Species]
    T. rumicivorus

Chionididae [Family]
  Chionis [Genus]
    C. alba [Species]
    C. minor

Stercorariidae [Family]
  Catharacta [Genus]
    C. skua [Species]

Stercorarius [Genus]
  S. pomarinus [Species]
  S. parasiticus
  S. longicaudus

Laridae [Family]
  Gabianus [Genus]
    G. pacificus [Species]
    G. scoresbii
  Pagophila [Genus]
    P. eburnea [Species]
  Larus [Genus]
    L. fuliginosus [Species]
    L. modestus
    L. heermanni
    L. leucophthalmus
    L. hemprichii
    L. belcheri
    L. crassirostris
    L. audouinii
    L. delawarensis
    L. canus
    L. argentatus
    L. fuscus
    L. californicus
    L. occidentalis
    L. dominicanus
    L. schistisagus
    L. marinus
    L. glaucescens
    L. hyperboreus
    L. leucopterus
    L. ichthyaetus
    L. atricilla
    L. brunnicephalus
    L. cirrocephalus
    L. serranus
    L. pipixcan
    L. novaehollandiae
    L. melanocephalus
    L. bulleri
    L. maculipennis
    L. ridibundus
    L. genei
    L. philadelphia
    L. minutus
    L. saundersi
  Rhodostethia [Genus]
    R. rosea [Species]
  Rissa [Genus]
    R. tridactyla [Species]
    R. brevirostris
  Creagrus [Genus]
    C. furcatus [Species]
  Xema [Genus]
    X. sabini [Species]
  Chlidonias [Genus]
    C. hybrida [Species]
    C. leucoptera
    C. nigra

Phaetusa [Genus]
  P. simplex [Species]
Gelochelidon [Genus]
  G. nilotica [Species]
Hydroprogne [Genus]
  H. tschegrava [Species]
Sterna [Genus]
  S. aurantia [Species]
  S. hirundinacea
  S. hirundo
  S. paradisaea
  S. vittata
  S. virgata
  S. forsteri
  S. trudeaui
  S. dougallii
  S. striata
  S. repressa
  S. sumatrana
  S. melanogaster
  S. aleutica
  S. lunata
  S. anaethetus
  S. fuscata
  S. nereis
  S. albistriata
  S. superciliaris
  S. balaenarum
  S. iorata
  S. albifrons
Thalasseus [Genus]
  T. bergii [Species]
  T. maximus
  T. bengalensis
  T. zimmermanni
  T. eurygnatha
  T. elegans
  T. sandvicensis
Larosterna [Genus]
  L. inca [Species]
Procelsterna [Genus]
  P. cerulea [Species]
Anous [Genus]
  A. stolidus [Species]
  A. tenuirostris
  A. minutus
Gygis [Genus]
  G. alba [Species]

Rynchopidae [Family]
  Rynchops [Genus]
    R. nigra [Species]
    R. flavirostris
    R. albicollis

Alcidae [Family]
  Plautus [Genus]
    P. alle [Species]
  Pinguinis [Genus]
    P. impennis [Species]
  Alca [Genus]
    A. torda [Species]

Uria [Genus]
  U. lomvia [Species]
  U. aalge
Cepphus [Genus]
  C. grylle [Species]
  C. columba
  C. carbo
Brachyramphus [Genus]
  B. marmoratus [Species]
  B. brevirostris
  B. hypoleucus
  B. craveri
Synthliboramphus [Genus]
  S. antiquus [Species]
  S. wumizusume
Ptychoramphus [Genus]
  P. aleuticus [Species]
Cyclorrhynchus [Genus]
  C. psittacula [Species]
Aethia [Genus]
  A. cristatella [Species]
  A. pusilla
  A. pygmaea
Cercorhinca [Genus]
  C. monocerata [Species]
Fratercula [Genus]
  F. arctica [Species]
  F. corniculata
Lunda [Genus]
  L. cirrhata [Species]

Columbiformes [Order]
Pteroclididae [Family]
  Syrrhaptes [Genus]
    S. tibetanus [Species]
    S. paradoxus
  Pterocles [Genus]
    P. alchata [Species]
    P. namaqua
    P. exustus
    P. senegallus
    P. orientalis
    P. coronatus
    P. gutturalis
    P. burchelli
    P. personatus
    P. decoratus
    P. lichtensteinii
    P. bicinctus
    P. indicus
    P. quadricinctus

Raphidae [Family]
  Raphus [Genus]
    R. cucullatus [Species]
    R. solitarius
  Pezophaps [Genus]
    P. solitaria [Species]

Columbidae [Family]
  Sphenurus [Genus]
    S. apicauda [Species]

  S. seimundi
  S. oxyura
  S. sphenurus
  S. korthalsi
  S. sieboldii
  S. farmosae
Butreron [Genus]
  B. capellei [Species]
Treron [Genus]
  T. curvirostra [Species]
  T. pompadora
  T. fulvicollis
  T. olax
  T. vernans
  T. bicincta
  T. s. thomae
  T. australis
  T. calva
  T. delalandii
  T. waalia
  T. phoenicoptera
Phapitreron [Genus]
  P. leucotis [Species]
  P. amethystina
Leucotreron [Genus]
  L. occipitalis [Species]
  L. fischeri
  L. merrilli
  L. marchei
  L. subgularis
  L. leclancheri
  L. cincta
  L. dohertyi
  L. porphyrea
Ptilinopus [Genus]
  P. dupetithouarsii [Species]
  P. regina
  P. mercierii
  P. purpuratus
  P. coralensis
  P. insularis
  P. rarotongensis
  P. huttoni
  P. porphyraceus
  P. greyii
  P. richardsii
  P. ponapensis
  P. pelewensis
  P. roseicapilla
  P. perousii
  P. superbus
  P. pulchellus
  P. coronulatus
  P. monacha
  P. iozonus
  P. insolitus
  P. rivoli
  P. miquelli
  P. bellus
  P. solomonensis

P. viridis
P. eugeniae
P. geelvinkiana
P. pectoralis
P. naina
P. hyogastra
P. granulifrons
P. melanospila
P. jambu
P. wallacii
P. aurantiifrons
P. ornatus
P. perlatus
P. tannensis
Chrysoena [Genus]
  C. victor [Species]
  C. viridis
  C. luteovirens
Alectroenas [Genus]
  A. pulcherrima [Species]
  A. sganzini
  A. madagascariensis
  A. nitidissima
Drepanoptila [Genus]
  D. holosericea [Species]
Megaloprepia [Genus]
  M. magnifica [Species]
  M. formosa
Ducula [Genus]
  D. galeata [Species]
  D. aurorae
  D. oceanica
  D. pacifica
  D. rubricera
  D. myristicivora
  D. concinna
  D. aenea
  D. oenothorax
  D. pistrinaria
  D. whartoni
  D. rosacea
  D. perspicillata
  D. pickeringii
  D. latrans
  D. bakeri
  D. brenchleyi
  D. goliath
  D. bicolor
  D. luctuosa
  D. melanura
  D. spilorrhoa
  D. cineracea
  D. lacernulata
  D. badia
  D. mullerii
  D. pinon
  D. melanochroa
  D. poliocephala
  D. forsteni
  D. mindorensis

  D. radiata
  D. rufigaster
  D. finschii
  D. chalconota
  D. zoeae
  D. carola
Cryptophaps [Genus]
  C. poecilorrhoa [Species]
Hemiphaga [Genus]
  H. novaeseelandiae [Species]
Lopholaimus [Genus]
  L. antarcticus [Species]
Gymnophaps [Genus]
  G. albertisii [Species]
  G. solomonensis
  G. mada
Columba [Genus]
  C. leuconota [Species]
  C. rupestris
  C. livia
  C. oenas
  C. eversmanni
  C. oliviae
  C. albitorques
  C. palumbus
  C. trocaz
  C. junoniae
  C. leucocephala
  C. picazuro
  C. gymnophtalmos
  C. squamosa
  C. maculosa
  C. unicincta
  C. guinea
  C. hodgsonii
  C. arquatrix
  C. thomensis
  C. albinucha
  C. flavirostris
  C. oenops
  C. inornata
  C. caribaea
  C. rufina
  C. fasciata
  C. albilinea
  C. araucana
  C. elphinstonii
  C. torringtoni
  C. pulchricollis
  C. punicea
  C. palumboides
  C. janthina
  C. versicolor
  C. jouyi
  C. vitiensis
  C. pallidiceps
  C. norfolciensis
  C. argentina
  C. pollenii
  C. speciosa

  C. nigriristris
  C. goodsoni
  C. subvinacea
  C. plumbea
  C. chiriquensis
  C. purpureotincta
  C. delegorguei
  C. iriditorques
  C. malherbii
Nesoenas [Genus]
  . mayeri [Species]
Turacoena [Genus]
  T. manadensis [Species]
  T. modesta
Macropygia [Genus]
  M. unchall [Species]
  M. amboinensis
  M. ruficeps
  M. magna
  M. phasianella
  M. rufipennis
  M. nigrirostris
  M. mackinlayi
Reinwardtoena [Genus]
  R. reinwardtsi [Species]
  R. browni
Coryphoenas [Genus]
  C. crassirostris [Species]
Ectopistes [Genus]
  E. migratoria [Species]
Zenaidura [Genus]
  Z. macroura [Species]
  Z. graysoni
  Z. auriculata
Zenaida [Genus]
  Z. aurita [Species]
  Z. asiatica
Nesopelia [Genus]
  N. galapagoensis [Species]
Streptopelia [Genus]
  S. turtur [Species]
  S. orientalis
  S. lugens
  S. picturata
  S. decaocto
  S. roseogrisea
  S. semitorquata
  S. decipiens
  S. capicola
  S. vinacea
  S. reichenowi
  S. fulvopectoralis
  S. bitorquata
  S. tranquebarica
  S. chinensis
  S. senegalensis
Geopelia [Genus]
  G. humeralis [Species]
  G. striata
  G. cuneata

Metriopelia [Genus]
  M. ceciliae [Species]
  M. morenoi
  M. melanoptera
  M. aymara
Scardafella [Genus]
  S. inca [Species]
  S. squammata
Uropelia [Genus]
  U. campestris [Species]
Columbina [Genus]
  C. picui [Species]
Columbigallina [Genus]
  C. passerina [Species]
  C. talpacoti
  C. minuta
  C. buckleyi
  C. cruziana
Oxypelia [Genus]
  O. cyanopis [Species]
Claravis [Genus]
  C. pretiosa [Species]
  C. mondetoura
  C. godefrida
Oena [Genus]
  O. capensis [Species]
Tympanistria [Genus]
  T. tympanistria [Species]
Turtur [Genus]
  T. afer [Species]
  T. abyssinicus
  T. chalcospilos
  T. brehmeri
Chalcophaps [Genus]
  C. indica [Species]
  C. stephani
Henicophaps [Genus]
  H. albifrons [Species]
  H. foersteri
Petrophassa [Genus]
  P. albipennis [Species]
  P. rufipennis
Phaps [Genus]
  P. chalcoptera [Species]
  P. elegans
Ocyphaps [Genus]
  O. lophotes [Species]
Lophophaps [Genus]
  L. plumifera [Species]
  L. ferruginea
Geophaps [Genus]
  G. scripta [Species]
  G. smithii
Histriophaps [Genus]
  H. histrionica [Species]
Aplopelia [Genus]
  A. larvata [Species]
  A. simplex
Leptotila [Genus]
  L. verreauxi [Species]

L. megalura
L. jamaicensis
L. plumbeiceps
L. rufaxilla
L. wellsi
L. cassini
L. ochraceiventris
Osculatia [Genus]
  O. saphirina [Species]
Oreopeleia [Genus]
  O. veraguensis [Species]
  O. lawrencii
  O. goldmani
  O. costaricensis
  O. chrysia
  O. mystacea
  O. martinica
  O. violacea
  O. montana
  O. caniceps
  O. albifacies
  O. chiriquensis
  O. linearis
  O. bourcieri
  O. erythropareia
Geotrygon [Genus]
  G. versicolor [Species]
Gallicolumba [Genus]
  G. luzonica [Species]
  G. platenae
  G. keayi
  G. criniger
  G. menagei
  G. rufigula
  G. tristigmata
  G. beccarii
  G. salamonis
  G. sanctaecrucis
  G. stairi
  G. canifrons
  G. xanthonura
  G. kubaryi
  G. jobiensis
  G. erythroptera
  G. rubescens
  G. hoedtii
Leucosarcia [Genus]
  L. melanoleuca [Species]
Trugon [Genus]
  T. terrestris [Species]
Microgoura [Genus]
  M. meeki [Species]
Starnoenas [Genus]
  S. cyanocephala [Species]
Otidiphaps [Genus]
  O. nobilis [Species]
Caloenas [Genus]
  C. nicobarica [Species]
Goura [Genus]
  G. cristata [Species]

G. scheepmakeri
G. victoria
Didunculus [Genus]
  D. strigirostris [Species]

Psittaciformes [Order]
Psittacidae [Family]
  Strigops [Genus]
    S. habroptilus [Species]
  Nestor [Genus]
    N. meridionalis [Species]
    N. notabilis
    N. productus
  Chalcopsitta [Genus]
    C. atra [Species]
    C. insignis
    C. sintillata
    C. duivenbodei
    C. cardinalis
  Eos [Genus]
    E. cyanogenia [Species]
    E. reticulata
    E. squamata
    E. histrio
    E. bornea
    E. semilarvata
    E. goodfellowi
  Trichoglossus [Genus]
    T. ornatus [Species]
    T. haematod
    T. rubiginosus
    T. chlorolepidotus
    T. euteles
  Psitteuteles [Genus]
    P. flavoviridis [Species]
    P. johnstoniae
    P. goldiei
    P. versicolor
    P. iris
  Pseudeos [Genus]
    P. fuscata [Species]
  Domicella [Genus]
    D. hypoinochroa [Species]
    D. amabilis
    D. lory
    D. domicella
    D. tibialis
    D. chlorocercus
    D. albidinucha
    D. garrula
  Phigys [Genus]
    P. solitarius [Species]
  Vini [Genus]
    V. australis [Species]
    V. kuhlii
    V. stepheni
    V. peruviana
    V. ultramarina
  Glossopsitta [Genus]
    G. concinna [Species]
    G. porphyrocephala
    G. pusilla

Charmosyna [Genus]
   C. palmarum [Species]
   C. meeki
   C. rubrigularis
   C. aureicincta
   C. diadema
   C. toxopei
   C. placentis
   C. rubronotata
   C. multistriata
   C. wilhelminae
   C. pulchella
   C. margarethae
   C. josefinae
   C. papou
Oreopsittacus [Genus]
   O. arfaki [Species]
Neopsittacus [Genus]
   N. musschenbroekii [Species]
   N. pullicauda
Psittaculirostris [Genus]
   P. desmaresti [Species]
   P. salvadorii
Opopsitta [Genus]
   P. gulielmitertii [Species]
   P. diophthalma
Lathamus [Genus]
   L. discolor [Species]
Micropsitta [Genus]
   M. bruijnii [Species]
   M. keiensis
   M. geelvinkiana
   M. pusio
   M. meeki
   M. finschii
Probosciger [Genus]
   P. aterrimus [Species]
Calyptorhynchus [Genus]
   C. baudinii [Species]
   C. funereus
   C. magnificus
   C. lathami
Callocephalon [Genus]
   C. fimbriatum [Species]
Kakatoe [Genus]
   K. galerita [Species]
   K. sulphurea
   K. alba
   K. moluccensis
   K. Haematuropygia
   K. leadbeateri
   K. ducrops
   K. sanguinea
   K. tenuirostris
   K. roseicapilla
Nymphicus [Genus]
   N. hollandicus [Species]
Anodorhynchus [Genus]
   A. hyacinthinus [Species]
   A. glaucus
   A. leari

Ara [Genus]
   A. ararauna [Species]
   A. caninde
   A. militaris
   A. ambigua
   A. macao
   A. chloroptera
   A. tricolor
   A. rubrogenys
   A. auricollis
   A. severa
   A. spixii
   A. manilata
   A. maracana
   A. couloni
   A. nobilis
Aratinga [Genus]
   A. acuticaudata [Species]
   A. guarouba
   A. holochlora
   A. strenua
   A. finschi
   A. wagleri
   A. mitrata
   A. erythrogenys
   A. leucophthalmus
   A. chloroptera
   A. euops
   A. auricapillus
   A. jandaya
   A. solstitialis
   A. weddellii
   A. astec
   A. nana
   A. canicularis
   A. pertinax
   A. cactorum
   A. aurea
Nandayus [Genus]
   N. nenday [Species]
Leptosittaca [Genus]
   L. branickii [Species]
Conuropsis [Genus]
   C. carolinensis [Species]
Rhynchopsitta [Genus]
   R. pachyrhyncha [Species]
Cyanoliseus [Genus]
   C. patagonus [Species]
   C. whitleyi
Ognorhynchus [Genus]
   O. icterotis [Species]
Pyrrhura [Genus]
   P. cruentata [Species]
   P. devillei
   P. frontalis
   P. perlata
   P. rhodogaster
   P. molinae
   P. hypoxantha
   P. hoematotis

   P. leucotis
   P. picta
   P. viridicata
   P. egregria
   P. melanura
   P. berlepschi
   P. rupicola
   P. albipectus
   P. calliptera
   P. rhodocephala
   P. hoffmanni
Microsittace [Genus]
   M. ferruginea [Species]
Enicognathus [Genus]
   E. leptorhynchus [Species]
Myiopsitta [Genus]
   M. monachus [Species]
Amoropsittaca [Genus]
   A. aymara [Species]
Psilopsaigon [Genus]
   P. aurifrons [Species]
Bolborhynchus [Genus]
   B. lineola [Species]
   B. ferrugineifrons
   B. andicolus
Forpus [Genus]
   F. cyanopygius [Species]
   F. passerinus
   F. conspicillatus
   F. sclateri
   F. coelestis
Brotogeris [Genus]
   B. tirica [Species]
   B. versicolurus
   B. pyrrhopterus
   B. jugularis
   B. gustavi
   B. chrysopterus
   B. sanctithomae
Nannopsittaca [Genus]
   N. panychlora [Species]
Touit [Genus]
   T. batavica [Species]
   T. purpurata
   T. melanonotus
   T. huetii
   T. dilectissima
   T. surda
   T. stictoptera
   T. emmae
Pionites [Genus]
   P. melanocephala [Species]
   P. leucogaster
Pionopsitta [Genus]
   P. pileata [Species]
   P. haematotis
   P. caica
   P. barrabandi
   P. pyrilia
Hapalopsittaca [Genus]
   H. melanotis [Species]

H. fuertesi
H. amazonina
H. pyrrhops
Gypopsitta [Genus]
  G. vulturina [Species]
Graydidascalus [Genus]
  G. brachyurus [Species]
Pionus [Genus]
  P. menstruus [Species]
  P. sordidus
  P. maximiliani
  P. tumultuosus
  P. seniloides
  P. senilis
  P. chalcopterus
  P. fuscus
Amazona [Genus]
  A. collaria [Species]
  A. leucocephala
  A. ventralis
  A. xantholora
  A. albifrons
  A. agilis
  A. vittata
  A. pretrei
  A. viridigenalis
  A. finschi
  A. autumnalis
  A. dufresniana
  A. brasiliensis
  A. arausiaca
  A. festiva
  A. xanthops
  A. barbadensis
  A. aestiva
  A. ochrocephala
  A. amazonica
  A. mercenaria
  A. farinosa
  A. vinacea
  A. guildingii
  A. versicolor
  A. imperialis
Deroptyus [Genus]
  D. accipitrinus [Species]
Triclaria [Genus]
  T. malachitacea [Species]
Poicephalus [Genus]
  P. robustus [Species]
  P. gulielmi
  P. flavifrons
  P. cryptoxanthus
  P. senegalus
  P. meyeri
  P. rufiventris
  P. ruppellii
Psittacus [Genus]
  P. erithacus [Species]
Coracopsis [Genus]
  C. vasa [Species]
  C. nigra

Psittrichas [Genus]
  P. fulgidus [Species]
Lorius [Genus]
  L. roratus [Species]
Geoffroyus [Genus]
  G. geoffroyi [Species]
  G. simplex
  G. heteroclitus
Prioniturus [Genus]
  P. luconensis [Species]
  P. discurus
  P. flavicans
  P. platurus
  P. mada
Tanygnathus [Genus]
  T. lucionensis [Species]
  T. mulleri
  T. gramineus
  T. heterurus
  T. megalorynchos
Mascarinus [Genus]
  M. mascarin [Species]
Psittacula [Genus]
  P. eupatria [Species]
  P. krameri
  P. alexandri
  P. caniceps
  P. exsul
  P. derbyana
  P. longicauda
  P. cyanocephala
  P. intermedia
  P. himalayana
  P. calthorpae
  P. columboides
Polytelis [Genus]
  P. swainsonii [Species]
  P. anthopeplus
  P. alexandrae
Aprosmictus [Genus]
  A. jonquillaceus [Species]
  A. erythropterus
Alisterus [Genus]
  A. amboinensis [Species]
  A. chloropterus
  A. scapularis
Prosopeia [Genus]
  P. tabuensis [Species]
  P. personata
Psittacella [Genus]
  P. brehmii [Species]
  P. picta
  P. modesta
Bolbopsittacus [Genus]
  B. lunulatus [Species]
Psittinus [Genus]
  P. cyanurus [Species]
Agapornis [Genus]
  A. cana [Species]
  A. pullaria
  A. roseicollis

A. taranta
A. swinderniana
A. fischeri
A. personata
A. lilianae
A. nigrigenis
Loriculus [Genus]
  L. vernalis [Species]
  L. beryllinus
  L. pusillus
  L. philippensis
  L. amabilis
  L. stigmatus
  L. galgulus
  L. exilis
  L. flosculus
  L. aurantiifrons
Platycercus [Genus]
  P. elegans [Species]
  P. caledonicus
  P. eximius
  P. icterotis
  P. adscitus
  P. venustus
  P. zonarius
Purpureicephalus [Genus]
  P. spurius [Species]
Northiella [Genus]
  N. haematogaster [Species]
Psephotus [Genus]
  P. haematonotus [Species]
  P. varius
  P. pulcherrimus
  P. chrysopterygius
Neophema [Genus]
  N. elegans [Species]
  N. chrysostomus
  N. chrysogaster
  N. petrophila
  N. pulchella
  N. splendida
  N. bourkii
Eunymphicus [Genus]
  E. cornutus [Species]
Cyanoramphus [Genus]
  C. unicolor [Species]
  C. novaezelandiae
  C. zealandicus
  C. auriceps
  C. malherbi
  C. ulietanus
Melopsittacus [Genus]
  M. undulatus [Species]
Pezoporus [Genus]
  P. wallicus [Species]
Geopsittacus [Genus]
  G. occidentalis [Species]

Cuculiformes [Order]
Musophagidae [Family]
Tauraco [Genus]
  T. persa [Species]

T. livingstonii
T. corythaix
T. schuttii
T. fischeri
T. erythrolophus
T. bannermani
T. ruspolii
T. leucotis
T. macrorhynchus
T. hartlaubi
T. leucolophus
Gallirex [Genus]
G. porphyreolophus [Species]
Ruwenzorornis [Genus]
R. johnstoni [Species]
Musophaga [Genus]
M. violacea [Species]
Corythaeola [Genus]
C. cristata [Species]
Crinifer [Genus]
C. leucogaster [Species]
C. africanus
C. concolor
C. personata

Cuculidae [Family]
Clamator [Genus]
C. glandarius [Species]
C. coromandus
C. serratus
C. jacobinus
C. cafer
Pachycoccyx [Genus]
P. audeberti [Species]
Cuculus [Genus]
C. crassirostris [Species]
C. sparverioides
C. varius
C. vagans
C. fugax
C. solitarius
C. clamosus
C. micropterus
C. canorus
C. saturatus
C. poliocephalus
C. pallidus
Cercococcyx [Genus]
C. mechowi [Species]
C. olivinus
C. montanus
Penthoceryx [Genus]
P. sonneratii [Species]
Cacomantis [Genus]
C. merulinus [Species]
C. variolosus
C. castaneiventris
C. heinrichi
C. pyrrhophanus
Rhamphomantis [Genus]
R. megarhynchus [Species]

Misocalius [Genus]
M. osculans [Species]
Chrysococcyx [Genus]
C. cupreus [Species]
C. flavigularis
C. klaas
C. caprius
Chalcites [Genus]
C. maculatus [Species]
C. xanthorhynchus
C. basalis
C. lucidus
C. malayanus
C. crassirostris
C. ruficollis
C. meyeri
Caliechthrus [Genus]
C. leucolophus [Species]
Surniculus [Genus]
S. lugubris [Species]
Microdynamis [Genus]
M. parva [Species]
Eudynamys [Genus]
E. scolopacea [Species]
Urodynamis [Genus]
U. taitensis [Species]
Scythrops [Genus]
S. novaehollandiae [Species]
Coccyzus [Genus]
C. pumilus [Species]
C. cinereus
C. erythropthalmus
C. americanus
C. euleri
C. minor
C. melacoryphus
C. lansbergi
Piaya [Genus]
P. rufigularis [Species]
P. pluvialis
P. cayana
P. melanogaster
P. minuta
Saurothera [Genus]
S. merlini [Species]
S. vetula
Ceuthmochares [Genus]
C. aereus [Species]
Rhopodytes [Genus]
R. diardi [Species]
R. sumatranus
R. tristis
R. viridirostris
Taccocua [Genus]
T. leschenaulti [Species]
Rhinortha [Genus]
R. chlorophaea [Species]
Zanclostomus [Genus]
Z. javanicus [Species]
Rhamphococcyx [Genus]

R. calyorhynchus [Species]
R. curvirostris
Phaenicophaeus [Genus]
P. pyrrhocephalus [Species]
Dasylophus [Genus]
D. superciliosus [Species]
Lepidogrammus [Genus]
L. cumingi [Species]
Crotophaga [Genus]
C. major [Species]
C. ani
C. sulcirostris
Guira [Genus]
G. guira [Species]
Tapera [Genus]
T. naevia [Species]
Morococcyx [Genus]
M. erythropygus [Species]
Dromococcyx [Genus]
D. phasianellus [Species]
D. pavoninus
Geococcyx [Genus]
G. californiana [Species]
G. velox
Neomorphus [Genus]
N. geoffroyi [Species]
N. squaminger
N. radiolosus
N. rufipennis
N. pucheranii
Carpococcyx [Genus]
C. radiceus [Species]
C. renauldi
Coua [Genus]
C. delalandei [Species]
C. gigas
C. coquereli
C. serriana
C. reynaudii
C. cursor
C. ruficeps
C. cristata
C. verreauxi
C. caerulea
Centropus [Genus]
C. milo [Species]
C. goliath
C. violaceus
C. menbeki
C. ateralbus
C. chalybeus
C. phasianinus
C. spilopterus
C. bernsteini
C. chlororhynchus
C. rectunguis
C. steerii
C. sinensis
C. andamanensis
C. nigrorufus
C. viridis

C. toulou
C. bengalensis
C. grillii
C. epomidis
C. leucogaster
C. anselli
C. monachus
C. senegalensis
C. superciliosus
C. melanops
C. celebensis
C. unirufus

Strigiformes [Order]
Tytonidae [Family]
Tyto [Genus]
T. soumagnei [Species]
T. alba
T. rosenbergii
T. inexpectata
T. novaehollandiae
T. aurantia
T. tenebricosa
T. capensis
T. longimembris
Phodilus [Genus]
P. badius [Species]

Strigidae [Family]
Otus [Genus]
O. sagittatus [Species]
O. rufescens
O. icterorhynchus
O. spilocephalus
O. vandewateri
O. balli
O. alfredi
O. brucei
O. scops
O. umbra
O. senegalensis
O. flammeolus
O. brookii
O. rutilus
O. manadensis
O. beccarii
O. silvicola
O. whiteheadi
O. insularis
O. bakkamoena
O. asio
O. trichopsis
O. barbarus
O. guatemalae
O. roboratus
O. cooperi
O. choliba
O. atricapillus
O. ingens
O. watsonii
O. nudipes

O. clarkii
O. albogularis
O. minimus
O. leucotis
O. hartlaubi
Pyrroglaux [Genus]
P. podargina [Species]
Mimizuku [Genus]
M. gurneyi [Species]
Jubula [Genus]
J. lettii [Species]
Lophostrix [Genus]
L. cristata [Species]
Bubo [Genus]
B. virginianus [Species]
B. bubo
B. capensis
B. africanus
B. poensis
B. nipalensis
B. sumatrana
B. shelleyi
B. lacteus
B. coromandus
B. leucostictus
Pseudoptynx [Genus]
P. philippensis [Species]
Ketupa [Genus]
K. blakstoni [Species]
K. zeylonensis
K. flavipes
K. ketupu
Scotopelia [Genus]
S. peli [Species]
S. ussheri
S. bouvieri
Pulsatrix [Genus]
P. perspicillata [Species]
P. koeniswaldiana
P. melanota
Nyctea [Genus]
N. scandiaca [Species]
Surnia [Genus]
S. ulula [Species]
Glaucidium [Genus]
G. passerinum [Species]
G. gnoma
G. siju
G. minutissimum
G. jardinii
G. brasilianum
G. perlatum
G. tephronotum
G. capense
G. brodiei
G. radiatum
G. cuculoides
G. sjostedti
Micrathene [Genus]
M. whitneyi [Species]

Uroglaux [Genus]
U. dimorpha [Species]
Ninox [Genus]
N. rufa [Species]
N. strenua
N. connivens
N. novaeseelandiae
N. scutulata
N. affinis
N. superciliaris
N. philippensis
N. spilonota
N. spilocephala
N. perversa
N. squamipila
N. theomacha
N. punctulata
N. meeki
N. solomonis
N. odiosa
N. jacquinoti
Gymnoglaux [Genus]
G. lawrencii [Species]
Sceloglaux [Genus]
S. albifacies [Species]
Athene [Genus]
A. noctua [Species]
A. brama
A. blewitti
Speotyto [Genus]
S. cunicularia [Species]
Ciccaba [Genus]
C. virgata [Species]
C. nigrolineata
C. huhula
C. albitarsus
C. woodfordii
Strix [Genus]
S. butleri [Species]
S. seloputo
S. ocellata
S. leptogrammica
S. aluco
S. occidentalis
S. varia
S. hylophila
S. rufipes
S. uralensis
S. davidi
S. nebulosa
Rhinoptynx [Genus]
R. clamator [Species]
Asio [Genus]
A. otus [Species]
A. stygius
A. abyssinicus
A. madagascariensis
A. flammeus
A. capensis
Pseudoscops [Genus]
P. grammicus [Species]

AVES SPECIES LIST

Nesasio [Genus]
   N. solomonensis [Species]
Aegolius [Genus]
   A. funereus [Species]
   A. acadicus
   A. ridgwayi
   A. harrisii

Caprimulgiformes [Order]
  Steatornithidae [Family]
    Steatornis [Genus]
      S. caripensis [Species]

  Podargidae [Family]
    Podargus [Genus]
      P. strigoides [Species]
      P. papuensis
      P. ocellatus
    Batrachostomus [Genus]
      B. auritus [Species]
      B. harteri
      B. septimus
      B. stellatus
      B. moniliger
      B. hodgsoni
      B. poliolophus
      B. javensis
      B. affinis

  Nyctibiidae [Family]
    Nyctibius [Genus]
      N. grandis [Species]
      N. aethereus
      N. griseus
      N. leucopterus
      N. bracteatus

  Aegothelidae [Family]
    Aegotheles [Genus]
      A. crinifrons [Species]
      A. insignis
      A. cristatus
      A. savesi
      A. bennettii
      A. wallacii
      A. albertisi

  Caprimulgidae [Family]
    Lurocalis [Genus]
      L. semitorquatus [Species]
    Chordeiles [Genus]
      C. pusillus [Species]
      C. rupestris
      C. acutipennis
      C. minor
    Nyctiprogne [Genus]
      N. leucopyga [Species]
    Podager [Genus]
      P. nacunda [Species]
    Eurostopodus [Genus]
      E. guttatus [Species]
      E. albogularis

      E. diabolicus
      E. papuensis
      E. archboldi
      E. temminckii
      E. macrotis
    Veles [Genus]
      V. binotatus [Species]
    Nyctidromus [Genus]
      N. albicollis [Species]
    Phalaenoptilus [Genus]
      P. nuttallii [Species]
    Siphonorhis [Genus]
      S. americanus [Species]
    Otophanes [Genus]
      O. mcleodii [Species]
      O. yucatanicus
    Nyctiphrynus [Genus]
      N. ocellatus [Species]
    Caprimulgus [Genus]
      C. carolinensis [Species]
      C. rufus
      C. cubanensis
      C. sericocaudatus
      C. ridgwayi
      C. vociferus
      C. saturatus
      C. longirostris
      C. cayennensis
      C. maculicaudus
      C. parvulus
      C. maculosus
      C. nigrescens
      C. hirundinaceus
      C. ruficollis
      C. indicus
      C. europaeus
      C. aegyptius
      C. mahrattensis
      C. nubicus
      C. eximius
      C. madagascariensis
      C. macrurus
      C. pectoralis
      C. rufigena
      C. donaldsoni
      C. poliocephalus
      C. asiaticus
      C. natalensis
      C. inornatus
      C. stellatus
      C. ludovicianus
      C. monticolus
      C. affinis
      C. tristigma
      C. concretus
      C. pulchellus
      C. enarratus
      C. batesi
    Scotornis [Genus]
      S. fossii [Species]
      S. climacurus

    Macrodipteryx [Genus]
      M. longipennis [Species]
    Semeiophorus [Genus]
      S. vexillarius [Species]
    Hydropsalis [Genus]
      H. climacocerca [Species]
      H. brasiliana
    Uropsalis [Genus]
      U. segmentata [Species]
      U. lyra
    Macropsalis [Genus]
      M. creagra [Species]
    Eleothreptus [Genus]
      E. anomalus [Species]

Apodiformes [Order]

  Apodidae [Family]
    Collocalia [Genus]
      C. gigas [Species]
      C. whiteheadi
      C. lowi
      C. fuciphaga
      C. brevirostris
      C. francica
      C. inexpectata
      C. inquieta
      C. vanikorensis
      C. leucophaea
      C. vestita
      C. spodiopygia
      C. hirundinacea
      C. troglodytes
      C. marginata
      C. esculenta
    Hirundapus [Genus]
      H. caudacutus [Species]
      H. giganteus
      H. ernsti
    Streptoprocne [Genus]
      S. zonaris [Species]
      S. biscutata
    Aerornis [Genus]
      A. senex [Species]
      A. semicollaris
    Chaetura [Genus]
      C. chapmani [Species]
      C. pelagica
      C. vauxi
      C. richmondi
      C. gaumeri
      C. leucopygialis
      C. sabini
      C. thomensis
      C. sylvatica
      C. nubicola
      C. cinereiventris
      C. spinicauda
      C. martinica
      C. rutila
      C. ussheri

C. andrei
C. melanopygia
C. brachyura
Zoonavena [Genus]
   Z. grandidieri [Species]
Mearnsia [Genus]
   M. picina [Species]
   M. novaeguineae
   M. cassini
   M. bohmi
Cypseloides [Genus]
   C. cherriei [Species]
   C. fumigatus
   C. major
Nephoecetes [Genus]
   N. niger [Species]
Apus [Genus]
   A. melba [Species]
   A. aequatorialis
   A. reichenowi
   A. apus
   A. sladeniae
   A. toulsoni
   A. pallidus
   A. acuticaudus
   A. pacificus
   A. unicolor
   A. myoptilus
   A. batesi
   A. caffer
   A. horus
   A. affinis
   A. andecolus
Aeronautes [Genus]
   A. saxatalis [Species]
   A. montivagus
Panyptila [Genus]
   P. sanctihieronymi [Species]
   P. cayennensis
Tachornis [Genus]
   T. phoenicobia [Species]
Micropanyptila [Genus]
   M. furcata [Species]
Reinarda [Genus]
   R. squamata [Species]
Cypsiurus [Genus]
   C. parvus [Species]

Hemiprocnidae [Family]
Hemiprocne [Genus]
   H. longipennis [Species]
   H. mystacea
   H. comata

Trochilidae [Family]
Doryfera [Genus]
   D. johannae [Species]
   D. ludovicae
Androdon [Genus]
   A. aequatorialis [Species]
Ramphodon [Genus]

R. naevius [Species]
R. dohrnii
Glaucis [Genus]
   G. hirsuta [Species]
Threnetes [Genus]
   T. niger [Species]
   T. leucurus
   T. ruckeri
Phaethornis [Genus]
   P. yaruqui [Species]
   P. guy
   P. syrmatophorus
   P. superciliosus
   P. malaris
   P. eurynome
   P. hispidus
   P. anthophilus
   P. bourcieri
   P. philippii
   P. squalidus
   P. augusti
   P. pretrei
   P. subochraceus
   P. nattereri
   P. gounellei
   P. rupurumii
   P. porcullae
   P. ruber
   P. griseogularis
   P. longuemareus
   P. zonura
Eutoxeres [Genus]
   E. aquila [Species]
   E. condamini
Phaeochroa [Genus]
   P. cuvierii [Species]
Campylopterus [Genus]
   C. curvipennis [Species]
   C. largipennis
   C. rufus
   C. hyperythrus
   C. hemileucurus
   C. ensipennis
   C. falcatus
   C. phainopeplus
   C. villaviscensio
Eupetomana [Genus]
   E. macroura [Species]
Florisuga [Genus]
   F. mellivora [Species]
Melanotrochilus [Genus]
   M. fuscus [Species]
Colibri [Genus]
   C. delphinae [Species]
   C. thalassinus
   C. coruscans
   C. serrirostris
Anthracothorax [Genus]
   A. viridigula [Species]
   A. prevostii

A. nigricollis
A. veraguensis
A. dominicus
A. viridis
A. mango
Avocettula [Genus]
   A. recurvirostris [Species]
Eulampis [Genus]
   E. jugularis [Species]
Sericotes [Genus]
   S. holosericeus [Species]
Chrysolampis [Genus]
   C. mosquitus [Species]
Orthorhyncus [Genus]
   O. cristatus [Species]
Klais [Genus]
   K. guimeti [Species]
Abeillia [Genus]
   A. albeillei [Species]
Stephanoxis [Genus]
   S. lalandi [Species]
Lophornis [Genus]
   L. ornata [Species]
   L. gouldii
   L. magnifica
   L. delattrei
   L. stictolopha
   L. melaniae
Polemistria [Genus]
   P. chalybea [Species]
   P. pavonina
Lithiophanes [Genus]
   L. insignibarbis [Species]
Paphosia [Genus]
   P. helenae [Species]
   P. adorabilis
Popelairia [Genus]
   P. popelairii [Species]
   P. langsdorffi
   P. letitiae
   P. conversii
Discosura [Genus]
   D. longicauda [Species]
Chlorestes [Genus]
   C. notatus [Species]
Chlorostilbon [Genus]
   C. prasinus [Species]
   C. vitticeps
   C. aureoventris
   C. canivetti
   C. ricordii
   C. swainsonii
   C. maugaeus
   C. russatus
   C. gibsoni
   C. inexpectatus
   C. stenura
   C. alice
   C. poortmani
   C. euchloris
   C. auratus

Cynanthus [Genus]
   C. sordidus [Species]
   C. latirostris
Ptochoptera [Genus]
   P. iolaima [Species]
Cyanophaia [Genus]
   C. bicolor [Species]
Thalurania [Genus]
   T. furcata [Species]
   T. watertonii
   T. glaucopis
   T. lerchi
Neolesbia [Genus]
   N. nehrkorni [Species]
Panterpe [Genus]
   P. insignis [Species]
Damophila [Genus]
   D. julie [Species]
Lepidopyga [Genus]
   L. coeruleogularis [Species]
   L. goudoti
   L. luminosa
Hylocharis [Genus]
   H. xantusii [Species]
   H. leucotis
   H. eliciae
   H. sapphirina
   H. cyanus
   H. chrysura
   H. grayi
Chrysuronia [Genus]
   C. oenone [Species]
Goldmania [Genus]
   G. violiceps [Species]
Goethalsia [Genus]
   G. bella [Species]
Trochilus [Genus]
   T. polytmus [Species]
Leucochloris [Genus]
   L. albicollis [Species]
Polytmus [Genus]
   P. guainumbi [Species]
Waldronia [Genus]
   W. milleri [Species]
Smaragdites [Genus]
   S. theresiae [Species]
Leucippus [Genus]
   L. fallax [Species]
   L. baeri
   L. chionogaster
   L. viridicauda
Talaphorus [Genus]
   T. hypostictus [Species]
   T. taczanowskii
   T. chlorocercus
Amazilia [Genus]
   A. candida [Species]
   A. chionopectus
   A. versicolor
   A. hollandi

A. luciae
A. fimbriata
A. lactea
A. amabilis
A. cyaneotincta
A. rosenbergi
A. boucardi
A. franciae
A. veneta
A. leucogaster
A. cyanocephala
A. microrhyncha
A. cyanifrons
A. beryllina
A. cyanura
A. saucerrottei
A. tobaci
A. viridigaster
A. edward
A. rutila
A. yucatanensis
A. tzacatl
A. castaneiventris
A. amazilia
A. violiceps
Eupherusa [Genus]
   E. eximia [Species]
   E. nigriventris
Elvira [Genus]
   E. chionura [Species]
   E. cupreiceps
Microchera [Genus]
   M. albocoronata [Species]
Chalybura [Genus]
   C. buffonii [Species]
   C. urochrysia
Aphantochroa [Genus]
   A. cirrochloris [Species]
Lampornis [Genus]
   L. clemenciae [Species]
   L. amethystinus
   L. viridipallens
   L. hemileucus
   L. castaneoventris
   L. cinereicauda
Lamprolaima [Genus]
   L. rhami [Species]
Adelomyia [Genus]
   A. melanogenys [Species]
Anthocephala [Genus]
   A. floriceps [Species]
Urosticte [Genus]
   U. ruficrissa [Species]
   U. benjamini
Phlogophilus [Genus]
   P. hemileucurus [Species]
   P. harterti
Clytolaema [Genus]
   C. rubricauda [Species]
Polyplancta [Genus]
   P. aurescens [Species]

Heliodoxa [Genus]
   H. rubinoides [Species]
   H. leadbeateri
   H. jacula
   H. xanthogonys
Ionolaima [Genus]
   I. schreibersii [Species]
Agapeta [Genus]
   A. gularis [Species]
Lampraster [Genus]
   L. branickii [Species]
Eugenia [Genus]
   E. imperatrix [Species]
Eugenes [Genus]
   E. fulgens [Species]
Hylonympha [Genus]
   H. macrocerca [Species]
Sternoclyta [Genus]
   S. cyanopectus [Species]
Topaza [Genus]
   T. pella [Species]
   T. pyra
Oreotrochilus [Genus]
   O. chimborazo [Species]
   O. stolzmanni
   O. melanogaster
   O. estella
   O. bolivianus
   O. leucopleurus
   O. adela
Urochroa [Genus]
   U. bougueri [Species]
Patagona [Genus]
   P. gigas [Species]
Aglaeactis [Genus]
   A. cupripennis [Species]
   A. aliciae
   A. castelnaudii
   A. pamela
Lafresnaya [Genus]
   L. lafresnayi [Species]
Pterophanes [Genus]
   P. cyanopterus [Species]
Coeligena [Genus]
   C. coeligena [Species]
   C. wilsoni
   C. prunellei
   C. torquata
   C. phalerata
   C. eos
   C. bonapartei
   C. helianthea
   C. lutetiae
   C. violifer
   C. iris
Ensifera [Genus]
   E. ensifera [Species]
Sephanoides [Genus]
   S. sephanoides [Species]
   S. fernandensis

Boissoneaua [Genus]
    B. flavescens [Species]
    B. matthewsii
    B. jardini
Heliangelus [Genus]
    H. mavors [Species]
    H. clarisse
    H. amethysticollis
    H. strophianus
    H. exortis
    H. viola
    H. micraster
    H. squamigularis
    H. speciosa
    H. rothschildi
    H. luminosus
Eriocnemis [Genus]
    E. nigrivestis [Species]
    E. soderstromi
    E. vestitus
    E. godini
    E. cupreoventris
    E. luciani
    E. isaacsonii
    E. mosquera
    E. glaucopoides
    E. alinae
    E. derbyi
Haplophaedia [Genus]
    H. aureliae [Species]
    H. lugens
Ocreatus [Genus]
    O. underwoodii [Species]
Lesbia [Genus]
    L. victoriae [Species]
    L. nuna
Sappho [Genus]
    S. sparganura [Species]
Polyonymus [Genus]
    P. caroli [Species]
Zodalia [Genus]
    Z. glyceria [Species]
Ramphomicron [Genus]
    R. microrhynchum [Species]
    R. dorsale
Metallura [Genus]
    M. phoebe [Species]
    M. theresiae
    M. purpureicauda
    M. aeneocauda
    M. melagae
    M. eupogon
    M. williami
    M. tyrianthina
    M. ruficeps
Chalcostigma [Genus]
    C. olivaceum [Species]
    C. stanleyi
    C. heteropogon
    C. herrani

Oxypogon [Genus]
    O. guerinii [Species]
Opisthoprora [Genus]
    O. euryptera [Species]
Taphrolesbia [Genus]
    T. griseiventris [Species]
Aglaiocercus [Genus]
    A. kingi [Species]
    A. emmae
    A. coelestis
Oreonympha [Genus]
    O. nobilis [Species]
Augastes [Genus]
    A. scutatus [Species]
    A. lumachellus
Schistes [Genus]
    S. geoffroyi [Species]
Heliothryx [Genus]
    H. barroti [Species]
    H. aurita
Heliactin [Genus]
    H. cornuta [Species]
Loddigesia [Genus]
    L. mirabilis [Species]
Heliomaster [Genus]
    H. constantii [Species]
    H. longirostris
    H. squamosus
    H. furcifer
Rhodopis [Genus]
    R. vesper [Species]
Thaumastura [Genus]
    T. cora [Species]
Philodice [Genus]
    P. evelynae [Species]
    P. bryantae
    P. mitchellii
Doricha [Genus]
    D. enicura [Species]
    D. eliza
Tilmatura [Genus]
    T. dupontii [Species]
Microstilbon [Genus]
    M. burmeisteri [Species]
Calothorax [Genus]
    C. lucifer [Species]
    C. pulcher
Archilochus [Genus]
    A. colubris [Species]
    A. alexandri
Calliphlox [Genus]
    C. amethystina [Species]
Mellisuga [Genus]
    M. minima [Species]
Calypte [Genus]
    C. anna [Species]
    C. costae
    C. helenae
Stellula [Genus]
    S. calliope [Species]

Atthis [Genus]
    A. heloisa [Species]
Myrtis [Genus]
    M. fanny [Species]
Eulidia [Genus]
    E. yarrellii [Species]
Myrmia [Genus]
    M. micrura [Species]
Acestrura [Genus]
    A. mulsanti [Species]
    A. decorata
    A. bombus
    A. heliodor
    A. berlepschi
    A. harteri
Chaetocercus [Genus]
    C. jourdanii [Species]
Selasphorus [Genus]
    S. platycercus [Species]
    S. rufus
    S. sasin
    S. flammula
    S. torridus
    S. simoni
    S. ardens
    S. scintilla

Coliiformes [Order]
Coliidae [Family]
    Colius [Genus]
        C. striatus [Species]
        C. castanotus
        C. colius
        C. leucocephalus
        C. indicus
        C. macrourus

Trogoniformes [Order]
Trogonidae [Family]
    Pharomachrus [Genus]
        P. mocinno [Species]
        P. fulgidus
        P. pavoninus
    Euptilotis [Genus]
        E. neoxenus [Species]
    Priotelus [Genus]
        P. temnurus [Species]
    Temnotrogon [Genus]
        T. roseigaster [Species]
    Trogon [Genus]
        T. massena [Species]
        T. clathratus
        T. melanurus
        T. strigilatus
        T. citreolus
        T. mexicanus
        T. elegans
        T. collaris
        T. aurantiiventris
        T. personatus
        T. rufus

T. surrucura
T. curucui
T. violaceus
Apaloderma [Genus]
  A. narina [Species]
  A. aequatoriale
Heterotrogon [Genus]
  H. vittatus [Species]
Harpactes [Genus]
  H. reinwardtii [Species]
  H. fasciatus
  H. kasumba
  H. diardii
  H. ardens
  H. whiteheadi
  H. orrhophaeus
  H. duvaucelii
  H. oreskios
  H. erythrocephalus
  H. wardi

Coraciiformes [Order]
Alcedinidae [Family]
  Ceryle [Genus]
    C. lugubris [Species]
    C. maxima
    C. torquata
    C. alcyon
    C. rudis
  Chloroceryle [Genus]
    C. amazona [Species]
    C. americana
    C. inda
    C. aenea
  Alcedo [Genus]
    A. hercules [Species]
    A. atthis
    A. semitorquata
    A. meninting
    A. quadribrachys
    A. euryzona
    A. coerulescens
    A. cristata
    A. leucogaster
  Myioceyx [Genus]
    M. lecontei [Species]
  Ispidina [Genus]
    I. picta [Species]
    I. madagascariensis
  Ceyx [Genus]
    C. cyanopectus [Species]
    C. argentatus
    C. goodfellowi
    C. lepidus
    C. azureus
    C. websteri
    C. pusillus
    C. erithacus
    C. rufidorsum
    C. melanurus
    C. fallax

Pelargopsis [Genus]
  P. amauroptera [Species]
  P. capensis
  P. melanorhyncha
Lacedo [Genus]
  L. pulchella [Species]
Dacelo [Genus]
  D. novaeguineae [Species]
  D. leachii
  D. tyro
  D. gaudichaud
Clytoceyx [Genus]
  C. rex [Species]
Melidora [Genus]
  M. macrorrhina [Species]
Cittura [Genus]
  C. cyanotis [Species]
Halcyon [Genus]
  H. coromanda [Species]
  H. badia
  H. smyrnensis
  H. pileata
  H. cyanoventris
  H. leucocephala
  H. senegalensis
  H. senegaloides
  H. malimbica
  H. albiventris
  H. chelicuti
  H. nigrocyanea
  H. winchelli
  H. diops
  H. macleayii
  H. albonotata
  H. leucopygia
  H. farquhari
  H. pyrrhopygia
  H. torotoro
  H. megarhyncha
  H. australasia
  H. sancta
  H. cinnamomina
  H. funebris
  H. chloris
  H. saurophaga
  H. recurvirostris
  H. venerata
  H. tuta
  H. gambieri
  H. godeffroyi
  H. miyakoensis
  H. bougainvillei
  H. concreta
  H. lindsayi
  H. fulgida
  H. monacha
  H. princeps
Tanysiptera [Genus]
  T. hydrocharis [Species]
  T. galatea

T. riedelii
T. carolinae
T. ellioti
T. nympha
T. danae
T. sylvia

Todidae [Family]
  Todus [Genus]
    T. multicolor [Species]
    T. angustirostris
    T. todus
    T. mexicanus
    T. subulatus

Momotidae [Family]
  Hylomanes [Genus]
    H. momotula [Species]
  Aspatha [Genus]
    A. gularis [Species]
  Electron [Genus]
    E. platyrhynchum [Species]
    E. carinatum
  Eumomota [Genus]
    E. superciliosa [Species]
  Baryphthengus [Genus]
    B. ruficapillus [Species]
  Momotus [Genus]
    M. mexicanus [Species]
    M. momota

Meropidae [Family]
  Dicrocercus [Genus]
    D. hirundineus [Species]
  Melittophagus [Genus]
    M. revoilii [Species]
    M. pusillus
    M. variegatus
    M. lafresnayii
    M. bullockoides
    M. bulocki
    M. gularis
    M. mulleri
  Aerops [Genus]
    A. albicollis [Species]
    A. boehmi
  Merops [Genus]
    M. leschenaulti [Species]
    M. apiaster
    M. superciliosus
    M. ornatus
    M. orientalis
    M. viridis
    M. malimbicus
    M. nubicus
    M. nubicoides
  Bombylonax [Genus]
    B. breweri [Species]
  Nyctyornis [Genus]
    N. amicta [Species]
    N. athertoni

Meropogon [Genus]
    M. forsteni [Species]

Leptosomatidae [Family]
  Leptosomus [Genus]
    L. discolor [Species]

Coraciidae [Family]
  Brachypteracias [Genus]
    B. leptosomus [Species]
    B. squamigera
  Atelornis [Genus]
    A. pittoides [Species]
    A. crossleyi
  Uratelornis [Genus]
    U. chimaera [Species]
  Coracias [Genus]
    C. garrulus [Species]
    C. abyssinica
    C. caudata
    C. spatulata
    C. noevia
    C. benghalensis
    C. temminckii
    C. cyanogaster
  Eurystomus [Genus]
    E. glaucurus [Species]
    E. gularis
    E. orientalis

Upupidae [Family]
  Upupa [Genus]
    U. epops [Species]

Phoeniculidae [Family]
  Phoeniculus [Genus]
    P. purpureus [Species]
    P. bollei
    P. castaneiceps
    P. aterrimus
  Rhinopomastus [Genus]
    R. minor [Species]
    R. cyanomelas
Bucerotidae [Family]
  Tockus [Genus]
    T. birostris [Species]
    T. fasciatus
    T. alboterminatus
    T. bradfieldi
    T. pallidirostris
    T. nasutus
    T. hemprichii
    T. monteiri
    T. griseus
    T. hartlaubi
    T. camurus
    T. erythrorhynchus
    T. flavirostris
    T. deckeni
    T. jacksoni
  Berenicornis [Genus]
    B. comatus [Species]
    B. albocristatus

Ptiloaemus [Genus]
    P. tickelli [Species]
Anorrhinus [Genus]
    A. galeritus [Species]
Penelopides [Genus]
    P. panini [Species]
    P. exarhatus
Aceros [Genus]
    A. nipalensis [Species]
    A. corrugatus
    A. leucocephalus
    A. cassidix
    A. undulatus
    A. plicatus
    A. everetti
    A. narcondami
Anthracoceros [Genus]
    A. malayanus [Species]
    A. malabaricus
    A. coronatus
    A. montani
    A. marchei
Bycanistes [Genus]
    B. bucinator [Species]
    B. cylindricus
    B. subcylindricus
    B. brevis
Ceratogymna [Genus]
    C. atrata [Species]
    C. elata
Buceros [Genus]
    B. rhinoceros [Species]
    B. bicornis
    B. hydrocorax
Rhinoplax [Genus]
    R. vigil [Species]
Bucorvus [Genus]
    B. abyssinicus [Species]
    B. leadbeateri

Piciformes [Order]
Galbulidae [Family]
  Galbalcyrhynchus [Genus]
    G. leucotis [Species]
  Brachygalba [Genus]
    B. lugubris [Species]
    B. phaeonota
    B. goeringi
    B. salmoni
    B. albogularis
  Jacamaralcyon [Genus]
    J. tridactyla [Species]
  Galbula [Genus]
    G. albirostris [Species]
    G. galbula
    G. tombacea
    G. cyanescens
    G. pastazae
    G. ruficauda
    G. leucogastra
    G. dea

Jacamerops [Genus]
    J. aurea [Species]

Bucconidae [Family]
  Notharchus [Genus]
    N. macrorhynchos [Species]
    N. pectoralis
    N. ordii
    N. tectus
  Bucco [Genus]
    B. macrodactylus [Species]
    B. tamatia
    B. noanamae
    B. capensis
  Nystalus [Genus]
    N. radiatus [Species]
    N. chacuru
    N. striolatus
    N. maculatus
  Hypnelus [Genus]
    H. ruficollis [Species]
    H. bicinctus
  Malacoptila [Genus]
    M. striata [Species]
    M. fusca
    M. fulvogularis
    M. rufa
    M. panamensis
    M. mystacalis
  Micromonacha [Genus]
    M. lanceolata [Species]
  Nonnula [Genus]
    N. rubecula [Species]
    N. sclateri
    N. brunnea
    N. frontalis
    N. ruficapilla
    N. amaurocephala
  Hapaloptila [Genus]
    H. castanea [Species]
  Monasa [Genus]
    M. atra [Species]
    M. nigrifrons
    M. morphoeus
    M. flavirostris
  Chelidoptera [Genus]
    C. tenebrosa [Species]

Capitonidae [Family]
  Capito [Genus]
    C. aurovirens [Species]
    C. maculicoronatus
    C. squamatus
    C. hypoleucus
    C. dayi
    C. quinticolor
    C. niger
  Eubucco [Genus]
    E. richardsoni [Species]
    E. bourcierii
    E. versicolor

Semnornis [Genus]
   S. frantzii [Species]
   S. ramphastinus
Psilopogon [Genus]
   P. pyrolophus [Species]
Megalaima [Genus]
   M. virens [Species]
   M. lagrandieri
   M. zeylanica
   M. viridis
   M. faiostricta
   M. corvina
   M. chrysopogon
   M. rafflesii
   M. mystacophanos
   M. javensis
   M. flavifrons
   M. franklinii
   M. oorti
   M. asiatica
   M. incognita
   M. henricii
   M. armillaris
   M. pulcherrima
   M. robustirostris
   M. australis
   M. eximia
   M. rubricapilla
   M. haemacephala
Calorhamphus [Genus]
   C. fuliginosus [Species]
Gymnobucco [Genus]
   G. calvus [Species]
   G. peli
   G. sladeni
   G. bonapartei
Smilorhis [Genus]
   S. leucotis [Species]
Stactolaema [Genus]
   S. olivacea [Species]
   S. anchietae
   S. whytii
Pogoniulus [Genus]
   P. duchaillui [Species]
   P. scolopaceus
   P. leucomystax
   P. simplex
   P. coryphaeus
   P. pusillus
   P. chrysoconus
   P. bilineatus
   P. subsulphureus
   P. atroflavus
Tricholaema [Genus]
   T. lacrymosum [Species]
   T. leucomelan
   T. diadematum
   T. melanocephalum
   T. flavibuccale
   T. hirsutum

Lybius [Genus]
   L. undatus [Species]
   L. vieilloti
   L. torquatus
   L. guifsobalito
   L. rubrifacies
   L. chaplini
   L. leucocephalus
   L. minor
   L. melanopterus
   L. bidentatus
   L. dubius
   L. rolleti
Trachyphonus [Genus]
   T. purpuratus [Species]
   T. vaillantii
   T. erythrocephalus
   T. darnaudii
   T. margaritatus

Indicatoridae [Family]
Prodotiscus [Genus]
   P. insignis [Species]
   P. regulus
Melignomon [Genus]
   M. zenkeri [Species]
indicator [Genus]
   I. exilis [Species]
   I. propinquus
   I. minor
   I. conirostris
   I. variegatus
   I. maculatus
   I. archipelagicus
   I. indicator
   I. xanthonotus
Melichneutes [Genus]
   M. robustus [Species]

Ramphastidae [Family]
Aulacorhynchus [Genus]
   A. sulcatus [Species]
   A. calorhynchus
   A. derbianus
   A. prasinus
   A. haematopygus
   A. coeruleicinctis
   A. huallagae
Pteroglossus [Genus]
   P. torquatus [Species]
   P. sanguineus
   P. erythropygius
   P. castanotis
   P. aracari
   P. pluricinctus
   P. viridis
   P. bitorquatus
   P. olallae
   P. flavirostris
   P. mariae
   P. beauharnaesii

Selenidera [Genus]
   S. spectabilis [Species]
   S. culik
   S. reinwardtii
   S. langsdorffi
   S. nattereri
   S. maculirostris
Andigena [Genus]
   A. bailloni [Species]
   A. laminirostris
   A. hypoglauca
   A. cucullata
   A. nigrirostris
Ramphastos [Genus]
   R. vitellinus [Species]
   R. dicolorus
   R. citreolaemus
   R. sulfuratus
   R. swainsonii
   R. ambiguus
   R. aurantiirostris
   R. tucanus
   R. cuvieri
   R. inca
   R. toco

Picidae [Family]
Jynx [Genus]
   J. torquilla [Species]
   J. ruficollis
Picumnus [Genus]
   P. cinnamomeus [Species]
   P. rufiventris
   P. fuscus
   P. castelnau
   P. leucogaster
   P. limae
   P. olivaceus
   P. granadensis
   P. nebulosus
   P. exilis
   P. borbae
   P. aurifrons
   P. temminckii
   P. cirratus
   P. sclateri
   P. steindachneri
   P. squamulatus
   P. minutissimus
   P. pallidus
   P. albosquamatus
   P. guttifer
   P. varzeae
   P. pygmaeus
   P. asterias
   P. pumilus
   P. innominatus
Nesoctites [Genus]
   N. micromegas [Species]
Verreauxia [Genus]
   V. africana [Species]

Sasia [Genus]
  S. ochracea [Species]
  S. abnormis
Geocolaptes [Genus]
  G. olivaceus [Species]
Colaptes [Genus]
  C. cafer [Species]
  C. auratus
  C. chrysoides
  C. rupicola
  C. pitius
  C. campestris
Nesoceleus [Genus]
  N. fernandinae [Species]
Chrysoptilus [Genus]
  C. melanochloros [Species]
  C. punctigula
  C. atricollis
Piculus [Genus]
  P. rivolii [Species]
  P. auricularis
  P. aeruginosus
  P. rubiginosus
  P. simplex
  P. flavigula
  P. leucolaemus
  P. aurulentus
  P. chrysochloros
Campethera [Genus]
  C. punctuligera [Species]
  C. nubica
  C. bennettii
  C. cailliautii
  C. notata
  C. abingoni
  C. taeniolaema
  C. tullbergi
  C. maculosa
  C. permista
  C. caroli
  C. nivosa
Celeus [Genus]
  C. flavescens [Species]
  C. spectabilis
  C. castaneus
  C. immaculatus
  C. elegans
  C. jumana
  C. grammicus
  C. loricatus
  C. undatus
  C. flavus
  C. torquatus
Micropternus [Genus]
  M. brachyurus [Species]
Picus [Genus]
  P. viridis [Species]
  P. vaillantii
  P. awokera
  P. squamatus

P. viridanus
P. vittatus
P. xanthopygaeus
P. canus
P. rabieri
P. erythropygius
P. flavinucha
P. puniceus
P. chlorolophus
P. mentalis
P. mineaceus
Dinopium [Genus]
  D. benghalense [Species]
  D. shorii
  D. javanense
  D. rafflesii
Gecinulus [Genus]
  G. grantia [Species]
  G. viridis
Meiglyptes [Genus]
  M. tristis [Species]
  M. jugularis
  M. tukki
Mulleripicus [Genus]
  M. pulverulentus [Species]
  M. funebris
  M. fuliginosus
  M. fulvus
Dryocopus [Genus]
  D. martius [Species]
  D. javensis
  D. pileatus
  D. lineatus
  D. erythrops
  D. schulzi
  D. galeatus
Asyndesmus [Genus]
  A. lewis [Species]
Melanerpes [Genus]
  M. erythrocephalus [Species]
  M. portoricensis
  M. herminieri
  M. formicivorus
  M. hypopolius
  M. carolinus
  M. aurifrons
  M. chrysogenys
  M. superciliaris
  M. caymanensis
  M. radiolatus
  M. striatus
  M. rubricapillus
  M. pucherani
  M. chrysauchen
  M. flavifrons
  M. cruentatus
  M. rubrifrons
Leuconerpes [Genus]
  L. candidus [Species]
Sphyrapicus [Genus]

S. varius [Species]
S. thyroideus
Trichopicus [Genus]
  T. cactorum [Species]
Veniliornis [Genus]
  V. fumigatus [Species]
  V. spilogaster
  V. passerinus
  V. frontalis
  V. maculifrons
  V. cassini
  V. affinis
  V. kirkii
  V. callonotus
  V. sanguineus
  V. dignus
  V. nigriceps
Dendropicos [Genus]
  D. fuscescens [Species]
  D. stierlingi
  D. elachus
  D. abyssinicus
  D. poecilolaemus
  D. gabonensis
  D. lugubris
Dendrocopos [Genus]
  D. major [Species]
  D. leucopterus
  D. syriacus
  D. assimilis
  D. himalayensis
  D. darjellensis
  D. medius
  D. leucotos
  D. cathpharius
  D. hyperythrus
  D. auriceps
  D. atratus
  D. macei
  D. mahrattensis
  D. minor
  D. canicapillus
  D. wattersi
  D. kizuki
  D. moluccensis
  D. maculatus
  D. temminckii
  D. obsoletus
  D. dorae
  D. albolarvatus
  D. villosus
  D. pubescens
  D. borealis
  D. nuttallii
  D. scalaris
  D. arizonae
  D. stricklandi
  D. mixtus
  D. lignarius
Picoides [Genus]

AVES SPECIES LIST

P. tridactylus [Species]
P. arcticus
Sapheopipo [Genus]
S. noguchii [Species]
Xiphidiopicus [Genus]
X. percussus [Species]
Polipicus [Genus]
P. johnstoni [Species]
P. elliotii
Mesopicos [Genus]
M. goertae [Species]
M. griseocephalus
Thripias [Genus]
T. namaquus [Species]
T. xantholophus
T. pyrrhogaster
Hemicircus [Genus]
H. concretus [Species]
H. canente
Blythipicus [Genus]
B. pyrrhotis [Species]
B. rubiginosus
Chrysocolaptes [Genus]
C. validus [Species]
C. festivus
C. lucidus
Phloeoceastes [Genus]
P. guatemalensis [Species]
P. melanoleucos
P. leucopogon
P. rubricollis
P. robustus
P. pollens
P. haematogaster
Campephilus [Genus]
C. principalis [Species]
C. imperialis
C. magellanicus

Passeriformes [Order]
Eurylaimidae [Family]
Smithornis [Genus]
S. capensis [Species]
S. rufolateralis
S. sharpei
Pseudocalyptomena [Genus]
P. graueri [Species]
Corydon [Genus]
C. sumatranus [Species]
Cymbirhynchus [Genus]
C. macrorhynchos [Species]
Eurylaimus [Genus]
E. javanicus [Species]
E. ochromalus
E. steerii
Serilophus [Genus]
S. lunatus [Species]
Psarisomus [Genus]
P. dalhousiae [Species]
Calyptomena [Genus]

C. viridis [Species]
C. hosii
C. whiteheadi

Dendrocolaptidae [Family]
Dendrocincla [Genus]
D. tyrannina [Species]
D. macrorhyncha
D. fuliginosa
D. anabatina
D. merula
D. homochroa
Deconychura [Genus]
D. longicauda [Species]
D. stictolaema
Sittasomus [Genus]
S. griseicapillus [Species]
Glyphorynchus [Genus]
G. spirurus [Species]
Drymornis [Genus]
D. bridgesii [Species]
Nasica [Genus]
N. longirostris [Species]
Dendrexetastes [Genus]
D. rufigula [Species]
Hylexetastes [Genus]
H. perrotii [Species]
H. stresemanni
Xiphocolaptes [Genus]
X. promeropirhynchus [Species]
X. albicollis
X. falcirostris
X. franciscanus
X. major
Dendrocolaptes [Genus]
D. certhia [Species]
D. concolor
D. hoffmannsi
D. picumnus
D. platyrostris
Xiphorhynchus [Genus]
X. picus [Species]
X. necopinus
X. obsoletus
X. ocellatus
X. spixii
X. elegans
X. pardalotus
X. guttatus
X. flavigaster
X. striatigularis
X. lachrymosus
X. erythropygius
X. triangularis
Lepidocolaptes [Genus]
L. leucogaster [Species]
L. souleyetii
L. angustirostris
L. affinis
L. squamatus
L. fuscus
L. albolineatus

Campylorhamphus [Genus]
C. pucherani [Species]
C. trochilirostris
C. pusillus
C. procurvoides

Furnariidae [Family]
Geobates [Genus]
G. poecilopterus [Species]
Geositta [Genus]
G. maritima [Species]
G. peruviana
G. saxicolina
G. isabellina
G. rufipennis
G. punensis
G. cunicularia
G. antarctica
G. tenuirostris
G. crassirostris
Upucerthia [Genus]
U. dumetaria [Species]
U. albigula
U. validirostris
U. serrana
U. andaecola
Ochetorhynchus [Genus]
O. ruficaudus [Species]
O. certhioides
O. harteri
Eremobius [Genus]
E. phoenicurus [Species]
Chilia [Genus]
C. melanura [Species]
Cinclodes [Genus]
C. antarcticus [Species]
C. patagonicus
C. oustaleti
C. fuscus
C. comechingonus
C. atacamensis
C. palliatus
C. taczanowskii
C. nigrofumosus
C. excelsior
Clibanornis [Genus]
C. dendrocolaptoides [Species]
Furnarius [Genus]
F. rufus [Species]
F. leucopus
F. torridus
F. minor
F. figulus
F. cristatus
Limnornis [Genus]
L. curvirostris [Species]
Sylviorthorhynchus [Genus]
S. desmursii [Species]
Aphrastura [Genus]
A. spinicauda [Species]
A. masafuerae

Phleocryptes [Genus]
 P. melanops [Species]
Leptasthenura [Genus]
 L. andicola [Species]
 L. striata
 L. pileata
 L. xenothorax
 L. striolata
 L. aegithaloides
 L. platensis
 L. fuliginiceps
 L. yanacensis
 L. setaria
Spartonoica [Genus]
 S. maluroides [Species]
Schizoeaca [Genus]
 S. coryi [Species]
 S. fuliginosa
 S. griseomurina
 S. palpebralis
 S. helleri
 S. harterti
Schoeniophylax [Genus]
 S. phryganophila [Species]
Oreophylax [Genus]
 O. moreirae [Species]
Synallaxis [Genus]
 S. ruficapilla [Species]
 S. superciliosa
 S. poliophrys
 S. azarae
 S. frontalis
 S. moesta
 S. cabanisi
 S. spixi
 S. hypospodia
 S. subpudica
 S. albescens
 S. brachyura
 S. albigularis
 S. gujanensis
 S. propinqua
 S. cinerascens
 S. tithys
 S. cinnamomea
 S. fuscorufa
 S. unirufa
 S. rutilans
 S. erythrothorax
 S. cherriei
 S. stictothorax
Hellmayrea [Genus]
 H. gularis [Species]
Gyalophylax [Genus]
 G. hellmayri [Species]
Certhiaxis [Genus]
 C. cinnamomea [Species]
 C. mustelina
Limnoctites [Genus]
 L. rectirostris [Species]

Poecilurus [Genus]
 P. candei [Species]
 P. kollari
 P. scutatus
Cranioleuca [Genus]
 C. sulphurifera [Species]
 C. semicinerea
 C. obsoleta
 C. pyrrhophia
 C. subcristata
 C. hellmayri
 C. curtata
 C. furcata
 C. demissa
 C. erythrops
 C. vulpina
 C. pallida
 C. antisiensis
 C. marcapatae
 C. albiceps
 C. baroni
 C. albicapilla
 C. mulleri
 C. gutturata
Siptornopsis [Genus]
 S. hypochondriacus [Species]
Asthenes [Genus]
 A. pyrrholeuca [Species]
 A. dorbignyi
 A. berlepschi
 A. baeri
 A. patagonica
 A. steinbachi
 A. humicola
 A. modesta
 A. pudibunda
 A. ottonis
 A. heterura
 A. wyatti
 A. humilis
 A. anthoides
 A. sclateri
 A. hudsoni
 A. virgata
 A. maculicauda
 A. flammulata
 A. urubambensis
Thripophaga [Genus]
 T. macroura [Species]
 T. cherriei
 T. fusciceps
 T. berlepschi
Phacellodomus [Genus]
 P. sibilatrix [Species]
 P. rufifrons
 P. striaticeps
 P. erythrophthalmus
 P. ruber
 P. striaticollis
 P. dorsalis

Coryphistera [Genus]
 C. alaudina [Species]
Anumbius [Genus]
 A. annumbi [Species]
Siptornis [Genus]
 S. striaticollis [Species]
Xenerpestes [Genus]
 X. minlosi [Species]
 X. singularis
Metopothrix [Genus]
 M. aurantiacus [Species]
Roraimia [Genus]
 R. adusta [Species]
Margarornis [Genus]
 M. squamiger [Species]
 M. bellulus
 M. rubiginosus
 M. stellatus
Premnornis [Genus]
 P. guttuligera [Species]
Premnoplex [Genus]
 P. brunnescens [Species]
Pseudocolaptes [Genus]
 P. lawrencii [Species]
 P. boissonneautii
Berlepschia [Genus]
 B. rikeri [Species]
Pseudoseisura [Genus]
 P. cristata [Species]
 P. lophotes
 P. gutturalis
Hyloctistes [Genus]
 H. subulatus [Species]
Ancistrops [Genus]
 A. strigilatus [Species]
Anabazenops [Genus]
 A. fuscus [Species]
Syndactyla [Genus]
 S. rufosuperciliata [Species]
 S. subalaris
 S. guttulata
 S. mirandae
Simoxenops [Genus]
 S. ucayalae [Species]
 S. striatus
Anabacerthia [Genus]
 A. striaticollis [Species]
 A. temporalis
 A. amaurotis
Philydor [Genus]
 P. atricapillus [Species]
 P. erythrocercus
 P. pyrrhodes
 P. dimidiatus
 P. baeri
 P. lichtensteini
 P. rufus
 P. erythropterus
 P. ruficaudatus
Automolus [Genus]

A. leucophthalmus [Species]
A. infuscatus
A. dorsalis
A. rubiginosus
A. albigularis
A. ochrolaemus
A. rufipileatus
A. ruficollis
A. melanopezus
Hylocryptus [Genus]
  H. erythrocephalus [Species]
  H. rectirostris
Cichlocolaptes [Genus]
  C. leucophrus [Species]
Heliobletus [Genus]
  H. contaminatus [Species]
Thripadectes [Genus]
  T. flammulatus [Species]
  T. holostictus
  T. melanorhynchus
  T. rufobrunneus
  T. virgaticeps
  T. scrutator
  T. ignobilis
Xenops [Genus]
  X. milleri [Species]
  X. tenuirostris
  X. rutilans
  X. minutus
Megaxenops [Genus]
  M. parnaguae [Species]
Pygarrhichas [Genus]
  P. albogularis [Species]
Sclerurus [Genus]
  S. scansor [Species]
  S. albigularis
  S. mexicanus
  S. rufigularis
  S. caudacutus
  S. guatemalensis
Lochmias [Genus]
  L. nematura [Species]

Formicariidae [Family]
  Cymbilaimus [Genus]
    C. lineatus [Species]
  Hypoedaleus [Genus]
    H. guttatus [Species]
  Batara [Genus]
    B. cinerea [Species]
  Mackenziaena [Genus]
    M. leachii [Species]
    M. severa
  Frederickena [Genus]
    F. viridis [Species]
    U. unduligera
  Taraba [Genus]
    T. major [Species]
  Sakesphorus [Genus]
    S. canadensis [Species]
    S. cristatus

S. bernardi
S. melanonotus
S. melanothorax
S. luctuosus
Biatas [Genus]
  B. nigropectus [Species]
Thamnophilus [Genus]
  T. doliatus [Species]
  T. multistriatus
  T. palliatus
  T. bridgesi
  T. nigriceps
  T. praecox
  T. nigrocinereus
  T. aethiops
  T. unicolor
  T. schistaceus
  T. murinus
  T. aroyae
  T. punctatus
  T. amazonicus
  T. insignis
  T. caerulescens
  T. torquatus
  T. ruficapillus
Pygiptila [Genus]
  P. stellaris [Species]
Megastictus [Genus]
  M. margaritatus [Species]
Neoctantes [Genus]
  N. niger [Species]
Clytoctantes [Genus]
  C. alixii [Species]
Xenornis [Genus]
  X. setifrons [Species]
Thamnistes [Genus]
  T. anabatinus [Species]
Dysithamnus [Genus]
  D. stictothorax [Species]
  D. mentalis
  D. striaticeps
  D. puncticeps
  D. xanthopterus
  D. ardesiacus
  D. saturninus
  D. occidentalis
  D. plumbeus
Thamnomanes [Genus]
  T. caesius [Species]
Myrmotherula [Genus]
  M. brachyura [Species]
  M. obscura
  M. sclateri
  M. klagesi
  M. surinamensis
  M. ambigua
  M. cherriei
  M. guttata
  M. longicauda
  M. hauxwelli

M. gularis
M. gutturalis
M. fulviventris
M. leucophthalma
M. haematonota
M. ornata
M. erythrura
M. erythronotos
M. axillaris
M. schisticolor
M. sunensis
M. longipennis
M. minor
M. iheringi
M. grisea
M. unicolor
M. behni
M. urosticta
M. menetriesii
M. assimilis
Dichrozona [Genus]
  D. cincta [Species]
Myrmorchilus [Genus]
  M. strigilatus [Species]
Herpsilochmus [Genus]
  H. pileatus [Species]
  H. sticturus
  H. stictocephalus
  H. dorsimaculatus
  H. roraimae
  H. pectoralis
  H. longirostris
  H. axillaris
  H. rufimarginatus
Microrhopias [Genus]
  M. quixensis [Species]
Formicivora [Genus]
  F. iheringi [Species]
  F. grisea
  F. serrana
  F. melanogaster
  F. rufa
Drymophila [Genus]
  D. ferruginea [Species]
  D. genei
  D. ochropyga
  D. devillei
  D. caudata
  D. malura
  D. squamata
Terenura [Genus]
  T. maculata [Species]
  T. callinota
  T. humeralis
  T. sharpei
  T. spodioptila
Cercomacra [Genus]
  C. cinerascens [Species]
  C. brasiliana
  C. tyrannina

C. nigriscens
C. serva
C. nigricans
C. carbonaria
C. melanaria
C. ferdinandi
Sipia [Genus]
    S. berlepschi [Species]
    S. rosenbergi
Pyriglena [Genus]
    P. leuconota [Species]
    P. atra
    P. leucoptera
Rhopornis [Genus]
    R. ardesiaca [Species]
Myrmoborus [Genus]
    M. leucophrys [Species]
    M. lugubris
    M. myotherinus
    M. melanurus
Hypocnemis [Genus]
    H. cantator [Species]
    H. hypoxantha
Hypocnemoides [Genus]
    H. melanopogon [Species]
    H. maculicauda
Myrmochanes [Genus]
    M. hemileucus [Species]
Gymnocichla [Genus]
    G. nudiceps [Species]
Sclateria [Genus]
    S. naevia [Species]
Percnostola [Genus]
    P. rufifrons [Species]
    P. schistacea
    P. leucostigma
    P. caurensis
    P. lophotes
Myrmeciza [Genus]
    M. longipes [Species]
    M. exsul
    M. ferruginea
    M. ruficauda
    M. laemosticta
    M. disjuncta
    M. pelzelni
    M. hemimelaena
    M. hyperythra
    M. goeldii
    M. melanoceps
    M. fortis
    M. immaculata
    M. griseiceps
Myrmoderus [Genus]
    M. loricatus [Species]
    M. squamosus
Myrmophylax [Genus]
    M. atrothorax [Species]
    M. stictothorax
Formicarius [Genus]

F. colma [Species]
F. analis
F. nigricapillus
F. rufipectus
Chamaeza [Genus]
    C. campanisona [Species]
    C. nobilis
    C. ruficauda
    C. mollissima
Pithys [Genus]
    P. albifrons [Species]
    P. castanea
Gymnopithys [Genus]
    G. rufigula [Species]
    G. salvini
    G. lunulata
    G. leucaspis
Rhegmatorhina [Genus]
    R. gymnops [Species]
    R. berlepschi
    R. cristata
    R. hoffmannsi
    R. melanosticta
Hylophylax [Genus]
    H. naevioides [Species]
    H. naevia
    H. punctulata
    H. poecilonota
Phlegopsis [Genus]
    P. nigromaculata [Species]
    P. erythroptera
    P. borbae
Phaenostictus [Genus]
    P. mcleannani [Species]
Myrmornis [Genus]
    M. torquata [Species]
Pittasoma [Genus]
    P. michleri [Species]
    P. rufopileatum
Grallaricula [Genus]
    G. flavirostris [Species]
    G. ferrugineipectus
    G. nana
    G. loricata
    G. peruviana
    G. lineifrons
    G. cucullata
Myrmothera [Genus]
    M. campanisona [Species]
    M. simplex
Thamnocharis [Genus]
    T. dignissima [Species]
Grallaria [Genus]
    G. squamigera [Species]
    G. excelsa
    G. gigantea
    G. guatimalensis
    G. varia
    G. alleni
    G. haplonota

G. milleri
G. bangsi
G. quitensis
G. erythrotis
G. hypoleuca
G. przewalskii
G. capitalis
G. nuchalis
G. albigula
G. ruficapilla
G. erythroleuca
G. rufocinerea
G. griseonucha
G. rufula
G. andicola
G. macularia
G. fulviventris
G. berlepschi
G. perspicillata
G. ochroleuca

Conopophagidae [Family]
    Conopophaga [Genus]
        C. lineata [Species]
        C. cearae
        C. aurita
        C. roberti
        C. peruviana
        C. ardesiaca
        C. castaneiceps
        C. melanops
        C. melanogaster
    Corythopis [Genus]
        C. delalandi [Species]
        C. torquata

Rhinocryptidae [Family]
    Pteroptochos [Genus]
        P. castaneus [Species]
        P. tarnii
        P. megapodius
    Scelorchilus [Genus]
        S. albicollis [Species]
        S. rubecula
    Rhinocrypta [Genus]
        R. lanceolata [Species]
    Teledromas [Genus]
        T. fuscus [Species]
    Liosceles [Genus]
        L. thoracicus [Species]
    Merulaxis [Genus]
        M. ater [Species]
    Melanopareia [Genus]
        M. torquata [Species]
        M. maximiliani
        M. maranonicus
        M. elegans
    Scytalopus [Genus]
        S. unicolor [Species]
        S. speluncae
        S. macropus

S. femoralis
S. argentifrons
S. chiriquensis
S. panamensis
S. latebricola
S. indigoticus
S. magellanicus
Psilorhamphus [Genus]
  P. guttatus [Species]
Myornis [Genus]
  M. senilis [Species]
Eugralla [Genus]
  E. paradoxa [Species]
Acropternis [Genus]
  A. orthonyx [Species]

Tyrannidae [Family]
Phyllomyias [Genus]
  P. fasciatus [Species]
  P. burmeisteri
  P. virescens
  P. sclateri
  P. griseocapilla
  P. griseiceps
  P. plumbeiceps
  P. nigrocapillus
  P. cinereiceps
  P. uropygialis
Zimmerius [Genus]
  Z. vilissimus [Species]
  Z. bolivianus
  Z. cinereicapillus
  Z. gracilipes
  Z. viridiflavus
Ornithion [Genus]
  O. inerme [Species]
  O. semiflavum
  O. brunneicapillum
Camptostoma [Genus]
  C. imberbe [Species]
  C. obsoletum
Phaeomyias [Genus]
  P. murina [Species]
Sublegatus [Genus]
  S. modestus [Species]
  S. obscurior
Suiriri [Genus]
  S. suiriri [Species]
Tyrannulus [Genus]
  T. elatus [Species]
Myiopagis [Genus]
  M. gaimardii [Species]
  M. caniceps
  M. subplacens
  M. flavivertex
  M. cotta
  M. viridicata
  M. leucospodia
Elaenia [Genus]
  E. martinica [Species]
  E. flavogaster

E. spectabilis
E. albiceps
E. parvirostris
E. mesoleuca
E. strepera
E. gigas
E. pelzelni
E. cristata
E. ruficeps
E. chiriquensis
E. frantzii
E. obscura
E. dayi
E. pallatangae
E. fallax
Mecocerculus [Genus]
  M. leucophrys [Species]
  M. poecilocercus
  M. hellmayri
  M. calopterus
  M. minor
  M. stictopterus
Serpophaga [Genus]
  S. cinerea [Species]
  S. hypoleuca
  S. nigricans
  S. araguayae
  S. subcristata
Inezia [Genus]
  I. inornata [Species]
  I. tenuirostris
  I. subflava
Stigmatura [Genus]
  S. napensis [Species]
  S. budytoides
Anairetes [Genus]
  A. alpinus [Species]
  A. agraphia
  A. agilis
  A. reguloides
  A. flavirostris
  A. fernandezianus
  A. parulus
Tachuris [Genus]
  T. rubrigastra [Species]
Culicivora [Genus]
  C. caudacuta [Species]
Polystictus [Genus]
  P. pectoralis [Species]
  P. superciliaris
Pseudocolopteryx [Genus]
  P. sclateri [Species]
  P. dinellianus
  P. acutipennis
  P. flaviventris
Euscarthmus [Genus]
  E. meloryphus [Species]
  E. rufomarginatus
Mionectes [Genus]
  M. striaticollis [Species]

M. oliveceus
M. oleagineus
M. macconnelli
M. rufiventris
Leptopogon [Genus]
  L. rufipectus [Species]
  L. taczanowskii
  L. amaurocephalus
  L. superciliaris
Phylloscartes [Genus]
  P. nigrifrons [Species]
  P. poecilotis
  P. chapmani
  P. ophthalmicus
  P. eximius
  P. gualaquizae
  P. flaviventris
  P. venezuelanus
  P. orbitalis
  P. flaveolus
  P. roquettei
  P. ventralis
  P. paulistus
  P. oustaleti
  P. difficilis
  P. flavovirens
  P. virescens
  P. superciliaris
  P. sylviolus
Pseudotriccus [Genus]
  P. pelzelni [Species]
  P. simplex
  P. ruficeps
Myiornis [Genus]
  M. auricularis [Species]
  M. albiventris
  M. ecaudatus
Lophotriccus [Genus]
  L. pileatus [Species]
  L. eulophotes
  L. vitiosus
  L. galeatus
Atalotriccus [Genus]
  A. pilaris [Species]
Poecilotriccus [Genus]
  P. ruficeps [Species]
  P. capitale
  P. tricolor
  P. andrei
Oncostoma [Genus]
  O. cinereigulare [Species]
  O. olivaceum
Hemitriccus [Genus]
  H. minor [Species]
  H. josephinae
  H. diops
  H. obsoletus
  H. flammulatus
  H. zosterops
  H. aenigma

H. orbitatus
H. iohannis
H. striaticollis
H. nidipendulus
H. spodiops
H. margaritaceiventer
H. inoratus
H. granadensis
H. mirandae
H. kaempferi
H. rufigularis
H. furcatus
Todirostrum [Genus]
  T. senex [Species]
  T. russatum
  T. plumbeiceps
  T. fumifrons
  T. latirostre
  T. sylvia
  T. maculatum
  T. poliocephalum
  T. cinereum
  T. pictum
  T. chrysocrotaphum
  T. nigriceps
  T. calopterum
Cnipodectes [Genus]
  C. subbrunneus [Species]
Ramphotrigon [Genus]
  R. megacephala [Species]
  R. fuscicauda
  R. ruficauda
Rhynchocyclus [Genus]
  R. brevirostris [Species]
  R. olivaceus
  R. fulvipectus
Tolmomyias [Genus]
  T. sulphurescens [Species]
  T. assimilis
  T. poliocephalus
  T. flaviventris
Platyrinchus [Genus]
  P. saturatus [Species]
  P. cancrominus
  P. mystaceus
  P. coronatus
  P. flavigularis
  P. platyrhynchos
  P. leucoryphus
Onychorhynchus [Genus]
  O. coronatus [Species]
Myiotriccus [Genus]
  M. ornatus [Species]
Terenotriccus [Genus]
  T. erythrurus [Species]
Myiobius [Genus]
  M. villosus [Species]
  M. barbatus
  M. atricaudus
Myiophobus [Genus]

M. flavicans [Species]
M. phoenicomitra
M. inornatus
M. roraimae
M. lintoni
M. pulcher
M. ochraceiventris
M. cryptoxanthus
M. fasciatus
Aphanotriccus [Genus]
  A. capitalis [Species]
  A. audax
Xenotriccus [Genus]
  X. callizonus [Species]
  X. mexicanus
Pyrrhomyias [Genus]
  P. cinnamomea [Species]
Mitrephanes [Genus]
  M. phaeocercus [Species]
  M. olivaceus
Contopus [Genus]
  C. borealis [Species]
  C. fumigatus
  C. ochraceus
  C. sordidulus
  C. virens
  C. cinereus
  C. nigrescens
  C. albogularis
  C. caribaeus
  C. latirostris
Empidonax [Genus]
  E. flaviventris [Species]
  E. virescens
  E. alnorum
  E. traillii
  E. albigularis
  E. euleri
  E. griseipectus
  E. minimus
  E. hammondii
  E. wrightii
  E. oberholseri
  E. affinis
  E. difficilis
  E. flavescens
  E. fulvifrons
  E. atriceps
Nesotriccus [Genus]
  N. ridgwayi [Species]
Cnemotriccus [Genus]
  C. fuscatus [Species]
Sayornis [Genus]
  S. phoebe [Species]
  S. saya
  S. nigricans
Pyrocephalus [Genus]
  P. rubinus [Species]
Ochthoeca [Genus]
  O. cinnamomeiventris [Species]
  O. diadema

O. frontalis
O. pulchella
O. rufipectoralis
O. fumicolor
O. oenanthoides
O. parvirostris
O. leucophrys
O. piurae
O. littoralis
Myiotheretes [Genus]
  M. striaticollis [Species]
  M. erythropygius
  M. rufipennis
  M. pernix
  M. fumigatus
  M. fuscorufus
Xolmis [Genus]
  X. pyrope [Species]
  X. cinerea
  X. coronata
  X. velata
  X. dominicana
  X. irupero
Neoxolmis [Genus]
  N. rubetra [Species]
  N. ruficentris
Agriornis [Genus]
  A. montana [Species]
  A. andicola
  A. livida
  A. microptera
  A. murina
Muscisaxicola [Genus]
  M. maculirostris [Species]
  M. fluviatilis
  M. macloviana
  M. capistrata
  M. rufivertex
  M. juninensis
  M. albilora
  M. alpina
  M. cinerea
  M. albifrons
  M. flavinucha
  M. frontalis
Lessonia [Genus]
  L. oreas [Species]
  L. rufa
Knipolegus [Genus]
  K. striaticeps [Species]
  K. hudsoni
  K. poecilocercus
  K. signatus
  K. cyanirostris
  K. poecilurus
  K. orenocensis
  K. aterrimus
  K. nigerrimus
  K. lophotes
Hymenops [Genus]
  H. perspicillata [Species]

AVES SPECIES LIST

Fluvicola [Genus]
  F. pica [Species]
  F. nengeta
  F. leucocephala
Colonia [Genus]
  C. colonus [Species]
Alectrurus [Genus]
  A. tricolor [Species]
  A. risora
Gubernetes [Genus]
  G. yetapa [Species]
Satrapa [Genus]
  S. icterophrys [Species]
Tumbezia [Genus]
  T. salvini [Species]
Muscigralla [Genus]
  M. brevicauda [Species]
Hirundinea [Genus]
  H. ferruginea [Species]
Machetornis [Genus]
  M. rixosus [Species]
Muscipipra [Genus]
  M. vetula [Species]
Attila [Genus]
  A. phoenicurus [Species]
  A. cinnamomeus
  A. torridus
  A. citriniventris
  A. bolivianus
  A. rufus
  A. spadiceus
Casiornis [Genus]
  C. rufa [Species]
  C. fusca
Rhytipterna [Genus]
  R. simplex [Species]
  R. holerythra
  R. immunda
Laniocera [Genus]
  L. hypopyrrha [Species]
  L. rufescens
Sirystes [Genus]
  S. sibilator [Species]
Myiarchus [Genus]
  M. semirufus [Species]
  M. yucatanensis
  M. barbirostris
  M. tuberculifer
  M. swainsoni
  M. venezuelensis
  M. panamensis
  M. ferox
  M. cephalotes
  M. phaeocephalus
  M. apicalis
  M. cinerascens
  M. nuttingi
  M. crinitus
  M. tyrannulus
  M. magnirostris

M. nugator
M. validus
M. sagrae
M. stolidus
M. antillarum
M. oberi
Deltarhynchus [Genus]
  D. flammulatus [Species]
Pitangus [Genus]
  P. lictor [Species]
  P. sulphuratus
Megarhynchus [Genus]
  M. pitangua [Species]
Myiozetetes [Genus]
  M. cayanensis [Species]
  M. similis
  M. granadensis
  M. luteiventris
Conopias [Genus]
  C. inornatus [Species]
  C. parva
  C. trivirgata
  C. cinchoneti
Myiodynastes [Genus]
  M. hemichrysus [Species]
  M. chrysocephalus
  M. bairdii
  M. maculatus
  M. luteiventris
Legatus [Genus]
  L. leucophaius [Species]
Empidonomus [Genus]
  E. varius [Species]
  E. aurantioatrocristatus
Tyrannopsis [Genus]
  T. sulphurea [Species]
Tyrannus [Genus]
  T. niveigularis [Species]
  T. albogularis
  T. melancholicus
  T. couchii
  T. vociferans
  T. crassirostris
  T. verticalis
  T. forficata
  T. savana
  T. tyrannus
  T. dominicensis
  T. caudifasciatus
  T. cubensis
Xenopsaris [Genus]
  X. albinucha [Species]
Pachyramphus [Genus]
  P. viridis [Species]
  P. versicolor
  P. spodiurus
  P. rufus
  P. castaneus
  P. cinnamomeus
  P. polychopterus

P. marginatus
P. albogriseus
P. major
P. surinamus
P. aglaiae
P. homochrous
P. minor
P. validus
P. niger
Tityra [Genus]
  T. cayana [Species]
  T. semifasciata
  T. inquisitor
  T. leucura

Pipridae [Family]
  Schiffornis [Genus]
    S. major [Species]
    S. turdinus
    S. virescens
  Sapayoa [Genus]
    S. aenigma [Species]
  Piprites [Genus]
    P. griseiceps [Species]
    P. chloris
    P. pileatus
  Neopipo [Genus]
    N. cinnamomea [Species]
  Chloropipo [Genus]
    C. flavicapilla [Species]
    C. holochlora
    C. uniformis
    C. unicolor
  Xenopipo [Genus]
    X. atronitens [Species]
  Antilophia [Genus]
    A. galeata [Species]
  Tyranneutes [Genus]
    T. stolzmanni [Species]
    T. virescens
  Neopelma [Genus]
    N. chrysocephalum [Species]
    N. pallescens
    N. aurifrons
    N. sulphureiventer
  Heterocercus [Genus]
    H. flavivertex [Species]
    H. aurantiivertex
    H. lineatus
  Machaeropterus [Genus]
    M. regulus [Species]
    M. pyrocephalus
    M. deliciosus
  Manacus [Genus]
    M. manacus [Species]
  Corapipo [Genus]
    C. leucorrhoa [Species]
    C. gutturalis
  Ilicura [Genus]
    I. militaris [Species]
  Masius [Genus]

M. chrysopterus [Species]
Chiroxiphia [Genus]
  C. linearis [Species]
  C. lanceolata
  C. pareola
  C. caudata
Pipra [Genus]
  P. pipra [Species]
  P. coronata
  P. isidorei
  P. coeruleocapilla
  P. nattereri
  P. vilasboasi
  P. iris
  P. serena
  P. aureola
  P. fasciicauda
  P. filicauda
  P. mentalis
  P. erythrocephala
  P. rubrocapilla
  P. chloromeros
  P. cornuta

Cotingidae [Family]
  Phoenicircus [Genus]
    P. carnifex [Species]
    P. nigricollis
  Laniisoma [Genus]
    L. elegans [Species]
  Phibalura [Genus]
    P. flavirostris [Species]
  Tijuca [Genus]
    T. atra [Species]
  Carpornis [Genus]
    C. cucullatus [Species]
    C. melanocephalus
  Ampelion [Genus]
    A. rubrocristatus [Species]
    A. rufaxilla
    A. sclateri
    A. stresemanni
  Pipreola [Genus]
    P. riefferii [Species]
    P. intermedia
    P. arcuata
    P. auroeopectus
    P. frontalis
    P. chlorolepidota
    P. formosa
    P. whitelyi
  Ampelioides [Genus]
    A. tschudii [Species]
  Iodopleura [Genus]
    I. pipra [Species]
    I. fusca
    I. isabellae
  Calyptura [Genus]
    C. cristata [Species]
  Lipaugus [Genus]
    L. subalaris [Species]

    L. cryptolophus
    L. fuscocinereus
    L. vociferans
    L. unirufus
    L. lanioides
    L. streptophorus
  Chirocylla [Genus]
    C. uropygialis [Species]
  Porphyrolaema [Genus]
    P. porphyrolaema [Species]
  Cotinga [Genus]
    C. amabilis [Species]
    C. ridgwayi
    C. nattererii
    C. maynana
    C. cotinga
    C. maculata
    C. cayana
  Xipholena [Genus]
    X. punicea [Species]
    X. lamellipennis
    X. atropurpurea
  Carpodectes [Genus]
    C. nitidus [Species]
    C. antoniae
    C. hopkei
  Conioptilon [Genus]
    C. mcilhennyi [Species]
  Gymnoderus [Genus]
    G. foetidus [Species]
  Haematoderus [Genus]
    H. militaris [Species]
  Querula [Genus]
    Q. purpurata [Species]
  Pyroderus [Genus]
    P. scutatus [Species]
  Cephalopterus [Genus]
    C. glabricollis [Species]
    C. penduliger
    C. ornatus
  Perissocephalus [Genus]
    P. tricolor [Species]
  Procnias [Genus]
    P. tricarunculata [Species]
    P. alba
    P. averano
    P. nudicollis
  Rupicola [Genus]
    R. rupicola [Species]
    R. peruviana

Oxyruncidae [Family]
  Oxyruncus [Genus]
    O. cristatus [Species]

Phytotomidae [Family]
  Phytotoma [Genus]
    P. raimondii [Species]
    P. rara
    P. rutila

Pittidae [Family]
  Pitta [Genus]
    P. phayrei [Species]
    P. nipalensis
    P. soror
    P. oatesi
    P. schneideri
    P. caerulea
    P. cyanea
    P. elliotii
    P. guajana
    P. gurneyi
    P. kochi
    P. erythrogaster
    P. arcuata
    P. granatina
    P. venusta
    P. baudii
    P. sordida
    P. brachyura
    P. nympha
    P. angolensis
    P. superba
    P. maxima
    P. steerii
    P. moluccensis
    P. versicolor
    P. anerythra

Philepittidae [Family]
  Philepitta [Genus]
    P. castanea [Species]
    P. schlegeli
  Neodrepanis [Genus]
    N. coruscans [Species]
    N. hypoxantha

Acanthisittidae [Family]
  Acanthisitta [Genus]
    A. chloris [Species]
  Xenicus [Genus]
    X. longipes [Species]
    X. gilviventris
    X. lyalli

Menuridae [Family]
  Menura [Genus]
    M. novaehollandiae [Species]
    M. alberti

Atrichornithidae [Family]
  Atrichornis [Genus]
    A. clamosus [Species]
    A. rufescens

Alaudidae [Family]
  Mirafra [Genus]
    M. javanica [Species]
    M. hova
    M. cordofanica
    M. williamsi
    M. cheniana

AVES SPECIES LIST

M. albicauda
M. passerina
M. candida
M. pulpa
M. hypermetra
M. somalica
M. africana
M. chuana
M. angolensis
M. rufocinnamomea
M. apiata
M. africanoides
M. collaris
M. assamica
M. rufa
M. gilleti
M. poecilosterna
M. sabota
M. erythroptera
M. nigricans
Heteromirafra [Genus]
H. ruddi [Species]
Certhilauda [Genus]
C. curvirostris [Species]
C. albescens
C. albofasciata
Eremopterix [Genus]
E. australis [Species]
E. leucotis
E. signata
E. verticalis
E. nigriceps
E. grisea
E. leucopareia
Ammomanes [Genus]
A. cincturus [Species]
A. phoenicurus
A. deserti
A. dunni
A. grayi
A. burrus
Alaemon [Genus]
A. alaudipes [Species]
A. hamertoni
Ramphocoris [Genus]
R. clotbey [Species]
Melanocorypha [Genus]
M. calandra [Species]
M. bimaculata
M. maxima
M. mongolica
M. leucoptera
M. yeltoniensis
Calandrella [Genus]
C. cinerea [Species]
C. blanfordi
C. acutirostris
C. raytal
C. rufescens
C. razae

C. conirostris
C. starki
C. sclateri
C. fringillaris
C. obbiensis
C. personata
Chersophilus [Genus]
C. duponti [Species]
Pseudalaemon [Genus]
P. fremantlii [Species]
Galerida [Genus]
G. cristata [Species]
G. theklae
G. malabarica
G. deva
G. modesta
G. magnirostris
Lullula [Genus]
L. arborea [Species]
Alauda [Genus]
A. arvensis [Species]
A. gulgula
Eremophila [Genus]
E. alpestris [Species]
E. bilopha

Hirundinidae [Family]
Pseudochelidon [Genus]
P. eurystomina [Species]
Tachycineta [Genus]
T. bicolor [Species]
T. albilinea
T. albiventer
T. leucorrhoa
T. leucopyga
T. thalassina
Callichelidon [Genus]
C. cyaneoviridis [Species]
Kalochelidon [Genus]
K. euchrysea [Species]
Progne [Genus]
P. tapera [Species]
P. subis
P. dominicensis
P. chalybea
P. modesta
Notiochelidon [Genus]
N. murina [Species]
N. cyanoleuca
N. flavipes
N. pileata
Atticora [Genus]
A. fasciata [Species]
A. melanoleuca
Neochelidon [Genus]
N. tibialis [Species]
Alopochelidon [Genus]
A. fucata [Species]
Stelgidopteryx [Genus]
S. ruficollis [Species]
Cheramoeca [Genus]

C. leucosternum [Species]
Pseudhirundo [Genus]
P. griseopyga [Species]
Riparia [Genus]
R. paludicola [Species]
R. congica
R. riparia
R. cincta
Phedina [Genus]
P. borbonica [Species]
P. brazzae
Ptyonoprogne [Genus]
P. rupestris [Species]
P. obsoleta
P. fuligula
P. concolor
Hirundo [Genus]
H. rustica [Species]
H. lucida
H. angolensis
H. tahitica
H. albigularis
H. aethiopica
H. smithii
H. atrocaerulea
H. nigrita
H. leucosoma
H. megaensis
H. nigrorufa
H. dimidiata
Cecropis [Genus]
C. cucullata [Species]
C. abyssinica
C. semirufa
C. senegalensis
C. daurica
C. striolata
Petrochelidon [Genus]
P. rufigula [Species]
P. preussi
P. andecola
P. nigricans
P. spilodera
P. pyrrhonota
P. fulva
P. fluvicola
P. ariel
P. fuliginosa
Delichon [Genus]
D. urbica [Species]
D. dasypus
D. nipalensis
Psalidoprocne [Genus]
P. nitens [Species]
P. fuliginosa
P. albiceps
P. pristoptera
P. oleaginea
P. antinorii
P. petiti

P. holomelaena
P. orientalis
P. mangebettorum
P. chalybea
P. obscura

Motacillidae [Family]
Dendronanthus [Genus]
D. indicus [Species]
Motacilla [Genus]
M. flava [Species]
M. citreola
M. cinerea
M. alba
M. grandis
M. madaraspatensis
M. aguimp
M. clara
M. capensis
M. flaviventris
Tmetothylacus [Genus]
T. tenellus [Species]
Macronyx [Genus]
M. capensis [Species]
M. croceus
M. fullebornii
M. sharpei
M. flavicollis
M. aurantiigula
M. ameliae
M. grimwoodi
Anthus [Genus]
A. novaeseelandiae [Species]
A. leucophrys
A. vaalensis
A. pallidiventris
A. melindae
A. campestris
A. godlewskii
A. berthelotii
A. similis
A. brachyurus
A. caffer
A. trivialis
A. nilghiriensis
A. hodgsoni
A. gustavi
A. pratensis
A. cervinus
A. roseatus
A. spinoletta
A. sylvanus
A. spragueii
A. furcatus
A. hellmayri
A. chacoensis
A. lutescens
A. correndera
A. nattereri
A. bogotensis
A. antarcticus

A. gutturalis
A. sokokensis
A. crenatus
A. lineiventris
A. chloris

Campephagidae [Family]
Pteropodocys [Genus]
P. maxima [Species]
Coracina [Genus]
C. novaehollandiae [Species]
C. fortis
C. atriceps
C. pollens
C. schistacea
C. caledonica
C. caeruleogrisea
C. temminckii
C. larvata
C. striata
C. bicolor
C. lineata
C. boyeri
C. leucopygia
C. papuensis
C. robusta
C. longicauda
C. parvula
C. abbotti
C. analis
C. caesia
C. pectoralis
C. graueri
C. cinerea
C. azurea
C. typica
C. newtoni
C. coerulescens
C. dohertyi
C. tenuirostris
C. morio
C. schisticeps
C. melaena
C. montana
C. holopolia
C. mcgregori
C. panayensis
C. polioptera
C. melaschistos
C. fimbriata
C. melanoptera
Campochaera [Genus]
C. sloetii [Species]
Chlamydochaera [Genus]
C. jefferyi [Species]
Lalage [Genus]
L. melanoleuca [Species]
L. nigra
L. sueurii
L. aurea
L. atrovirens

L. leucomela
L. maculosa
L. sharpei
L. leucopygia
Campephaga [Genus]
C. phoenicea [Species]
C. quiscalina
C. lobata
Pericrocotus [Genus]
P. roseus [Species]
P. divaricus
P. cinnamomeus
P. lansbergei
P. erythropygius
P. solaris
P. ethologus
P. brevirostris
P. miniatus
P. flammeus
Hemipus [Genus]
H. picatus [Species]
H. hirundinaceus
Tephrodornis [Genus]
T. gularis [Species]
T. pondicerianus

Pycnonotidae [Family]
Spizixos [Genus]
S. canifrons [Species]
S. semitorques
Pycnonotus [Genus]
P. zeylanicus [Species]
P. striatus
P. leucogrammicus
P. tympanistrigus
P. melanoleucos
P. priocephalus
P. atriceps
P. melanicterus
P. squamatus
P. cyaniventris
P. jocosus
P. xanthorrhous
P. sinensis
P. taivanus
P. leucogenys
P. cafer
P. aurigaster
P. xanthopygos
P. nigricans
P. capensis
P. barbatus
P. eutilotus
P. nieuwenhuisii
P. urostictus
P. bimaculatus
P. finlaysoni
P. xantholaemus
P. penicillatus
P. flavescens
P. goiavier

P. luteolus
P. plumosus
P. blanfordi
P. simplex
P. brunneus
P. erythropthalmos
P. masukuensis
P. montanus
P. virens
P. gracilis
P. ansorgei
P. curvirostris
P. importunus
P. latirostris
P. gracilirostris
P. tephrolaemus
P. milanjensis
Calyptocichla [Genus]
C. serina [Species]
Baeopogon [Genus]
B. indicator [Species]
B. clamans
Ixonotus [Genus]
I. guttatus [Species]
Chlorocichla [Genus]
C. falkensteini [Species]
C. simplex
C. flavicollis
C. flaviventris
C. laetissima
Thescelocichla [Genus]
T. leucopleura [Species]
Phyllastrephus [Genus]
P. scandens [Species]
P. terrestris
P. strepitans
P. cerviniventris
P. fulviventris
P. poensis
P. hypochloris
P. baumanni
P. poliocephalus
P. flavostriatus
P. debilis
P. lorenzi
P. albigularis
P. fischeri
P. orostruthus
P. icterinus
P. xavieri
P. madagascariensis
P. zosterops
P. tenebrosus
P. xanthophrys
P. cinereiceps
Bleda [Genus]
B. syndactyla [Species]
B. eximia
B. canicapilla
Nicator [Genus]

N. chloris [Species]
N. gularis
N. vireo
Criniger [Genus]
C. barbatus [Species]
C. calurus
C. ndussumensis
C. olivaceus
C. finschii
C. flaveolus
C. pallidus
C. ochraceus
C. bres
C. phaeocephalus
Setornis [Genus]
S. criniger [Species]
Hypsipetes [Genus]
H. viridescens [Species]
H. propinquus
H. charlottae
H. palawanensis
H. criniger
H. philippinus
H. siquijorensis
H. everetti
H. affinis
H. indicus
H. mcclellandii
H. malaccensis
H. virescens
H. flavala
H. amaurotis
H. crassirostris
H. borbonicus
H. madagascariensis
H. nicobariensis
H. thompsoni
Neolestes [Genus]
N. torquatus [Species]
Tylas [Genus]
T. eduardi [Species]

Irenidae [Family]
Aegithina [Genus]
A. tiphia [Species]
A. nigrolutea
A. viridissima
A. lafresnayei
Chloropsis [Genus]
C. flavipennis [Species]
C. palawanensis
C. sonnerati
C. cyanopogon
C. cochinchinensis
C. aurifrons
C. hardwickei
C. venusta
Irena [Genus]
I. puella [Species]
I. cyanogaster
Eurocephalus [Genus]

E. ruppelli [Species]
E. anguitimens
Laniidae [Family]
Prionops [Genus]
P. plumata [Species]
P. poliolopha
P. caniceps
P. alberti
P. retzii
P. gabela
P. scopifrons
Lanioturdus [Genus]
L. torquatus [Species]
Nilaus [Genus]
N. afer [Species]
Dryoscopus [Genus]
D. pringlii [Species]
D. gambensis
D. cubla
D. senegalensis
D. angolensis
D. sabini
Tchagra [Genus]
T. minuta [Species]
T. senegala
T. tchagra
T. australis
T. jamesi
T. cruenta
Laniarius [Genus]
L. ruficeps [Species]
L. luhderi
L. ferrugineus
L. barbarus
L. mufumbiri
L. atrococcineus
L. atroflavus
L. fulleborni
L. funebris
L. leucorhynchus
Telophorus [Genus]
T. bocagei [Species]
T. sulfureopectus
T. olivaceus
T. nigrifrons
T. multicolor
T. kupeensis
T. zeylonus
T. viridis
T. quadricolor
T. dohertyi
Malaconotus [Genus]
M. cruentus [Species]
M. lagdeni
M. gladiator
M. blanchoti
M. alius
Corvinella [Genus]
C. corvina [Species]
C. melanoleuca

Lanius [Genus]
   L. tigrinus [Species]
   L. souzae
   L. bucephalus
   L. cristatus
   L. collurio
   L. collueioides
   L. gubernator
   L. vittatus
   L. schach
   L. validirostris
   L. mackinnoni
   L. minor
   L. ludovicianus
   L. excubitor
   L. excubitoroides
   L. sphenocercus
   L. cabanisi
   L. dorsalis
   L. somalicus
   L. collaris
   L. newtoni
   L. senator
   L. nubicus
Pityriasis [Genus]
   P. gymnocephala [Species]

Vangidae [Family]
  Calicalicus [Genus]
   C. madagascariensis [Species]
  Schetba [Genus]
   S. rufa [Species]
  Vanga [Genus]
   V. curvirostris [Species]
  Xenopirostris [Genus]
   X. xenopirostris [Species]
   X. damii
   X. polleni
  Falculea [Genus]
   F. palliata [Species]
  Leptopterus [Genus]
   L. viridis [Species]
   L. chabert
   L. madagascarinus
  Oriolia [Genus]
   O. bernieri [Species]
  Euryceros [Genus]
   E. prevostrii [Species]

Bombycillidae [Family]
  Bombycilla [Genus]
   B. garrulus [Species]
   B. japonica
   B. cedrorum
  Ptilogonys [Genus]
   P. cinereus [Species]
   P. caudatus
  Phainopepla [Genus]
   P. nitens [Species]
  Phainoptila [Genus]
   P. melanoxantha [Species]

Hypocolius [Genus]
   H. ampelinus [Species]

Dulidae [Family]
  Dulus [Genus]
   D. dominicus [Species]

Cinclidae [Family]
  Cinclus [Genus]
   C. cinclus [Species]
   C. pallasii
   C. mexicanus
   C. leucocephalus

Troglodytidae [Family]
  Campylorhynchus [Genus]
   C. jocosus [Species]
   C. gularis
   C. yucatanicus
   C. brunneicapillus
   C. griseus
   C. rufinucha
   C. turdinus
   C. nuchalis
   C. fasciatus
   C. zonatus
   C. megalopterus
  Odontorchilus [Genus]
   O. cinereus [Species]
   O. branickii
  Salpinctes [Genus]
   S. obsoletus [Species]
   S. mexicanus
  Hylorchilus [Genus]
   H. sumichrasti [Species]
  Cinnycerthia [Genus]
   C. unirufa [Species]
   C. peruana
  Cistothorus [Genus]
   C. platensis [Species]
   C. meridae
   C. apolinari
   C. palustris
  Thryomanes [Genus]
   T. bewickii [Species]
   T. sissonii
  Ferminia [Genus]
   F. cerverai [Species]
  Thryothorus [Genus]
   T. atrogularis [Species]
   T. fasciatoventris
   T. euophrys
   T. genibarbis
   T. coraya
   T. felix
   T. maculipectus
   T. rutilus
   T. nigricapillus
   T. thoracicus
   T. pleurostictus
   T. ludovicianus

   T. rufalbus
   T. nicefori
   T. sinaloa
   T. modestus
   T. leucotis
   T. superciliaris
   T. guarayanus
   T. longirostris
   T. griseus
  Troglodytes [Genus]
   T. troglodytes [Species]
   T. aedon
   T. solstitialis
   T. rufulus
   T. browni
  Uropsila [Genus]
   U. leucogastra [Species]
  Henicorhina [Genus]
   H. leucosticta [Species]
   H. leucophrys
  Microcerculus [Genus]
   M. marginatus [Species]
   M. ustulatus
   M. bambla
  Cyphorhinus [Genus]
   C. thoracicus [Species]
   C. aradus

Mimidae [Family]
  Dumetalla [Genus]
   D. carolinensis [Species]
  Melanoptila [Genus]
   M. glabrirostris [Species]
  Melanotis [Genus]
   M. caerulescens [Species]
   M. hypoleucus
  Mimus [Genus]
   M. polyglottos [Species]
   M. gilvus
   M. gundlachii
   M. thenca
   M. longicaudatus
   M. saturninus
   M. patagonicus
   M. triurus
   M. dorsalis
  Nesomimus [Genus]
   N. trifasciatus [Species]
  Mimodes [Genus]
   M. graysoni [Species]
  Oreoscoptes [Genus]
   O. montanus [Species]
  Toxostoma [Genus]
   T. rufum [Species]
   T. longirostre
   T. guttatum
   T. cinereum
   T. bendirei
   T. ocellatum
   T. curvirostre
   T. lecontei

T. redivivum
T. dorsale
Cinclocerthia [Genus]
  C. ruficauda [Species]
Ramphocinclus [Genus]
  R. brachyurus [Species]
Donacobius [Genus]
  D. atricapillus [Species]
Allenia [Genus]
  A. fusca [Species]
Margarops [Genus]
  M. fuscatus [Species]

Prunellidae [Family]
  Prunella [Genus]
    P. collaris [Species]
    P. himalayana
    P. rubeculoides
    P. strophiata
    P. montanella
    P. fulvescens
    P. ocularis
    P. atrogularis
    P. koslowi
    P. modularis
    P. rubida
    P. immaculata

Turdidae [Family]
  Brachypteryx [Genus]
    B. stellata [Species]
    B. hyperythra
    B. major
    B. calligyna
    B. leucophrys
    B. montana
  Zeledonia [Genus]
    Z. coronata [Species]
  Erythropygia [Genus]
    E. coryphaeus [Species]
    E. leucophrys
    E. hartlaubi
    E. galactotes
    E. paena
    E. leucosticta
    E. quadrivirgata
    E. barbata
    E. signata
  Namibornis [Genus]
    N. herero [Species]
  Cercotrichas [Genus]
    C. podobe [Species]
  Pinarornis [Genus]
    P. plumosus [Species]
  Chaetops [Genus]
    C. frenatus [Species]
  Drymodes [Genus]
    D. brunneopygia [Species]
    D. superciliaris
  Pogonocichla [Genus]
    P. stellata [Species]

P. swynnertoni
Erithacus [Genus]
  E. gabela [Species]
  E. cyornithopsis
  E. aequatorialis
  E. erythrothorax
  E. sharpei
  E. gunningi
  E. rubecula
  E. akahige
  E. komadori
  E. sibilans
  E. luscinia
  E. megarhynchos
  E. calliope
  E. svecicus
  E. pectoralis
  E. ruficeps
  E. obscurus
  E. pectardens
  E. brunneus
  E. cyane
  E. cyanurus
  E. chrysaeus
  E. indicus
  E. hyperythrus
  E. johnstoniae
Cossypha [Genus]
  C. roberti [Species]
  C. bocagei
  C. polioptera
  C. archeri
  C. isabellae
  C. natalensis
  C. dichroa
  C. semirufa
  C. heuglini
  C. cyanocampter
  C. caffra
  C. anomala
  C. humeralis
  C. ansorgei
  C. niveicapilla
  C. heinrichi
  C. albicapilla
Modulatrix [Genus]
  M. stictigula [Species]
Cichladusa [Genus]
  C. guttata [Species]
  C. arquata
  C. ruficauda
Alethe [Genus]
  A. diademata [Species]
  A. poliophrys
  A. fuelleborni
  A. montana
  A. lowei
  A. poliocephala
  A. choloensis
Copsychus [Genus]

C. saularis [Species]
C. sechellarum
C. albospecularis
C. malabaricus
C. stricklandii
C. luzoniensis
C. niger
C. pyrropygus
Irania [Genus]
  I. gutturalis [Species]
Phoenicurus [Genus]
  P. alaschanicus [Species]
  P. erythronotus
  P. caeruleocephalus
  P. ochruros
  P. phoenicurus
  P. hodgsoni
  P. frontalis
  P. schisticeps
  P. auroreus
  P. moussieri
  P. erythrogaster
Rhyacornis [Genus]
  R. bicolor [Species]
  R. fuliginosus
Hodgsonius [Genus]
  H. phaenicuroides [Species]
Cinclidium [Genus]
  C. leucurum [Species]
  C. diana
  C. frontale
Grandala [Genus]
  G. coelicolor [Species]
Sialia [Genus]
  S. sialis [Species]
  S. mexicana
  S. currucoides
Enicurus [Genus]
  E. scouleri [Species]
  E. velatus
  E. ruficapillus
  E. immaculatus
  E. schistaceus
  E. leschenaulti
  E. maculatus
Cochoa [Genus]
  C. purpurea [Species]
  C. viridis
  C. azurea
Myadestes [Genus]
  M. townsendi [Species]
  M. obscurus
  M. elisabeth
  M. genibarbis
  M. ralloides
  M. unicolor
  M. leucogenys
Entomodestes [Genus]
  E. leucotis [Species]
  E. coracinus

Stizorhina [Genus]
  S. fraseri [Species]
  S. finschii
Neocossyphus [Genus]
  N. rufus [Species]
  N. poensis
Cercomela [Genus]
  C. sinuata [Species]
  C. familiaris
  C. tractrac
  C. schlegelii
  C. fusca
  C. dubia
  C. melanura
  C. scotocerca
  C. sordida
Saxicola [Genus]
  S. rubetra [Species]
  S. macrorhyncha
  S. insignis
  S. dacotiae
  S. torquata
  S. leucura
  S. caprata
  S. jerdoni
  S. ferrea
  S. gutturalis
Myrmecocichla [Genus]
  M. tholloni [Species]
  M. aethiops
  M. formicivora
  M. nigra
  M. arnotti
  M. albifrons
  M. melaena
Thamnolaea [Genus]
  T. cinnamomeiventris [Species]
  T. coronata
  T. semirufa
Oenanthe [Genus]
  O. bifasciata [Species]
  O. isabellina
  O. bottae
  O. xanthoprymna
  O. oenanthe
  O. deserti
  O. hispanica
  O. finschii
  O. picata
  O. lugens
  O. monacha
  O. alboniger
  O. pleschanka
  O. leucopyga
  O. leucura
  O. monticola
  O. moesta
  O. pileata
Chaimarrornis [Genus]
  C. leucocephalus [Species]

Saxicoloides [Genus]
  S. fulicata [Species]
Pseudocossyphus [Genus]
  P. imerinus [Species]
Monticola [Genus]
  M. rupestris [Species]
  M. explorator
  M. brevipes
  M. rufocinereus
  M. angolensis
  M. saxatilis
  M. cinclorhynchus
  M. rufiventris
  M. solitarius
Myophonus [Genus]
  M. blighi [Species]
  M. melanurus
  M. glaucinus
  M. robinsoni
  M. horsfieldii
  M. insularis
  M. caeruleus
Geomalia [Genus]
  G. heinrichi [Species]
Zoothera [Genus]
  Z. schistacea [Species]
  Z. dumasi
  Z. interpres
  Z. erythronota
  Z. wardii
  Z. cinerea
  Z. peronii
  Z. citrina
  Z. everetti
  Z. sibirica
  Z. naevia
  Z. pinicola
  Z. piaggiae
  Z. oberlaenderi
  Z. gurneyi
  Z. cameronensis
  Z. princei
  Z. crossleyi
  Z. guttata
  Z. spiloptera
  Z. andromedae
  Z. mollissima
  Z. dixoni
  Z. dauma
  Z. talaseae
  Z. margaretae
  Z. monticola
  Z. marginata
  Z. terrestris
Amalocichla [Genus]
  A. sclateriana [Species]
  A. incerta
Cataponera [Genus]
  C. turdoides [Species]
Nesocichla [Genus]

  N. eremita [Species]
Cichlherminia [Genus]
  C. lherminieri [Species]
Phaeornis [Genus]
  P. obscurus [Species]
  P. palmeri
Catharus [Genus]
  C. gracilirostris [Species]
  C. aurantiirostris
  C. fuscater
  C. occidentalis
  C. mexicanus
  C. dryas
  C. fuscescens
  C. minimus
  C. ustulatus
  C. guttatus
Hylocichla [Genus]
  H. mustelina [Species]
Platycichla [Genus]
  P. flavipes [Species]
  P. leucops
Turdus [Genus]
  T. bewsheri [Species]
  T. olivaceofuscus
  T. olivaceus
  T. abyssinicus
  T. helleri
  T. libonyanus
  T. tephronotus
  T. menachensis
  T. ludoviciae
  T. litsipsirupa
  T. dissimilis
  T. unicolor
  T. cardis
  T. albocinctus
  T. torquatus
  T. boulboul
  T. merula
  T. poliocephalus
  T. chrysolaus
  T. celaenops
  T. rubrocanus
  T. kessleri
  T. feae
  T. pallidus
  T. obscurus
  T. ruficollis
  T. naumanni
  T. pilaris
  T. iliacus
  T. philomelos
  T. mupinensis
  T. viscivorus
  T. aurantius
  T. ravidus
  T. plumbeus
  T. chiguanco
  T. nigriscens

T. fuscater
T. serranus
T. nigriceps
T. reevei
T. olivater
T. maranonicus
T. fulviventris
T. rufiventris
T. falcklandii
T. leucomelas
T. amaurochalinus
T. plebejus
T. ignobilis
T. lawrencii
T. fumigatus
T. hauxwelli
T. haplochrous
T. grayi
T. nudigenis
T. jamaicensis
T. albicollis
T. rufopalliatus
T. swalesi
T. rufitorques
T. migratorius

Orthonychidae [Family]
  Orthonyx [Genus]
    O. temminckii [Species]
    O. spaldingii
  Androphobus [Genus]
    A. viridis [Species]
  Psophodes [Genus]
    P. olivaceus [Species]
    P. nigrogularis
  Sphenostoma [Genus]
    S. cristatum [Species]
  Cinclostoma [Genus]
    C. punctatum [Species]
    C. castanotum
    C. cinnamomeum
    C. ajax
  Ptilorrhoa [Genus]
    P. leucosticta [Species]
    P. caerulescens
    P. castanonota
  Eupetes [Genus]
    E. macrocercus [Species]
  Melampitta [Genus]
    M. lugubris [Species]
    M. gigantea
  Ifrita [Genus]
    I. kowaldi [Species]

Timaliidae [Family]
  Pellorneum [Genus]
    P. ruficeps [Species]
    P. palustre
    P. fuscocapillum
    P. capistratum
    P. albiventre

Trichastoma [Genus]
  T. tickelli [Species]
  T. pyrrogenys
  T. malaccense
  T. cinereiceps
  T. rostratum
  T. bicolor
  T. separium
  T. celebense
  T. abbotti
  T. perspicillatum
  T. vanderbilti
  T. pyrrhopterum
  T. cleaveri
  T. albipectus
  T. rufescens
  T. rufipenne
  T. fulvescens
  T. puveli
  T. poliothorax
Leonardina [Genus]
  L. woodi [Species]
Ptyrticus [Genus]
  P. turdinus [Species]
Malacopteron [Genus]
  M. magnirostre [Species]
  M. affine
  M. cinereum
  M. magnum
  M. palawanense
  M. albogulare
Pomatorhinus [Genus]
  P. hypoleucos [Species]
  P. erythrogenys
  P. horsfieldii
  P. schisticeps
  P. montanus
  P. ruficollis
  P. ochraceiceps
  P. ferruginosus
Garritornis [Genus]
  G. isidorei [Species]
Pomatostomus [Genus]
  P. temporalis [Species]
  P. superciliosus
  P. ruficeps
Xiphirhynchus [Genus]
  X. superciliaris [Species]
Jabouilleia [Genus]
  J. danjoui [Species]
Rimator [Genus]
  R. malacoptilus [Species]
Ptilocichla [Genus]
  P. leucogrammica [Species]
  P. mindanensis
  P. falcata
Kenopia [Genus]
  K. striata [Species]
Napothera [Genus]
  N. rufipectus [Species]

N. atrigularis
N. macrodactyla
N. marmorata
N. crispifrons
N. brevicaudata
N. crassa
N. rabori
N. epilepidota
Pnoepyga [Genus]
  P. albiventer [Species]
  P. pusilla
Spelaeornis [Genus]
  S. caudatus [Species]
  S. troglodytoides
  S. formosus
  S. chocolatinus
  S. longicaudatus
Sphenocichla [Genus]
  S. humei [Species]
Neomixis [Genus]
  N. tenella [Species]
  N. viridis
  N. striatigula
  N. flavoviridis
Stachyris [Genus]
  S. rodolphei [Species]
  S. rufifrons
  S. ambigua
  S. ruficeps
  S. pyrrhops
  S. chrysaea
  S. plateni
  S. capitalis
  S. speciosa
  S. whiteheadi
  S. striata
  S. nigrorum
  S. hypogrammica
  S. grammiceps
  S. herberti
  S. nigriceps
  S. poliocephala
  S. striolata
  S. oglei
  S. maculata
  S. leucotis
  S. nigricollis
  S. thoracica
  S. erythroptera
  S. melanothorax
Dumetia [Genus]
  D. hyperythra [Species]
Rhopocichla [Genus]
  R. atriceps [Species]
Macronous [Genus]
  M. flavicollis [Species]
  M. gularis
  M. kelleyi
  M. striaticeps
  M. ptilosus

Micromacronus [Genus]
    M. leytensis [Species]
Timalia [Genus]
    T. pileata [Species]
Chrysomma [Genus]
    C. sinense [Species]
Moupinia [Genus]
    M. altirostris [Species]
    M. poecilotis
Chamaea [Genus]
    C. fasciata [Species]
Turdoides [Genus]
    T. nipalensis [Species]
    T. altirostris
    T. caudatus
    T. earlei
    T. gularis
    T. longirostris
    T. malcolmi
    T. squamiceps
    T. fulvus
    T. aylmeri
    T. rubiginosus
    T. subrufus
    T. striatus
    T. affinis
    T. melanops
    T. tenebrosus
    T. reinwardtii
    T. plebejus
    T. jardineii
    T. squamulatus
    T. leucopygius
    T. hindei
    T. hypoleucus
    T. bicolor
    T. gymnogenys
Babax [Genus]
    B. lanceolatus [Species]
    B. waddelli
    B. koslowi
Garrulax [Genus]
    G. cinereifrons [Species]
    G. palliatus
    G. rufifrons
    G. perspicillatus
    G. albogularis
    G. leucolophus
    G. monileger
    G. pectoralis
    G. lugubris
    G. striatus
    G. strepitans
    G. milleti
    G. maesi
    G. chinensis
    G. vassali
    G. galbanus
    G. delesserti
    G. variegatus

G. davidi
G. sukatschewi
G. cineraceus
G. rufogularis
G. lunulatus
G. maximus
G. ocellatus
G. caerulatus
G. mitratus
G. ruficollis
G. merulinus
G. canorus
G. sannio
G. cachinnans
G. lineatus
G. virgatus
G. austeni
G. squamatus
G. subunicolor
G. elliotii
G. henrici
G. affinis
G. erythrocephalus
G. yersini
G. formosus
G. milnei
Liocichla [Genus]
    L. phoenicea [Species]
    L. steerii
Leiothrix [Genus]
    L. argentauris [Species]
    L. lutea
Cutia [Genus]
    C. nipalensis [Species]
Pteruthius [Genus]
    P. rufiventer [Species]
    P. flaviscapis
    P. xanthochlorus
    P. melanotis
    P. aenobarbus
Gampsorhynchus [Genus]
    G. rufulus [Species]
Actinodura [Genus]
    A. egertoni [Species]
    A. ramsayi
    A. nipalensis
    A. waldeni
    A. souliei
    A. morrisoniana
Minla [Genus]
    M. cyanouroptera [Species]
    M. strigula
    M. ignotincta
Alcippe [Genus]
    A. chrysotis [Species]
    A. variegaticeps
    A. cinerea
    A. castaneceps
    A. vinipectus
    A. striaticollis

A. ruficapilla
A. cinereiceps
A. rufogularis
A. brunnea
A. brunneicauda
A. poioicephala
A. pyrrhoptera
A. peracensis
A. morrisonia
A. nipalensis
A. abyssinica
A. atriceps
Lioptilus [Genus]
    L. nigricapillus [Species]
    L. gilberti
    L. rufocinctus
    L. chapini
Parophasma [Genus]
    P. galinieri [Species]
Phyllanthus [Genus]
    P. atripennis [Species]
Crocias [Genus]
    C. langbianis [Species]
    C. albonotatus
Heterophasia [Genus]
    H. annectens [Species]
    H. capistrata
    H. gracilis
    H. melanoleuca
    H. auricularis
    H. pulchella
    H. picaoides
Yuhina [Genus]
    Y. castaniceps [Species]
    Y. bakeri
    Y. flavicollis
    Y. gularis
    Y. diademata
    Y. occipitalis
    Y. brunneiceps
    Y. nigrimenta
    Y. zantholeuca
Malia [Genus]
    M. grata [Species]
Myzornis [Genus]
    M. pyrrhoura [Species]
Horizorhinus [Genus]
    H. dohrni [Species]
Oxylabes [Genus]
    O. madagascariensis [Species]
Mystacornis [Genus]
    M. crossleyi [Species]

Panuridae [Family]
Panurus [Genus]
    P. biarmicus [Species]
Conostoma [Genus]
    C. oemodium [Species]
Paradoxornis [Genus]
    P. paradoxus [Species]
    P. unicolor

P. flavirostris
P. guttaticollis
P. conspicillatus
P. ricketti
P. webbianus
P. alphonsianus
P. zappeyi
P. przewalskii
P. fulvifrons
P. nipalensis
P. davidianus
P. atrosuperciliaris
P. ruficeps
P. gularis
P. heudei

Picathartidae [Family]
  Picathartes [Genus]
    P. gymnocephalus [Species]
    P. oreas

Polioptilidae [Family]
  Microbates [Genus]
    M. collaris [Species]
    M. cinereiventris
  Ramphocaenus [Genus]
    R. melanurus [Species]
  Polioptila [Genus]
    P. caerulea [Species]
    P. melanura
    P. lembeyei
    P. albiloris
    P. plumbea
    P. lactea
    P. guianensis
    P. schistaceigula
    P. dumicola

Sylviidae [Family]
  Oligura [Genus]
    O. castaneocoronata [Species]
  Tesia [Genus]
    T. superciliaris [Species]
    T. olivea
    T. cyaniventer
  Urosphena [Genus]
    U. subulata [Species]
    U. whiteheadi
    U. squameiceps
    U. pallidipes
  Cettia [Genus]
    C. diphone [Species]
    C. annae
    C. parens
    C. ruficapilla
    C. fortipes
    C. vulcania
    C. major
    C. flavolivacea
    C. robustipes
    C. brunnifrons
    C. cetti

Bradypterus [Genus]
    B. baboecala [Species]
    B. graueri
    B. grandis
    B. carpalis
    B. alfredi
    B. sylvaticus
    B. barratti
    B. victorini
    B. cinnamomeus
    B. thoracicus
    B. major
    B. tacsanowskius
    B. luteoventris
    B. palliseri
    B. seebohmi
    B. caudatus
    B. accentor
    B. castaneus
  Bathmocercus [Genus]
    B. cerviniventris [Species]
    B. rufus
    B. winifredae
  Dromaeocercus [Genus]
    D. brunneus [Species]
    D. seeboehmi
  Nesillas [Genus]
    N. typica [Species]
    N. aldabranus
    N. mariae
  Thamnornis [Genus]
    T. chloropetoides [Species]
  Melocichla [Genus]
    M. mentalis [Species]
  Achaetops [Genus]
    A. pycnopygius [Species]
  Sphenoeacus [Genus]
    S. afer [Species]
  Megalurus [Genus]
    M. pryeri [Species]
    M. timoriensis
    M. palustris
    M. albolimbatus
    M. gramineus
    M. punctatus
  Cincloramphus [Genus]
    C. cruralis [Species]
    C. mathewsi
  Eremiornis [Genus]
    E. carteri [Species]
  Megalurulus [Genus]
    M. bivittata [Species]
    M. mariei
  Cichlornis [Genus]
    C. whitneyi [Species]
    C. llaneae
    C. grosvenori
  Ortygocichla [Genus]
    O. rubiginosa [Species]
    O. rufa

Chaetornis [Genus]
    C. striatus [Species]
  Graminicola [Genus]
    G. bengalensis [Species]
  Schoenicola [Genus]
    S. platyura [Species]
  Locustella [Genus]
    L. lanceolata [Species]
    L. naevia
    L. certhiola
    L. ochotensis
    L. pleskei
    L. fluvialtilis
    L. luscinioides
    L. fasciolata
    L. amnicola
  Acrocephalus [Genus]
    A. melanopogon [Species]
    A. paludicola
    A. schoenobaenus
    A. sorghophilus
    A. bistrigiceps
    A. agricola
    A. concinens
    A. scirpaceus
    A. cinnamomeus
    A. baeticatus
    A. palustris
    A. dumetorum
    A. arundinaceus
    A. stentoreus
    A. orinus
    A. orientalis
    A. luscinia
    A. familiaris
    A. aequinoctialis
    A. caffer
    A. atyphus
    A. vaughani
    A. rufescens
    A. brevipennis
    A. gracilirostris
    A. newtoni
    A. aedon
  Bebrornis [Genus]
    B. rodericanus [Species]
    B. sechellensis
  Hippolais [Genus]
    H. caligata [Species]
    H. pallida
    H. languida
    H. olivetorum
    H. polyglotta
    H. icterina
  Chloropeta [Genus]
    C. natalensis [Species]
    C. similis
    C. gracilirostris
  Cisticola [Genus]
    C. erythrops [Species]
    C. lepe

C. cantans
C. lateralis
C. woosnami
C. anonyma
C. bulliens
C. chubbi
C. hunteri
C. nigriloris
C. aberrans
C. bodessa
C. chiniana
C. cinereola
C. ruficeps
C. rufilata
C. subruficapilla
C. lais
C. restricta
C. njombe
C. galactotes
C. pipiens
C. carruthersi
C. tinniens
C. robusta
C. aberdare
C. natalensis
C. fulvicapilla
C. angusticauda
C. melanura
C. brachyptera
C. rufa
C. troglodytes
C. nana
C. incana
C. juncidis
C. cherina
C. haesitata
C. aridula
C. textrix
C. eximia
C. dambo
C. brunnescens
C. ayresii
C. exilis
Scotocerca [Genus]
S. inquieta [Species]
Rhopophilus [Genus]
R. pekinensis [Species]
Prinia [Genus]
P. burnesi [Species]
P. criniger
P. polychroa
P. atrogularis
P. cinereocapilla
P. buchanani
P. rufescens
P. hodgsoni
P. gracilis
P. sylvatica
P. familiaris
P. flaviventris

P. socialis
P. subflava
P. somalica
P. fluviatilis
P. maculosa
P. flavicans
P. substriata
P. molleri
P. robertsi
P. leucopogon
P. leontica
P. bairdii
P. erythroptera
P. pectoralis
Drymocichla [Genus]
D. incana [Species]
Urolais [Genus]
U. epichlora [Species]
Spiloptila [Genus]
S. clamans [Species]
Apalis [Genus]
A. thoracica [Species]
A. pulchra
A. ruwenzori
A. nigriceps
A. jacksoni
A. chariessa
A. binotata
A. flavida
A. ruddi
A. rufogularis
A. sharpii
A. goslingi
A. bamendae
A. porphyrolaema
A. melanocephala
A. chirindensis
A. cinerea
A. alticola
A. karamojae
A. rufifrons
Stenostira [Genus]
S. scita [Species]
Phyllolais [Genus]
P. pulchella [Species]
Orthotomus [Genus]
O. metopias [Species]
O. moreaui
O. cucullatus
O. sutorius
O. atrogularis
O. derbianus
O. sericeus
O. ruficeps
O. sepium
O. cinereiceps
O. nigriceps
O. samaransis
Camaroptera [Genus]
C. brachyura [Species]

C. brevicauda
C. harterti
C. superciliaris
C. chloronota
Calamonastes [Genus]
C. simplex [Species]
C. stierlingi
C. fasciolatus
Euryptila [Genus]
E. subcinnamomea [Species]
Poliolais [Genus]
P. lopesi [Species]
Graueria [Genus]
G. vittata [Species]
Eremomela [Genus]
E. icteropygialis [Species]
E. flavocrissalis
E. scotops
E. pusilla
E. canescens
E. gregalis
E. badiceps
E. turneri
E. atricollis
E. usticollis
Randia [Genus]
R. pseudozosterops [Species]
Newtonia [Genus]
N. brunneicauda [Species]
N. amphichroa
N. archboldi
N. fanovanae
Sylvietta [Genus]
S. virens [Species]
S. denti
S. leucophrys
S. brachyura
S. philippae
S. whytii
S. ruficapilla
S. rufescens
S. isabellina
Hemitesia [Genus]
H. neumanni [Species]
Macrosphenus [Genus]
M. kempi [Species]
M. flavicans
M. concolor
M. pulitzeri
M. kretschmeri
Amaurocichla [Genus]
A. bocagei [Species]
Hypergerus [Genus]
H. atriceps [Species]
H. lepidus
Hyliota [Genus]
H. flavigaster [Species]
H. australis
H. violacea
Hylia [Genus]

H. prasina [Species]
Phylloscopus [Genus]
  P. ruficapilla [Species]
  P. laurae
  P. laetus
  P. herberti
  P. budongoensis
  P. umbrovirens
  P. trochilus
  P. collybita
  P. sindianus
  P. neglectus
  P. bonelli
  P. sibilatrix
  P. fuscatus
  P. fuligiventer
  P. affinis
  P. griseolus
  P. armandii
  P. schwarzi
  P. pulcher
  P. maculipennis
  P. proregulus
  P. subviridis
  P. inornatus
  P. borealis
  P. trochiloides
  P. nitidus
  P. plumbeitarsus
  P. tenellipes
  P. magnirostris
  P. tytleri
  P. occipitalis
  P. coronatus
  P. ijimae
  P. reguloides
  P. davisoni
  P. cantator
  P. ricketti
  P. olivaceus
  P. cebuensis
  P. trivirgatus
  P. sarasinorum
  P. presbytes
  P. poliocephalus
  P. makirensis
  P. amoenus
Seicercus [Genus]
  S. burkii [Species]
  S. xanthoschistos
  S. affinis
  S. poliogenys
  S. castaniceps
  S. montis
  S. grammiceps
Tickellia [Genus]
  T. hodgsoni [Species]
Abroscopus [Genus]
  A. albogularis [Species]
  A. schisticeps
  A. superciliaris

Parisoma [Genus]
  P. buryi [Species]
  P. lugens
  P. boehmi
  P. layardi
  P. subcaeruleum
Sylvia [Genus]
  S. atricapilla [Species]
  S. borin
  S. communis
  S. curruca
  S. nana
  S. nisoria
  S. hortensis
  S. leucomelaena
  S. rueppelli
  S. melanocephala
  S. melanothorax
  S. mystacea
  S. cantillans
  S. conspicillata
  S. deserticola
  S. undata
  S. sarda
Regulus [Genus]
  R. ignicapillus [Species]
  R. regulus
  R. goodfellowi
  R. satrapa
  R. calendula
Leptopoecile [Genus]
  L. sophiae [Species]
  L. elegans

Muscicapidae [Family]
  Melaenornis [Genus]
    M. semipartitus [Species]
    M. pallidus
    M. infuscatus
    M. mariquensis
    M. microrhynchus
    M. chocolatinus
    M. fischeri
    M. brunneus
    M. edolioides
    M. pammelaina
    M. ardesiacus
    M. annamarulae
    M. ocreatus
    M. cinerascens
    M. silens
  Rhinomyias [Genus]
    R. addita [Species]
    R. oscillans
    R. brunneata
    R. olivacea
    R. umbratilis
    R. ruficauda
    R. colonus
    R. gularis
    R. insignis
    R. goodfellowi

Muscicapa [Genus]
  M. striata [Species]
  M. gambagae
  M. griseisticta
  M. sibirica
  M. dauurica
  M. ruficauda
  M. muttui
  M. ferruginea
  M. sordida
  M. thalassina
  M. panayensis
  M. albicaudata
  M. indigo
  M. infuscata
  M. ussheri
  M. boehmi
  M. aquatica
  M. olivascens
  M. lendu
  M. adusta
  M. epulata
  M. sethsmithii
  M. comitata
  M. tessmanni
  M. cassini
  M. caerulescens
  M. griseigularis
Myioparus [Genus]
  M. plumbeus [Species]
Humblotia [Genus]
  H. flavirostris [Species]
Ficedula [Genus]
  F. hypoleuca [Species]
  F. albicollis
  F. zanthopygia
  F. narcissina
  F. mugimaki
  F. hodgsonii
  F. dumetoria
  F. strophiata
  F. parva
  F. subruba
  F. monileger
  F. solitaris
  F. hyperythra
  F. basilanica
  F. rufigula
  F. buruensis
  F. henrici
  F. harterti
  F. platenae
  F. bonthaina
  F. westermanni
  F. superciliaris
  F. tricolor
  F. sapphira
  F. nigrorufa
  F. timorensis
  F. cyanomelana

Niltava [Genus]
  N. grandis [Species]
  N. macgrigoriae
  N. davidi
  N. sundara
  N. sumatrana
  N. vivida
  N. hyacinthina
  N. hoevelli
  N. sanfordi
  N. concreta
  N. ruecki
  N. herioti
  N. hainana
  N. pallipes
  N. poliogenys
  N. unicolor
  N. rubeculoides
  N. banyumas
  N. superba
  N. caerulata
  N. turcosa
  N. tickelliae
  N. rufigastra
  N. hodgsoni
Culicicapa [Genus]
  C. ceylonensis [Species]
  C. helianthea

Platysteiridae [Family]
  Bias [Genus]
    B. flammulatus [Species]
    B. musicus
  Pseudobias [Genus]
    P. wardi [Species]
  Batis [Genus]
    B. diops [Species]
    B. margaritae
    B. mixta
    B. dimorpha
    B. capensis
    B. fratrum
    B. molitor
    B. soror
    B. pririt
    B. senegalensis
    B. orientalis
    B. minor
    B. perkeo
    B. minulla
    B. minima
    B. ituriensis
    B. poensis
  Platysteira [Genus]
    P. cyanea [Species]
    P. albifrons
    P. peltata
    P. laticincta
    P. castanea
    P. tonsa
    P. blissetti

P. chalybea
P. jamesoni
P. concreta

Maluridae [Family]
  Clytomyias [Genus]
    C. insignis [Species]
  malurus [Genus]
    M. wallacii [Species]
    M. grayi
    M. alboscapulatus
    M. melanocephalus
    M. leucopterus
    M. cyaneus
    M. splendens
    M. lamberti
    M. amabilis
    M. pulcherrimus
    M. elegans
    M. coronatus
    M. cyanocephalus
  Stipiturus [Genus]
    S. malachurus [Species]
    S. mallee
    M. ruficeps
  Amytornis [Genus]
    A. textilis [Species]
    A. purnelli
    A. housei
    A. woodwardi
    A. dorotheae
    A. striatus
    A. barbatus
    A. goyderi

Acanthizidae [Family]
  Dasyornis [Genus]
    D. brachypterus [Species]
    D. broadbenti
  Pycnoptilus [Genus]
    P. floccosus [Species]
  Origma [Genus]
    O. solitaria [Species]
  Crateroscelis [Genus]
    C. gutturalis [Species]
    C. murina
    C. nigrorufa
    C. robusta
  Sericornis [Genus]
    S. citreogularis [Species]
    S. maculatus
    S. humilis
    S. frontalis
    S. beccarii
    S. nouhuysi
    S. magnirostris
    S. keri
    S. spilodera
    S. perspicillatus
    S. rufescens
    S. papuensis

S. arfakianus
S. magnus
Pyrrholaemus [Genus]
  P. brunneus [Species]
Chthonicola [Genus]
  C. sagittatus [Species]
Calamanthus [Genus]
  C. fuliginosus [Species]
  C. campestris
Hylacola [Genus]
  H. pyrrhopygius [Species]
  H. cautus
Acanthiza [Genus]
  A. murina [Species]
  A. inornata
  A. reguloides
  A. iredalei
  A. katherina
  A. pusilla
  A. apicalis
  A. ewingii
  A. chrysorrhoa
  A. uropygialis
  A. robustirostris
  A. nana
  A. lineata
Smicrornis [Genus]
  S. brevirostris [Species]
Gerygone [Genus]
  G. cinerea [Species]
  G. chloronota
  G. palpebrosa
  G. olivacea
  G. dorsalis
  G. chrysogaster
  G. ruficauda
  G. magnirostris
  G. sulphurea
  G. inornata
  G. ruficollis
  G. fusca
  G. tenebrosa
  G. laevigaster
  G. flavolateralis
  G. insularis
  G. mouki
  G. modesta
  G. igata
  G. albofrontata
Aphelocephala [Genus]
  A. leucopsis [Species]
  A. pectoralis
  A. nigricincta
Mohoua [Genus]
  M. ochrocephala [Species]
Finschia [Genus]
  F. novaeseelandiae [Species]
Epthianura [Genus]
  E. albifrons [Species]
  E. tricolor

E. aurifrons
E. crocea
Ashbyia [Genus]
A. lovensis [Species]

Monarchidae [Family]
Erythrocercus [Genus]
E. mccallii [Species]
E. holochlorus
E. livingstonei
Elminia [Genus]
E. longicauda [Species]
E. albicauda
Trochocercus [Genus]
T. nigromitratus [Species]
T. albiventris
T. albonotatus
T. cyanomelas
T. nitens
Philentoma [Genus]
P. pyrhopterum [Species]
P. velatum
Hypothymis [Genus]
H. azurea [Species]
H. helenae
H. coelestris
Eutrichomyias [Genus]
E. rowleyi [Species]
Terpsiphone [Genus]
T. rufiventer [Species]
T. bedfordi
T. rufocinerea
T. viridis
T. paradisi
T. atrocaudata
T. cyanescens
T. cinnamomea
T. atrochalybeia
T. mutata
T. corvina
T. bourbonnensis
Chasiempis [Genus]
C. sandwichensis [Species]
Pomarea [Genus]
P. dimidiata [Species]
P. nigra
P. mendozae
P. iphis
P. whitneyi
Mayrornis [Genus]
M. versicolor [Species]
M. lessoni
M. schistaceus
Neolalage [Genus]
N. banksiana [Species]
Clytorhynchus [Genus]
C. pachycephaloides [Species]
C. vitiensis
C. nigrogularis
C. hamlini
Metabolus [Genus]

M. rugensis [Species]
Monarcha [Genus]
M. axillaris [Species]
M. rubiensis
M. cinerascens
M. melanopsis
M. frater
M. erythrostictus
M. castaneiventris
M. richardsii
M. leucotis
M. guttulus
M. mundus
M. sacerdotum
M. trivirgatus
M. leucurus
M. julianae
M. manadensis
M. brehmii
M. infelix
M. menckei
M. verticalis
M. barbatus
M. browni
M. viduus
M. godeffroyi
M. takatsukasae
M. chrysomela
Arses [Genus]
A. insularis [Species]
A. telescophthalmus
A. kaupi
Myiagra [Genus]
M. oceanica [Species]
M. galeata
M. atra
M. rubecula
M. ferrocyanea
M. cervinicauda
M. caledonica
M. vanikorensis
M. albiventris
M. azureocapilla
M. ruficollis
M. cyanoleuca
M. alecto
M. hebetior
M. inquieta
Lamprolia [Genus]
L. victoriae [Species]
Machaerirhynchus [Genus]
M. flaviventer [Species]
M. nigripectus
Peltops [Genus]
P. blainvillii [Species]
P. montanus
Rhipidura [Genus]
R. hypoxantha [Species]
R. superciliaris
R. cyaniceps

R. phoenicura
R. nigrocinnamomea
R. albicollis
R. euryura
R. aureola
R. javanica
R. perlata
R. leucophrys
R. rufiventris
R. cockerelli
R. albolimbata
R. hyperythra
R. threnothorax
R. maculipectus
R. leucothorax
R. atra
R. fuliginosa
R. drownei
R. tenebrosa
R. rennelliana
R. spilodera
R. nebulosa
R. brachyrhyncha
R. personata
R. dedemi
R. superflua
R. teysmanni
R. lepida
R. opistherythra
R. rufidorsa
R. dahli
R. matthiae
R. malaitae
R. rufifrons

Eopsaltriidae [Family]
Monachella [Genus]
M. muelleriana [Species]
Microeca [Genus]
M. leucophaea [Species]
M. flavigaster
M. hemixantha
M. griseoceps
M. flavovirescens
M. papuana
Eugerygone [Genus]
E. rubra [Species]
Petroica [Genus]
P. bivittata [Species]
P. archboldi
P. multicolor
P. goodenovii
P. phoenicea
P. rosea
P. rodinogaster
P. cucullata
P. vittata
P. macrocephala
P. australis
P. traversi
Tregellasia [Genus]

T. capito [Species]
T. leucops
Eopsaltria [Genus]
   E. australis [Species]
   E. flaviventris
   E. georgiana
Peneoenanthe [Genus]
   P. pulverulenta [Species]
Peocilodryas [Genus]
   P. brachyura [Species]
   P. hypoleuca
   P. placens
   P. albonotata
   P. superciliosa
Peneothello [Genus]
   P. sigillatus [Species]
   P. cryptoleucus
   P. cyanus
   P. bimaculatus
Heteromyias [Genus]
   H. cinereifrons [Species]
   H. albispecularis
Pachycephalopsis [Genus]
   P. hattamensis [Species]
   P. poliosoma

Pachycephalidae [Family]
Eulacestoma [Genus]
   E. nigropectus [Species]
Falcunculus [Genus]
   F. frontatus [Species]
Oreoica [Genus]
   O. gutturalis [Species]
Pachycare [Genus]
   P. flavogrisea [Species]
Rhagologus [Genus]
   R. leucostigma [Species]
Hylocitrea [Genus]
   H. bonensis [Species]
Pachycephala [Genus]
   P. raveni [Species]
   P. rufinucha
   P. tenebrosa
   P. olivacea
   P. rufogularis
   P. inornata
   P. hypoxantha
   P. cinerea
   P. phaionota
   P. hyperythra
   P. modesta
   P. philippensis
   P. sulfuriventer
   P. meyeri
   P. soror
   P. simplex
   P. orpheus
   P. pectoralis
   P. flavifrons
   P. caledonica
   P. implicata

P. nudigula
P. lorentzi
P. schlegelii
P. aurea
P. rufiventris
P. lanioides
Colluricincla [Genus]
   C. megarhyncha [Species]
   C. parvula
   C. boweri
   C. harmonica
   C. woodwardi
Pitohui [Genus]
   P. kirhocephalus [Species]
   P. dichrous
   P. incertus
   P. ferrugineus
   P. cristatus
   P. nigrescens
   P. tenebrosus
Turnagra [Genus]
   T. capensis [Species]

Aegithalidae [Family]
Aegithalos [Genus]
   A. caudatus [Species]
   A. leucogenys
   A. concinnus
   A. iouschistos
   A. fuliginosus
Psaltria [Genus]
   P. exilis [Species]
Psaltriparus [Genus]
   P. minimus [Species]
   P. melanotis

Remizidae [Family]
Remiz [Genus]
   R. pendulinus [Species]
Anthoscopus [Genus]
   A. punctifrons [Species]
   A. parvulus
   A. musculus
   A. flavifrons
   A. caroli
   A. sylviella
   A. minutus
Auriparus [Genus]
   A. flaviceps [Species]
Cephalopyrus [Genus]
   C. flammiceps [Species]

Paridae [Family]
Parus [Genus]
   P. palustris [Species]
   P. lugubris
   P. montanus
   P. atricapillus
   P. carolinensis
   P. sclateri
   P. gambeli

P. superciliosus
P. davidi
P. cinctus
P. hudsonicus
P. rufescens
P. wollweberi
P. rubidiventris
P. melanolophus
P. ater
P. venustulus
P. elegans
P. amabilis
P. cristatus
P. dichrous
P. afer
P. griseiventris
P. niger
P. leucomelas
P. albiventris
P. leuconotus
P. funereus
P. fasciiventer
P. fringillinus
P. rufiventris
P. major
P. bokharensis
P. monticolus
P. nuchalis
P. xanthogenys
P. spilonotus
P. holsti
P. caeruleus
P. cyanus
P. varius
P. semilarvatus
P. inornatus
P. bicolor
Melanochlora [Genus]
   M. sultanea [Species]
Sylviparus [Genus]
   S. modestus [Species]
Hypositta [Genus]
   H. corallirostris [Species]

Sittidae [Family]
Sitta [Genus]
   S. europaea [Species]
   S. nagaensis
   S. castanea
   S. himalayensis
   S. victoriae
   S. pygmaea
   S. pusilla
   S. whiteheadi
   S. yunnanensis
   S. canadensis
   S. villosa
   S. leucopsis
   S. carolinensis
   S. krueperi
   S. neumayer

S. tephronota
S. frontalis
S. solangiae
S. azurea
S. magna
S. formosa
Neositta [Genus]
    N. chrysoptera [Species]
    N. papuensis
Daphoenositta [Genus]
    D. miranda [Species]
Tichodroma [Genus]
    T. muraria [Species]

Certhiidae [Family]
    Certhia [Genus]
        F. familiaris [Species]
        F. brachydactyla
        F. himalayana
        F. nipalensis
        F. discolor
    Salpornis [Genus]
        S. spilonotus [Species]

Rhabdornithidae [Family]
    Rhabdornis [Genus]
        R. mysticalis [Species]
        R. inornatus

Climacteridae [Family]
    Climacteris [Genus]
        C. erythrops [Species]
        C. affinis
        C. picumnus
        C. rufa
        C. melanura
        C. leucophaea

Dicaeidae [Family]
    Melanocharis [Genus]
        M. arfakiana [Species]
        M. nigra
        M. longicauda
        M. versteri
        M. striativentris
    Rhamphocharis [Genus]
        R. crassirostris [Species]
    Prionochilus [Genus]
        P. olivaceus [Species]
        P. maculatus
        P. percussus
        P. plateni
        P. xanthopygius
        P. thoracicus
    Dicaeum [Genus]
        D. annae [Species]
        D. agile
        D. everetti
        D. aeruginosum
        D. proprium
        D. chrysorrheum
        D. melanoxanthum

D. vincens
D. aureolimbatum
D. nigrilore
D. anthonyi
D. bicolor
D. quadricolor
D. australe
D. retrocinctum
D. trigonostigma
D. hypoleucum
D. erythrorhynchos
D. concolor
D. pygmaeum
D. nehrkorni
D. vulneratum
D. erythrothorax
D. pectorale
D. eximium
D. aeneum
D. tristrami
D. igniferum
D. maugei
D. sanguinolentum
D. hirundinaceum
D. celebicum
D. monticolum
D. ignipectus
D. cruentatum
D. trochileum
Oreocharis [Genus]
    O. arfaki [Species]
Paramythia [Genus]
    P. montium [Species]
Pardalotus [Genus]
    P. quadragintus [Species]
    P. punctatus
    P. xanthopygus
    P. rubricatus
    P. striatus
    P. ornatus
    P. substriatus
    P. melanocephalus

Nectariniidae [Family]
    Anthreptes [Genus]
        A. gabonicus [Species]
        A. fraseri
        A. reichenowi
        A. anchietae
        A. simplex
        A. malacensis
        A. rhodolaema
        A. singalensis
        A. longuemarei
        A. orientalis
        A. neglectus
        A. aurantium
        A. pallidogaster
        A. pujoli
        A. rectirostris
        A. collaris
        A. platurus

Hypogramma [Genus]
    H. hypogrammicum [Species]
Nectarina [Genus]
    N. seimundi [Species]
    N. batesi
    N. olivacea
    N. ursulae
    N. veroxii
    N. balfouri
    N. reichenbachii
    N. hartlaubii
    N. newtonii
    N. thomensis
    N. oritis
    N. alinae
    N. bannermani
    N. verticalis
    N. cyanolaema
    N. fuliginosa
    N. rubescens
    N. amethystina
    N. senegalensis
    N. adelberti
    N. zeylonica
    N. minima
    N. sperata
    N. sericea
    N. calcostetha
    N. dussumeiri
    N. lotenia
    N. jugularis
    N. buettikoferi
    N. solaris
    N. asiatica
    N. souimanga
    N. humbloti
    N. comorensis
    N. coquerellii
    N. venusta
    N. talatala
    N. oustaleti
    N. fusca
    N. chalybea
    N. afra
    N. mediocris
    N. preussi
    N. neergaardi
    N. chloropygia
    N. minulla
    N. regia
    N. loveridgei
    N. rockefelleri
    N. violacea
    N. habessinica
    N. bouvieri
    N. osea
    N. cuprea
    N. tacazze
    N. bocagii
    N. purpureiventris

N. shelleyi
N. mariquensis
N. bifasciata
N. pembae
N. chalcomelas
N. coccinigastra
N. erythrocerca
N. congensis
N. pulchella
N. nectarinioides
N. famosa
N. johnstoni
N. notata
N. johannae
N. superba
N. kilimensis
N. reichenowi
Aethopyga [Genus]
A. primigenius [Species]
A. boltoni
A. flagrans
A. pulcherrima
A. duyvenbodei
A. shelleyi
A. gouldiae
A. nipalensis
A. eximia
A. christinae
A. saturata
A. siparaja
A. mysticalis
A. ignicauda
Arachnothera [Genus]
A. longirostra [Species]
A. crassirostris
A. robusta
A. flavigaster
A. chrysogenys
A. clarae
A. affinis
A. magna
A. everetti
A. juliae

Zosteropidae [Family]
Zosterops [Genus]
Z. erythropleura [Species]
Z. japonica
Z. palpebrosa
Z. ceylonensis
Z. conspicillata
Z. salvadorii
Z. atricapilla
Z. everetti
Z. nigrorum
Z. montana
Z. wallacei
Z. flava
Z. chloris
Z. consibrinorum
Z. grayi

Z. uropygialis
Z. anomala
Z. atriceps
Z. atrifrons
Z. mysorensis
Z. fuscicapilla
Z. buruensis
Z. kuehni
Z. novaeguineae
Z. metcalfi
Z. natalis
Z. lutea
Z. griseotincta
Z. rennelliana
Z. vellalavella
Z. luteirostris
Z. rendovae
Z. murphyi
Z. ugiensis
Z. stresemanni
Z. sanctaecrucis
Z. samoensis
Z. explorator
Z. flavifrons
Z. minuta
Z. xanthochroa
Z. lateralis
Z. strenua
Z. tenuirostris
Z. albogularis
Z. inornata
Z. cinerea
Z. abyssinica
Z. pallida
Z. senegalensis
Z. virens
Z. borbonica
Z. ficedulina
Z. griseovirescens
Z. maderaspatana
Z. mayottensis
Z. modesta
Z. mouroniensis
Z. olivacea
Z. vaughani
Woodfordia [Genus]
W. superciliosa [Species]
W. lacertosa
Rukia [Genus]
R. palauensis [Species]
R. oleaginea
R. ruki
R. longirostra
Tephrozosterops [Genus]
T. stalkeri [Species]
Madanga [Genus]
M. ruficollis [Species]
Lophozosterops [Genus]
L. pinaiae [Species]
L. goodfellowi

L. squamiceps
L. javanica
L. superciliaris
L. dohertyi
Oculocincta [Genus]
O. squamifrons [Species]
Heleia [Genus]
H. muelleri [Species]
H. crassirostris
Chlorocharis [Genus]
C. emiliae [Species]
Hypocryptadius [Genus]
H. cinnamomeus [Species]
Speirops [Genus]
S. brunnea [Species]
S. leucophoeca
S. lugubris

Meliphagidae [Family]
Timeliopsis [Genus]
T. fulvigula [Species]
T. griseigula
Melilestes [Genus]
M. megarhynchus [Species]
M. bouganvillei
Toxorhamphus [Genus]
T. novaeguineae [Species]
T. poliopterus
Oedistoma [Genus]
O. iliolophum [Species]
O. pygmaeum
Glycichaera [Genus]
G. fallax [Species]
Lichmera [Genus]
L. lombokia [Species]
L. argentauris
L. indistincta
L. incana
L. alboauricularis
L. squamata
L. deningeri
L. monticola
L. flavicans
L. notabilis
L. cockerelli
Myzomela [Genus]
M. blasii [Species]
M. albigula
M. cineracea
M. eques
M. obscura
M. cruentata
M. nigrita
M. pulchella
M. kuehni
M. erythrocephala
M. adolphinae
M. sanguinolenta
M. cardinalis
M. chermesina
M. sclateri

M. lafargei
M. melanocephala
M. eichhorni
M. malaitae
M. tristrami
M. jugularis
M. erythromelas
M. vulnerata
M. rosenbergii
Certhionyx [Genus]
   C. niger [Species]
   C. variegatus
Meliphaga [Genus]
   M. mimikae [Species]
   M. montana
   M. orientalis
   M. albonotata
   M. aruensis
   M. analoga
   M. vicina
   M. gracilis
   M. notata
   M. flavirictus
   M. lewinii
   M. flava
   M. albilineata
   M. virescens
   M. versicolor
   M. fasciogularis
   M. inexpectata
   M. fusca
   M. plumula
   M. chrysops
   M. cratitia
   M. keartlandi
   M. penicillata
   M. ornata
   M. reticulata
   M. leucotis
   M. flavicollis
   M. melanops
   M. cassidix
   M. unicolor
   M. flaviventer
   M. polygramma
   M. macleayana
   M. frenata
   M. subfrenata
   M. obscura
Oreornis [Genus]
   O. chrysogenys [Species]
Foulehaio [Genus]
   F. carunculata [Species]
   F. provocator
Cleptornis [Genus]
   C. marchei [Species]
Apalopteron [Genus]
   A. familiare [Species]
Melithreptus [Genus]
   M. brevirostris [Species]

M. lunatus
M. albogularis
M. affinis
M. gularis
M. laetior
M. validirostris
Entomyzon [Genus]
   E. cyanotis [Species]
Notiomystis [Genus]
   N. cincta [Species]
Pycnopygius [Genus]
   P. ixoides [Species]
   P. cinereus
   P. stictocephalus
Philemon [Genus]
   P. meyeri [Species]
   P. brassi
   P. citreogularis
   P. inornatus
   P. gilolensis
   P. fuscicapillus
   P. subcorniculatus
   P. moluccensis
   P. buceroides
   P. novaeguineae
   P. cockerelli
   P. eichhorni
   P. albitorques
   P. argenticeps
   P. corniculatus
   P. diemenensis
Ptiloprora [Genus]
   P. plumbea [Species]
   P. meekiana
   P. erythropleura
   P. guisei
   P. perstriata
Melidectes [Genus]
   M. fuscus [Species]
   M. princeps
   M. nouhuysi
   M. ochromelas
   M. leucostephes
   M. belfordi
   M. torquatus
Melipotes [Genus]
   M. gymnops [Species]
   M. fumigatus
   M. ater
Vosea [Genus]
   V. whitemanensis [Species]
Myza [Genus]
   M. celebensis [Species]
   M. sarasinorum
Meliarchus [Genus]
   M. sclateri [Species]
Gymnomyza [Genus]
   G. viridis [Species]
   G. samoensis
   G. aubryana

Moho [Genus]
   M. braccatus [Species]
   M. bishopi
   M. apicalis
   M. nobilis
Chaetoptila [Genus]
   C. angustipluma [Species]
Phylidonyris [Genus]
   P. pyrrhoptera [Species]
   P. novaehollandiae
   P. nigra
   P. albifrons
   P. melanops
   P. undulata
   P. notabilis
Ramsayornis [Genus]
   R. fasciatus [Species]
   R. modestus
Plectorhyncha [Genus]
   P. lanceolata [Species]
Conopophila [Genus]
   C. whitei [Species]
   C. albogularis
   C. rufogularis
   C. picta
Xanthomyza [Genus]
   X. phrygia [Species]
Cissomela [Genus]
   C. pectoralis [Species]
Acanthorhynchus [Genus]
   A. tenuirostris [Species]
   A. superciliosus
Manorina [Genus]
   M. melanophrys [Species]
   M. melanocephala
   M. flavigula
   M. melanotis
Anthornis [Genus]
   A. melanura [Species]
Anthochaera [Genus]
   A. rufogularis [Species]
   A. chrysoptera
   A. carunculata
   A. paradoxa
Prosthemadera [Genus]
   P. novaeseelandiae [Species]
Promerops [Genus]
   P. cafer [Species]
   P. gurneyi

Emberizidae [Family]
Melophus [Genus]
   M. lathami [Species]
Latoucheornis [Genus]
   L. siemsseni [Species]
Emberiza [Genus]
   E. calandra [Species]
   E. citrinella
   E. leucocephala
   E. cia
   E. cioides

E. jankowskii
E. buchanani
E. stewarti
E. cineracea
E. hortulana
E. caesia
E. cirlus
E. striolata
E. impetuani
E. tahapisi
E. socotrana
E. capensis
E. yessoensis
E. tristami
E. fucata
E. pusilla
E. chrysophrys
E. rustica
E. elegans
E. aureola
E. poliopleura
E. flaviventris
E. affinis
E. cabanisi
E. rutila
E. koslowi
E. melanocephala
E. bruniceps
E. sulphurata
E. spodocephala
E. variabilis
E. pallasi
E. schoeniclus
Calcarius [Genus]
  C. mccownii [Species]
  C. lapponicus
  C. pictus
  C. ornatus
Plectrophenax [Genus]
  P. nivalis [Species]
Calamospiza [Genus]
  C. melanocorys [Species]
Zonotrichia [Genus]
  Z. iliaca [Species]
  Z. melodia
  Z. lincolnii
  Z. georgiana
  Z. capensis
  Z. querula
  Z. leucophrys
  Z. albicollis
  Z. atricapilla
Junco [Genus]
  J. vulcani [Species]
  J. hyemalis
  J. phaeonotus
Ammodramus [Genus]
  A. sandwichensis [Species]
  A. maritimus
  A. caudacutus

A. leconteii
A. bairdii
A. baileyi
A. henslowii
A. savannarum
A. humeralis
A. aurifrons
Spizella [Genus]
  S. arborea [Species]
  S. passerina
  S. pusilla
  S. atrogularis
  S. pallida
  S. breweri
Pooecetes [Genus]
  P. gramineus [Species]
Chondestes [Genus]
  C. grammacus [Species]
Amphispiza [Genus]
  A. bilineata [Species]
  A. belli
Aimophila [Genus]
  A. mystacalis [Species]
  A. humeralis
  A. ruficauda
  A. sumichrasti
  A. stolzmanni
  A. strigiceps
  A. aestivalis
  A. botterii
  A. cassinii
  A. quinquestriata
  A. carpalis
  A. ruficeps
  A. notosticta
  A. rufescens
Torreornis [Genus]
  T. inexpectata [Species]
Oriturus [Genus]
  O. superciliosus [Species]
Phrygilus [Genus]
  P. atriceps [Species]
  P. gayi
  P. patagonicus
  P. fruticeti
  P. unicolor
  P. dorsalis
  P. erythronotus
  P. plebejus
  P. carbonarius
  P. alaudinus
Melanodera [Genus]
  M. melanodera [Species]
  M. xanthogramma
Haplospiza [Genus]
  H. rustica [Species]
  H. unicolor
Acanthidops [Genus]
  A. bairdii [Species]
Lophospingus [Genus]

L. pusillus [Species]
L. griseocristatus
Donacospiza [Genus]
  D. albifrons [Species]
Rowettia [Genus]
  R. goughensis [Species]
Nesospiza [Genus]
  N. acunhae [Species]
  N. wilkinsi
Diuca [Genus]
  D. speculifera [Species]
  D. diuca
Idiopsar [Genus]
  I. brachyurus [Species]
Piezorhina [Genus]
  P. cinerea [Species]
Xenospingus [Genus]
  X. concolor [Species]
Incaspiza [Genus]
  I. pulchra [Species]
  I. ortizi
  I. laeta
  I. watkinsi
Poospiza [Genus]
  P. thoracica [Species]
  P. boliviana
  P. alticola
  P. hypochondria
  P. erythrophrys
  P. ornata
  P. nigrorufa
  P. lateralis
  P. rubecula
  P. garleppi
  P. baeri
  P. caesar
  P. hispaniolensis
  P. torquata
  P. cinerea
Sicalis [Genus]
  S. citrina [Species]
  S. lutea
  S. uropygialis
  S. luteocephala
  S. auriventris
  S. olivascens
  S. columbiana
  S. flaveola
  S. luteola
  S. raimondii
  S. taczanowskii
Emberizoides [Genus]
  E. herbicola [Species]
Embernagra [Genus]
  E. platensis [Species]
  E. longicauda
Volatinia [Genus]
  V. jacarina [Species]
Sporophila [Genus]
  S. frontalis [Species]
  S. falcirostris

S. schistacea
S. intermedia
S. plumbea
S. americana
S. torqueola
S. collaris
S. lineola
S. luctuosa
S. nigricollis
S. ardesiaca
S. melanops
S. obscura
S. caerulescens
S. albogularis
S. leucoptera
S. peruviana
S. simplex
S. nigrorufa
S. bouvreuil
S. insulata
S. minuta
S. hypoxantha
S. hypochroma
S. ruficollis
S. palustris
S. castaneiventris
S. cinnamomea
S. melanogaster
S. telasco
Oryzoborus [Genus]
  O. crassirostris [Species]
  O. angolensis
Amaurospiza [Genus]
  A. concolor [Species]
  A. moesta
Melopyrrha [Genus]
  M. nigra [Species]
Dolospingus [Genus]
  D. fringilloides [Species]
Catamenia [Genus]
  C. analis [Species]
  C. inornata
  C. homochroa
  C. oreophila
Tiaris [Genus]
  T. canora [Species]
  T. olivacea
  T. bicolor
  T. fuliginosa
Loxipasser [Genus]
  L. anoxanthus [Species]
Loxigilla [Genus]
  L. portorocensis [Species]
  L. violacea
  L. noctis
Melanospiza [Genus]
  M. richardsoni [Species]
Geospiza [Genus]
  G. magnirostris [Species]
  G. fortis

G. fuliginosa
G. difficilis
G. scandens
G. conirostris
Camarhynchus [Genus]
  C. crassirostris [Species]
  C. psittacula
  C. pauper
  C. parvulus
  C. pallidus
  C. heliobates
Certhidea [Genus]
  C. olivacea [Species]
Pinaroloxias [Genus]
  P. inornata [Species]
Pipilo [Genus]
  P. chlorurus [Species]
  P. ocai
  P. erythrophthalmus
  P. socorroensis
  P. fuscus
  P. aberti
  P. albicollis
Melozone [Genus]
  M. kieneri [Species]
  M. biarcuatum
  M. leucotis
Arremon [Genus]
  A. taciturnus [Species]
  A. flavirostris
  A. aurantiirostris
  A. schlegeli
  A. abeillei
Arremonops [Genus]
  A. rufivirgatus [Species]
  A. tocuyensis
  A. chlorinotus
  A. conirostris
Atlapetes [Genus]
  A. albinucha [Species]
  A. pallidinucha
  A. rufinucha
  A. leucopis
  A. pileatus
  A. melanocephalus
  A. flaviceps
  A. fuscoolivaceus
  A. tricolor
  A. albofrenatus
  A. schistaceus
  A. nationi
  A. leucopterus
  A. albiceps
  A. pallidiceps
  A. rufigenis
  A. semirufus
  A. personatus
  A. fulviceps
  A. citrinellus
  A. brunneinucha
  A. torquatus

Pezopetes [Genus]
  P. capitalis [Species]
Oreothraupis [Genus]
  O. arremonops [Species]
Pselliophorus [Genus]
  P. tibialis [Species]
  P. luteoviridis
Lysurus [Genus]
  L. castaneiceps [Species]
Urothraupis [Genus]
  U. stolzmanni [Species]
Charitospiza [Genus]
  C. eucosma [Species]
Coryphaspiza [Genus]
  C. melanotis [Species]
Saltatricula [Genus]
  S. multicolor [Species]
Gubernatrix [Genus]
  G. cristata [Species]
Coryphospingus [Genus]
  C. pileatus [Species]
  C. cucullatus
Rhodospingus [Genus]
  R. cruentus [Species]
Paroaria [Genus]
  P. coronata [Species]
  P. dominicana
  P. gularis
  P. baeri
  P. capitata
Catamblyrhynchus [Genus]
  C. diadema [Species]
Spiza [Genus]
  S. americana [Species]
Pheucticus [Genus]
  P. chrysopeplus [Species]
  P. aureoventris
  P. ludovicianus
  P. melanocephalus
Cardinalis [Genus]
  C. cardinalis [Species]
  C. phoeniceus
  C. sinuatus
Caryothraustes [Genus]
  C. canadensis [Species]
  C. humeralis
Rhodothraupis [Genus]
  R. celaeno [Species]
Periporphyrus [Genus]
  P. erythromelas [Species]
Pitylus [Genus]
  P. grossus [Species]
Saltator [Genus]
  S. atriceps [Species]
  S. maximus
  S. atripennis
  S. similis
  S. coerulescens
  S. orenocensis
  S. maxillosus

S. aurantiirostris
S. cinctus
S. atricollis
S. rufiventris
S. albicollis
Passerina [Genus]
P. glaucocaerulea [Species]
P. cyanoides
P. brissonii
P. parellina
P. caerulea
P. cyanea
P. amoena
P. versicolor
P. ciris
P. rositae
P. leclancherii
P. caerulescens
Orchesticus [Genus]
O. albeillei [Species]
Schistochlamys [Genus]
S. ruficapillus [Species]
S. melanopis
Neothraupis [Genus]
N. fasciata [Species]
Cypsnagra [Genus]
C. hirundinacea [Species]
Conothraupis [Genus]
C. speculigera [Species]
C. mesoleuca
Lamprospiza [Genus]
L. melanoleuca [Species]
Cissopis [Genus]
C. leveriana [Species]
Chlorornis [Genus]
C. riefferii [Species]
Compsothraupis [Genus]
C. loricata [Species]
Sericossypha [Genus]
S. albocristata [Species]
Nesospingus [Genus]
N. speculiferus [Species]
Chlorospingus [Genus]
C. ophthalmicus [Species]
C. tacarcunae
C. inornatus
C. punctulatus
C. semifuscus
C. zeledoni
C. pileatus
C. parvirostris
C. flavigularis
C. flavovirens
C. canigularis
Cnemoscopus [Genus]
C. rubrirostris [Species]
Hemispingus [Genus]
H. atropileus [Species]
H. superciliaris
H. reyi

H. frontalis
H. melanotis
H. goeringi
H. verticalis
H. xanthophthalmus
H. trifasciatus
Pyrrhocoma [Genus]
P. ruficeps [Species]
Thlypopsis [Genus]
T. fulviceps [Species]
T. ornata
T. pectoralis
T. sordida
T. inornata
T. ruficeps
Hemithraupis [Genus]
H. guira [Species]
H. ruficapilla
H. flavicollis
Chrysothlypis [Genus]
C. chrysomelas [Species]
C. salmoni
Nemosia [Genus]
N. pileata [Species]
N. rourei
Phaenicophilus [Genus]
P. palmarum [Species]
P. poliocephalus
Calyptophilus [Genus]
C. frugivorus [Species]
Rhodinocichla [Genus]
R. rosea [Species]
Mitrospingus [Genus]
M. cassinii [Species]
M. oleagineus
Chlorothraupis [Genus]
C. carmioli [Species]
C. olivacea
C. stolzmanni
Orthogonys [Genus]
O. chloricterus [Species]
Eucometis [Genus]
E. penicillata [Species]
Lanio [Genus]
L. fulvus [Species]
L. versicolor
L. aurantius
L. leucothorax
Creurgops [Genus]
C. verticalis [Species]
C. dentata
Heterospingus [Genus]
H. xanthopygius [Species]
Tachyphonus [Genus]
T. cristatus [Species]
T. rufiventer
T. surinamus
T. luctuosus
T. delatrii
T. coronatus

T. rufus
T. phoenicius
Trichothraupis [Genus]
T. melanops [Species]
Habia [Genus]
H. rubica [Species]
H. fuscicauda
H. atrimaxillaris
H. gutturalis
H. cristata
Piranga [Genus]
P. bidentata [Species]
P. flava
P. rubra
P. roseogularis
P. olivacea
P. ludoviciana
P. leucoptera
P. erythrocephala
P. rubriceps
Calochaetes [Genus]
C. coccineus [Species]
Ramphocelus [Genus]
R. sanguinolentus [Species]
R. nigrogularis
R. dimidiatus
R. melanogaster
R. carbo
R. bresilius
R. passerinii
R. flammigerus
Spindalis [Genus]
S. zena [Species]
Thraupis [Genus]
T. episcopus [Species]
T. sayaca
T. cyanoptera
T. ornata
T. abbas
T. palmarum
T. cyanocephala
T. bonariensis
Cyanicterus [Genus]
C. cyanicterus [Species]
Buthraupis [Genus]
B. arcaei [Species]
B. melanochlamys
B. rothschildi
B. edwardsi
B. aureocincta
B. montana
B. eximia
B. wetmorei
Wetmorethraupis [Genus]
W. sterrhopteron [Species]
Anisognathus [Genus]
A. lacrymosus [Species]
A. igniventris
A. flavinuchus
A. notabilis

Stephanophorus [Genus]
  S. diadematus [Species]
Iridosornis [Genus]
  I. porphyrocephala [Species]
  I. analis
  I. jelskii
  I. rufivertex
Dubusia [Genus]
  D. taeniata [Species]
Delothraupis [Genus]
  D. castaneoventris [Species]
Pipraeidea [Genus]
  P. melanonota [Species]
Euphonia [Genus]
  E. jamaica [Species]
  E. plumbea
  E. affinis
  E. luteicapilla
  E. chlorotica
  E. trinitatis
  E. concinna
  E. saturata
  E. finschi
  E. violacea
  E. laniirostris
  E. hirundinacea
  E. chalybea
  E. musica
  E. fulvicrissa
  E. imitans
  E. gouldi
  E. chrysopasta
  E. mesochrysa
  E. minuta
  E. anneae
  E. xanthogaster
  E. rufiventris
  E. pectoralis
  E. cayennensis
Chlorophonia [Genus]
  C. flavirostris [Species]
  C. cyanea
  C. pyrrhophrys
  C. occipitalis
Chlorochrysa [Genus]
  C. phoenicotis [Species]
  C. calliparaea
  C. nitidissima
Tangara [Genus]
  T. inornata [Species]
  T. cabanisi
  T. palmeri
  T. mexicana
  T. chilensis
  T. fastuosa
  T. seledon
  T. cyanocephala
  T. desmaresti
  T. cyanoventris
  T. johannae

T. schrankii
T. florida
T. arthus
T. icterocephala
T. xanthocephala
T. chrysotis
T. parzudakii
T. xanthogastra
T. punctata
T. guttata
T. varia
T. rufigula
T. gyrola
T. lavinia
T. cayana
T. cucullata
T. peruviana
T. preciosa
T. vitriolina
T. rufigenis
T. ruficervix
T. labradorides
T. cyanotis
T. cyanicollis
T. larvata
T. nigrocincta
T. dowii
T. nigroviridis
T. vassorii
T. heinei
T. viridicollis
T. argyrofenges
T. cyanoptera
T. pulcherrima
T. velia
T. callophrys
Dacnis [Genus]
  D. albiventris [Species]
  D. lineata
  D. flaviventer
  D. hartlaubi
  D. nigripes
  D. venusta
  D. cayana
  D. viguieri
  D. berlepschi
Chlorophanes [Genus]
  C. spiza [Species]
Cyanerpes [Genus]
  C. nitidus [Species]
  C. lucidus
  C. caeruleus
  C. cyaneus
Xenodacnis [Genus]
  X. parina [Species]
Oreomanes [Genus]
  O. fraseri [Species]
Diglossa [Genus]
  D. baritula [Species]
  D. lafresnayii

D. carbonaria
D. venezuelensis
D. albilatera
D. duidae
D. major
D. indigotica
D. glauca
D. caerulescens
D. cyanea
Euneornis [Genus]
  E. campestris [Species]
Tersina [Genus]
  T. viridis [Species]

Parulidae [Family]
Mniotilta [Genus]
  M. varia [Species]
Vermivora [Genus]
  V. bachmanii [Species]
  V. chrysoptera
  V. pinus
  V. peregrina
  V. celata
  V. ruficapilla
  V. virginiae
  V. crissalis
  V. luciae
  V. gutturalis
  V. superciliosa
Parula [Genus]
  P. americana [Species]
  P. pitiayumi
Dendroica [Genus]
  D. petechia [Species]
  D. pensylvanica
  D. cerulea
  D. caerulescens
  D. plumbea
  D. pharetra
  D. pinus
  D. graciae
  D. adelaidae
  D. pityophila
  D. dominica
  D. nigrescens
  D. townsendi
  D. occidentalis
  D. chrysoparia
  D. virens
  D. discolor
  D. vitellina
  D. tigrina
  D. fusca
  D. magnolia
  D. coronata
  D. palmarum
  D. kirtlandii
  D. striata
  D. castanea
Catharopeza [Genus]
  C. bishopi [Species]

Setophaga [Genus]
　S. ruticilla [Species]
Seiurus [Genus]
　S. aurocapillus [Species]
　S. noveboracensis
　S. motacilla
Limnothlypis [Genus]
　L. swainsonii [Species]
Helmitheros [Genus]
　H. vermivorus [Species]
Protonotaria [Genus]
　P. citrea [Species]
Geothlypis [Genus]
　G. trichas [Species]
　G. beldingi
　G. flavovelata
　G. rostrata
　G. semiflava
　G. speciosa
　G. nelsoni
　G. chiriquensis
　G. aequinoctialis
　G. poliocephala
　G. formosa
　G. agilis
　G. philadelphia
　G. tolmiei
Microligea [Genus]
　M. palustris [Species]
Teretistris [Genus]
　T. fernandinae [Species]
　T. fornsi
Leucopeza [Genus]
　L. semperi [Species]
Wilsonia [Genus]
　W. citrina [Species]
　W. pusilla
　W. canadensis
Cardellina [Genus]
　C. rubrifrons [Species]
Ergaticus [Genus]
　E. ruber [Species]
　E. versicolor
Myioborus [Genus]
　M. pictus [Species]
　M. miniatus
　M. brunniceps
　M. pariae
　M. cardonai
　M. torquatus
　M. ornatus
　M. melanocephalus
　M. albifrons
　M. flavivertex
　M. albifacies
Euthlypis [Genus]
　E. lachrymosa [Species]
Basileuterus [Genus]
　B. fraseri [Species]
　B. bivittatus

B. chrysogaster
B. flaveolus
B. luteoviridis
B. signatus
B. nigrocristatus
B. griseiceps
B. basilicus
B. cinereicollis
B. conspicillatus
B. coronatus
B. culicivorus
B. rufifrons
B. belli
B. melanogenys
B. tristriatus
B. trifasciatus
B. hypoleucus
B. leucoblepharus
B. leucophrys
Phaeothlypis [Genus]
　P. fulvicauda [Species]
　P. rivularis
Peucedramus [Genus]
　P. taeniatus [Species]
Xenoligea [Genus]
　X. montana [Species]
Granatellus [Genus]
　G. venustus [Species]
　G. sallaei
　G. pelzelni
Icteria [Genus]
　I. virens [Species]
Conirostrum [Genus]
　C. speciosum [Species]
　C. leucogenys
　C. bicolor
　C. margaritae
　C. cinereum
　C. ferrugineiventre
　C. rufum
　C. sitticolor
　C. albifrons
Coereba [Genus]
　C. flaveola [Species]

Drepanididae [Family]
　Himatione [Genus]
　　H. sanguinea [Species]
　Palmeria [Genus]
　　P. dolei [Species]
　Vestiaria [Genus]
　　V. coccinea [Species]
　Drepanis [Genus]
　　D. funerea [Species]
　　D. pacifica
　Ciridops [Genus]
　　C. anna [Species]
　Viridonia [Genus]
　　V. virens [Species]
　　V. parva
　　V. sagittirostris

Hemignathus [Genus]
　H. obscurus [Species]
　H. lucidus
　H. wilsoni
Loxops [Genus]
　L. coccinea [Species]
Paroreomyza [Genus]
　P. maculata [Species]
Pseudonester [Genus]
　P. xanthophrys [Species]
Psittirostra [Genus]
　P. psittacea [Species]
Loxioides [Genus]
　L. cantans [Species]
　L. palmeri
　L. flaviceps
　L. bailleui
　L. kona

Vireonidae [Family]
　Cyclarhis [Genus]
　　C. gujanensis [Species]
　　C. nigrirostris
　Vireolanius [Genus]
　　V. melitophrys [Species]
　　V. pulchellus
　　V. leucotis
　Vireo [Genus]
　　V. brevipennis [Species]
　　V. huttoni
　　V. atricapillus
　　V. griseus
　　V. pallens
　　V. caribaeus
　　V. bairdi
　　V. gundlachii
　　V. crassirostris
　　V. bellii
　　V. vicinior
　　V. nelsoni
　　V. hypochryseus
　　V. modestus
　　V. nanus
　　V. latimeri
　　V. osburni
　　V. carmioli
　　V. solitarius
　　V. flavifrons
　　V. philadelphicus
　　V. olivaceus
　　V. magister
　　V. altiloquus
　　V. gilvus
　Hylophilus [Genus]
　　H. poicilotis [Species]
　　H. thoracicus
　　H. semicinereus
　　H. pectoralis
　　H. sclateri
　　H. muscicapinus
　　H. brunneiceps

H. semibrunneus
H. aurantifrons
H. hypoxanthus
H. flavipes
H. ochraceiceps
H. decurtatus

Icteridae [Family]
Psarocolius [Genus]
  P. oseryi [Species]
  P. latirostris
  P. decumanus
  P. viridis
  P. atrovirens
  P. angustifrons
  P. wagleri
  P. montezuma
  P. cassini
  P. bifasciatus
  P. guatimozinus
  P. yuracares
Cacicus [Genus]
  C. cela [Species]
  C. uropygialis
  C. chrysopterus
  C. koepckeae
  C. leucoramphus
  C. chrysonotus
  C. sclateri
  C. solitarius
  C. melanicterus
  C. holosericeus
Icterus [Genus]
  I. cayanensis [Species]
  I. chrysater
  I. nigrogularis
  I. leucopteryx
  I. auratus
  I. mesomelas
  I. auricapillus
  I. graceannae
  I. xantholemus
  I. pectoralis
  I. gularis
  I. pustulatus
  I. cucullatus
  I. icterus
  I. galbula
  I. spurius
  I. dominicensis
  I. wagleri
  I. laudabilis
  I. bonana
  I. oberi
  I. graduacauda
  I. maculialatus
  I. parisorum
Nesopsar [Genus]
  N. nigerrimus [Species]
Xanthopsar [Genus]
  X. flavus [Species]

Gymnomystax [Genus]
  G. mexicanus [Species]
Xanthocephalus [Genus]
  X. xanthocephalus [Species]
Agelaius [Genus]
  A. thilius [Species]
  A. phoeniceus
  A. tricolor
  A. icterocephalus
  A. humeralis
  A. xanthomus
  A. cyanopus
  A. ruficapillus
Leistes [Genus]
  L. militaris [Species]
Pezites [Genus]
  P. militaris [Species]
Sturnella [Genus]
  S. magna [Species]
  S. neglecta
Pseudoleistes [Genus]
  P. guirahuro [Species]
  P. virescens
Amblyramphus [Genus]
  A. holosericeus [Species]
Hypopyrrhus [Genus]
  H. pyrohypogaster [Species]
Curaeus [Genus]
  C. curaeus [Species]
  C. forbesi
Gnorimopsar [Genus]
  G. chopi [Species]
Oreopsar [Genus]
  O. bolivianus [Species]
Lampropsar [Genus]
  L. tanagrinus [Species]
Macroagelaius [Genus]
  M. subalaris [Species]
Dives [Genus]
  D. atroviolacea [Species]
  D. dives
Quiscalus [Genus]
  Q. mexicanus [Species]
  Q. major
  Q. palustris
  Q. nicaraguensis
  Q. quiscula
  Q. niger
  Q. lugubris
Euphagus [Genus]
  E. carolinus [Species]
  E. cyanocephalus
Molothrus [Genus]
  M. badius [Species]
  M. rufoaxillaris
  M. bonariensis
  M. aeneus
  M. ater
Scaphidura [Genus]
  S. oryzivorus [Species]

Fringillidae [Family]
Fringilla [Genus]
  F. coelebs [Species]
  F. teydea
  F. montifringilla
Serinus [Genus]
  S. pusillus [Species]
  S. serinus
  S. syriacus
  S. canaria
  S. citrinella
  S. thibetanus
  S. canicollis
  S. nigriceps
  S. citrinelloides
  S. frontalis
  S. capistratus
  S. koliensis
  S. scotops
  S. leucopygius
  S. atrogularis
  S. citrinipectus
  S. mozambicus
  S. donaldsoni
  S. flaviventris
  S. sulphuratus
  S. albogularis
  S. gularis
  S. mennelli
  S. tristriatus
  S. menschensis
  S. striolatus
  S. burtoni
  S. rufobrunneus
  S. leucopterus
  S. totta
  S. alario
  S. estherae
Neospiza [Genus]
  N. concolor [Species]
Linurgus [Genus]
  L. olivaceus [Species]
Rhynchostruthus [Genus]
  R. socotranus [Species]
Carduelis [Genus]
  C. chloris [Species]
  C. sinica
  C. spinoides
  C. ambigua
  C. spinus
  C. pinus
  C. atriceps
  C. spinescens
  C. yarrellii
  C. cucullata
  C. crassirostris
  C. magellanica
  C. dominicensis
  C. siemiradzkii
  C. olivacea

C. notata
C. xanthogastra
C. atrata
C. uropygialis
C. barbata
C. tristis
C. psaltria
C. lawrencei
C. carduelis
Acanthis [Genus]
  A. flammea [Species]
  A. hornemanni
  A. flavirostris
  A. cannabina
  A. yemenensis
  A. johannis
Leucosticte [Genus]
  L. nemoricola [Species]
  L. brandti
  L. arctoa
Callacanthis [Genus]
  C. burtoni [Species]
Rhodopechys [Genus]
  R. sanguinea [Species]
  R. githaginea
  R. mongolica
  R. obsoleta
Uragus [Genus]
  U. sibiricus [Species]
Urocynchramus [Genus]
  U. pylzowi [Species]
Carpodacus [Genus]
  C. rubescens [Species]
  C. nipalensis
  C. erythrinus
  C. purpureus
  C. cassinii
  C. mexicanus
  C. pulcherrimus
  C. eos
  C. rhodochrous
  C. vinaceus
  C. edwardsii
  C. synoicus
  C. roseus
  C. trifasciatus
  C. rhodopeplus
  C. thura
  C. rhodochlamys
  C. rubicilloides
  C. rubicilla
  C. puniceus
  C. roborowskii
Chaunoproctus [Genus]
  C. ferreorostris [Species]
Pinicola [Genus]
  P. enucleator [Species]
  P. subhimachalus
Haematospiza [Genus]
  H. sipahi [Species]

Loxia [Genus]
  L. pytyopsittacus [Species]
  L. curvirostra
  L. leucoptera
Pyrrhula [Genus]
  P. nipalensis [Species]
  P. leucogenys
  P. aurantiaca
  P. erythrocephala
  P. erythaca
  P. pyrrhula
Coccothraustes [Genus]
  C. coccothraustes [Species]
  C. migratorius
  C. personatus
  C. icterioides
  C. affinis
  C. melanozanthos
  C. carnipes
  C. vespertinus
  C. abeillei
Pyrrhoplectes [Genus]
  P. epauletta [Species]

Estrildidae [Family]
Parmoptila [Genus]
  P. woodhousei [Species]
Nigrita [Genus]
  N. fusconota [Species]
  N. bicolor
  N. luteifrons
  N. canicapilla
Nesocharis [Genus]
  N. shelleyi [Species]
  N. ansorgei
  N. capistrata
Pytilia [Genus]
  P. phoenicoptera [Species]
  P. hypogrammica
  P. afra
  P. melba
Mandingoa [Genus]
  M. nitidula [Species]
Cryptospiza [Genus]
  C. reichenovii [Species]
  C. salvadorii
  C. jacksoni
  C. shelleyi
Pyrenestes [Genus]
  P. sanguineus [Species]
  P. ostrinus
  P. minor
Spermophaga [Genus]
  P. poliogenys [Species]
  P. haematina
  P. ruficapilla
Clytospiza [Genus]
  C. monteiri [Species]
Hypargos [Genus]
  H. margaritatus [Species]
  H. niveoguttatus

Euschistospiza [Genus]
  E. dybowskii [Species]
  E. cinereovinacea
Lagonosticta [Genus]
  L. rara [Species]
  L. rufopicta
  L. nitidula
  L. senegala
  L. rubricata
  L. landanae
  L. rhodopareia
  L. larvata
Uraeginthus [Genus]
  U. angolensis [Species]
  U. bengalus
  U. cyanocephala
  U. granatina
  U. ianthinogaster
Estrilda [Genus]
  E. caerulescens [Species]
  E. perreini
  E. thomensis
  E. melanotis
  E. paludicola
  E. melpoda
  E. rhodopyga
  E. rufibarba
  E. troglodytes
  E. astrild
  E. nigriloris
  E. nonnula
  E. atricapilla
  E. erythronotos
  E. charmosyna
Amandava [Genus]
  A. amandava [Species]
  A. formosa
  A. subflava
Ortygospiza [Genus]
  O. atricollis [Species]
  O. gabonensis
  O. locustella
Aegintha [Genus]
  A. temporalis [Species]
Emblema [Genus]
  E. picta [Species]
  E. bella
  E. oculata
  E. guttata
Oreostruthus [Genus]
  O. fuliginosus [Species]
Neochmia [Genus]
  N. phaeton [Species]
  N. ruficauda
Poephila [Genus]
  P. guttata [Species]
  P. bichenovii
  P. personata
  P. acuticauda
  P. cincta

Erythrura [Genus]
    E. hyperythra [Species]
    E. prasina
    E. viridifacies
    E. tricolor
    E. coloria
    E. trichroa
    E. papuana
    E. psittacea
    E. cyaneovirens
    E. kleinschmidti
Chloebia [Genus]
    C. gouldiae [Species]
Aidemosyne [Genus]
    A. modesta [Species]
Lonchura [Genus]
    L. malabarica [Species]
    L. griseicapilla
    L. nana
    L. cucullata
    L. bicolor
    L. fringilloides
    L. striata
    L. leucogastroides
    L. fuscans
    L. molucca
    L. punctulata
    L. kelaarti
    L. leucogastra
    L. tristissima
    L. leucosticta
    L. quinticolor
    L. malacca
    L. maja
    L. pallida
    L. grandis
    L. vana
    L. caniceps
    L. nevermanni
    L. spectabilis
    L. forbesi
    L. hunsteini
    L. flaviprymna
    L. castaneothorax
    L. stygia
    L. teerinki
    L. monticola
    L. montana
    L. melaena
    L. pectoralis
Padda [Genus]
    P. fuscata [Species]
    P. oryzivora
Amadina [Genus]
    A. erythrocephala [Species]
    A. fasciata
Pholidornis [Genus]
    P. rushiae [Species]

Ploceidae [Family]
    Vidua [Genus]

V. chalybeata [Species]
V. funerea
V. wilsoni
V. hypocherina
V. fischeri
V. regia
V. macroura
V. paradisaea
V. orientalis
Bubalornis [Genus]
    B. albirostris [Species]
Dinemellia [Genus]
    D. dinemelli [Species]
Plocepasser [Genus]
    P. mahali [Species]
    P. superciliosus
    P. donaldsoni
    P. rufoscapulatus
Histurgops [Genus]
    H. ruficauda [Species]
Pseudonigrita [Genus]
    P. arnaudi [Species]
    P. cabanisi
Philetairus [Genus]
    P. socius [Species]
Passer [Genus]
    P. ammodendri [Species]
    P. domesticus
    P. hispaniolensis
    P. pyrrhonotus
    P. castanopterus
    P. rutilans
    P. flaveolus
    P. moabiticus
    P. iagoensis
    P. melanurus
    P. griseus
    P. simplex
    P. montanus
    P. luteus
    P. eminibey
Petronia [Genus]
    P. brachydactyla [Species]
    P. xanthocollis
    P. petronia
    P. superciliaris
    P. dentata
Montifringilla [Genus]
    M. nivalis [Species]
    M. adamsi
    M. taczanowskii
    M. davidiana
    M. ruficollis
    M. blanfordi
    M. theresae
Sporopipes [Genus]
    S. squamifrons [Species]
    S. frontalis
Amblyospiza [Genus]
    A. albifrons [Species]

Ploceus [Genus]
    P. baglafecht [Species]
    P. bannermani
    P. batesi
    P. nigrimentum
    P. bertrandi
    P. pelzelni
    P. subpersonatus
    P. luteolus
    P. ocularis
    P. nigricollis
    P. alienus
    P. melanogaster
    P. capensis
    P. subaureus
    P. xanthops
    P. aurantius
    P. heuglini
    P. bojeri
    P. castaneiceps
    P. princeps
    P. xanthopterus
    P. castanops
    P. galbula
    P. taeniopterus
    P. intermedius
    P. velatus
    P. spekei
    P. spekeoides
    P. cucullatus
    P. grandis
    P. nigerrimus
    P. weynsi
    P. golandi
    P. dicrocephalus
    P. melanocephalus
    P. jacksoni
    P. badius
    P. rubiginosus
    P. aureonucha
    P. tricolor
    P. albinucha
    P. nelicourvi
    P. sakalava
    P. hypoxanthus
    P. superciliosus
    P. benghalensis
    P. manyar
    P. philippinus
    P. megarhynchus
    P. bicolor
    P. flavipes
    P. preussi
    P. dorsomaculatus
    P. olivaceiceps
    P. insignis
    P. angolensis
    P. sanctithomae
Malimbus [Genus]
    M. coronatus [Species]
    M. cassini

M. scutatus
M. racheliae
M. ibadanensis
M. nitens
M. rubricollis
M. erythrogaster
M. malimbicus
M. rubriceps
Quelea [Genus]
  Q. cardinalis [Species]
  Q. erythrops
  Q. quelea
Foudia [Genus]
  F. madagascariensis [Species]
  F. eminentissima
  F. rubra
  F. bruante
  F. sechellarum
  F. flavicans
Euplectes [Genus]
  E. anomalus [Species]
  E. afer
  E. diadematus
  E. gierowii
  E. nigroventris
  E. hordeaceus
  E. orix
  E. aureus
  E. capensis
  E. axillaris
  E. macrourus
  E. hartlaubi
  E. albonotatus
  E. ardens
  E. progne
  E. jacksoni
Anomalospiza [Genus]
  A. imberbis [Species]

Sturnidae [Family]
Aplonis [Genus]
  A. zelandica [Species]
  A. santovestris
  A. pelzelni
  A. atrifusca
  A. corvina
  A. mavornata
  A. cinerascens
  A. tabuensis
  A. striata
  A. fusca
  A. opaca
  A. cantoroides
  A. crassa
  A. feadensis
  A. insularis
  A. dichroa
  A. mysolensis
  A. magna
  A. minor
  A. panayensis

A. metallica
A. mystacea
A. brunneicapilla
Poeoptera [Genus]
  P. kenricki [Species]
  P. stuhlmanni
  P. lugubris
Grafisia [Genus]
  G. torquata [Species]
Onychognathus [Genus]
  O. walleri [Species]
  O. nabouroup
  O. morio
  O. blythii
  O. frater
  O. tristramii
  O. fulgidus
  O. tenuirostris
  O. albirostris
  O. salvadorii
Lamprotornis [Genus]
  L. iris [Species]
  L. cupreocauda
  L. purpureiceps
  L. curruscus
  L. purpureus
  L. nitens
  L. chalcurus
  L. chalybaeus
  L. chloropterus
  L. acuticaudus
  L. splendidus
  L. ornatus
  L. australis
  L. mevesii
  L. purpuropterus
  L. caudatus
Cinnyricinclus [Genus]
  C. femoralis [Species]
  C. sharpii
  C. leucogaster
Speculipastor [Genus]
  S. bicolor [Species]
Neocichla [Genus]
  N. gutturalis [Species]
Spreo [Genus]
  S. fischeri [Species]
  S. bicolor
  S. albicapillus
  S. superbus
  S. pulcher
  S. hildebrandti
Cosmopsarus [Genus]
  C. regius [Species]
  C. unicolor
Saroglossa [Genus]
  S. aurata [Species]
  S. spiloptera
Creatophora [Genus]
  C. cinerea [Species]

Necropsar [Genus]
  N. leguati [Species]
Fregilupus [Genus]
  F. varius [Species]
Sturnus [Genus]
  S. senex [Species]
  S. malabaricus
  S. erythropygius
  S. pagodarum
  S. sericeus
  S. philippensis
  S. sturninus
  S. roseus
  S. vulgaris
  S. unicolor
  S. cinerascens
  S. contra
  S. nigricollis
  S. burmannicus
  S. melanopterus
  S. sinensis
Leucopsar [Genus]
  L. rothschildi [Species]
Acridotheres [Genus]
  A. tristis [Species]
  A. ginginianus
  A. fuscus
  A. grandis
  A. albocinctus
  A. cristatellus
Ampeliceps [Genus]
  A. coronatus [Species]
Mino [Genus]
  M. anais [Species]
  M. dumontii
Basilornis [Genus]
  B. celebensis [Species]
  B. galeatus
  B. corythaix
  B. miranda
Streptocitta [Genus]
  S. albicollis [Species]
  S. albertinae
Sarcops [Genus]
  S. calvus [Species]
Gracula [Genus]
  G. ptilogenys [Species]
  G. religiosa
Enodes [Genus]
  E. erythrophris [Species]
Scissirostrum [Genus]
  S. dubium [Species]
Buphagus [Genus]
  B. africanus [Species]
  B. erythrorhynchus

Oriolidae [Family]
Oriolus [Genus]
  O. szalayi [Species]
  O. phaeochromus
  O. forsteni

O. bouroensis
O. viridifuscus
O. sagittatus
O. flavocinctus
O. xanthonotus
O. albiloris
O. isabellae
O. oriolus
O. auratus
O. chinensis
O. chlorocephalus
O. crassirostris
O. brachyrhynchus
O. monacha
O. larvatus
O. nigripennis
O. xanthornus
O. hosii
O. crentus
O. traillii
O. mellianus
Sphecotheres [Genus]
S. vieilloti [Species]
S. flaviventris
S. viridis
S. hypoleucus

Dicruridae [Family]
Chaetorhynchus [Genus]
C. papuensis [Species]
Dicrurus [Genus]
D. ludwigii [Species]
D. atripennis
D. adsimilis
D. fuscipennis
D. aldabranus
D. forficatus
D. waldenii
D. macrocercus
D. leucophaeus
D. caerulescens
D. annectans
D. aeneus
D. remifer
D. balicassius
D. hottentottus
D. megarhynchus
D. montanus
D. andamanensis
D. paradiseus

Callaeidae [Family]
Callaeas [Genus]
C. cinerea [Species]
Creadion [Genus]
C. carunculatus [Species]
Heterolocha [Genus]
H. acutirostris [Species]

Grallinidae [Family]
Grallina [Genus]

G. cyanoleuca [Species]
G. brujini
Corcorax [Genus]
C. melanorhamphos [Species]
Struthidea [Genus]
S. cinerea [Species]

Artamidae [Family]
Artamus [Genus]
A. fuscus [Species]
A. leucorhynchus
A. monachus
A. maximus
A. insignis
A. personatus
A. superciliosus
A. cinereus
A. cyanopterus
A. minor

Cracticidae [Family]
Cracticus [Genus]
C. mentalis [Species]
C. torquatus
C. cassicus
C. louisiadensis
C. nigrogularis
C. quoyi
Gymnorhina [Genus]
G. tibicen [Species]
Strepera [Genus]
S. graculina [Species]
S. fuliginosa
S. versicolor

Ptilonorhynchidae [Family]
Ailuroedus [Genus]
A. buccoides [Species]
A. crassirostris
Scenopoeetes [Genus]
S. dentirostris [Species]
Archboldia [Genus]
A. papuensis [Species]
Amblyornis [Genus]
A. inornatus [Species]
A. macgregoriae
A. subalaris
A. flavifrons
Prionodura [Genus]
P. newtoniana [Species]
Sericulus [Genus]
S. aureus [Species]
S. bakeri
S. chrysocephalus
Ptilonorhynchus [Genus]
P. violaceus [Species]
Chlamydera [Genus]
C. maculata [Species]
C. nuchalis
C. lauterbachi
C. cerviniventris

Paradisaeidae [Family]
Loria [Genus]
L. loriae [Species]
Loboparadisea [Genus]
L. sericea [Species]
Cnemophilus [Genus]
C. macgregorii [Species]
Macgregoria [Genus]
M. pulchra [Species]
Lycocorax [Genus]
L. pyrrhopterus [Species]
Manucodia [Genus]
M. ater [Species]
M. jobiensis
M. chalybatus
M. comrii
Phonygammus [Genus]
P. keraudrenii [Species]
Ptiloris [Genus]
P. paradiseus [Species]
P. victoriae
P. magnificus
Semioptera [Genus]
S. wallacei [Species]
Seleucidis [Genus]
S. melanuleuca [Species]
Paradigalla [Genus]
P. carunculata [Species]
Drepanornis [Genus]
D. albertisi [Species]
D. brujini
Epimachus [Genus]
E. fastuosus [Species]
E. meyeri
Astrapia [Genus]
A. nigra [Species]
A. splendidissima
A. mayeri
A. stephaniae
A. rothschildi
Lophorina [Genus]
L. superba [Species]
Parotia [Genus]
P. sefilata [Species]
P. carolae
P. lawesii
P. wahnesi
Pteridophora [Genus]
P. alberti [Species]
Cicinnurus [Genus]
C. regius [Species]
Diphyllodes [Genus]
D. magnificus [Species]
D. respublica
Paradisaea [Genus]
P. apoda [Species]
P. minor
P. decora
P. rubra
P. guilielmi
P. rudolphi

Corvidae [Family]
  Platylophus [Genus]
    P. galericulatus [Species]
  Platysmurus [Genus]
    P. leucopterus [Species]
  Gymnorhinus [Genus]
    G. cyanocephala [Species]
  Cyanocitta [Genus]
    C. cristata [Species]
    C. stelleri
  Aphelocoma [Genus]
    A. coerulescens [Species]
    A. ultramarina
    A. unicolor
  Cyanolyca [Genus]
    C. viridicyana [Species]
    C. pulchra
    C. cucullata
    C. pumilo
    C. nana
    C. mirabilis
    C. argentigula
  Cissilopha [Genus]
    C. melanocyanea [Species]
    C. sanblasiana
    C. beecheii
  Cyanocorax [Genus]
    C. caeruleus [Species]
    C. cyanomelas
    C. violaceus
    C. cristatellus
    C. heilprini
    C. cayanus
    C. affinis
    C. chrysops
    C. mysticalis
    C. dickeyi
    C. yncas
  Psilorhinus [Genus]
    P. morio [Species]
  Calocitta [Genus]
    C. formosa [Species]
  Garrulus [Genus]

    G. glandarius [Species]
    G. lanceolatus
    G. lidthi
  Perisoreus [Genus]
    P. canadensis [Species]
    P. infaustus
    P. internigrans
  Urocissa [Genus]
    U. ornata [Species]
    U. caerulea
    U. flavirostris
    U. erythrorhyncha
    U. whiteheadi
  Cissa [Genus]
    C. chinensis [Species]
    C. thalassina
  Cyanopica [Genus]
    C. cyana [Species]
  Dendrocitta [Genus]
    D. vagabunda [Species]
    D. occipitalis
    D. formosae
    D. leucogastra
    D. frontalis
    D. baileyi
  Crypsirina [Genus]
    C. temia [Species]
    C. cucullata
  Temnurus [Genus]
    T. temnurus [Species]
  Pica [Genus]
    P. pica [Species]
    P. nuttali
  Zavattariornis [Genus]
    Z. stresemanni [Species]
  Podoces [Genus]
    P. hendersoni [Species]
    P. biddulphi
    P. panderi
    P. pleskei
  Pseudopodoces [Genus]
    P. humilis [Species]
  Nucifraga [Genus]

    N. columbiana [Species]
    N. caryocatactes
  Pyrrhocorax [Genus]
    P. pyrrhocorax [Species]
    P. graculus
  Ptilostomus [Genus]
    P. afer [Species]
  Corvus [Genus]
    C. monedula [Species]
    C. dauuricus
    C. splendens
    C. moneduloides
    C. enca
    C. typicus
    C. florensis
    C. kubaryi
    C. validus
    C. woodfordi
    C. fuscicapillus
    C. tristis
    C. capensis
    C. frugilegus
    C. brachyrhynchos
    C. caurinus
    C. imparatus
    C. ossifragus
    C. palmarum
    C. jamaicensis
    C. nasicus
    C. leucognaphalus
    C. corone
    C. macrorhynchos
    C. orru
    C. bennetti
    C. coronoides
    C. torquatus
    C. albus
    C. tropicus
    C. cryptoleucus
    C. ruficollis
    C. corax
    C. rhipidurus
    C. albicollis
    C. crassirostris

• • • • •

A brief geologic history of animal life

A note about geologic time scales: A cursory look will reveal that the timing of various geological periods differs among textbooks. Is one right and the others wrong? Not necessarily. Scientists use different methods to estimate geological time—methods with a precision sometimes measured in tens of millions of years. There is, however, a general agreement on the magnitude and relative timing associated with modern time scales. The closer in geological time one comes to the present, the more accurate science can be—and sometimes the more disagreement there seems to be. The following account was compiled using the more widely accepted boundaries from a diverse selection of reputable scientific resources.

Geologic time scale

Era	Period	Epoch	Dates	Life forms
Proterozoic			2,500-544 mya*	First single-celled organisms, simple plants, and invertebrates (such as algae, amoebas, and jellyfish)
Paleozoic	Cambrian		544-490 mya	First crustaceans, mollusks, sponges, nautiloids, and annelids (worms)
	Ordovician		490-438 mya	Trilobites dominant. Also first fungi, jawless vertebrates, starfish, sea scorpions, and urchins
	Silurian		438-408 mya	First terrestrial plants, sharks, and bony fish
	Devonian		408-360 mya	First insects, arachnids (scorpions), and tetrapods
	Carboniferous	Mississippian	360-325 mya	Amphibians abundant. Also first spiders, land snails
		Pennsylvanian	325-286 mya	First reptiles and synapsids
	Permian		286-248 mya	Reptiles abundant. Extinction of trilobytes
Mesozoic	Triassic		248-205 mya	Diversification of reptiles: turtles, crocodiles, therapsids (mammal-like reptiles), first dinosaurs
	Jurassic		205-145 mya	Insects abundant, dinosaurs dominant in later stage. First mammals, lizards, frogs, and birds
	Cretaceous		145-65 mya	First snakes and modern fish. Extinction of dinosaurs, rise and fall of toothed birds
Cenozoic	Tertiary	Paleocene	65-55.5 mya	Diversification of mammals
		Eocene	55.5-33.7 mya	First horses, whales, and monkeys
		Oligocene	33.7-23.8 mya	Diversification of birds. First anthropoids (higher primates)
		Miocene	23.8-5.6 mya	First hominids
		Pliocene	5.6-1.8 mya	First australopithecines
	Quaternary	Pleistocene	1.8 mya-8,000 ya	Mammoths, mastodons, and Neanderthals
		Holocene	8,000 ya-present	First modern humans

*Millions of years ago (mya)

· · · · ·

Index

Bold page numbers indicate the primary discussion of a topic; page numbers in italics indicate illustrations.

INDEX

INDEX

INDEX

INDEX

INDEX

INDEX

INDEX

INDEX

INDEX

INDEX

INDEX

INDEX

INDEX

Z